# 22nd Topical Conference on Radio-Frequency Power in Plasmas (RFPPC 2017)



EPJ Web of Conferences Volume 157 (2017)

Aix-en-Provence, France
30 May - 2 June 2017

**Editor:**

**Julien Hillairet**

ISBN: 978-1-5108-5011-8

**Printed from e-media with permission by:**

Curran Associates, Inc.
57 Morehouse Lane
Red Hook, NY 12571

**Some format issues inherent in the e-media version may also appear in this print version.**



Printed by Curran Associates, Inc. (2017)

For additional information, please contact EDP Sciences – Web of Conferences at the address below.

EDP Sciences – Web of Conferences
17, Avenue du Hoggar
Parc d'Activité de Courtabœuf
BP 112
F-91944 Les Ulis Cedex A
France

Phone: +33 (0) 1 69 18 75 75
Fax: +33 (0) 1 69 28 84 91

contact-edps@webofconferences.org

**Additional copies of this publication are available from:**

Curran Associates, Inc.
57 Morehouse Lane
Red Hook, NY 12571 USA
Phone: 845-758-0400
Fax: 845-758-2633
Email: curran@proceedings.com
Web: www.proceedings.com

# TABLE OF CONTENTS

# 22 Topical Conference on
# RADIOFREQUENCY POWER IN PLASMAS

## 30 May - 2 June 2017 - Centre de Congrès - Aix en Provence – France

**Scope and topics**

This conference is the Twenty-Second Edition of the International Conference on electromagnetic waves in plasmas; including RF applications in magnetic fusion devices, RF plasma sources, and Basic RF wave Physics. This conference will also offer an opportunity to discuss advances of both Technology and Physical aspects of the ITER Heating and Current Drive systems.

Among the topics of the joint conference, recent developments and research results in the following fields are addressed:

1- R&D for ITER and Reactor-grade devices;
2- Plasma wave interactions, such as heating, current generation, other phase, space engineering, diagnostics and plasma control;
3- Applications of RF power in fusion devices such as tokamaks, stellarators, spherical tori, reversed-field pinches and other confinement concepts;
4- Basic plasma wave physics;
5- Applications to plasma sources for material processing, space propulsion, ionospheric and space plasmas.

**Scientific Editor**: Julien Hillairet

22nd Topical Conference on

# RADIOFREQUENCY POWER IN PLASMAS

PLATINUM SPONSOR

THALES

SILVER SPONSOR

ANSYS

AIX EN PROVENCE LA VILLE

MÉTROPOLE AIX-MARSEILLE PROVENCE | PAYS D'AIX

cea

BRONZE SPONSOR

iter

edp sciences

EPJ Web of Conferences

# Statement of Peer review

In submitting conference proceedings to EPJ Web of Conferences, I certify to the Publisher that I adhere to the **Policy on Publishing Integrity** of the journal in order to safeguard good scientific practice in publishing.

1. All articles have been subjected to peer review administered by the proceedings editors.
2. Reviews have been conducted by expert referees, who have been requested to provide unbiased and constructive comments aimed, whenever possible, at improving the work.
3. Proceedings editors have taken all reasonable steps to ensure the quality of the materials they publish and their decision to accept or reject a paper for publication has been based only on the merits of the work and the relevance to the journal.

Title, date and place of the conference

22 Topical Conference on Radio-Frequency Power in Plasmas

30 May – 2 June 2017

Aix-En-Provence, France

Proceedings editor(s):

Julien Hillairet

Date and editor's signature

25/9/2017

 DOI: 10.1051/epjconf/201715701001

# SOL RF physics modelling in Europe, in support of ICRF experiments

*Laurent* Colas[1,*], *LingFeng* Lu[1,2,3,†], *Jonathan* Jacquot[4], *Wouter* Tierens[4], *Alena* Křivská[5], *Stéphane* Heuraux[2], *Eric* Faudot[2], *Patrick* Tamain[1], *Bruno* Després[6], *Dirk* Van Eester[5], *Kristel* Crombé[3,5], *Fabrice* Louche[5], *Julien* Hillairet[1], *Walid* Helou[1], and *Marc* Goniche[1]

[1]CEA, IRFM, F-13108 Saint Paul Lez Durance, France.
[2]Institut Jean Lamour, UMR 7198, CNRS-University of Lorraine, F-54506 Vandoeuvre Cedex, France.
[3]Department of Applied Physics, Ghent University, Belgium
[4]Max-Planck-Institut für Plasmaphysik, Garching, Germany.
[5]LPP-ERM-KMS, TEC partner, Brussels, Belgium.
[6]Laboratoire Jacques Louis Lions, UPMC-Paris VI, CNRS UMR 7598, Paris, France.
[†] present address: TianQin Research Center for Gravitational Physics, School of Physics and Astronomy, Sun Yat-Sen University, Zhuhai 519082, P. R. China

**Abstract.** A European project was undertaken to improve the available SOL ICRF physics simulation tools and confront them with measurements. This paper first reviews code upgrades within the project. Using the multi-physics finite element solver COMSOL, the SSWICH code couples RF full-wave propagation with DC plasma biasing over "antenna-scale" 2D (toroidal/radial) domains, *via* non-linear RF and DC sheath boundary conditions (SBCs) applied at shaped plasma-facing boundaries. For the different modules and associated SBCs, more elaborate basic research in RF-sheath physics, SOL turbulent transport and applied mathematics, generally over smaller spatial scales, guides code improvement. The available simulation tools were applied to interpret experimental observations on various tokamaks. We focus on robust qualitative results common to several devices: the spatial distribution of RF-induced DC bias; left-right asymmetries over strap power unbalance; parametric dependence and antenna electrical tuning; DC SOL biasing far from the antennas, and RF-induced density modifications. From these results we try to identify the relevant physical ingredients necessary to reproduce the measurements, *e.g.* accurate radiated field maps from 3D antenna codes, spatial proximity effects from wave evanescence in the near RF field, or DC current transport. Pending issues towards quantitative predictions are also outlined.

## 1 ICRF antennas as active Plasma-Facing Components

The phased strap arrays used to launch Ion Cyclotron Range of Frequency (ICRF, 30-80MHz) waves into magnetic fusion devices poorly radiate in vacuum: efficient coupling of the fast wave to the main plasma necessitates minimizing the distance from the straps to a critical peripheral density: the *R*-cutoff layer. In the Scrape-Off Layer (SOL) the antennas then behave as Plasma-facing Components (PFCs), subject to Plasma-Surface Interactions (PSI). As RF-field-emitting structures, ICRF antennas are active PFCs able to create RF-specific PSI and to modify locally their environment: enhanced ion energies, heat loads, erosion and density modifications have been observed locally. These phenomena are critical in the prospect of long-pulse machines with high-*Z* plasma-facing materials. Predicting the magnitude and spatial location of these processes, in relation with plasma parameters, launcher design and electrical settings, remains challenging. So far, realistic tokamak predictions relied on very simple models of oscillating double probes. A European project was undertaken to improve the available SOL RF physics simulation tools and confront them with measurements. This paper first reviews code improvements and associated basic research within the project. The available simulation tools are then applied to interpret experimental observations on various machines. From these results we exhibit qualitative properties common to the various devices, identify relevant physical ingredients necessary to reproduce the measurements, as well as pending issues towards quantitative predictions.

## 2 Improvements in RF+DC sheaths simulations, associated basic research

The intense time-harmonic RF electric fields **E** emitted in the SOL at frequency $\omega_0$ are generally suspected to cause RF oscillations $V_{RF}\exp(-i\omega_0 t)$ of the sheath voltage at plasma-wall interfaces. Non-linear rectification of these oscillations then produces a Direct-Current (DC) self-biasing of the SOL plasma. Ion acceleration across the larger DC potential $V_{DC}$ is suspected to enhance the PSI locally. Sheaths also modify the RF wave reflection at material boundaries, in a way depending on $V_{DC}$. Our simulations over spatial scales comparable to antenna dimensions couple simple models of RF wave

* Corresponding author: laurent.colas@cea.fr

propagation for **E** and DC SOL biasing for $V_{DC}$ *via* non-linear sheath electrical properties. In Europe, these are implemented in the Self-consistent Sheaths and Waves for Ion Cyclotron Heating (SSWICH) code [1], [2], [3], using the Newton-Raphson scheme in the multiphysics solver COMSOL. As an intermediate step the code also computes the complex oscillating voltages $V_{RF}$ at sheath boundaries. Rectification likely generates harmonics of the wave frequency $\omega_0$ but these are neglected in our modelling. We examine below the different modules successively. In several key areas, more elaborate basic research in physics and applied mathematics, generally over smaller spatial scales, guides code improvement.

## 2.1 RF and DC sheath boundary conditions.

In "antenna-scale" modelling, limited spatial resolution presently imposes treating the sheaths as non-linear sheath boundary conditions (SBCs) at plasma/sheath interfaces, applied at plasma-facing material boundaries. SSWICH presently incorporates simple RF and DC SBCs.

DC SBCs in the DC biasing module reflect the rectifying properties of the sheaths. Current-voltage ($j/V$) characteristics of sheaths are generally non-linear. Consider for simplicity the case of a static sheath

$$j = j^{+}\left[1-\exp\left[-e\left(V-V_{f}\right)/kT_{e}\right]\right] \quad (1)$$

With $T_e$ the electron temperature, $V_f$ the floating potential in the absence of RF sheaths, and $j^+$ the ion saturation current. When the confinement magnetic field $\mathbf{B_0}$ is tilted by angle $\alpha$ with respect to the wall, $j+$ is supposed to scale as $\sin(\alpha)$ [3]. Due to the non-linearity in (1), adding a sinusoidal RF oscillations $V_{RF}$ of $V$ on top of a DC component $V_{DC}$, and time averaging $j(t)$ shifts the $j/V$ characteristics

$$j_{DC} = j^{+}\left[1-\exp\left[-e\left(V_{DC}-V_{b}-V_{f}\right)/kT_{e}\right]\right] \quad (2)$$

Oscillating sheaths behave as if the (grounded) wall was locally biased to a DC voltage $V_b$ expressed as

$$V_{b} = kT_{e}\log\left[I_{0}\left(e\left|V_{RF}/kT_{e}\right|\right)\right]/e \quad (3)$$

With $I_0$ the modified Bessel function. DC SBCs (2) therefore couple the DC biasing to RF wave-propagation *via* $V_{RF}$. Conversely the RF SBCs couple the wave propagation to the DC bias. At the RF timescale, RF sheaths are often treated as thin dielectric layers of width $\delta$ between the wall and the quasi-neutral plasma. The associated RF SBC was first formulated in [4] and reads

$$V_{RF}=D_n\delta/\varepsilon_{sh} \quad (4)$$

where $D_n$ is the component of the electric displacement normal to the boundary, $\varepsilon_{sh}$ is a dielectric constant of order 1 and the sheath width $\delta$ depends on the DC sheath voltage $V_{DC}$ *via* the Child-Langmuir law.

$$\delta \sim \lambda_{De}(eV_{DC}/kT_e)^{3/4} \quad (5)$$

where $\lambda_{De}$ is the electron Debye length at the entrance of the Debye sheath. Consistent with the electrostatic sheath approximation, the RF field $\mathbf{E_t}$ tangent to the sheath/plasma interface is not null

$$\mathbf{E_t}=-\nabla_t V_{RF} \quad (6)$$

The electrostatic approximation also imposes that

$$B_n=-i(\nabla\times\mathbf{E})_n/\omega_0= i\nabla\times(\nabla_t V_{RF})/\omega_0=0 \quad (7)$$

To start the iterations between RF and DC modules, RF wave propagation is first run using asymptotic RF SBCs valid for large sheath widths and independent of $V_{DC}$. Together with relation (7), the general form of the asymptotic RF SBC reads [1], [5].

$$D_n=0 \quad (8)$$

Since their first formulation, SBCs have been improved using more basic models of spatially-resolved magnetized RF sheaths over scales much smaller than antenna dimensions [6], [7]. The structure of SSWICH is modular and can incorporate these improvements progressively. Several basic models were tested in Europe. Paper [8] investigated the linear reflection of cold plane waves on a flat boundary across a prescribed time-independent 1D density profile characteristic of a sheath. Such plane waves exist with Fast Wave or Slow Wave (SW) polarization. As already shown by [9] such incoming waves are generally depolarized upon reflection. The process is described *via* a 2×2 matrix relation between incident and outgoing Fast and SW, depending on plasma profiles, $\alpha$ and tangent wavevector. The parametric sensitivity of the matrix components was stressed in [8]. In [10] the above model was made more sophisticated: the intense RF fields change the DC density profile *via* ponderomotive forces, in addition to the usual electric and pressure forces in the DC sheath. This in turn modifies the wave propagation. Self-consistent simulations were performed. Similar to [11], kinetic 1d3v PIC simulations of an oscillating double Langmuir probe are used to refine the RF impedance (4), $j_{DC}/V_{DC}$ characteristics (2) (3), and sheath widths (5) of magnetized RF sheaths, as a function of $\alpha$, relevant frequencies ($\omega_0$, plasma frequency $\omega_{pi}$, cyclotron frequency $\Omega_{ci}$,...) and temperatures. First runs suggest that $V_b$ in equation (3) should depend on $\alpha$: it tends to vanish below the Chodura angle where DC sheaths disappear: $\sin(\alpha_c)=\left[2\pi(1+T_i/T_e)m_e/m_i\right]^{1/2}$. Other valuable output includes the time-averaged ion energy distribution at the wall as well as higher harmonics of the wave frequency $\omega_0$.

## 2.2 RF wave propagation module for tokamaks

RF wave propagation is solved using the Finite Element Method (FEM) to allow versatile wall shapes and plasma in contact with material boundaries for sheath excitation. The original SSWICH-SW version solves for the scalar field $E_{//}$ representing the time harmonic cold SW [2]

DOI: 10.1051/epjconf/201715701001

$$\varepsilon_{//}\Delta_{//}E_{//} + \varepsilon_{\perp}\Delta_{\perp}E_{//} + k_0^2\varepsilon_{//}\varepsilon_{\perp} = 0 \quad (9)$$

With $k_0=\omega_0/c$ and ($\varepsilon_{//}$, $\varepsilon_{\perp}$) the cold-plasma dielectric constants [12]. Walls are either parallel or normal to $\mathbf{B_0}$. Excitation is provided by a map of $E_{//}$ from an external antenna code, and prescribed at the antenna mouth (see Figure 1). Both fully coupled versions and asymptotic RF-SBCs exist, in a 2D (radial, toroidal) plane.

Within the project SSWICH was upgraded to full-wave polarization to solve the vectorial wave equation

$$\nabla \times \nabla \times \mathbf{E} - k_0^2 \boldsymbol{\varepsilon}\mathbf{E} = 0 \quad (10)$$

excited either by straps or vectorial input field maps (see Figure 1). A 2D version was tested in [3] including shaped walls and magnetized RF sheaths. Eq. (6) was used to integrate $V_{RF}$ along the curved 1D boundary, assuming $V_{RF}$=0 far from the antenna. A standalone 3D version of SSWICH-FW with metallic BCs and embedded antenna was developed as RAPLICASOL [13]. It is presently used as an antenna code in [14], [15]. An iterative solver (GM-RES) reduces the huge RAM requirements of 3D computations.

Figure 1 sketches the simulation domain for the RF module of SSWICH-FW. Focus is put on the antenna vicinity at the Low Field Side (LFS) of a straightened tokamak and does not cover the whole radial and toroidal extension of the vessel. It cannot investigate RF-sheaths likely present at the high-field side (HFS). The magnetic connection lengths at LFS are also reduced compared to reality. In order to fairly reproduce the power coupling in presence of inhomogeneous plasma, the radial extent of the simulation domain needs to include all the relevant $R$-cutoffs in the radiated $k_{//}$ spectrum.

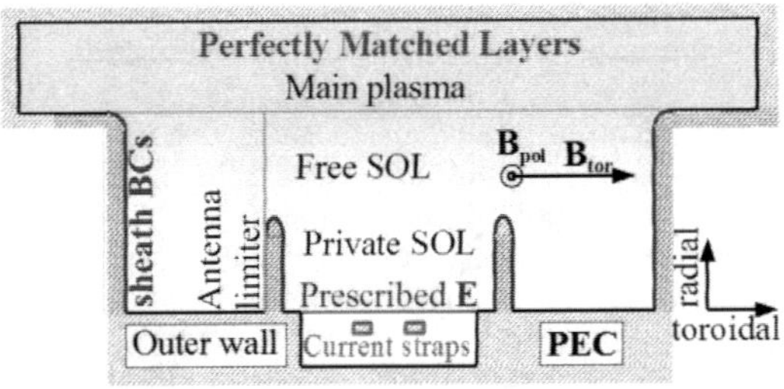


**Fig. 1.** Sketch of 2D (toroidal/radial) simulation domain for the RF wave module of SSWICH-FW in tokamaks (not to scale). Confinement magnetic field $\mathbf{B_0}$ includes in-plane ($B_{tor}$) and out-of-plane ($B_{pol}$) components (see Figure 3). Colour intensities are indicative of plasma local density. RF waves are excited either by prescribed RF currents on poloidal straps or prescribed RF electric field tangent to the magenta boundary (cut through "antenna aperture" on Figure 3). RF SBCs are implemented at blue boundaries. Orange line: cut through "reference plane" on Figure 3.

At the inner part of the domain, SSWICH-FW and RAPLICASOL emulate radiation to infinity using Perfectly Matched Layers (PMLs). Inside PMLs, spatial coordinates are stretched into the complex plane, turning propagative waves into evanescent ones. The only restriction is to avoid coexisting forward and backward waves at PML location. PMLs are implemented as lossy inhomogeneous materials with artificial dielectric and magnetic tensors depending on plasma parameters, the stretching function and stretched coordinates. For gyrotropic media like cold plasmas, PMLs were formulated in slab [17] and cylindrical geometries [13].

The FEM allows versatile spatial distributions of the plasma parameters. In the radial direction this includes realistic density profiles from tokamak measurements. Inhomogeneity can also be implemented in the other directions. Paper [2] defined a private SOL surrounded by the two antenna limiters on figure 1. References [16], [18] investigated waves in presence of RF-induced density modifications.

As one such application, the impact of a tenuous plasma inside a realistic ICRF antenna box was explored in [20], using RAPLICASOL in 2D. The tenuous plasma introduces the Lower-Hybrid (LH) resonance in the box when $\varepsilon_{\perp}$=0. The SW excited parasiticallybecomes propagative at densities below LH resonance. Local RF fields in this region evolved upon mesh refinement. This is attributed to the short wavelength of the SW, combined with the box geometry, a domain without volume losses closed by metallic walls on three sides, and coupled to the straps *via* the LH resonance. This complicates near field studies in the box. Yet, the radiated power was found independent of the mesh: the sloshing power carried by the SW is transferred to the fast wave. The radiated power spectrum was weakly dependent on the density in the box, except for low $k_{//}$ components, whose $R$-cutoff moved closer to the straps in presence of tenuous plasma.

Crossing the cold LH resonance causes specific numerical issues for the fast wave in perpendicular propagation: the radial electric field exhibits a singularity there, causing finite power damping in the collisionless limit [21]. Regularization relies frequently on artificial friction, *i.e.* an imaginary part $\nu$ added to $\omega_0$, *e.g.* in [13]. Numerical tests however show that a subtle trade-off between the choices of $\nu$ and mesh size is necessary to recover the collisionless limit $\nu \rightarrow 0^+$. An alternative formulation was proposed in [22], [23]: at the resonance position, the wave equation is replaced by an integral relation involving only the non-singular field components, as well as collisionless "manufactured solutions". Numerical tests show the good behaviour of 1D FEM using this scheme against analytical solutions of the Budden problem. Extension to 2D is under study.

### 2.3. Full-wave RF wave propagation in ALINE

Europe hosts two RF-sheath testbeds: IShTAR [24] and ALINE [25], [26]. ALINE is a cylindrical device (length 1m×diameter 30cm) producing partially ionized Ar and He plasmas with longitudinal magnetic fields up to 0.1T and electron densities in the range of $10^{16}$m$^{-3}$. Plasma and waves (frequencies 10kHz-250MHz) are produced by a cylindrical RF electrode located at the centre of the vessel. Waves are damped in the plasma volume by

DOI: 10.1051/epjconf/201715701001

collisions. The ALINE plasma can be mapped in 3D by a RF-compensated moveable Langmuir probe.

SSWICH was recently adapted to ALINE. The simulation domain for this device, sketched on figure 2, features a 2D cut across the closed vacuum vessel, filled with cold collisional magnetized plasma. RF excitation is provided by imposing an oscillating voltage between the vessel body and the coaxial feeder of the RF electrode, surrounded by a ceramic tube. Asymptotic RF SBCs are implemented at vessel boundaries normal to $\mathbf{B_0}$ while perfect conductors are assumed on walls parallel to $\mathbf{B_0}$

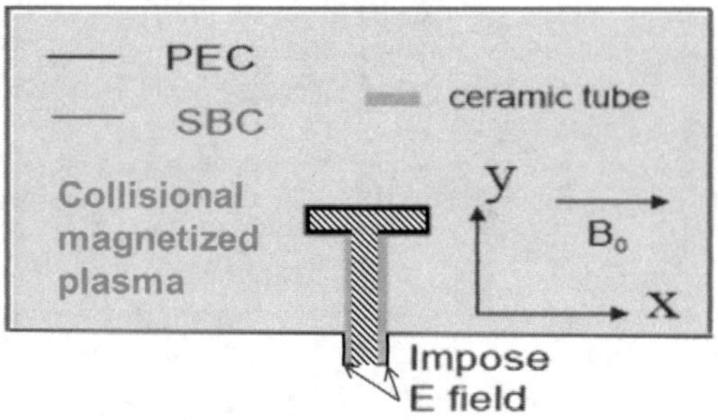


**Fig. 2.** 2D (axial/longtitudinal) simulation domain for the RF wave module of SSWICH-FW in ALINE. Not to scale.

### 2.4. DC plasma biasing module of SSWICH

The DC plasma biasing module of SSWICH solves the conservation of DC charge over the SOL.

$$\mathrm{div}(\mathbf{j_{DC}})=0 \tag{11}$$

At the separatrix the plasma is supposed to be floating. DC SBCs implement $j_{DC}/V_{DC}$ characteristics (2) with artificial DC wall bias $V_b$ (3) driving enhanced DC plasma potentials $V_{DC}$. In the volume the conduction is supposed to follow a highly anisotropic Ohm's law

$$\mathbf{j_{DC}} = -\boldsymbol{\sigma}_{\mathbf{DC}}\nabla\mathbf{V_{DC}} \tag{12}$$

In the parallel direction the Spitzer DC conductivity is taken, while a phenomenological conductivity is introduced transverse to $\mathbf{B_0}$. In references [2] and [15] typical quantitative values in the range $\sigma_{\perp DC}$~$10^{-6}\sigma_{//DC}$ in the antenna private SOL and $\sigma_{\perp DC}$~$10^{-8}\sigma_{//DC}$ in the main SOL were found to reasonably fit the measurements. Relation (12) is strictly valid only for charged particles colliding with neutrals. This process is however quite weak in the tokamak SOL. The transverse transport of DC current is more likely turbulent, but SSWICH cannot afford modelling it. Therefore $\sigma_{\perp DC}$ is a loosely constrained parameter.

The above DC current transport process was assessed numerically against 3D SOL turbulence simulations using the TOKAM3X fluid code [27]. As a proxy to the self-biasing by sheath rectification, a local DC bias was imposed on material boundaries protruding into the outer half of the SOL, similar to the antenna side limiters on Figure 1. From these biased wall elements, the DC potential spread along connected magnetic flux tubes over long distances, with parallel attenuation of the peaks and broadening to adjacent field lines. An Ohm's law like (12) was found inadequate to describe the local transverse conduction. No linear relation exists between local DC currents and DC electric fields. Besides, the simulated conduction was different in the radial and poloidal directions. The local shear of the DC electric field $\mathbf{E_{DC}}=-\nabla_\perp V_{DC}$ in the biased region likely plays a role in the turbulent conduction. Deeper studies will however be needed to determine a more accurate mean-field description of transverse anomalous charge transport.

### 2.5. Brief panorama of antenna-scale RF-sheath modelling

SSWICH is a unique tool in Europe to model RF-sheath effects over antenna-scale spatial domains in tokamaks. Similar tools however exist worldwide. Their main characteristics are compared in table 1. Although all these tools make use of RF-SBCs, each one features specificities. Vsim [28] uses the Finite Difference Time Domain (FDTD) method. Although this is not yet implemented, it can potentially handle altogether all the harmonics of the wave frequency and the full non-linearity of the initial RF-sheath problem, as well as ponderomotive effects. rfSOL [5] models an entire poloidal cross-section of the machine. It was used to investigate qualitatively RF-sheaths at the HFS walls, due to residual fast waves crossing the plasma core [29]. The DC plasma biasing module allowing DC current transport is a specificity of SSWICH. Its necessity will be further discussed in the context of tokamak experiments. All alternative codes implicitly assume that the sheaths float over time-scales much longer than the RF period, so that no DC current crosses them. The local DC potential $V_{DC}$ at one sheath then depends only on the local $V_{RF}$. It is independent of the sheaths at the opposite extremity of the same field line or on neighbouring flux tubes.

**Table 1.** Existing "antenna-scale" RF models with RF-sheath boundary conditions for tokamaks

| | Method | Dimensions | RF module | DC module | Excitation |
|---|---|---|---|---|---|
| **Vsim** [28] | FDTD | 3D | Time-resolved floating sheaths | | CAD-based antenna model |
| **rfSOL** [5] | FEM | 2D cross-section (radial/poloidal) | Full Wave | Floating sheaths | Current sheet |
| **SSWICH** | FEM | Multi-2D, LFS (radial/toroidal) | Slow Wave or Full Wave | DC currents (Ohm's law) | Straps/realistic field maps |

EPJ Web of Conferences **157**, 01001 (2017)    DOI: 10.1051/epjconf/201715701001
*22 Topical Conference on Radio-Frequency Power in Plasmas*

## 3 Application of SSWICH to qualitatively interpret tokamak experiments.

The SSWICH-SW code was applied for the first time in 2012 to interpret observations on Tore Supra (TS) [2]. Predictions were also made for WEST and ITER [30]. The project revisited the TS simulations with the upgraded SSWICH-FW tool [3], and extended the tokamak database to ASDEX upgrade (AUG) 2-strap and 3-strap antennas [14], [15] as well as JET ITER-like Antenna (ILA) [31].

Mostly run was the SSWICH-SW version using asymptotic RF-SBCs (7) and (8), with no attempt at self-consistency. The fully-coupled SSWICH-SW version was run in a few TS cases, showing that the asymptotic approximation, corresponding to the first turn around the self-consistent loop, was a fair approximation of the final result in these cases [2]. In [3] the TS simulations were also revisited using the SSWICH-FW code with asymptotic RF-SBCs. Specific effects related to the Fast Wave were investigated. In all cases a multi-2D procedure was followed and is sketched on Figure 3 for SSWICH-FW: using 2D (toroidal/poloidal) input field maps from 3D antenna codes with plasma load but without sheaths (either TOPICA [32] or RAPLICASOL [13]), independent 2D (radial/toroidal) simulations were performed at each altitude along the antenna mouth, assuming negligible poloidal gradients. For SSWICH-SW simulations, this multi-2D procedure was assessed in [33]: it is valid as long as the input RF field maps do not exhibit small-scale poloidal structures below the electron skin depth $c/\omega_{pe}$. The multi-2D approach will be further discussed below for Fast Wave propagation.

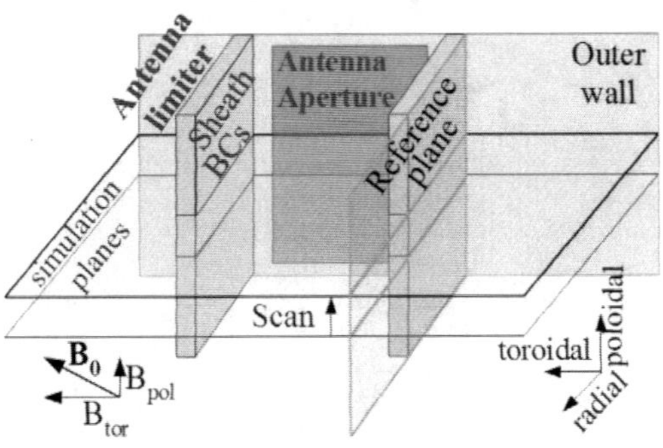


**Fig. 3.** 3D sketch of multi-2D simulation procedure with SSWICH-FW code excited using antenna codes. Independent 2D runs are successively performed in each (toroidal/radial) simulation plane, detailed on Figure 1 and visualized here from plasma side. They take as excitation poloidal cuts through a 2D (toroidal/poloidal) input field map prescribed at antenna aperture. From the simulation output, $V_{DC}$ is mapped in section 3.1 in the orange (radial/poloidal) reference plane.

Quantitative sensitivity of the simulations on loosely constrained parameters is frequently reported [20], [15], [2]. We concentrate here on relative comparisons and robust qualitative results reproducible over many runs, on several machines, in relation with experimental observations.

### 3.1. $V_{DC}$ spatial structure near antenna limiters.

The RF-induced PSI does not affect the SOL uniformly, but rather concerns a limited number of flux tubes. 2D (radial/poloidal) maps of the DC plasma potential $V_{DC}$ were simulated on the external face of antenna side limiters (reference plane on Figure 3) for active TS antennas [2], [3], AUG 2-strap and 3-strap antennas [15], [14]. All these maps exhibit similar spatial structures. $V_{DC}$ in the free SOL is peaked radially at the leading edge of the limiters. A 2-hump poloidal variation appears, with local maxima near the poloidal extremities of the antenna box, and local minima near the equatorial plane. Such spatial structure was observed experimentally on TS [34] and AUG [35] for ion parallel energies. A 2-hump poloidal structure was also present for heat loads along TS antenna limiters [36] and W sputtering yield on AUG [37]. A 2-hump structure was also visualized on the LAPD testbed [38].

The JET ILA consists of two strap arrays on top of each other that can be powered independently. When the upper ILA runs in dipole, numerical $V_{DC}$ peaks are present near the poloidal extremities of the upper ILA [31]. The peaks move to the poloidal extremities of the lower array when it is powered, while smaller peaks exist at the poloidal ends of the whole box when the full array radiates. Correlation of the poloidal structure with BeII line intensity on a limiter magnetically connected to the ILA is shown in [39] over a change of feeding in CD phasing.

The robustness of the simulated $V_{DC}$ poloidal structure, despite different RF designs, feeding schemes and Faraday screens, points to a possible rôle of the antenna box. Correlation of the poloidal structures of $V_{DC}$ and of $E_{//}$ on the antenna frame in the input field map was noticed in [14].

For TS, 2D (radial/poloidal) maps of $V_{DC}$ were also produced using the SSWICH-FW asymptotic code [3]. In order to evidence a possible role of the Fast Wave, the code was run using either $E_{//}$ only (dominant SW) or $E_{poloidal}$ only (dominant Fast Wave), or their combination in the vectorial input field map from RAPLICASOL. The experimental poloidal distribution of $V_{DC}$ was best reproduced with SW only. Even in the antenna vicinity the presence of the Fast Wave changed $V_{RF}$ and $V_{DC}$ magnitudes and poloidal shapes: contrary to experiments, small-scale poloidal modulations appeared in the simulations, correlated with the poloidal spacing of FS bars. It is suspected that the multi-2D approach over-estimates these poloidal modulations: it under-estimates the Fast Wave radial evanescence for large poloidal wavenumbers.

### 3.2. Left-right unbalance of 2-strap antennas

The RF-induced PSI observed experimentally is weaker with two balanced straps whose toroidal phasing is

* Corresponding author: laurent.colas@cea.fr

dipole ($\Delta\varphi=\pi$) than monopole ($\Delta\varphi=0$), (see *e.g.* [40]). Yet it is never totally suppressed, despite the anti-symmetry of the radiated field map. In a series of TS experiments, the left-right ratio of strap voltage amplitudes was varied [36]. Over this scan, the antenna side limiter near the strap with higher voltage heated up, while the limiter at the opposite toroidal side cooled down. A similar unbalance on AUG produced opposite variations of RF currents measured at the surface of toroidally opposite antenna limiters [40]. In this experiment with $\Delta\varphi=\pi$, to minimize the collected RF current, the RF voltage imposed on the remote strap was nearly twice the RF voltage on the strap nearer to the limiter. The "optimal" voltage ratios were approximately inverse at the two toroidal sides. These trends can hardly be explained using a single physical parameter simultaneously relevant at both extremities of the same field line. They rather suggest that the toroidal distance between radiating elements and the observed walls might play a role in the RF-sheath excitation.

Within the SSWICH model these observations can be interpreted as follows. For fixed sheath widths the RF part of SSWICH is linear. For multi-port antenna excitation, it is then possible to represent the oscillating sheath voltage $V_{RF}(\mathbf{r})$ in every point $\mathbf{r}$ of the wall as a linear combination of port contributions ($G_1(\mathbf{r})$, $G_2(\mathbf{r})$, …), weighted by the complex port voltages ($V_1$, $V_2$, …),

$$V_{RF}(\mathbf{r})=V_1G_1(\mathbf{r})+V_2G_2(\mathbf{r})+\ldots \tag{13}$$

For a given wall location $\mathbf{r}$, a combination of 2 port voltages always exists allowing local $V_{RF}(\mathbf{r})$ cancellation

$$V_2/V_1(\mathbf{r})=-G_1(\mathbf{r})/G_2(\mathbf{r}) \tag{14}$$

If $G_1(\mathbf{r})$ and $G_2(\mathbf{r})$ are real positive, the optimal setting is for dipole phasing (minus sign). Although a local $V_{RF}$ cancellation is always possible the optimal voltage ratio in (14) is location-dependant. Considering two toroidally opposite points $\mathbf{r}=(x,z)$ and $\mathbf{r}=(x,-z)$ on a toroidally symmetric antenna, one expects for symmetry reasons that $G_1(x,-z)=G_2(x,z)$. Relation (14) then implies that the optimal voltage ratios at the two opposite sides are inverse of each other, consistent with AUG measurements. Over left-right power unbalance, $V_{RF}(x,z)$ and $V_{RF}(x,-z)$ can vary in opposite ways, even if the two points lie on the same open magnetic field line [33]. Formula (13) also provides a general form for a scan of $\Delta\varphi$:

$$\begin{aligned}|V_{RF}(\mathbf{r})|^2 &= V_1^2|G_1(\mathbf{r})|^2+V_2^2|G_2(\mathbf{r})|^2+\\ &\ldots 2V_1V_2\,\mathrm{Re}\left[G_1(\mathbf{r})G_2^*(\mathbf{r})\exp(i\Delta\varphi)\right]\end{aligned} \tag{15}$$

The detailed expressions of port contributions $G_i(\mathbf{r})$ depend on antenna geometry and plasma parameters. Another perspective to interpret the experiments in relation with antenna geometry, is to consider the SSWICH-SW code excited by a prescribed $E_{//}$ field map, still assuming fixed sheath widths [33]. The linearity of the RF model then implies that $V_{RF}(\mathbf{r})$ can be re-expressed formally as a linear combination of individual contributions by every emitting point in the field map.

$$V_{RF}(\mathbf{r})=\int_{aperture}E_{//}(\mathbf{r_0})G(\mathbf{r},\mathbf{r_0})d\mathbf{r_0} \tag{16}$$

Expression (16) is a weighted integral of $E_{//}$ and replaces the line integral $\tilde{V}=E_{//}.dl$ often used to assess RF sheaths. Symmetry arguments imply $G(x,-z,z_0)=G(x,z,-z_0)$ for toroidally opposite limiters. SW evanescence makes individual weights $G(\mathbf{r},\mathbf{r_0})$ all the larger as the wave emission point $\mathbf{r_0}$ is located closer to the observation point $\mathbf{r}$, both in the parallel and transverse directions. For realistic geometries and target SOL plasmas, poloidal decay occurs over a few centimetres. Typical parallel decay lengths for $G(\mathbf{r},\mathbf{r_0})$ are found to be smaller than antenna parallel extension. $V_{RF}$ at antenna side limiters are therefore mainly sensitive to $E_{//}$ emission by metallic elements near these limiters, as suggested by experimental observations. Parallel proximity effects could also explain why sheath oscillations persist with dipole phasing, despite the parallel anti-symmetry of the radiated field map. They could finally justify reducing the RF fields induced near antenna boxes to attenuate sheath oscillations in their vicinity [40].

Paper [3] extended the previous study using full-wave calculations. In the case of Tore Supra, it showed that spatial proximity effects persist in the antenna near field in presence of both evanescent SW and Fast Waves.

### 3.3. Electrical tuning of AUG 3-strap antenna

Although the linearity makes $V_{RF}$ easier to study, deleterious RF-induced PSI arises from the DC plasma biasing. Consider one open magnetic field line of length $L_{//}$, exchanging DC currents with its neighbours. The sheaths at its two extremities oscillate with amplitudes $V_{RFl}$ and $V_{RFr}$. Assume that $V_{DC}$ is homogeneous along the field lines. Combining eq. (2), (3), and (11) yields

$$\begin{aligned}V_{DC} &= V_f+\ldots\\ &\frac{kT_e}{e}\log\left[\frac{I_0(e|V_{RFl}|/kT_e)+I_0(e|V_{RFr}|/kT_e)}{2}\right]\\ &\ldots+\frac{kT_e}{e}\log\left[1+\frac{L_{//}\,\mathrm{div}_\perp\,\mathbf{j}_{\perp\mathbf{DC}}}{j^+}\right]\end{aligned} \tag{17}$$

Where from (12) $\mathbf{j}_{\perp\mathbf{DC}}=-\sigma_{\perp DC}\nabla_\perp V_{DC}$ in the case of Ohm's law. From relation (17) one deduces that $V_{DC}$ mitigation on the considered open field line imposes cancelling $V_{RF}$ simultaneously at its two extremities, as well as on the neighbouring field lines. Otherwise DC current flows from the high-$|V_{RF}|$ boundary to the low-$|V_{RF}|$ one, without significant attenuation of $V_{DC}$.

Symmetry arguments in the previous subsection imply that the $V_{DC}$ reduction can only be partial with toroidally symmetric 2-strap antennas, over a scan of the ratio $V_2/V_1$. This was confirmed numerically on AUG using SSWICH-SW asymptotic and RAPLICASOL [14]. At fixed coupled power, the minimal $V_{DC}$ was obtained for nearly balanced antenna phased $\Delta\varphi=\pi$.

Similar symmetry considerations show that a simultaneous reduction of $V_{RFl}$ and $V_{RFr}$ is possible with the 3-strap AUG antenna [41] or with more straps [42]. Over a scan of $V_2/V_1$, the AUG experiment found an optimal electrical setting for which the measured tungsten (W) sputtering yield is reduced simultaneously on both toroidal sides of the AUG 3-strap antenna. This reduction was correlated with a local minimum in the RF currents collected on the frame. W production was mitigated with $\Delta\varphi=\pi$ phasing and a power ratio of the order of 2. This experiment was simulated using the SSWICH-SW asymptotic code [14], [15]. Over the parametric scan, a local minimum of the simulated $V_{DC}$ was obtained for electrical settings close to the experimental optimum. The minimal $V_{DC}$ for the 3-strap antenna was significantly smaller than the one for the 2-strap antenna at the same power on the same plasma. On the RAPLICASOL input field map, the optimal setting corresponded to low values of the local $E_{//}$ near the antenna frame, consistent with expected spatial proximity effects. Yet, complete sheath cancellation could not be achieved simultaneously everywhere over the 3-strap antenna structure: a residual RF-induced DC bias still persisted even with the optimal RF setting.

### 3.4. DC plasma biasing far from the antennas

RF-induced DC modifications of the SOL are frequently observed very far away from the active ICRF antennas, on field lines magnetically connected near the antenna limiters [43], [34], [44]. These large parallel distances (*e.g.* 12m in [34]) contrast with the much smaller evanescence lengths for the SW invoked above to interpret the power unbalance experiments.

To explain this apparent paradox, the SSWICH-FW model proposes two possible mechanisms able to produce DC plasma biasing far from the active antenna. In the first one, propagative Fast Waves induce RF fields at remote boundaries. These RF far fields are subsequently rectified by "far-field" sheaths [9], [29]. An alternative mechanism was proposed in [2]: it relies on a rectification of the near RF fields in the antenna private SOL by "near-field" sheaths, and a subsequent spread of the local DC bias to remote areas by DC current transport. Using the SSWICH-FW asymptotic code, the two processes were compared numerically in [3].

In SSWICH-SW simulations of TS antennas, intense sheath oscillations were present only in the immediate vicinity of the antenna mouth [2]. In SSWICH-FW simulations using the same wall geometry, $V_{RF}$ values much larger than $kT_e/e$ could also be obtained on remote sheath boundaries at the LFS of the vacuum vessel, that are accessible only to the propagative Fast Wave [3]. These simulated $V_{RF}$ values were still lower than the ones obtained in the private SOL.

Over changes of the wall geometry, it was observed that the amplitude of the simulated $V_{RF}$ decreased with the parallel distance from the antenna to the observed remote sheath boundary. In particular efficient generation of $V_{RF}$ by propagative Fast Waves seems unlikely 12m away from the active launcher. The simulated $V_{RF}$ did also increase with larger radial distance from the wall to the antenna. In comparison with the spatial proximity effects in the antenna near field, this result is counter-intuitive and calls for a re-assessment of the "spatial proximity effects" in regions where the RF waves are propagative. One possible interpretation is that over this geometrical change, the observed sheath wall moved closer to the propagation lobe of the emitted Fast Wave.

The role of DC current transport in remote DC plasma biasing was investigated by scanning the plasma transverse DC conductivity $\sigma_{\perp DC}$ in SSWICH-FW simulations [3]. Even in presence of propagating Fast Waves, this DC plasma conductivity was found necessary to produce a $V_{DC}$ spatial distribution peaked radially at the leading edge of the side limiters, as measured experimentally [34], [43]. The amplitude and radial width of the peak was sensitive to the badly constrained value of $\sigma_{\perp DC}$. A parametric scaling was proposed in [2]. The simulated $V_{DC}$ peak did propagate very far away along the magnetic field lines connected to the side limiter.

The present DC current transport model in SSWICH, despite its caveats, is yet necessary to interpret the experimental observations. TOKAM3X simulations also predict efficient parallel spread of the DC bias. The existence of DC current transport, mainly parallel to $\mathbf{B_0}$, is also attested by several experimental observations. Biased electrodes (*e.g.* Langmuir probes) create DC current circuits *via* the plasma. DC current flows were measured experimentally from active ICRF antennas to passive objects [45], [37]. On JET, the regions of RF-induced PSI far from the antenna did move poloidally over a scan of the field line pitch angle *via* the safety factor $q_{95}$ [44], [39].

### 3.5. RF-induced local density modifications

Spatially inhomogeneous SOL density modifications were measured in several devices during ICRH, *e.g.* in [46], [47], [48]. Local RF-induced density $\mathbf{E_{DC}}\times\mathbf{B_0}$ convection is suspected in the intense DC field $\mathbf{E_{DC}}=-\nabla_\perp V_{DC}$. The density changes in turn alter the RF wave propagation and subsequent sheath rectification. To model this process self-consistently for AUG antennas, the SSWICH-SW asymptotic code was iterated with RAPLICASOL and EMC3-EIRENE 3D SOL code for density transport [18], [19]. In EMC3, $V_{DC}$ was assumed homogeneous along the open field lines. The obtained self-consistent density patterns were compared with local measurements from reflectometers embedded in the AUG 3-strap antenna, showing reasonable agreement.

## 4 Conclusions and outlook.

Within the reported project, important upgrades were realized in the SSWICH code, coupling RF wave propagation and DC SOL biasing *via* sheath boundary conditions (SBCs). Full-wave RF wave propagation was implemented in 2D (radial/toroidal) geometry, together with PMLs and magnetized sheaths at shaped walls, for tokamaks and the RF-sheath testbed ALINE.

Even after these upgrades, present « antenna-scale » models remain relatively simple. The geometry is simpler than a real tokamak vessel and focuses on the antenna vicinity. Most tokamak simulations accounted for the Slow Wave (SW) only. The multi-2D approach followed, although well-suited for the SW, appears questionable to reproduce the Fast Wave evanescence for small-scale poloidal modulations in the input RF field maps. The implemented SBCs are simpler than more detailed fluid or PIC models of RF sheaths. Most tokamak simulations used asymptotic RF-SBCs that are even simpler. DC current transport relies on a phenomenological Ohm's law that cannot account for the turbulent DC current conduction likely at play. In the prospect of quantitative prediction, further improvement is needed from more basic research into RF-sheath physics, SOL turbulence and applied mathematics.

In its simplest form the SSWICH code gets now widely used in Europe. Despite its relative simplicity, it was already able to reproduce qualitative experimental observations: the poloidal distributions of RF-induced near-field PSI over many antenna configurations; the radial position of $V_{DC}$ peaks; left-right asymmetries upon strap power unbalance; electrical tuning of various antennas; spread of DC bias over long parallel distances and DC current flows. Earlier models, *e.g.* the oscillating double probe analogy with line integral $\tilde{V}=E_{//}.dl$, failed to interpret some of these observations. From these studies, some key physical ingredients have emerged as essential to reproduce the measurement results: accurate input RF field maps from 3D antenna codes; spatial proximity effects in the near field due to SW and possibly fast wave evanescence; parallel and transverse DC current transport.

Detailed comparisons with local measurements, especially 2D mappings, were essential for assessment. Of particular interest for future investigation would be local measurements in the antenna private SOL, where the rectification is predicted most intense. Measurements might be easier in RF-sheath testbeds, motivating more extensive simulation of these devices.

Presently SSWICH-SW can be used to assess parametric dependences and compare antenna designs. Spatial proximity effects provide guidelines for future antenna optimization. Yet, aside from the numerous approximations made, sensitivity of the results to loosely constrained parameters prevents good quantitative prediction. The amplitude and radial width of $V_{DC}$ peaks were found sensitive to uncertainties in the measured profiles, private SOL poorly accessible to diagnostics [20], RF-induced local SOL modifications [15], and transverse DC conductivity $\sigma_{\perp DC}$ [2]. Predicting "far field sheaths" is probably even more critical.

**Acknowledgements.** This work has been carried out within the framework of the EUROfusion Consortium and has received funding from the European research and training programme under grant agreement N° 633053. The views and opinions expressed herein do not necessarily reflect those of the European Commission.

## References

1. L. Colas et al., *Phys. Plasmas* **19** 092505 (2012)
2. J. Jacquot et al. *Phys. Plasmas* **21**, 061509 (2014)
3. L. Lu, et al., "Non-linear RF wave-sheath interaction in magnetized plasma edge: the role of the fast wave", ECA Vol. 40A, P2.068, http://ocs.ciemat.es/EPS2016PAP/pdf/P2.068.pdf, to be submitted to *PPCF*
4. D.A. D'Ippolito & J.R. Myra, *Physics of Plasmas* **13** 102508 (2006)
5. H. Kohno et al. *Computer Physics Communications* **183** (2012) p. 2116
6. J. R. Myra and D. A. D'Ippolito, *Phys. Plasmas* **22**, 062507 (2015)
7. J.R. Myra, *Physics of Plasmas* **24**, 072507 (2017)
8. D. Van Eester and K. Crombé, *Plasma Physics and Controlled Fusion* **55** (2013) 055001
9. D. A. D'Ippolito et al., *Physics of Plasmas* **15**, 102501 (2008)
10. D. Van Eester and K. Crombé, *Phys. Plasmas* **22**, 122505 (2015)
11. T.G. Jenkins and D.N. Smithe, *Plasma Sources Sci. Technol.* **24** (2015) 015020
12. T.H. Stix, "Waves in plasmas", 1992, AIP Press
13. J. Jacquot et al., *AIP Conference Proceedings* **1689**, 050008-1 050008-4 (2015)
14. J. Jacquot et al., this conference poster B-39
15. Tierens W. et al, *Nucl. Fusion* **57** (2017) 116034
16. W. Tierens, Poster A-25, this conference
17. J. Jacquot et al., *Plasma Phys. Control. Fusion* **55** (2013) 115004
18. Zhang. W et al., *Nucl. Fusion* **57** (2017) 116048
19. W. Zhang et al., Inv-07, this conference
20. L. Lu et al., *PPCF* **58** (2016) 055001
21. K.G. Budden, "Radio Waves in the Ionosphere", Cambridge University Press 1961
22. M. Campos-Pinto and B. Després, "Constructive formulations of resonant Maxwell's equations", to appear in *Siam Journal Math. Anal.* 2017, https://hal.archives-ouvertes.fr/hal-01278860/document.
23. B. Després et al., poster C-32, this conference
24. Crombé, K. et al., *AIP Conference Proceedings*, **1689**, 030006 (2015)
25. E. Faudot et al., *Review of Scientific Instruments*, **86** 063502 (2015).
26. E. Faudot et al. poster A-33, this conference
27. P. Tamain, et al., *Nucl. Materials and Energy* (2017), http://dx.doi.org/10.1016/j.nme.2016.12.022
28. T.G. Jenkins et al., poster B-30, this conf.
29. H. Kohno et al., *Phys. Plasmas* **22**, 072504 (2015), ibid. **23** (2016) 089901.

30. L. Colas et al., Proc. 21st IAEA FEC conference, St Petersburg (Russia) 2014, TH/P6-9
31. A. Křivská et al., poster C-44 , this conference
32. D. Milanesio et al., *Nuclear Fusion* **49** (2009) 115019
33. L. Colas et al., *Plasma Phys. Control. Fusion* **59** (2017) 025014
34. M. Kubič et al., *Journal of Nuclear Materials* **438** (2013) S509–S512
35. L. Colas et al. *AIP conf. proc.* **1580** p.259 (2014)
36. L. Colas et al., *JNM* **438** 2013 S330–3
37. V.V. Bobkov et al. *Nuclear Fusion* **50** (2010) 035004
38. B. VanCompernolle et al., poster C-26, this conf.
39. C.C. Klepper et al., poster A-12, this conference
40. V. Bobkov et al., *AIP Conf. Proc.* **1689**, 030004-1 030004-8 (2015)
41. V. Bobkov et al., *Plasma Phys. Control. Fusion* **59** (2017) 014022
42. S.J. Wukitch et al., Inv-08 this conference
43. Cziegler I. et al., *Plasma Physics and Controlled Fusion* **54** (2012) 105019
44. C.C. Klepper et al. *Journal of Nuclear Materials* **438** (2013) S594
45. R. Van Nieuwenhove and G. Van Oost, *PPCF* **34** (4), pp.525–532 (1992)
46. Bécoulet M. *et al.*, *Physics of Plasmas* **9** (2002) pp. 2619-32
47. Lau C. et al., *PPCF* **55** (2013) 095003
48. Colas L. *et al.*, (2015) *JNM* **463** pp.735-8

EPJ Web of Conferences **157**, 01002 (2017) DOI: 10.1051/epjconf/201715701002
*22 Topical Conference on Radio-Frequency Power in Plasmas*

# Progress and prospect of true steady state operation with RF

*Jean* Jacquinot[1,*]

[1]IRFM CEA Cadarache 13108 Saint Paul lez Durance, France

**Abstract.** Operation of fusion confinement experiments in full steady state is a major challenge for the development towards fusion energy. Critical to achieving this goal is the availability of actively cooled plasma facing components and auxiliary systems withstanding the very harsh plasma environment. Equally challenging are physics issues related to achieving plasma conditions and current drive efficiency required by reactor plasmas. RF heating and current drive systems have been key instruments for obtaining the progress made until today towards steady state. They hold all the records of long pulse plasma operation both in tokamaks and in stellarators. Nevertheless much progress remains to be made in particular for integrating all the requirements necessary for maintaining in steady state the density and plasma pressure conditions of a reactor. This is an important stated aim of ITER and of devices equipped with superconducting magnets. After considering the present state of the art, this review will address the key issues which remain to be solved both in physics and technology for reaching this goal. They constitute very active subjects of research which will require much dedicated experimentation in the new generation of superconducting devices which are now in operation or becoming close to it.

## 1 Introduction

In the past decades, emphasis of fusion research has been placed on demonstrating, albeit during short pulses, plasma parameters required, at least in dimensionless form, by future fusion reactors. In particular, the world tokamak community has adopted a "wind tunnel" approach which has led to dimensioning the ITER tokamak for achieving, with high confidence, a fusion gain Q = 10 lasting in excess of 300 seconds. This approach left the goal of running in true steady state or at least during very long pulses as an objective to be achieved later in a second phase. This steady state objective has now become a first priority and accordingly the new generation of national machines is conceived with super conducting magnets with the stated aim to be operating in steady state. These new devices are already operating (EAST, KSTAR, SST1, W7X and very recently WEST) or will operate in the near future (JT60-SA).

Several excellent review papers address the physics and technical challenges of steady state operation in reactor relevant conditions. An early paper draws the attention [1] that the baseline tokamak "Elmy H-mode" would require far too much external current drive power which would result in too much recirculating power in a reactor. A large fraction of self-generated bootstrap current is mandatory for a steady state tokamak reactor. A comprehensive physics basis is provided in a recent review [2]. It puts the emphasis on the profound coupling between MHD stability, bootstrap current, transport and edge plasma conditions and shows the narrow way forward towards a steady state tokamak. Chapter 6 of the Progress in the ITER physics basis [3] reviews the situation for ITER. It considers two target scenarios in addition to the baseline Elmy H-mode for ITER operating either in a hybrid mode (meaning with some remaining inductive drive) of 1000 s or in a non-inductive steady state mode lasting more than 3000 s.

In this paper we first consider recent integrated attempts to steady state achieved to date by the fusion community. We then review the progress made by the actuators based on waves in the electron cyclotron (EC), the ion cyclotron (IC) and the lower hybrid (LH) range of frequencies both in their technical realization and for their properties for driving plasma current. We include using combined operation of neutral beam (NB) and RF systems. Finally we outline the major issues which need to be resolved in line with the ITER programme.

## 2 Attempts to steady state operation

Experiments attempting to run fusion devices in steady state can fall in 2 categories. The first one is to optimize the pulse duration generally at the expense of the equivalent fusion performance. The second category does the opposite, trying to lengthen as much as possible a discharge with dimensionless parameters relevant to ITER or fusion reactors. The preferred figures of merit of the first category are the pulse duration and the total

* Corresponding author: jean.jacquinot@cea.fr

 DOI: 10.1051/epjconf/201715701002

energy extracted from the discharge. For the second category one can choose, for tokamaks, a fusion figure of merit, G, to be maintained for at least several current diffusion times. Here we use G defined as [2, 3]:

$$G=H_{89}\beta_N/(q_{95})^2 \quad (1)$$

With $$\beta_N = \beta/(I_p/aB_T);\ \beta = <P>/(B^2/2\mu_0) \quad (2)$$

A discharge running in a small or medium size tokamak with a high value of G would extrapolate to a high fusion gain in a tokamak scaled up to the size of ITER or DEMO and operating with the same dimensionless parameters except ρ*. For instance G = 0.3 for the Q = 5 steady state scenario foreseen in ITER.

## 2.1. The long duration approach

Figure 1 shows some experimental results of steady state operation of fusion devices in terms of energy extracted versus pulse length and figure 2 represents the same data in terms of G versus pulse duration.

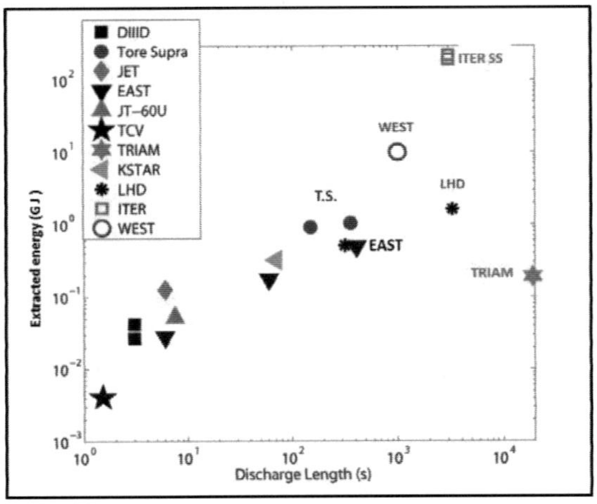


**Fig. 1.** Energy versus duration for steady state operation in recent fusion devices operating with zero loop voltage. The TCV point (black star) refers to a 100% bootstrap current discharge. Open symbols refer to targets set for WEST and the ITER steady state scenarios.

The largest extracted energy of 1.6 GJ has been reached in the stellarator LHD and durations in excess of 5 hours was obtained in the tokamak TRIAM. Tore-Supra holds the record of the highest energy extracted in a tokamak. These records have been achieved in devices equipped with super-conducting magnets potentially allowing of CW plasma operation but, interestingly, in these cases the duration was not limited by the capability of some components of the machine but by unexpected events such as a radiation collapse triggered by detached flakes (LHD), very low frequency oscillations (TRIAM) or MHD events (Tore-Supra).

The ability to operate with $V_{loop}$ strictly zero and without any core particle source using RF systems opens the possibility to reveal new physics. For example the neo-classical Ware particle pinch vanishes in these conditions. Nevertheless the $V_{loop}$ = 0 Tore Supra experiments [4] reveal moderately peaked density profiles corresponding to a turbulent inward pinch which would, if present in ITER, provide a significant gain in Q values on the order of 30%.

## 2.2 The fusion performance approach

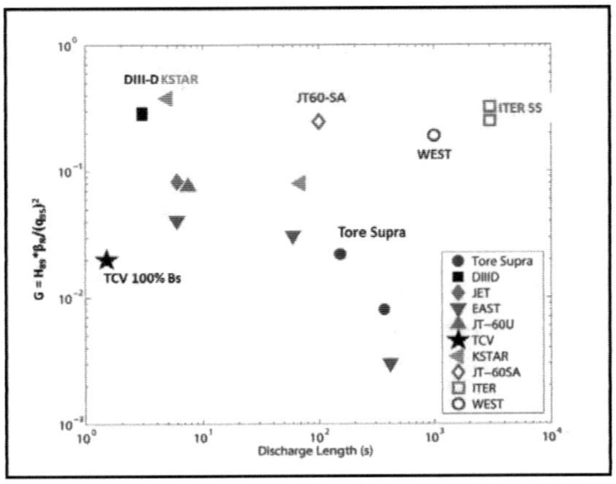


**Fig. 2.** The fusion figure of merit G versus duration for steady state operation in recent fusion devices operating with zero loop voltage. The TCV point (black star) refers to the 100% bootstrap current discharge. Open symbols refer to targets set for JT60-SA, WEST and the ITER steady state scenarios.

When the data of figure 1 is now plotted with respect to the fusion figure of merit G, the high performance steady state discharges (DIII-D, KSTAR) are now seen to match the G requirement for ITER but the duration is short by 3 orders of magnitude. Nevertheless, as discussed below, they stand as the most promising approach to steady state for tokamaks. Also quite remarkable but for a different reason, is the TCV point [5] which corresponds to a fully non inductive plasma despite avoiding the use of any external current drive. It shows that auto generated plasma currents (the bootstrap current) can reach 100% of the plasma current and maintain the configuration in steady state. Furthermore, it demonstrates that the bootstrap current profile can be exactly and stably aligned with the high gradient region it engenders; this auto regulation process is remarkable and will also be seen in the section 2.3 dealing with the hybrid discharges.

Let us now recall that the bootstrap current fraction $\alpha_{BS}$ is proportional to the poloidal pressure $\beta_p$ and can written as:

$$\alpha_{BS} \propto \beta_p \propto \beta_N B/I = C_{BS}\beta_N q_{95} \quad (3)$$

As already mentioned in the introduction $\alpha_{BS}$ should be large in a steady state tokamak reactor (say > 0.6) in order to avoid excessive external current drive power. Comparing equations 1 and 3 and wanting to realize large values of both G and $\alpha_{BS}$, we are faced with the challenge to create good confinement discharges at fairly large values of $q_{95}$ which is contradictory at first sight!

The 100% TCV discharge does realize good normalized confinement at very high $q_{95}$ thanks to an ITB (Internal Transport Barrier) resulting from a reverse magnetic shear configuration. However the $q_{95}$ is far too high for achieving a reactor relevant fusion performance. The next section shows how to remedy this at the expense of a lower $\alpha_{BS}$ which is still quite acceptable.

### 2.3 Realizing both high G and high $\alpha_{BS}$

Reactor/ITER relevant steady state plasmas require high values of both G and $\alpha_{BS}$. Modern tokamaks (Asdex-Upgrade, DIII-D, JET, JT60-U and KSTAR) have demonstrated this by realizing simultaneously high values of $H_{89}$ and $\beta_N$ while keeping $q_{95}$ rather large in order to keep at a low level the requirement for external current drive.

The limit of $\beta_N$ is set by ideal MHD stability. Neglecting the influence of the walls and with an ITER like configuration this limit is close to 3.5 (the 'no-wall limit'). With broad plasmas close enough to the walls and/or with feedback stabilization this can be extended to values approaching 5, the 'wall-limit'. $H_{89}$, the confinement H-mode factor, which reaches a value of 2 in the standard elmy H-mode thanks to a confinement barrier forming in the plasma periphery can be increased further by proper tailoring of the plasma current density profile.

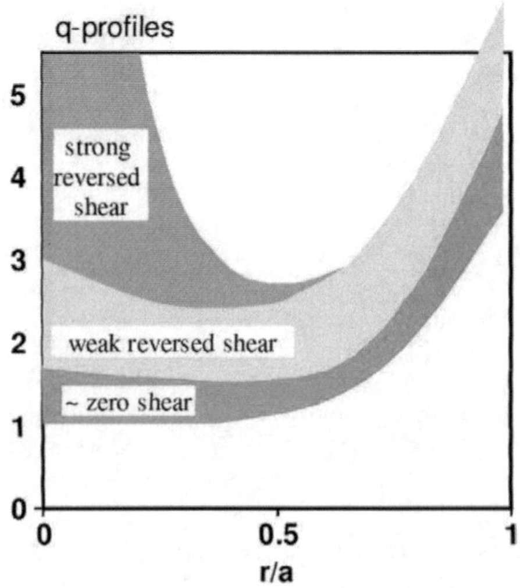


**Fig. 3.** Profiles of the safety factor q leading to improved confinement in the plasma core. Reproduced from [3]

An important consideration for improving the core confinement is to avoid the existence in the plasma of low rational numbers of q which can sustain major MHD instabilities such as 1/1 leading to sawteeth or 3/2 leading to NTM (neo classical tearing modes). When this is achieved, it is observed that reversed, weak or nearly zero central shear profiles generates high core confinement.

Highly reversed shear produces a very strong localized confinement barrier which in turn creates a significant bootstrap current. A spectacular example of such a regime is the JET D/T shot [6] where the ion temperature reached 40 keV and the fusion power 8 MW. However MHD stabilities with such a steep pressure gradient do not allow reaching a high value of $\beta_N$ which was limited to 1.9 in the JET case before going unstable. Therefore such a scenario does not seem particularly well suited to steady-state operation for fusion energy.

On the contrary stable plasmas with $\beta_N$ values reaching 3.5 were obtained in weak central shear regimes were the minimum value of q, $q_{min}$, is raised either slightly above 1 or even above 1.5.

#### *2.3.1 The 'hybrid': a scenario with weak central shear and $q_{min}$ slightly above 1*

The so-called hybrid scenario involves a weak central shear and $q_{min}$ slightly above 1. It has been pioneered by the Asdex-Upgrade team [7], reproduced with variants in most other tokamaks and developed on DIII-D [8] to be shown compatible with the planned Q=5 ITER steady state scenario.

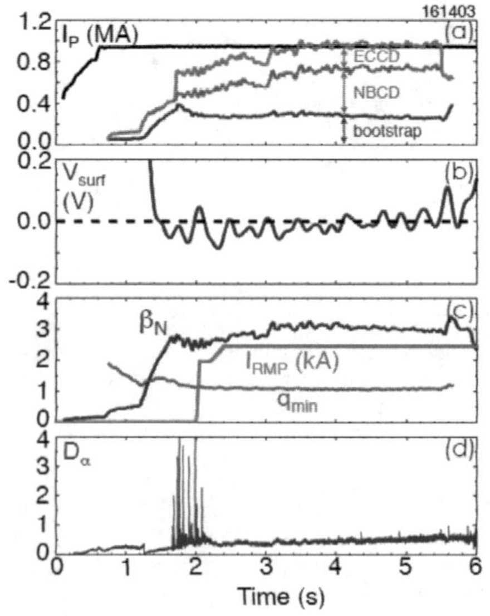


**Fig. 4.** Traces of a fully non-inductive hybrid scenario in DIII-D which also tested ELM mitigation with resonant magnetic perturbation coils (RMP). Reproduced from [8]

DOI: 10.1051/epjconf/201715701002

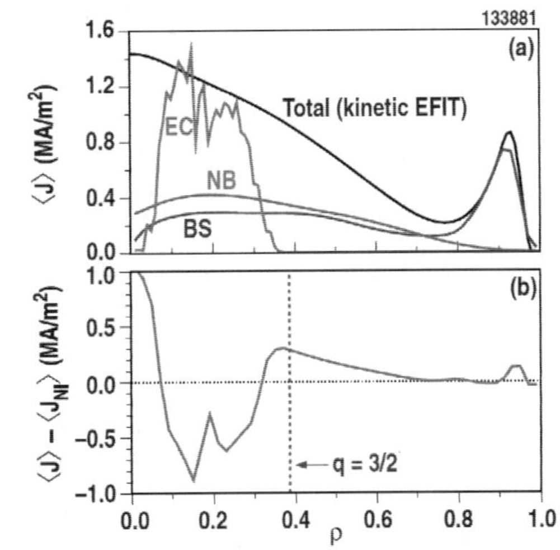


**Fig. 5.** Current drive sources in a hybrid discharge in DIII-D. ECCD overdrives the central part of the current but poloidal magnetic flux pumping manages to distribute the current across the profile in a stable way. Reproduced from [8].

Initially, the term 'hybrid' was given to suggest that inductive discharges could, thanks to a large bootstrap fraction, be lengthened significantly whilst not being fully steady state. It now turns out that, with proper optimisation, they could well become fully non-inductive as shown on figures 4 and 5 which give their peculiar characteristics. The heating and current drive systems are started before the ohmic current has fully penetrated in the plasma centre. Therefore the safety factor q is above 1 everywhere, sawteeth are absent and the current profile is broad which is favourable for MHD stability. This configuration becomes frozen by programming the H&CD systems for maintaining null the loop voltage (figure 4). Note that ECCD overdrives the central current density. Nevertheless $q_{min}$ keeps hovering slightly and stably above 1. A weak 3/2 NTM is detected. This mode has a minute effect on energy confinement but seems to be responsible for maintaining a stable current profile and energy confinement. The use of ECCD near the centre offers the benefit of good current drive efficiency and there is no need to align the current sources.

This mode is resilient to both ELM mitigation using RMP coils (figure 4) and to enhancing radiation in the divertor using argon injection [8] as would be required in ITER. It has also been demonstrated recently in Asdex-Upgrade equipped with tungsten plasma facing components (figure 6, [9]).

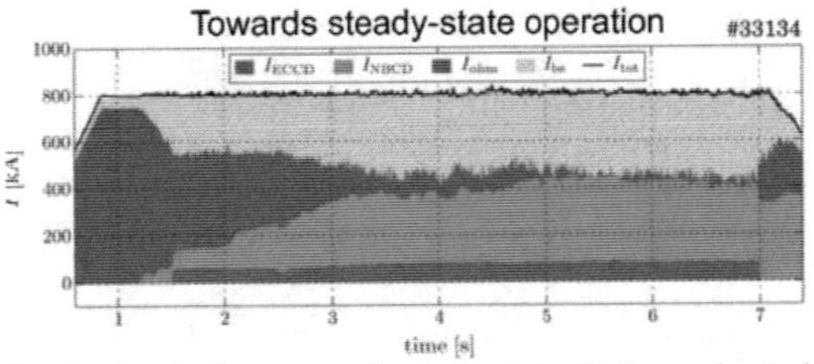


**Fig. 6.** A raised $q_{min}$ quasi steady state discharge in Asdex-Upgrade equipped with tungsten PFC. The contributions to the plasma current are in grey for ohmic, yellow from bootstrap red from the beams and blue from EC. Courtesy of S. Bock

### *2.3.2 The raised $q_{min}$ around 1.4*

It is also possible [10] to achieve similar results without having to rely on the somewhat miraculous action of the weak 3/2 mode which also takes its toll by reducing the global energy confinement time by some 10%. In this case, significant external off-axis current drive needs to be available and is programmed to maintain $q_{min}$ around 1.4 where optimum results are obtained. A broad pressure profile is obtained which is favourable to MHD stability thanks to the proximity of the walls. Therefore this mode offers more stability margin than the previous one but requires off-axis current drive power. ECCD and NBCD (neutral beam current drive) become notoriously inefficient the further you go off-axis. The use of LHCD (Lower Hybrid Current Drive) for the off-axis drive would be well adapted to this (section 3.3).

## 3 The RF actuators

If all the longest discharges have been obtained with RF systems only, the discharges with the highest values of G values have all been obtained with neutral beams as a work horse complemented by some RF power. In present day machines, the beams have the advantage of heating mainly the ions and of injecting torque and particles. However these 3 properties will not extrapolate to ITER conditions and even less to a reactor. One has to wonder if the nice results just reported will carry over in ITER when fusion born alpha particles will be the main heat source and with a much increased role of RF systems. It is therefore a priority to demonstrate equivalent scenarios using mainly RF systems. Let us consider them in turn.

### 3.1. ECRH and ECCD

The electron cyclotron resonance heating (ECRH) and current drive (ECCD) methods have impressive qualities. The waves can propagate freely from the launcher to the plasma core and the energy of the narrow microwave beams of millimetric wavelength can be precisely deposited where needed. The theory of plasma-wave interactions is well verified by experiments. ECRH and ECCD are used quite successfully for plasma

DOI: 10.1051/epjconf/201715701002

breakdown, highly localised heating and current drive and control of sawteeth and NTMs. No wonder, they are tools highly sought after in fusion research. However they involve advanced hardware (gyrotrons, diamond windows, long transmission lines) which can be delicate and costly.

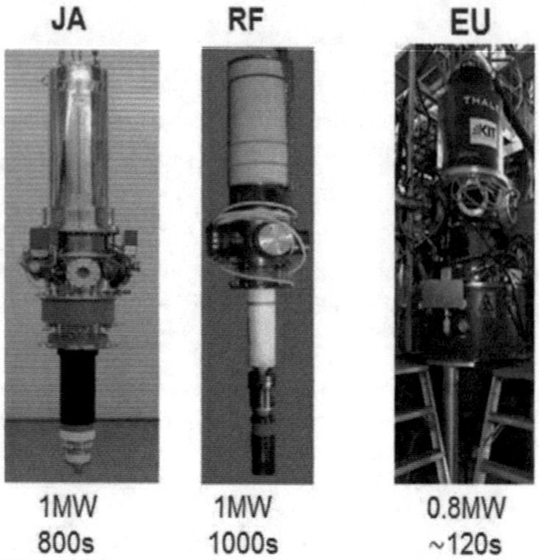


**Fig. 7.** The three 170 GHz gyrotrons developed for ITER by Russia, Japan and EU. The maximum power and pulse duration obtained during tests are indicated. Reproduced from [11].

Figure 7 shows the three gyrotrons [12, 13, and 14] which have been developed for ITER and the performance in duration and power reached so far. They are well on the way to reaching the ITER specifications (pulse length of 3600 s, 1MW at window with ≥95% $TEM_{00}$ mode purity, LHe free cryogenic magnets, >50% efficiency ($P_{out}/P_{in}$)). Some designs still have to face the challenges of mass production, high reliability, higher power (≥1.0MW), long life (≥5 years) and resilience to external magnetic fields (0.1mT < $|B_r|$ < 0.25mT). Research in the field of gyrotrons is very active and making progress towards higher power units and the capability of operating at several different frequencies with the same tube [13].

The W7-X stellarators is not yet ready for true steady state operation with RF but its 140 GHz 1000 s system is fully commissioned for operating in this way when required [15]. It includes actively cooled steerable mirrors and interlocked systems for protecting against stray microwave radiation which could damage in-vessel components such as cryo-pumps. It is also worth noting the development of a remote steering capability of the EC beams [16] which offers a simple and compact solution.

### 3.2. ICRH and ICCD

Despite obvious constraints of having to couple through an evanescent zone at the plasma edge, ICRH and ICCD (Ion cyclotron Resonance Heating and Ion Cyclotron Curent Drive)) are toolsLHCD of choice in modern tokamak research. Used with good success on the D/T programs of both JET and TFTR [17], the method extrapolates well to large D/T plasmas such as ITER where temperature and size guarantee single–pass damping in several scenarios and the broad frequency bandwidth of the entire system allows interacting with electrons or a chosen ion species.

This property can be used, not only for localised heating but also in steady state tokamak scenarios to drive current either close to the plasma axis with an equatorial launcher or around $\rho$ = 0.3 with a launcher located on the top of the machine (figure 8). Although modest, the current drive efficiency is the best one can do with any external central CD system for instance with NBCD at 1.5MeV [18] because the plasma electrons receive directly parallel energy from TTMP and Landau damping. We will see later that LHCD can also give an even more competitive CD efficiency but its use in ITER-size machines is restricted to a far off axis contribution.

The IC power sources use high power tetrodes or diacrodes [14]. The ITER sources are specified to be 1.5MW per module with a 35-65 MHz bandwidth and to withstand a VSWR of 2 for all phases. Such a module has already been tested successfully by the Indian domestic agency [19] and demonstrated in excess of 1.5 MW during 3600 s with an impressive RF power efficiency of 65% in a well matched load. The weaker part of the system is the antenna which has to operate at high voltage and has to maintain coupling on variable loads in particular when ELMs are present. Modern antenna designs are now capable of operating at 40 kV in CW and include an ELM resilient circuit (see for instance [20]) and maintaining coupling during ELMs is no longer an issue. However, the question of impurity generation from the edge RF fields via the sheath rectification mechanism [21, 22] still raises some issues. Impurity generation became negligible when dipole phasing was used and when the antenna components exposed to the plasma were made of a low Z material such as beryllium [23], boron or carbon. However, the recent generalisation of using tungsten as plasma facing components is reopening this issue. It has recently been addressed by the Asdex-Upgrade team [24] proposing and testing an elegant solution based on using a three strap antenna and adjusting the amplitude and phase of the strap currents so that the edge fields are minimized. The tungsten impurity release is indeed minimized in line with simulation of the sheath rectification mechanisms with the SSWICH-SW code [22]. It is worth noting that the ITER antenna design [11] is equivalent to having 4 straps distributed toroidally. They can be adjusted in current amplitude and phase in the same way minimizing the edge fields. It is highly prudent that such a mode of operation be prepared.

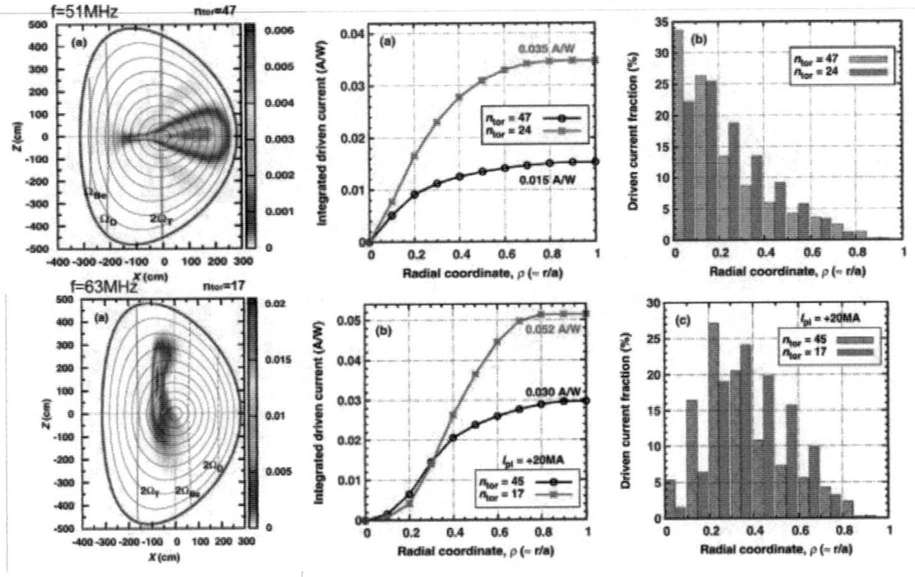

**Fig. 8.** ICCD for DEMO plasmas with 2 different locations of IC antennas. The 3 graphs on the top row corresponds to the equatorial launcher and compares the efficiency and the current profile of 2 toroidal mode numbers 47 and 24. The 3 graphs on the bottom row corresponds to the top launcher and compares the efficiency of the toroidal mode numbers 45 and 17 The upper launcher gives the best CD efficiency of 0.052 A/W. Reproduced from [18].

### 3.3. LHCD

Historically, the first demonstration of steady state operation of a tokamak without any inductive drive for more than 3 s was made with lower hybrid waves launched in PLT [25]. In small or medium size devices LHCD has the clear advantage over all the other methods that the waves can damp most their energy to super-thermal electrons which, because of their reduced drag, provides a very effective way to drive current. LHCD antennas are made of multiple waveguides (figure 9) each being phased to launch, with high directivity, traveling waves with a parallel index of refraction chosen to be high enough for providing accessibility ($N_{\|} > N_{\|a} \approx \omega_{pe}/\Omega_{ce} \propto \sqrt{n_e}/B$) but small enough to couple to fast electrons carrying the current. In small or medium size devices, typically $N_{//}$ equals 1.4 to 3 and supra-thermal electrons are in the range of 30 to 120keV. The waves can travel a long way in the plasma before being damped and during this process undergo an up-shift of the index of refraction (figure 10). A quasi-linear plateau in the electron distribution function is created leading to the formation of the super-thermal population. Detailed comparisons between theory, simulation and experiments [27] show good agreement provided the most accurate description of the wave ray-tracing propagation and the relativistic effects in the quasi-linear interaction are taken into account. It also seems necessary [27] to assume that a tail in the $N_{//}$ spectrum is already formed at, or very close to, the antenna (figure 10). The physics origin of this tail formation near or at the antenna remains to be clarified.

**Fig. 9.** Two types of LHCD launchers used in Tore-Supra and installed in WEST: Left: FAM, all wave guides are powered and phased using a multijunction. Right: PAM, waveguide are alternately active and passive to ease active cooling. These modules are rated for a total of 7 MW 1000s operation [28, 29].

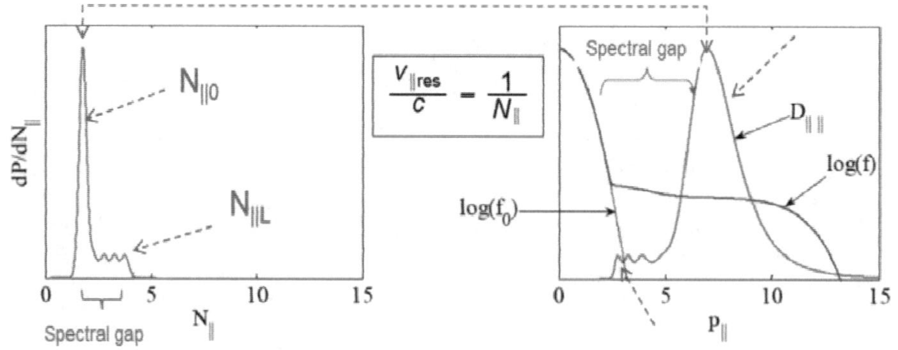


**Fig. 10.** Illustration of the $N_{||}$ upshift, occurring during propagation in the plasma or near the antenna. It closes the spectral gap so that the waves launched at $N_{//0}$ can couple to fast electrons. Left: The launched wave spectrum $N_{//0}$ brodenned to $N_{//L}$. Right and in blue: The electron distribution functions before application of the RF (log($f_0$) and after (log (f). Reproduced from [26].

In large, dense and hot plasmas such as ITER, $N_{||a}$, the minimum refraction index of waves which can access to the plasma, increases and $N_{||L}$ ($6.5/\sqrt{T_e}$ [keV]), the refraction index coupling to thermal electrons, becomes close to $N_{//0}$ the index launched by the antenna. We are now in a different situation of single pass damping driving current far off-axis without creation of very fast electrons; best results from simulations of the current drive efficiency are obtained [30] with an index close to $N_{//} = 1.9$.

Figure 9 illustrates actively cooled LHCD launchers capable of launching in Tore Supra 3 to 4 MW per port for 1000 s. The LH waves do not propagate in vacuum and have to tunnel through a cut-off layer. Nevertheless, it is possible to couple at about 10 cm from the plasma using gas puffing in the vicinity of the launcher. Klystrons and sometimes gyrotrons provide the power sources in frequencies ranging from 2.45 GHz to 8 GHz. For instance, Tore-Supra/WEST can use 16 klystrons at 3.7 GHz capable of delivering 9.2 MW for 1000 s. These klystrons are not, at present, using a depressed collector and their efficiency is only 37.5%. Recent developments have led to availability of 4.6 GHz systems used on Alcator C-mod and EAST and 5 GHz systems intended for KSTAR and ITER [31, 32].

## 4 Steady state operation of ITER

The ITER design basis foresees 3 scenarios (table 1). The inductive operation scenario (table 1) aims at the highest fusion power and power gain (500 MW, Q = 10) but the burn time will be limited to 300 – 500 s.

* Corresponding author: jean.jacquinot@cea.fr

DOI: 10.1051/epjconf/201715701002

**Table 1.** The ITER design basis scenarios.

| Parameter | Inductive Operation | Hybrid Operation | Steady-State Operation |
|---|---|---|---|
| Plasma Current, $I_p$ (MA) | 15 | 13.8 | 9 |
| Safety Factor, $q_{95}$ | 3.0 | 3.3 | 5.3 |
| Confinement Time, $\tau_E$ (s) | 3.4 | 2.7 | 3.1 |
| Fusion Power, $P_{fus}$ (MW) | 500 | 400 | 360 |
| Power Multiplication, Q | 10 | 5.4 | 6 |
| Burn time (s) | 300 – 500 | 1000 | 3000 |

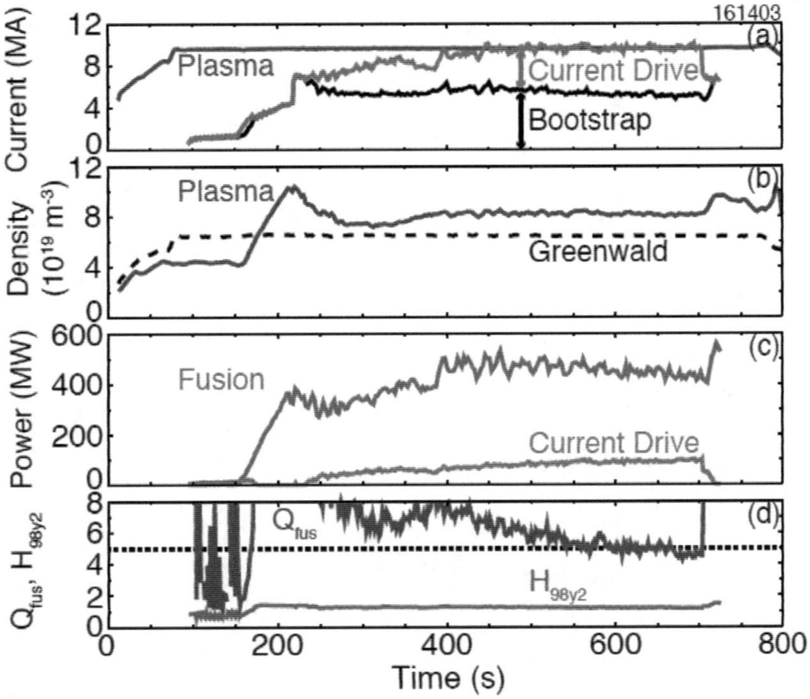


**Fig. 11.** Scaling of the DIII-D discharge (figure 4) to ITER's R and magnetic field with fixed dimensionless parameters except ρ*. Note that the current relaxation time $\tau_R$ extrapolates from 1.9 s to 245 s in ITER. Reproduced from [8].

The hybrid scenario plans is based on the hybrid mode of operation described in section 2.2.1. Long burn times of 1000 s are expected from a discharge sustained by a high bootstrap current supplemented by the external current supplied by the ITER H&CD systems not far from the axis and by some remaining inductive current. Figure 11 shows that the high G discharges discussed in figure 4 would indeed extrapolate well to the ITER conditions with the same dimensionless parameters except ρ*. The extrapolation matches the gain Q and the fusion power and because of the very long current diffusion time, the discharge duration is nearly adequate. However, it would require 85 MW of H&CD power which is larger than foreseen in the ITER in its initial phase of operation. We should also stress, as already done in section 2, that no experiment has so far managed a discharge with both a large value of G and a duration requiring active cooling of all the in-vessel components. This is a challenging target for future research.

* Corresponding author: jean.jacquinot@cea.fr

 DOI: 10.1051/epjconf/201715701002

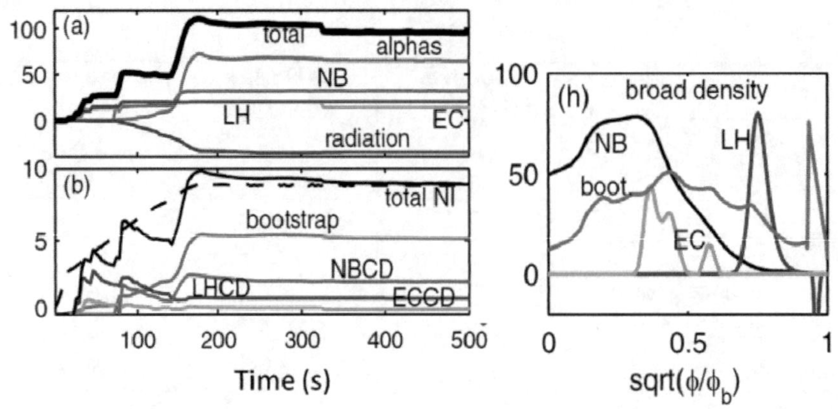


**Fig. 12.** Simulation of the full steady state operation in ITER using 33 MW of NB, 20 MW of both EC and LH. Left: power and plasma current traces during the setting-up phase (first 500s). Right: profiles of the currents generated by bootstrap, NB, EC and LH. It is found that the 20 MW of LH is necessary to broaden the profile ensuring both high $\beta_N$ stability and high confinement.

The third scenario, steady state operation, reduces further the plasma current and the fusion power to ensure that full non inductive operation is possible. However the target gain (Q = 6) remains significant which requires an excellent normalized confinement at high $\beta_N$. We have seen in section 2.2.2 that this can be achieved with $q_{min}$ raised at about 1.4 which requires off-axis current drive in order to broaden the current profile. Recently, the current diffusive ballooning mode (CDBM) model has been used to simulate steady state operation of ITER [33]. This model simulates, in particular, the formation of ITBs resulting from a vanishing local shear. Figure 12 gives the results of a simulation for ITER. It is shown that the combination of EC and LH waves is an effective combination for this mode of operation. ECCD at mid radius prevents the bootstrap current from peaking and LHCD ensures that the current profile remains broad. These very interesting results should however be seen from the perspective that CDBM has not yet been benchmarked against the international ITB data base and that there are some doubts that the GENRAY code used in this simulation accounts correctly to the LH physics at high density.

## 5 Conclusions

Operation in true steady state has now become a major research area in fusion research. Although no experiments have yet been able to combine simultaneously ITER-long pulses and high performance, it is apparent from this review that considerable progress has been made on the many aspects, sometimes conflictual, which have to be integrated for reaching the goal.

New modern devices equipped with CW components (super-conducting magnets, active cooling of all in vessel components, RF power sources and controls) are now operational or coming into operation. New steady state records are around the corner. In what area would they matter most?

We have noted the gap between the high G discharges dominated by NB but only capable so far of relatively short pulses and the very long pulse discharges dominated by RF with low G values. Bridging this gap should receive a high priority; it would demonstrate the successful integration of the core/edge plasma constraints with the plasma/wall interactions operating in complete active cooling conditions. Doing it with RF only, with low torque and no direct ion heating, would be particularly ITER and reactor relevant.

A major recent advance is without doubt the experimental demonstration by nearly all large tokamaks of stable high performance discharges with $q_{min}$ raised above 1 and with low central shear. They provide high performance and a high bootstrap current fraction. They extrapolate well to the ITER secondary goal of extended burn and provide a path towards a reactor running in steady state with a tolerable recirculating power devoted to external current drive. The documentation of this scenario by an international data base is of great value.

Of major interest are also the advances in the development of specific RF systems. LHCD is the most efficient current driver. Modelling and simulation continue to document the question of the spectral gap. Experimental clarification of the launched spectrum at the antenna would comfort the prediction of the role of LHCD for the ITER steady state scenarios. We also note that EC is a high precision tool required in most applications. It is ready to demonstrate CW operation in advanced stellarators and the reliability of its various components in delivering continuously high power. Finally IC has again demonstrated with ITER sources its long pulse advantages. It has also made major progress in ELM resilient coupling and in controlling impurity generation using current/phase shaping of the elements of a multi strap (>2) antenna.

* Corresponding author: jean.jacquinot@cea.fr

The author would like to acknowledge gratefully the information and advices provided by many colleagues especially from: B. Beaumont, V. Bobkov, G. Denisov, A. Ekedahl, R. Ikeda, E. Joffrin, M. Goniche, Ph. Lamalle, T. Luce, A. Mukherjee, J.-M. Noterdaeme, Y. Peysson, D. Van Ester, W. Zhang.

## References

1. J. Jacquinot, V.P. Bhatnagar, C. Gormezano and the JET Team, Plasma Phys. Control. Fusion **35** A35 (1993)
2. T.C. Luce, Physics of Plasmas **18**, 030501 (2011); doi: 10.1063/1.3551571
3. C. Gormezano et al. Nucl. Fusion **47** S285–S336 (2007)
4. G.T. Hoang, Phys. Rev. Lett. **90** 1555507 (2003)
5. S. Coda, O. Sauter, M. A. Henderson, and T. P. Goodman, *Proceedings of the IAEA Fusion Energy Conference* (2008)
6. C. Gormezano et al. Phys. Rev. Lett. **80**, 5544 (1998)
7. A. C. C. Sips, Plasma Phys. Control. Fusion 44 B69–B83 (2002)
8. C. C. Petty et al. Nucl. Fusion **56** (2016) and *Proceedings of the IAEA Fusion Energy Conference* (2016)
9. S. Bock, submitted to Nucl. Fusion
10. C.T. Holtcomb, Nucl. Fusion **54** 093009 (2014)
11. B. Beaumont, ITER Business Forum, March (2017)
12. G.G. Denisov and A.G. Litvak FIP/1-6Ra, *Proceedings of the IAEA Fusion Energy Conference* (2016)
13. R. Ikeda et al. FIP/1-6Rb *Proceedings of the IAEA Fusion Energy Conference* (2016); see also this conference (2017)
14. S. Bethuys, this conference (2017)
15. H.-S. Bosch and H. Laqua, private communication; T. Stange, *this conference* (2017)
16. B. Plaum et al. J. Phys. Conference Series **25** 84-91(2005)
17. J. Jacquinot and the JET team, Plasma Phys. Control. Fusion **41** A13 (1999)
18. Y. Kazakov, D. Van Ester et al. Plasma Phys. Control. Fusion **57** 025014(2014)
19. A. Mukherjee et al, *Proceedings of the IAEA Fusion Energy Conference* (2016)
20. C. Bourdelle et al. Nucl. Fusion **55** 063017 (2015)
21. D. D'Ippolito, J. Myra, M. Bures and J. Jacquinot, Plasma Phys. and Control. Fusion **33** 607 (1991).
22. L. Colas et al. Phys. Plasmas **19** 052505 (2012); L. Colas *review paper at this conference* (2017)
23. M. Bures, J. Jacquinot, M. Stamp, D. Summers, D. Start, T. Wade, D.A. D'Ippolito and J.R. Myra, Nucl. Fusion **32** 1139 (1992)
24. V. Bobkov et al, Plasma Phys. Control. Fusion **59** 014022 (2017)
25. S. Bernabei *et al.*, Phys. Rev. Lett. **49**, 1255 (1982)
26. Y. Peysson et al. 42$^{nd}$ EPS conf. Lisbon (2015)
27. Y. Peysson et al. Plasma Phys. Control. Fusion **58** 044008 (2016); see also Y. Peysson et al. *invited paper at this conference* (2017)
28. A. Ekedahl et al., Nucl. Fusion **50** 112002 (2010) and A. Ekedahl et al. *Proceedings of the IAEA Fusion Energy Conference* (2016)
29. M. Goniche et al. Physics of Plasmas **21** 061515 (2014)
30. J. Decker et al.,Nucl. Fusion, 073025 (2011)
31. J. Hillairet et al. Fusion Engineering and Design **94** 22-30 (2015)
32. G.S. Lee et al. Nucl. Fusion **48** 087001 (2008)
33. F.M. Poli et al Nucl. Fusion **54** 073007 (2014)

EPJ Web of Conferences **157**, 02001 (2017) DOI: 10.1051/epjconf/201715702001
*22 Topical Conference on Radio-Frequency Power in Plasmas*

# Development and Application of 3.7GHz LHCD system on HL-2A and Development of RF Heating system on HL-2M

*XingYu* Bai[1,*], *Bo* Lu[1], *Hao* Zeng[1] , *Jun* Liang[1], *Chao* Wang[1], *GuoLiang* Xiao[1,3], *He* Wang[1], *YaLi* Chen[1], *JieQiong* Wang[1], *Kun* Feng[1], *ShaoDong* Song[1], *Mei* Huang[1], *Jun* Rao[1], *XianMing* Song[1], *Rui* Mao[1], *Jun* Cheng[1], *YiPo* Zhang[1], *DeLiang* Yu[1], *Yan* Zhou[1], *ZhongBing* Shi[1], *Annika* Ekedahl[2], *Julien* Hillairet[2], *Emmanuel* Bertrand[2], *Lena* Delpech[2], *Gerardo* Giruzzi[2], *Tuong* Hoang[2], *Roland* Magne[2], *Didier* Mazon[2], *Yves* Peysson[2], *XiaoLan* Zou[2] and the HL-2A Team[1]

[1]Southwestern Institute of Physics (SWIP), P.O. Box 432, Chengdu, China.

[2]CEA, IRFM, F-13108 Saint-Paul-lez-Durance, France

[3]Department of Engineering Physics, Tsinghua University, Beijing, China.

**Abstract.** The first Lower Hybrid (LH) experiments were carried out with a Passive-Active Multijunction (PAM) launcher in H-mode plasmas. The experiments were performed on the HL-2A tokamak with the new 3.7 GHz LHCD system, installed and tested by SWIP in collaboration with CEA/IRFM. The ELMs and local gas impact on LH power coupling was studied in the experiments. The coupled LH power in HL-2A was 200-500kW at large gap at the first experiments and reaches 900 kW now in H-mode, while it reaches 1MW in L-mode. The LH experiments on HL-2A show that the PAM launcher is a viable concept for high performance scenarios. The LH power can be coupled at large plasma-launcher gap, and assist in triggering and sustaining H-modes. Finally, an overview of the RF heating systems for the tokamak HL-2M is given. HL-2M will dispose of a 4 MW LH system and a 8 MW ECRH system, both of which are currently under installation at SWIP.

## 1 Introduction

Lower hybrid current drive (LHCD) is a well known method of non-inductive current drive in tokamaks, which has been used successfully in tokamaks around the world since the early 1980s. A fraction of the plasma current is replaced by the non-inductively LH driven current in the tokamak by injecting LHW into one toroidal direction, which can increase the duration of the plasma discharge[1].

In view of foreseeing a LH system for the second phase of ITER [2], a relevant LH launcher design, the passive-active multijunction (PAM), has been developed [3]. The first experimental test of a PAM module was carried out in the FTU tokamak [4, 5]. Following this, a PAM launcher was designed and constructed for Tore Supra [6], demonstrating long pulse operation at high coupled power (2.7 MW, 80 s) and ITER-relevant power density [7]. So far, the PAM experiments in FTU and Tore Supra have shown excellent results of power handling and coupling at large plasma-launcher distances, making it an attractive launcher design for operation in high performance scenarios and in long pulses.

However, this has so far only been demonstrated in L-mode plasmas. Now, as a part of the 3.7 GHz LHCD system constructed on HL-2A tokamak, the third PAM launcher is developed and applied, making it possible to carry out LH experiments in H-mode plasmas by PAM antenna.

## 2 LHCD system description

The PAM launcher was designed by SWIP (Southwestern Institute of Physics) in Chengdu, in cooperation with IRFM (Institut de Recherche sur la Fusion par confinement Magnetique) at CEA. The layout of the LHCD system is depicted in Fig. 1 [8].The four klystrons are located on the ground floor and the launcher in the torus hall on the first floor. Eight transmission lines (~ 20 m each) connect the klystrons to the launcher.

* Corresponding author: baixy@swip.ac.cn

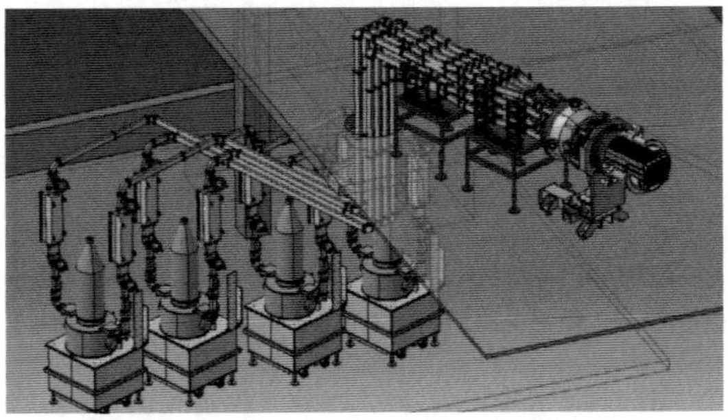

**Fig. 1.** Schematic layout of the 3.7 GHz LHCD system.

### 2.1. PAM launcher

The PAM launcher consists of 16 modules, with four active and four passive grills each, mounted in four rows and four columns. One passive grill is added at the end of each toroidal row, resulting in four rows with 16 active and 17 passive grills (see photo in Fig. 2). The size parameters of the PAM mouth are width = 365 mm; height = 325 mm.

**Fig. 2.** Photo of the PAM launcher in HL-2A

A dedicated gas puffing system with three poloidal injection points is installed next to the launcher mouth, on the left of the launcher, signed by the arrows in Fig 2. Local gas puffing near LH launchers has been used routinely, for example in JET [9] and EAST [10], as a method to increase the local density and improve the LH coupling, in particular during H-mode operation. A set of Langmuir probes, consisting of a total of eight probes mounted at two radial locations, is installed on the right hand side of the launcher mouth. The probes allow measuring the electron density at two radial locations and will thus also allow deducing the density and temperature decay lengths at the launcher location.

The ALOHA coupling code [11] was used for defining the design parameters for the PAM launcher. The peak parallel refractive index was chosen to $n_{//0}$ = 2.75, in order to allow accessibility of the LH wave in HL-2A plasmas. The width and height of each grill are 8.8 mm and 72.14 mm, respectively, with a wall thickness of 2.2 mm. The depth of the passive grill is 24.5 mm, i.e. a quarter of a wavelength. According to the thermal and electromagnetic force analysis, the launcher is made of copper to avoid local overheating in arcing and covered by stainless steel in order to protect the antenna from the twisting in plasma disruption. As the pulse duration is less than 2.0 s, no active cooling is installed on the launcher. The launcher can be moved in the radial direction as much as 100 mm in order to have a good coupling in different discharge conditions. Graphite protection tiles are installed around the front of the launcher, protruding by 2 mm in front of the launcher mouth. ALOHA-code analysis shows that the power reflection coefficient, averaged over the 16 modules, is less than 3% when the electron density at the launcher mouth is larger than $2\times10^{17}$ $m^{-3}$, the cut-off density for the LH wave being $n_{co} = 1.7\times10^{17}$ $m^{-3}$ at f = 3.7 GHz.

### 2.2. Wave source

The PAM launcher is fed by four klystrons, giving out 500 kW RF power maximum via two ports each. The klystrons are powered by a high-voltage power supply (HVPS) and several auxiliary power supplies including a filament heater power supply, three electromagnet power supplies, two ion pump power supplies and an anode power supply modulator for each klystron, as suggested. But the klystrons are operated in totally a new way for the TH2013A type. A set of divider resistance is chosen to obtain the anode voltage from the cathode HVPS, replacing the anode high voltage modulator, making the HVPS system much simpler and cheaper. In this way, 500 kW/2s output power was successfully obtained from every klystron in the separately test, indicating the new way effective.

The maximum output power of the RF exciter is 10 W. Besides the four channels for power amplification, there is an additional output channel with a power level of 0 dBm as a phase reference for the phase feedback control system, by which the $n_{//}$ spectrum can range between 2.55 and 2.95, with low reflection coefficient. The protection system can switch off the HVPS in 10μs if arcing occurs or backward power exceeds the threshold, keeping the system safe.

### 2.3. Transmission lines

The output RF power from each klystron is fed via two ports, and then divided equally to four channels by two 3-dB power dividers. Each of the four columns of the launcher is fed by one klystron via the transmission lines, with a transmission efficiency more than 80%. Since each klystron has two output arms, there are eight transmission lines in total. Eight 4-port ferrite circulators are specified near the klystrions' output windows. The circulator is rated to handle the full load reflection at full power over the full duration. Therefor, the experiment regime is extended with the help of the circulator. Each transmission line is about 20 meters length, based on the WR284 waveguide with a propagation mode of TE10. The transmission lines are pressurized with two bars of nitrogen in order to prevent arcing. The forward and backward power at the output windows near the klystron (8+8=16 channels) and the input ports of the antenna (16+16=32 channels) is measured by bidirectional couplers. As a result, the LH power mentioned in this paper refers to the pure power injected into the tokamak, reflected power subtracted.

## 3 LH coupling experiments

Prior to the experiments in H-mode plasmas, the klystrons and the PAM launcher had been commissioned in L-mode plasmas, both limiter and divertor configuration, up to 500 kW coupled power in 400 ms length pulses on HL-2A[12]. With the commissioning, the reflection coefficient is decreased much from 80% at the first experiments.

In the following experiments, coupling optimization in H-mode plasmas was carried out. Coupling effect of ELMs and local gas puffing at the antenna mouth was investigated. 200-500 kW LH power was successfully coupled during type I ELMs in H-mode plasmas triggered and sustained by Neutral Beam Injection (NBI) combined with LH waves, with a large gap (more than 10 cm) between LCFS and the antenna mouth. The coupled power reached 900kW in H-mode and 1MW in L-mode by the optimization. The experimental results indicate that the PAM launcher is a viable concept for high performance scenarios. The LH power can be coupled at large plasma-launcher gap, and assist in triggering and sustaining H-modes.

### 3.1. Coupling effect on ELMs

The LH coupling experiments with the PAM in H-mode were carried out in a lower single null (LSN) configuration with plasma current Ip = 160 kA, toroidal magnetic field Bt = 1.4 T and edge safety factor $q_{95} = 4$. In shot #26862, the distance from the last closed flux surface (LCFS) to the PAM launcher mouth was 11.5 cm at the mid-plane. Fig. 3 is the time evolution of the coupled LH power, the average RC, the Dα signal that shows the ELM behavior and the line integrated density. H-mode is triggered 20 ms after the LH power is applied, when the density reaches $2.1 \times 10^{19}$ m$^{-3}$. 700 kW of NBI power is injected between t = 500 ms and t = 1000 ms.

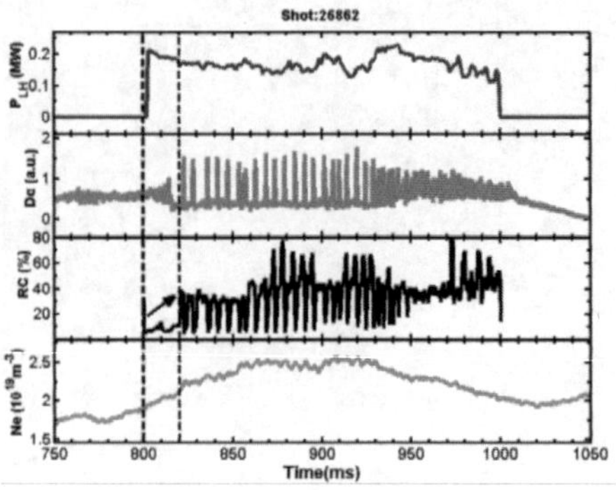


**Fig. 3.** LH coupling of shot #26862.

Fig. 3 also shows the evolution of the average reflection coefficient during the ELMs. As can be expected, RC is highest during the phase between ELMs, as the density in the scrape-off-layer (SOL) is lowest in this phase. During the ELM-bursts, the electron density increases and RC decreases significantly, as expected by ALOHA modeling. RC varies between 5% and 35% during an ELM exciting. In order to show the RC behavior during an ELM more clearly, RC is plotted against the Dα signal in Fig. 4a, while Fig. 4b shows RC on the electron density given by Langmuir probes compared with the calculation results using ALOHA code, in the same duration. Fig. 4b gives the computed RC for the PAM at different electron density decay lengths in front of the launcher mouth, namely 2 mm, 5 mm and 10 mm. The experimental RC has the same trend of density but a little higher than that of the calculation. The exact value of the density in front of the launcher, as measured by the Langmuir probes, can be quite difficult to determine. If one assumes that the density at the launcher mouth is somewhat smaller than the value given by the probe, since the launcher mouth is retracted behind its protection limiters, the experimental curve would agree better with the ALOHA code. It can be noted that in Fig. 4a, the RC behaves differently during the ELM-rise and after the ELM-crash. One plausible hypothesis is enhanced SOL ionization by the LH power during an ELM, as was investigated and modeled in JET [13]. During the ELM-crash, the SOL temperature increases and so the ionization by the LH power is enhanced. The ionization will increase the electron density locally in front of the powered grills, leading to a reduction in reflected power. However, this local increase in electron density may not necessarily be measured by the Langmuir probes, since they are located outside of the grill region.

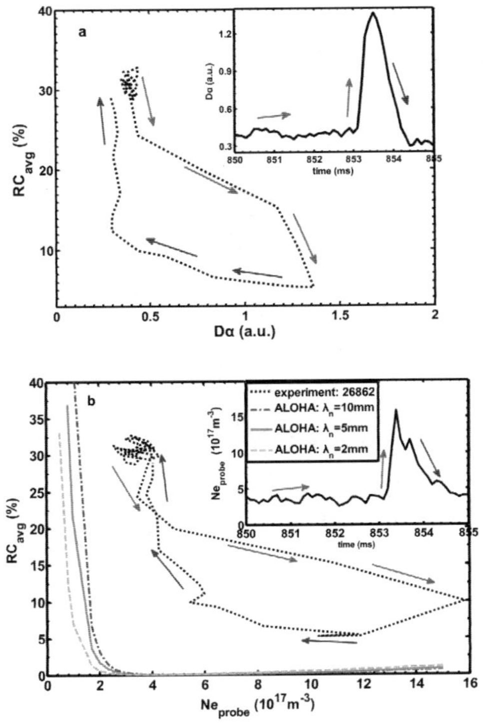


**Fig. 4.** RC versus the amplitude of the Dα signal during one ELM (a) and versus the electron density measured on Langmuir probes on the launcher compared with calculation of ALOHA code for three different density decay lengths (b).

### 3.2. Coupling effect on local gas puffing

The local gas injection was found to be crucial in this experiment for increasing the local electron density at the launcher, since the plasma-launcher gap was always very large (usually more than 10 cm). The gas dosing was controlled by opening the valve during short intervals. This timing of the gas puffs is indicated in Fig. 5. The coupled LH power, RC, the Dα signal showing the ELM behavior and the density measured by a Langmuir probe at the antenna mouth are shown in the figure. The timing of the gas puffs are indicated by dashed vertical lines.

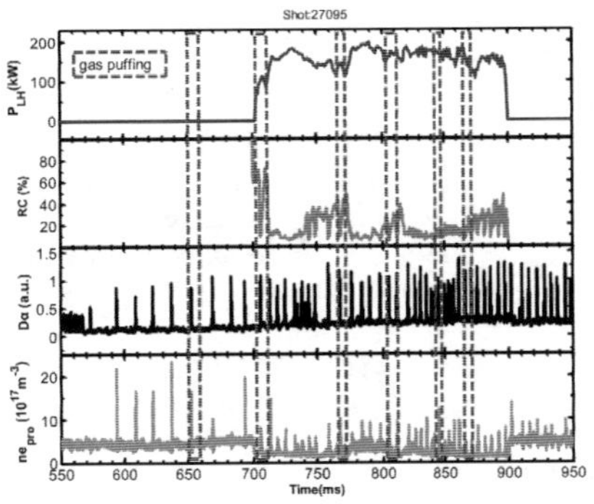


**Fig. 5.** LH coupling on gas injection, shot #27095

According to Fig. 5, an increase in ELM-frequency (at around t = 740 ms and 850 ms) can be observed. The reflection coefficient decreases typically 10 ms after the gas puff is switched on. In this discharge, the average RC decreases to 8% following the gas puffs.

### 3.3. Coupling optimization

In the first coupling experiments, 200-500kW LH power was coupled to H-mode plasmas via PAM antenna. Coupling optimization such as local gas puffing, plasma position control, antenna position set, and also plasma parameters changing were carried out in the experiments, decreasing RC to 10%.

Following the first experimental campaign in 2015, a number of modifications on the LH system and on the plasma control system were carried out in order to increase the coupled LH power. A piece of shorter flexible waveguide was changed in order to reduce transmission line losses and the RF measurements at the directional couplers were improved to get better reliability of forward and reflected power measurements in each module. On the plasma control system side, the divertor coil currents could be better controlled so as to maintain a smaller plasma-launcher gap during the H-mode phase. These modifications all had a significant effect on the LH power injection capability.

In the experiments carried out in June 2016 and May 2017, the plasma-launcher gap could be controlled at a smaller value during the H-mode phase, allowing lower fraction of reflected power to the LH launcher. The coupled LH power has thus reached 900 kW in H-mode and 1 MW in L-mode plasmas without any local gas injection [14].

One H-mode discharge with NBI + LH is shown in Fig. 6 (a). The H-mode is triggered when the LH power is applied, and the H-mode is then sustained throughout the duration of the LH power. Fig. 6 (b) shows that the coupled power reached 1MW/1s with really low average RC. The density was constant during the LH power injection. The RC was a little high for the first several ms (~30%). This is mainly because of the plasma radial drift impacted by the LH power injection. Then RC maintained a low level about 7%, and dropped to 4% after Electron Cyclotron Resonance Heating (ECRH) injected, indicating that the ECRH injection can benefit LH coupling. Local gas injection was not systematically used in these discharges, since the reflected power level was sufficiently low event at a smaller gap (~2.5 cm) without any local gas puffing.

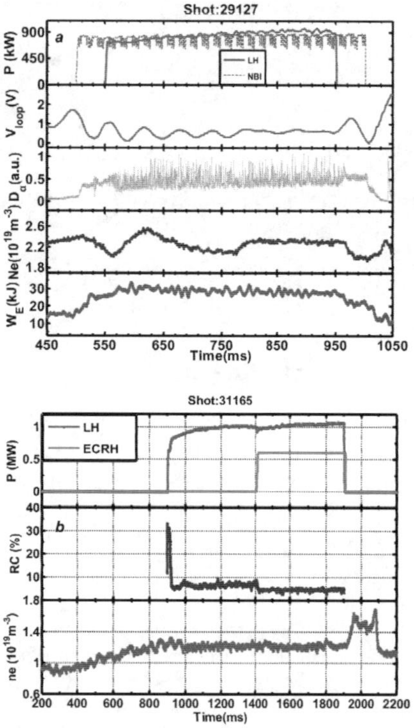


**Fig. 6.** 900kW LH power coupled in H-mode plasmas, shot #29127 (a) and 1MW in L-mode, shot # 31165 (b)

### 3.4. ELMs mitigation by LH power

The effect of LH power on the H-mode pedestal was investigated in these experiments, in particular the effect on the ELM frequency and ELM amplitude [15]. The results indicate an effect on the ELM behavior, i.e. increase in ELM frequency and decrease in ELM amplitude during LH power modulations [14], when the LH power is larger than ~ 300 kW. Fig. 7 is the time evolution of shot #27141, showing ELM modification during LH injection, including coupled LH power (a), local gas puffing signal (b), ELM behavior, as seen on ion saturation current (c) and plasma radiation (d), ELM frequency (e). The mitigation effect with LH waves was found to be sensitive to plasma parameters and there appears to be a minimum density, ne ~ $2.5\times10^{19}$ m$^{-3}$, above which the mitigation effect occurs. An increase of the pedestal turbulence measured by Doppler reflectometry could also be observed, which suggests that an enhancement of the particle transport due to pedestal turbulence could be the cause of the ELM mitigation [14]. Similar effect was previously observed during supersonic molecular beam injection (SMBI) in HL-2A [16]. It appears that LH power injection can produce a comparable increase of the ELM frequency, as was also observed in EAST [17].

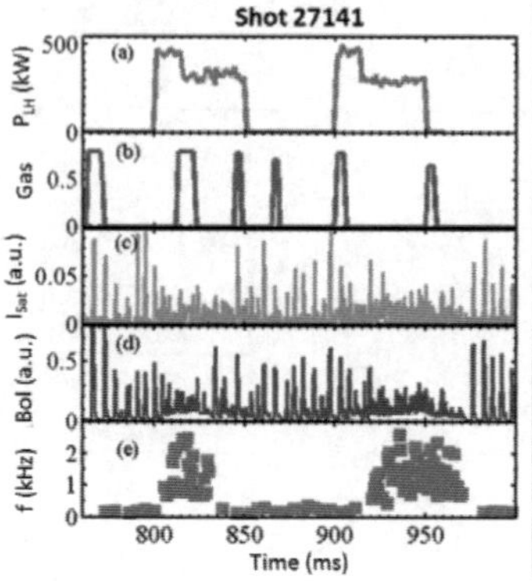


**Fig. 7.** ELMs mitigation by LH power injection, shot #27141.

## 4 RF heating system of HL-2M

A new tokamak device called HL-2M is being constructed in SWIP. The RF heating system including LHCD and ECRH system are developed. HL-2M will dispose of a 4 MW LHCD system and an 8 MW ECRH system, both of which are currently under installation.

### 4.1. LHCD system

The main parameters of the LHCD system on HL-2M are 3.7 GHz frequency, 4 MW power and 3 s duration. 2 MW wave source would be constructed for the first step with a 4 MW power capacity Full Active Multijunction (FAM) antenna. For the wave source, Four TH2103C1 klystrons are ready located in the heating hall of HL-2M. All of the auxiliary equipments are also ready for the klystron operation test, which will be started at early Jun 2017. There are 8 transmission lines in the system, about 30 m length each. Part of the transmission components such as power combiner, absorption load and the ceramic window have been tested since the high power test bed was built in 2015. The antenna is designed in FAM structure and the peak parallel refractive index $n_{//0}$ is 2.25. TE10 to TE30 mode converters are designed for the antenna to divide the power in three parts equally in ploidal direction. As a result, There are 192 active grills (6×32) facing to the plasma, with passive ones at each side.

**Fig. 8.** LHCD system layout of HL-2M.

Fig. 8 shows the layout of the LHCD system. The smaller klystrons in the figure have not been installed yet and the bigger ones are ready.

### 4.2. ECRH system

The ECRH system takes 105/140 GHz frequency, second harmonic injection to the plasma on HL-2M. the ECRH system includes eight gyrotrons (1 MW /3 s each), two of them can be operated at dual-frequency (140 GHz and 105 GHz), other six gyrotrons operated at single-frequency 105HGz. Fig. 9 shows the layout of the ECRH system. There are eight transmission lines, about 40 m length each. Corrugated circular waveguides (Ø = 63.5 mm ) are applied in the transmission lines to keep the wave $HE_{11}$ propagation mode. The transmission lines are pumped to $10^{-3}$ Pa level to avoid arcing. Three ports in the east of HL-2M, one equatorial port and two upper ports situated at the same vacuum segment, were allocated for ECRH launchers, corresponding to three different structure launchers, one 2×3 structure 6 MW equatorial launcher, one 1×2 structure 2 MW upper launcher and one 1×1 structure 1 MW upper launcher. All of them can realize real-time feedback control in poloidal direction. For example, the response time of the equatorial launcher, which has been manufactured and tested in 2016, could reach 45ms for full scan range. One can note that the total power is less than the power for launching. It is because the three antennas would not be operated at the same time. A switch will be installed in the transmission line, guiding the power to different line in different shot as wish. Up to now, six gyrotrons, four transmission lines and one antenna is ready, including the auxiliary equipments.

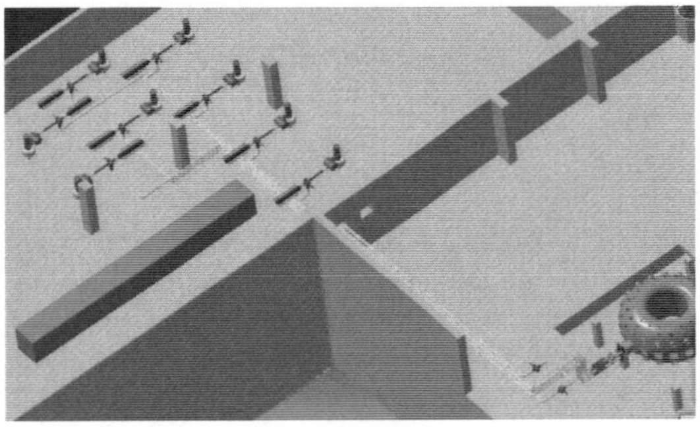

**Fig. 9.** ECRH system layout of HL-2M.

## 5 Summary and conclusions

A new 3.7 GHz LHCD system was developed and applied on HL-2A, in which a new passive-active multijunction (PAM) launcher was successfully used in H-mode experiments for the first time. The LH power was generated from 4 TH2103A klystrons, operated in a new way, and transmitted to the antenna by eight transmission lines, with a transmission efficiency more than 80%. In the applying experiments, coupling optimization in H-mode plasmas was carried out. Coupling effect of ELMs and local gas puffing at the antenna mouth was investigated. It was found that the RC was higher during the ELM-rise than that of after the ELM-crash in the experiments, with same trend but higher RC than that of calculation result via ALOHA code. The local gas puffing at the antenna mouth was found beneficial for the LH coupling in H-mode plasmas with large gap. The optimization was successful since the coupled LH power reached 900 kW/400 ms in H-mode and 1 MW/ 1 s in L-mode plasmas. The LH power was found triggering and maintaining H-mode when the plasma was heated by 800 kW of Neutral Beam Injection (NBI) power. It was also found that the LH power can mitigate ELMs in the H-mode plasmas. The experiments indicates that the PAM launcher is a viable concept for high performance scenarios, the LH power can be coupled at large plasma-launcher gap, and assist in triggering and sustaining H-modes. In addition, LHCD and ECRH system of HL-2M tokamak was briefly introduced. There would be a 4 MW LHCD system and an 8 MW ECRH system on HL-2M. Parts of the two systems have been installed and tested.

## Acknowledgments

The authors gratefully acknowledge the support of the HL-2A Team at SWIP, in particular the Control & Operation group, the Vacuum & Gas puffing group, the Diagnostic group and also the High Voltage Power

Supply group. This work was carried out within the framework of the Associated Laboratory in the field of fusion between CEA/IRFM and SWIP.

This work is also supported by the domestic ITER project of China (No. 2011GB104000, No. 2014GB102000), funded by Ministry of Science and Technology (MOST) and the Natural Science Foundation of China (NSFC) project (No. 11575054, No. 11261140327), funded by NSFC.

## References

1. P. Bonoli, Physics of Plasmas **21** (2014) 061508
2. G.T. Hoang et al, Nucl. Fusion **49** (2009) 075001
3. Ph. Bibet et al, Nucl. Fusion **35** (1995) 1213
4. F. Mirizzi et al, Fusion Eng. Des. **74** (2005) 237
5. V. Pericoli Ridolfini et al, Nucl. Fusion **45** (2005) 1085
6. D. Guilhem et al, Fusion Eng. Des. **86** (2011) 279
7. A. Ekedahl et al., Nucl. Fusion **50** (2010) 112002
8. Lu Bo et al., Journal of the Korean Physical Society **65** (2014) 1243
9. A. Ekedahl et al., Plasma Phys. Control. Fusion **51** (2009) 044001
10. F.K. Liu et al., Nucl. Fusion **55** (2015) 123022
11. J. Hillairet et al, Nucl. Fusion **50** (2010) 125010
12. X.Y. Bai et al., Proc. 42nd EPS Conference on Plasma Physics, Lisbon (2015), paper P5.137
13. V. Petrzilka et al., Plasma Phys. Control. Fusion **54** (2012) 074005
14. X.R. Duan et al., 26th IAEA Fusion Energy Conference (2016), paper OV/4-4
15. G.L. Xiao et al. This conference poster, A-24,
16. W.W. Xiao et al., Nucl. Fusion **52** (2012) 114027
17. Y. Liang et al., Phys. Rev. Lett. **110** (2013) 235002

DOI: 10.1051/epjconf/201715702002

# Status of the ITER Ion Cyclotron H&CD

*Bertrand* Beaumont[a,*], *Rohit* Agarwal[b], *Tania* Alonzo Montemayor[e], *Rohit* Anand[b], *P* Ajesh[b], *Francois* Calarco[a], *Hrushikesh* Dalicha[b], *Craig* Deibele[c], *Nicolas* Ferrigno[f], *Thibault* Gassmann[a], *Davide* Kleiner[d], *Richard* Hanks[c], *JVS* Hari[b], *Richard* Goulding[c], *Margaret* Graham[a], *Akhil* Jha[b], *Fabienne* Kazarian[a], *Philippe* Lamalle[a], *Harsha* Machchhar[b], *Mike* McCarthy[c], *Mark* Middendorf[c], *Kartik* Mohan[b], *Aparajita* Mukherjee[b], *Amit* Patel[b], *Manoj* Patel[b], *Hriday* Patel[b], *Kumar* Rajnish[b], *David* Rasmussen[c], *Dipal* Soni[b], *Manoj* Singh[b], *Narinder Pal* Singh[b], *Raghuraj* Singh[b], *Gajendra* Suthar[b], *Rajesh* Trivedi[b], *Roberto* Sanabria[c], *Roberta* Sartori[d], *Michael* Smith[c], *David* Swain[c], *Pareshkumar* Vasava[b], *Sriprakash* Verma[b], *Kurt* Vetter[c], *Paul* Wright[c]

[a] *ITER Organization, Route de Vinon sur Verdon, CS 90 046, 13067 St Paul Lez Durance, France*
[b] *ITER-India, Institute for Plasma Research, Bhat, Gandhinagar – 382428, India*
[c] *US ITER Project Office, ORNL, 055 Commerce Park, PO Box 2008, Oak Ridge, TN 37831, USA*
[d] *Fusion for Energy, C/ Josep Pla 2, Torres Diagonal Litoral-B3,E-08019 Barcelona – Spain*
[e] *Solution F, Allée du Verdon, 13770 Venelles, France*
[f] *Ametra, Z.I. Les Cabassols, 13770 Venelles, France*

**Abstract.** The ITER Ion Cyclotron Heating and Current Drive system (IC H&CD) is designed to deliver 20MW to a broad range of plasma scenarios between 40 and 55MHz, during very long pulses. It consists of two broadband equatorial port plug antennas, their pre-matching and matching systems, transmission lines, Radio Frequency (RF) Sources and High Voltage Power Supplies. The overall project schedule has been revised and agreed by ITER Council; it re-integrates the second antenna and its power supplies in construction baseline and sets the dates for progressive installation with DT phase planned in 2035. Recent progress on ICRF subsystems is reported, covering design evolution, qualification of test articles and specific R&D results in domestic agencies, suppliers, associated laboratories and IO.

## 1 Introduction

A revised overall project schedule has been presented at the ITER Council in November 2016. This schedule is defining the steps between now and the start of DT operation in 2035, with progressive installation of the basic machine and its equipment. The specific targets for the ICRF system, compatible with the development of the ITER Research Plan have been presented. The Council has also requested to evaluate how to anticipate installation and commissioning of part of the ICRF system.

Meanwhile, the R&D is progressing well for the RF sources and the transmission equipment. The antenna design has been adapted to the most recent evolution of the interfaces, which have been modified in the last 3 years. Specific R&D is in progress on key technologies that are required in the antenna design: joining technique between Titanium Alloy and stainless steel, copper coating method, RF contacts.

## 2 Schedule and Staged Approach

A realistic and detailed timetable up to the First Plasma, together with the associated costs, was submitted to the ITER Council in June 2016. At this $18^{th}$ session, the members of the Council have validated the elements of principle of the updated calendar submitted by ITER Organization. As a result of two years of efforts by ITER Organization and the seven domestic agencies, an integrated calendar sets the date of the first Plasma in December 2025. In November 2016, the ITER Council adopted the current complete updated calendar covering the period from now on, via the First Plasma, until the start of deuterium-tritium operations[1].

The calendar details a 4-stage approach as shown in fig. 1. First plasma is planned in December 2025 that will allow performing technical tests on the basic components of the machine: magnets, vacuum, gas, etc… At this stage, the torus will be essentially empty with very limited set of diagnostics and only part of the ECRH system (max 8 MW), mainly to assist plasma breakdown. It will be followed by an assembly and commissioning phase leading to an 18 months experimental session (Pre-Fusion-Power-Operation-1). At this stage, the machine will be equipped with inner components, and the ECRH system will be completed to its nominal power, 20 MW. During PFPO1, planned during 2029 and first part of 2030, experiments will take place in H and/or He plasmas, at various levels of toroidal field (third, half and nominal). From mid-2030

* Corresponding author: bertrand.beaumont@iter.org

DOI: 10.1051/epjconf/201715702002

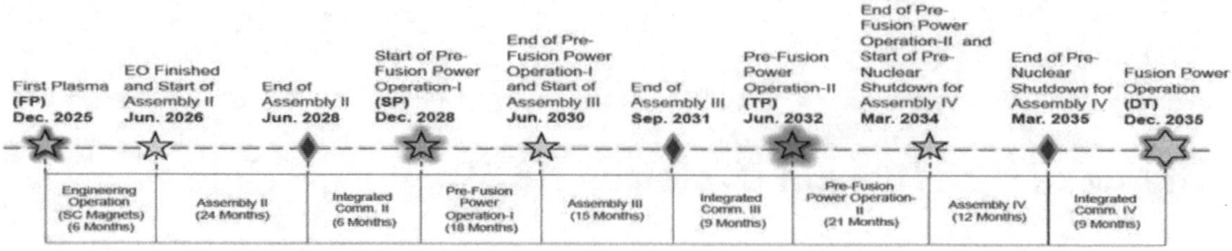


**Fig. 1.** Four Stages Approach from First Plasma to DT Operation

to mid-2032, the heating systems will be completed to provide the nominal heating power: 33 MW of NNBI and 20 MW of ICRH, together with the already existing 20 MW of ECRH power. Experiments will follow during PFPO2 with full capacity of heating systems until March 2034, aiming to prepare for the DT operation.

It is important to note that the Council approved that the second ICRF antenna be re-integrated in the construction baseline and budget, together with the second set of power supplies that were previously deferred to the operation budget. This is a very significant and positive step: the ICRF construction baseline is now restored to its full performance, based on all components of the 20 MW system initially planned in the initial baseline.

In addition, the November 2016 Council requested that the IO, Members and their Domestic Agencies (DAs) accelerate the Neutral Beam Injection (NBI) and Ion Cyclotron Resonance Heating (ICRH) systems, making use of existing expertise in the Members and their DAs and of progress that will be achieved in the coming 2-3 years, with a view to enabling the use of the ICRH system for the PFPO1 in the interests of shortening the period from First Plasma to DT Operation. This acceleration would allow to better use the PFPO1 experimental session, and perform an earlier commissioning of the ICRF system [2].

### 2.1 ICRF acceleration:

As a consequence, the IO and respective DAs are investigating the feasibility of accelerating the ICRH system with a view to providing 10 MW of ICRH power in the PFPO-1 experimental campaign.

The 4 stage approach has been developed with the inputs of the configuration workshop and the cost constraints of the DAs for the relevant procurements. Issues have been identified with advancing the delivery of some of the transmission line components, as currently the USDA cost constraints and schedule baseline are not compliant with an accelerated program. However the main critical path, to ensure 10 MW for PFPO-1, is the antenna which is not anticipated to be ready until 2030.

To accelerate 10 MW of ICRH power with one antenna, so that it is available for assembly phase II, would imply that the ICRH antenna would need to be delivered to the IO site by October 2027 thus allowing for six months of testing in the Port plug test facility (PPTF) and three months for the installation, i.e. this would mean that the antenna would need to be installed and ready for the machine integrated commissioning in July 2028. Therefore, three year acceleration on the current antenna schedule (current delivery foreseen in September 2030) would be required. The evaluation of an accelerated path is in progress and will be reported at the next IC-21 and include the possible solutions for a three year accelerated schedule. A final report will then be made available at the IC-22 in June 2018 and will include the full risk assessment and required additional budget and resources for the acceleration.

A late installation during a short shutdown during PFPO1 is also an alternative to be considered.

## 3. ICRF system progress

### 3.1 HVPS and RF Sources

ITER India is responsible for the procurement of the RF Sources and part of the corresponding power supplies to be installed in the RF building. ITER India has initiated a voluntary R&D program for establishing the technology prior to prototype and series production. This R&D is motivated by the demanding performance requirements set upon the ITER RF sources when compared to the state of the art. Only two potential suppliers are able to provide high power RF tubes that could be suitable for ITER: CPI in the USA and TED in Europe. Table 1 provides typical demonstrated capabilities against ITER needs.

**Table 1.** Comparison between ITER specifications and typical values from Industry

| ITER Specifications | Tetrode CPI | Diacrode TED |
|---|---|---|
| 2.5 MW; VSWR 2:1; 2000s ; 35-65 MHz; | ~1.9MW; VSWR 1:1 | 1 MW; VSWR 1:1 |
| 3.0 MW; VSWR 1.5:1; 2000s ; 40-55 MHz | 300s; 30-60 MHz | 1000s; 200 MHz |
| $\eta \sim 50\% - 60\%$ | > 60% | > 60% |
| BW +/- 1 MHz @ 1 dB | High Q cavity | High Q cavity |

One of the important specifications is the ability to deliver a constant output power even with VSWR up to 2:1 for any phase of the reflection coefficient, corresponding to 11% of reflected power. ITER will then be equipped with a set of robust RF sources.

The ITER RF source layout consists of two parallel three-stage amplifier chains, with a combiner circuit on the output side. This kind of RF source is unique in terms of its stringent specifications. The R&D program in India is aiming to select the supplier for the driver and end-stage of the three-stage amplifier chain. In this program, both technical solutions proposed by suppliers will be installed in a test facility in IPR: the suppliers have designed and fabricated driver and final stage amplifiers, and tests have been performed according to the possibilities of the test equipment in factories. In the ITER India test facility, these are powered by a solid state pre-driver (10 kW typ.) provided by ITER India, and deliver the RF power in a 3 MW dummy load (Fig. 2), that can be detuned for testing on high VSWR (fig. 3). Full power and long duration tests can be performed on various loading conditions [3].

**Fig. 2.** High Power 3 MW matched load at IPR

**Fig. 3.** Stub and phase shifter for high VSWR tests at IPR

For the Diacrode-based system, the entire system has been stabilized against parasitic oscillation before powering the amplifiers up to the 1.5MW level at various frequencies from 35 to 65 MHz for 2000s pulses [3]. A further test has been performed for 1.7MW/3600s at 36 MHz as extended operation (Fig. 4).

The tetrode based system is now also installed in the RF test facility and tests are performed at different power levels, frequencies, and VSWR in preparation of final site acceptance tests.

Both proposals experience instabilities on the band edges near 35 and 65 MHz, but these frequencies are well out of the ITER required range: 40 to 55 MHz.

The tests will conclude if one or both proposals meet the specifications required for ITER, and will determine the voltage requirement for the anode driver, which is fed by an intermediate tap point in the high voltage power supply.

The tests in ITER India are also used to validate the HVPS design and its integration, and interaction, with the RF source: a fast switch off and a low stored energy helps to keep the dissipated energy low to protect the vacuum tube in case of arcing.

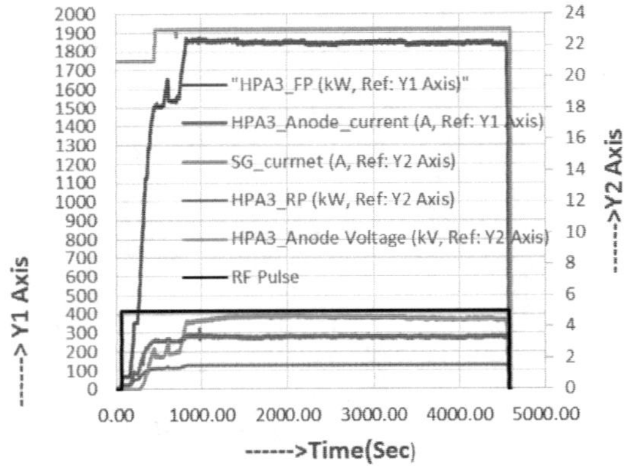


**Fig. 4.** 3600s high power shot with Diacrode based system; HPA3 = end stage.

The combiner prototype circuit is also being tested at various levels of power and duration.

Once the tests are completed and the final series supplier is selected, the preliminary design of the RF sources will be reviewed to IO, and the contract will be placed by ITER India for the procurement of the series of amplifiers, auxiliaries and high voltage power supplies.

### 3.2 Transmission lines and matching system

The transmission lines and matching system are procured by USIPO (the US Domestic Agency) and will be installed in the RF, Assembly, Tokamak and Hot Cell buildings. It is a large network with around 1100 m of 300 mm diameter rigid coaxial lines and approximately 300 90º elbows.

The layout in the RF building allows interconnecting the RF sources with the 8 main transmission lines, the 2 power dummy loads and the test line connected to the Port Plug Test Facility in the Hot Cell Facility. By means of 4 port coaxial switches (Fig. 5 & Fig. 6), the source #9 can replace any faulty source among #1 to #8, while that faulty source is connected to a power dummy load for maintenance.

The port cell equipment layout is still a matter of concerns: Remote handling have imposed large permanent structures in the cells that are not compatible with the initial proposal of a 3 stubs pre-matching system. A revised configuration based on 2 stubs for

DOI: 10.1051/epjconf/201715702002

each line is considered, with reduced frequency range [5].
USIPO is developing a large test program to validate technical choices as well as potential suppliers for each type of components. Standard coaxial support using 3 quartz rods has been already chosen [5], and the most recent tests focus on hybrid splitters and 4 port switches (Fig. 5 & Fig. 6).
Test rating is corresponding to levels up to 6 MW power handling (typically 37 kV with 42 kV transients in 50 Ω components, with pulse duration of one hour. These specifications cover the ICRF upgrade case: transmission equipment in the tokamak building is then compatible with the increase to 40 MW of the ICRH system, and thus do not require modifying the transmission system installed in the nuclear building.

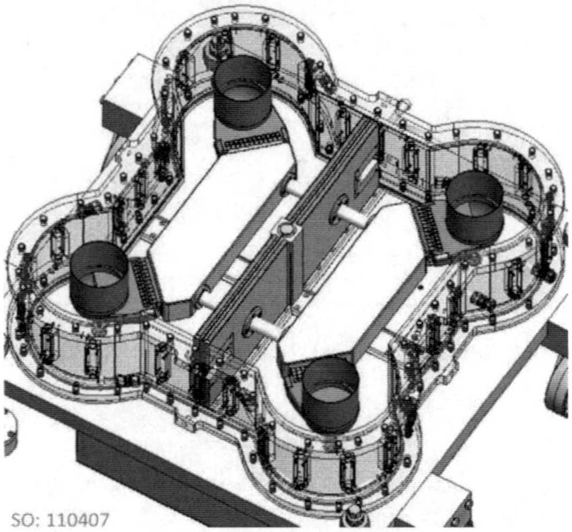


**Fig. 5.** Cut-away view of 4-port switch enabling connections 2 by 2

**Fig. 6.** Test article of 4-ports switch

Structural analysis of current lines layout is used to establish loads for main line components. These components will be reviewed by IO in September 2017 in order to prepare the first bulk orders for series of straights, elbows, 3MW load and auxiliary equipment. The first deliveries are needed for the RF source testing that should take place in RF building in 2020-2025.

## 3.3 Antenna system

The antennas are installed in equatorial ports #13 and #15. The environment has evolved in the past 3 years and the antenna design that was approved at the Preliminary Design Review in 2013 needs to be modified.

The front face cross section is reduced to accommodate the blanket modules final design and provide realistic gaps. The port plug structure is modified to allow proper tolerance with port extension and wider gaps during installation, to provide location for skids used during installation, to accommodate the sealing flange system and to provide gripping points for the remote handling interface.

In addition, the back flange of the antenna is simplified by moving some components away from the antenna or by reconsidering the shimming requirements, and the rear bulkhead position has been shifted toward the plasma to allow more space at the back of the antenna.

The design evolution ensures compatibility with interfaces and simplifies the shimming concept which allows adjusting the radial position of the RF modules (and thus the strap/plasma edge distance).

Cover plates to protect the antenna back flange from contamination during cask transfer and hot cell refurbishment provide provision for access to Removable Vacuum Transmission Lines (Fig. 7). Each individual RVTL can be accessed and replaced through a hatch, while the rest of the antenna back flange stays protected from the hot cell workshop environment.

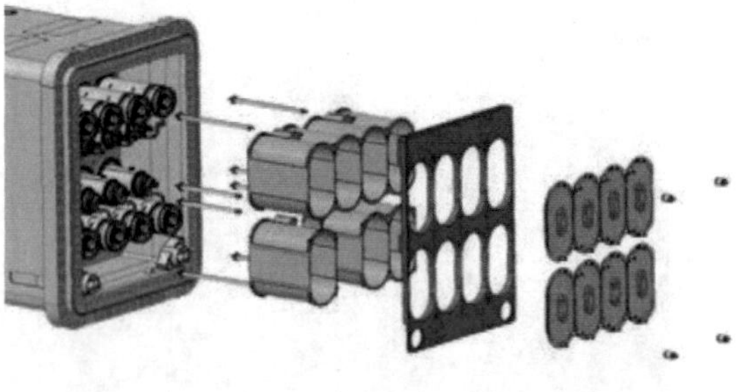

**Fig. 7.** Cover plate system to limit contamination of antenna backplate; individual plates provided to access each RF input

The design of a set of hot cell tools is progressing: a main frame (Fig. 8) holds the antenna itself and provide handling points for lifting and tilting. A lifting tool is used to catch the front of the RF module and extract it from the plug structure, and guiding tools are used during assembly phases in the hot cell workshop.

DOI: 10.1051/epjconf/201715702002

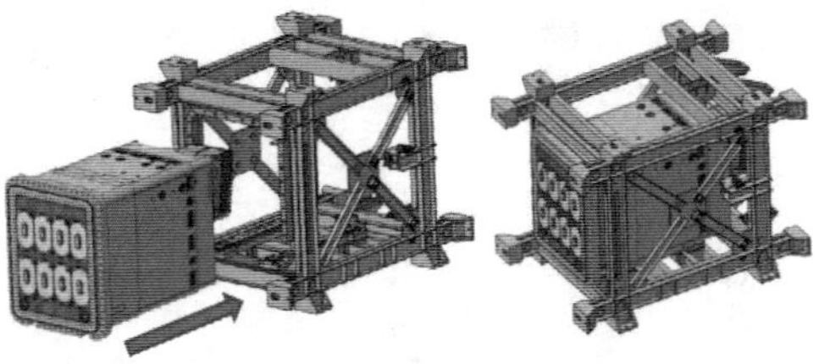

**Fig. 8**: Handling frame (14 t) for IC antenna Port Plug (45 t)

A numerical model for structural analysis has been developed to assess the driving load cases that will be used to develop the final design. Future analysis work will be completed by using a master model and sub structuring approach (Fig. 9 & Fig. 10).

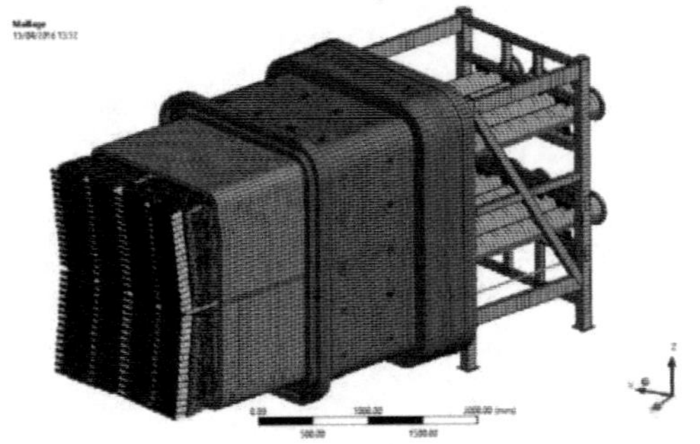

**Fig. 9** Antenna model for structural analysis (>3 million nodes)

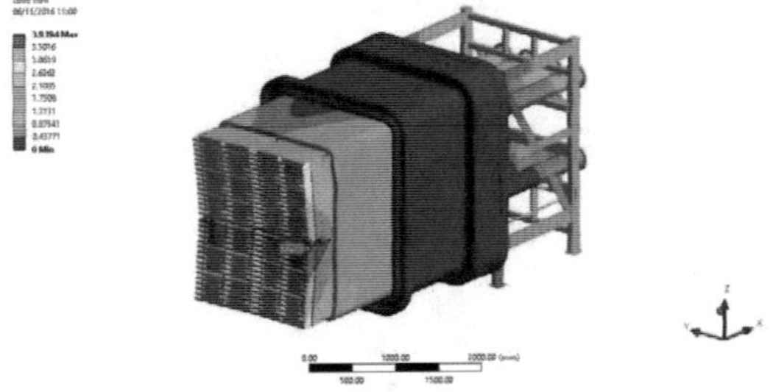

**Fig. 10**. Typical load case (VDEII): 4 mm max.displacement.

The design work is performed under a task agreement with F4E and is reactivated from this new antenna configuration.

In parallel, R&D activities are in progress.

High heat flux tests of Faraday Shield elements have been reported earlier [5]. A real size Faraday shield mock up is planned to be built to test manufacturing process.

Copper coating is used to reduce RF losses on antenna transmission lines. Two methods have been considered: 1) Galvanic coating on the titanium alloy used in the vacuum transmission lines has exhibited issues of adherence and reproducibility: composition, purity of bath, process control do not guarantee stable results. 2) Cold spraying has been investigated and gives promising results: adherence to substrate is good as well as compatibility with the high temperature brazing cycle. The method uses a robotic arm driving a spraying gun (Fig. 11). The surface is prepared by thorough cleaning and sand blasting using alumina grit, then the copper powder is sprayed at high velocity on the titanium. The gas pressure and gas temperature have a large influence on the deposition efficiency, coating thickness and quality of the deposited coating. The resulting coating can be thick, up to few tenths of a mm, and reworked (machining or polishing). Coating purity is high and the electrical conductivity is in the range of 85 to 90% IACS. Successful tests have been performed for outer surface of inner conductors as well as inner surface of outer conductor. Samples submitted to Ticusil brazing cycle have confirmed no significant loss of adhesion and no creation of blisters.

**Fig. 11.** Robotic arm for cold spray; micrographic view of coating

The RF window is a critical element of the antenna; it is part of the confinement boundary. The design uses titanium alloy rings brazed to ceramic cones. Differential expansion between metal and ceramic is used to induce a compressive stress on the ceramic during cooling down of the brazing process. The initial proposal at PDR is to build all vacuum transmission line in titanium alloys, but the transition to the stainless steel cooling circuit was not defined in that initial proposal. It is now foreseen to keep Ti alloy only were it is necessary, i.e. to join the ceramic, and to perform a transition to stainless steel parts as near as possible to the ceramic windows. This strategy allows using conventional welding techniques to build the transmission lines, but requires joining stainless steel to Ti alloy. The joint will have to survive high temperature ceramic brazing cycle, and be compatible with water cooling near the ceramic edges. Two techniques are being investigated: rotary friction weld and uniaxial diffusion bonding. Both joints are performed without fusion of the metals, a solid state process, to avoid degrading the material properties.

Metallic interlayers are used to facilitate bonding and prevent the formation of brittle phases between dissimilar materials (Fig. 12). Niobium, Copper, Nickel, Vanadium and Zirconium are investigated in test samples. Tests are in progress for friction weld and will start soon for uniaxial diffusion bonding. Figure 11 shows a cut through the joint for test between 316L and Niobium. In that test, the weld was not successful. The other example shows a good weld between Zirconium and Ti grade 5.

DOI: 10.1051/epjconf/201715702002

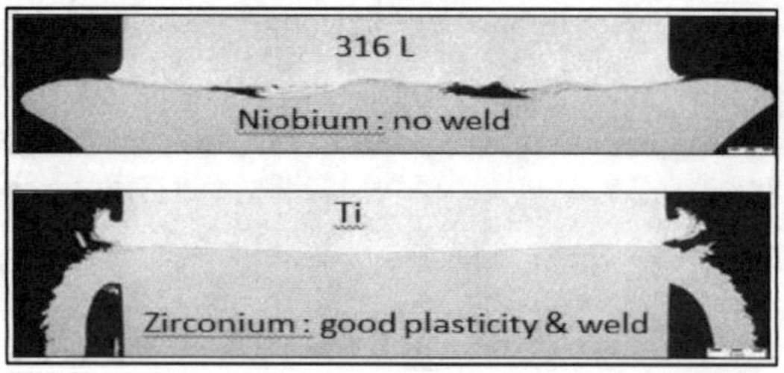


**Fig. 12**. Early tests of rotary friction welds with interlayer

The tests have discarded Nb and V, and now concentrate on Zr to optimize the friction weld parameters. In parallel, the uniaxial diffusion bonding contract is about to be launched.

RF contacts are used in the antenna to ensure electrical continuity while providing mechanical compliance. They also allow assembly of vacuum transmission lines to antenna structure and RF modules. Contacts are critical components of the antenna design: they need to withstand high currents up to 6A/mm, be resistant to long baking cycles, and ideally to the full life cycle of the antenna representing 30000 mechanical cycles. Several designs have been proposed and tests are performed in CEA/IRFM laboratory for RF vacuum tests and in CEA/SDTC for sliding tests in relevant conditions [7]. As of today, the Multi-Contact® LA-CUT/0,25/0 have been tested up to 50% of the required current. It is a special version in a CuCrZr alloy that would resist creeping during long baking period at 240°C. Further tests are necessary with optimized geometry and improved cooling in order to raise the current limit.

## 3.4 Plant Control system

IO team is in charge of the ICRF plant control system that will interface CODAC with the local control units controlling HVPS, RF sources, transmission lines and antennas, delivered by each DA.

This plant control system is in final design phase. A prototype is installed and tested on real hardware within the CODAC lab in IO, with 2 PLCs (CPU-400 station and SM-300 IO Modules) and 2 fast controllers with PXI components (Chassis & IO Modules). IC-PSC function configuration and communication tests are performed to check data exchange between the different levels, and to validate the best architecture.

## 4. Conclusion

The overall schedule recently agreed by the ITER Council gives the path for staged installation of ITER components. The ICRF system is due to deliver power in plasmas in 2032, but an earlier installation is considered for plasma experiments from 2029. The subsystem development is progressing well with good results of HVPS and RF source testing in India, of RF transmission line and layout studies in US, and antenna design and R&D in EU.

## References

1. ITER Council Press release : http://www.iter.org/doc/www/content/com/Lists/list_items/Attachments/708/2016_11_IC-19.pdf
2. "ICRF scenarios for the ITER non-active phase ", Mireille Schneider, this conference (to be published)
3. "Progress in High Power Test of R&D Source for ITER ICRF system", A. Mukherjee, 26th IAEA Fusion Energy Conference (FEC-2016), 17-22 OCT 2016, Kyoto, Japan
4. "THALES Electron Devices activities for ITER and satellites projects in plasma heating solutions - Recent results", Stéphane Bethuys, this conference
5. "Status of the ITER Ion Cyclotron Heating and Current Drive System", P.Lamalle et al., AIP Conference Proceedings;2015, Vol. 1689 Issue 1
6. "Developments in the Testing of ITER Ion Cyclotron Transmission Line Components", R.G. Goulding, this conference
7. "RF Characterisations of High Current Steady-State Sliding Contacts for the ITER ICRH Antenna", J.Hillairet, this conference

## Disclaimer

The views and opinions expressed herein do not necessarily reflect those of the ITER Organization.

# Development of slow and fast wave coupling and heating from the C-Stellarator to NSTX

*Joel* Hosea[1,*] and the PPPL ICRF Teams

[1]Princeton Plasma Physics Laboratory, Princeton University, Princeton, NJ 08543, USA

**Abstract.** A historical perspective on key discoveries which contributed to understanding the properties of coupling both slow and fast waves and the effects on plasma heating and current drive will be presented. Important steps made include the demonstration that the Alfven resonance was in fact a mode conversion on the C-stellarator, that toroidal m = -1 eigenmodes were excited in toroidal geometry and impurity influx caused the Z mode on the ST tokamak, that the H minority regime provided strong heating and that $^3$He minority could be used as well on PLT, that the 2$^{nd}$ harmonic majority tritium regime was viable on TFTR, and that high harmonic fast wave heating was efficient when the SOL losses were avoided on NSTX.

## 1 Introduction

Understanding of ICRF coupling and heating physics has evolved through the combination of experiments, modeling, and technology over several decades. At PPPL this has involved experiments and modeling for a line of experimental devices including – C stellarator, ST tokamak, PLT tokamak, TFTR tokamak, and NSTX/NSTX-U spherical tokamak – and employing critical hardware and diagnostics for achieving high power operation and observing the RF effects on the core and scrape off layer (SOL) plasmas. It is informative to track the historical evolution of advances in understanding on these devices and a selection of these advances is given here.

## 2 C stellarator: Alfven "resonance" shown to be an Alfven mode conversion

The C stellarator was configured in a race track geometry with the ICRF four strap antenna located in one straight section and a toroidal limiter located in the other straight section [1]. Thus the waves were launched in a purely cylindrical region, propagated away from the antenna, and damped in the magnetic beaches [2].

Early ICRF experiments on the C stellarator showed that the antenna loading was up to ~ 6 times larger than predicted with an approximate model for the loading [2], and this mysterious loading was designated as mode X [3]. The technology addition of a faraday shield between the antenna straps and the plasma removed this mode X loading [3] and opened the way to comparing experimental loading predictions to quantitative predictions from a cylindrical model utilizing the full

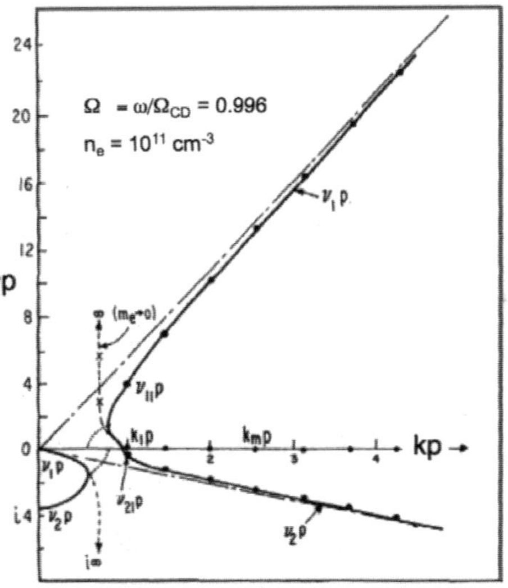


**Figure 1.** Alfven mode conversion of perpendicular wavenumber vp versus parallel wavenumber kp for Model-C parameters for Ω < 1. (Plasma radius p = 6.1 cm.) (Fig. 1 of Ref. 4.)

cold plasma relation. The first step for developing this model was recognizing that the "Alfven resonance" was in fact an "Alfven mode conversion" even in the absence of finite temperature as shown in Fig. 1 [4,5]. When it was assumed earlier that the electron mass could be set to zero and the essential wave physics would be preserved, an artificial resonance resulted. But keeping the electron mass in the plasma dispersion indeed gives the mode conversion at $n^2_{\parallel} = S$ (T. Stix notation [2]) for this regime with $\omega \lesssim \Omega_{ci}$. The orthogonal eigenmodes

* Corresponding author: jhosea@pppl.gov

 DOI: 10.1051/epjconf/201715702003

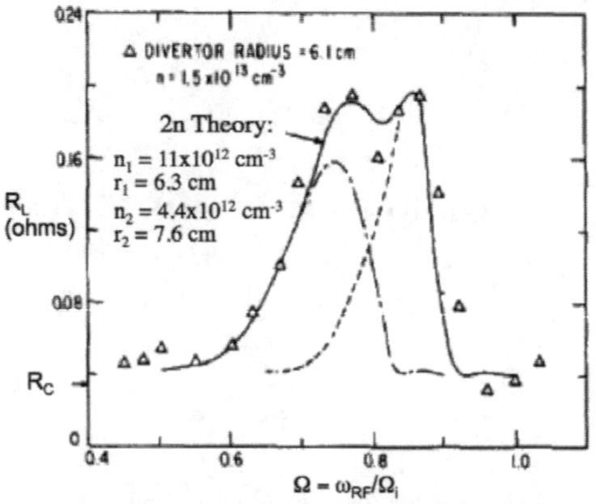


**Figure 2.** Experimental loading of Stix coil versus Ω, and its prediction by the two-density cold plasma theory. (Fig. 3 of Ref. 6 with AIP Publishing permission)

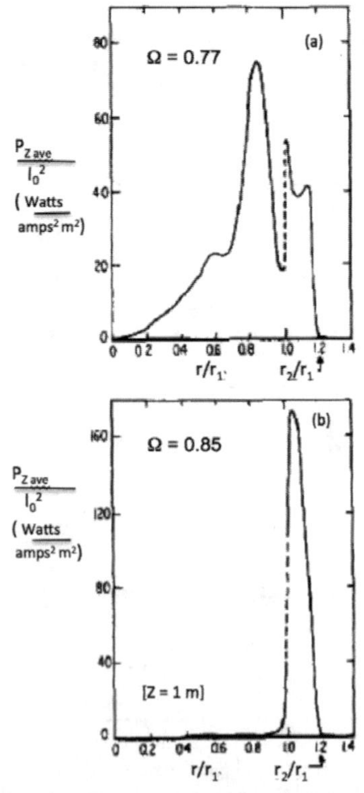


**Figure 3.** The average Poynting flux in the axial direction 1 m along the axis from the center of the Stix coil for the two peaks in Fig, 2. (Fig. 4 of Ref.6 with AIP Publishing permission)

which fit in the cylindrical model straight section have two radial mode numbers for each parallel wavenumber - one on the "electrostatic" arm and one on the "electromagnetic" arm – on the right branch in Fig. 1. It should be noted that all of the modes fall on the slow wave branch since fast wave modes were not excited under the $\omega/\Omega_{ci} \lesssim 1$ conditions on the C stellarator. Subsequently, the loading for each of the modes could be calculated with the residue method described in Ref. 5 and summed over the antenna spectrum, which peaked at $k_{||}\rho \sim 1$, to compare to the measured loading.

To obtain sharper loading profiles in $\Omega = \omega/\Omega_{ci}$ a second technology step was taken to make the magnetic field very close to constant over the antenna region [4]. The theoretical $R_L$ loading profiles versus Ω followed the experimental profiles quantitatively over the entire range of densities studied – from 1.5 x $10^{18}$ $m^{-3}$ to 2 x $10^{16}$ $m^{-3}$ [4]. Importantly, the maximum of $R_L$ falls off as $\Omega = 1$ is approached in contrast to the approximate prediction with $m_e = 0$. This comparison clearly verifies the cold plasma theory with the Alfven mode conversion. Subsequently, this modeling was extended to a two step density profile that accounted for the density drop-off in the C stellarator toroidal divertor and used to explain the two peak loading profile versus Ω observed at higher density (Fig. 2) [6]. This comparison further revealed that the peak closest to $\Omega = 1$ was due to power coupled to the SOL as shown in Fig.3, and such coupling can be a major effect in the fast-wave regime on NSTX as will be discussed later.

## 3 ST tokamak: Transition to toroidal geometry and fast wave excitation

The C stellarator was converted to the ST tokamak in 1970 with the toroidal configuration being provided by removing its straight sections. This forced us to rethink the wave launching and wave damping which overlap in the same cross-section in a torus. We adopted the strategy of launching slow and fast waves from the inside of the plasma torus with a half-turn current strap. We charged ahead with a cylindrical plasma model to provide an approximate analysis of both the slow wave and fast wave excitation [7]. Eigenmodes were predicted for both slow and fast waves with poloidal m values of 0, +1, and –1. One question at the time was do distinct toroidal eigenmodes exist in toroidal geometry or are the modes coupled and therefore not discrete. Discrete eigenmodes were indeed observed [8], and the loading from these varied in time as plasma density changed during the discharge as predicted by the fast wave eigenmode analysis.

Clear ion heating was observed for short (1.5 ms) RF pulses at $\Omega_d \approx 2$ with charge exchange systems: directed radially near the antenna and tangentially on the other side of the torus from the antenna (Fig. 4) [8]. The perpendicular energy distribution was bi-Maxwellian (Fig. 5) whereas the tangential distribution was a single Maxwellian. It was not clear at the time what caused the bi-Maxwellian distribution since the charge exchange analyzer did not have mass discrimination and the minority heating effect was not yet known.

For longer RF pulses (5.5 ms) the parallel temperature time response was rather slow during the fall off after the pulse due to the effect of impurities induced by the RF energy [8]. This effect was reproduced approximately with a neon gas pulse set to produce a Z heating mode with the same density

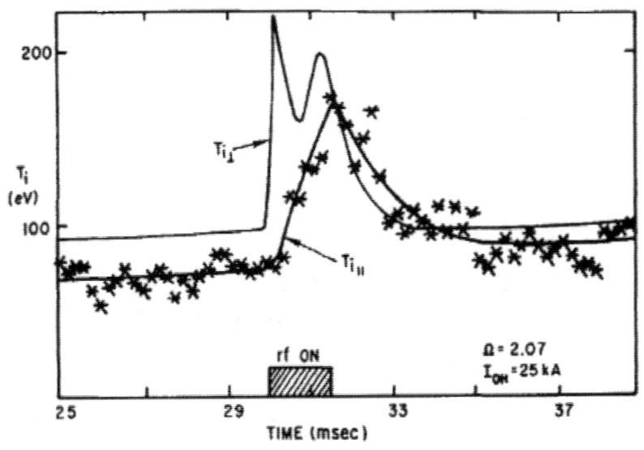


**Figure 4.** $T_{i\parallel}$ and $T_{i\perp}$ versus time from charge exchange measurements in $D_2$. ($P_{RF}$ ~ 70 kW, $n_e = 6 \times 10^{12}$ cm$^{-3}$, $\Omega$ = 2.07). (Fig. 10 of Ref. 8 with IAEA permission)

perturbation as the RF pulse [8]. An attempt was made to reduce the RF produced density by sealing off a Thomson scattering window leak and enclosing it in helium thinking that water vapor might be contributing to the observed impurity effect on density. In hindsight, the energetic minority ions were very poorly confined in ST, and even more so as the hydrogen concentration was reduced with the water vapor leak at the Thomson scattering window displaced with helium. It was noticed that the RF heating improved substantially immediately following a vessel opening, presumably due to a greater hydrogen concentration resulting in a lower energy tail on the hydrogen energy distribution.

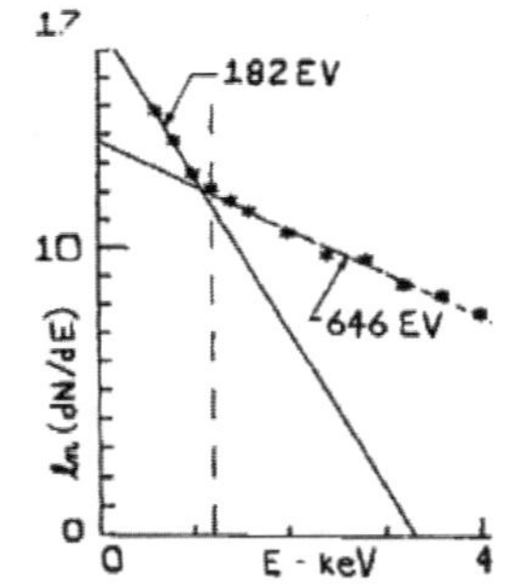


**Figure 5.** Perpendicular energy distribution induced at $\Omega$ = 2. (Fig. 9 of Ref. 8 with IAEA permission)

## 4 PLT tokamak: Clear demonstration of the minority H regime, minority $^3$He regime established, and second harmonic regime shown to be viable in hydrogen

Prior to the start of ICRF experiments on PLT an important technology advance was made with the installation of a mass sensitive charge exchange system being developed for TFTR. The initial ICRF experiment was conducted with a modest RF power (~ 40 kW) and the deuterium channel showed a relatively small temperature rise. An RF team member suggested looking also at the hydrogen channel, since we could, and it was discovered that the hydrogen energy distribution was greatly enhanced by the ICRF power and exhibited a very energetic tail. It was suddenly clear that the confusing heating results obtained on the ST tokamak were due to operating in the "minority" hydrogen regime. On PLT the energetic hydrogen confinement was no longer affected by severe banana-orbit loss cones as on the ST tokamak and the measured hydrogen energy extended to ~ 40 keV at $P_{RF}$ of only ~ 40 kW (Fig. 6) [9]. The theoretical curve is for $Z_{eff}$ = 2.2, $E_j$ = 1.8 keV, and $\xi$ =13.8 [9]. As $P_{RF}$ was subsequently increased to ~ 350 kW, $T_d$ was increased substantially to ~ 1.21 keV from the ohmic level of ~ 0.54 keV [10], thereby establishing that the minority H regime was an effective heating method when the energetic H ions were adequately confined.

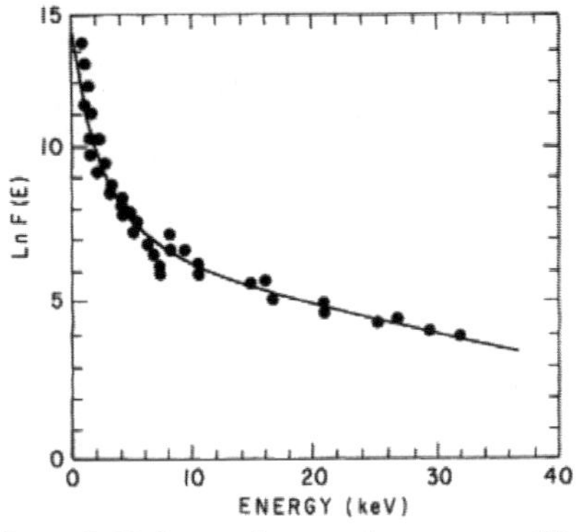


**Figure 6.** Hydrogen charge exchange spectra 20 ms into a ~ 40 kW RF pulse for initial ICRF operation at 25 MHz on PLT. (Fig. 4 of Ref. 9)

A technological event occurred – an internal turn-to-turn short in a toroidal field coil - and PLT machine operations were stopped for several months. We used this time to explore possible minority species that could be used to optimize minority heating near the maximum $B_T$ available after the coil repair - ~ 3 T. This search resulted in the selection of helium-3 ($^3$He) as the minority species for which our 30 MHz system frequency matched the cyclotron frequency at 3 T. This $^3$He minority regime was analyzed to be ~ 65% more efficient at heating deuterium ions over the electrons than for the H minority regime [10]. This was born out experimentally by comparing the scaling of $\Delta T_d$ with $P_{RF}/n_e$ for the two minority regimes [10].

In order to push to higher ICRF power on PLT, it was critical to develop a high power alumina cone feedthrough which eliminated multipacting and allowed

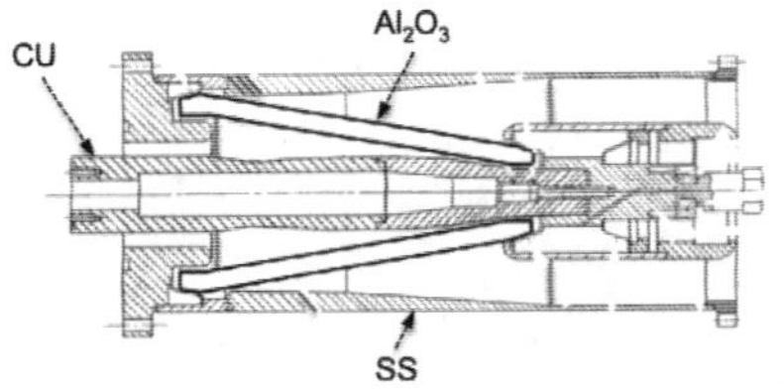


**Figure 7.** Alumina cone feedthrough developed to support high power operation on PLT.

DOI: 10.1051/epjconf/201715702003

high voltage operation without multi-pacting (Fig. 7) [11]. With this feedthrough installed, the RF power was increased to 4.3 MW and very large increases in $T_d$ and $T_e$ were produced in the D/$^3$He minority regime (Fig. 8) [12].

The second harmonic heating regime was also shown to be viable on PLT. Second harmonic heating of a hydrogen plasma with $P_{RF}$ at the 1 MW level was successful in increasing the hydrogen effective temperature [$T_{eff}$ = 2/3⟨E⟩] from ~ 0.5 keV to ~ 3 keV (Fig. 9) [13]. This is an important method that can be used in the low-B startup phase of ITER and for the second harmonic T regime in a D-T plasma.

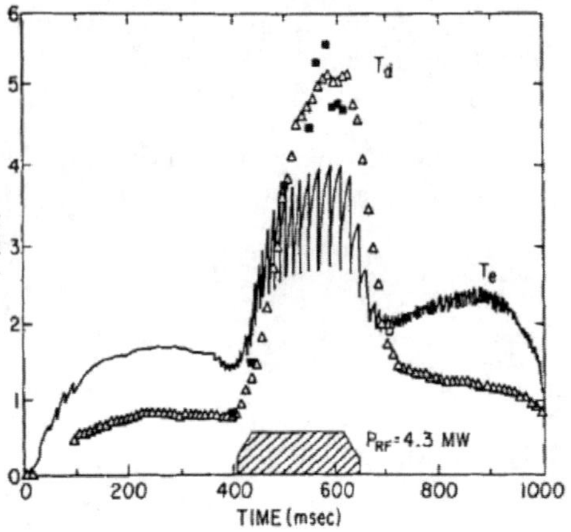


**Figure 8.** Time evolution of $T_e(0)$ (solid line) and bulk $T_d$ (charge exchange Δ and neutron emission ■) for a D/$^3$He case [$P_{RF}$ = 4.3 MW at 30 MHz, $B_T$ = 3.3 T, $n_e$ = 3.7 x $10^{13}$ cm$^{-3}$, $\eta_{3He}$ = 5 – 10%]. (Fig.1 of Ref. 12 with AIP Publishing permission)

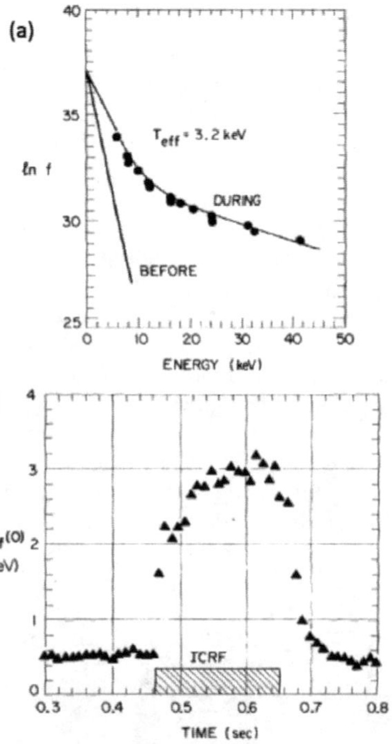


**Figure 9.** a) Hydrogen ion energy distribution before and during ICRF heating ($P_{RF}$ ~ 1 MW, $n_e$ = 3.8 x $10^{13}$ cm$^{-3}$), and b) time evolution of average energy, $T_{eff}$ = 2/3⟨E⟩. (Fig. 1 of Ref. 13)

## 5 TFTR tokamak: Second harmonic tritium regime demonstrated, and efficient mode conversion heating and current drive established

ICRF experiments were carried out on TFTR in the minority hydrogen regime prior to the introduction of tritium for $P_{RF}$ up to 8.4 MW [14]. Electron heating resulted in a large increase in $T_e(0)$ from ~ 2.4 keV to 6 keV. Also, above 3 MW the sawteeth were stabilized or significantly lengthened during the RF pulse. The energetic H distribution was measured out to 1 MeV energy and found to have a high effective temperature. A large increase in TAE mode amplitude was observed versus $P_{RF}$ and this later led to a vacuum failure during the D-T phase of operation due to the loss of energetic ions via banana trapping which bombarded exposed welds at the bottom ports on TFTR.

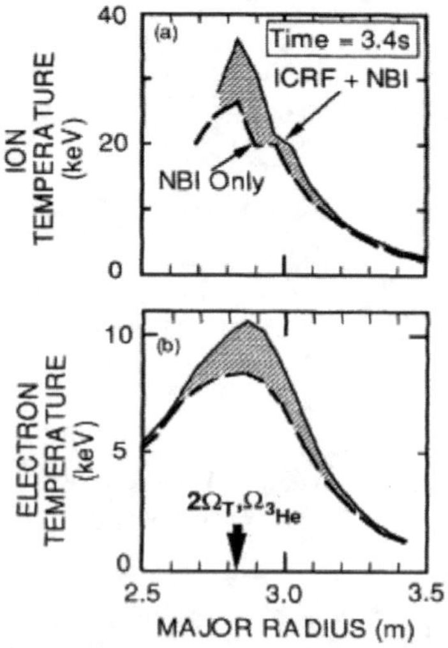


**Figure 10.** Ion and electron temperature profiles showing strong ICRF central ion and electron heating at $2\Omega_{CT}$ on TFTR. (Fig. 6 of Ref. 15)

In the first D-T ICRF experiments on a tokamak, substantial ion and electron temperature increments were obtained with $2\Omega_{ct}$ damping in hot ion mode discharges as shown in Fig. 10 [15,16]. For this case of $P_{RF}$ = 5.5 MW with $P_{NBI}$ = 23.5 MW (60% T), the ICRF majority tritium heating at $2\Omega_{ct}$ resulted in temperature increments of $\Delta T_i(0)$ = 9 keV and $\Delta T_e(0)$ = 2.6 keV. It is to be noted that a small amount of $^3$He minority, which has a cyclotron frequency equal to $2\Omega_{cT}$, was also present in the discharge. The majority $2\Omega_{cT}$/minority $^3$He heating regime is ideal for heating in the reactor D-T-$^3$He plasma

DOI: 10.1051/epjconf/201715702003

since damping on the $^3$He can serve to heat the tritium to a temperature level at which the $2\Omega_{cT}$ heating will take over even in the absence of neutral beam ion heating.

Mode conversion heating and current drive with outside launch were also demonstrated on TFTR [17,16]. The electron heating for a $^3$He – $^4$He case with $P_{RF}$ = 4.4 MW resulted in $T_e(0)$ doubling to 6 keV and $T_e$ approximately doubling across the profile as well. The measured electron power deposition profile peaks off axis at the mode conversion layer as predicted.

## 6 NSTX tokamak: Transition to high harmonic fast wave (HHFW) regime and close examination of SOL losses

The AORSA and GENRAY codes for HHFW launched at the plasma separatrix predicted very strong RF power absorption in the plasma core on NSTX with the wave power damping being ~ 80% in approximately ¼ turn around the torus [18]. This strong "single pass" absorption serves to accentuate the competition between core RF power deposition and RF power loss in the SOL. It was immediately observed that the HHFW heating efficiency was optimum when the density in front of the antenna in the SOL was below the fast wave onset density for the k value launched by the antenna (set by phase between antenna straps) [18]. The heating efficiency ($W_T$ at constant $P_{RF}$) falls off for phasing below 90° for the conditions studied [18]. This result suggests strong RF power loss occurs in the SOL for k values that put the density in front of the antenna above the fast wave cut-off density, $n_{co} \propto B^*k^2/\omega$.

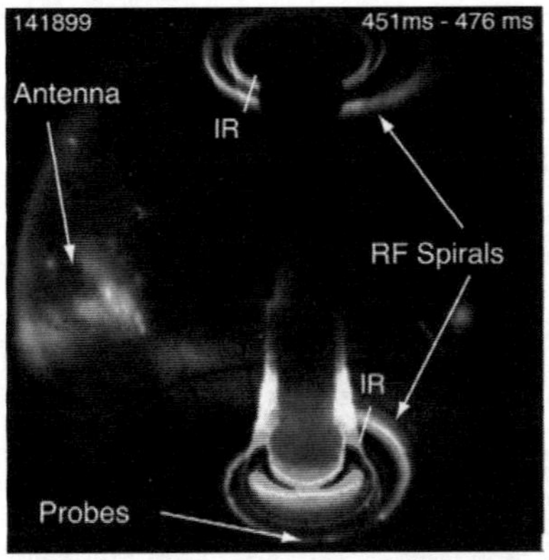


**Figure 11.** RF spirals produced on the bottom and top divertor plates of NSTX with HHFW heating ($P_{RF}$ = 1.3 MW, $P_{NB}$ = 2MW, D). (Fig. 1 of Ref. 21 with AIP Publishing permission)

HHFW heating in L-mode plasmas was strong when the edge density was low relative to the cut off density, producing $T_e(0)$ values of ~ 5 keV in both He and D plasmas [19]. HHFW heating in ELM free H-mode plasmas showed significant heating toward the periphery of the plasma core for antenna strap phasings of – 150° ($k_\phi$ ~ 14 m$^{-1}$) and – 90° (~ - 8 m$^{-1}$) [20]. Here the core heating efficiencies ($\eta = P_{dep}/P_{applied}$, with $P_{dep}$ measured from $\Delta W_T/\tau$ measurements) were 66% and 40% for -150° and -90°, respectively. This indicates a large RF power loss in the SOL that increases inversely with $k_\phi$, as might be expected since the cut off density decreases as $k_\phi^2$.

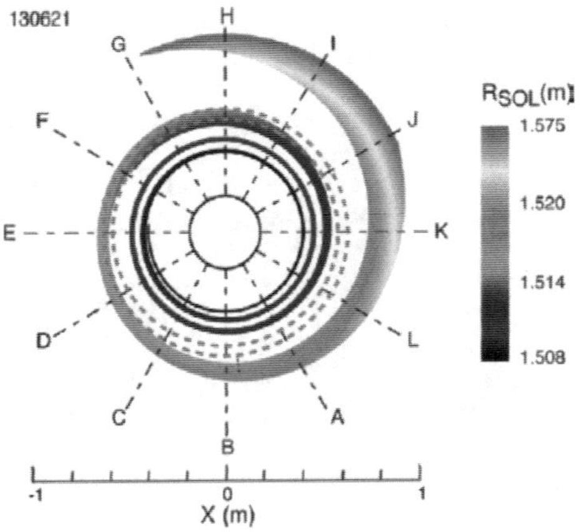


**Figure 12.** Magnetic field strike points on the lower divertor versus midplane field line $R_{SOL}$ location in front of the antenna. (Fig. 4 of Ref. 23 with IAEA permission)

It came as a surprise to us that much of the power loss to the SOL ended up on the top and bottom divertors in the form of "hot" spirals as shown in Fig. 11 [21]. This edge loss is reminiscent of the power loss to the SOL on the C stellarator. Considerable effort has been applied to understanding these spiral patterns. The SPIRAL code was used to map the magnetic field lines passing in the SOL in front of the antenna at the midplane to their strike points on the divertors, and indeed these strikepoints form a spiral (Fig. 12) [22,23]. The passes of this spiral line up with the IR camera peaks observed along a major radius view of the divertor floor [23]. Furthermore, the RF effects for the lower spiral are observed on probe floating potentials and instrumented tile currents [22]. The negative shift of the probe floating potential for the probe under the spiral is consistent with RF rectification for the far RF fields

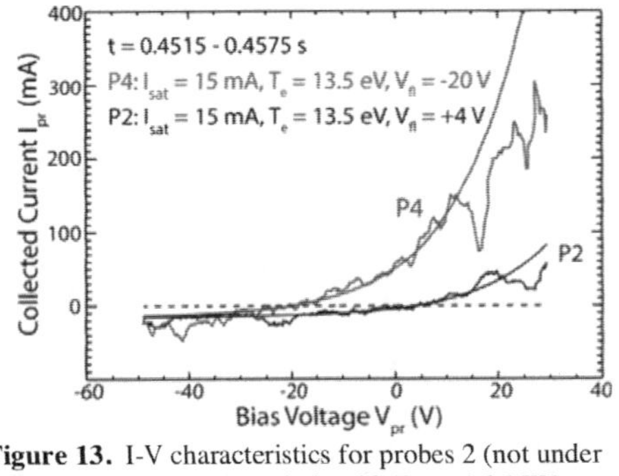


**Figure 13.** I-V characteristics for probes 2 (not under spiral) and 4 (under the spiral) with $P_{RF}$ = 1.3 MW, averaged over 6 $V_{pr}$ sweeps. (Fig. 6 of Ref. 21 with AIP Publishing permission)

arriving at the divertor along the magnetic field lines [21]. Also, for the tiles under the spiral the RF induced currents are caused by electron collection which is also consistent with RF rectification. Upon averaging the probe characteristics over six consecutive voltage sweeps, the negative shift of the I-V characteristic under the spiral is as expected for RF rectification as well (Fig. 13) [21]. For Fig. 13, the RF voltage across the probe sheath consistent with the $V_{fl}$ shift is $V_{RF}$ = 43.7 V.

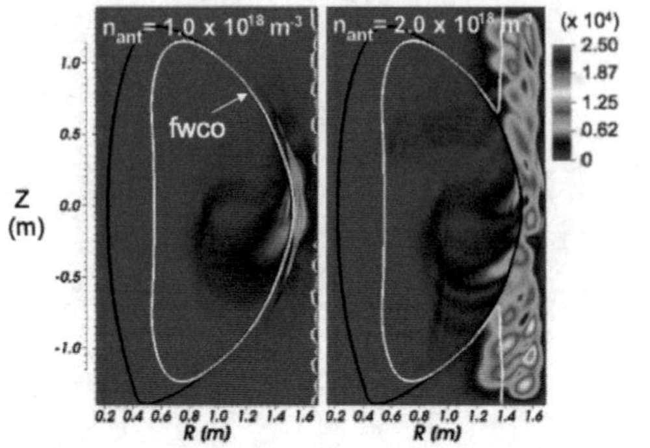


**Figure 14.** AORSA prediction of RF electric field increase in the SOL versus density at the antenna for $n_\phi$ = -21. The enhanced SOL field occurs when the FW cutoff (fwco) begins to open in front of antenna. (Fig. 2 of Ref. 24 with IAEA permission)

Finally, the AORSA code has been extended to include waves in the SOL and compared to the results obtained on NSTX [24]. Strong standing wave RF electric field amplitudes in the SOL are predicted for edge densities in front of the antenna above the fast wave cutoff density. This is illustrated clearly in Fig. 14 where the strong fields are predicted when the density at the antenna is raised above the FW cutoff density in front of the antenna causing the fast wave cutoff to open in front of the antenna. These enhanced fields are thought to be responsible for the observed far field rectification at the divertor.

## 7 Summary

The evolution of understanding of ICRF regimes on the C stellarator through to the NSTX tokamak with experiments, modeling and technology advances has been summarized. Much is now understood from this research and research on other devices in the world fusion program. It is generally recognized that far field wave fields away from the antenna and near field wave fields in the vicinity of the antenna can cause considerable RF power loss and result in unwanted impurity production as well. Extensive efforts have been and are underway to quantify the near field and far field effects, and efforts are also underway to modify antenna designs to reduce these fields. Particularly noteworthy recent antenna advances include the field aligned antenna on C-mod [25] and the 3-strap antenna on ASDEX Upgrade [26].

The authors wish to acknowledge the support of the several fusion machine project teams, and especially the machine, RF, and neutral beam operations groups. This work is supported by USDOE Contract No. DE-AC02-09CH11466.

## References

1. M.A. Rothman, R.M. Sinclair, I.G. Brown, and J.C. Hosea, Phys. Fluids **12**, 2211 (1969)
2. T.H. Stix, *Theory of Plasma Waves* (McGraw-Hill Book Company, New York, 1962)
3. M.A. Rothman, R.M. Sinclair, and S. Yoshikawa, J. Nucl. Ener. Part C_Plasma Physics **8**, 241 (1966)
4. J.C. Hosea and R.M. Sinclair, PRL **23**, 3 (1969)
5. J.C. Hosea and R.M. Sinclair, Phys Fluids **13,** 701 (1970)
6. J.C. Hosea and R.M. Sinclair, Phys. Fluids **16**, 1268 (1973)
7. J.C. Hosea and W.M. Hooke, PRL **31**, 150 (1973)
8. J. Adam, M. Chance, H. Eubank, et al., 5th IAEA Fusion Energy Conf **Vol 1**, 65 (Tokyo, November 1974)
9. J. Hosea, S. Bernabei, P. Colestock, et al., PRL **43**, 1802 (1979)
10. J. Hosea, V. Arunasalam, S. Bernabei, et al., Proc. Course on Physics of Plasmas Close to Thermonuclear Conditions **Vol 2**, 571 (Varenna, August-September 1979)
11. D.Q. Hwang, G. Grotz, and J.C. Hosea, J. Vac. Sci. and Technol. **28**, 1273 (1982)
12. J.R. Wilson and the PLT Group, RF Conf. Proc. AIP **129**, 28 (1985)
13. D.Q. Hwang, J. Hosea, H. Thompson, et al., PRL **51**, 1865 (1983)
14. C.K. Phillips, J.R. Wilson, M. Bell, et al., RF Conf. Proc. AIP **289**, 44 (1994)
15. J.R. Wilson, C.E. Bush, D. Darrow, et al., PRL **75**, 842 (1995)
16. C.K. Phillips, M.G. Bell, R. Bell, et al., Phys. Plasmas **2**, 2427 (1995)
17. R. Majeski, J.H. Rodgers, S.H. Batha, et al., PRL **76**, 764 (1996)
18. J. Hosea, R.E. Bell, B.P. LeBlanc, et al., Phys. Plasmas **15**, 056104 (2008)
19. G. Taylor et al., RF Conf. Proc. AIP **1187**, 113 (2009)
20. J.C. Hosea, R.E. Bell, E. Feibush, et al., RF Conf. Proc. AIP **1187**, 105 (2009)
21. R.J. Perkins, J.C. Hosea, M.A. Jaworski, et al., Phys. Plasmas **22**, 042506 (2015)
22. R.J. Perkins, J.C. Hosea, G.J. Kramer, et al., PRL **109**, 045001 (2012)
23. R.J. Perkins, J.-W. Ahn, R.E. Bell, et al., Nucl. Fusion **53**, 083025 (2013)
24. N. Bertelli, E.F. Jaeger, J.C. Hosea, et al., Nucl. Fusion **54**, 083004 (2014)
25. S.J. Wukitch, M.L. Garrett, R. Ochoukov, et al., Phys. Plasmas **20**, 056117 (2013)
26. V. Bobkov, F. Braun, R. Dux, et al., Nucl. Fusion **56,** 084001 (2016)

EPJ Web of Conferences **157**, 02004 (2017)
*22 Topical Conference on Radio-Frequency Power in Plasmas* DOI: 10.1051/epjconf/201715702004

# ICRH physics and technology achievements in JET-ILW

P. Jacquet[1,]*, D. Van Eester[2], E. Lerche[2,1], V. Bobkov[3], T. Blackman[1], L. Colas[4], C. Challis[1], A. Czarnecka[5], P. Dumortier[2,1], D. Frigione[6], F. Durodié[2], L. Garzotti[1], M. Goniche[4], J. Graves[7], Y. Kazakov[2], K. Kirov[1], C. C. Klepper[8], N. Krawczyk[5], A. Krivska[2], M. Mantsinen[9,10], I. Monakhov[1], I. Nunes[11], J. Ongena[2], M. Reinke[8], F. Rimini[1], W. Zhang[3, 12] and JET contributors[13]

EUROfusion Consortium, JET, Culham Science Centre, Abingdon, OX14 3DB, UK.
[1]CCFE, Culham Science Centre, Abingdon, OX14 3DB, UK.
[2]LPP-ERM/KMS, EUROfusion Consortium Member, TEC Partner, Brussels, Belgium.
[3]Max-Planck-Institut für Plasmaphysik, Boltzmannstr.2, D-85748, Germany.
[4]CEA, IRFM, F-13108 Saint-Paul-lez-Durance, France.
[5]IPPLM, Hery 23 Str., 01-497 Warsaw, Poland.
[6]Associazione ENEA, Fusion and Nuclear Safety Department, C. R. Frascati, Via E. Fermi 45, 00044 Frascati (Roma), Italy.
[7]Centre de Recherches en Physique des Plasmas, EPFL, 1015 Lausanne, Switzerland.
[8]Oak Ridge National Laboratory, PO Box 2008, Oak Ridge, USA.
[9]Barcelona Supercomputing Center (BSC), 08034 Barcelona, Spain.
[10]ICREA, Pg. Lluis Companys 23, 08010 Barcelona, Spain.
[11]Instituto de Plasmas e Fusão Nuclear, IST, Universidade de Lisboa, Portugal.
[12]Department of Applied Physics, University Ghent, 9000 Gent, Belgium
[13]See the author list of "Overview of the JET results in support to ITER" by X. Litaudon et al. to be published in Nuclear Fusion Special issue: overview and summary reports from the 26th Fusion Energy Conference (Kyoto, Japan, 17-22 October 2016).

**Abstract.** ICRH was extensively used in the 2015-16 JET-ILW (ITER like wall) experimental campaign; bulk heating together with high-Z impurity chase-out from plasma centre importantly contributed to the good DD fusion performance obtained recently in JET. Power up to 6 MW was launched in H-mode deuterium plasmas and 8 MW during the hydrogen campaign. The ILA was re-installed and contributed positively to the availability of ICRH power. The ILA produces slightly less high-Z impurities than the A2's and the PWI measured via Be line emission on limiters is in the same ballpark. Specific experiments were conducted to optimise ICRH scenarios in preparation for DT in particular the dual frequency scheme, (H)D and (He)D were tested. In addition, it was confirmed that the (D)H scenario is accessible in a ILW environment and the novel 3-ions ICRH scheme was validated experimentally.

## 1 Introduction

In 2015-16, JET experiments were run to develop plasma scenarios in preparation for the DT campaign. ICRH was routinely used in plasma scenario development. ICRH provides bulk plasma heating and a localised central heat source for high-Z core impurity screening. The paper describes the progress to reliably and efficiently deliver high ICRH power to the plasma and describes experiments aiming at optimizing ICRH heating scenarios in preparation for the JET DT campaign and ITER non-active phase operation.

## 2 Operational aspects

The JET ICRH system [1, 2] includes 4 A2 antennas and the ITER-Like Antenna (ILA) which, after its initial operation in 2008-2009 [3] was re-installed and re-commissioned in 2015 [4]. The frequency coverage is 23-57 MHz for the A2's and 29-51 MHz for the ILA. Each A2 antenna is a phased array of 4 poloidal straps; controlling the phase between straps allows waves to be

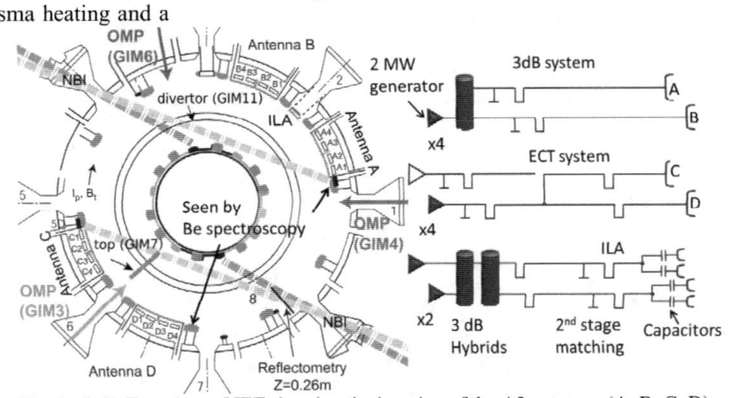


Fig. 1. (left) Top view of JET showing the location of the A2 antennas (A, B, C, D) and the ILA. (right) Configuration of the transmission lines feeding the antennas.

* Corresponding author: philippe.jacquet@ukaea.uk

DOI: 10.1051/epjconf/201715702004

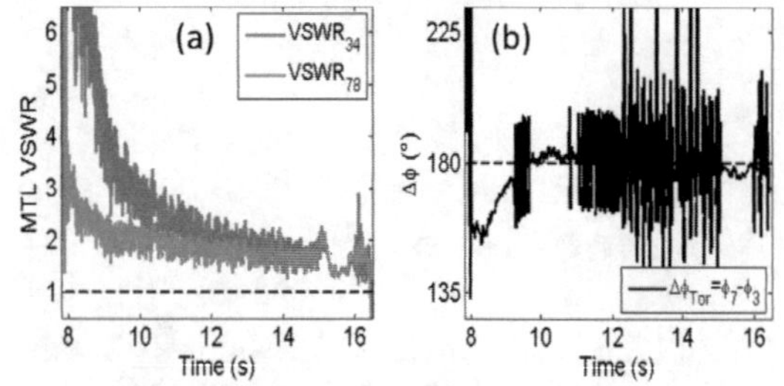

Fig. 2. Reproduced from [4], illustration of ILA real time matching, JPN 91916 (ELM-y H-mode), bottom row only is used. (a) VSWR in the main transmission lines feeding bottom straps 34 and 78 (2nd stage matching); (b) phase between toroidally adjacent straps 7 and 3.

launched with different $k_{//}$ spectra. Usually [0, π, 0, π] (π phasing), [0, π/2, π, -π/2] (+π/2 phasing) or [0, -π/2, π, π/2] (-π/2 phasing) phasing is applied to the straps [1, 2, 3, 4]. ELM resilience for the A2 antennas is achieved by pairing the antennas using 3 dB hybrid couplers (A&B antenna system) or external conjugate-T's (ECT, C&D antenna system) (see Fig. 1). The ILA is an array of 8 straps (4 in the poloidal direction × 2 in the toroidal direction); neighbour straps in the poloidal direction are paired as Resonant Double Loops (RDL) using in-vacuum capacitors and T-junctions, which, with the matching system, provide ELM resilience.

The implementation of real-time antenna matching algorithms greatly improved the ILA system availability [4]. Now automated are: the impedance at the RDL's T-junction, the stub/trombone impedance matching (2$^{nd}$ stage matching) behind the RDL's, the phase between toroidally adjacent RDL's, and the phase between top/bottom rows. Fig. 2 shows an example of real-time matching in an H-mode pulse. In most of the campaign, only one row of the ILA was used (either top or bottom); a fluid leak in a capacitor actuator developed in summer 2016 (stopping then operation of the top array) before all controls for operation of the full array were fully commissioned. Despite this limitation, power levels of up to 2.8 MW were coupled from the ILA (~1.5 MW in H-mode) and in general the ILA contributed to ~1/3 of the total ICRH power.

Operational aspects for the A2 antennas and the ICRH plant are discussed in details in [5]. When the plasma conditions are such that antenna coupling resistance ($R_c$) is low ($R_c$<1.0 Ω, typical of H-mode plasmas with low gas fuelling) the power is bounded by the maximum voltage in the transmission lines of ~30 kV. Operating above this voltage increases the risk of arcing. However, improving the power balance between straps and between generators is a way to effectively increase the launched power in these low coupling conditions. Statistically, most of the ICRH pulses in JET-ILW were performed with antenna coupling resistance $R_c$>1.0 Ω. In this case, the capability of the A2 system to deliver power to the plasma is effectively bounded by the available power from the generators and the ohmic losses. ICRH power gain can be obtained from ICRH generator tuning and maintenance to ensure they operate reliably close to their specifications.

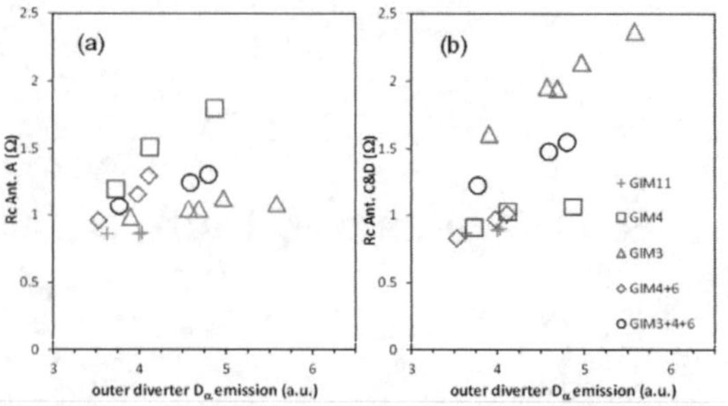

Fig. 3. $R_c$ vs divertor $D_\alpha$, with divertor or OMP gas injection. (a) antenna A, (b) antennas C&D. Experimental conditions: ELMy H-mode plasma with N=1 (H)D ICRH, $f_{ICRH}$=42 MHz, $B_t$=2.7T, $I_P$=2.5MA, $P_{NBI}$=15 MW, $P_{ICRH}$=3MW.

## 3 SOL density tailoring to increase ICRH coupling

As explained in the previous section, increasing $R_c$ allows operation away from transmission line voltage limits and maximisation of the ICRH power. The fast wave can only propagate beyond the evanescent layer in the Scrape-Of Layer (SOL). Typically the cut-off density is a few $10^{-18}$ $m^{-3}$ in tokamaks with $B_T$ = a few Tesla. Since 2007, experiments are carried-out in various tokamaks, including JET, to tailor the SOL density in front of the ICRH antennas and reduce the fast wave evanescence gap [6, 7]. In 2016, a new Outer-Mid Plane (OMP) Gas Injection Module (GIM), was installed between antenna C and antenna D (GIM 3, see Fig. 1), allowing to develop further the gas injection technique to maximise the ICRH power. Fig. 3 illustrates the gain in coupling resistance on antenna A and antennas C&D when using gas from GIM 4 (resp. GIM 3) in H-mode plasmas. It is also important to note that using a combination of outer GIMs (for example GIM 3+4+6) allows a substantial increase of $R_c$ on all JET antennas simultaneously. The effect of local gas injection on JET ICRH antenna coupling was modelled [8] using the 3D EMC3-Eirene SOL modelling. A 3D JET model was built as a realistic simulation model. This model includes all 5 ICRH antennas, the poloidal limiters, the pumped divertor, and all major obstacles for gas motion in the JET chamber. First, a reference pulse from the experiment with symmetrical divertor gas puffing was simulated and in particular, transport parameters were adjusted to match the measured profiles of $n_e$, $T_e$ and ion saturation current in the divertor. This validated plasma was then used in the simulations to study the changes in the SOL when puffing gas from a specific GIM location at the OMP or at the top of the vessel, while other parameters were kept constant. No specific ICRH-SOL interaction was taken into account in the simulations; the changes in the SOL are exclusively from the change to the location of the ionisation via plasma electron impact of the neutrals. Simulations in these conditions show that localised OMP gas injection increases the SOL density close to the injection point which results in a shift of few centimetres of the cut-off layer position. Using a simple 1D fast wave coupling code [7], the changes in SOL

 DOI: 10.1051/epjconf/201715702004

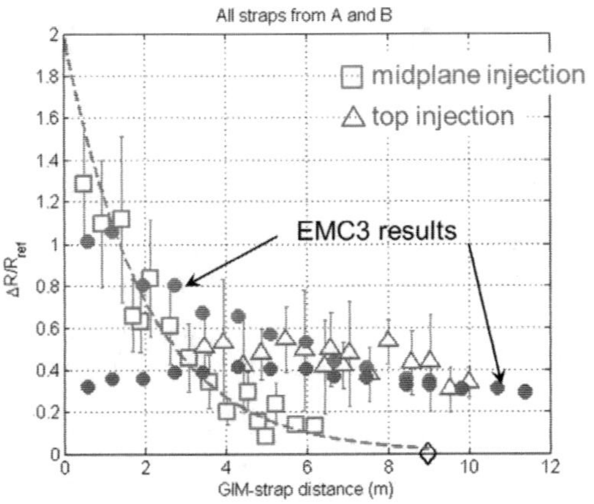


Fig. 4. Reproduced from [8], $R_c$ improvement w.r.t. divertor gas, measurements and simulations. Open symbols: experimental data. The simulation results are shown with the plain circles (red: OMP injection; green: top injection).

density profiles at different toroidal positions (averaged poloidally over the height of an A2 antenna) were translated in terms of relative $R_c$ increase when changing from divertor to top or OMP gas injection. The result is shown in Fig. 4. The simulations can reproduce qualitatively the $\Delta R_c/Rc$ trend with injection-strap distance indicating that ionisation of the neutrals by the plasma is the main physical process at play in the local increase of density in front of the antennas when using localised gas injection. The JET benchmarking of the EMC3-Eirene simulations against the experimental results also gives confidence that this simulation tool can confidently be used on ITER to assess the effect of localised gas injection on ICRH antenna performance.

## 4 Characterisation of ICRH enhanced PWI

Although ICRH is the main tool used to prevent W accumulation in the centre of JET plasmas, application of ICRH usually leads to an overall increase of the plasma impurity content, and in particular in JET-ILW, tungsten (W) and nickel (Ni). This is in general attributed to an enhanced Plasma Wall Interaction (PWI) and to sputtering of the Plasma Facing Components (PFC) when applying ICRH. It is also known that the rapid variation of the Radio-Frequency (RF) rectified potential across the equilibrium magnetic field can cause significant convective transport (E × B drifts) in the scrape-off layer (SOL) [9, 10]. One could argue that this RF driven SOL transport, can enhance the penetration of impurities from the SOL to the plasma core. Experiments were carried out in JET to assess this latter effect [11]. By injecting a fixed amount of the $N_2$ molecules from single location in L-mode discharges and toggling the ICRF power between the antennas, the influence of the proximity of the $N_2$ injection to the active antennas on the core N content was characterized via measurement of the N VII spectral line intensity. No significant influence of proximity of the active antenna on the core N content was observed when toggling the ICRF power between antennas A+B (close to $N_2$ injection) and D (away). This is independent of the order of the antennas. A similar result is obtained when using π, -π/2, or +π/2 strap phasing. These observations show that proximity of a powered antenna close to an impurity source does not significantly affect penetration of these impurities, ruling out the hypothesis that RF driven changes in edge transport plays a significant role in the enhanced impurity content when using ICRH.

The exact location of the RF driven impurity sources is still unknown, as is the exact release mechanism(s). In particular, there is no tungsten or nickel PFCs close to the ICRH antennas; the W-I line emission on tungsten PFCs (for example, divertor baffles, inner wall NBI shine-through protection) [12], did not show evidence of enhanced W sputtering when using ICRH. However RF sheath rectification [13] is suspected to play a significant role. In the past, SOL density modifications at locations magnetically connected few meters away from active ICRH antennas have been evidenced [14]. It is well possible that given the size of the antennas, a large area of tungsten (for example divertor baffles) is experiencing enhanced sputtering, however below spectroscopy diagnostic sensibility level. Experiments were recently conducted to address the question of the impact of the antenna geometry on RF induced PWI. First, an experiment was conducted to compare Be sputtering on objects connected either to a conventional A2 antenna or connected to the ILA antenna [15]. In L-mode plasmas, Be II line emission signal at the outer poloidal limiter locations 'D14' and 'K14' was monitored when the ICRH power was toggled between different antennas:

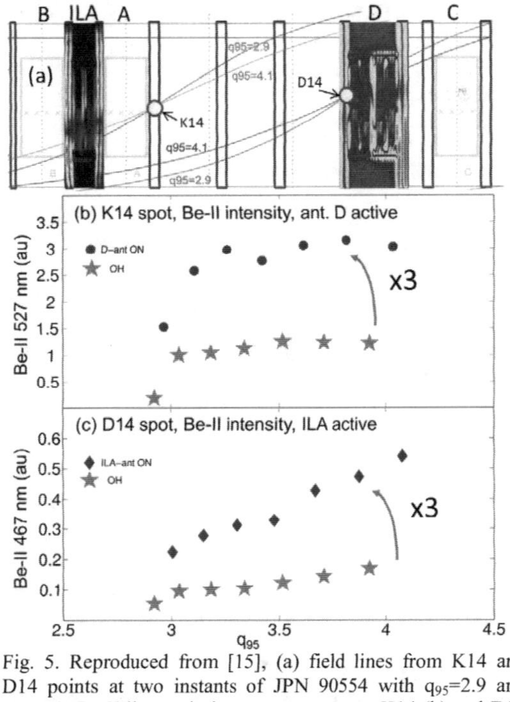


Fig. 5. Reproduced from [15], (a) field lines from K14 and D14 points at two instants of JPN 90554 with $q_{95}$=2.9 and $q_{95}$=4.1. Be II line emission measurement at K14 (b) and D14 (c) with resp. D and ILA active. 0.5 MW per antenna and –π/2 strap phasing was used.

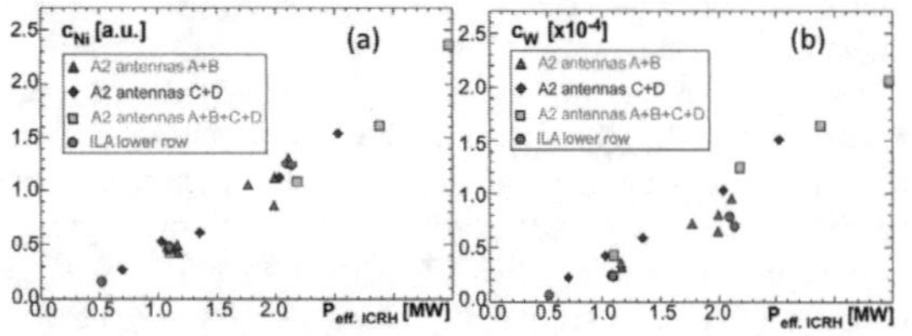


Fig. 6. JET plasma impurity content computed at plasma mid-radius from VUV spectroscopy vs ICRH power when using the ILA, A&B, C&D or all the A2 antennas, (a) Nickel and (b) Tungsten.

A&B, D, or ILA. D14 intercepts field lines passing few centimetres in front of the bottom of the ILA limiter while K14 intercepts field lines passing few centimetres in front of the top of antenna D limiter. In these pulses, $q_{95}$ was changed so as to scan the altitude of the field lines in front of the ILA (resp. D) limiters probed by D14 (resp. K14) measurement. A typical result is on Fig. 5. The same level of Be line emission increase (w.r.t. the ohmic level) is observed when probing in front of the active ILA or active antenna D limiters. Further analysis are ongoing to verify if these observations are coherent with sputtering models (using the ERO code [16]) and combined antenna /RF sheath rectification models [17] (using the TOPICA antenna code [18] and the SSWICH code [19]).

Secondly, the impact of using antennas with different geometries on core impurity content (W and Ni) was studied in a series of ICRH L-mode pulses where different combinations of antennas were energised [11]. Fig. 6 summarises the results. The Ni and W concentration are in the same ballpark for all combinations of antennas used, the ILA producing slightly less W and radiation when looking at a larger database. As a next step in the analysis, we will compare $E_{//}$ maps calculated by TOPICA and the rectified potentials modelled by SSWICH for the two antennas.

## 5 ICRH scenarios for JET DTE2

Recently, record DD fusion performances for JET-ILW plasmas were obtained; ICRH was extensively used in the baseline (proxy for the Q=10 ITER reference scenario with $I_p$ = 15MA at $q_{95}$~3) and hybrid (proxy for the Q = 5 t > 1000s $I_p$ = 12MA at $q_{95}$~4.3 ITER scenario) experiments aiming at preparing plasma scenarios for the future (2018-19) JET D-T campaign; this is illustrated for baseline pulses in Fig. 7. Reference [20] also reports on the use of ICRH in the hybrid scenario. Several factors contributed to the recent record performances, in particular, the availability of high power NBI and ICRH, and operation with reduced gas dosing. This latest condition has two implications: (i) discharges with lower gas dosing are more prone to impurity accumulation, then ICRH became a key ingredient, providing central heating and tungsten chase-out from the centre, to ensure stability of these discharges [21,22]; (ii) as the total gas dosing was reduced, OMP GIMs must be used to maximise ICRH power (see previous section). However some questions remain to be addressed in the context of D-T plasmas. The 'reference' ICRH scenario for D-T is

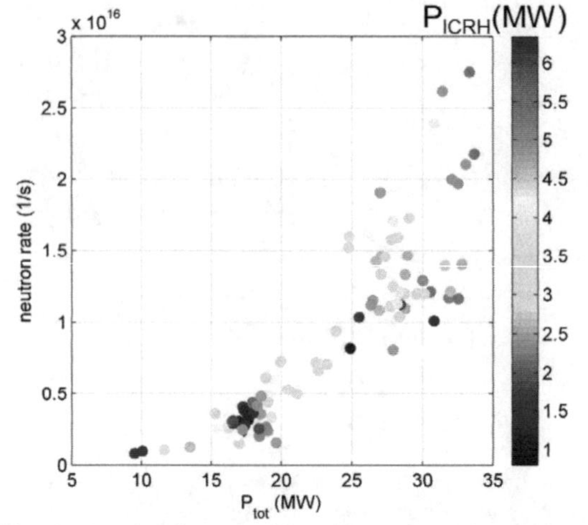


Fig. 7. For baseline scenario plasmas, neutron yield vs $P_{tot}=P_{NBI}+P_{ICRH}$ power. The data points are colour coded with the ICRH power, the N=1 (H)D scenario was used. The data base includes JET campaign C36 pulses with Ip>2.5 MA.

N=2 T heating assisted by N=1 minority $^3He$ heating; this scenario maximises ion heating and hence has the potential to boost fusion power [23]. However, some authors have emphasized the role of central electron heating that, aside from providing peaked temperature profiles, increases turbulent transport to prevent tungsten accumulation [24]. In this context the effectiveness of minority $^3He$ heating to chase-out W from plasma centre needs to be characterised. One can also envisage a scenario where hydrogen minority and $^3He$ minority are combined to act on W accumulation and fusion performance at the same time. Some experiments were recently carried out to compare different ICRH schemes in similar 2.7T/2MA H-mode discharges with $P_{NBI}$~15 MW and $P_{ICRH}$ in excess of 6 MW: (H)D heating with

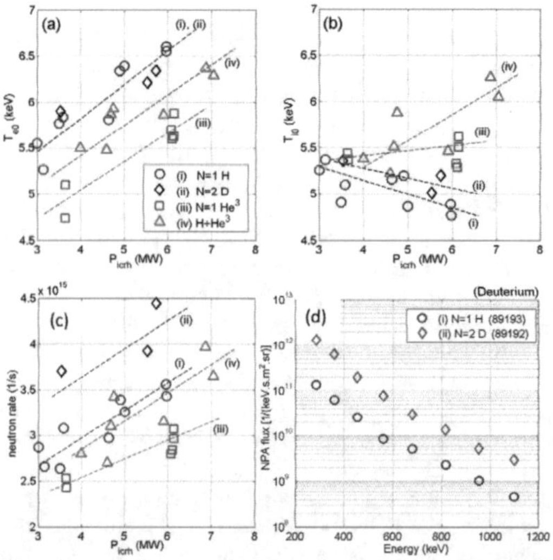


Fig. 8. Summary of experiments to compare different ICRH scenarios: N=1 H minority (and N=2 D heating when %H is low), N=1 $^3He$ minority, and combined heating. (a) Te, (b) Ti from CXS, (c) neutron yield. On (d) NPA measurement for JPN 89192 (%H<1%) and 89193 (%H=2.5%) are compared.

$f_{ICRH}$=51 MHz, ($^3$He)D heating with $f_{ICRH}$=32.5 MHz, and the combined heating scheme where both minorities are present in the plasma and the ICRH power is split between 51 MHz and 32.5 MHz. This was the first time this combined heating scenario was characterised, in particular we could demonstrate safe operation of (H)D ICRH with the presence of the N=1 $^3$He resonance at the JET inner wall.

A summary of these experiments is shown in Fig. 8. As expected, electron temperature was maximised when using (H)D ICRH; ion temperature was maximised when using the combined heating scheme. It must be emphasized that in the ($^3$He)D heating pulses the $^3$He concentration was < 2 % which is too low since the optimum RF power absorption is expected to take place for %$^3$He~7% [25]. The combined heating pulses which show a better ion heating were run with %$^3$He~3.5%. Because of the non-optimal conditions for %$^3$He in the pure ($^3$He)D heating pulses, no conclusion could be drawn about the effectiveness of this scenario to prevent W accumulation. Also, future similar experiments should ideally be done in conditions where, without ICRH, plasma is unstable due to W accumulation.

Also noticeable in Fig. 8 is the family of points where all the ICRH power was used with $f_{ICRH}$=51 MHz and with the hydrogen concentration reduced from ~2.5% (○) to less than 1% (◊) In this case, N=2 D heating plays a significant role, creating a population of energetic D ions leading to a 25% increase of the neutron yield w.r.t the (H)D standard case. This is supported by D fast ions measurements from the Neutral Particle Analyser and neutron time of flight spectroscopy (TOFOR) [25]. Similar observations were made in recent hybrid scenario pulses [20].

## 6 ICRH scenarios for JET H campaign and ITER non-active phase

The path towards the JET D-T campaign includes

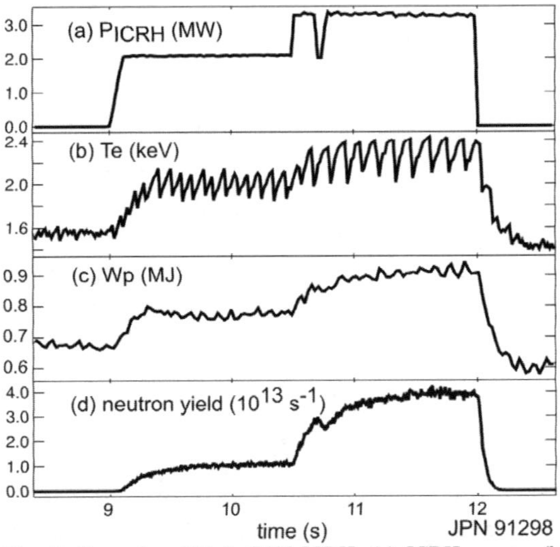


Fig. 9. Example of N=1 (D)H ICRH. (a) ICRH power, (b) central electron temperature, (c) MHD plasma energy, (d) neutron yield. The D concentration is this pulse was ~1.5%

Hydrogen and Tritium campaigns to characterise isotope effects on confinement properties and H-mode threshold in which ICRH is needed. It is also important to test the ICRH scenarios considered for the ITER non-active phase [26] in the ILW environment, and validate the modelling codes used to predict ICRH absorption in ITER. Some of these scenarios were never tested experimentally (the 3-ion scheme) or were tested in un-favourable conditions (N=1 (D)H in JET-C). The following summarises the ICRH scenarios used or tested during the JET 2016 hydrogen campaign.

- N=2 H majority, with $B_T$~1.8T/$f_{ICRH}$=51MHz, applicable for ITER 1/3 field non-active phase with the same $B_T/f_{ICRH}$ parameters: this scenario was extensively used in H-mode access experiments at low field and to characterize low activation p+Be nuclear reactions [27]; the ICRH aspects are reported in [28].

- N=1 ($^3$He)H with $B_T$~3.2T/$f_{ICRH}$=32.5MHz, applicable for the ITER full-field non-active phase with $B_T$=5.3T/$f_{ICRH}$=53MHz: this scenario was extensively used in the H-mode access experiments at higher field and in transport studies [29]. A recent report on this ICRH scheme is given in [30]. The heating efficiency evaluated using the brake-in slope analysis technique was $P_{abs}/P_{in}$~70% and heating performances of ~0.14 MJ/MW ($I_P$ = 2.5 MA) to ~0.11 MJ/MW ($I_P$ = 1.8 MA) were observed (note that the heating performance critically depends on plasma energy transport properties and hence on details of the experimental conditions). In this so called 'inverted scenario' it is important that the minority concentration is controlled to the optimal value of ~2%. At too small minority concentration, the heating degrades while at larger concentrations the mode conversion regime is entered and direct electron heating comes more significantly into play.

- N=1 (D)H with $B_T$~3.3T/$f_{ICRH}$=25MHz; the parent ($^4$He)H is applicable for the ITER full-field non-active phase with $B_T$=5.3T/$f_{ICRH}$=41MHz. This heating scenario was not accessible on JET-C because the concentration of carbon (2-3%) (carbon has the same Z/A=6/12 as deuterium) was such that a mode-conversion layer was located on the low field side of the D resonance layer [31]. In JET-ILW the heating scheme becomes available as the carbon concentration is a fraction of percent and Be, the main low-Z impurity has Z/A=4/9. Fig. 9 summarises the results of preliminary tests for this heating scheme performed in the last campaign. Plasma heating is indeed now observed; the heating efficiency was ~45% and the plasma performance was ~0.1 MJ/MW ($I_p$ = 2MA). The increase of the neutron yield and the detection of a fast deuterium tail by the TOFOR diagnostic (not shown) are other indications of D minority heating. Further experiments are planned to optimise this scheme, as the condition for maximum wave absorption, expected for %D~3%, was not reached.

- The 3-ion scheme [32] D-($^3$He)-H with $B_T$~3.2T and $f_{ICRH}$=32.5MHz where %D~20-30%; the equivalent $^4$He-($^3$He)-H scenario is applicable for the ITER full field non-active phase with $B_T$=5.3T/$f_{ICRH}$=53MHz: in this scenario, the D concentration is adjusted so that the D/H

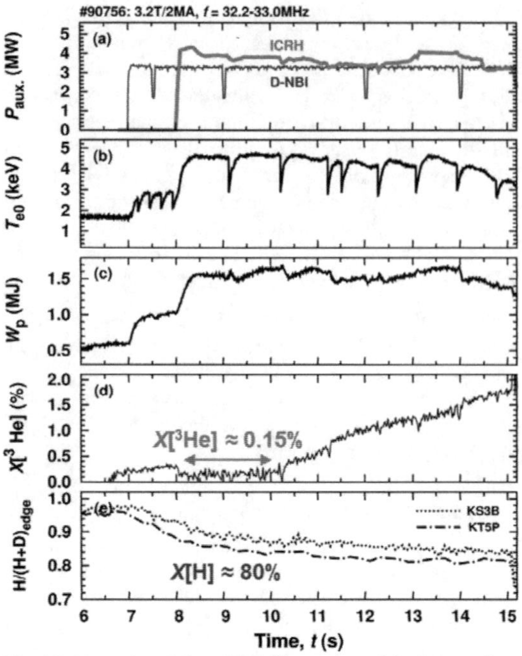


Fig. 10. Reproduced from [32], illustration of the 3 ion scheme heating (a) NBI and ICRH power waveforms, (b) Te, (c) MHD plasma energy, (d) $^3$He concentration and (e) D concentration. In this pulse $+\pi/2$ strap phasing was used.

hybrid resonance sits close to the N=1 $^3$He cyclotron resonance layer which, with this combination of $B_T$ and $f_{ICRH}$ is close to the plasma centre. In these conditions, the amplitude and polarisation of the fast wave electric field is favourable for efficient power transfer to $^3$He ions in tiny concentration (fraction of percent). Fig. 10 shows an example where the 3-ion scheme was used. The heating efficiency was ~70% and the plasma performance was ~0.15 MJ/MW ($I_p$ = 1.8MA and 2 MA). This heating scheme allows creation of a $^3$He fast population with MeV energies as was measured with the gamma-ray emission spectroscopy. Using $+\pi/2$ strap phasing enhanced the efficiency of fast ions generation and plasma heating; this is attributed to a lower $k_{//}$ and the inward pinch effect which results in a better confinement of the fast $^3$He population. We note that the 3-ion heating scheme can also potentially be applied to JET or ITER DT plasmas by choosing an appropriate D-T mix and adjusting $f_{ICRH}$ for resonant Be or fast T heating [32, 33].

## 7 Conclusions and prospects

ICRH played an important role in the high DD performances recently obtained in JET. Not only does it provide bulk plasma heating but central ICRH is also key for tungsten chase-out and stationarity of discharges. Optimisation of plasma operation to maximize the benefit of ICRH is now part of the scenario development, for example local OMP gas injection to increase antenna coupling is routinely used. Specific experiments were also carried-out to test/tune the ICRH scenarios in preparation for DT operation, or to test some of the heating schemes envisaged for the non-active ITER phase. In particular the dual frequency scheme (H)D and ($^3$He)D was tested. In addition, it was confirmed that the (D)H scenario is accessible in an ILW environment and the novel 3-ions ICRH scheme was validated experimentally. One open issue related to the use of ICRH in a metallic environment is the RF enhanced PWI. Experiments were conducted to compare impurity production using antennas with completely different geometries. The ILA produces slightly less high-Z impurities than the A2's and the PWI measured via Be line emission on limiters is in the same ballpark. This will be further studied using antenna and RF-sheath rectification models applied to JET. The availability of the ICRH system and the re-installation of the ILA are two factors that contributed to this progress. Work is on-going to further improve the RF generators availability and performance, and a Real-Time RF power management system is being implemented to maximise the launched ICRH power when operating close to voltage limits [5]. Remedial work is also in progress to operate the full ILA array, despite the hydraulic fluid leak that developed in the antenna. Ironically, this issue related to the ILA in-vessel matching capacitors validate the choice of external matching elements for ITER.

*This work has been carried out within the framework of the EUROfusion Consortium and has received funding from the Euratom research and training programme 2014-2018 under grant agreement No 633053 and from the RCUK Energy Programme [grant number EP/P012450/1]. The views and opinions expressed herein do not necessarily reflect those of the European Commission 'To obtain further information on the data and models underlying this paper please contact PublicationsManager@ukaea.uk.*

## References

1. M. Graham et al., Plasma Phys. Control. Fusion **54** 074011 (2012)
2. I. Monakhov et al., Nucl. Fusion **53** 083013 (2013)
3. F. Durodié et al., Plasma Phys. Control. Fusion **54** 074012 (2012)
4. P. Dumortier et al., these proceedings
5. I. Monakhov et al., these proceedings
6. P. Jacquet et al., Nucl. Fusion **56** 046001 (2016)
7. E. Lerche et al., J. Nucl. Mater. **463** 634–639 (2015)
8. W. Zhang et al., Nucl. Fusion **57** 056042 (2017)
9. M. Bécoulet et al., Phys. Plasmas **9** 2619–32 (2002)
10. W. Zhang et al., Plasma Phys. Control. Fusion **58** 095005 (2016)
11. V. Bobkov et al., Nuclear Materials and Energy, https://doi.org/10.1016/j.nme.2016.10.026
12. Vl. Bobkov et al., J. Nucl. Mater. **438** S160–S165 (2013)
13. F. W. Perkins et al., Nucl. Fusion **29** 583–92 (1989)
14. L. Colas et al., J. Nucl. Mater. **463** 735–738 (2015)
15. C.C Klepper et al., these proceedings
16. C. C Klepper et al., Phys. Scr. **T167** 014035 (2016)

DOI: 10.1051/epjconf/201715702004

17. A. Krivska et al., these proceedings
18. D. Milanesio et al., Nucl. Fusion **49** 115019 (2009)
19. L. Colas et al., Phys. Plasmas **19** 092505 (2012)
20. M. Mantsinen et al., these proceedings
21. E. Lerche et al., Nucl. Fusion **56** 036022 (2016)
22. M. Goniche et al., Plasma Phys. Control. Fusion **59** 055001 (2017)
23. D. Gallart et al., 43rd EPS Conf. on Plasma Physics (Leuven, Belgium, 4–8 July) P2.003 (2016)
24. C. Angioni et al., Nucl. Fusion **57** 056015 (2017)
25. D. Van Eester et al., Proc. 26th IAEA-FEC Conference, Kyoto, EX/P6-10
26. M. Schneider et al., these proceedings
27. A. Krasilnikov et al., submitted to Nucl. Fusion
28. E. Lerche et al., AIP Conf. Proc. **1689** 040003 (2015)
29. P. Mantica et al., Proc. 26th IAEA-FEC Conference, Kyoto, EX/P6-14 (2016)
30. D. Van Eester et al., these proceedings
31. M.-L. Mayoral et al., Nucl. Fusion **46**, S550–S563 (2006).
32. Y. Kazakov et al., these proceedings
33. J. Ongena et al., these proceedings

EPJ Web of Conferences **157**, 02005 (2017) DOI: 10.1051/epjconf/201715702005
*22 Topical Conference on Radio-Frequency Power in Plasmas*

# 2D full-wave simulation of waves in space and tokamak plasmas

*Eun-Hwa* Kim[1,*], *Nicola* Bertelli[1], *Jay* Johnson[2], *Ernest* Valeo[1], and *Joel* Hosea[1]

[1] Princeton Plasma Physics Laboratory, Princeton University, Princeton NJ 08540, USA
[2] Department of Engineering and Computer Science, Andrews University, Berrien Springs, MI 49103, USA

**Abstract.** Simulation results using a 2D full-wave code (FW2D) for space and NSTX fusion plasmas are presented. The FW2D code solves the cold plasma wave equations using the finite element method. The wave code has been successfully applied to describe low frequency waves in planetary magnetospheres (i.e., dipole geometry) and the results include generation and propagation of externally driven ultra-low frequency waves via mode conversion at Mercury and mode coupling, refraction and reflection of internally driven field-aligned propagating left-handed electromagnetic ion cyclotron (EMIC) waves at Earth. In this paper, global structure of linearly polarized EMIC waves is examined and the result shows such resonant wave modes can be localized near the equatorial plane. We also adopt the FW2D code to tokamak geometry and examine radio frequency (RF) waves in the scape-off layer (SOL) of tokamaks. By adopting the rectangular and limiter boundary, we compare the results with existing AORSA simulations. The FW2D code results for the high harmonic fast wave heating case on NSTX with a rectangular vessel boundary shows excellent agreement with the AORSA code.

## 1 Introduction

In the solar system, many planets including Mercury, Earth, Jupiter, Saturn, Uranus, and Neptune, and their moons are magnetized. The solar wind acts to confine the planetary magnetic field in a region known as the magnetosphere and interaction of the solar wind with these magnetospheres leads to the formation of large scale field aligned currents along with various electrostatic and electromagnetic waves.

Ultra-low frequency (ULF) waves are one example of such magnetospheric waves that exist in the lowest observed frequency range (mHz-Hz). One very common ULF wave occurs near the ion cyclotron frequency ($\Omega_{ion}$) and is called the ElectroMagnetic Ion Cyclotron (EMIC) wave.

ULF waves can lead to significant acceleration, transport, and loss of electrons during magnetic storms [1]. They can be a source for plasma transport at the magnetopause and have been shown to be responsible for ion outflows [2-3]. Such outflows may dominate the energy of the ring current during magnetic storms and lead to significant modifications of magnetospheric dynamics. EMIC waves are also considered to be a dominant radiation belt loss mechanism leading to rapid pitch angle scattering radiation belt electrons of almost all energies on the time-scale of seconds with subsequent loss to the ionosphere [4]. In addition, these waves have also been used as a diagnostic tool to measure plasma densities [5-6].

There are many numerical efforts to examine ULF waves in planetary magnetospheres. To date, many models for waves in the magnetosphere are based primarily on WKB-based methods, such as local dispersion relation analysis or ray tracing models [7-8]. While these models can follow the flow of energy due to a particular excited wave, this approach will fail if the wave enters a region in space in which it can couple power to other propagating modes via tunneling and /or mode conversion. One the other hand, time-dependent fluid wave models have been used for MHD waves [9] and EMIC waves [10]. Although they successfully demonstrate mode conversion process for MHD and EMIC waves, since MHD wave models are only valid for waves with $\omega << \Omega_{ion}$, where $\omega$ is angular frequency, and multi-ion fluid wave code adopted a slab coordinate so that the magnetic field curvature effect on waves could not be captured.

A recently developed 2D full-wave (FW2D) code using finite element method [11] overcomes these shortcomings using an approach that describes wave propagation, mode effects for arbitrary plasma and magnetic field configurations. The FW2D wave code has been successfully used to describe low frequency waves in Earth and Mercury's multi-ion magnetospheres (i.e., dipole geometry). The results include generation and propagation of externally driven ULF waves via mode conversion at Mercury [11]. Ref. [11] showed that efficient mode conversion from the fast compressional waves to the ion-ion hybrid (IIH) resonance occurs at Mercury and such mode-converted waves globally oscillate similar to field-line resonance at Earth. Mode coupling, refraction and reflection of internally driven field-aligned propagating left-handed EMIC waves at

* Corresponding author: ehkim@pppl.gov

DOI: 10.1051/epjconf/201715702005

Earth [12] has also been demonstrated. Ref. [12] suggested that polarization reversal and mode conversion occurs at the crossover frequency, where the left-hand (LHP) and right-hand (RHP) polarization modes cross each other in the dispersion relation, from LHP to RHP polarization and such RHP EMIC waves can propagate to the inner magnetosphere, which is consistent with 1D full-wave calculations [13, 14].

Furthermore, we have adapted the FW2D code to tokamak geometry and have used it to examine radio frequency (RF) waves in the scape-off layer (SOL) of tokamaks, the region of the plasma between the last closed flux surface (LCFS) and the tokamak vessel. Taking into account the SOL region is an important aspect because the interaction between fast waves and the SOL region can be significant. More details, specifically, for National Spherical Torus eXperiment (NSTX) can be found in Refs. [15-21]. This code is ideal for waves in SOL plasma, because realistic boundary shapes and arbitrary density profiles can be easily adopted in the code and the SOL plasma can be approximated as cold plasma.

In this paper, we use the FW2D code to examine EMIC waves at Earth's magnetosphere and RF waves in the SOL region of a tokamak plasma. It is found that (a) linearly polarized EMIC waves generated via mode conversion are localized near the magnetic equator; (2) FW2D and AORSA results show an excellent agreement in the rectangular boundary; and (3) FW2D results with a realistic limiter boundary are significantly different to results with the rectangular vessel boundary.

## 2 FW2D Model Description

We have developed a 2D full wave code, so called the FW2D code, appropriate for general geometries [7], which solves the cold plasma wave equations in two-dimensions. Assuming time dependence, $\exp(-i\omega t)$, the linear, cold plasma wave equation takes the form:

$$\nabla\times\nabla\times\mathbf{E}-\frac{\omega^2}{c^2}\boldsymbol{\varepsilon}\cdot\mathbf{E}=4\pi i\frac{\omega}{c^2}\mathbf{j}_e, \quad (1)$$

where $\mathbf{E}$ is the perturbed electric field, $c$ is a speed of light, $\boldsymbol{\varepsilon}$ is the dielectric tensor, and $\mathbf{j}_e$ is the (localized) external current source, which generates the waves within our model. In space plasma, the source could be instability in the equatorial region excited, for example, by free energy in velocity space, and its structure could be specified from kinetic simulations in the equatorial magnetosphere. Alternatively, the source could be external compressions in the outer magnetosphere that propagate energy into the inner magnetosphere.

The dielectric tensor $\boldsymbol{\varepsilon}$ is naturally expressed in coordinates aligned along and across the local ambient magnetic field direction $\mathbf{b}=\mathbf{B}/|\mathbf{B}|$. For an axisymmetric magnetospheric and tokamak model, Eq. (1) is most naturally expressed in cylindrical ($\mathbf{r}$, $\mathbf{z}$, $\boldsymbol{\phi}$) coordinates. The field is represented in terms of its projections along ($\mathbf{b}$) and perpendicular ($\boldsymbol{\xi}$, $\boldsymbol{\eta}$) to $\mathbf{B}$, thus we pick a local and orthogonal basis ($\mathbf{b}$, $\boldsymbol{\xi}$, $\boldsymbol{\eta}=\mathbf{b}\times\boldsymbol{\xi}$). For magnetospheric plasma, we define $\boldsymbol{\xi}=\boldsymbol{\phi}$ with the assumption that $\mathbf{B}\cdot\boldsymbol{\phi}=0$ while for tokamak geometry, we define $\boldsymbol{\xi}=\mathbf{r}\times\mathbf{b}$.

The wave equation in Equation (1) has been solved on an unstructured triangular mesh using the finite element method. One advantage of using the finite element method is that the local basis functions that are employed can be readily adapted to boundary shapes and can be packed in such a way as to provide higher resolution in regions where solutions may exhibit singular behavior.

## 3 Global structure of EMIC waves at Earth

We demonstrate the linearly polarized EMIC waves at Earth using the FW2D code. EMIC waves are in the Pc 1-2 (0.1-5Hz) frequency range, which are excited below the proton gyrofrequency and commonly observed in the magnetosphere. The polarization is one of the key characteristics that can be used to identify wave properties of EMIC waves. Their polarization is generally reported to be LHP, however, RHP and linear polarization have also been reported. Recent statistical studies showed that the occurrence of linearly polarized waves is comparable to LHP waves even in the source region [22]. Furthermore, these waves predominantly occur in the inner magnetosphere independent of magnetic local time (MLT) [23].

The dispersion relation of heavy ions indicates that any branch of the wave mode should be predominantly LHP or RHP. Since linear polarization occurs only near the crossover frequency or at oblique propagation near the multi-ion hybrid resonances, the linearly polarized EMIC waves are suggested to occur at such critical frequencies [6, 24-25].

Ref. [6, 24] calculated the efficiency of mode conversion of a radially propagating compressional wave at the IIH resonance using a simplified 1D slab cold plasma model that captures the essential features of the IIH resonance, however, global structure at Earth's magnetosphere has not been investigated.

By adopting realistic plasma density model, we examine global structure of such resonant waves using the FW2D code. In the simulation, we adopt a dipole magnetic field. We also use the global core plasma model (GCPM) [26], which provides an empirically derived plasma density as a function of geomagnetic and solar conditions throughout the inner magnetosphere. Figure 1 shows the plasma densities calculated by the GCPM at the MLT=12, where is on the magnetic meridian (magnetic line of longitude) facing the Sun, on 1/22/2000 for Kp = 1. Here, $r$ and $z$ are the radial and south-north directions normalized to Earth's radius ($R_E$). We also show contour of the Alfvén speed ($V_A$) in Figure 2. Because the electron density profile has an asymmetry structure between northern and southern hemispheres, the Alfvén speed profile in Figure 2 also shows asymmetry. In the inner magnetosphere the minimum value of Alfvén speed occurs at the magnetic equator near $r\sim 3.2$ $R_E$ (i.e., in the plasmasphere).

The waves are initially launched by specification of $\mathbf{j}_e = (0, 0, j_\phi)$ at $r = 6.4$ $R_E$ near the magnetic equator as a compressional fast waves (FW) and the frequency is $f$ = 0.4 Hz, which is between local oxygen ($\Omega_O$) and helium ($\Omega_{He}$) cyclotron frequencies. Reflecting boundary conditions are prescribed at the Earth's surface, while absorbing boundary conditions are used at all other boundaries.

Figure 3 shows two-dimensional wave solutions of the electric field components (a) $|E_\eta|$ and (b) $|E_\phi|$, respectively. In this figure, white dotted lines represent where the wave frequency matches $\Omega_O$ or $\Omega_{He}$, and the wave amplitude of $|E_\eta|$ and $|E_\phi|$ is normalized to the maximum value of the $|E_\eta|$.

The FW mode with a long wavelength along the $z$ direction appears in the $E_\phi$ component in Figure 3(b). For $\omega << \Omega_e$, $\omega_{pe}$, where $\omega_{pe}$ and $\Omega_e$ are electron plasma and cyclotron frequencies, dispersion relation of the FW mode can be written as

$$n_\perp^2 = \frac{(\varepsilon_R - n_\parallel^2)(\varepsilon_L - n_\parallel^2)}{(\varepsilon_S - n_\parallel^2)} \quad , \qquad (2)$$

where $n_\parallel$ ($n_\perp$) is refractive index parallel (perpendicular) to the ambient magnetic field, and $\varepsilon_R$, $\varepsilon_L$, and $\varepsilon_S$ are the components of the cold dielectric tensor [27]. Equation (2) exhibits two cutoffs at $n_\parallel^2 \sim \varepsilon_{R(L)}$. FW mode launched near $r$ = 6.4 $R_E$ propagates toward Earth quasi-perpendicular to the ambient magnetic field. These waves reach the cutoff location near $r \sim 4$ $R_E$ where $\varepsilon_L$ = 0 is satisfied and are partially reflected. The rest of the FW energy tunnels through the stopgap between the cutoff and resonance ($n_\parallel^2 \sim \varepsilon_S \sim 0$) and finally encounters the ionosphere. The FW amplitude in the inner magnetosphere is concentrated/trapped near the equatorial plane where Alfvén speed has minimum value as shown in Figure 3.

The IIH resonant waves appear only in $E_\eta$ showing field-aligned structure (See Figure 3a). Because the waves are initially launched as a pure FW mode, the waves exhibiting field-aligned structure in $E_\eta$ provide evidence of mode conversion [11]. The IIH resonant waves reach higher magnetic latitude, but the waves reflect before reaching the altitude where the frequency matches the oxygen gyrofrequency, which is unlike to the IIH resonant waves at Mercury [11]. Such localization of the EMIC waves, which is predicted by Ref. [28], is similar to the LHP EMIC waves [12]. Ref. [12] also showed that the EMIC waves generated in the LHP propagate along the field line and reach the heavier ion cyclotron frequency. However, the waves cannot tunnel through the cut-off and heavier ion cyclotron frequency even though the heavy ion concentration ratio is 5% (because the stop gap is too large compared with the wavelength). Because EMIC waves are simultaneously detected at the ground as well as space on conjugate field lines [29-30], it is particularly confounding that the full wave calculations show that the waves cannot reach higher magnetic latitude, and the resolution of this descrepancy likely involves kinetic effects.

# 4 HHFW propagation in the SOL of tokamak

Recent experimental studies of high harmonic fast wave (HHFW) heating on the NSTX [31] have demonstrated that up to 60% of the coupled HHFW power can be lost in the SOL (i.e., the region of the plasma between LCFS and tokamak vessel, where the magnetic field lines are open and the plasma density has a sharp profile) [15-16].

Numerical efforts to examine waves in the SOL have been performed using the AORSA code [32] and they showed a large amplitude electric field occurs in the SOL when edge densities are high enough that the fast waves can propagate in the SOL [20-21]. However, they adopted a rectangular boundary, which differs from the vacuum vessel structures. For comparison we adopt (1) a rectangular boundary to benchmark with the AORSA results and (2) a limiter boundary to examine boundary effects on HHFW propagation in the SOL of NSTX.

## 4.1 Comparisoin between FW2D and AORSA

In order to verify the FW2D, we have performed a benchmark between the AORSA and FW2D codes for a rectangular boundary case as shown in Ref. [20].

Figure 4 show the wave total amplitude of the wave electric field (normalized to its maximum value) calculated by AORSA (Figs. a and c) and FW2D (Figs. b and d) for a single toroidal mode $n_\phi$ = -21 (Figs. a and b) and $n_\phi$ =-12 (Figs. c and d) for NSTX shot 130608. In this figure, we define that $r$ and $z$ are radial and vertical directions of the tokamak, respectively, while $\phi$ is a toroidal direction. For the FW2D simulation, we adopt radial direction between 0.8m < $r$ < 1.7m to focus on the SOL and reduce computational time. We also assumed a strong collisions at the inner boundary near $r$ = 0.8m to avoid the wave reflection at left boundary. The wave frequency is $f$ = 30 MHz and the antenna is assumed to be located at $r$ = 1.575m. An electron density is assumed to have exponential profile as used in AORSA simulations [20-21] and electron density in front of the antenna is $N_{ant} = 1.5 \times 10^{18}$ m$^{-3}$ in electron-deuterium plasma.

In Figure 4, AORSA and FW2D simulations show excellent agreement. Even in the core plasmas, although the cold plasma model in FW2D, the FW2D solution shows a wave field behavior similar to the AORSA results. Therefore, these results confirm that the FW2D code is successfully verified.

## 4.2 Limiter boundary vs. rectangular boundary

Although previous numerical studies using AORSA clearly demonstrated that substantial HHFW power loss occurs in the SOL, the rectangular boundary differs from physical vacuum vessel structure and could affect the wave solutions.

In order to overcome this limitation, we adopt a limiter boundary to the FW2D code similar to Ref. [33] for various $N_{ant}$ as shown in Figure 5. In this case, we assume that the antenna is located along the limiter boundary for -0.5m < $z$ < 0.5m. Similar to rectangular boundary cases as shown in Figure 4, we adopt an artificial inner boundary at $r$ = 0.7m with strong collision. However we assume the limiter to be a perfect conductor, thus waves totally reflect at the boundary. For comparison, wave solutions from the AORSA code are also presented in Figure 5(d)-(f).

For $N_{ant} = 1 \times 10$ m$^{-3}$ in Figure 5(a) and 5(d), because the FW cut-off is 'closed', the FW field is strongly localized in front of antenna and waves are evanescent. On the other hand, for $N_{ant} = 2.0 \times 10^{18}$ m$^{-3}$ and $4.0 \times 10^{18}$ m$^{-3}$, the electron density is sufficiently large to 'open' the FW cut-off and thus standing waves clearly appear in the SOL.

The FW2D simulation shows that the maximum amplitude of the wave electric field for $N_{ant} = 4.0 \times 10^{18}$ m$^{-3}$ seems to be stronger than for $N_{ant} = 2.0 \times 10^{18}$ m$^{-3}$. On the other hand the AORSA shows the maximum amplitude of standing wave modes occur for $N_{ant} = 2.0 \times 10^{18}$ m$^{-3}$ and the peak of the wave electric field in the SOL for the AORSA simulation occurs outside region of the limiter boundary. In addition, for $N_{ant} = 4 \times 10^{18}$ m$^{-3}$, standing wave amplitude for $z > 0$ is stronger than for $z < 0$, which is different with respect to the AORSA showing stronger wave electric field appear at $z < 0$.

Therefore, the preliminary results in this paper clearly show that the boundary shape can significantly affect the FW propagation in the SOL and it is definitely necessary to adopt a realistic boundary shape of tokamak to investigate RF wave propagation.

## 5 Summary

In this paper, we present wave simulations using 2D full-wave FW2D code in Earth's magnetosphere and in the NSTX spherical tokamak. For the Earth case, we show strong mode conversion from incoming FW to linearly polarized IIH resonant waves occurs and such mode-converted waves are localized near the equatorial plane. Preliminary simulations for HHFW in NSTX show that the wave electic field in the SOL plasma is sensitive to the boundary shape (rectangular vs limiter boundaries) and suggest that the realistic boundary should be considered to determine wave power loss in the SOL.

The work at the Princeton University was supported by NASA grants (NNH13AV37I and NNH14AY20I) and DOE contract DE-AC02-09CH11466.

## References

1. S. R. Elkington, M. K. Hudson, and A. A. Chan, J. Geophys. Res. **108**, 1116 (2003)
2. L. C. Lee, J. R. Johnson, and Z. W. Ma, J. Geophys. Res. **99**, 17405, (1994)
3. J. R. Johnson, J. R., C. Z. Cheng, and P. Song, Geophys. Res. Lett. **28**, 227 (2001).
4. A. Y. Ukhorskiy et al., Geophys. Res. Lett. **37**, L09101 (2010)
5. K. Takahashi, R. E. Denton, R. R. Anderson and W. J. Hughes, J. Geophys. Res. **111**, 1201 (2006)
6. E.-H. Kim, J. R. Johnson, H. Kim, and D.-H. Lee, J. Geophys. Res. **120**, 6464 (2015)
7. J. L. Rauch, J. L. and A. Roux, J. Geophys. Res. **87**, 8191 (1982)
8. K. Rönnmark and M. André, J. Geophys. Res. **96**, 17573 (1991)
9. D.-H. Lee and R. L. Lysak, J. Geophys. Res. **94**, 17097 (1989)
10. E.-H. Kim, J. R. Johnson and D. -H. Lee, J. Geophys. Res. **113**, A11207 (2008)
11. E.-H. Kim, J. R. Johnson, E. Valeo, and C. K. Phillips, Geophys. Res. Lett. **42**, 5147–5154 (2015)
12. E.-H. Kim and J. R. Johnson, Geophys. Res. Lett., **43**, 13-21(2016)
13. J. R. Johnson, J. R., T. Chang, and G. B. Crew, Phys. Plasmas **2**, 1274 (1995)
14. J. R. Johnson and C. Z. Cheng, Geophys. Res. Lett. **26**, 671 (1999)
15. J. Hosea et al., Phys. Plasmas **15**, 056104 (2008)
16. Perkins et al., Phys. Rev. Lett. **109**, 045001 (2012)
17. Perkins et al., Nucl. Fusion **53**, 083025 (2013)
18. Perkins et al., Phys. Plasmas **19**, 042501 (2012)
19. Perkins et al., Phys. Plasmas **23**, 070702 (2016)
20. N. Bertelli et al., Nucl. Fusion **54**, 083004 (2014)
21. N. Bertelli et al., Nucl. Fusion **56**, 016019 (2016)
22. R. C. Allen el al., J. Geophys. Res. **120**, 5574 (2015)
23. A. A. Saikin et al., J. Geophys. Res. **120**, 7477 (2015)
24. D.-H. Lee et al., J. Geophys. Res. **113**, A11212 (2008)
25. B. J. Fraser, H. J. Singer, M. L. Andrian, D. L. Gallagher and M. F. Thomsen, in Inner Magnetosphere Interactions: New Perspectives from Imaging (edited by J. L. Burch, 366 M. Schulz, and H. Spence), American Geophysical Union, Washington, D. C., p55 (2005)
26. D. L. Gallagher, P. D. Craven and R. H. Comfort, J. Geophys. Res. **105**, 18819 (2000)
27. T. H. Stix, Waves in plasmas, Springer (1992)
28. E.-H. Kim, J. R. Johnson, and D.-H. Lee, J. Astron. Space Sci. **32**, 289-295, (2015)
29. S. Perraut et al., J. Geophys. Res. **89**, 195 (1984)
30. B. J. Anderson, R. E. Erlandson and L. J. Zanetti, J. Geophys. Res. **97**, 3075 (1992)
31. S. M. Kaye et al., Nucl. Fusion **55**, 104002 (2015)
32. E. F. Jaeger et al., Phys. Plasmas **8,** 1573 (2001)
33. D. L. Green et al., Phys. Rev. Lett. **107**, 145001 (2011)

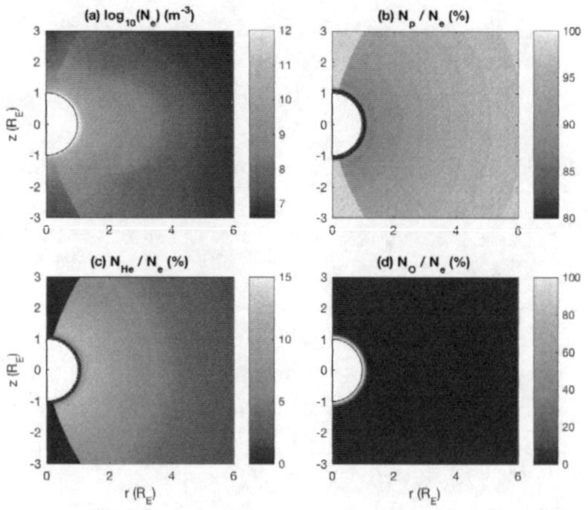


**Fig. 1.** Plasma density profle calculated using the global core plasma model (GCPM) on 1/22/2000 for MLT = 12 and Kp = 1 ; (a) Electron density ($N_e$) in logarizm scale and (b-d) proton ($N_p$), helium ($N_{He}$), and oxygen ($N_O$) density ratio to the electron density, respectively.

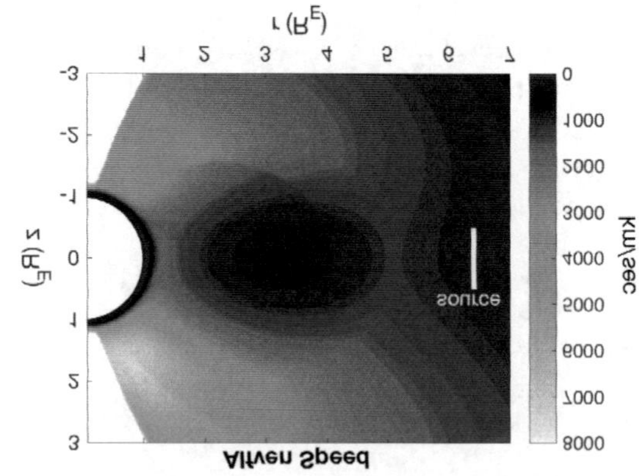

**Fig. 2.** Alfven speed profile calculated using density profiles in Figure 1 and dipole magnetic field configuration. Yellow bar near $r/R_E$ = 6.4 is where the initial FW waves are launched.

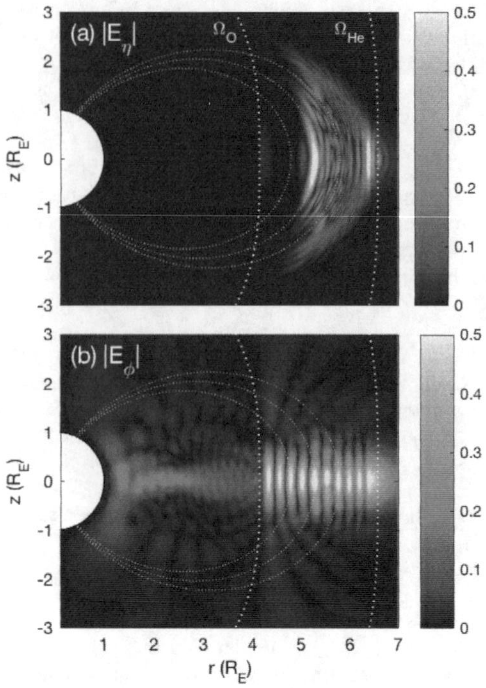


**Fig. 3.** Wave solutions using the FW2D; (a) $|E_\eta|$, electric field amplitude normal direction to local magnetic field line, (b) $|E_\phi|$, electric field amplitude azimuthal (toroidal) component, respectively. Here each power has been normalized to the maximum value of $|E_\eta|$. Bold white dotted lines indicate the oxygen ($\Omega_O$) and helium ($\Omega_{He}$) cyclotron freuqencies, and light dotted lines are magnetic field line at $r_{equator}/R_E$ = 4.8, 5.3, and 5.8, respectively.

 DOI: 10.1051/epjconf/201715702005

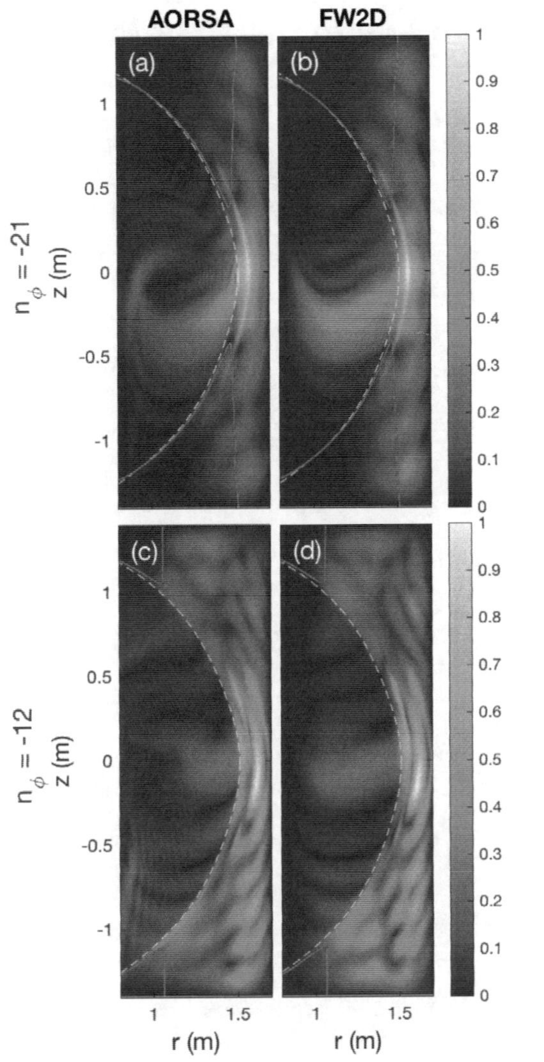


**Fig. 4.** Two-dimensional simulation results of the total electric field amplitude using the AORSA and FW2D codes for (a-b) $n_\phi$ = -21 and (c-d) $n_\phi$ = -12 for $N_{ant}$ = 1.5 × $10^{18}$ $m^{-3}$. Each electric field amplitude has been normalized to its maximum value. White and red solid lines are LCFS and FW cut-off, respectively.

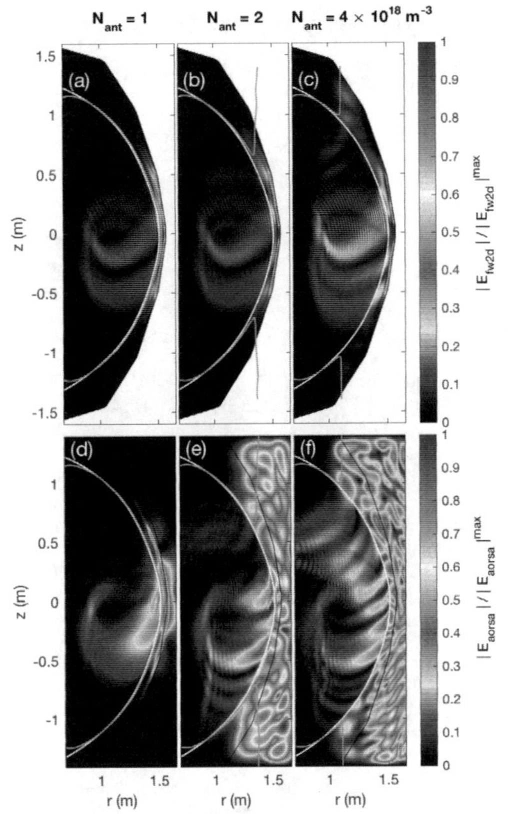


**Fig. 5.** Two-dimensional electric field amplitude from (a-c) FW2D simulation with a limiter boundary and (d-f) AORSA simulation assuming a rectangular boundary for $N_{ant}$ = 1, 2, and 4 × $10^{18}$ $m^{-3}$, respectively. In this figure, white, black and light red solid lines are the LCFS, the limiter boundary, and the FW cut-off, respectively.

 DOI: 10.1051/epjconf/201715702006

# Synergetic heating of D-NBI ions in the vicinity of the mode conversion layer in H-D plasmas in JET with the ITER like wall.

*J. Ongena*[1,*], *Ye.O. Kazakov*[1], *Y. Baranov*[2], *C. Hellesen*[3], *J. Eriksson*[3], *T. Johnson*[4], *V.G. Kiptily*[2], *M.J. Mantsinen*[5,6], *M. Nocente*[7,8], *R. Bilato*[9], *A. Cardinali*[10], *C. Castaldo*[10], *K. Crombé*[1,11], *A. Czarnecka*[12], *R. Dumont*[13], *J. Faustin*[14]. *L. Giacomelli*[8], *V. Goloborodko*[15], *J. Graves*[16], *Ph. Jacquet*[2], *N. Krawczyk*[12], *E. Lerche*[1,2], *L. Meneses*[17], *M.F.F. Nave*[17], *H. Patten*[16], *M. Schneider*[18], *D. Van Eester*[1], *H. Weisen*[17], *J.C. Wright*[19] *and JET Contributors*[†]

EUROfusion Consortium, JET, Culham Science Centre, Abingdon, OX14 3DB, UK.

[1] Laboratory for Plasma Physics, LPP-ERM/KMS, TEC Partner, 1000 Brussels, Belgium.
[2] Culham Centre for Fusion Energy (CCFE), Culham Science Centre, Abingdon, OX14 3DB, UK.
[3] Department of Physics and Astronomy, Uppsala University, Sweden.
[4] KTH Royal Institute of Technology, 114 28 Stockholm, Sweden.
[5] Barcelona Supercomputing Center (BSC), 08034 Barcelona, Spain.
[6] ICREA, Pg. Lluis Companys 23, 08010 Barcelona, Spain.
[7] Dipartimento di Fisica, Università di Milano-Bicocca, 20126, Milan, Italy.
[8] Istituto di Fisica del Plasma, CNR, 20125 Milan, Italy.
[9] Max-Planck-Institut für Plasmaphysik, 85748 Garching, Germany.
[10] ENEA Centro Ricerche, Frascati, Italy.
[11] Department of Applied Physics, Ghent University, 9000 Gent, Belgium.
[12] Institute of Plasma Physics and Laser Microfusion, 01-497 Warsaw, Poland.
[13] CEA, IRFM, 13108 Saint-Paul-Lez-Durance, France.
[14] Max-Planck-Institut für Plasmaphysik, 17491 Greifswald, Germany
[15] Institute for Nuclear Research, Kyiv 03680, Ukraine.
[16] EPFL, Swiss Plasma Center (SPC), 1015 Lausanne, Switzerland.
[17] Instituto de Plasmas e Fusão Nuclear, IST, Universidade de Lisboa, 1049-001 Lisboa, Portugal.
[18] ITER Organization, Route de Vignon-sur-Verdon, 13067 St Paul-lez-Durance Cedex, France.
[19] Massachusetts Institute of Technology, Plasma Science and Fusion Center, Cambridge, MA 02139, USA.
[†] See the author list of "Overview of the JET results in support of ITER" by X. Litaudon et al., *Nucl. Fusion* 57, 102001 (2017)

**Abstract.** This paper discusses the extension of the 'three-ion' species ICRF technique for heating mixture plasmas using fast injected NBI ions as resonant 'third' species. In this scenario the ICRF power is absorbed by the fast beam ions in the vicinity of the mode conversion layer where the left-hand polarized RF electric field $E_+$ is strongly enhanced. The ions in the beam velocity distribution that have a Doppler-shifted resonance close to the mode conversion layer efficiently absorb RF power and undergo acceleration. We show first experimental observations of ICRF heating of D-NBI ions in H-D plasmas in JET with the ITER-like wall. In agreement with theoretical predictions and numerical modelling, acceleration of the D-NBI ions in this D-($D_{NBI}$)-H scenario is confirmed by several fast-ion measurements. An extension of the heating scheme discussed here is acceleration of T-NBI and D-NBI ions in D-T plasmas, offering the potential to further boost the *Q*-value in future D-T campaigns in JET.

## 1. Introduction

Plasma heating in the ion cyclotron range of frequencies (ICRF) is widely used in magnetic confinement fusion research. The fast magnetosonic waves excited in the plasma by the RF currents on the antenna strap(s) can undergo a large variety of collisionless interactions, both with ions and electrons [1]. Strong wave-particle interaction is possible when the wave frequency matches the Doppler-shifted cyclotron frequency or one of its harmonics of the plasma ions:

$$\omega = n\,\omega_{cs} + k_{||}v_{||}\ (n = 0, \pm1, \pm2, \ldots). \quad (1)$$

Here, $\omega$ is the angular frequency of the magnetosonic wave, $\omega_{cs} = |q_s|\,B/m_s$ the local cyclotron frequency of

* Corresponding author: j.ongena@fz-juelich.de

particle species '*s*' with charge $q_s$ and mass $m_s$, $k_{\parallel}$ is the wavenumber and $v_{\parallel}$ the particle velocity parallel to the confining magnetic field. A review of traditionally used ICRF scenarios is given in [2-4], and recent advances in physics and technology of ICRF are summarized in [5].

The electric field associated with the fast magnetosonic waves in the plasma can be written as the sum of a component rotating in the sense of the ions (left-hand polarized $E_+$), and one rotating in the opposite direction (right-hand polarized $E_-$). The RF power $P_{abs}$ absorbed by the ions at the $n^{th}$ harmonic is given by

$$P_{abs} \propto |E_+ J_{n-1}(k_\perp \rho_L) + E_- J_{n+1}(k_\perp \rho_L)|^2 \qquad (2)$$

with $k_\perp$ the component of the wave vector perpendicular to the magnetic field, $\rho_L$ the Larmor radius of the resonant ion and $n$ the harmonic number of the cyclotron frequency. $J_m$ is the Bessel function of the first kind and $m^{th}$ order. For fundamental heating ($n = 1$), wave absorption by thermal and moderately energetic particles (energies of a few hundred keV) is to first order determined by the left-hand polarized component $E_+$, i.e. $P_{abs} \propto |E_+|^2$, while the contribution to power absorption due to $E_-$ is negligibly small. Indeed, for these ions $k_\perp \rho_L \ll 1$ resulting for $n = 1$ in $J_{n-1}(k_\perp \rho_L) = J_0(0) \simeq 1$ and $J_{n+1}(k_\perp \rho_L) = J_2(0) \simeq 0$. For more energetic ions, typically in the MeV-energy range (such that $k_\perp \rho_L$ is no longer negligibly small), the situation is more complicated and both the RF-field components $E_+$ and $E_-$ contribute to wave absorption.

The parallel velocity needed for resonant interaction by ions with waves with an angular frequency $\omega$ follows from Eq. (1):

$$v_{\parallel} = (\omega - n\,\omega_{cs})/k_{\parallel} \qquad (3)$$

This condition can be locally fulfilled for thermal ions (low $v_{\parallel}$) if the wave frequency closely matches with one of the ion cyclotron frequencies $\omega = n\,\omega_{ci}$ ($n = 1, 2, \ldots$). Larger values of the difference $\omega - n\,\omega_{ci}$, require higher values for $v_{\parallel}$. This condition alone is not sufficient for efficient power absorption by the ions. Indeed, as follows from Eq. (2), the local $E_+$ component needs also to be sufficiently large and this is the case at the ion-ion hybrid resonance / L-cutoff pair located in between the resonance layers of the two main plasma species. In what follows we will call this pair of layers the Mode Conversion (MC) layer (see Fig. 1).

In the case of minority heating, where the concentration of one of the two ion species is typically a few percent, this MC layer is located rather close to the resonance of the minority species, and thermal minority ions can then profit from the large $E_+$ associated with this layer to efficiently absorb power from the injected magnetosonic waves. However, if the concentration of the minority species becomes larger (typically ~ 10-20%), the MC layer moves radially outward, away from the minority cyclotron resonance. As a result, thermal ions can then no longer reach this layer and instead dominant electron absorption takes over.

There is an alternative way to achieve dominant ion absorption under such conditions: extend the plasma composition beyond two thermal ion species only and provide an additional population of ions that resonantly absorb RF power close to the MC layer. Two possibilities to do so follow immediately from Eq. (1). First, by adding a third ion species to the plasma with a different $Z/A$ value than that of the two main ions such that $(Z/A)_2 < (Z/A)_3 < (Z/A)_1$. Such 'three-ion' ICRF scenarios have been theoretically developed in [6, 7], and later confirmed in dedicated JET and Alcator C-Mod experiments [8].

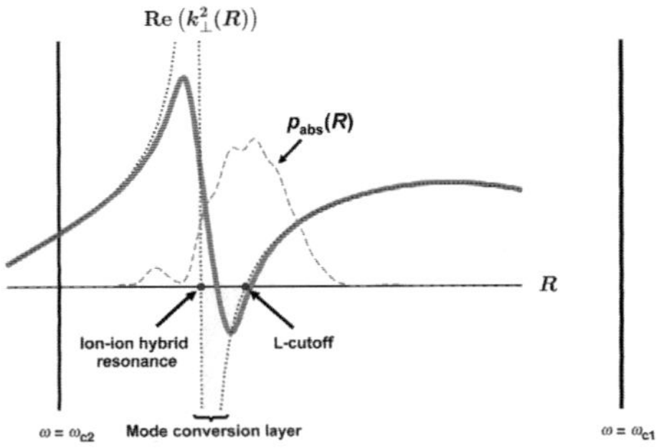


Fig. 1: Typical dispersion curve (hot-plasma approximation) for the fast magnetosonic wave in a mixture plasma, without (blue dotted line) and with a population of fast beam ions (red solid line). The position of the mode conversion layer between the ion cyclotron layers $\omega = \omega_{c1}$ and $\omega = \omega_{c2}$ can be controlled by adjusting the plasma composition. In the presence of fast ions (injected NBI ions, fusion-born alphas, etc.), efficient absorption of ICRH power occurs in the immediate neighbourhood of the mode conversion layer (red dashed line).

The second option relies on using fast ions with sufficiently large parallel velocities to make use of the Doppler-shift term $k_{\parallel}v_{\parallel}$ in Eq. (1). Indeed, for such fast ions the condition for resonant wave-particle interaction, Eq. (1), is fulfilled at a different radial location than for thermal ions. The radial separation between the resonant layer for the thermal ions $\omega \approx \omega_{ci}$ and that for the fast ions $\omega = \omega_{ci} + k_{\parallel}v_{\parallel}$ is given by

$$\Delta R_{fast} \approx n_{tor} \times v_{\parallel,fast}/\omega, \qquad (4)$$

where $n_{tor}$ is the toroidal wave number in the antenna spectrum ($k_{\parallel} \approx n_{tor}/R$). In this case, the $Z/A$ value of resonant fast particles does not necessarily have to differ from that of the main plasma ions, $(Z/A)_1$ and $(Z/A)_2$. Fast ions provided by the neutral beam injection (NBI) system are a natural choice for such particles. Remark that the same physics applies to energetic ions born in various fusion reactions (e.g. $^4$He ions from D + T and D + $^3$He reactions). They can therefore also effectively absorb RF power in the vicinity of the MC layer in

DOI: 10.1051/epjconf/201715702006

mixture plasmas. Absorption of ICRF power by alpha particles in D-T plasmas is well-known and has been studied earlier, see e.g. Fig. 4 in Ref. [9].

Note that the Doppler-shifted ion cyclotron absorption by D-beam ions in D plasmas was earlier reported on JET [10, 11]. In those experiments there was no MC layer present in the plasma, and the double-pass absorption was computed to be 5–15% [11]. Although upon crude inspection the two techniques look rather similar, the heating scenario reported here is in two aspects fundamentally different, as (i) a mixture plasma is used with a well defined plasma composition such that (ii) the fast beam ions can profit from the large $E_+$ located close to the MC layer resulting in a greatly enhanced double-pass absorption (> 80%).

## 2. Three-ion ICRH experiments using fast beam ions as resonant particles: experimental conditions

To test the possibility of ICRH absorption by fast beam particles in the vicinity of the MC layer, a number of dedicated pulses were performed in H-D plasmas with various values for the isotopic H/D ratio in JET with the ITER-like wall (ILW).

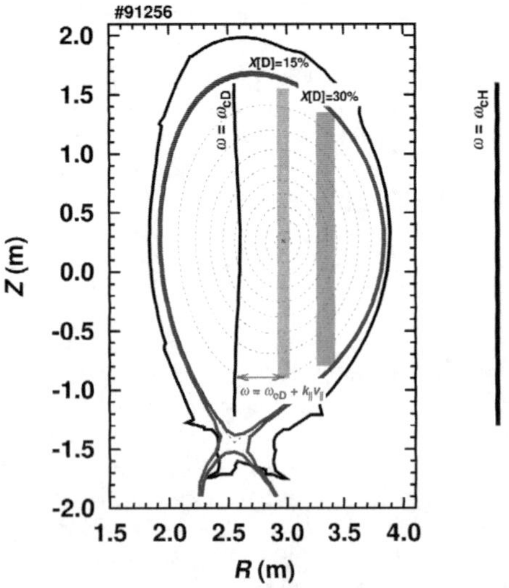


Fig. 2: Tokamak poloidal cross-section with the location of the ion cyclotron resonance layers for thermal D ions for JET pulse #91256 ($B_0 \approx 2.9$T, $I_p = 2$MA, $f = 25$MHz). The computed locations of the mode conversion layer in H-D plasmas are also shown for $X$[D] = 15% and $X$[D] = 30% ($n_{tor} = 27$ and $X$[$^9$Be] = 0.5% assumed). The reported synergetic ICRF +NBI scenario relies on wave absorption in the vicinity of the mode conversion layer by NBI-injected fast D ions through the Doppler-shifted fundamental ion cyclotron interaction $\omega = \omega_{cD} + k_{//}v_{//}$.

The operational conditions for the experiments described here were as follows: H-D plasma with $X$[D] ≈ 10–15%, $B_0 \approx 2.9$T, $I_p = 2$MA, $n_{e0} \approx 4\times10^{19}$ m$^{-3}$, dipole antenna phasing at $f = 25$MHz. The location of the ion cyclotron resonances for those operational parameters is shown in Fig. 2.

The JET tangential D-NBI injectors ($v_{||}/v \approx 0.62$) were used at an acceleration voltage of $E_0 \approx 100$ keV. This defines the maximum parallel velocity $v_{\parallel}^* = 2\times10^6$ m/s in the slowing-down distribution of the injected beam ions, and beam particle velocities range thus from $v_{\parallel} = v_{\parallel}^*$ to thermal ion velocities. Taking into account that the spectrum for dipole phasing of the ICRH antennas in JET shows a maximum for the toroidal wavenumber $n_{tor} = 27$ and that the fastest D-NBI ions have a characteristic Doppler shift $\Delta R_{fast} \approx$ 35-40 cm, (Eq. (4)) we thus choose the value for the toroidal magnetic field and ICRH frequency such that the ion cyclotron resonance layer for thermal D ions is located at $R_{ic,D} \approx 2.6$ m, i.e. about 40 cm from the magnetic axis towards the high field side (HFS). The cyclotron resonance of H ions is then located at $R_{ic,H} \approx 5.2$ m, i.e. outside the plasma at the low field side (LFS).

H-D mixture plasmas with various isotopic compositions can be used to realize the D-($D_{NBI}$)-H scenario. We recall that in our experiments D concentrations can be used up that value for which the outward shifted position $R_{MC}$ of the MC layer is 'overshooting' the Doppler shifted resonance position of the fastest beam ions in the distribution. In other words, the useful D concentrations in the plasma have to be such that the following condition is fulfilled:

$$R_{MC} \lesssim R_{ic,D} + \Delta R_{fast} \quad (5)$$

For every percent increase in the D concentration in the plasma, the L-cutoff and the ion-ion hybrid resonance layer shift to the LFS by about 2.8 cm and 2.2 cm, respectively. The position of the MC layer in H-D plasmas can thus be approximated as $R_{MC} \approx R_{ic,D} \times (1 + X[\mathrm{D}])$. Combining this with Eq. (5) then leads immediately to the following estimate of the upper limit of the concentration for the D ions:

$$X[\mathrm{D}] \lesssim \Delta R_{fast}/R_{ic,D} \quad (6)$$

For the experimental parameters in the experiments reported here, this yields $X[\mathrm{D}]_{max} \approx 15\%$. At higher D concentrations $X[\mathrm{D}] > X[\mathrm{D}]_{max}$ (e.g. 30% as also indicated in Fig. 2), the MC layer shifts further to the LFS and the separation between the $\omega = \omega_{cD}$ and the MC layer becomes too large. Since for such conditions no beam ions have parallel velocities that allow them to resonate at the MC layer, electron damping through mode conversion becomes dominant. Maximizing absorption by fast NBI ions (and thus minimizing electron absorption) under those conditions is still possible, but requires higher NBI injection energies and/or ICRH operation with larger dominant $k_{||}$.

For the calculation of the location of the MC layer and optimal plasma conditions we have to take into account the total concentration of D ions (thermal and beam ions). For the described experiments, the fraction of fast D-NBI ions in the plasma core was ranging between 2% and 4%. The presence of those ions modifies the fast wave dispersion as shown by the red solid curve in Fig. 1. The corresponding radial profile for the ICRF power absorption (red dashed line in Fig. 1) shows that RF power is absorbed by the fast ions in the vicinity of the MC layer. It is important to realize that the D-NBI injectors not only provide a seed of resonant fast ions, but also contribute to the H-D plasma ratio, via thermalized D-ions. This means that the proposed ICRF + NBI heating technique can also be applied for heating nearly pure H plasmas, where the necessary D ion concentration is provided by the thermalized fast ions from D-NBI injection.

## 3. Experimental results

Figure 3 shows the waveforms for JET pulse #91256: (a) NBI and ICRF power, (b) the temporal evolution of the central electron temperature $T_{e0}$, (c) plasma stored energy $W_p$, (d) the neutron rate and (e) the number of gamma-ray counts with $E_\gamma = 3.37$ MeV from one of the channels of D + $^9$Be nuclear reactions.

The edge isotopic ratio H/(H+D) in this pulse varied between 0.92 and 0.90. At $t = 7.5$ s, about 3.5 MW of D-NBI power (obtained using two tangential PINIs) is injected into the plasma, resulting in central electron temperatures $T_{e0} \approx 2.0$–$2.2$ keV, and plasma stored energy $W_p \approx 1.05$ MJ ($\Delta W_p/\Delta P_{NBI} \approx 0.13$ MJ/MW). At $t = 9.0$ s, 2.5 MW ICRF power (dipole phasing at $f = 25$ MHz) is coupled to the plasma.

As soon as the ICRH system is switched on, the central electron temperature increases up to $T_{e0} \approx 4.0$ keV and the plasma stored energy reaches $W_p \approx 1.45$ MJ ($\Delta W_p/\Delta P_{ICRH} \approx 0.15$ MJ/MW). While the increase in plasma stored energy per MW of injected NBI or ICRF power is similar, the ratio $\Delta T_{e0}/\Delta P_{aux}$ is significantly higher for the ICRH +NBI heating phase, namely 0.64 keV/MW vs. 0.27 keV/MW.

Figure 4 shows the electron temperature profile during various heating phases of the pulse and shows clearly the difference between the $T_e$ profiles obtained during application of NBI-only (rather broad power deposition) and during the combined ICRH +NBI heating phase. It clearly shows that the ICRF scenario discussed here provides a rather localized core plasma heating.

The stabilization of the sawtooth oscillations is a first clear indication for the presence of fast particles due to synergetic ICRH+NBI acceleration: from sawteeth with periods $T_{saw} \approx 0.11$s during the NBI-only heating phase of the pulse ($t = 7.5$–$9.0$ s) to $T_{saw} \approx 0.53$s during the combined 2.5MW ICRH + 3.5MW NBI heating phase ($t = 9.0$–$11.5$ s). Further confirmation of synergetic ICRF+NBI acceleration of the D-beam ions follows from the large increase in the neutron yield (panel (d) in Fig. 3) and from measurements of gamma rays with energy $E_\gamma = 3.37$ MeV, resulting from nuclear reactions between $^9$Be nuclei and D particles (panel (e) in Fig. 3). The minimal fast D energy needed to observe gamma rays from the $^9$Be+D reaction is ~500keV [12]. The modulation in both signals is due to the expulsion of the fast particles from the centre associated with the sawtooth crashes.

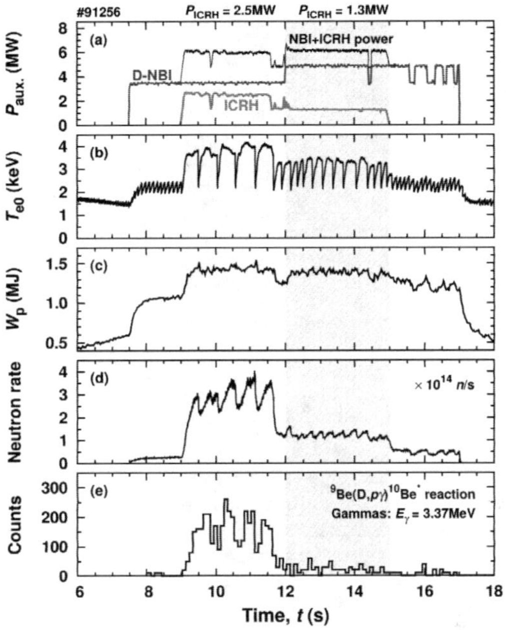


Fig. 3: Evolution of various parameters in JET pulse #91256 ($B_0 \approx 2.9$T, $I_p = 2.0$ MA, $f = 25$ MHz, dipole phasing, H/(H + D) ≈ 0.90): (a) auxiliary heating power (NBI, ICRH and NBI+ICRH); (b) central electron temperature $T_{e0}$; (c) plasma stored energy $W_p$; (d) neutron rate; (e) number of gamma-ray counts with $E_\gamma = 3.37$ MeV, produced in one of the channels of the D + $^9$Be nuclear reaction. With 2.5 MW of coupled ICRH power a population of energetic D ions is generated in the plasma as confirmed by the sawtooth stabilization and the increase in the neutron yield and gamma-ray measurements.

The determining role of ICRH for accelerating D particles to high energies is further illustrated in the second phase of the discharge ($t = 12$–$15$s), where the coupled ICRH power was reduced to 1.3MW and the NBI power was increased to 4.9MW, in order to keep the total heating power approximately constant. This led to a shorter sawtooth period ($T_{saw} \approx 0.25$s) and a reduced number of gamma ray counts.

## 4. Comparison with earlier results obtained in D-$^3$He plasmas

Acceleration of fast D beam particles in plasma mixtures was already seen before in D-$^3$He discharges on TFTR and JET [13–15]. In JET pulse #55712 the plasma consisted of ~25% $^3$He and correspondingly 40-50% D [14]. The applied magnetic field was 3.45 T and the ICRH frequency $f = 33$ MHz. As a result, the distance between the ion cyclotron resonance layers of thermal D and $^3$He ions is about 80 cm. The MC layer in this D-$^3$He plasma shifts towards the HFS over about 1.6cm for every percent of added $^3$He. A concentration of about 25% $^3$He thus locates the MC layer close to the plasma centre, at nearly the same distance of about 40 cm from both the D and $^3$He cyclotron layers.

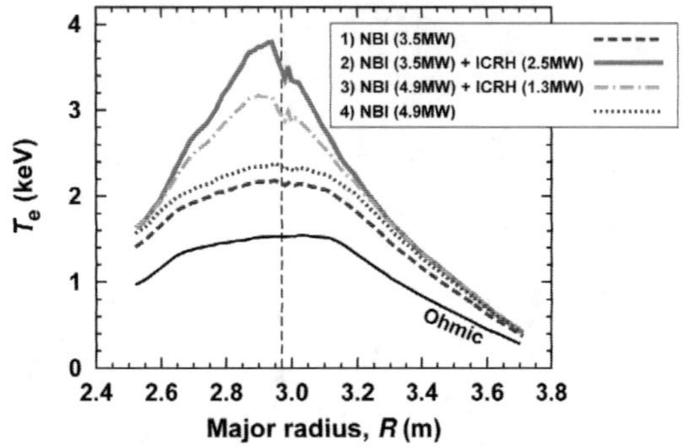


Fig. 4: Electron temperature profiles for JET pulse #91256 during the ohmic, NBI-only and ICRH+NBI heating phases, averaged over a 1s time interval. The dashed vertical line corresponds to the location of the magnetic axis. With 3.5 MW of D-NBI power $T_{e0}$ increases from 1.5 keV to 2.2 keV ($\Delta T_{e0}/\Delta P_{NBI} \approx$ 0.2keV/MW). By adding 2.5MW of coupled ICRH power, $T_{e0}$ increases further up to 3.8keV ($\Delta T_{e0}/\Delta P_{ICRH} \approx$ 0.65keV/MW) and centrally peaked temperature profiles are achieved.

Taking into account the beam injection parameters ($E_0 \approx 120$ keV) and dominant $k_{\parallel}$ value in the antenna spectrum (dipole phasing) of the earlier experiments, we find that the resonance layer for the fast D-NBI particles is shifted over about 40cm towards the LFS. Thus, it coincides with the MC layer, creating the necessary conditions for efficient synergetic acceleration of D-NBI ions with ICRH. That this is indeed the case is illustrated by the strong increase in the neutron yield and long neutron decay times when a blip of D-NBI injection is applied (see Fig. 5, red lines in the bottom panel).

In pulse #55714, for otherwise similar conditions, $^3$He ions were replaced by the same amount of $^4$He ions. As illustrated by the blue dotted line in the bottom panel of Fig. 5, the D-$^4$He plasma shows a strongly reduced neutron rate and much faster neutron decay. The decay time of the neutron rate in #55714 is consistent with the slowing down time for 120keV D-NBI ions [14], indicating no extra acceleration of D-NBI ions with ICRH in D-$^4$He plasmas, in contrast to the D-$^3$He case (#55712). Indeed, D and $^4$He ions have the same $(Z/A) = 1/2$ and thus there is no MC layer in the plasma and no local amplification of the $E_+$ field. In fact, pulse #55714 is very similar to the discharges studied in [10, 11], and highlights that fundamental ICRH heating of beam ions can be made significantly much more efficient in plasmas with a purposely chosen particle composition.

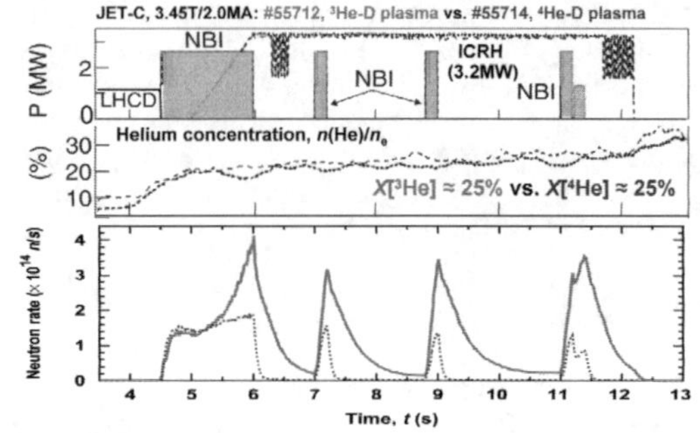


Fig. 5: Time evolution and comparison of various plasma signals from JET pulse #55712 and #55714 (adapted from Fig. 21 of Ref. [14]). The plasma in pulse #55712 consisted of a mixture of D with ~25% of $^3$He; in pulse #55714 $^3$He ions were replaced by $^4$He for otherwise similar conditions. The presence of a MC layer in pulse #55712 leads to synergetic acceleration of D-NBI ions to high energies.

This comparison nicely illustrates the key role for the presence of the MC layer and choosing the optimal plasma mix: the same ICRH power absorption mechanism is at work in both the H-D discharges reported here and the D-$^3$He discharges discussed in [14].

## 5. Relevance for D-T operation

The insights gained from the experiments reported here leads to a direct extension for use in D-T plasmas. In this section, we briefly discuss how to extend the scenario discussed here towards D-T plasmas in a future DTE2 campaign on JET.

The foreseen parameters of the JET T-NBI system after the recent upgrade are as follows: $E_{max}$ = 118 keV, power fractions for full-, half- and third-energy components 63%, 26%, 11%, respectively [16]. This results in an average energy for the injected beam ions $\langle E_{T-NBI} \rangle \approx$ 80keV and an upper estimate for the effective temperature $T_{eff,T} = 2/3 \langle E_{T-NBI} \rangle \approx 53$ keV (the average energy and $T_{eff}$ for the actual slowing-down distribution of beam ions will be smaller). From Ref. [17] follows that the fusion reactivity in D-T plasmas is maximized at $3T_D + 2T_T \approx 330$ keV. Thus for plasma temperatures in the range $T_i \approx 5$–15keV, the optimal temperature for

DOI: 10.1051/epjconf/201715702006

T and D ions to maximize the fusion reactivity should be $T_T \approx 150$keV and $T_D \approx 110$keV, respectively. This can be realized with appropriate T-($T_{NBI}$)-D or T-($D_{NBI}$)-D heating scenarios in D-T plasmas, as explained below.

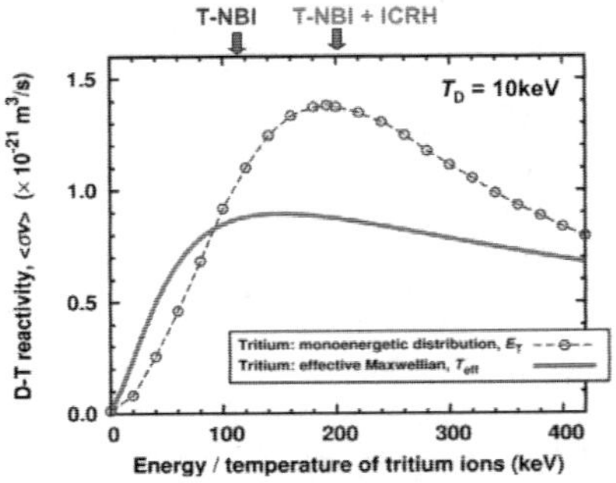


Fig.6: D-T reaction rate $<\sigma$ v$>$ in a Maxwellian D plasma ($T_D$= 10keV) resulting (i) from the injection of a mono-energetic T beam (blue dashed line), and (ii) from thermonuclear reactions with a Maxwellian T ion population with an effective temperature $T_{eff} > T_D$ (red full line).

Figure 6 illustrates D-T $<\sigma$ v$>$ values computed using as an approximation for the accelerated tritium beam particle distribution (i) a mono-energetic $T_{NBI}$ beam (using the injection energy as a characterizing parameter) and (ii) a Maxwellian distribution (using the tritium effective temperature $T_{eff}$ as characterizing parameter). The maximum fusion yield $<\sigma$ v$> \approx 1.4 \times 10^{-21}$ m$^3$/s is achieved at $E_T \approx 200$ keV in the first case and $<\sigma$ v$> \approx 0.9 \times 10^{-21}$ m$^3$/s at $T_T \approx 150$ keV in the second case. The real optimum $<\sigma$ v$>$ value should be located somewhere between these two extremes.

From Fig. 6 follows that increasing the energy of the injected T-NBI ions by a factor of ~2 should maximize the fusion output in such plasmas. Moderate acceleration of T-NBI ions can be done in D-T plasmas with T concentrations up to ~ 30% at $f$ = 25 MHz and at a toroidal magnetic field $B_0 \approx$ 3.7 T or higher. The advantage of the proposal here compared to earlier tritium minority ICRH experiments in JET [18] is that we now make use of more optimal frequencies for the JET ICRH system (25 MHz vs. 23MHz) and that the plasma mix is closer to the optimal D-T mixture.

Figure 7 shows 2D TORIC [19] calculations for the ICRF power deposition in a D-T plasma with the following operational parameters in JET-ILW: $B_0$ = 3.7T, $f$=25 MHz, $n_{tor}$ = 27, $n_{e0} = 6\times10^{19}$m$^{-3}$, $T_0$ = 5 keV. The concentration of the various ions is $X$[D] = 68%, $X$[T] = 30%, $X$[$T_{NBI}$] = 5% and $X$[$^9$Be] = 0.5%. The energy distribution of the injected T-NBI ions is approximated by a Maxwellian with an effective temperature $T_{eff}$ = 55 keV. For these conditions, TORIC predicts that most of coupled ICRF power is absorbed by the fast T-NBI ions ($p_{abs}$($T_{NBI}$) = 55%) at $R \approx 2.6$ m, in the close vicinity of the MC layer in the D-T plasma.

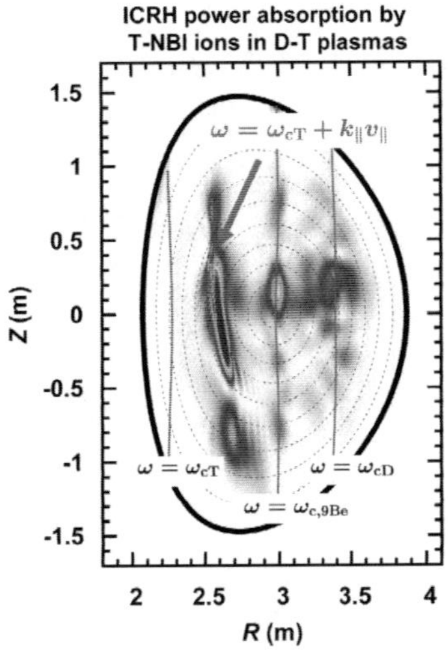


Fig. 7: 2D TORIC computations of the ICRF power absorption for the T-($T_{NBI}$)-D three-ion scenario, for a D-T plasma containing 30% tritium ions. Off-axis absorption of RF power in the vicinity of the MC layer will moderate the acceleration of the T-NBI ions, favourable to maximize the D-T fusion yield.

In contrast to harmonic ICRH absorption scenarios such as $\omega = 2\omega_{ci}$ and $\omega = 3\omega_{ci}$ that naturally produce MeV-range energetic ions, the proposed ICRH technique allows a more modest acceleration of the NBI ions in D-T plasmas required to maximize the fusion reactivity. Moreover, similar to T-minority experiments reported in [18], off-axis synergetic ICRF acceleration of $T_{NBI}$ ions can also be used. This has the additional advantage to reduce the local absorbed RF power density, and is a further possibility to prevent excessive acceleration of the T-NBI ions in order to avoid 'overshooting' the maximum in $<\sigma$ v$>$. As discussed in Ref. [20], additional actuators for controlling fast-ion energies are varying the plasma density ($T_{eff} \propto 1/n_e^2$) and concentration of resonant ions (e.g. by varying NBI power). Similarly, D-NBI ions can also be used, accelerating them in D-T plasmas through the T-($D_{NBI}$)-D heating scenario. The proposed ICRF + NBI scenarios offer thus an additional tool to maximize the $Q$-value in the forthcoming DTE2 campaign on JET and are compatible with the existing technological constraints.

## 6. Conclusions

Absorption of ICRH power by fast D-NBI ions in H-D plasmas can be made very effective by arranging the plasma conditions such that the MC layer is located in the close vicinity of the Doppler-shifted resonance layer of the injected fast ions. This is an extension of the three-ion species ICRH technique reported earlier [6–8] to convert traditional electron heating scenarios (through mode conversion) to effective ion absorption scenarios. Synergetic ICRH+NBI acceleration of D-NBI ions was also at work in earlier D-$^3$He plasmas [13–15]. A promising extension of this ion cyclotron heating scheme is to accelerate T-NBI and D-NBI ions, which has the potential to further boost the Q-value in future D-T campaigns on JET.

## Acknowledgements

The support from the JET team is warmly acknowledged. This work has been carried out within the framework of the EUROfusion Consortium and has received funding from the Euratom research and training programme 2014-2018 under grant agreement No 633053. The views and opinions expressed herein do not necessarily reflect those of the European Commission and the ITER Organization.

## References

1. ITER Physics Basis, Chapter 6: Plasma auxiliary heating and current drive. Nucl. Fusion, Vol. 39, No. 12, 2496-2539 (1999).
2. Adam, J. Review of tokamak plasma heating by wave damping in the ion cyclotron range of frequency. Plasma Phys. Control. Fusion **29**, 443–472 (1987).
3. Porkolab, M. et al.. Recent progress in ICRF physics. Plasma Phys. Control. Fusion **40**, A35–A52 (1998).
4. Noterdaeme, J.-M. et al.. Physics studies with the additional heating systems in JET. Fusion Sci. Tech. **53**, 1103–1151 (2008).
5. Ongena, J. et al., Recent advances in physics and technology of ion cyclotron resonance heating in view of future fusion reactors. Plasma Phys. Contr. Fusion **59**, 054002 (2017).
6. Kazakov, Ye.O. et al, On resonant ICRF absorption in three-ion component plasmas: a new promising tool for fast ion generation. Nucl. Fusion **55**, 032001 (2015).
7. Kazakov, Ye.O. et al, A new ion-cyclotron range of frequency scenario for bluk ion heating in deuterium-tritium plasmas: How to utilize intrinsic impurities in our favour. Phys. Plasmas **22**, 082511 (2015).
8. Kazakov, Ye.O. et al, Efficient generation of energetic ions in multi-ion plasmas by radio-frequency heating, Nature Physics (2017), advance online publication; http://dx.doi.org/10.1038/nphys4167
9. Kazakov, Ye.O. et al., Study of ICRH scenarios for thermal ion heating in D-T plasmas, Nucl. Fusion 52, 094012 (2012).
10. Krasilnikov, A.V. et al., Fundamental ion cyclotron resonance heating of JET deuterium plasmas, Plasma Phys. Control. Fusion **51,** 044005 (2009).
11. Lerche, E.A. et al., Modelling of D majority ICRH at JET: impact of absorption at the Doppler-shifted resonance. Plasma Phys. Control. Fusion **51**, 044006 (2009).
12. Kiptily, V.G. et al., Gamma ray diagnostics of high temperature magnetically confined fusion plasmas. Plasma Phys. Contr. Fusion **48**, R59-R82 (2006).
13. Darrow, D.S. et al., Enhanced loss of fast ions during mode conversion ion Bernstein wave heating in TFTR, Nucl. Fusion **36**, 509–513 (1996).
14. Mantsinen, M.J. et al., Localized bulk electron heating with ICRF mode conversion in the JET tokamak. Nucl. Fusion **44**, 33–46 (2004).
15. Van Eester, D. et al., JET ($^3$He)-D scenarios relying on RF heating: survey of selected recent experiments. Plasma Phys. Contr. Fusion **51**, 044007 (2009).
16. Ciric, D. et al., Performance of upgraded JET neutral beam injectors. Fusion Eng. Design **86**, 509-512 (2011).
17. Nath, D. et al., Thermonuclear Fusion Reactivities for Drifting Tri-Maxwellian Ion Velocity Distributions. J Fusion Energy **32**, 457–463 (2013).
18. Lamalle, P.U. et al., Expanding the operating space of ICRF on JET with a view to ITER. Nucl. Fusion **46**, 391–400 (2006).
19. Bilato, R. et al., Simulations of combined neutral beam injection and ion cyclotron heating with the TORIC-SSFPQL package. Nucl. Fusion **51**, 103034 (2011).
20. Stix, T.H., Fast-wave heating of a two-component plasma. Nucl. Fusion **15**, 737–754 (1975).

# Current Challenges in the First Principle Quantitative Modelling of the Lower Hybrid Current Drive in Tokamaks

Y. Peysson[1*], P. T. Bonoli[2], J. Chen[3], A Garofalo[3], J. Hillairet[1], M. Li[4], J. Qian[4], S. Shiraiwa[2], J. Decker[5], B. J. Ding[4], A. Ekedahl[1], M. Goniche[1], X. Zhai[3]

[1]*CEA, IRFM, F-13108, Saint-Paul-lez-Durance, France*

[2]*PSFC, MIT, MA-02139, Cambridge, USA*

[3]*General Atomics, CA-92186-5606, San Diego, USA*

[4]*ASIPP, 230031, Hefei, China*

[5]*EPFL, CRPP, F-13108, Lausanne*
**yves.peysson@cea.fr*

**Abstract.**

The Lower Hybrid (LH) wave is widely used in existing tokamaks for tailoring current density profile or extending pulse duration to steady-state regimes. Its high efficiency makes it particularly attractive for a fusion reactor, leading to consider it for this purpose in ITER tokamak. Nevertheless, if basics of the LH wave in tokamak plasma are well known, quantitative modeling of experimental observations based on first principles remains a highly challenging exercise, despite considerable numerical efforts achieved so far. In this context, a rigorous methodology must be carried out in the simulations to identify the minimum number of physical mechanisms that must be considered to reproduce experimental shot to shot observations and also scalings (density, power spectrum). Based on recent simulations carried out for EAST, Alcator C-Mod and Tore Supra tokamaks, the state of the art in LH modeling is reviewed. The capability of fast electron bremsstrahlung, internal inductance $l_i$ and LH driven current at zero loop voltage to constrain all together LH simulations is discussed, as well as the needs of further improvements (diagnostics, codes, LH model), for robust interpretative and predictive simulations.

## 1. Introduction

With the highest known current drive (CD) efficiency in tokamaks and off-axis power absorption, the rf wave at the Lower Hybrid (LH) frequency is particularly attractive for current profile shaping or steady-state operation [1]. In this context, first principle modeling has started long time ago and is still active after more than 30 years of research. Thanks to the challenges concerning ITER physics performances by appropriate current profile tailoring, an active international modeling activity has been carried out during the last five years, involving quantitative comparisons between simulations and experiments for several tokamaks (Tore Supra, Alcator C-Mod, EAST) [2,3]. In this paper, a review of main achievements is presented, which highlights the state of the art in this field of physics. While modeling tools are able to reproduce quantitatively observations in low density plasmas, simulation becomes challenging when LH wave absorption becomes weak, i.e. for plasmas at high density and low temperature. In this regime characterized by a rather large spectral gap, the level of agreement between simulations and experimental results may be significantly improved by a broadening of the launched power spectrum at the separatrix. The potential influence of edge plasma physics on core absorption is one the many challenges to be tackled concerning LH current drive in tokamak, in particular for ITER.

## 2. Numerical tools used by LH current drive first principle modeling

The aim of first principles modeling is to build a consistent interpretative framework of LH experiments with reliable predictive capabilities for existing tokamaks and ITER reactor. Such an objective represents a considerable challenge because of the multiple physical mechanisms involved concerning propagation and absorption. To this day, there is still a remarkable contrast between the reproducibility of LH experiments and the sensitivity of LH modelling. Multi-machine modeling using well benchmarked numerical tools with quantitative comparison against experimental observations is the reference path to reach this ambitious goal.

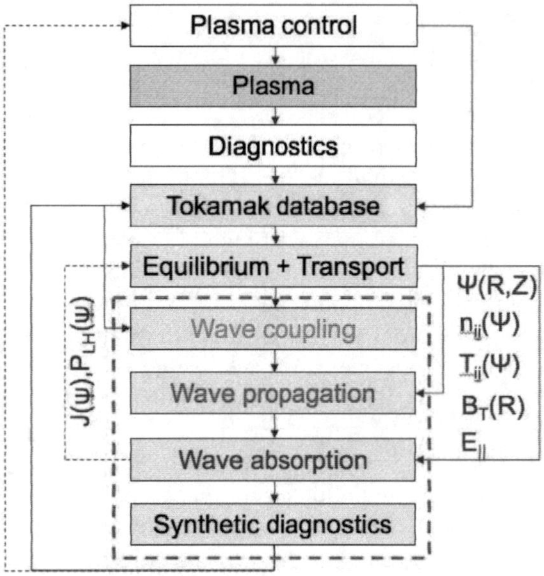


*Figure 1. General scheme for first principle modelling for rf heating and current drive in tokamaks.*

As shown in Fig. 1, the general scheme for first principle modeling consists to couple different models describing successively wave propagation from the antenna in the Scrape-Off Layer (SOL), while wave propagation and absorption of the LH wave inside the separatrix are determined by solving numerically Maxwell and Boltzman equations in the geometrical optics and Fokker-Planck approximations respectively [2]. The standard tools used for this purpose are C3PO/LUKE or GENRAYCQL3D ray-tracing/Fokker-Planck codes [4,5,6,7]. In presence of a tail of fast electrons that are pulled out from the thermal bulk by resonant Landau interaction with the LH wave to high kinetic energies, non-thermal bremsstrahlung in the hard X-ray (HXR) photon energy range provides a natural diagnostic with an almost direct insight on the electron momentum distribution function built-up at different radial positions, an invaluable diagnostic to validate models and numerical tools. The R5-X2 code coupled to LUKE calculates the fast electron bremsstrahlung (FEB) and an analogous calculation is performed inside CQL3D itself, using the same numerical kernel for the quantum-relativistic cross-sections [8].

All the concepts for first principle modeling LH wave current drive were already established about 30 years ago. Since, progresses have concerned principally more realistic descriptions of the LH wave and electron dynamics in momentum and radial spaces, allowing systematic quantitative comparisons between simulations and experiments, which represents itself a significant achievement. In the meantime, powerful toroidal full-wave codes have been developed (TORLH, LHEAF) [9,10,11]. Although both are the same full-wave code for LH waves, the two codes took different sort of complimentary approaches. TORLH started from TORIC, which uses a spectral basis and solved a dense matrix problem, while LHEAF uses the finite element method and solves a sparse matrix problem. Both codes have advantages and issues. For the LH wave, the electron Landau damping is the only term which requires a dense matrix and even TORLH simplifies a dispersion relation to eliminate hot plasma branch. This fact has been exploited in LHEAF, making the code becomes more efficient in terms of computer resource. However, the approach in LHEAF eliminates some of physics such as a mode conversion at the LH resonance, which could be of interest to investigate in some specific situations. On the other hand, the spectral approach does not allow a realistic description of the SOL while some further improvements must be considered for the launcher implementation in TORLH. It is important to recall that TORLH has been used to validate a posteriori some critical assumptions considered in ray-tracing calculations, which is particularly important for present day simulations [9]. A systematic use of full-wave codes for first principle modeling is likely one the challenges in this domain, thanks to the very large computational effort, whatever the considered method.

The first step in LH current drive modeling is to calculate the power spectrum that must be considered at the plasma separatrix, by performing of FFT of the wave electric pattern along the toroidal direction (z). This is done by wave coupling codes like ALOHA [12], considering the antenna structure as arrays of square waveguides along the toroidal and poloidal directions. Depending upon the waveguide phasing along the toroidal direction, the LH power spectrum is made of a set of lobes characterized by a Gaussian-like shape, each of them centered at a toroidal refraction number $N_z$, with a width $\Delta N_z$ resulting from a FFT of the antenna size in the toroidal direction. Uniform variations of density in front of the antenna by any process (ponderomotive effects at high input power, gas feed,...) do not modify spectral lobe positions, but their relative heights, changing the overall directivity, i.e. the relative fraction of power that generates co- or counter-current in the plasma [13]. The determination of the directivity has a large impact in code predictions of the total LH current, especially for antennas who turns out to be very sensitive to edge density variations like at 2.45 GHz in EAST tokamak. In this case, the consistency between edge density deduced from reflection coefficient (RC) and measured directly by probes in front of the antenna is essential to guarantee the robustness of the launched power spectrum. For the antenna at 2.45 GHz in EAST tokamak, a large discrepancy between RC and density measurement is observed as shown in Fig. 2, a problem already encountered in Tore Supra LH experiments [14].

If the SOL is turbulent, depending upon the spatial characteristics of the density fluctuations, lobe positions may change randomly, as well as their amplitude [15]. This approach is providing a *a posteriori* interpretation of the tail LH model, which was introduced to explain successfully HXR profiles in Tore Supra [16]. The transfer of power from the main lobe to the tail is a free parameter, as illustrated in Fig. 3, and significant effects on the predicted current density and HXR profiles require at least that $P_{tail}/P_{tot} > 0.5$. Beyond this threshold, predicted results are weakly sensitive to this ratio, making the model rather robust. Ray-tracing calculations are used for describing LH wave propagation that is separated from the absorption process, as weak damping approximation is always satisfied. The well collimated beam and WKB approximations are applicable to a large domain in the plasma, except at cut-off and caustics, which may represent an issue when ray length is long in weak absorption regime. However, thanks to full-wave calculations, errors are likely benign for cut-off near the plasma edge, while, at caustics, the impact of this problem on overall simulations remains more open [17]. Since ray-tracing captures toroidal refraction from B-field inhomogeneity, a natural mechanism in tokamaks due to toroidal curvature, it is an appropriate tool for bridging the spectral gap leading to wave absorption, while for $N_{\|0}$ corresponding to the main lobe, the plasma is in principle transparent. However, this mechanism is not necessarily the single one that could contribute to LH wave absorption, and the difficulty in first principle modeling is to identify the possible existence of additional external physical mechanisms for bridging the spectral gap.

DOI: 10.1051/epjconf/201715702007

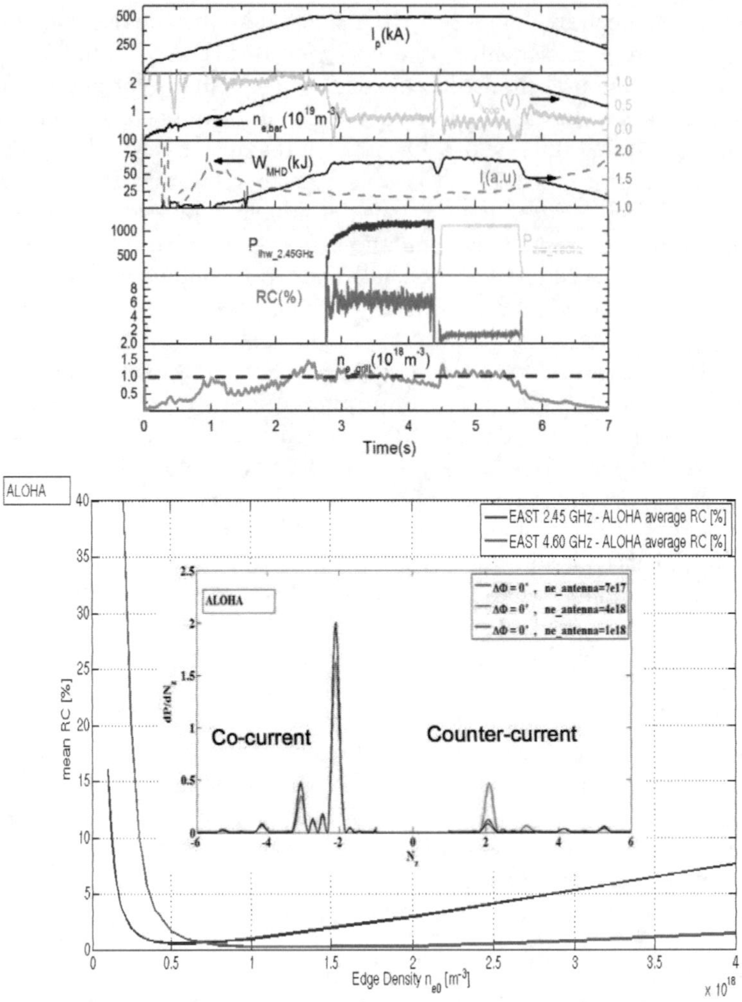


*Figure 2. (Top) Waveform of the reflection coefficient (RC) and the measured density in front of the 2.45 and 4.6 GHz antennas of the EAST tokamak; (down) RC variation with density in front of the antenna and corresponding LH power spectra (insert) as calculated by the wave coupling code ALOHA [12]. For the 2.45 GHz antenna, the measured RC level is corresponding to a density of ~$4x10^{18}m^{-3}$ while direct density measurement is 4 times lower. Furthermore, it is shown that the power spectrum of the 2.45 GHz antenna is significantly modified by changing the density (red curve in the insert: ~$4x10^{18}m^{-3}$, blue and black curves: ~$7x10^{17}$-$1x10^{18}m^{-3}$). For the 4.6 GHz antenna, RC level and density measurement are consistent, leading to a robust estimate of the launched power spectrum.*

The condition of absorption corresponds to large $N_{||}$ upshift combined with inward propagation, which is difficult to satisfy for plasmas with circular poloidal cross-section or large aspect ratio tokamaks, while small aspect ratio diverted plasmas are much more favorable. This is an important difference between magnetic configurations. Indeed, if ray stochasticity develops before absorption, LH simulations are becoming very sensitive to any small variations of the toroidal MHD equilibrium or initial ray condition, a problem never observed experimentally [16].

This suggests that additional mechanisms may come into play to complement weak toroidal refraction, and bridge the spectral gap before wave penetration in the plasma core. To tackle this difficulty, a large number of rays with a small numerical width is considered using GENRAY/CQL3D calculations, assuming that numerical solution converges towards the physical one as the number of rays is large [18]. For C3PO/LUKE, each lobe in the power spectrum is considered locally as a plane wave, even for the tail LH model, with a spectral width given by the Fourier transform of the initial wave electric field pattern. With this approach, all rays are well separated from each other in phase space, a condition for a correct calculation the LH diffusion coefficient by simply adding all ray contributions [2,16].

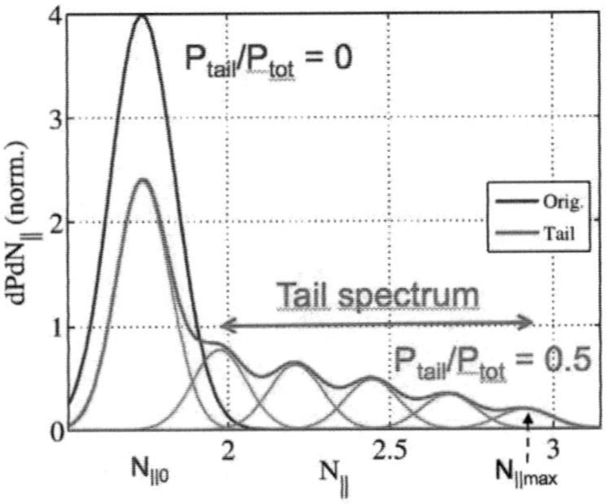


*Figure 3. Standard and tail power spectra for a main lobe at $N_{\|0}$. The tail spectrum is made of 5 secondary lobes up to $N_{\|max}$, which is generally chosen as $6.5/\sqrt{T_{e0}}$ [16].*

The LH wave absorption is calculated by a 3-D bounce-averaged relativistic Fokker-Planck solver using a linearized collision operator (either in LUKE or CQL3D) that determines the distortion of the electron distribution functions under the influence of the LH wave electric field. The corresponding rf diffusion operator describing the resonant Landau absorption (RLA) is determined in LUKE and CQL3D codes from the relativistic bounce-averaged Kennel-Engelman-Lerche formula for plane waves, taking into account of the exact LH wave polarization (slow and fast modes). Concerning the power decrement for each ray, both CQL3D and LUKE have similar approaches, taking into account of multiple passes of the same ray through an incremental plasma volume between neighboring magnetic flux surfaces. The self-consistency between the electron distribution and the rf diffusion is achieved until all LH power is transferred to electrons for each ray. The quasilinear calculation is performed while other processes may contribute, like parasitic non-resonant collisional absorption (NRCA) which may be important in very cold regions of the plasma, and anomalous fast electron transport, an ad-hoc free parameter in simulations [19].

It is important to recall that RLA calculations are valid for circulating and trapped electrons far from low order rational magnetic flux surfaces (not valid for CD in islands). The calculation of the power transferred from the waves to the plasma is based on the narrow beam approximation in the poloidal direction, i.e. plasma must be almost homogeneous transverse to the beam, a condition which is sometimes not well fulfilled. Besides, ray-tracing does not provide any prescription for $\Delta N_{\|}$ evolution in the plasma. But it turns out that results are weakly sensitive to $\Delta N_{\|}$ along the ray trajectory as long as $\Delta N_{\|} << N_{\|}$ corresponding to the well collimated beam approximation [6].

Finally, the ultimate step in first principle modelling consists in determining moments of the electron distribution function, the so-called synthetic diagnostics. They allow direct comparison between simulations and experiments, and are crucial steps in the simulation process. The convolution of the physical signal with the detector's response leads to an additional complexity in the physical interpretation. So, it is important to recall that HXR

DOI: 10.1051/epjconf/201715702007

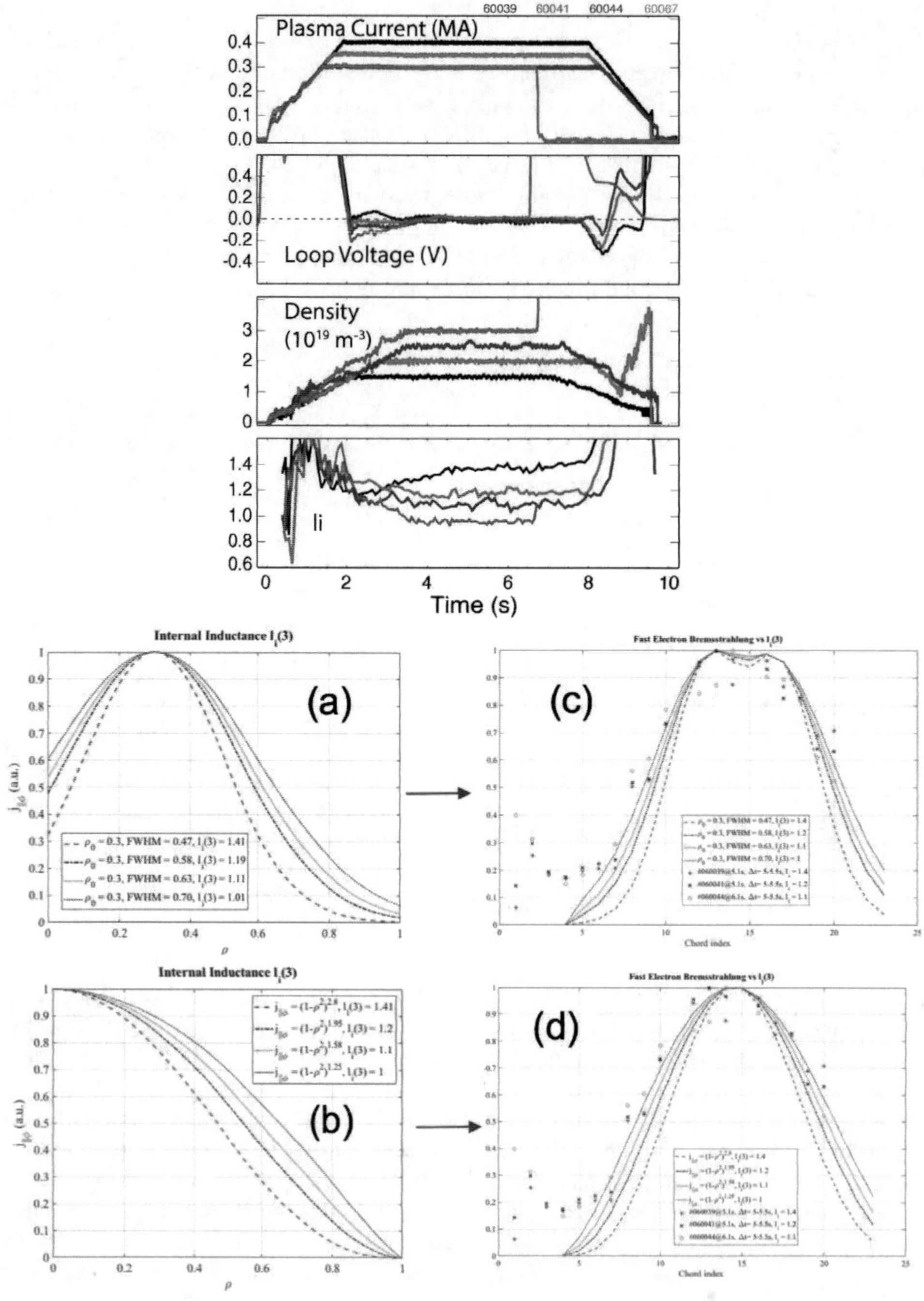


*Figure 4. (Top) waveforms of LH-driven EAST discharges at $V_{loop}$ = 0 V, where the internal inductance $l_i$ is ranging from 1.0 (green), 1.1 (blue), 1.2 (red) to 1.4 (black); (Down) (a) and (b) are peaked and off-axis current density profiles which reproduce both the range of observed $l_i$ : 1.0 (black), 1.1 (green), 1.2 (blue) to 1.4 (red). Curves in (c) and (d) are corresponding normalized line-integrated HXR profiles assuming proportionality to the current density, while dots are experimental measurements. Because of their statistical noises, both hollow (a) and peaked (b) current density profiles are consistent the normalized HXR measurements, whatever the $l_i$ value.*

tomography in a poloidal cross-section of the plasma is indicative of the LH power absorption profile instead of LH driven current density profile [14]. Furthermore, the standard effective charge $Z_{eff}$ cannot be used readily for HXR calculations, except if the plasma contains only fully stripped low Z impurities. For coming experiments with tungsten walls, the use of impurity transport codes for HXR calculations must be systematic [8]. Besides HXR measurements, internal inductance, which is a normalized integral moment of the current density profile, is supposed to give also valuable indications on the current density profile [3]. However, combined with the plasma current, it cannot constrain unambiguously simulations as illustrated in Fig.4 for a set of EAST tokamak discharges. Only quantitative measurements that combine simultaneously dynamics in momentum and configuration spaces like HXR are more selective between models [2]. Time modulation of the LH power or the antenna phasing may represent also a challenging constraint for first principle modeling.

## 3. A selection of results

Quantitative "snapshot" simulations of a single shot in almost steady-state conditions thanks to the resistive current diffusion have been performed several times for Tore Supra, Alcator C-Mod and EAST tokamaks. For low density discharges, a good quantitative agreement is found between simulations and observations concerning HXR line-integrated profiles and the plasma current, as shown for Tore Supra in Fig. 5 [14]. This represents an important validation of the whole chain of calculations. For plasmas of higher densities, large discrepancies are observed between FEB profiles and simulations, using the standard LH model (power spectrum with a quiescent SOL). In order to recover a good agreement, the introduction of a fast fluctuating power spectrum (tail LH model) with respect to the fast electron slowing-down time must be introduced in LH-driven Tore Supra plasmas [16]. This is similar for Alcator C-Mod as shown in Fig.6, the tail LH model allowing to recover simultaneously the bell-shape HXR line-integrated profiles observed up to $n_e \leq 10^{+20}$ $m^{-3}$ and the total plasma current [2].

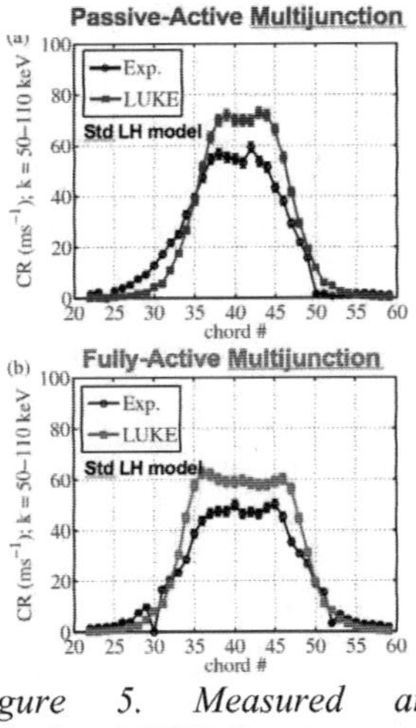


*Figure 5. Measured and calculated HXR line-integrated profiles in Tore Supra for PAM and FAM antennas at 2MW with $N_{\parallel 0}$ = 1.7. Standard LH model, low density [14].*

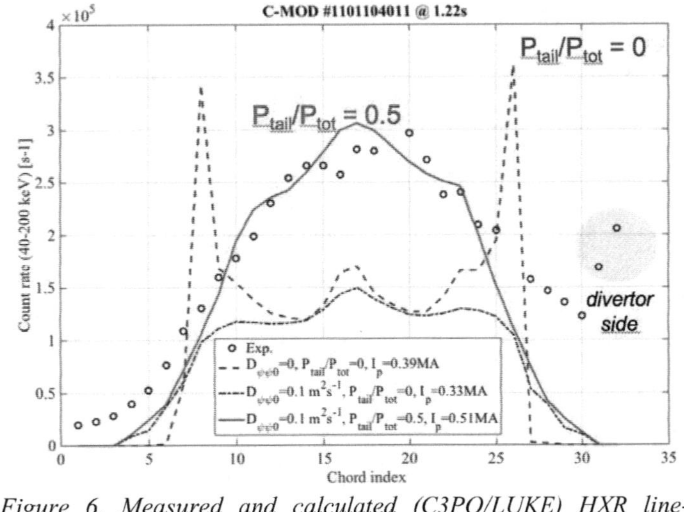


*Figure 6. Measured and calculated (C3PO/LUKE) HXR line-integrated profiles in a high density ($n_e$ = $4.5x10^{+19}m^{-3}$) Alcator C-Mod LH-driven discharge at 0.53 MA. The overall quantitative agreement is obtained using the tail LH model only since the spectral gap is large, $N_{\parallel 0}$ = 1.55. [2]*

Attempt to reproduce the decrease of the internal inductance as the density rises at $V_{loop} = 0V$ is a challenging exercise. This study has been carried out for several sets of EAST tokamak LH-driven discharges, using interfero-polarimetry to enhance quality of the toroidal MHD equilibrium reconstruction by EFIT code [3]. As shown in Fig. 7, the decrease of $l_i$ is well predicted using C3PO/LUKE using the standard LH model, but the variation of the internal inductance is too large as compared to observations, indicating that the calculated current profile is likely too off-axis. The relative agreement between C3PO/LUKE and GENRAY/CQL3D predictions is remarkable with the standard LH model, except that the current predicted by GENRAY/CQL3D is higher than measurements, while for C3PO/LUKE, its absolute level is very close to the experimental value in the lowest density case. The tail LH model gives conversely the correct current level at all densities, but profiles are too peaked, leading to $l_i$ values higher than those deduced from magnetic measurements. From calculations, it could be deduced that LH power absorption is less off-axis than the standard LH model, and more off-axis for the tail one, which highlights the importance of the LH power spectrum considered for ray-tracing calculations in prediction of the current density profile.

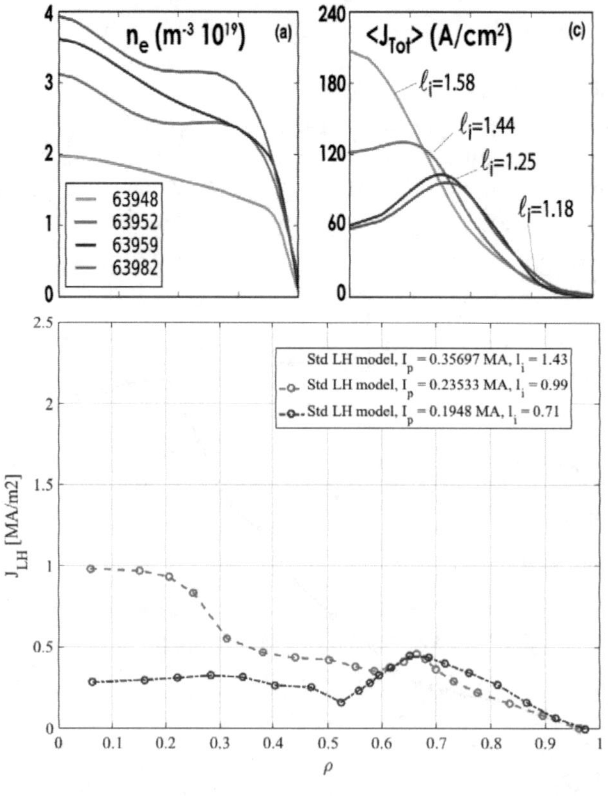


*Figure 7. (Left, a and c) density profiles of LH-driven EAST discharges at $V_{loop} = 0$ V ($I_p \sim 330$ kA), where the internal inductance $l_i$ is ranging from 1.18 (green, #63982), 1.25 (blue, #63959), 1.44 (red, #63952) to 1.58 (orange, #63948). Equilibrium reconstruction is constrained by interfero-polarimetry [3]; (Right) Current density profiles as calculated by the standard LH model (C3PO/LUKE). Predicted $l_i/I_p$(kA) is 0.71/195 (blue, #63959), 0.99/235 (red, #63952) to 1.43/356 (yellow, #63948) For high density cases, only the tail LH model gives the correct plasma current, but the predicted $l_i$ is always too high.*

Since $l_i$ is very sensitive to small variations of the current density at mid-radius and outside, the use of diagnostics that provide additional constraints in this region of the plasma is essential. The multi-code analysis illustrates the challenge of current density calculation from first principles, and the importance of the launched power spectrum at the separatrix on general conclusions. Its direct experimental determination will be likely one of the most important challenge to decide on the relative influence between core and edge LH physics on the overall simulations. This problem is crucial for ITER LH current drive predictions, as already pointed out [2].

Scaling of the current drive efficiency with electron density has been found to be highly anomalous in Alcator C-Mod, considering the HXR intensity of central chords as a proxy [20]. From GENRAY/CQL3D calculations, the sharp decrease above $10^{+20}m^{-3}$ is ascribed to parasitic absorption of the LH wave in the cold and dense SOL. The LH power available for current drive in the core plasma is then dramatically reduced. Simulations of HXR emission at $1.3x10^{+20}m^{-3}$ with C3PO/LUKE/R5-X2 codes have shown a good agreement with observations, without invoking NRCA at the plasma edge as show in Fig.8 [2].

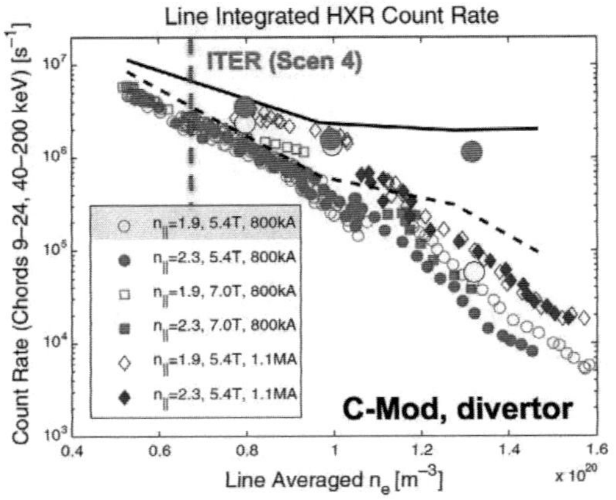


*Figure 8. Scaling of the central chord HXR intensity as function of density for Alcator C-Mod. Full line: GENRAY-CQL3D (standard LH model without NRCA) dashed line: GENRAY-CQL3D (standard LH model with NRCA), yellow points: C3PO/LUKE (standard LH model without NRCA), red points: C3PO/LUKE (tail LH model without NRCA). Other points are experiments [20,2]*

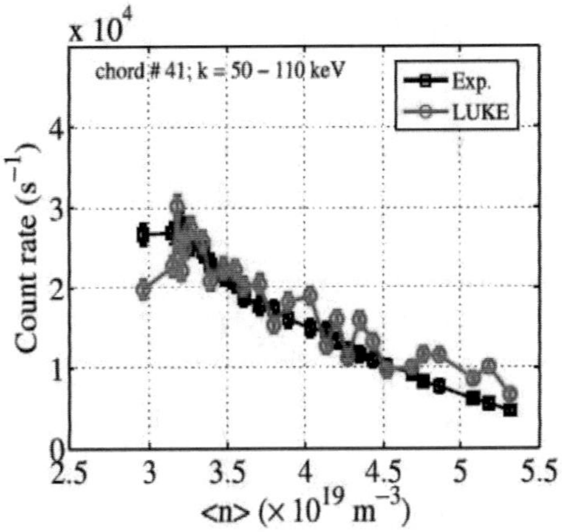


*Figure 9. Scaling of the central chord HXR intensity as function of density for Tore Supra, using C3PO/LUKE with tail LH model without NRCA. With the standard LH model, the tendency is similar, but predictions are noisy, because of the ray stochasticity [16].*

For this calculation, only 5% of the total current at this density is driven by the LH wave, which leads to questioning the non-inductive nature of the discharge [2]. For densities lower than $10^{+20}m^{-3}$, as foreseen in ITER, the experimental HXR scaling with $n_e$ is grossly reproduced with the standard or tail LH model, either by C3PO/LUKE and GENRAY/CQL3D codes, as shown in Fig.8. This scaling is also well reproduced in Tore Supra up to $5.5x10^{+19}$ $m^{-3}$ as shown in Fig. 9, but the quality of the agreement between modelling and experiment is greatly improved using the tail LH model [16].

## 4. **Conclusions**

For small spectral gap regimes, the existing first principle modeling tools are able to describe observations quantitatively, which represents a remarkable achievement. However, large spectral gap regimes remain still challenging to model, and it is difficult to decide if the discrepancy arises from a failure in the chain of modeling or from physical processes that could be missed in LH current drive modeling at the edge or in the core of the plasma. In this context, one of the most important challenge is to determine experimentally the characteristics of the power spectrum as the LH propagates through the SOL, considering the large impact for LH current drive in ITER. But, numerous other investigations must be carried out to describe simultaneously HXR profiles and $l_i$ at $V_{loop}$ = 0V and to improve the consistency between measured density in front of the antenna and the RC level. Besides, simulations must be performed on a much larger database, to estimate the robustness of LH simulations. These short terms objectives must not hide the numerous studies that must be revisited, like Ohmic/LH synergy, each of them being able to put larger constraints on modeling, in particular when quasilinear effects become very important. Finally, on a long term, the full-wave calculations must likely become a standard, in replacement of ray-tracing calculations, in such a way all intrinsic limitations may be removed and new synthetic diagnostics must be developed.

This work was supported in part by US DoE Contract Nos. DE-SC-0010492 and DE-FC02-99ER54512-CMOD.

[1] P. T. Bonoli, et al., Phys. Plasma 21 (2014) 061508.
[2] Y. Peysson, et al., Plasma. Phys. Contr. Fusion 58 (2016) 044008.
[3] A. Garofalo, et al., Nucl. Fusion 57 (2017) 076037.
[4] Y. Peysson, et al., Plasma. Phys. Contr. Fusion 54 (2012) 045003.
[5] A.P. Smirnov and R. W. Harvey, Report CompX-2000-01 (2003).
[6] Y. Peysson and J. Decker, Fusion Sci. Tech., 65 (2014) 22.
[7] R.W. Harvey and M.G. McCoy, IAEA Conf. Proc. (1992) 489.
[8] Y. Peysson and J. Decker, Phys. Plasmas, 15 (2008) 092509.
[9] J. C. Wright, et al., Phys. Plasmas 16 (2009) 072502.
[10] O. Meneghini, et al., Phys. Plasmas 16 (2009) 090701
[11] S. Shiraiwa , et al., Phys. Plasmas 18 (2011) 080705.
[12] J. Hillairet, et al., Nucl. Fusion, 50 (2010) 125010.
[13] M. Preynas, et al., Nucl. Fusion, 51 (2011) 023001.
[14] E. Nilsson, et al. , Nucl. Fusion, 53 (2013) 083018.
[15] M. Madi, et al., Plasma. Phys. Contr. Fusion, 57 (2016) 125001.
[16] J. Decker, et al. Phys. Plasmas, 21 (2014) 092504.
[17] A. S. Richardson, et al., Phys. Plasmas, 17 (2010) 052107.
[18] J. P. Bizarro and D. Moreau, Phys. Fluids B, 5 (1993) 1227.
[19] P. T. Bonoli, et al., Phys. Fluids, 29 (1986), 2937.
[20] G.M. Wallace, et al., Nucl. Fusion, 51 (2011) 083032.

DOI: 10.1051/epjconf/201715702008

# Advanced electron cyclotron heating and current drive experiments on the stellarator Wendelstein 7-X

*Torsten* Stange[1,*], *Heinrich Peter* Laqua[1], *Marc* Beurskens[1], *Hans-Stephan* Bosch[1], *Sergey* Bozhenkov[1], *Rudolf* Brakel[1], *Harald* Braune[1], *Kai Jakob* Brunner[1], *Alvaro* Cappa[2], *Andreas* Dinklage[1], *Volker* Erckmann[1], *Golo* Fuchert[1], *Gerd* Gantenbein[4], *Florian* Gellert[1], *Olaf* Grulke[1], *Dirk* Hartmann[1], *Matthias* Hirsch[1], *Udo* Höfel[1], *Walter* Kasparek[3], *Jens* Knauer[1], *Andreas* Langenberg[1], *Stefan* Marsen[1], *Nikolai* Marushchenko[1], *Dmitry* Moseev[1], *Novomir* Pablant[5], *Ekkehard* Pasch[1], *Kian* Rahbarnia[1], *Humberto* Trimino Mora[1], *Toru* Tsujimura[6], *Yuriy* Turkin[1], *Tom* Wauters[7], *Robert* Wolf[1], & the W7-X-Team

[1]Max-Planck-Institute for Plasma Physics, Greifswald, Germany
[2]Laboratorio Nacional de Fusión, CIEMAT, Madrid, Spain
[3]Institute for Interfacial Process Engineering and Plasma Technology, University Stuttgart, Germany
[4]Institute of Pulsed Power and Microwave Technology, Karlsruhe Institute of Technology (KIT), Germany
[5]Princeton Plasma Physics Laboratory, Princeton, New Jersey 08543, USA
[6]National Institute for Fusion Science, Toki 509-5292, Japan
[7]Laboratory for Plasma Physics, ERM/KMS, Brussels, Belgium

**Abstract.** During the first operational phase (OP 1.1) of Wendelstein 7-X (W7-X) electron cyclotron resonance heating (ECRH) was the exclusive heating method and provided plasma start-up, wall conditioning, heating and current drive. Six gyrotrons were commissioned for OP1.1 and used in parallel for plasma operation with a power of up to 4.3 MW. During standard X2-heating the spatially localized power deposition with high power density allowed controlling the radial profiles of the electron temperature and the rotational transform. Even though W7-X was not fully equipped with first wall tiles and operated with a graphite limiter instead of a divertor, electron densities of $n_e > 3 \cdot 10^{19}$ m$^{-3}$ could be achieved at electron temperatures of several keV and ion temperatures above 2 keV. These plasma parameters allowed the first demonstration of a multipath O2-heating scenario, which is envisaged for safe operation near the X-cutoff-density of $1.2 \cdot 10^{20}$ m$^{-3}$ after full commissioning of the ECRH system in the next operation phase OP1.2.

## 1 INTRODUCTION

On 10th December 2015 the first plasma was created in Wendelstein 7-X (W7-X). It is the world's largest optimized stellarator (R = 5.5 m, a = 0.5 m) with 3-D shaped superconducting modular coils and a five-fold symmetry. W7-X aims to achieve reactor relevant plasma parameters in quasi steady state operation [1]. For this reason, the device will be equipped with a 10 MW electron cyclotron resonance heating (ECRH) system allowing at least 30 minutes continuous operation. The design frequency of the 10 gyrotrons already installed is 140 GHz corresponding to second harmonic heating at the magnetic axis with a local magnetic field of 2.5 T [2]. As a very versatile technique, ECRH can provide plasma start-up, bulk plasma heating, current drive as well as wall conditioning. Even though, the confinement of a stellarator is only given by the external field coils, electron cyclotron current drive (ECCD) can be used at W7-X to compensate a finite bootstrap current of up to tens of kilo-Ampéres in some magnetic field configurations, which can cause a shift of the strike line on the divertor. Furthermore, the steady state operating magnetic field prevents any re-conditioning of the wall by glow discharges, requiring the development of adapted ECRH wall conditioning scenarios.

ECRH was the exclusive heating system in the first operational phase (OP1.1) which ended in March 2016 after 10 weeks of operation and was primarily dedicated to an integral commissioning of the heating system, the diagnostics and especially the components of the device itself, e.g. vacuum vessel and coil system. The machine was not fully equipped with wall protecting graphite tiles leaving a larger surface of the CuCrZr-cooling structure unprotected. Furthermore, instead of the foreseen 10 divertor elements, a graphite limiter was installed on the high field side (HFS) in each bean plane of the five modules preventing a direct contact of the plasma with the unprotected regions of the plasma vessel [3]. Therefore, the plasma volume was limited to an effective minor radius of a = 0.49 m. In parallel, the flexible coil system of W7-X was used to operate a magnetic field configuration at a lower rotational transform developing

* Corresponding author: torsten.stange@ipp.mpg.de

DOI: 10.1051/epjconf/201715702008

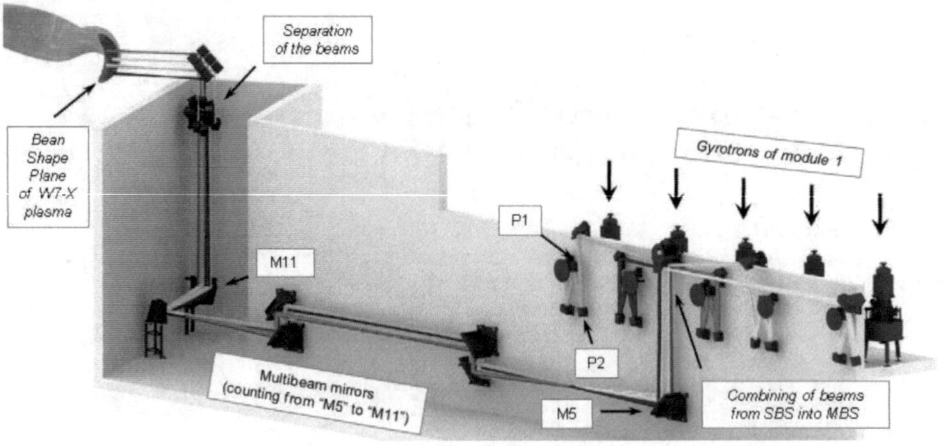


**Fig. 1.** Fully quasi-optical transmission line to the torus of Wendelstein 7-X

from $\iota_0 = 0.79$ in the center to $\iota_a = 0.87$ at the scrape off layer to push the 5/6 island chain inside the plasma volume.

## 2 Setup of the ECRH system

The heating beams are provided by 10 gyrotrons, which are aligned in two symmetrical rows. The gyrotron outputs of each row face each other into a beam channel with two identical quasi-optical multi-beam transmission lines, one for each gyrotron row. Fig. 1 shows one half of the whole installation excluding periphery like cooling and high voltage supplies. The beam line is designed such that an additional gyrotron can be installed on each side during a later upgrade.

### 2.1. Gyrotrons

The gyrotrons were developed within the PMW-project ("Projekt Mikrowellenheizung für W7-X") at KIT and have a nominal power of 1 MW at 140 GHz [4]. Nine gyrotrons were built by THALES in France and another one by CPI in the USA. The overall power of all gyrotrons is up to 9 MW in short pulse operation and up to 8 MW for steady state. The power can be modulated with arbitrary waveforms at frequencies up to 10 kHz to determine the power deposition zone within the plasma and conduct advanced heat wave experiments. In OP1.1 only 6 gyrotrons were in operation. This was sufficient, as the 4 MJ per discharge given by the finite heat capacity of the limiter elements in the plasma vessel did not require more gyrotrons.

### 2.2. Transmission Line

Both quasi-optical transmission lines to the plasma vessel were part of the PMW-project, as well, and are first-of-its-kind systems. Each of them consists of five single beam sections (SBS) matching the non-perfect Gaussian beam output of the gyrotrons to the subsequent 40 m long multi-beam section (MBS), which is intrinsically broadband. The whole MBS consists of 7 large-area multi-beam mirrors being able to transmit 7 gyrotron beams in parallel (5 gyrotron beams + 1 optional beam line for switching one gyrotron to another launcher + 1 beam line of the spare box). Finally, the MBS separates the beams and directs them to their respective vacuum window in a single beam line. The windows interface to the four equatorial launchers installed on W7-X with three beam lines each (see Fig. 2). The plasma facing mirror of each beam line is steerable in the poloidal and toroidal direction. This way, the radial heat deposition in the plasma and the plasma current drive can be varied, respectively. The polarization of the beams can be arbitrarily adjusted by two grooved mirrors, the polarizer 1 (P1) and polarizer 2 (P2), generating a 90°- and 180°-phaseshift, respectively.

The beam power of each beam line is measured by a calorimetric load before entering the MBS, as depicted in Fig. 3. The measured value has to be corrected by the theoretical transmission losses up to the vessel window to determine the final plasma heating power. However, the multi-beam transmission technique gives the opportunity to introduce a retroreflector at the end of the

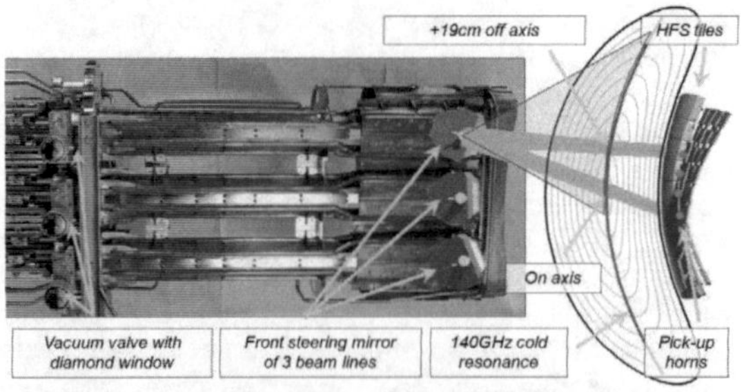


**Fig. 2.** Cross section of one of the 4 ECRH launchers in relation to the vacuum flux surfaces of the limiter configuration in OP1.1. The beam with a diameter of about 80mm in the plasma center is shown for two examples of the upper beam line. Its whole steering range is visualized by the green shaded area.

 DOI: 10.1051/epjconf/201715702008

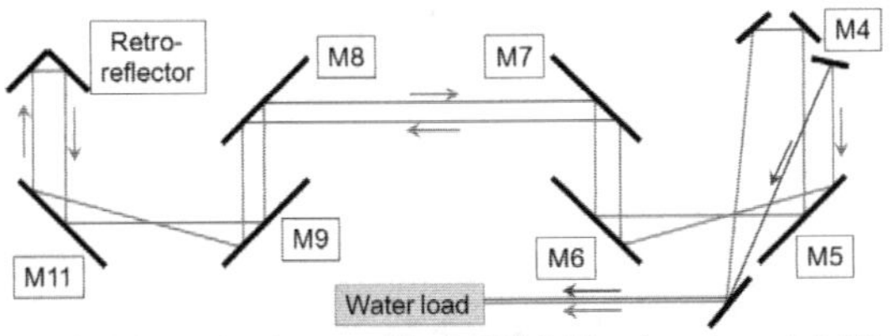


**Fig. 3.** Schematic picture of the MBS fed by the mirror 4 (M4) which is the last mirror of the SBS and is switchable between the water load (blue beam line) or the MBS to W7-X (red). Optional use of a retroreflector allows the additional power measurement after passing the MBS twice.

MBS to redirect a beam to the same calorimetric load for measuring the transmission losses after passing the MBS two times (see Fig. 3). Using the same load has the advantage, that uncertainties of the single power measurement cancel out. The time traces of the two separate power measurements are overlaid in Fig. 4. After 300 seconds the gyrotron as well as the load is fully equilibrated allowing an accurate comparison of both power levels. The transmission efficiency was determined for three different power levels with an averaged value of (94.4 ± 0.4) % being in very good agreement with the theoretical value of 94.2 %. Therefore, the losses of the MBS can be specified to 0.06 %/m including atmospheric absorption of 0.02 %/m for 140 GHz. Finally, the overall transmission efficiency from the gyrotron window up to the launcher window can be calculated to 94% including diffraction, beam truncation, misalignment and absorption by the mirrors and by the atmosphere.

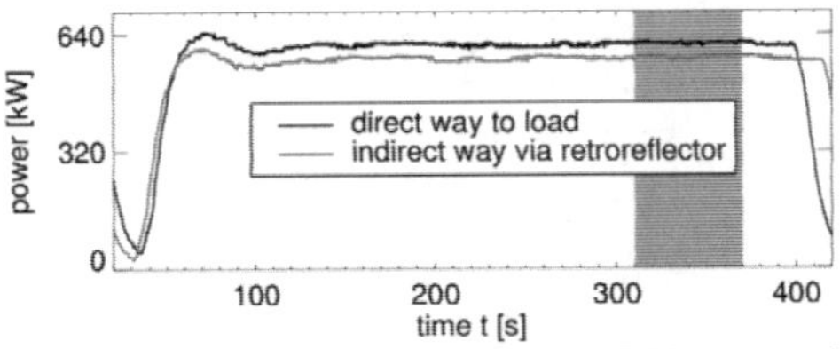


**Fig. 4.** Evaluation of the efficiency of the MBS by comparison of the gyrotron beam power on a direct way to the load with the indirect way via a retroflector at the end of the MBS (the evaluated time windows are shaded)

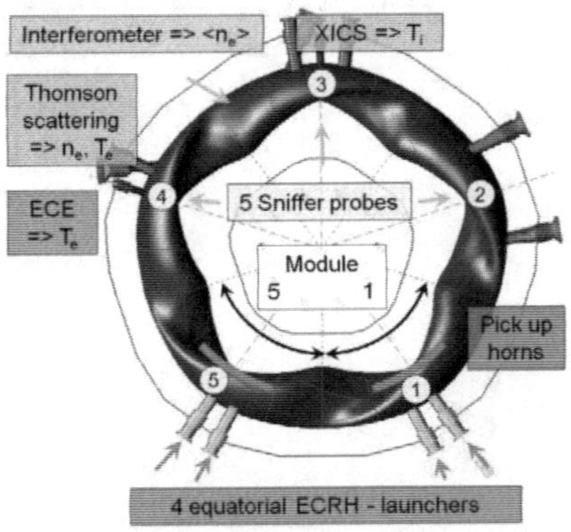


**Fig. 5.** Top view on the W7-X torus with the positions of the main diagnostics and the ECRH launchers in module 1 and 5.

## 3 Diagnostics for plasma operation

Fig. 5 shows an overview of the W7-X torus indicating the position of the ECRH launchers and some important diagnostics, which were commissioned during OP1.1. Safety relevant diagnostics like the stray radiation probes (sniffer) [5], indicating the ECRH absorption, and the electron cyclotron emission diagnostic (ECE) [6], measuring the electron temperature, were available from the first day of operation. A detailed description of all diagnostics is given in [7].

## 4 ECRH scenarios during OP1.1

The standard heating scenario at W7-X is X2-heating at 140 GHz with an electron cutoff-density of $n_{e,X\text{-cut}} = 1.2 \cdot 10^{20}\ m^{-3}$. However, for electron densities above $0.8 \cdot 10^{20}\ m^{-3}$, which are envisaged in the future, a completely different O2-heating-scenario ($n_{e,O\text{-cut}} = 2.4 \cdot 10^{20}\ m^{-3}$) should be used to prevent beam refraction or reflection in case of losing density control during the discharge. Both scenarios were successfully demonstrated up to electron densities of $n_e = 4 \cdot 10^{19}\ m^{-3}$. The subsequent presentation follows the timeline of progress during OP1.1. The conducted heating scenarios including the expected absorption were optimized in advance with the aid of the ray tracing code TRAVIS [8] developed at IPP.

### 4.1 X2-heating

To similarly fulfil the on-axis resonance condition for thermal and suprathermal electrons during the plasma breakdown an optimal magnetic field with $B_{00} = 2.52$ T on axis was forecast and proved in a magnetic field scan (2.48-2.54 T) on the first day of OP1.1 However, poor vacuum conditions necessitated the use of ECRH for wall conditioning during the first days of operation.

#### *4.1.1 Wall Conditioning*

In preparation of plasma operation, the vacuum vessel was baked one week at 150°C to remove most of the water. Due to possible copper sputtering and other problems with the glow discharge system, the plasma operation was started in helium without glow discharge cleaning (GDC). For this reason, the first days of plasma discharges were governed by strong outgassing of the walls, resulting in incomplete ECRH absorption and a radiative collapse immediately after breakdown. The wall conditions could be drastically improved by repeating 50 ms shots hundreds of times. Fig. 6 shows the full discharge behaviour after 110 pulses and in comparison the stray radiation time trace after 220 further pulses. The time period with full absorption could be doubled. All available gyrotrons were used during the cleaning pulses to bring as much energy as possible into the plasma and thus onto the plasma wetted limiter elements. Furthermore, energy is transferred to the walls by soft X-ray and charge exchange processes, thus releasing further impurities.

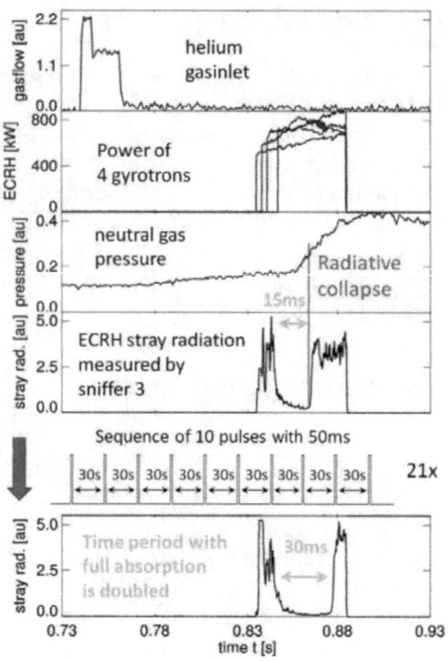


**Fig. 6.** Temporal evolution of a cleaning discharge @ 2.7 MW. Stray radiation after 0.1s cumulated discharge time compared with 3s cumulated discharge time (as published in [12])

After 3 days of W7-X operation, the glow discharge system was operated with helium for periods of 20 min - 40 min in advance of each experiment day. Fig. 7 shows the value of the neutral pressure peak after a helium discharge normalized to the absorbed energy over the cumulated discharge time in which almost full absorption was achieved. After GDC the outgassing is decreased because of hydrogen desaturation of the whole first wall, which re-absorbs in the successive discharges and partially allows density control. However, the overall trend was recovered at the end of an experiment day indicating that the vacuum conditions are mainly determined by the energy carried into the plasma-wetted areas leading to a depletion of impurities from deeper trapped states. For this reason, only GDC with hydrogen or ion cyclotron resonance heating could speed up the time consuming ECRH wall conditioning. All details of the analysis can be found in [9].

Nevertheless, desaturation of the walls after massive hydrogen operation was successfully demonstrated by means of an intermediate helium ECRH discharge being

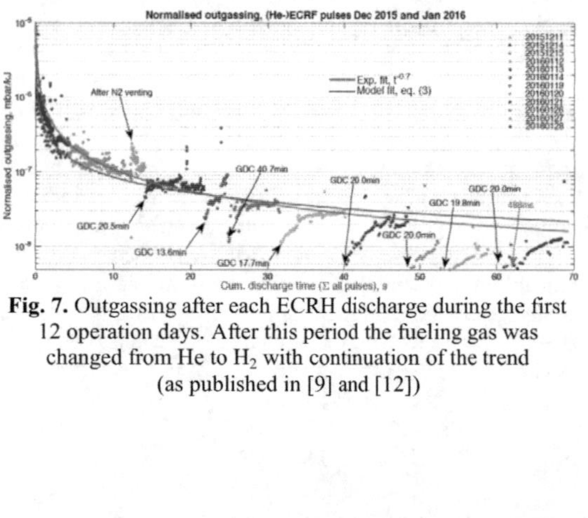


**Fig. 7.** Outgassing after each ECRH discharge during the first 12 operation days. After this period the fueling gas was changed from He to $H_2$ with continuation of the trend (as published in [9] and [12])

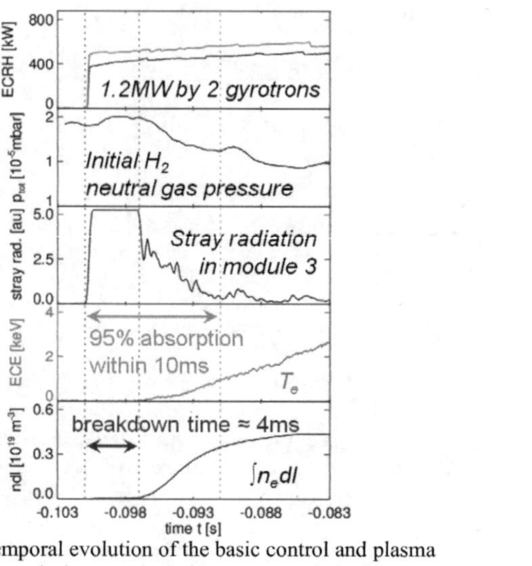


**Fig. 8.** Temporal evolution of the basic control and plasma parameters during a typical plasma startup in hydrogen

the only possibility to recover density control during the experiment day.

### *4.1.2 Startup behaviour & bulk plasma heating*

The breakdown of W7-X-plasmas happens on a considerably fast time scale. Fig. 8 shows the temporal evolution for a typical startup scenario using 2 gyrotrons with an overall power of 1.2 MW. Nearly full absorption can be achieved within 10 ms after injecting microwaves including a breakdown time of 4 ms. A minimum breakdown time of 2 ms can be achieved at about 2 MW as can be seen in Fig. 9. In general, the shown breakdown scenario was 100% reliable, if the used gyrotrons and the gasinlet were activated as programmed.

The ECR heated W7-X plasmas are governed in general by central electron root confinement leading to peaked electron temperature profiles of several keV and low ion temperatures of about 1 keV at the beginning of a discharge. Further gyrotrons were added after 30 ms including further gas puffs to increase the density as shown in Fig. 10. Exhausting the energy limit of 4 MJ at 4.3 MW allowed to increase the density up to $n_e = 4 \cdot 10^{19}\ m^{-3}$ at central $T_e = 7$ keV. Because of the improved heat transfer from electrons to ions with increasing density, the ion temperature rose up to $T_i \approx 2$ keV.

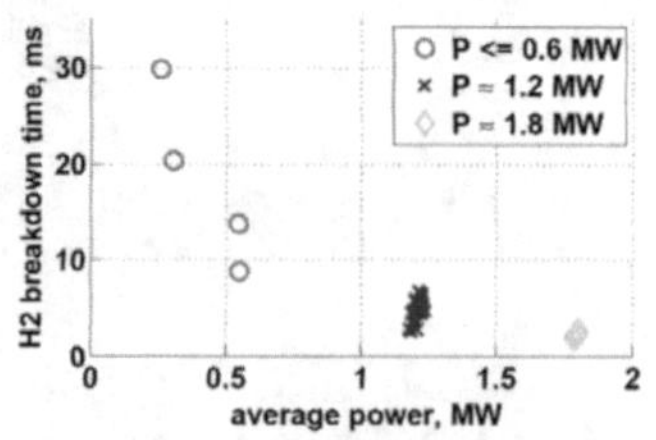


**Fig. 9.** Breakdown time in H2-plasmas in dependence on the used ECRH power

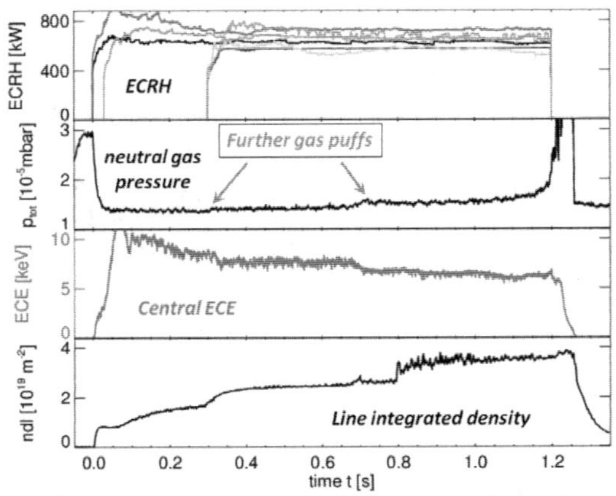


**Fig. 10.** Temporal evolution of a discharge with the maximum ECRH power of 4.3 MW (after the second gas puff the spike in the density time trace indicates an higher density between 0.7s and 0.8s due to an error in the interferometer measurement)

### *4.1.3 Control of electron temperature & iota profile*

The spatially localized ECRH power deposition ($\Delta r/a < 0.1$) with high power density typically allows the control of non-stiff plasma profiles. For this purpose, the front steering mirrors of the ECRH-beam lines can be varied in the vertical direction. In this case the beams have to be corrected in the horizontal direction to sustain the vertical incidence on the magnetic field lines because of the 3D-plasma shape of W7-X.

Fig. 11 shows a conducted scenario, where the plasma was ignited by 2 on-axis gyrotrons followed by a switch to 3 off-axis gyrotrons with a deposition radius of $r/a = 0.4$. Additionally, the power of each gyrotron was square wave modulated with a frequency of 17 Hz to analyze the heat transport and compare it with a similar on axis discharge. A first analysis can be found in [11] casting an electron heat diffusivity of expected magnitude, albeit with error bars too high for an detailed analysis. However, a closer look onto the Thomson profiles (see Fig. 12) of an on- and off-axis discharge with a power level of 0.6 MW shows non-stiff temperature profiles and an increase of the central

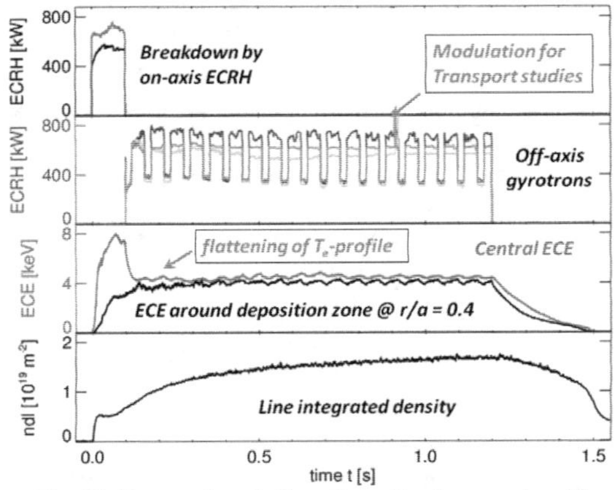


**Fig. 11.** Temporal evolution of an off-axis scenario with modulated gyrotrons @ 1.9 MW for transport studies

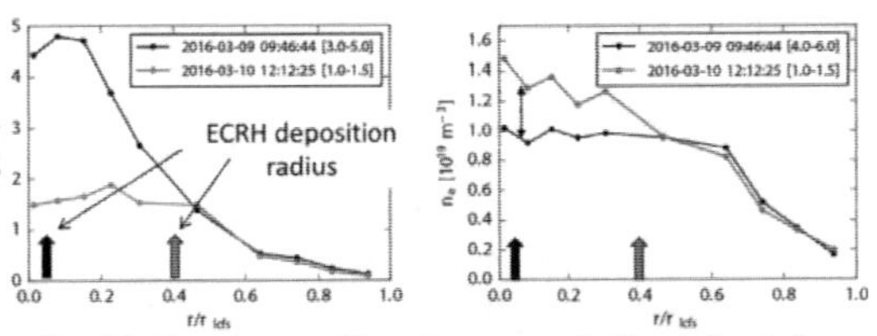


**Fig. 12.** $T_e$ and $n_e$ profiles of an on- and off-axis heated discharge @ 0.6 MW heating power (as published in [12])

density by about 30 % going from on-axis to off-axis. This could indicate a reduction of an outward particle diffusion, which is predicted by neoclassical theory in presence of a strong electron temperature gradient [10]. A more detailed investigation on this topic can be conducted with the availability of more advanced diagnostics in the next operation phase (OP1.2) allowing a more accurate determination of the central radial electric field, which plays a fundamental role in the present long mean-free-path regime. A summary of the actual state of knowledge can be found in [12].

A much more drastic change of the central confinement properties can be triggered by varying the rotational transform by ECCD. Fig. 13 shows the comparison of 2 discharges with a centrally driven current co- and counter-acting to the rotational transform with a central value of $\iota = 0.79$ for the vacuum configuration. The central electron temperature crashes in both cases on a timescale of μs similar to sawtooth crashes in tokamaks. The reason could be an instability on a rational flux surface, which forms a reconnection of magnetic field lines, because of the central iota changing to the resonant value of 1.0 (co-ECCD) and 0.5 (counter-ECCD), respectively. The resulting flattening of the central profile would also short-wire the induced central current peaks, so that the confinement can be recovered leading to a rise of the profiles until the next crash. These fast internal current changes are not visible by the rogowski coils, because of slowly varying shielding currents with a L/R-time of several seconds [3].

A possible application in later operation phases could be a controlled reduction of a central impurity contamination, because the typically flat density profile would not be affected by a central profile short circuit. A further possibility would be the flattening of hollow density profiles which are predicted for high density operation [10].

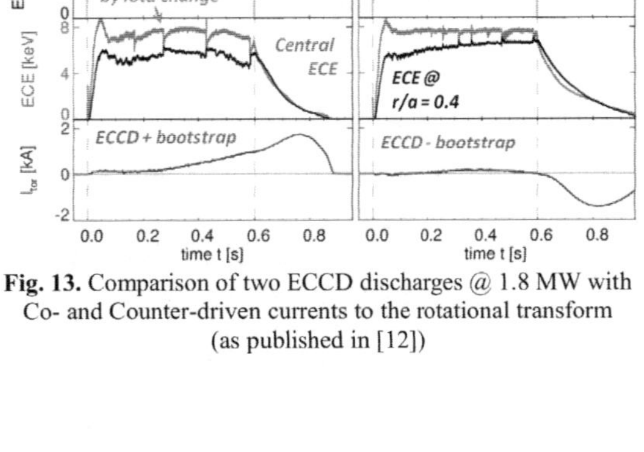


**Fig. 13.** Comparison of two ECCD discharges @ 1.8 MW with Co- and Counter-driven currents to the rotational transform (as published in [12])

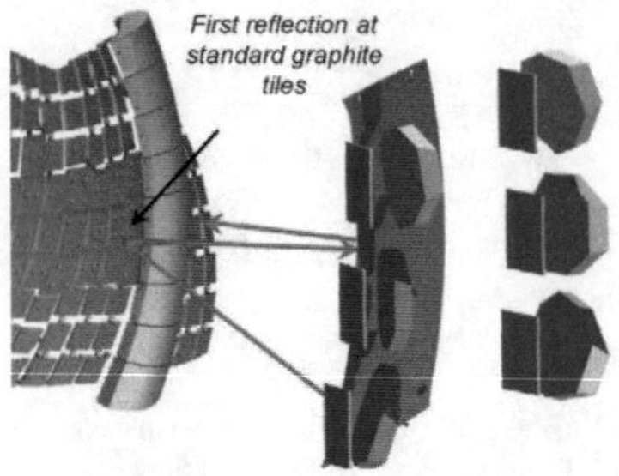


**Fig. 14.** 3-beampath-scenario used in OP1.1 to increase the ECRH-absorption in O2-heating scenarios

### 4.2 O2-heating

Even though the achieved electron densities in OP1.1 were far away from the cut-off for X2-heating, an advanced O2-heating scenario, foreseen for later operation phases, was proven. In contrast to X2-heating, full absorption of the O2-mode is not possible for the expected plasma parameters of $n_e \approx 10^{20}\,m^{-3}$ and $T_e < 5$ keV at a heating power of up to 10 MW. The shinethrough power is of the order 30 % depending on the particular beam path which is with an incidence angle of about 78° already optimized to maximum. For this reason, the shinethrough beam should be reflected at an HFS-tile, redirecting the beam again through the plasma center onto a polished stainless steel panel in between two neighbouring ECRH-launchers (see Fig. 14). The second reflection on the panel should pass the plasma again centrally, so that the non-absorbed power will be less than 5%.

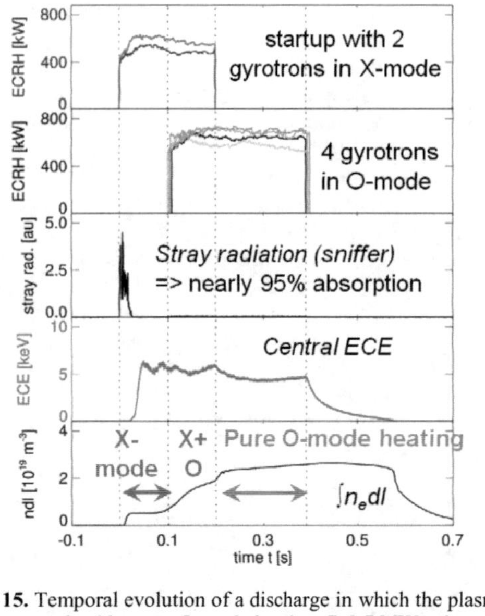


**Fig. 15.** Temporal evolution of a discharge in which the plasma was sustained by pure O-mode heating @ 2.7 MW using a 3-path-scenario (as published in [12])

The experiment (shown in Fig. 15) was conducted at higher magnetic field of $B_{00} = 2.57$ T to centralize the broad O2 deposition zone. The plasma breakdown was lower axis magnetic field. After a further gas puff the O-mode gyrotrons were activated at 100 ms leading to a clear increase of the density. The X-mode gyrotrons were deactivated at 200 ms with a slight drop in the electron temperature. However, the density increases further and the plasma can be fully sustained by O2-heating.

## 5 Conclusions and outlook to OP1.2

Even though the plasmas of OP1.1 were typically influenced by many impurities due to the restricted use of glow discharge cleaning and the limited input energy allowed per pulse, the feasibility of the basic ECRH scenarios important for OP1.2 could be proven. However, the O2-heating scenario in OP1.2 will make use of holographic tungsten reflection tiles at the HFS, as already demonstrated at ASDEX Upgrade [13], to provide all 10 gyrotrons with an efficient triple-pass path through the plasma center. In addition, two so-called remote steering launchers have been installed in a toroidal position where the magnetic field gradient is low. These enable advanced current drive scenarios with beneficial ECRH coupling to suprathermal passing electrons [14].

This work has been carried out within the framework of the EUROfusion Consortium and has received funding from the Euratom research and training programme 2014-2018 under grant agreement No 633053. The views and opinions expressed herein do not necessarily reflect those of the European Commission.

## References

1. G. Grieger, et al., Phys. Fluids B **4** 2081 (1992)
2. V. Erckmann, et al., Fusion Sci. Technol. **52** 291 (2007)
3. T. Sunn Pedersen, et al., Nucl. Fusion **55** 126001 (2015)
4. G. Gantenbein, et al., J Infrared Milli Terahz Waves **32** 320 (2011)
5. D. Moseev, et al., Rev. Sci. Instr. **87** 083505 (2016)
6. S. Schmuck, et al., Fusion Eng. Des. **84** 1739 (2009)
7. R. König, et al., J. Instrum. **10** P10002 (2015)
8. N. B. Marushchenko, et al., Comp. Phys. Commun. **185** 165 (2014)
9. T. Wauters, et al., 43rd EPS Conference on Plasma Physics, Leuven, Belgium, P4.047 (2016)
10. H. Maasberg, et al., Plasma Phys. Control. Fusion **41** 1135 (1999)
11. U. Höfel, et al., 43rd EPS Conference on Plasma Physics, Leuven, Belgium, P4.008 (2016)
12. R. Wolf, et al., Nucl. Fusion **57** 102020, 2017
13. D. Wagner, et al, J Infrared Milli Terahz Waves 37:45-54 (2016)
14. N. B. Marushchenko, et al., AIP Conference Proceedings **1580** 518 (2014)

DOI: 10.1051/epjconf/201715702009

# Fully non-inductive plasma start-up with lower-hybrid waves using the outboard-launch and top-launch antennas on the TST-2 spherical tokamak

*Naoto* Tsujii[1,*], *Yuichi* Takase[1], *Akira* Ejiri[1], *Takahiro* Shinya[2], *Satoru* Yajima[1], *Hibiki* Yamazaki[1], *Hiro* Togashi[1], *Charles P.* Moeller[3], *Benedikt* Roidl[1], *Wataru* Takahashi[1], *Kazuya* Toida[1], and *Yusuke* Yoshida[1]

[1]The University of Tokyo, Kashiwa 277-8561, Japan
[2]National Institute for Quantum and Radiological Science and Technology, Rokkasho 039-3212, Japan
[3]General Atomics, San Diego, CA 92186, USA

**Abstract.** Removal of the central solenoid is essential to realize an economical spherical tokamak fusion reactor, but non-inductive plasma start-up is a challenge. On the TST-2 spherical tokamak, non-inductive plasma start-up using lower-hybrid (LH) waves has been investigated. Using the capacitively-coupled combline (CCC) antenna installed at the outboard midplane, fully non-inductive plasma current ramp-up up to a quarter of that of the typical Ohmic discharges has been achieved. Although it was desirable to keep the density low during the plasma current ramp-up to avoid the LH density limit, it was recognized that there was a maximum current density that could be carried by a given electron density. Since the density needed to increase as the plasma current was ramped-up, the achievable plasma current was limited by the maximum operational toroidal field of TST-2. The top-launch CCC antenna was installed to access higher density with up-shift of the parallel index of refraction. Numerical analysis of LH current drive with the outboard-launch and top-launch antennas was performed and the results were qualitatively consistent with the experimental observations.

## 1 Introduction

Spherical tokamaks (STs) are able to operate at high beta [1] and have the potential to realize an economical fusion reactor with low magnetic field. However, it is essential that the central solenoid (CS) be removed to achieve the optimum performance [2-3]. Steady-state plasma current can be carried mostly by the bootstrap current with small assistance from neutral beam current drive [2]. On the other hand, establishment of an efficient and reliable non-inductive plasma start-up method is an active area of research [4-11]. Since neutral beam current drive requires finite plasma current for confinement of fast ions, plasma current ramp-up from no current needs to be performed by other means such as coaxial helicity injection [4] or radio frequency (RF) waves [5-11].

Lower-hybrid (LH) current drive has the highest efficiency among all the known RF current drive methods under optimized conditions. Non-inductive plasma current ramp-up using LH waves has been demonstrated on JT-60U [12]. Formation of an advanced tokamak by switch-over to neutral beam injection has also been observed [12]. The possibility of plasma start-up with LH waves on STs has been investigated on TST-2 [8-11]. Formation of stable ST equilibria only with LH current drive has been demonstrated [8]. Although there is no fundamental reason to prohibit application of LH waves to STs, operation at high density is more challenging than conventional tokamaks due to the lower toroidal field. For parameters of present day STs, the parallel index of refraction of the LH waves needs to be relatively large and high current drive efficiency observed in conventional tokamaks [13-15] may be hard to realize. However, future ST reactors [2-3] will have a toroidal field of a few tesla, and efficient current drive can be possible using LH waves especially for plasma start-up when the density is low.

A greater concern may be the uncertainty of the impact of the strong toroidicity of the low-aspect ratio machines. The strong gradient of the toroidal field causes large up-shift or down-shift of the parallel index of refraction, and the wave propagation and damping are strongly altered from what they would be in the absence of up/down-shift. Although difference in the wave propagation for different launch scenarios are qualitatively similar between STs and conventional tokamaks, application of unconventional scenarios such as inboard-launch, top-launch, etc., will likely have greater impact on the current drive efficiency on STs. Therefore, testing different launch scenarios on ST is also a good code validation exercise for the predicted effectiveness of those unique launch scenarios. In this paper, we present our recent numerical analysis of the LH driven fully non-inductive discharges on TST-2.

* Corresponding author: tsujii@k.u-tokyo.ac.jp

 DOI: 10.1051/epjconf/201715702009

## 2 Experimental setup

TST-2 is a spherical tokamak ($R_0$ = 0.36 m, a = 0.23 m, $B_{t0}$ < 0.3 T, $I_p$ < 120 kA) [16] located at the University of Tokyo in Japan. The magnets are powered by capacitor banks and the on axis toroidal field is limited to 0.16 T for LH start-up discharges due to longer pulse length. The maximum plasma current achieved for LH start-up discharges is 25 kA which is about a quarter of that of the typical Ohmic discharges.

Four 100 kW sources at 200 MHz are available for LH current drive. Presently, two capacitively-coupled combline (CCC) antennas are installed [9, 11]. Compared to the previous antennas, the CCC antennas have good plasma coupling, directional wavenumber spectrum, and high power handling capability, so that most of the available source power can be delivered to the plasma at the desired spectral range. The outboard-launch antenna has a peak parallel index of refraction $n_{\parallel}$ of 6.0 and 200 kW source power. The top-launch antenna has a peak $n_{\parallel}$ of 4.7 at the center of the straps and 100 kW source power. The $n_{\parallel}$ needs to be greater than ~3 for core accessibility. Absorption on the bulk (thermal) distribution occurs at about $n_{\parallel}$ = 20 for the present TST-2 temperature. The highest plasma current was achieved using the outboard-launch CCC antenna.

A 5 kW electron cyclotron heating (ECH) system at 2.45 GHz is also available for reliable breakdown of the plasma. ECH is turned off during the plasma current ramp-up for the results shown in this paper.

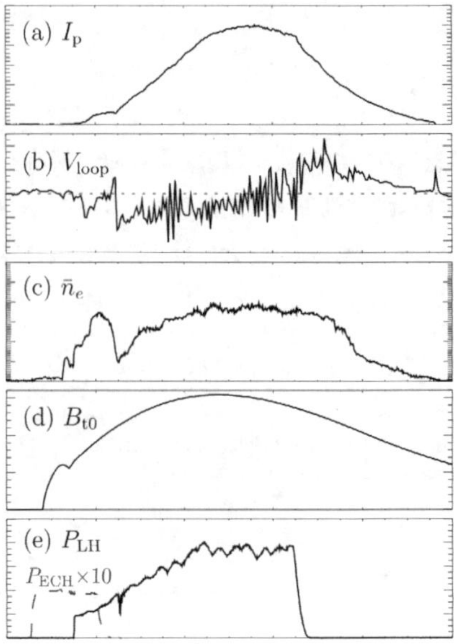


**Fig. 1.** Time traces of a fully non-inductive plasma start-up discharge by LH waves. The plotted parameters are (a) plasma current, (b) loop voltage, (c) line-averaged electron density, (d) toroidal field, and (e) LH power. ECH power (×10) is also shown with the blue dashed curve.

## 3 Numerical simulations

A ray-tracing code GENRAY [17] and a Fokker-Planck solver CQL3D [18] was used to simulate LH current drive in this work. The ray trajectories calculated in GENRAY are given to CQL3D for self-consistent simulation of LH wave damping and the electron distribution function. CQL3D solves the bounce-averaged Fokker-Planck equation in the limit of zero orbit width (electrons tied to the flux surfaces). The electron distribution function is evolved in time in the presence of LH wave quasilinear diffusion until it reaches a near steady-state under the experimentally given density, temperature and current profiles. In LH current drive experiments, waves with highly superthermal phase velocity are usually launched for efficiency. The absorption is weak at LH power turn on, but the fast electron plateau quickly develops, resulting in much stronger absorption and efficient plasma current sustainment. Therefore, it is essential that the wave damping and evolution of the electron distribution function be simulated self-consistently.

## 4 LH start-up experimental results

### 4.1 Outboard-launch experimental results

Figure 1 shows the time evolution of the parameters of a typical non-inductive discharge on TST-2 driven solely by LH waves. The plasma current reaches a maximum of 25 kA about 40 ms after the LH power turn on, and sustained for about 10 ms. Note the negative loop voltage (b) which indicates that this is fully non-inductive plasma current ramp-up. The line-averaged density is about $4\times10^{17}$ $m^{-3}$ which is only a few percent of the typical TST-2 Ohmic discharges. The maximum on axis toroidal field is 0.16 T for non-inductive discharges and peaks at about 35 ms after the LH power turn on. The LH power required to sustain this discharge is about 80 kW (e) which is less than half of the available source power. It was not possible to increase the plasma current by increasing the LH power. This point will be discussed further later.

The electron density and temperature profiles were measured by the Thomson scattering diagnostic (Fig. 2). Because of the low density, the same discharge was repeated and the measurements were ensemble averaged to bring the Thomson scattering signal out of the noise floor [19]. In fully non-inductive plasma current ramp-up, current holes due to back EMF with little particle and energy confinement have been observed elsewhere [20]. The density and temperature profiles were flat inside the current hole (a "box-type" profile) in these discharges. This does not seem to be the case for TST-2 since the density profile is peaked. The temperature profile is hollow, indicating that the LH power is deposited near the edge. This seems to indicate that there is a substantial radially inward transport of the current density.

The plasma current at the flat-top of the discharge is plotted versus the line-averaged density in Fig. 3. The

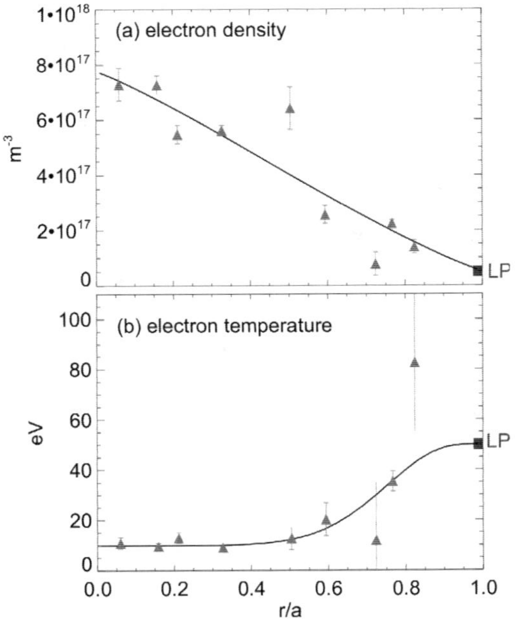


**Fig. 2.** Density (a) and temperature (b) profiles measured by the Thomson scattering diagnostic plotted versus the normalized minor radius (red triangles). The Langmuir probe measurement is also shown with the blue squares. Fits to the measured points used in the simulation are shown with the solid curves.

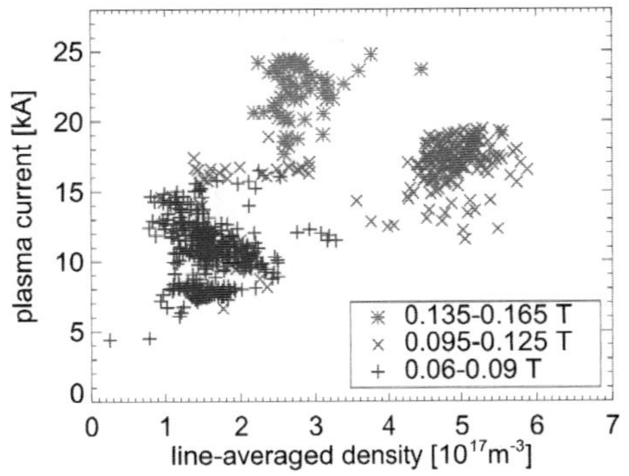


**Fig. 3.** The observed plasma current versus the line-averaged density for three different ranges of magnetic field; red asterisks: 0.135-0.165 T, green crosses: 0.095-0.125 T, blue plus symbols: 0.06-0.09 T.

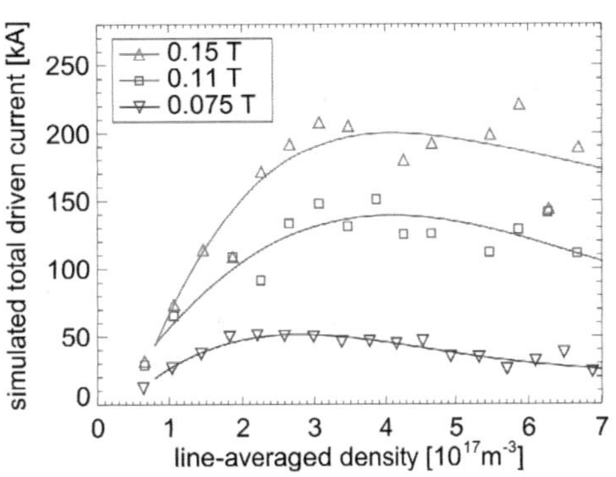


**Fig. 4.** Simulated total driven current versus the line-averaged density for three different values of toroidal field; red upward triangles: 0.15 T, green squares: 0.11 T, blue downward triangles: 0.075 T. The solid curves are the fits to the simulated points.

plasma current increased at higher toroidal field, which is consistent with the improved accessibility of low $n_{\|}$ component of the LH waves. On the other hand, the achievable plasma current decreased slightly with decreasing density, which was the opposite of what was expected from previous LH current drive experiments in conventional tokamaks [13-15]. This is a potential limitation for using LH on STs, since operation at as low density as possible is desired. If the density needs to increase with the plasma current, the plasma current will be limited at the mode conversion limit under fixed toroidal field.

To understand the observed parametric dependence of the plasma current, LH current drive was modelled using the GENRAY-CQL3D simulation. The density was scanned in the simulation for discharges obtained in TST-2 for three different values of toroidal fields. The results are shown in Fig. 4. In the simulation, the scrape-off-layer (SOL) was assumed to have a constant density of $5\times10^{16}$ m$^{-3}$ and a constant temperature of 50 eV which are the typical values measured by the Langmuir probe installed on the limiter. The temperature profile was fixed to the profile shown in Fig. 2 (b) since little variation of temperature was observed in the experiment. About 60 % of the power was predicted to be absorbed through electron Landau damping in the SOL. A smaller fraction of power (<10 %) was absorbed in the cold core of the plasma. The simulated total driven current decreased at higher density due to deterioration of the LH wave accessibility to the core as was expected.

On the other hand, reduction of the driven current at low density is not usually seen. The simulated parallel electron distribution function at 0.11 T together with the radial profile of the driven current is shown in Fig. 5 for the optimum and the low density (see Fig. 4, green square symbols). In LH current drive, the wave generates a plateau in the electron parallel velocity distribution function which carries the current [21]. The low-velocity end of the plateau is 2-3 times the thermal velocity below which the absorption is strong and no wave power exists. The high-velocity end, i.e., the maximum phase velocity (corresponding to the minimum $n_{\|}$) of the LH waves, is more important regarding the maximum current density. This limit corresponds to the wave accessibility limit determined by the mode conversion condition. For the cases shown in Fig. 5, it can be seen that the fast electron plateau is completely flat and saturated. Under this condition, the plateau level

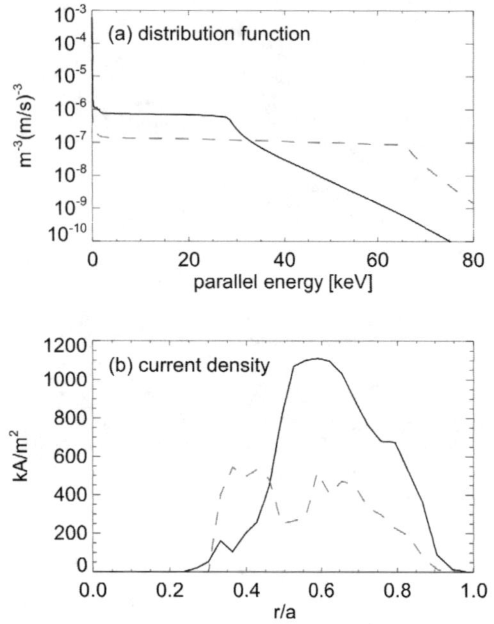


**Fig. 5.** The parallel electron distribution function at r/a = 0.59 (a) and the radial profile of the driven current (b) at the optimum and the low density, 0.11 T. Black solid curves: $3.9\times10^{17}$ m$^{-3}$, red dashed curves: $1.1\times10^{17}$ m$^{-3}$.

decreases in proportion to the density. Although wave spectrum with higher phase velocity become accessible at lower density so that the electrons are accelerated to higher energy, the net change is a factor of two reduction in the driven current density (Fig. 5 (b)). That is, our density is so low that the current density is at its theoretical maximum which is proportional to the electron density (Equation (4) in [21]). Increasing the LH power does not help increase the driven current under this circumstance since the fast electron plateau is already saturated.

In such a low density regime, there is an optimum density for current drive determined by the balance of a sufficient current carrier (electron) density and good LH wave accessibility to the plasma core which requires low density. According to the GENRAY-CQL3D simulation, the optimum density is $2\text{-}6\times10^{17}$ m$^{-3}$ for $B_{t0}$ = 0.11 T (Fig. 4, green squares). This is consistent with the operational density range in the actual experiment (Fig. 3, green crosses). The simulated total driven current starts to fall off as a function of density above $4.0\times10^{17}$ m$^{-3}$ at 0.11 T. The fall-off density decreases to $2.5\times10^{17}$ m$^{-3}$ at 0.075 T due to worse accessibility. The operational density range in the actual experiment also decreases by a factor of two from ~0.11 T to ~0.075 T, consistently with the simulation.

Appreciable increase in the total driven current was predicted when the toroidal field was increased from 0.11 T to 0.15 T, which was observed in the experiment as well. On the other hand, the predicted increase in the driven current from 0.076 T to 0.11 T was not observed in the experiment. In the previous experiments on TST-2, the plasma current ramp-up up to about 4 kA was possible with counter-current drive LH antenna phasing [8], indicating substantial contribution of the bootstrap current for the lower current discharges. Since bootstrap current is not taken into account in the present analysis, this may explain the discrepancy.

The overall value of the simulated driven current (Fig. 4) was up to an order of magnitude greater than the overall plasma current observed in the actual experiment (Fig. 3). The typical fast electron energy for the LH driven non-inductive discharges on TST-2 is about 20 keV (Fig. 5 (a)). At the present level of plasma current, the poloidal field is at most ~10 mT, and the trapped particle orbit widths for the 20 keV electrons are 0.07 m, which is one-third of the plasma minor radius. Under such a condition, acceleration of electrons through RF quasilinear diffusion is balanced not only by the collisional slowing down, but also with collisional radial diffusion (neoclassical diffusion). Additional losses of fast electrons associated with the radial diffusion will likely lead to smaller driven current and may bring the predicted value closer to the experimental observation. Either Monte-Carlo approach [22] or Fokker-Planck solver valid for finite orbit width [23] is necessary to account for these effects and also to evaluate bootstrap current in the presence of fast electrons.

## 4.2 Top-launch experimental results

For further current ramp-up under the present toroidal field, we have developed a top-launch antenna [11]. LH waves launched from the top of the plasma undergo up-shift of the parallel index of refraction and higher and higher density becomes accessible as the waves propagate radially inward. Although current drive efficiency decreases when the parallel index of refraction increases, it is hoped that higher operational density will allow us to achieve higher plasma current. When the rays are launched from the outboard side, they diffract outward at higher density. On the other hand, the ray trajectory for the waves launched from the top of the plasma barely changes (Fig. 5 in [11]).

Non-inductive plasma start-up experiment using the top-launch antenna started at the beginning of 2016 [11]. Although we were able to develop an operational scenario using only the top-launch antenna, plasma current ramp-up was most successful when the outboard-launch antenna was used together with the top-launch antenna. For demonstration of the improvement of the wave accessibility, we show the comparison of the discharges with and without the top-launch antenna power during the ramp-down of the toroidal field since it was difficult to perform accurate density control with the present system implemented on TST-2. The time traces of the parameters are shown in Fig. 6. After the plasma current reached a flat-top value of about 10 kA, the density (b) and the vertical field (d) was kept steady while the toroidal field (c) ramped down. The black curves show the traces of the discharge with only the outboard-launch antenna power of 60 kW. The red

DOI: 10.1051/epjconf/201715702009

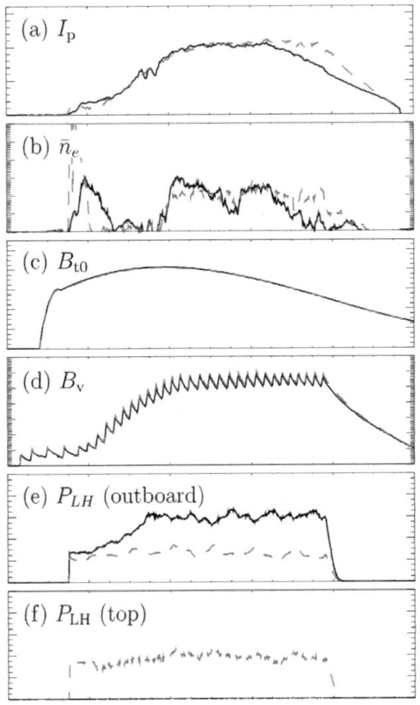


**Fig. 6.** Time traces of the fully non-inductive plasma start-up discharge by LH wave with only the outboard-launch antenna (black solid curves) and with both the top-launch and the outboard-launch antennas (red dashed curves). The plotted parameters are (a) plasma current, (b) line-averaged electron density, (c) toroidal field, (d) vertical field, (e) outboard-launch antenna power, and (f) top-launch antenna power.

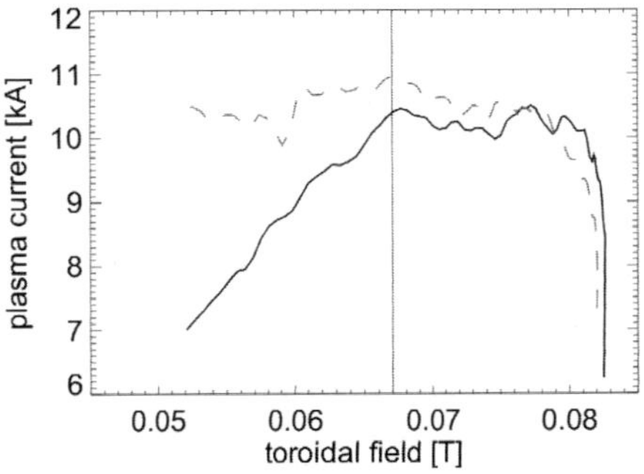


**Fig. 7.** The measured plasma current plotted versus the toroidal field. Black solid curve: outboard-launch only, red dashed curve: both top-launch and outboard-launch.

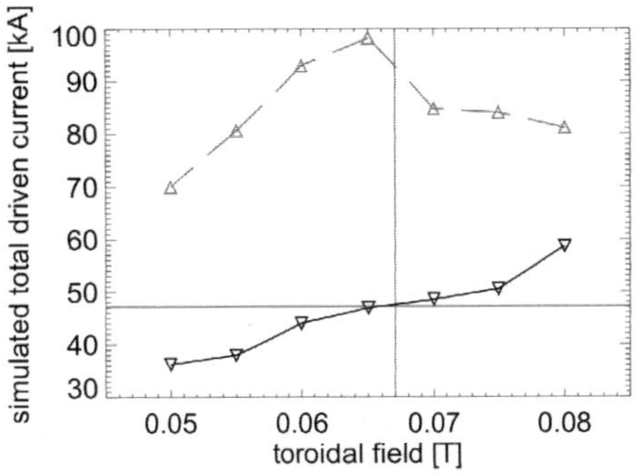


**Fig. 8.** The simulated total driven current plotted versus the toroidal field. Black solid curve: outboard-launch only, red dashed curve: both top-launch and outboard-launch.

curves show those with the same total LH power but half the LH power replaced by the top-launch antenna power.

The measured plasma current (a) is plotted versus the toroidal field in Fig. 7. When only the outboard-launch antenna was used, plasma current starts to fall off below 0.067 T. On the other hand, when the top-launch antenna was used together with the outboard-launch antenna, the plasma current was sustained until the end of the discharge despite the same total LH power. The simulated total driven current for the two discharges is shown in Fig. 8. The simulated driven current with the top-launch antenna was always higher than that at 0.067 T with the outboard-launch antenna only, consistently with the experimental observation. Note that the simulated total driven current shows how much current can potentially be driven for the given parameters. In the actual experiment, the plasma current is limited by the vertical field and is held constant as long as sufficient current drive power is provided.

## 5 Optimization of LH current drive with poloidal phasing

The dependence of the experimentally observed driven current on the toroidal field and different launch locations could be qualitatively reproduced with the present GENRAY-CQL3D model. If the present plasma current limit is indeed do to saturation of the fast electron plateau, higher current could be obtained only by optimizing the wave spectrum in the core plasma.

The theoretical maximum current density is achieved when the $n_{\|}$ spectrum is such that the parallel phase velocity extends all the way from electron thermal velocity up to the maximum phase velocity given by the mode conversion condition. As mentioned in the introduction, waves with highly superthermal parallel phase velocity are usually launched, and the gap between this velocity and the electron thermal velocity is filled by up-shift of $n_{\|}$ as the waves travel in and out of the plasma core. Such a strategy is empirically known to work under the ideal circumstance with minimal SOL losses.

The optimum launch condition can be obtained by targeting the rays such that they mode convert at the location where current drive is desired. The condition was evaluated for TST-2 by launching the rays at r/a = 0.2 on the high-field side midplane with mode conversion condition, and propagating them back in time. This gave a launch location slightly above the midplane with strong poloidal phasing in addition to the toroidal phasing. Fig. 9 shows the comparison of the driven current profile with the present outboard-launch antenna phasing (toroidal index of refraction $n_{tor}$ = 6.1) and with

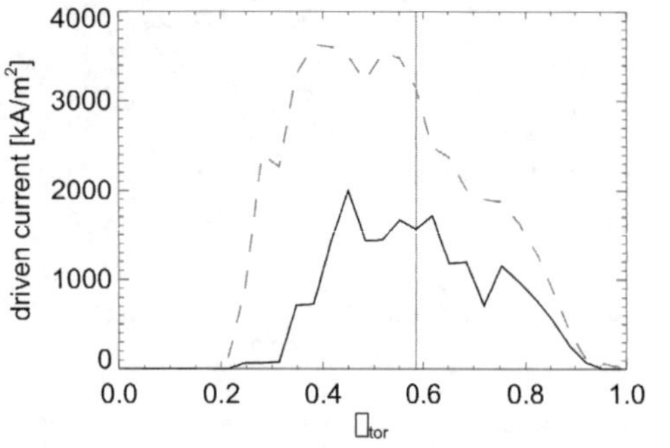


**Fig. 9.** The simulated total driven current plotted versus the normalized minor radius. Black solid curve: the present outboard-launch antenna, $n_{tor}$ = 6.1 (216 kA), red dashed curve: with poloidal phasing, $n_{tor}$ = 1.8, $n_{pol}$ = -17 (442 kA).

strong poloidal phasing ($n_{tor}$ = 1.8 and poloidal index of refraction $n_{pol}$ = -17). Increase of the total driven current by a factor of two was predicted. With the poloidal phasing, the launched rays propagate below the midplane while the imposed poloidal phasing is down-shifted and $n_{\parallel}$ is maintained nearly constant until they transform to fast waves around the targeted location. Further analysis is still required to find the optimum launch condition for various toroidal field and density, but imposing certain degree of poloidal phasing appears to have a beneficial impact in general.

## 6 Conclusions and future work

Non-inductive plasma start-up with the LH current drive on a ST has been demonstrated on TST-2. With the newly available electron density and temperature profiles, realistic numerical modeling of LH current drive was performed using the GENRAY-CQL3D simulation. The plasma current was observed to increase slightly with increasing density, which is the trend opposite to what was expected from previous LH current drive experiments in conventional tokamaks. The numerical analysis of the TST-2 discharges showed that the driven current indeed decreases at lower density due to shortage of current carriers (electrons). This parameter regime has never been accessed in other devices which have much higher density and and much lower LH power per electron. When the top-launch antenna was used, the density limit was higher than the discharge driven only by the outboard-launch antenna as was expected theoretically.

For future work, the cause of the overprediction of the simulated driven current needs to be investigated. The present analysis assumes zero electron orbit width. Finite orbit model is probably necessary to evaluate quantitatively accurately the LH driven current and the bootstrap current on TST-2. An improved antenna model may also be necessary to accurately describe the high $n_{\parallel}$ part of the wave spectrum. The internal current profile is also not accurately evaluated in the experiment. The impact of current profile on the wave propagation needs to be investigated. The MEGA code [24] is being extended to investigate the impact of fast electrons on the MHD equilibrium and instabilities.

Finally, poloidal phasing was theoretically shown to be beneficial for current drive. Feasibility of such an antenna design will be investigated.

This work was supported by JSPS KAKENHI Grant number 21226021, NIFS Collaboration Research Program under NIFSKOCR001, NIFS12KUTR078, and NIFSKNWR001, and Japan/US Cooperation in Fusion Research and Development. Work at General Atomics is supported by US DoE contract DE-AC03-97ER-54411. Collaborations with P. Bonoli, J. Wright, and J. Lee at MIT on GENRAY/CQL3D calculations are gratefully acknowledged.

## References

1. Y.K. Peng, and D. Strickler, Nucl. Fusion **26**, 769 (1986)
2. F. Najmabadi, Fusion Eng. Des. **65**, 143 (2003)
3. S. Nishio, et al., 20th International Conference on Fusion Energy, FT/P7-35 (2004)
4. R. Raman, et al., Phys. Rev. Lett. **104**, 095003 (2010)
5. V.F. Shevchenko, et al., Nucl. Fusion **50**, 022004 (2010)
6. M. Uchida, et al., Phys. Rev. Lett. **104**, 065001 (2010)
7. H. Miura, et al., Plasma Fus. Res. **10**, 3402066 (2015)
8. Y. Takase, et al., Nucl. Fusion **51**, 063017 (2011)
9. Y. Takase, et al., Nucl. Fusion **53**, 063006 (2013)
10. T. Wakatsuki, et al., Nucl. Fusion **54**, 093014 (2014)
11. T. Shinya, et al., Nucl. Fusion **57**, 036006 (2017)
12. S. Shiraiwa, et al., Phys. Rev. Lett **92**, 035001 (2004)
13. S. Shiraiwa, et al., Nucl. Fusion **53**, 113028 (2013)
14. K.K. Kirov, et al., Plasma Phys. Control. Fusion **55**, 115008 (2013)
15. M. Goniche, et al., Nucl. Fusion **53**, 033010 (2013)
16. Y. Takase, et al., Nucl. Fusion **41**, 1543 (2001)
17. A. Smirnov, and R. Harvey, *Bull. Am. Phys. Soc.* **39**, 1626 (1994)
18. R.W. Harvey, and M.G. McCoy, *Proceedings of the IAEA Technical Committee Meeting on Advances in Simulation and Modeling of Thermonuclear Plasmas*, 489 (1992)
19. H. Togashi, et al., J. Inst. **10**, C12020 (2015)
20. T. Fujita, Nucl. Fusion **50**, 113001 (2010)
21. N.J. Fisch, Phys. Rev. Lett. **41**, 873 (1978)
22. S. Murakami, et al., Nucl. Fusion **46**, S425 (2006)
23. Y.V. Petrov, and R.W. Harvey, Plasma Phys. Control. Fusion **58**, 115001 (2016)

DOI: 10.1051/epjconf/201715702009

24. Y. Todo, and A. Bierwage, Plasma Fus. Res. **9**, 3403068 (2014)

EPJ Web of Conferences **157**, 02010 (2017) DOI: 10.1051/epjconf/201715702010
*22 Topical Conference on Radio-Frequency Power in Plasmas*

# Recent progress of RF-dominated experiments on EAST

F.K. Liu,*, Y.P. Zhao, J.F. Shan, X.J. Zhang, B.J. Ding, X.J. Wang, M. Wang, H.D. Xu, C.M. Qin, M.H. Li, X.Z. Gong, L.Q. Hu, B.N. Wan, Y.T. Song, and J.G. Li for the EAST Team and Collaborators

Institute of Plasma Physics, Chinese Academy of Sciences, Hefei 230031, China

**Abstract.** The research of EAST program is mostly focused on the development of high performance steady state scenario with ITER-like poloidal configuration and RF-dominated heating schemes. With the enhanced ITER-relevant auxiliary heating and current drive systems, the plasma profile control by coupling/integration of various combinations has been investigated, including lower hybrid current drive (LHCD), electron cyclotron resonance heating (ECRH) and ion cyclotron resonance heating (ICRH). The 12 MW ICRH system has been installed on EAST. Heating and confinement studies using the Hydrogen Minority Heating scheme have been investigated. One of the importance challenges for EAST is coupling higher power into the core plasma, experiments including changing plasma position, electron density, local gas puffing and antenna phasing scanning were performed to improve ICRF coupling efficiency on EAST. Results show that local gas injection and reducing the k|| can improve the coupling efficiency directly. By means of the 4.6 GHz and 2.45 GHz LHCD systems, H-mode can be obtained and sustained at relatively high density, even up to ne ~ $4.5 \times 10^{19}$ $m^{-3}$, where a current drive effect is still observed. Meanwhile, effect of source frequency (2.45GHz and 4.6GHz) on LHCD characteristic has been studied on EAST, showing that higher frequency improves penetration of the coupled LH (lower hybrid) power into the plasma core and leads to a better effect on plasma characteristics. Studies demonstrate the role of parasitic effects of edge plasma in LHCD and the mitigation by increasing source frequency. Experiments of effect of LH spectrum and plasma density on plasma characteristics are performed, suggesting the possibility of plasma control for high performance. The development of a 4MW ECRH system is in progress for the purpose of plasma heating and MHD control. The built ECRH system with 1MW source power has been successfully put into use on EAST in 2015. H-mode discharges with L-H transition triggered by ECRH injection were obtained and its effects on the electron temperature, particle confinement and the core MHD stabilities were observed. By further exploring and optimizing the RF combination for the sole RF heating and current drive regime, fully non-inductive H-mode discharges with $V_{loop}$~0V has progressed steadily in the 2016 campaign. The overview of the significant progress of RF dominated experiments is presented in this paper.

## 1 Introduction

Experimental Advanced Superconducting Tokamak (EAST) [1-3] is dedicated to provide a unique platform to address plasma physics and technology issues relevant for ITER under long-pulse operation conditions. Recently, great efforts [4] have been made on both physics and technology fronts to accomplish this mission with all ITER-relevant auxiliary heating and current drive systems, enabling the investigation of plasma profile control by coupling/integration of various combinations. By means of the 4.6 GHz and 2.45 GHz lower hybrid current drive (LHCD) systems, H-mode can be obtained and maintained at relatively high density, even up to ne ~ $4.5 \times 10^{19}$ $m^{-3}$, where a current drive effect is still observed; A new small/no ELM H-mode regime has been demonstrated to be a good solution to the problem of divertor transient heat load, which facilitate the exploration of long-pulse high-performance H-mode operation on EAST. In addition, significant progress has been achieved on EAST, including: i) Demonstration of a steady-state scenario (fully non-inductive with $V_{loop}$ ~ 0.0V at high $\beta_P$ ~ 1.8 and high performance ($H_{98,y2}$> 1.0) in upper single-null ($\varepsilon$ ~ 1.6) configuration with the tungsten divertor; ii)Discovery of a stationary ELM-stable H-mode regime with 4.6 GHz LHCD.This overview will report the main advances in RF dominated steady-state H mode scenarios. The progress on RF physics will also be presented.

## 2 RF Heating and Current Drive systems

To fulfil the physical objectives of EAST and increase the relevance to ITER, over 30 MW CW heating and current-drive (H&CD) power have been installed on EAST. Figure 1 shows the distribution of auxiliary heating systems on EAST. Besides a new LHCD system (at 4.6 GHz) and NBI system, an ECH system, which would offer greater flexibility with regards to plasma

* Corresponding author: fkliu@ipp.ac.cn

shape and plasma stabilization, is under developing for pressure and current density profile control on EAST.

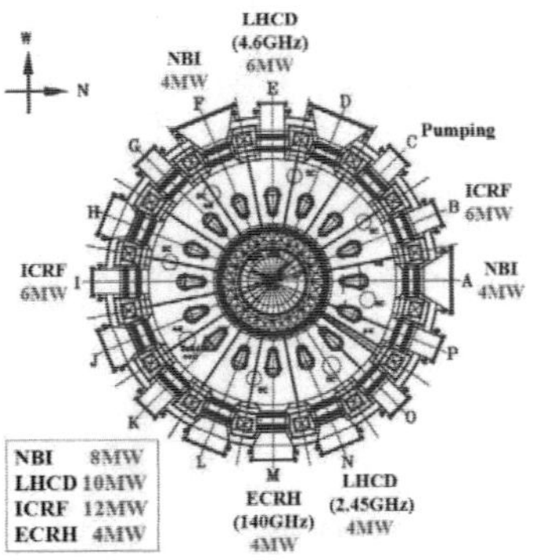


**Fig.1.** The distribution of auxiliary heating systems on EAST

### 2.1 ICRH system

The EAST ICRH system [5] has been developed to support long-pulse advanced tokamak fusion physics experiments. The source power of the ICRH system is 12.0 MW in total. The range of ICRH frequency is from 25 MHz to 70 MHz. The 8 ICRH sources at 25-70 MHz have been successfully commissioned at full power on water dummy load. An averaged maximum RF output power of 1.5 MW for each source has been achieved in a frequency range of 25-65 MHz with an efficiency varying from 60% to 70%.

The ICRH system consists of two antennas, each delivering 6.0 MW source power. One is a two toroidal, two poloidal strap (B-port) antenna and the other is a four-strap, folded strap (I-port) antenna. For B-port antenna [5], each strap is powered by one transmitter. Each poloidal pair at one toroidal location are phased at 180 and each toroidal pair is typically operated at [0, $\pi$ ] phasing. For the I-port antenna, each toroidal strap is powered by a single transmitter and can be phased relative to the other straps. During the 2016 experimental campaign, the ICRH system has been operated at total forward power up to 800 kW for 60s on long pulse H mode plasma on EAST.

### 2.2 LHW system

The long pulse operation of EAST relies exclusively on LHCD. Two CW LHCD systems are installed and operated in EAST campaign [6]. There are a 4.0 MW / 2.45 GHz system and a new 6.0 MW/ 4.6 GHz system utilizing 2 individual multi-junction phase antennas. Sustained plasmas over 400 s has been demonstrated by non-inductive current driven by LHCD systems. High current drive efficiency and nearly full non-inductive plasmas maintained by the new 4.6 GHz LHCD system. Technical solutions for long pulse high power system commissioning have been successfully demonstrated during the R&D activities.

### 2.3 ECRH system

As one of the most promising external heating methods, a long pulse electron cyclotron resonance heating (ECRH) system [7, 8] with 4MW power is under development at the frequency of 140 GHz (2nd harmonic X mode injection at a field of 2.5 Tesla), it will be used to provide central heating, optimize the plasma profiles for high pressure and good confinement as well as suppression of magneto-hydrodynamic (MHD) instabilities by a flexible two-mirror steerable antenna. At present, 1MW system was put into operation since 2015.

## 3 RF experiments on EAST

### 3.1 LHCD experiments at high density on EAST

LHCD experiments at high density in EAST aim at fully assessing conditions useful for enabling the LHCD effect into dense plasma core. To explore long pulse and high performance with LHCD, as shown in Fig.2, the capability of controlling current profile by optimizing lower hybrid (LH) spectrum with 4.6GHz LHCD system was demonstrated[9] on EAST. Results indicate the highest current drive (CD) efficiency and the most peaked current density profile occurs with $N_{//}^{peak}$=2.04, suggesting the possibility of profile control by changing the wave spectrum. The effect of LH frequency, with $N_{//}^{peak}$=2.1 for 2.45GHz and $N_{//}^{peak}$=2.04 for 4.6GHz and comparable directivities of the spectrum (0.74 vs. 0.76), on LHCD characteristics (ne =$2.0\times10^{19}m^{-3}$) with a same coupled power (1.05MW) in a lower single null (LSN) configuration was investigated in EAST (see Fig. 3) [10]. It shows the residual voltages ($V_{loop}$) are 0.27V and 0.15V, respectively, during 2.45GHz and 4.6GHz application, implying a better CD efficiency for 4.6GHz LH waves. Better plasma heating effect for 4.6GHz can be inferred from the time evolution of plasma stored energy (WMHD ~ 68.3kJ and 74.8kJ, respectively for 2.45GHz and 4.6GHz). The internal inductance (li) is higher with the 4.6GHz LH wave injection, meaning a more peaked current profile.

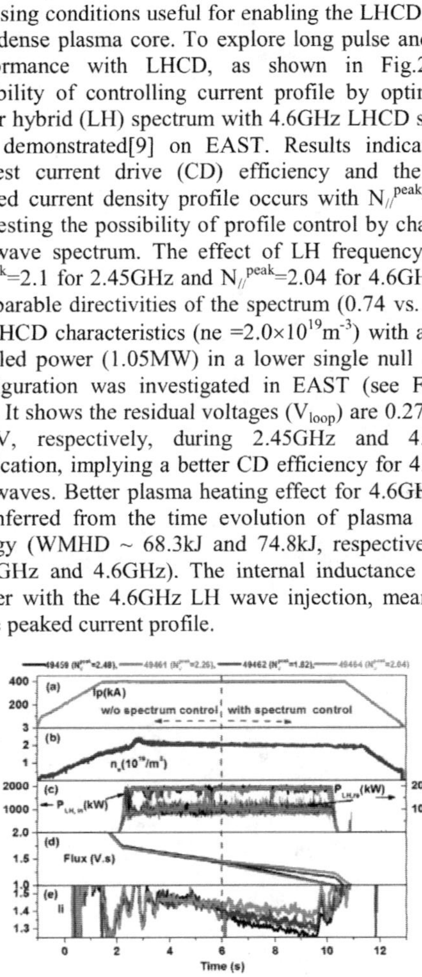


**Fig. 2.** Demonstration of LH spectrum on controlling current profile (4.6GHz)

Above results show that LH wave at 4.6 GHz exhibit stronger CD capability than at 2.45 GHz. As indicated by the parametric instability (PI) measurement by a radio frequency (RF) loop antenna (Fig. 4), it is speculated to be mainly ascribed to the less PI behavior with 4.6GHz LH wave. The effect of LH frequency (2.45 GHz and 4.60 GHz) on PI has been analysed by LHPI and MIT code in Ref. 6, both showing that PI is stronger with 2.45GHz wave. Based on the PI modelling, the LHstar suite of numerical codes [11] has been further utilised for calculating the driven current profile [6]. Due to the PI effect, the power deposited in the edge region, which cannot contribute to useful LHCD mechanism, is different, leading to a different LH current profile. Such result is qualitatively consistent with the much weaker LHCD effects that are routinely observed on EAST when operating with the 2.45 GHz frequency, in comparison with the 4.60 GHz case.

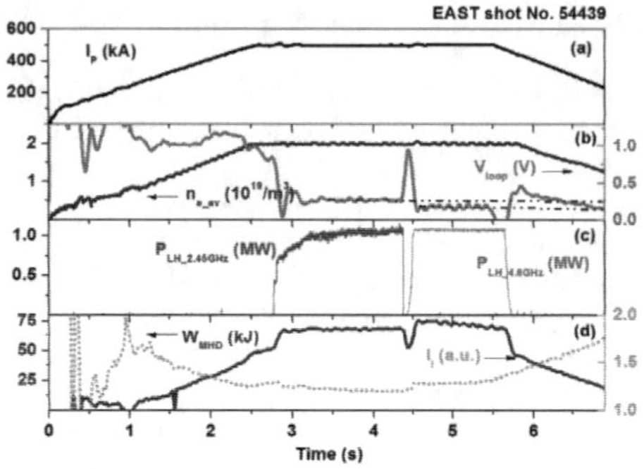


**Fig. 3.** Typical waveform of LH frequency effect on plasma characteristics

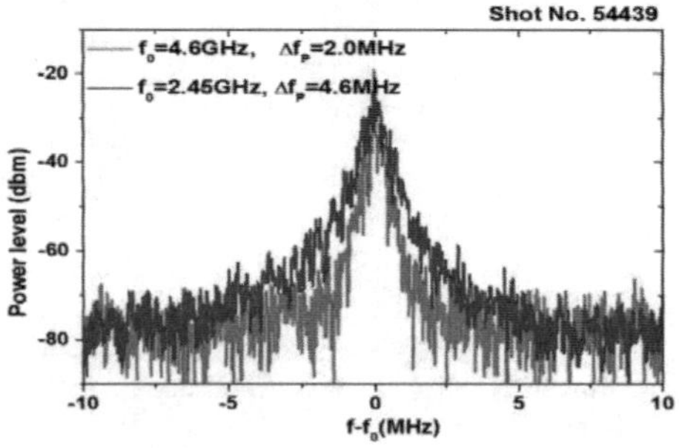


**Fig. 4.** PI signal by RF probe

With the 2.45GHz and 4.6GHz LHCD systems, high density H-mode is obtained and the typical waveforms are shown in Fig. 5, demonstrating the CD capability at high density. The electron cyclotron emission (ECE) signal drops quickly at the L-H transition, implying the decrease of LH driven current, possibly due to the concomitant density increase. Seen from the loop voltage and ECE signal, it is inferred that part of current is driven by LHW even if at the density of 4.5×1019m-3. Such density is higher than that of 1.5-2.5×1019m-3 obtained by 2.45GHz wave alone[12].

When injecting high LHW power at 4.6GHz, strong hot spots were often observed on the guard limiter of LHCD antenna by visible CCD (Fig. 6), leading to a sudden increase of impurity influx such as carbon and cooper, and often ending with disruptions. To identify the operation regime with hot spot occurring, a careful scan of global parameters such as plasma density, outer gap and LHW antenna phase difference was performed and the threshold power for the hot spots was around 2.5~3.0MW. The graphite tiles are damaged with the severest region located above the mid-plane. When using the RMP with rotating mode, hot spots could be slightly lightened by tuning the particle flux hitting on the guard limiter. More testing is still ongoing with a new guard limiter design to avoid strong particle flux directly hitting on the limiter.

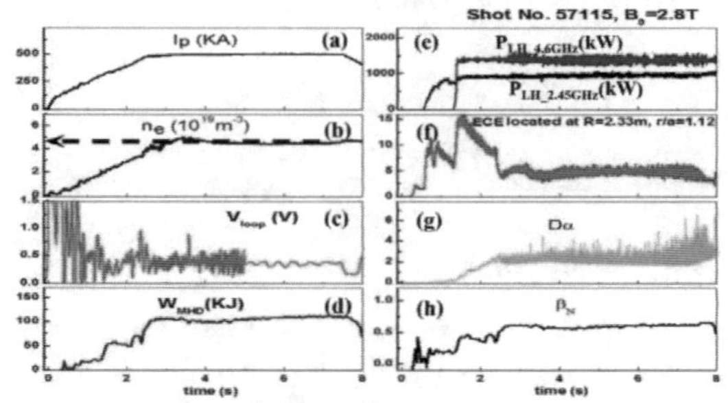


**Fig. 5.** H-mode with LHCD at high density

**Fig.6.** Hot spot observed by CCD and damage found in the 4.6GHz LHW antenna

## 3.2 ICRF Coupling and Heating on EAST

One of the importance challenges for EAST ICRH experiments is coupling higher power into the core plasma. The coupling of ICRF waves relies on the tunnelling of the fast wave between the antenna structure and the cut-off layer within the plasma. The experimental antenna loading is calculated from the input ICRF power measured by directional couplers and the maximum of the voltage in the transmission line measured by voltage probes. The antenna loading are calculated separately for each strap of each antenna. Here the mutual coupling between the straps is neglecting. Local gas puffing and antenna phasing scanning have been tested to improve ICRF coupling efficiency on EAST. The experimental results show that local gas injection and reducing the k parallel improves the coupling efficiency directly.

In order to study the ICRF coupling, some important experiments [13] including changing plasma position and electron density scanning have been performed. We change the gap between the limiter and the plasma last closed flux surface. The antenna is fixed in dipole phasing. As shown in Fig.7, the coupling resistance decreases with the gap. The results show that widening the gap will increase the evanescent layer width and reduce the coupling efficiency. This is also confirmed by the electron density measured by microwave reflectometry. ICRF antenna loading increase with the line averaged density of the core plasma. This is because increasing the central electron density is equivalent to pushing the position of the cutoff density to close to the antenna and reducing evanescent layer.

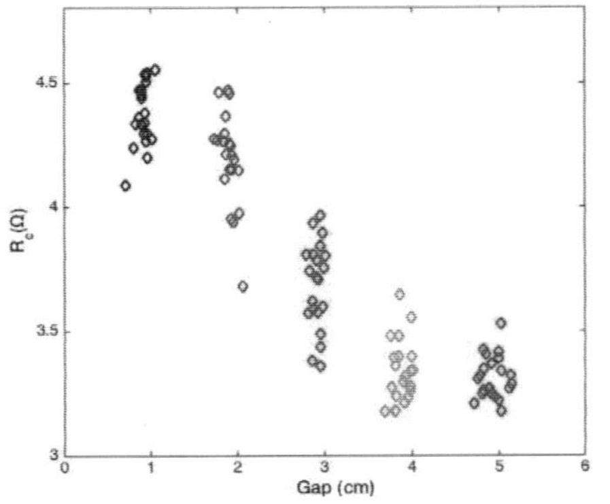


**Fig.7.** ICRF antenna loading versus gap between the limiter and the plasmas last closed flux surface.

Heating and confinement studies using H minority heating regime show a typical L-mode behavior. The ICRF heating efficiency is averaged at 35%. The heating efficiency can be improved when combined with the lower hybrid wave. The confinement of purely ICRF heated plasmas has been studied on EAST. The result of calculation shows that there is a good agreement with the result of the ITER89-P empirical scaling. And we found that the energy confinement of ICRH essentially follow an L-mode scaling.

### 3.3 Preliminary results from ECRH

It is generally known that ECRH is a highly controllable heating tool and the power deposition profiles can be changed by adjustment of the launcher in poloidal and toroidal directions. Fig. 8 shows the range of EC power deposition. The experimental locations were obtained by power modulation method and the simulated results were from TORAY code. It is seen that the locations can be varied in the region of ρ< 0.8 and good agreement between the experiments and simulations are achieved, which validates the steerable performance of the launcher.

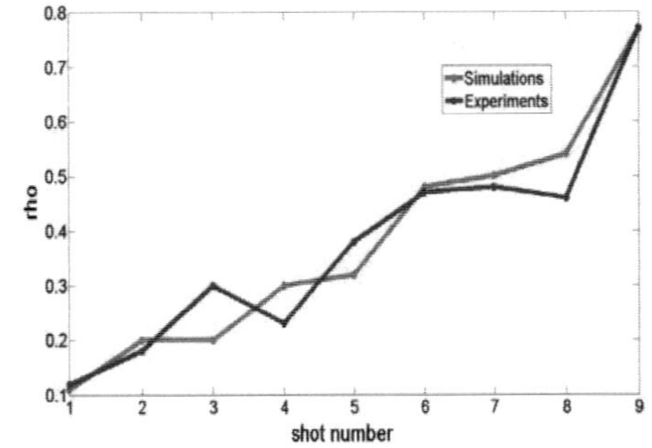


**Fig.8.** Comparison of EC power locations between experiments and simulations.

Good plasma heating effects were obtained during ohmic target plasma [8] and H-mode phase by 400 kW EC waves injection. As shown in Fig.9, the electron temperature measured by electron-cyclotron emission (ECE) radiometer increases globally during ECRH phase.

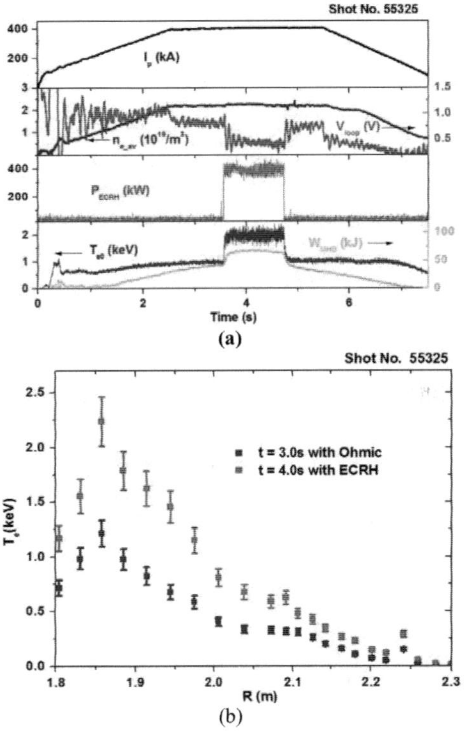


**Fig. 9.** (a) Typical waveform of ECRH power injected into Ohmic plasma, and (b) comparison on Te profiles measured by the electron-cyclotron emission (ECE) radiometer between Ohmic heating and ECRH phases.

An H-mode discharge with L-H transition was triggered by ECRH as illustrated in Fig.10 characterized by a sharp drop in Dα emission, an increase in line-averaged density (ne_av) and plasma stored energy ($\Delta W_{MHD} \sim 78$ kJ).

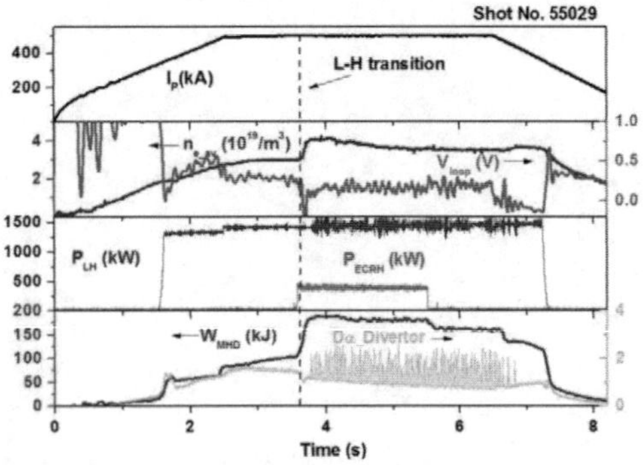


**Fig. 10.** Time histories from an ELMy H-mode discharge with L-H transition triggered by ECRH

The continued EC beam injection has been applied to stabilize the m/n=2/1 tearing mode on EAST tokomak. The stabilization effect is investigated as shown in the Fig.11 (a). It is found that the island width can be reduced up to 60% of the initial size before the EC beam on, with the gyrotron power of 320 kW and deposited at 0.5. As the beam deposition position moves away from the original position, the stabilization effect gradually reduces and the mode is even enhanced by the heating in the plasma core. The m/n=2/1 mode is also stabilized when co-ECCD is applied, even though its stabilization effect is less significant than ECRH. However, the counter-ECCD seems to have few effects on the mode stabilization.

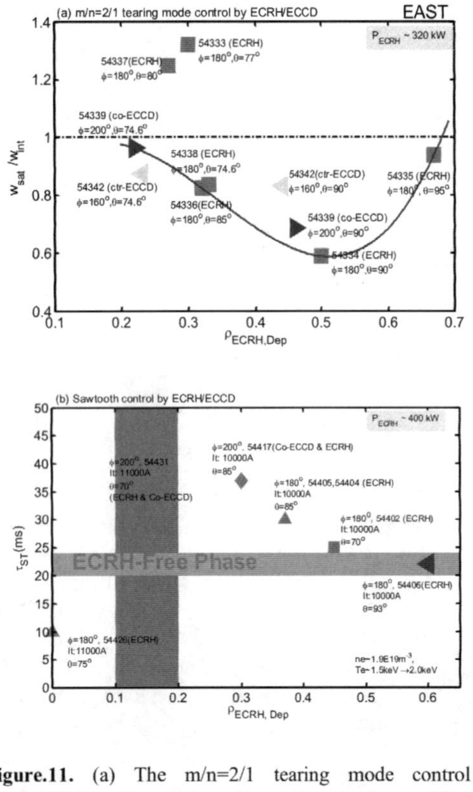


**Figure.11.** (a) The m/n=2/1 tearing mode control by ECRH/ECCD. The dashed line indicates free stabilization effect, above which the mode is enhanced and below which the mode is stabilized. (b) Sawtooth control by ECRH/ECCD. Sawtooth period reaches its maximum at about q = 1 surface.

The sawtooth period is found to be minimized in EAST experiments [14] when the ECRH resonance location is deposited to be just inside the q =1 surface, as shown in the Fig. 11(b). When ECRH deposition is outside the q = 1 surface, the sawtooth oscillation is stabilized (characterized by prolonged sawtooth period), and the sawtooth periods gradually decrease as ECRH deposition position sweeps away from q = 1 surface.

# 4 Long pulse discharge by H&CD systems

## 4.1 High electron temperature discharge

Stationary long pulse plasma of high electron temperature was produced on EAST for the first time through an integrated control of plasma shape, divertor heat flux, particle exhaust, wall conditioning, impurity management, and the coupling of multiple heating and current drive power. As shown in Fig.12, a discharge[15] with a lower single null divertor configuration was maintained for 103 s at a plasma current of 0.4 MA, q95 ≈7.0, a peak electron temperature of >4.5 keV, and a central density ne(0)~$2.5\times10^{19}$ $m^{-3}$. This scenario allows the further work on the synergy of electron heating by LHW and ECH and electron thermal transport in the near future. It further introduced an example of integrated "hybrid" operating scenario of interest to ITER and CFETR.

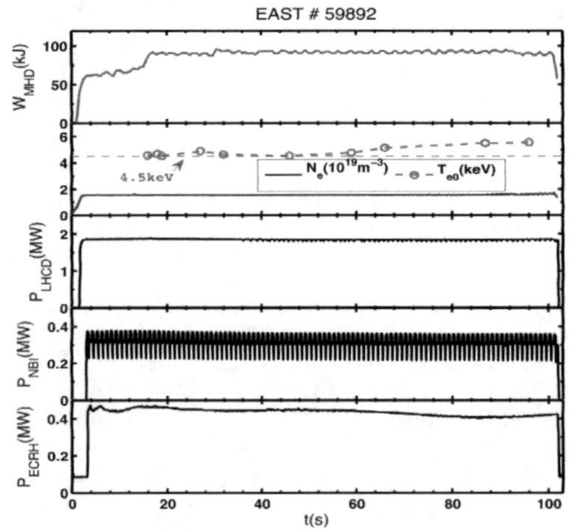


**Fig.12.** 102-sec long pulse discharge with Te> 4.5keV

## 4.2 long pulse H mode

To further explore the coupling or integration of several key physics process issues, long-pulse operation [16] has been explored on EAST. Fig. 13 shows a discharge of duration over one minute with the multi-RF power combination, i.e. 0.4 MW LHW at 2.45 GHz, 2.1 MW LHW at 4.6 GHz, 0.4 MW ECH and 0.8 MW ICRF. The

DOI: 10.1051/epjconf/201715702010

plasma parameters are as follows: Ip ~ 0.45 MA, toroidal magnetic field BT = 2.5 T, elongation k~1.6, and safety factor q95~6. The loop voltage was well controlled to be slightly negative, which indicates a fully non-inductive current drive condition. It is also worth pointing out that the loop voltage increases and crosses over zero after 50 s. Small edge localized modes (ELMs) are obtained in this long-pulse H-mode discharge, with low energy ejection per ELM event. These small ELMs facilitate the RF power coupling in the H-mode phase and the avoidance of shape oscillation caused by large ELMs. A confinement enhancement factor relative to the standard H-mode, H98y2 of 1.1~1.2, was achieved and maintained constant during the discharge, with a line averaged electron density ~3.0 × $10^{19}$ $m^{-3}$.

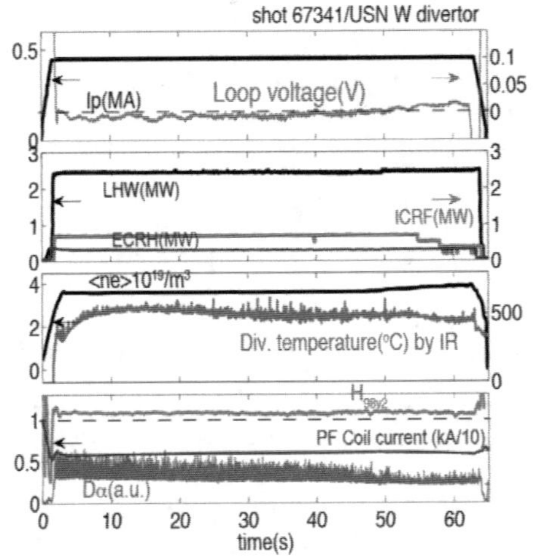


**Fig. 13.** Overview of >60s long-pulse RF-heated Steady-state H-mode discharge. From top to bottom panel: 1). Plasma current (Ip) and Loop voltage (V); 2). Line averaged density and divertor temperature; 3). LHW, ECRH and ICRF power; 4). H98y2 factor, PF current and Dα signal

## 5 Summary

On EAST, a series of long-pulse H-modes have been successfully achieved towards more ITER-relevant condition, such as low torque injection, RF-dominated heating and current drive with an ITER-like tungsten divertor. These newly achieved H-mode scenarios may offer a promising regime for long-pulse high-performance operation. To accomplish this, great efforts have been made to control both steady state and transient ELM induced divertor heat fluxes. EAST could be in operation with more than 30 MW CW heating and current drive power (LHCD, ICRH, NBI and ECRH), enhanced diagnostic capabilities and full actively-cooled metal wall from 2015. The upgraded EAST capabilities provide very high potential to investigate the critical issue for long pulse H mode operation with dominant electron heating and low torque input, which is essential for ITER and future reactors.

## Acknowledgements

This work was supported partly by National key research and development program (grant nos 2016YFA0400600, 2016YFA0400601 and 2016YFA0400602). This work was supported partly by National Magnetic confinement Fusion Science Programme (grant nos 2015GB101001, 2015GB103002, 2015GB102003, 2013GB106001B, 2013GB112003). This work was also supported partly by the National Natural Science Foundation of China under grant nos11675213, 11375235, 11175206, 11305211, 11275233 and 11375236.

## References

1. Y. X. Wan et al., 2006 Overview progress and future plan of EAST Project Proc. 21th IAEA FEC 2006 (Chengdu, China 2006) (Vienna: IAEA) OV/1-1and http://www.naweb.iaea.org/napc/physics/fec/fec2006/html/index.htm
2. B. N. Wan et al., 2012 Progress of long pulse and H-mode experiments in EAST Proc. 24th IAEA FEC (San Diego, US 2012) (Vienna: IAEA) OV2-5 and http://www-naweb.iaea.org/napc/physics/FEC/FEC2012/papers/278_OV25.pdf
3. J. G. Li et al., 2013 Nature Phys. 9 817
4. B.N. Wan et al., 2015 Nucl. Fusion 55 104015
5. Y. P. Zhao et al., 2014 Fusion Engineering and Design 89 2642-2646
6. F. K. Liu et al., 2016 Fusion Engineering and Design 113, 131-138
7. X. J. Wang et al., 2015 Fusion Engineering and Design, 96-97, 181-186
8. H. D. Xu et al., 2016 Plasma Sci. Technol.18 442
9. F. K. Liu et al., 2015 Nucl. Fusion 55 123022.
10. B. J. Ding et al., 2017 Nucl. Fusion 57 022022
11. B. J. Ding et al., Recent experimental and modeling advances in the understanding of lower hybrid current drive in ITER-relevant regimes, 26th IAEA FEC
12. B. J. Ding et al., 2013 Nucl. Fusion 55 113027
13. J. H. Zhang et al., 2017 Nucl. Fusion 57 066030
14. Y. Yuan et al., 2016 Phys. Plasmas 23, 062503
15. Y. Zhang et al., 2016 Fusion Science Technology, Vol 70 , 62–72
16. X. Z. Gong et al., 2017 Plasma Sci. Technol. 19 032001

DOI: 10.1051/epjconf/201715702011

# Antenna to Core: A New Approach to RF Modelling

*J. Wright and S. Shiraiwa*

Plasma Science and Fusion, Massachusetts Institute of Technology, Cambridge, MA, USA

**Abstract.** A new technique for the calculation of RF waves in toroidal geometry enables the simultaneous incorporation of antenna geometry, plasma facing components (PFCs), the scrape off-layer (SOL) and core propagation [Shiraiwa et al, "HIS-TORIC: Extending core ICRF wave simulation to include realistic SOL plasmas", Nucl. Fusion in press **10** (2017)]. The calculation of antenna coupling has been traditionally performed separately from those of core propagation whereas this method allows them to be combined permitting the modelling of such phenomena as waves reflected from plasma-vacuum boundary or cyclotron resonances back to the antenna. Calculations with this technique also naturally capture wave propagation in the SOL and its interactions with non-conforming PFCs permitting self-consistent calculation of core absorption and edge power loss.

## 1 Introduction

In modelling and theory of radio frequency (RF) fields in different frequency regimes, the core and edge plasmas have been treated in a separate manner. High fidelity core ion cyclotron and lower hybrid (LH) codes use sophisticated dielectric models for the confined plasma but model the scrape off layer and plasma facing components more simply often with uniform cold plasmas and conducting walls or simple vacuum regions [1-4]. Antenna coupling models will use very accurate models of the antenna structures but calculate the plasma impedance with one dimensional half infinite models [5,6] and typically do not model wave propagation into the plasma. There is a growing realization in the community of the importance of the connection between these two regimes [7-12].

The finite element method (FEM) and the spectral approaches to simulation of ion cyclotron (IC) waves in toroidal plasmas each have strengths and weaknesses. For example, the spectral approach (e.g. TORIC[1] and AORSA [2]]) has a natural algebraic representation of the parallel wavenumber and hence the wave dispersion model but does not easily represent complex geometries present outside the last closed flux surface, whereas the FEM approach (e.g. LHEAF [13,14]) naturally represents arbitrary geometries but does not easily represent thermal kinetic corrections to the plasma dispersion. The two domains: thermal core with flux surfaces and cold edge plasma with open field lines may be combined in such a way that each approach is used where it works naturally. Among the possible ways of doing this, we demonstrate the method of mode matching. This method provides an easy way of combining the two linear systems without significant modifications to the separate codes. We will present proof of principal cases and initial applications to minority heating.

The main motivating insight is that the core plasma region that is assumed to have closed flux surfaces requires a hot plasma dielectric while the open field line region in the scrape-off layer needs only a cold plasma dielectric. Spectral approaches work well for the former and finite elements work well for the latter. The two approaches are combined with a domain decomposition technique that constructs the solution from Green's functions obtained from a core spectral code, TORIC [1], and an edge finite element code using COMSOL [15] or MFEM [16], operating in their respective regions. The validity of this process follows directly from the superposition principle of the linear Maxwell's equations making this technique exact. The method is independent of the codes or representations used and works for any frequency regime.

The outline of this paper is as follows: we will briefly review the method of domain decomposition to separate the solution of wave propagation in the scrape off layer (SOL) from the core as presented by Shiraiwa and Wright [17-20]. We will then discuss extensions of this model to include resistive dissipation in the SOL to calculate core heating efficiency. In the next section, we will demonstrate applications to minority heating in Alcator C-Mod and high harmonic heating in NSTX-U will be presented in single pass and multi-pass regimes. Finally, we will discuss extensions of the method to three dimensions and future applications.

## 2 Description of the method

In the frequency domain Maxwell's equations are given by:

* Corresponding author: jcwright@mit.edu

DOI: 10.1051/epjconf/201715702011

$$\nabla \times \nabla \times \boldsymbol{E} = \frac{\omega^2}{c^2}\left\{\boldsymbol{E} + \frac{4\pi i}{\omega}(\boldsymbol{J}^P + \boldsymbol{J}^A)\right\} \quad (1)$$

$$\boldsymbol{J}^P = \overleftrightarrow{\sigma}[f_0(\boldsymbol{x}, \boldsymbol{k}, \boldsymbol{v}_\perp, \boldsymbol{v}_\parallel)] \cdot \boldsymbol{E}, \quad (2)$$

where $\boldsymbol{J}^P$ is the plasma current response to the electric RF field, $\boldsymbol{E}$, as determined by the plasma conductivity, $\sigma$. The antenna current, $\boldsymbol{J}^A$, gives the source of RF wave energy. In typical RF full wave codes, it is an idealized current strap without detailed antenna structures or limiters [1, 2]. The Vorpal code [8,9], this publication] is an exception that uses a finite difference mesh to model the complex antenna geometry and includes cold plasma wave propagation from the antenna into a partial region of the core plasma. Other recently developed codes also focus on the cold plasma wave propagation near the antenna with perfectly matched layers (PML) at the domain boundaries such as RAPLICASOL [21,22]. Such outgoing boundary conditions as PML or absorbing boundary layers preclude the study of possibly important physics effects such as the reflection of waves back towards the antenna, far field sheath effects [23,24], and the modelling of scrape-off layer (SOL) propagation and collisional dissipation of coupled wave energy.

Here, we present a method to seamlessly connect the physics of hot plasma core propagation codes with the detailed antenna models such as used by TOPICA [6] including SOL propagation and dissipation. Maxwell's Equation's as typically used in plasma physics with a linear dielectric response support the superposition principle. A consequence of this property is that an RF (radio-frequency) problem described by Eq. (1) along with boundary conditions (BCs) may be decomposed into separate solvable systems with BCs that while differing from the original system are equivalent to it. This leads to a domain decomposition technique that is described in detail in Wright [17] and Shiraiwa [18-20].

The tokamak plasma is separated into two domains: inside the last closed flux surface a hot plasma dielectric is used with a spectral decomposition and outside a finite element representation with a cold plasma dielectric is used. The two solution domains are connected by an admittance condition at the matching boundary – the last closed flux surface – with the approximation that no significant kinetic flux is present there. Solutions in the two domains are built up from a family of Green's functions resulting from a source electric field on the boundary that is a pure poloidal mode, $E \sim \exp(im\theta)$, for each tangential component. Continuity of the tangential electric field requires that the modal amplitudes of the inner and outer solutions are the same. Continuity of the tangential magnetic field is given by an admittance equation that determines the amplitudes of the solution. Mathematically, this is given by Eqns (3) and (4):

$$\sum_m a_m E_m^{\text{core}} = \sum_m a'_m E_m^{\text{edge}} \quad (3)$$

$$\sum_m a_m B_m^{\text{core}} = \left(\sum_m a'_m B_m^{\text{edge}} + B_m^{\text{ant}}\right), \quad (4)$$

where $B_m^{\text{ant}}$ is the contribution of the antenna magnetic field's $m^{\text{th}}$ poloidal mode to the matching boundary. For unit drive electric field, $E_m$, the resulting magnetic field, $\sum_{m'} B_{m'}$, is the admittance response (cf. Fig. 5 in Shiraiwa [19]).

There are several advantages to this approach. We may use an existing core code with a kinetic dielectric that has been well verified and validated such as TORIC [1]. The edge finite element model may represent complicated limiter structures, even the antenna itself, and straightforwardly supports other boundary conditions such as a rectified sheath [7]. The admittance equation solved to couple the two systems has only a degree of freedom equal to the number of poloidal modes used in the coupling while the core and edge solutions are of one dimension higher (in three dimensions the order of both is raised by one more dimension). And the core and edge solutions are only coupled via this admittance equation. A consequence of this is that the core and edge responses may be recalculated independently; for example, if the antenna structure or phasing was changed it would only require recalculation of the edge response of the coupled system.

The overall computational cost of the entire system is comparable to a normal standalone solution from a core code. All the mode responses are computed in a single simulation by replacing the normal boundary condition associated with an antenna strap by an array of boundary conditions each one associated with one of the driving poloidal modes. The consequence of this is to add a matrix-matrix product in addition to the normal matrix inverse in place of a matrix-vector product. This approximately doubles the calculation time. If we include the degrees of freedom (DoF) of the edge solver, the total DoF is approximately doubled compared to the original core solver for only ½ the computational requirement one would expect because of the domain partitioning. This is because for a direct solve scaling as $(\text{DoF})^3$ we would expect a factor of 8 increase in computation for doubling the DoF but instead have two regions with the same as the original DoF but requiring approximately twice the computation.

## 3 Application to edge losses

An important capability of this model is the capture of edge dissipation in the scrape of layer (SOL) or rf sheath rectification on plasma facing components. In this section, we demonstrate the role of Spitzer resistivity in dissipating wave energy in the SOL. In Alcator C-Mod, minority heating in Deuterium plasmas may use either a Hydrogen or a Helium-3 minority species. Observed heating efficiencies for these two cases differ significantly [25]. Including the SOL propagation and dissipation explains the difference. In NSTX, surface wave excitation has been shown [10] to be an issue in current drive efficiency by high harmonic fast waves when the SOL density is above the fast wave cutoff. We show that significant amounts of wave power are lost in the SOL and that the wave propagation is more limited with realistic geometry and SOL densities than shown previously [10].

DOI: 10.1051/epjconf/201715702011

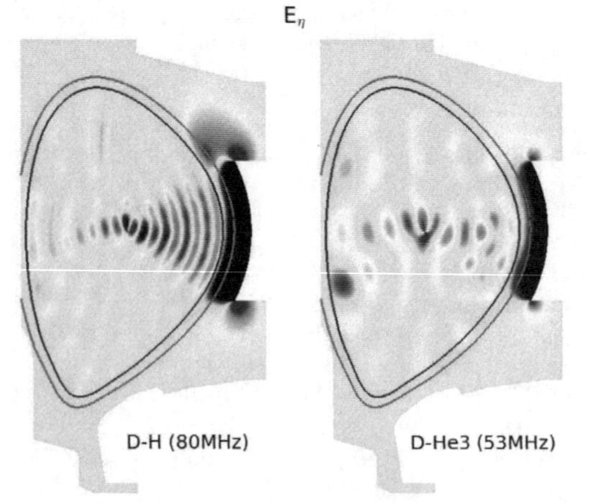


**Fig. 1.** Real part of the poloidal (η) component of the wave electric field. A comparison of strong D-(H) (left panel) and weak D-(He3) (right panel) absorption in Alcator C-Mod minority heating scenarios. Minority concentration in both cases is 5%. Simulations used resolution of 127 poloidal modes and 240 radial cubic finite elements. The light line border around the core plasma is the matching layer for the inner and outer solutions and the thick red line is the plasma – vacuum boundary.

## 3.1. Strong and Weak core absorption

In Fig. 1, D-(He3) in the left panel demonstrates weak single pass absorption. The waves reflect off the walls multiple times setting up standing wave patterns and fill the core plasma volume. In contrast, the D-(H) plasma shows strong absorption and focussing of the ICRF waves in the center of the plasma cross section. The antenna loading for both scenarios is comparable (14.5Ω and 16Ω respectively from model calculations) but experimentally, the D-(H) scenario is much more efficient at heating the plasma. A competing loss mechanism provided by resistive dissipation in the SOL enables the calculation of a core heating efficiency.

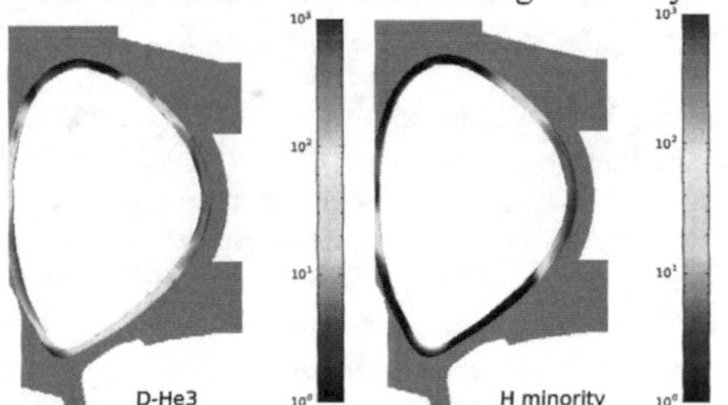


**Fig. 2.** Highlight of SOL collisional power loss for two minority heating cases in Alcator C-Mod on a logarithmic scale. 95% majority D, 5% minority He3 or H, B=5.254 T, $n_{e0}$=5x10$^{14}$, $T_{e0}$=4 keV, $T_{minority}$=120 keV, f=80 MHz, nphi=10.

To the edge FEM model surrounding the core plasma we add a 2.5 cm SOL with a cold plasma dielectric. A finite electron temperature of 15 eV is retained in a collision frequency calculated from Spitzer resistivity. The density is 4x10$^{19}$cm$^{-3}$ at the top of the pedestal and has a minimum of 2x10$^{19}$cm$^{-3}$ at the bottom. The density in the SOL everywhere is above the lower hybrid resonance value and the fast wave cutoff. The D-(H) minority case predicts that 10% of the wave power is dissipated in the scrape-off layer and 90% in the core. For D-(He3), 50% is lost in the scrape-off layer. These two heating efficiencies of 90% for D-(H) and 50% for D-(He3) are in the same range as experimental observations [25] for these minority heating scenarios. Figure 2 shows the collisional power deposition in the SOL in two dimensions. In both cases, there is power deposited in front of the antenna. In the D-(He3) case we see SOL power loss on the high field side of the cross-section indicating that some of the ICRF power passes through the plasma and is deposited in the SOL near the inner walls. This could indicate a possibility of far field sheath power losses and localized plasma facing component (PFC) heating as well.

## 3.2 Surface waves

For some plasma conditions in high harmonic fast wave current drive experiments on NSTX, much of the coupled wave power does not appear in the plasma core [10]. When the density in the SOL is above the fast wave cutoff density, surface waves may propagate between the antenna and the last closed flux surface (LCFS). In Fig. 13 from Bertelli [10], as the density is raised above the fast wave cutoff in AORSA full wave simulations, the region outside the LCFS is filled with a standing wave pattern. These simulations used a constant density everywhere outside the LCFS in a rectangular wall. We reproduce this case in Fig.3 using the actual limiter geometry and constrain the density to be constant on field lines outside the LCFS with a 3 cm SOL at the midplane.

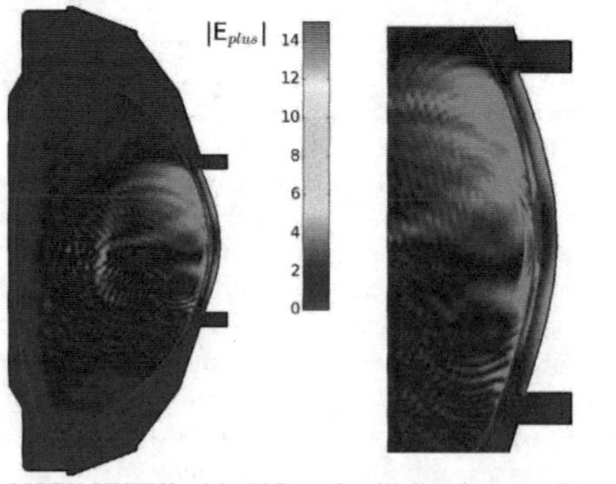


**Fig. 3.** NSTX HHFW with SOL and collisional losses. Density has 5 cm SOL gradient. Nphi=-12, f=30 MHz, $n_{e0}$=3.7x10$^{13}$, $T_{e0}$=1.1 keV, and B=0.55 T. Deuterium plasma with Deuterium beams.

The enlargement of the fields around the antenna show a limited area of fast wave propagation in the SOL. This is a consequence of the wave being cutoff by low in the shadow of the antenna limiters. The density in the SOL is modelled by diffusing density along open field lines from the last closed flux surface. The waves in this

NSTX case are in a single pass absorption regime but the presence of relatively high plasma density outside the LCFS permits some fast wave propagation in the immediate area of the antenna. This is sufficient for 20% of the coupled power to be lost in the SOL to collisional damping. In contrast to the results shown in Bertelli [10], the wave propagation is fairly localized due to the higher fidelity modelling of the SOL and limiter surfaces. This could be important in experimental comparisons with diagnostics measuring RF fields in the SOL, for example.

## 4 Extending to three dimensions

The domain decomposition technique used in these two dimensional simulations trivially extends to full three dimensional simulations. Instead of constructing an admittance system of poloidal mode responses for a single toroidal mode, the admittance matrix includes a spectrum of toroidal modes as well. In this case, the size in $N_m x N_n$ where $N_m$ is the number of poloidal modes and $N_n$ is the number of toroidal modes. The matrix then has the form of dense blocks representing poloidal coupling along the diagonal and more spare lower magnitude blocks off the diagonals representing toroidal coupling as seen in Fig. 4.

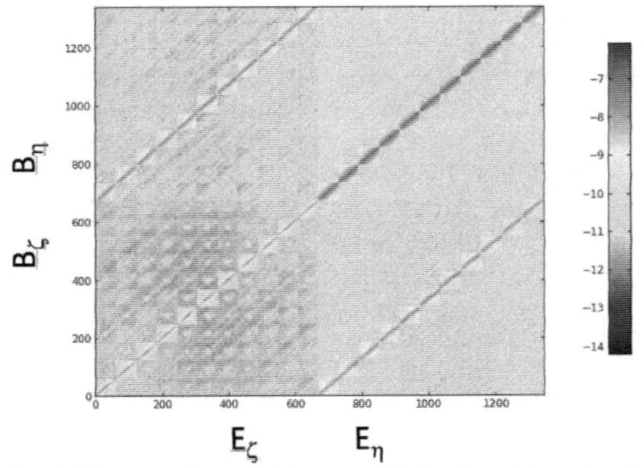


**Fig. 4.** Edge coupling admittance matrix for three dimensional ICRF simulation of Alcator C-Mod. 1342 independent full wave solutions by the each of the full wave codes for the core and edge plasma response are used to solve the full admittance system.

If we may assume the core plasma is axisymmetric for a tokamak, then the core solutions may be generated one toroidal mode at a time. The edge solution is fully three dimensional and any non-axisymmetric structures in the edge plasma such as the antenna result in finite cross coupling between toroidal modes. As a result, when solving the matching for the core – edge interface, the core toroidal modes are also coupled together by edge asymmetries in the final solution.

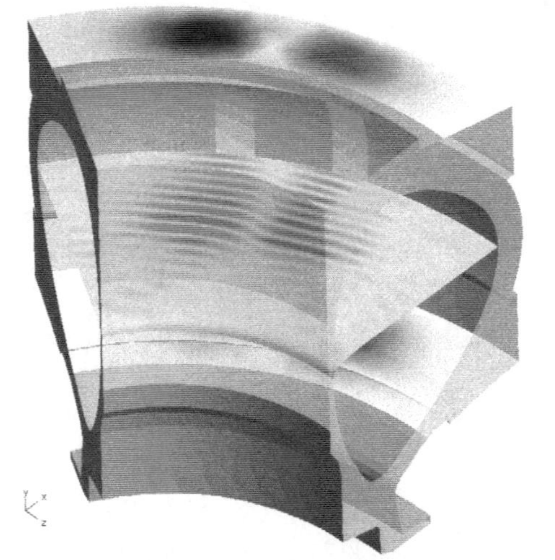

**Fig. 5.** A 60 degree toroidal slice of a full three dimensional simulation of Alcator C-Mod showing the left hand polarized component. 95% majority D, 5% minority He3 or H, B=5.254 T, $n_{e0}=5x10^{14}$, $T_{e0}=4$ keV, $T_{minority}=120$ keV, f=80 MHz, nphi=[-30, -24 … +24, +30].

For the case shown in Fig. 5, we choose to break the edge response calculation into six 60 degree toroidal wedges. As a consequence, only the sixth toroidal mode harmonics (-30, -24 … +24, +30) need to be calculated for the core response. This is the same symmetry technique used in previous stellarator full wave calculations [3]. Figure 5 shows a 60 degree slice of the reconstructed full three dimensional simulation including antenna straps and edge propagation. A midplane slice of the RF fields shows the toroidal extent of the waves launched by the two straps The two straps are phased 180 degrees apart for heating and this is evident in the RF field perpendicular to the wall of the straps shown at their endpoints. The only asymmetry in the edge plasma for this simulation is from the antenna straps. Stellarator simulations are also possible using the algorithm presented in this paper. To capture the toroidally periodicity of stellarators, a code such as AORSA [2] or SCENIC [26] could be used for the core response.

## 5 Conclusions

In this paper, we have presented a core to edge RF model that is capable for the first time of including an antenna model, scrape-off layer propagation and collisional damping, and core hot plasma propagation and kinetic damping in a single integrated simulation. The method is general and has been applied to ICRF and HHFW regimes. The model can be used in two dimensional or three dimensional configurations and is extensible to stellarators given an appropriate core wave solver. The inclusion of damping mechanisms in the SOL permits the simultaneous calculations of antenna loading and core heating efficiency. Results shown for Alcator C-Mod are in agreement with experimental

DOI: 10.1051/epjconf/201715702011

observations for two different minority heating scenarios. The combination of core and edge models permits the future exploration of many RF modelling questions previously unavailable such as the differences between antenna loading and core heating efficiencies, two and three dimensional effects of the plasma on antenna loading and heat deposition of RF power on plasma facing components.

The simulations presented in this paper were performed on the MIT-PSFC partition on the Engaging cluster at the MGHPCC (www.mghpcc.org). Funding for this research was provided under USDoE Contract No. DE-FC02-01ER54648 and supported by the U.S. Department of Energy, Office of Science, Office of Fusion Energy Sciences, using User Facility Alcator C-Mod, under Award Number DE-FC02-99ER54512.

## References

1. M. Brambilla, Plasma Phys. Controlled Fusion, **41**, 1–34 (1999).
2. E.F. Jaeger, D.B. Batchelor, H. Weitzner, J.H. Whealton, Computer Phys. Comm. **40** (1986). http://dx.doi.org/10.1016/0010-4655(86)90147-5
3. E.F. Jaeger, L. A. Berry, E. D'Azevedo, D. B. Batchelor, M. D. Carter, K. F. White and H. Weitzner, Phys. Plasmas **9**, 1873 (2002). http://dx.doi.org/10.1063/1.1455001
4. J.C. Wright, E.J. Valeo, C.K. Phillips, P.T. Bonoli and M. Brambilla, Commun. Comput. Phys., **4**, 545 (2008).
5. Lancellotti, et al, "Prediction of plasma-facing ICRH antenna behavior via a Finite-Element solution of coupled Integral Equations," in *RADIO FREQUENCY POWER IN PLASMAS: 16th Topical Conference on Radio Frequency Power in Plasmas*, edited by S. J. Wukitch and P. T. Bonoli, AIP Conference Proceedings, **787** (2005).
6. Lancellotti, V., Milanesio, D., Maggiora, R., Vecchi, G., and Kyrytsyabern, V., Nucl. Fusion, **46**, S476–S499 (2006).
7. D.A. D'Ippolito, et al, Plasma Phys. and Controlled Fusion **55**, 085001 (2013).
8. T. G. Jenkins and D. N. Smithe, Plasma Sources Sci. Technol. **24**, 015020 (2015).
9. D. N. Smithe, Phys. Plasmas **14**, 056104 (2007).
10. N Bertelli, et al, Nuclear Fusion **56** (1), 016019 (2016).
11. G.M. Wallace, et al, Phys Plasmas **17** (8), 082508 (2010).
12. D.L. Green, L.A. Berry, G. Chen, P.M. Ryan, J.M. Canik, and E.F. Jaeger, Phys. Rev. Lett. **107** 145001 (2011).
13. O. Meneghini, S. Shiraiwa and R. Parker, Phys. Plasmas **16**, 090701 (2009). http://dx.doi.org/10.1063/1.3216548
14. S. Shiraiwa, O. Meneghini, R. Parker, P. Bonoli, M. Garrett, J.C. Wright, and S. Wukitch, Phys. Plasmas **17**, 056119 (2010).
15. "*COMSOL Multiphysics Modeling Software". COMSOL.com. Comsol, Inc. Retrieved 20 November* (2016).
16. http://www.mfem.org
17. J. Wright and S. Shiraiwa, "Coupling an ICRF core spectral solver to an edge FEM code" in *21st Topical Conference on Radio Frequency Power in Plasmas*, AIP Conference Proceedings, edited by R. Pinsker, New York, **1689**, 05007-1 (2015).
18. S. Shiraiwa, J. Wright, J.P Lee, and P. Bonoli, "Integration of core/edge plasmas in fullwave RF simulation", in *RADIO FREQUENCY POWER IN PLASMAS: Proc. of 26th Fusion Energy Conf.* (TH/P4-27) (Kyoto, 2016).
19. S. Shiraiwa, J. Wright, J.P Lee, and P. Bonoli, "HIS-TORIC: Extending core ICRF wave simulation to include realistic SOL plasmas", Nucl. Fusion in press **10** (2017).
20. S. Shiraiwa, J.C. Wright, P.T. Bonoli, T. Kolev, and M. Stowell, "RF wave simulation forcold edge plasmas using the MFEM library" in *22nd Topical Conference on Radio Frequency Power in Plasmas*, European Phys. J. Web of Conferences, **this issue**, B-31 (2017).
21. J. Jacquot, V. Bobkov, L. Colas, S. Heuraux, A. Krivska, L. Lu, J.M. Noterdaeme, and the Tore Supra Team and ASDEX Upgrade Team, "Full Wave Propagation Modelling in View to Integrated ICRH Wave Coupling/RF Sheaths Modelling" in *21st Topical Conference on Radio Frequency Power in Plasmas*, AIP Conference Proceedings, edited by R. Pinsker, **1689**, 050008 (2015).
22. J. Jacquot, W. Tierens, W. Zhang, V. Bobkov, L. Colas, J.M. Noterdaeme, and ASDEX Upgrade Team, "Sequential modelling of ICRF wave coupling and self-consistent RF sheaths description for AUG ICRF antennas" " in *22nd Topical Conference on Radio Frequency Power in Plasmas*, European Phys. J. Web of Conferences, **this issue**, B-39 (2017).
23. J.R. Myra, D.A. D'Ippolito, and M. Bures, Phys. Plasmas, **1**, 2890 (1994).
24. J. R. Myra and D. A. D'Ippolito, "Radio frequency sheaths in an oblique magnetic field," Phys. Plasmas **22**, 062507 (2015).
25. S.J. Wukitch et al, "Recent ICRF Results in Alcator C-Mod", *35th EPS Plasma Physics Conf. (Hersonissos, Crete, Greece, June 2008)* P5.085 (2008).
26. M. Jucker, J.P. Graves, W.A. Cooper, N. Mellet, T. Johnson, and S. Brunner, Comp. Phys. Comm. **182**, 912 (2011).

EPJ Web of Conferences **157**, 02012 (2017) DOI: 10.1051/epjconf/201715702012
*22 Topical Conference on Radio-Frequency Power in Plasmas*

# High Field Side Lower Hybrid Current Drive Simulations for Off-axis Current Drive in DIII-D

*S.J.* Wukitch[1,*], *S.* Shiraiwa[1], *G.M.* Wallace[1], *P.T.* Bonoli[1], *C.* Holcomb[2], and *R.I.* Pinsker[3]

[1]Massachusetts Institute of Technology, Plasma Science and Fusion Center, Cambridge, Massachusetts 02139, USA
[2]Lawrence Livermore National Laboratory, Livermore, California 94551, USA
[3]General Atomics, P.O. Box 85608, San Diego, California 92186-5608, USA

**Abstract.** Efficient off-axis current drive scalable to reactors is a key enabling technology for developing economical, steady state tokamak. Previous studies have focussed on high field side (HFS) launch of lower hybrid current drive (LHCD) in double null configurations in reactor grade plasmas and found improved wave penetration and high current drive efficiency with driven current profile peaked near a normalized radius, ρ, of 0.6-0.8, consistent with advanced tokamak scenarios. Further, HFS launch potentially mitigates plasma material interaction and coupling issues. For this work, we sought credible HFS LHCD scenario for DIII-D advanced tokamak discharges through utilizing advanced ray tracing and Fokker Planck simulation tools (GENRAY+CQL3D) constrained by experimental considerations. For a model and existing discharge, HFS LHCD scenarios with excellent wave penetration and current drive were identified. The LHCD is peaked off axis, ρ~0.6-0.8, with FWHM Δρ=0.2 and driven current up to 0.37 MA/MW coupled. For HFS near mid plane launch, wave penetration is excellent and have access to single pass absorption scenarios for variety of plasmas for $n_{||}$=2.6-3.4. These DIII-D discharge simulations indicate that HFS LHCD has potential to demonstrate efficient off axis current drive and current profile control in DIII-D existing and model discharge.

## Introduction

Efficient off-axis current drive (OACD) scalable to reactors is a key enabling technology for economical, steady state tokamak. Efficient OACD scalable to reactors is yet to be demonstrated both from efficiency and survivability perspective. Studies of reactor grade, double null plasmas utilizing high field side (HFS) launch lower hybrid current drive (LHCD) have found improved wave penetration enabling efficient OACD.[1] Further, HFS launch potentially mitigates plasma material interaction and coupling issues due to quiescent HFS scrape off layer (SOL)[2], lower plasma particle[3] and neutron flux, and improved impurity screening[4]. HFS LHCD in DIII-D discharges offer an opportunity to validate HFS wave physics and SOL benefits for RF actuators.

In addition to primarily RF constraints, the analysis presented here optimizes for scenarios compatible with DIII-D advanced tokamak and scenario development research.[5] A key ingredient is current profile optimization.[6] Given the existing and planned DIII-D off axis current drive actuators (neutral beam[7], electron cyclotron[8], and helicon[9]), the HFS LHCD is to drive current at normalized minor radius, ρ, in the range of 0.6-0.8 with current density approaching 0.4 MA/m$^2$ for fields down to 1.6 T with central densities approaching $1x10^{20}$ m$^{-3}$. The goal of this work is to identify a credible HFS LHCD scenario for DIII-D. The simulation tool utilized in this study is GENRAY+CQL3D (advanced ray tracing and Fokker Planck solver[10,11]) for which existing and model DIII-D discharges were investigated.

The motivation to investigate HFS LHCD in DIII-D derives from the possibility to significantly improve the wave penetration and absorption compared to low field side (LFS) launch. For lower hybrid (LH) waves, wave access is a set by wave accessibility[12]and the condition for electron Landau damping[13]. For a given wave frequency, the minimum index of refraction, $n_{||}$, that will penetrate into plasma is proportional to $n_e^{1/2}/B$ where $n_e$ is plasma density and B is magnetic field. Ideally, the wave penetrates into the plasma until electron Landau damped which can be expressed as $n_{||}\geq(30/T_e)^{1/2}$ where $T_e$ is electron temperature in keV[14]. On the HFS, the toroidal field is higher and allows launch of lower $n_{||}$ waves that penetrate farther into the plasma core before damping. A further benefit to lower $n_{||}$ launch is that the waves are absorbed at higher $T_e$ yielding a higher current drive efficiency, $\sim 1/n_{||}^2$, because wave momentum is transferred to less collisional electrons.[15] Looking towards reactor concepts where current drive efficiency is the primary parameter to be optimized, the poloidal position is selected to balance the effects of toroidicity

* Corresponding author: wukitch@psfc.mit.edu

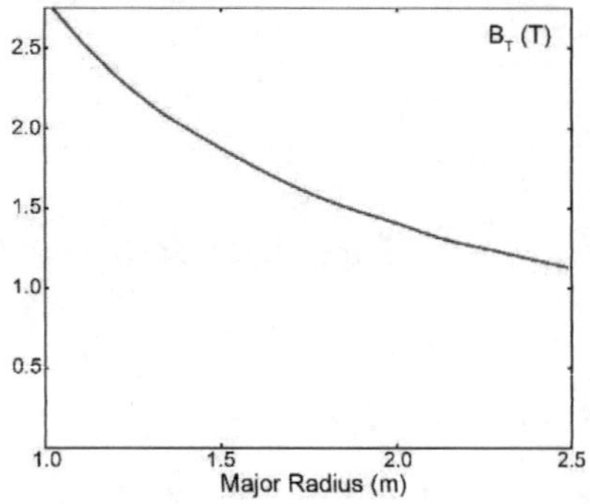


**Fig. 1** Magnetic field profile for 147634 (1.66 T discharge) shows that the HFS local magnetic field is 2.5x the LFS value.

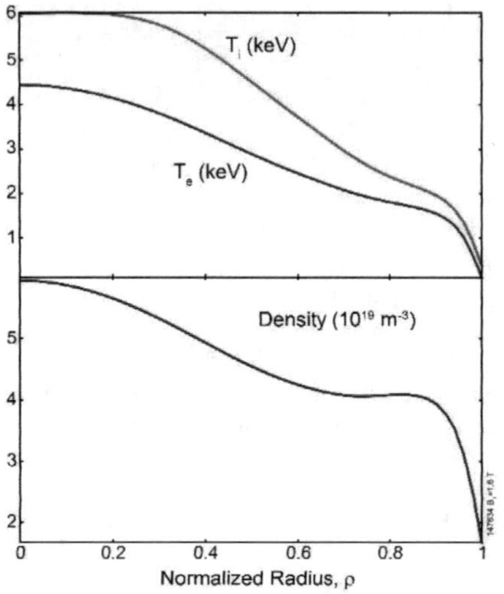


**Fig. 2** Electron density, ion and electron temperature profiles are peaked.

and poloidal field to maintain the smallest $n_{||}$ possible. In DIII-D, the primary consideration is the damping location. Thus, proper poloidal positioning allows for waves to penetrate and damp in region ρ~0.6-0.8.

## HFS LHCD Scenario Identification

As mentioned above, LFS LHCD is unattractive in DIII-D and reactor plasmas due to wave penetration into the plasma core.[1] Using a typical low $q_{min}$ (q is safety factor) discharge (147634), the mid plane B-field, density electron and ion temperature profiles are shown in figure 1 and 2. The waves launched with $n_{||}$=2.7 at 4.6 GHz from the outboard mid plane fail to penetrate into the plasma core and pass through the divertor region as shown in figure 3. The discharge profiles are representative of profiles in other DIII-D advanced tokamak (AT) and model discharges investigated here. The study examined a variety of discharges with different launcher locations to identify launch positions to optimize wave penetration, absorption location, and current drive efficiency. The waves were launched assuming four poloidal locations to represent an

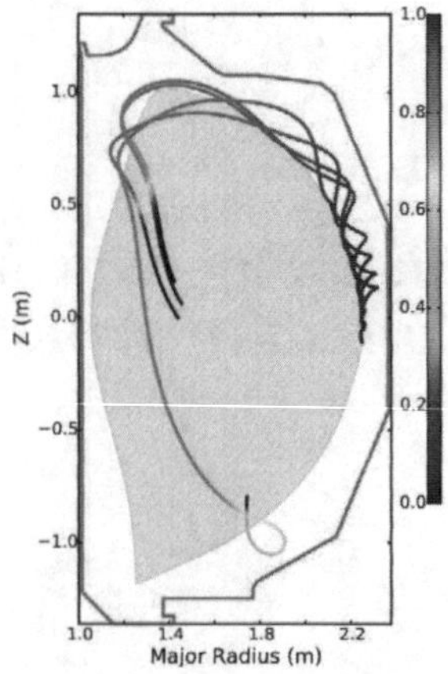


**Fig. 3** LH waves launched from LFS mid plane remain in the plasma periphery for 147634 due to the density and low local B field.

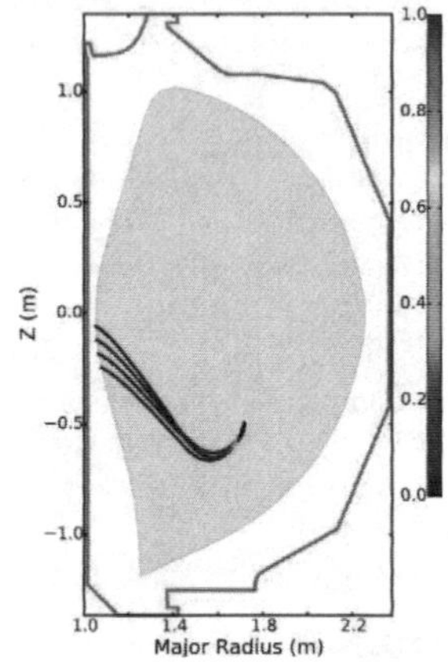


**Fig. 4** LH waves launched from HFS near mid plane penetrate and single pass damp near ρ~0.6 for 147634.

expected coupler with a spectrum peaked near 2.7 and width of +/-0.2 which is consistent with 32 toroidal apertures. Two HFS launch positions showed the greatest promise: near the upper divertor and near the mid plane.

An example of near mid plane HFS launch position is shown in figure 4. The rays are launched at poloidal positions 3-24 cm below the mid plane. The launch spectrum has a peak at $n_{||}$=2.7+/-0.2. The rays penetrate and drive ~150 kA/MW peaked near ρ~0.6 with a corresponding efficiency $0.12 \times 10^{20}$ $A{*}W^{-1}m^{-2}$. As shown in Figure 5, the simulation suggests with 1 MW of coupled power can satisfy the required OACD in the range of ρ~0.6-0.8 with current density approaching 0.4 $MA/m^2$ for AT discharges. The ray penetration and trajectory are sensitive to $n_{||}$ and poloidal launch position. The rays from -3 cm propagate to the largest Z values with a few rays, $n_{||}$~2.5, refracting back to the plasma edge at the top of the device. Launching the rays at or above the mid plane results in many more rays becoming trapped in the plasma edge.

An example of near upper divertor HFS launch position is shown in figure 6. The rays are launched at poloidal

DOI: 10.1051/epjconf/201715702012

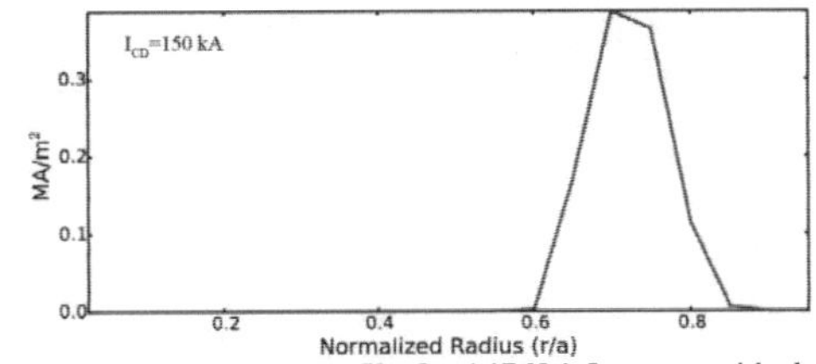


**Fig. 5** Driven current profile for 147634 for near mid plane launch is peaked near ρ~0.6 and approaching the required 0.4 MA/m$^2$ with 1 MW coupled.

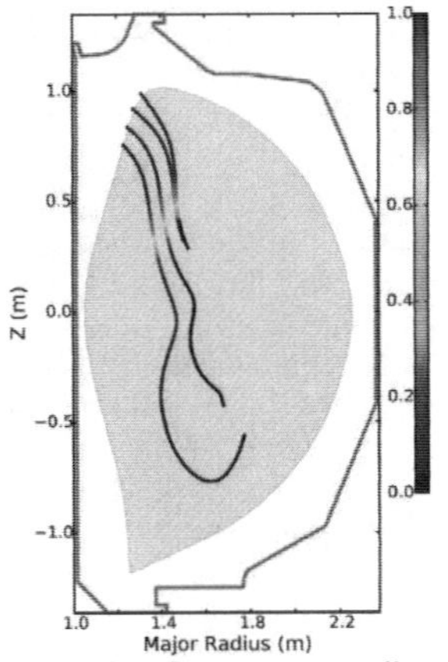


**Fig. 6** Ray propagation for near upper divertor LH wave launch shows good wave penetration and absorption.

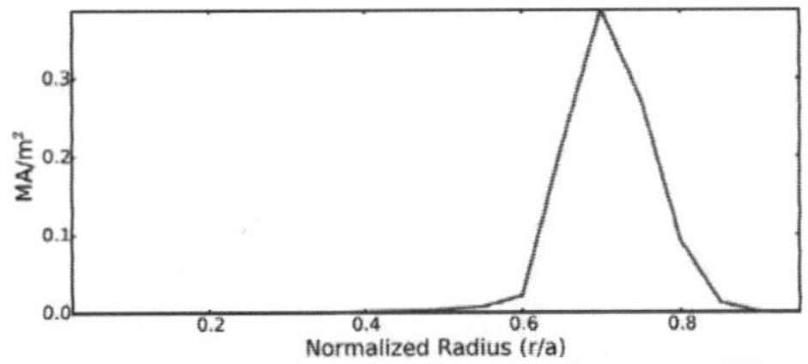


**Fig. 7** Driven current profile is peaked near ρ~0.7 with ~140 kA driven current.

positions 0.85 to 1.0 m above the mid plane. The launch spectrum is peaked at $n_{||}$=2.7+/-0.2. The rays penetrate and damp near ρ~0.7 and drive ~140 kA per MW coupled peaked with a corresponding efficiency 0.1x10$^{20}$ A*W$^{-1}$m$^{-2}$ as shown in Fig. 7. Moving the launch location to closer to the mid plane results in decreased penetration and multiple reflections. Although similar to the near mid plane launch in penetration and efficiency, the near upper divertor launch has practical considerations that make it less ideal. For upper discharges with upper X-points, the launcher would be in a zone of high heat flux and high density. The former would make a structure difficult to design and the latter may prevent the waves from penetrating into the plasma. Further, long distance coupling would be very challenging. The anticipated long distance coupling, the distance to the last closed flux surface (LCFS), is >10 cm.

For near mid plane launch, the sensitivity to magnetic field, electron temperature, and density were examined. The general trend with B-field is $n_{||}$ required for damping at ρ~0.6-0.8 increased to 3 as B-field decreased to 1.6 T.

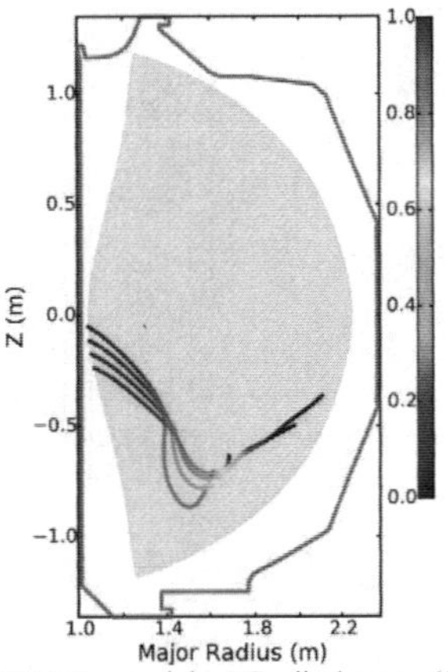


**Fig. 8** For DIII-D model AT discharge (1.6 T), HFS launched near mid plane penetrate and has single pass damping.

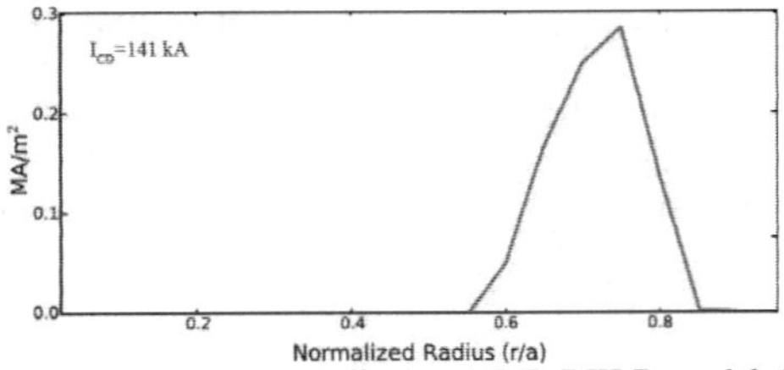


**Fig. 9** Driven current profile for 1.6 T, DIII-D model AT discharge is peaked near ρ~0.7 for near mid plane launch.

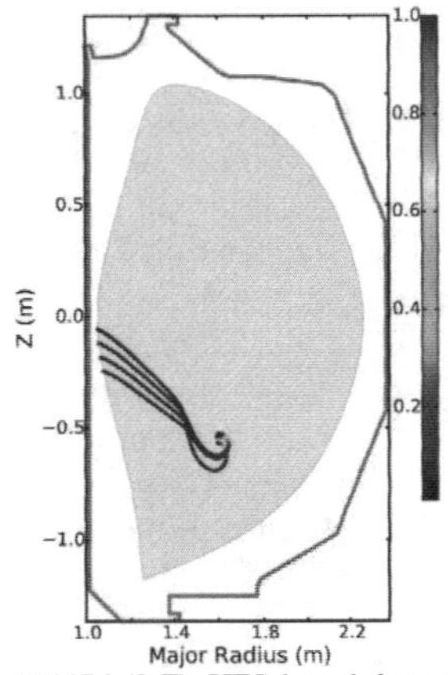


**Fig. 10** For 144476 (2 T), HFS launch has excellent wave penetration and single pass damp near r~0.6.

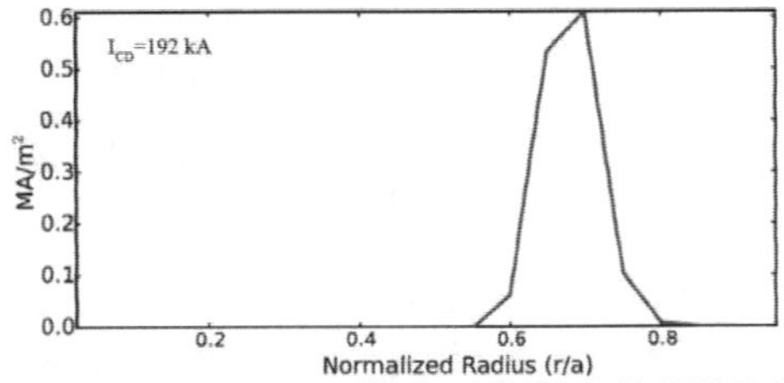


**Fig. 11** Driven current profile for 144476 (2 T) DIII-D AT discharge for n||=2.6+/-0.2 is peaked near r~0.7 and drives ~190 kA/MW with current density >0.4 MA/m$^2$.

Fig 8 and 9 show the wave propagation and driven current for a DIII-D 1.6 T AT discharge. For $n_{||}$=3+/-0.2,

DOI: 10.1051/epjconf/201715702012

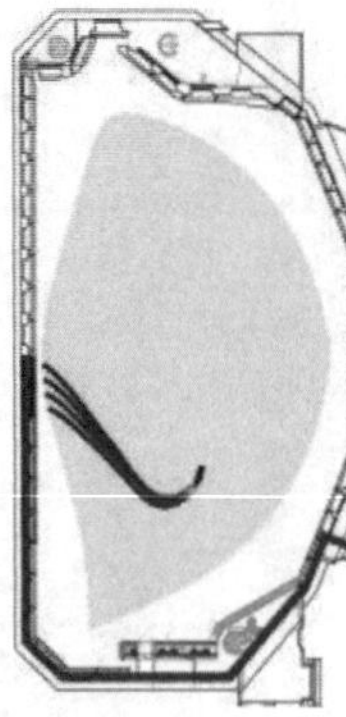

**Fig. 12** The wave guide route fits within the in-vessel profile except on the inner wall and baffle plate.

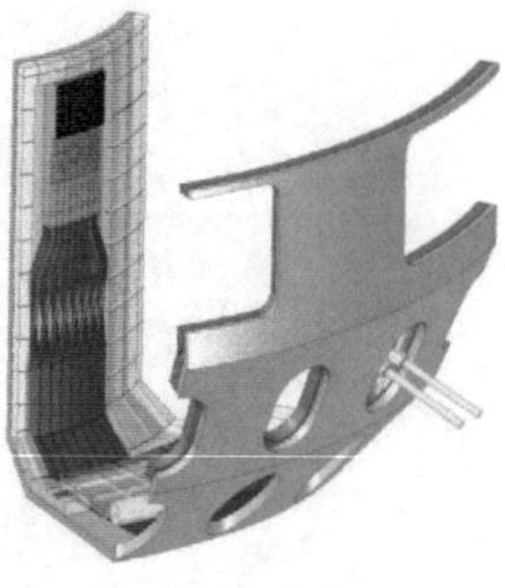

**Fig. 13** Placement of HFS LHCD coupler and wave guide network in the DIII-D model vacuum vessel is challenging but appears plausible assuming one can increase the radius of the center column tiles 2.5 cm.

the waves propagate and drive ~150 kA/MW current between ρ~0.6-0.8. To obtain 0.4 MA/m$^2$ for this discharge, the HFS LHCD power would need to couple ~1.3 MW. At ~2 T, the higher field allows a lower $n_{||}$ to be launched and the corresponding driven current increases to 190 kA/MW. At this level, the current density is sufficient to allow current profile control from a simulation perspective. Similar studies were done for density and temperature. As expected increasing the density required increasing $n_{||}$ to 3.4 for central density ~9x10$^{20}$ m$^{-3}$ at 1.6 T and the driven current falls to ~85 kA/MW. At higher B-fields, the density window and driven current levels improve. For temperature, the primary impact was on single pass damping. Decreasing the temperature by 25% from existing discharges resulted in waves having multiple reflections where ~85% of the power is damped on the first pass.

## Practical Outlook for HFS LHCD at DIII-D

The conceptual launcher and wave guide network design is based on existing technology and is shown in Fig 12 and Fig 13. One WR-187 wave guide from a 250 kW klystron will be split into 16 apertures using two bi-junctions[16] toroidally and a 4 way poloidal splitter[17]. The conceptual launcher is eight WR-187 split into 32 toroidal apertures. This results in an expected power density of <30 MW/m$^2$ assuming 200 kW delivered and similar to power densities obtained in Tore Supra and planned for ITER.[16] The aperture dimension is 6 cm tall and 0.5 cm wide with a 1 mm septa and results in a peak $n_{||}$=2.7 for $\Delta\phi$=90° excitation. With the passive elements, the directivity is ~62% and can access 2.4≤$n_{||}$≤3 with >55% directivity. The wave guide enters from the LFS, passes under the cryopump and divertor and up the inner wall with the long dimension along the vessel. Above the inner strike point, the wave guide twists to orient the wave guide's long dimension perpendicular to the inner column. To accommodate the twist, the inner column radius would need to increase by 2.54 cm. For the AT plasmas simulated here, the smallest inner gap is 3.8 cm suggesting there is may be room to expand the inner column without impacting the target AT plasmas. In addition to the inner column tile modifications, a number of protective tiles and baffle plate would need to be modified to allow routing of the wave guides to the launcher. From an operational perspective, impurities from a metal surface on the inner wall needs to be assessed.

## Summary of HFS LHCD at DIII-D

The simulations suggest efficient HFS LHCD off-axis current drive is possible for DIII-D existing and target AT discharges. An experiment based on these simulations could provide an opportunity to demonstrate q profile control for $B_T$≥1.6 T, $n_e$<9x10$^{19}$m$^{-3}$ with high current drive efficiency and deposition peaked between ρ=0.6-0.8 compatible with high temperature plasmas and scalable to reactors. Further, such an experiment would allow investigation of HFS wave propagation and SOL benefits for RF actuators. The primary in-vessel challenge is increase of the central column radius.

Work supported by the U.S. Department of Energy, Office of Science, Office of Fusion Energy Sciences, using User Facility DIII-D, under Award Number DE-FC02-04ER54698 and by US DoE Contract No. DE-FC02-01ER54648 under a Scientific Discovery through Advanced Computing Initiative.

## References

[1]P.T. Bonoli et al., "Novel Reactor Relevant RF Actuator Schemes for the Lower Hybrid and the Ion Cyclotron Range of Frequencies". Preprint: 2016 IAEA Fusion Energy Conference, Kyoto [TH/5-1].
[2] N. Smick, B. Labombard, and I.H. Hutchinson, Nuclear Fusion **53**, 023001 (2013).
[3] B. LaBombard, J.E. Rice, A.E. Hubbard, J.W. Hughes et al., Nuclear Fusion **44**, 1047 (2004).
[4] B. Labombard,A. Kuang, D. Brunner, I. Faust et al., Nuclear Fusion **57**, -76021 (2017).
[5] T. C. Luce, Fusion Sci. Tech. **48**, 1212–1225 (2005).
[6] C. T. Holcomb, J. R. Ferron, T. C. Luce, T. W. Petrie, et al. Nuclear Fusion **54**, 093009 (2014).
[7] J. M. Park, M. Murakami, C. C. Petty, W. W. Heidbrink et al., Phys. Plasmas **16**, 092508 (2009).
[8] R. Prater and C. C. Petty, Fusion Sci. Tech. **48**, 1141–1148 (2005).
[9] R. Prater, C. P. Moeller, R. I. Pinsker, M. Porkolab et al., Nucl. Fusion **54**, 083024 (2014).
[10] A.P. Smirnov and R.W. Harvey, Bull. Am. Physical Society **40**, 1837 (1995).
[11] R. W. Harvey, and M. McCoy, in Proceedings of the IAEA Technical Committee Meeting on Simulation and Modeling of Thermonuclear Plasmas, Montreal, Canada, 1992 (IAEA, Vienna, 1992), US DOC NTIS Document No. DE93002962.
[12] M. Brambilla Nuc. Fusion **19**, 1357 (1979).
[13] M. Brambilla Physics of Plasma Close to Thermonuclear Conditions (Brussels, 1980) 291.
[14] Y. Peysson and Tore Supra Team, Nuclear Fusion **41**, 1703 (2001). doi:10.1088/0029-5515/41/11/320.
[15] N. J. Fisch, Phys. Rev. Lett. **41**, 873 (1978).

DOI: 10.1051/epjconf/201715702012

[16] A. Bécoulet, Ph. Bibet, F. Mirizzi, M.L. Apicella et al., Fus. Eng. Design **86**, 490–496 (2011).
[17] S. Shiraiwa O. Meneghini, R.R. Parker, G. Wallace et al., Nucl. Fusion **51**, 103024 (2011).

EPJ Web of Conferences **157**, 02013 (2017)
DOI: 10.1051/epjconf/201715702013
*22 Topical Conference on Radio-Frequency Power in Plasmas*

# Recent progress on improving ICRF coupling and reducing RF-specific impurities in ASDEX Upgrade

*Wei Zhang*[1,2,3*], *Volodymyr Bobkov*[1], *Jean-Marie Noterdaeme*[1,2], *Wouter Tierens*[1], *Diogo Aguiam*[4], *Roberto Bilato*[1], *David Coster*[1], *Laurent Colas*[5], *Kristel Crombé*[2,6], *Helmut Fuenfgelder*[1], *Helmut Faugel*[1], *Yuhe Feng*[7], *Jonathan Jacquot*[1], *Philippe Jacquet*[8], *Arne Kallenbach*[1], *Ana Kostic*[1,2], *Tilmann Lunt*[1], *Riccardo Maggiora*[9], *Roman Ochoukov*[1], *Antonio Silva*[4], *Guillermo Suárez*[1], *Angelo A. Tuccilo*[10], *Onofrio Tudisco*[10], *Mariia Usoltceva*[1,2], *Dirk Van Eester*[6], *Yongsheng Wang*[3], *Qingxi Yang*[3], *the ASDEX Upgrade Team*[1] *and the EUROfusion MST1 Team*[11]

[1]Max-Planck-Institut für Plasmaphysik, Garching, Germany
[2]Applied Physics Department, University of Ghent, Ghent, Belgium
[3]Institute of Plasma Physics, Chinese Academy of Sciences, Hefei, P. R. China
[4]Instituto de Plasmas e Fusão Nuclear, IST, Universidade de Lisboa, Portugal
[5]CEA, IRFM, F-13108 Saint-Paul-Lez-Durance, France
[6]Laboratory for Plasma Physics, ERM/KMS, Brussels, Belgium
[7]Max-Planck-Institut für Plasmaphysik, Greifswald, Germany
[8]CCFE, Culham Science Centre, Abingdon, UK
[9]Politecnico di Torino, Torino, Italy
[10]ENEA, Fusion Physics Division, Frascati, Italy

**Abstract.** The recent scientific research on ASDEX Upgrade (AUG) has greatly advanced solutions to two issues of Radio Frequency (RF) heating in the Ion Cyclotron Range of Frequencies (ICRF): (a) the coupling of ICRF power to the plasma is significantly improved by density tailoring with local gas puffing; (b) the release of RF-specific impurities is significantly reduced by minimizing the RF near field with 3-strap antennas. This paper summarizes the applied methods and reviews the associated achievements.

## 1 Introduction

The coupling of Radio Frequency (RF) power in the Ion Cyclotron Range of Frequencies (ICRF) to the plasma depends sensitively on the Scrape-Off Layer (SOL) density because the fast wave is evanescent below the cut-off density (typically in the order of $10^{18}\ m^{-3}$ in nowadays tokamaks). The ICRF power coupling can be greatly enhanced by increasing the density in front of the antennas with local gas puffing. In ASDEX Upgrade (AUG) tokamak, it is found that compared to divertor gas puffing, midplane gas puffing close to the antennas increases the ICRF power coupling most significantly and top gas puffing increases it in a smaller level. The 3D edge plasma fluid and neutral transport code EMC3-EIRENE [1] is used to model the SOL density during different gas puffing scenarios in AUG. Antenna codes such as FELICE [2], TOPICA [3] and RAPLICASOL [4] are then used to calculate the antenna coupling resistance. Good agreement is found between the simulation and experimental results [5, 6].

Among the effects responsible for RF-specific impurity generation, the RF sheaths induced by the near-field effects have been found to be dominant in AUG so far [7, 8]. The RF enhanced sheath potential can be attributed to the image current driven on the antenna box. The novel 3-strap antennas, recently installed in AUG [9], has been designed to reduce those undesired currents through adjusting properly the phase and amplitude of the current in the straps. The recent results have shown that compared to 2-strap antennas, the use of 3-strap antennas reduces the RF-specific tungsten release significantly [10]. The SOL density convection due to the inhomogeneous RF-sheath potential was measured with poloidally distributed reflectometers embedded in one 3-strap antenna [11]. The measured density convection can clearly be reproduced in the self-consistent simulations by running the EMC3-EIRENE, RAPLICASOL and SSWICH [12] codes in an iterative way [13]. Progress of these studies are reviewed in the paper.

## 2 Improvement of ICRF coupling

The basic strategy of improving the ICRF power coupling is to decrease the width of the evanescence layer between the antenna strap and the position of the fast wave cut-off density by increasing the local density in the SOL with main chamber deuterium (D2) gas puffing (or so-called local gas puffing). Both experiments and simulations in AUG [5, 6, 14-16] showed tht significant increases of the effective coupling resistances were achieved by moving the gas source

* Corresponding author: wei.zhang@ipp.mpg.de
[11]See the author list of IAEA FEC 2016 OV/P-12 by H. Meyer et al., to be published in Nucl. Fusion

from the divertor to the main chamber. The core plasma propertites were kept close to identical during different gas puffing schemes.

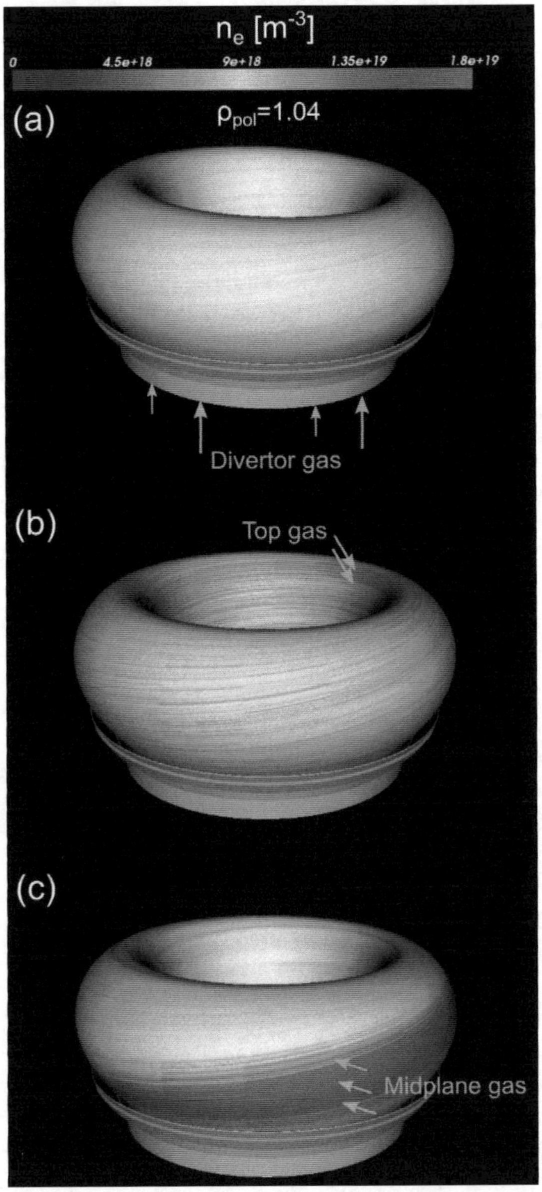


**Fig. 1.** 3D electron density in AUG during (a) divertor gas puffing; (b) top gas puffing; (c) midplane gas puffing. The densities are plotted on the same flux surface. The same gas puff rate is used ( $1.2 \times 10^{22}$ $el/s$ ) and the gas puffing locations and directions are indicated by the arrows.

Here, the EMC3-EIRENE code is used to calculate the SOL plasma density in the presence of different gas sources. A toroidal 360° grid with high resolution in the SOL is used. All the plasma facing components as well as the realistic gas source positions are taken into account. Moreover, the SOL plasma parameters, such as the upstream density and temperature profiles and the downstream power and particle fluxes, are fitted to the experimental ones by choosing appropriate particle and energy transport coefficients [5, 6]. SOL boundary conditions, such as the separatrix density and power across the separatrix, are set the same as those in the experiments. No impurity is considered in the simulations and the part of power radiated by impurities is subtracted from the total heating power.

An example of the simulated SOL electron densities during divertor, top and midplane gas puffing is shown in Fig. 1. Compared to divertor gas puffing (reference case), top gas puffing increases the SOL density almost uniformly toroidally, but to a small extent; midplane gas puffing increases the SOL density very significantly and locally. The density profiles measured by the reflectometry at particular toroidal positions are in good agreement with those in simulations [5, 6]. The increase of the local density leads to a shift of the fast wave cut-off density layer towards the vessel. This shift is ~0.7cm during top gas puffing and can locally be up to about 2.0cm close to the gas valve during midplane gas puffing. The shift of the fast wave cut-off density layer decreases the width of evanescence layer and increases the ICRF power coupling.

The mechanisms of the local density increase by local gas puffing are summarized as follows. Firstly, a localized neutral gas cloud developes in front of the top or midplane gas valve. Due to ionization ($T_e$>2eV), a localized high plasma density cloud develops in the SOL. Charged particles in this high density cloud transport very quickly along the magnetic field lines with a consequent increase of the density in the regions magnetically connected with the valves. Field lines starting from the top density cloud towards the midplane spread widely and lead to an evenly increased SOL density. Field lines starting from the midplane density cloud are much more concentrated and result in a localized density increase. Additionally, the plasma temperature at the locations of the density cloud are lowered, because the outward plasma flows (particle convection) and ionization provide energy sinks. The plasma pressure (density×temperature) along the same flux tubes are the same, thus density gradients develop along the same flux tubes. The largest density is expected to be at location where the ionization is largest.

Based on the 3D density calculated with EMC3-EIRENE, antenna codes such as FELICE (1D) [2] or RAPLICASOL (3D) [4] are then used to calculate the effective antenna coupling resistance. For FELICE calculations, the 3D density has to be averaged into a 1D radial profile, and this 1D averaging can bring additional errors to the final results. Nevertheless the calculated resistances with FELICE are in qualitative agreement with the experimental ones [5]. They indicate that compared to divertor gas puffing in H-mode plasmas, top gas puffing increases the coupling resistance by ~25% for all the antennas; midplane gas puffing increases this value by ~120% for the antenna close to the gas valve and this increase decays exponentially

DOI: 10.1051/epjconf/201715702013

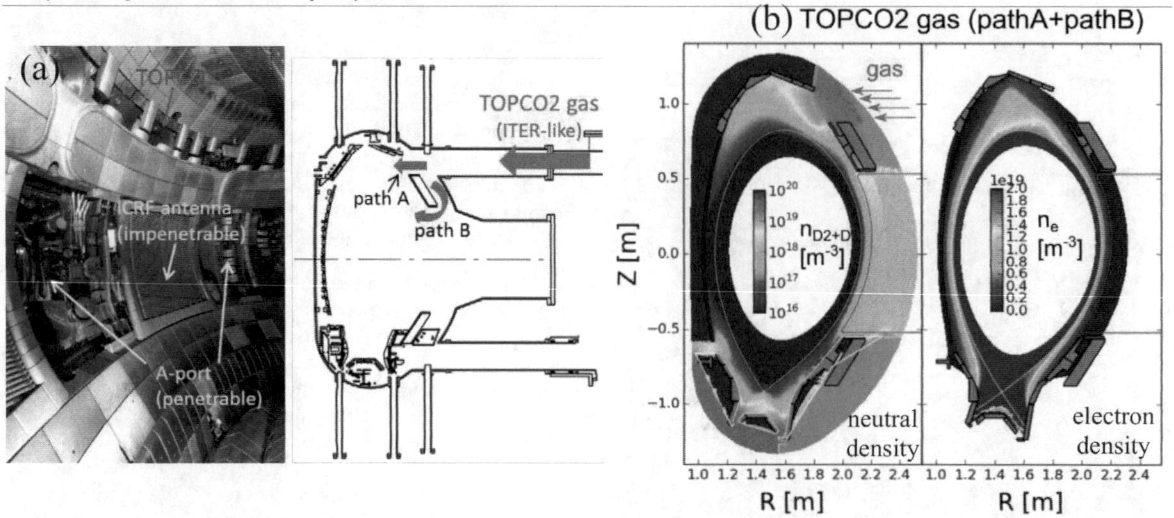


**Fig 2.** (a) Outer top gas puffing (TOPCO2, ITER relevant) and the surrounding wall structures in AUG; (b) neutral and electron density during outer top gas puffing. Figure reproduced with permission from [6].

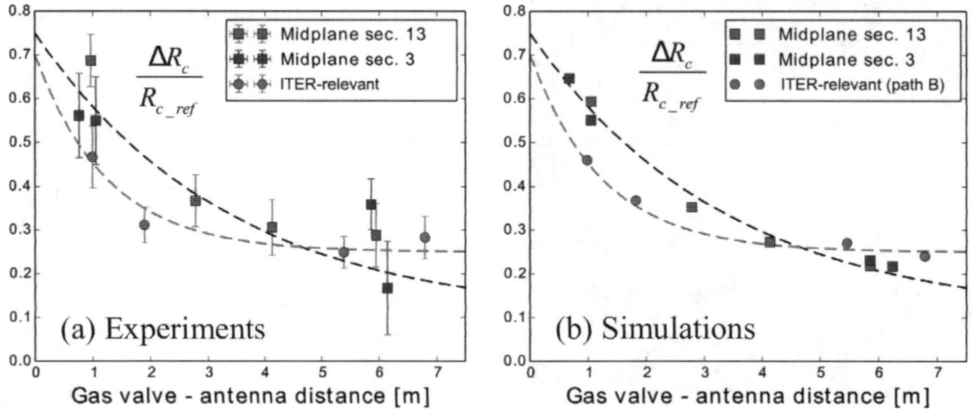


**Fig. 3.** Relative increase of coupling resistance $\Delta R_c / R_{c\_ref}$ obtained from (a) experiments; (b) simulations with the EMC3-EIRENE and RAPLICASOL codes. Figure reproduced with permission from [6]. The whole AUG torus has 16 sectors toroidally. "Midplane sec. 13" and "Midplane sec. 3" represent midplane gas puffing in sector 13 and sector 3, respectively.

(decay length ~3.7m) as the distance between the antenna and the gas valve becomes larger.

The previously discussed top gas sources are located at the inner top of the vessel. In addition, we have studied the outer top gas puffing (TOPCO2) in AUG L-mode plasmas [6]. This gas puffing is more ITER relevant, because in ITER the fueling gas is planned to be puffed from the outer top positions. The distribution of this gas depends on the surrounding wall structures and is different in different machines. In AUG, the outer top gas enters the plasma either through the top windows (path A) or through the A-ports in the midplane (path B), Fig. 2(a). It is shown that a localized neutral cloud is developed in the top while a more homogenous one is formed in the midplane, Fig. 2(b).

The calculated relative increase of coupling resistances with the RAPLICASOL code is in quantitative agreement with the experimental values shown in Fig 3. The largest increase of coupling resistance is ~70% in L-mode plasmas with midplane gas puffing (gas puff rate $\sim 6.5 \times 10^{21}\ el/s$). As for the outer top gas puffing (ITER releveant), the experimental results are in better agreement with simulations when only path B is assumed to be open (i.e. with the part of gas clouds only in the midplane). The antenna closest to the gas valve has the largest coupling resistance increase and this increase decays with the gas valve - antenna distance. Considering the fact that about half of the injected gas passes through path A and the corresponding gas cloud in the top is magnetically connected to antenna in sector 10 (with a gas valve – antenna distance ~5.4m), the effect of the localized high neutral density on the local increase of plasma density plays a more important role than the effect of field line connections. Indeed, in the simulations the top gas cloud associated with path A increases the SOL density in a much more toroidal uniform way [6].

From our previous studies, we conclude that the poloidal distributed gas valves close to the ICRF

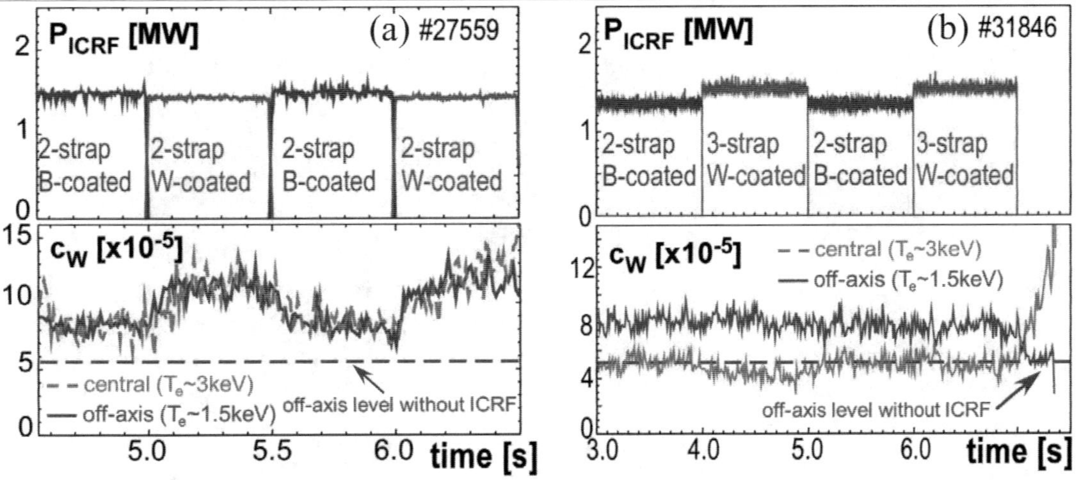


**Fig. 4.** Central and edge W concentrations during ICRF heating with (a) B-coated and W-coated 2-strap antennas; (b) B-coated 2-strap and W-coated 3-strap antennas. Figure reproduced with permission from [7, 10].

antennas are the best option to maximize the ICRF power coupling. This conclusion is also expected to be valid for future machines such as ITER and DEMO.

An additional benefit of local gas puffing is the reduction of total outer wall sputtering and core W concentration [17]. This is attributed to the local decrease of plasma temperature in the far SOL and the decrease of sheath potential in front of the antennas during local gas puffing. Future investigations are underway to understand the mechanisms. Furthermore, the influence of magnetic perturbation fields and MHD modes on the SOL density and thus on the ICRF coupling properties is now being studied [22]. Possible synergies with gas puffing will also be addressed in the future.

## 3 Reduce RF-specific impurities

ICRF antennas excite fast wave via oscillating currents on the radiating straps. However, the oscillating currents induce image currents on the antenna box accompanied by $E_{\parallel}$ fields (parallel to the equilibrium magnetic field) in the vicinity of the antennas. Subsequently, an increase of sheath potential is driven by the $E_{\parallel}$ fields. The large potential drop across the sheath accelerates ions towards the wall, leading to considerable physical sputtering and even to hot spots if the impacting ions are energetic enough. The hot spots can damage the wall components and the impurities induced by sputtering can affect the plasma performances. In AUG (full tungsten (W) wall), substantial impurity radiation losses can be generated if the W concentration ($c_W$) in the plasma core exceeds $5 \times 10^{-5}$ [7].

The recent research activities in AUG have made significant progress in reducing the ICRF-induced impurities. The first strategy is to coat the antenna limiters with a low-Z material boron (B) instead of W since experiments suggest that the major ICRF-induced W sources are from the antenna limiters [18]. In 2012, limiters of two antennas were coated with boron. The experimental results (Fig. 4(a)) clearly show that a reduction of RF-induced fraction of W concentration by a factor of two both in the edge and central plasma has been achieved with B-coated 2-strap antennas. The antenna limiters account for more than half of the increase in W concentration during ICRF heating. In these experiments, a standard H minority scheme with dipole phasing was used. The shifting of sputtering from tungsten to boron is less problematic since the boron content in the core plasma is more tolerable and causes less power radiation. The B-coated antennas allow the ICRF heating to operate in a wider range of plasma conditions. However, the B-coating is only a remedy to the high-Z impurity production induced by ICRF and not a solution of the problem associated with the RF fields close to the antenna.

The long-term strategy to reduce the ICRF-specific impurities involves optimizing the antenna design so that the generated image currents and $E_{\parallel}$ fields at antenna limiters are minimized. One of the strategies was to use broad-limiter antennas. The release of W was reduced by up to ~40% in a single antenna operation in experiments operated with broad-limiter antennas [7]. The other more advanced approach is the design and use of 3-strap antennas in AUG. The basic idea of this approach is to minimize the RF image currents on the antenna limiters by balancing the dipole contributions of the central strap with that of the outer straps [10]. In 2015, a pair of W-coated 3-strap antennas has been installed in ASDEX Upgrade. The experimental results (Fig. 4(b)) show that the W-coated 3-strap antennas lead to almost the same values of edge and core W concentration as those for the B-coated 2-strap antennas. The heating effectiveness of the W-coated 3-strap antenna is not lower than that of B-coated 2-strap [10] while the ICRF heating power, plasma current, total input and radiated power are almost the same in the studied discharge. Together with the previous results (Fig4. (a)), we can conclude that the use of W-coated 3-strap antennas reduces the RF-induced

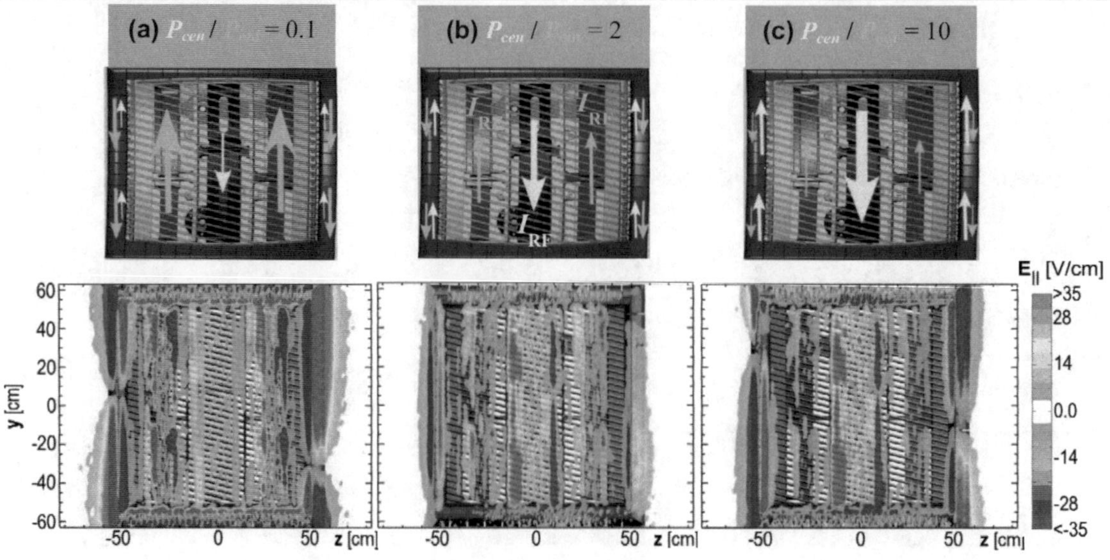

**Fig. 5.** Schematic image currents and parallel electric field calculated by TOPICA [3] in the vacuum gap 2 mm in front of limiters of 3-strap antenna, with power balance of (a) $P_{cen}/P_{out} = 0.1$; (b) $P_{cen}/P_{out} = 2$; (c) $P_{cen}/P_{out} = 10$. In these cases dipole phasing is used. Figure reproduced with permission from [8].

fraction of W concentration also by at least a factor of two compared to the W-coated 2-strap antennas.

A prominent rise of core W concentration after switching the ICRF power off (Fig. 4 (b)) is due to changes in the neoclassical transport of W. During ICRF heating, the core plasma density profiles are flattened and the core plasma temperature profiles are peaked. The flattening of density leads to reduced density gradients and thus reduced inward W transport while the peaking of temperature leads to increased temperature gradients and thus increased outward W transport [19]. As a result the core W concentration during ICRF heating is much lower than that without.

The final W concentration and W sputtering yield depend on the 3-strap antenna feeding configurations, including the phasing and power ratio between the central and outer straps ($P_{cen}/P_{out}$, where $P_{cen}$ is the power on the central strap and $P_{out}$ is the total power on the two outer straps, respectively). The lowest W concentration and the lowest W sputtering yield on the antenna limiters were observed with dipole phasing and $P_{cen}/P_{out}$ in the range (1.5, 2.0) [10]. Non-optimized feeding configurations cause high image currents and W sputtering yield. An example of this is shown in Fig. 5, in which large unbalanced image currents on the antenna limiters and $E_{\|}$ fields in front of the antenna limiters (calculated by TOPICA [3]) are generated when $P_{cen}/P_{out}$ is too small (case (a)) or too large (case (c), see [23] for more details). Numerical simulations show that the 3-strap antenna even with non-optimized settings (i.e. dipole phasing, $P_{cen}/P_{out} \sim 9$) generates $E_{\|}$ fields and sheath potential which are smaller than those due to the 2-strap antenna with typical feeding configurations (i.e. dipole phasing, $P_{cen}/P_{out} = 1$) [13].

Further experiments show that the local measurements of RF voltages at the shunts at the side limiters of one 3-strap antenna correlate with the measured W sputtering yield [8]. The calculated local $E_{\|}$ fields by TOPICA at the same locations are qualitatively consistent with the measured RF voltages [8]. These findings provide evidence for links between the $E_{\|}$ fields, RF voltages and W sputtering yield.

The number of sputtered impurities not only depend on the sputtering yield but also on the number of impacting particles, i.e. on the local plasma density in the far SOL. E×B drifts driven by the inhomogenous sheath potential will however influence this local plasma density. Recently, progress has been made on understanding the ICRF induced density convection in the SOL from two numerical approaches [13, 20]: (a) EMC3-EIRENE simulations based on the measured potential; (b) Self-consistent simulations with several codes including EMC3-EIRENE, RAPLICASOL and SSWICH. In the first approach, the E×B drifts are calculated from the measured potentials and are then used as prescribed drifts in EMC3-EIRENE. In the second approach, the output of EMC3-EIRENE (plasma density) is used as inputs in RAPLICASOL and SSWICH, the output of RAPLICASOL (the $E_{\|}$ fields) are used as inputs in SSWICH and the output of SSWICH (sheath potential) is used to calculate the E×B drifts, which are then used as inputs in the EMC3-EIRENE code. The obtained results from these simulations are in agreement [20] or consistent [13] with the corresponding measurements.

## 4 Conclusions

This paper summarizes the recent progress of ICRF heating in ASDEX Upgrade tokamak on two aspects:

improving the ICRF power coupling and reducing the ICRF-specific impurities.

The ICRF power coupling is improved by increasing the local density in front of the antennas with local deuterium gas puffing. Compared to divertor gas puffing, midplane gas puffing close to the ICRF antenna increases the ICRF power coupling most significantly (~120%) while top gas puffing increases it to a much smaller level (~25%) [5]. The effects of local gas puffing on local plasma density are attributed to the local high neutral density, magnetic filed line connections and the local decrease of temperature [5, 21]. It can be extrapolated that for future machines, such as ITER and DEMO, local gas valves close to the antennas are the best option in terms of maximizing ICRF power coupling.

The tungsten concentration in the central and edge plasma was reduced by using one of the three antenna modifications: (1) the broad-limiter antenna (up to ~40% for single antenna operation) [7]; (2) the boron-coated antennas (by about a factor of 2) [7]; (3) the 3-strap antennas (by more than a factor of 2) [10]. The tungsten sources, especially those from the antenna limiters are reduced either by replacing the tungsten sputtering by sputtering of a low-Z material (boron-coated antennas) or by decreasing the image currents on the antenna limiters (broad-limiter or 3-strap antennas). The latter solution is the most promising one, because it reduces the image currents and the parallel electric field most significantly. When operating the 3-strap antenna with dipole phasing and a power ratio between the central and outer straps in the range (1.5, 2.0), tungsten sputtering at the antenna limiters is minimized [10]. Further experiments and simulations show that the RF voltages measured on the antenna limiters correlate with the measured tungsten effective sputtering yield and are consistent with the calculated local parallel electric field at the antenna limiters [8].

Following the previous achievement, EMC3-EIRENE simulations for ITER and DEMO are being carried out to find the optimized gas puff locations and gas puff rate for the best ICRF power coupling. Theoretical approaches [24-26] to model sheaths are being developed and will be validated against measurements on ASDEX Upgrade [11, 27] and on IShTAR [28, 29]. Theoretical insight of experimental results will provide key guidelines to design antennas with reduced sheaths and RF-induced impurities.

*This work has been carried out within the framework of the EUROfusion Consortium and has received funding from the Euratom research and training programme 2014-2018 under grant agreement No 633053. The views and opinions expressed herein do not necessarily reflect those of the European Commission.*

## References

[1] Feng Y. *et al* 2004 *Contributions to Plasma Physics* **44** 57-69
[2] Brambilla M. 1989 *Plasma Physics and Controlled Fusion* **31** 723-57
[3] Lancellotti V. *et al* 2006 *Nuclear Fusion* **46** S476-S99
[4] Jacquot J. *et al* 2015 *AIP Conference Proceedings* **1689** 050008
[5] Zhang W. *et al* 2016 *Nuclear Fusion* **56** 036007
[6] Zhang W. *et al* 2017 *Plasma Phys. Control. Fusion* **59** 075004
[7] Bobkov V. *et al* 2013 *Nuclear Fusion* **53** 093018
[8] Bobkov V. *et al* 2017 *Plasma Phys. Control. Fusion* **59** 014022
[9] Fuenfgelder H. *et al* 2017 *Fusion Engineering and Design,* "A double success story: The international cooperation to built the new ICRF antennas on ASDEX Upgrade and the results obtained", http://dx.doi.org/10.1016/j.fusengdes.2017.03.114
[10] Bobkov V. *et al* 2016 *Nuclear Fusion* **56** 084001
[11] Aguiam D. E. *et al* 2016 *Review of Scientific Instruments* **87** 11E722
[12] Jacquot J. *et al* 2014 *Physics of Plasmas* **21** 061509
[13] Zhang W. *et al* 2017 *Nuclear Fusion* **57** 116048
[14] Bobkov V. *et al* 2014 *AIP Conference Proceedings* **1580** 271
[15] Zhang W. *et al* 2015 *AIP Conference Proceedings* **1689** 050006
[16] Jacquet P. *et al* 2016 *Nuclear Fusion* **56** 046001
[17] Bobkov V. *et al* 2015 *AIP Conference Proceedings* **1689** 030004
[18] Bobkov V. *et al* 2011 *Journal of Nuclear Materials* **415** S1005-S8
[19] Neu R. *et al* 2007 *Plasma Physics and Controlled Fusion* **49** B59-B70
[20] Zhang W. *et al* 2016 *Plasma Phys. Control. Fusion* **58** 095005
[21] Zhang W. *et al* 2017 *Nuclear Fusion* **57** 056042
[22] Suárez G. *et al*, " Investigation of the coupling properties of the Ion Cyclotron Fast Wave under applied Magnetic Perturbations and MHD Phenomena in ASDEX Upgrade", *this conference*, C-28
[23] Bobkov V. *et al*, "Characterization of 3-strap antennas in ASDEX Upgrade", *this conference*, C-29
[24] Tierens W. *et al*, "3-Dimensional density profiles in edge plasma simulations for ICRF heating", *this conference*, A-25
[25] Jacquot J. *et al*, "Sequential modelling of ICRF wave coupling and self-consistent RF sheaths description for AUG ICRF antennas", *this conference*, B-39
[26] Zhang W. *et al*, "Plasma edge modelling with ICRF coupling", *this conference*, C-05
[27] Ochoukov R. *et al*, "Latest upgrades of the high frequency B-dot probe diagnostic suite on ASDEX Upgrade", *this conference*, A-38
[28] Kostic A. *et al*, "Feasibility study of Passive Optical Emission Spectroscopy for the electric field measurements in IShTAR", *this conference*, C-42
[29] Usoltceva M. *et al*, "IShTAR ICRF antenna field characterization in vacuum and plasma by using probe diagnostic", *this conference*, B-24

# Recent progress in the upgrade of the TCV EC-system with two 1MW/2s dual-frequency (84/126GHz) gyrotrons

*Stefano* Alberti[1,*], *Jérémy* Genoud[1], *Timothy* Goodman[1], *Jean-Philippe* Hogge[1], *Laurie* Porte[1], *Miguel* Silva[1], *Trach-Minh* Tran[1], *Minh-Quang* Tran[1], *Konstantinos* Avramidis[2], *Ioannis* Pagonakis[2], *Jianbo* Jin[2], *Stefan* Illy[2], *Gerd* Gantenbein[2], *John* Jelonnek[2], *Manfred* Thumm[2], *William* Bin[3], *Alex* Bruschi[3], *Saul* Garavaglia[3], *Alessandro* Moro[3], *Walter* Kasparek[4], *François* Legrand[5], *Etienne* Perial[5], *Yoan* Rozier, *Fabio* Cismondi[6], *Niek* Doelman[7]

[1]Ecole Polytechnique Fédérale de Lausanne (EPFL), SPC, CH-1015, Lausanne, Switzerland
[2]IHM, Karlsruhe Institute of Technology (KIT), Kaiserstrasse 12, D-76131, Karlsruhe, Germany
[3]Institute of Plasma Physics "P. Caldirola", National Research Council of Italy, Milano, Italy
[4]Institute for Interfacial Process Engineering and Plasma Technology, University of Stuttgart, D-70569 Stuttgart, Germany
[5]Thales Electron Devices, Vélizy-Villacoublay, France
[6]PPPT Power Plant Physics and Technology EUROfusion, D-85748 Garching, Germany
[7]Department of OptoMechatronics, TNO Technical Sciences, NL-2600 AD, Delft, The Netherlands

**Abstract.** The upgrade of the EC-system of the TCV tokamak has entered in its realization phase and is part of a broader upgrade of TCV. The MW-class dual-frequency gyrotrons (84 or 126GHz/2s/1MW) are presently being manufactured by Thales Electron Devices with the first gyrotron foreseen to be delivered at SPC by the end of 2017. In parallel to the gyrotron development, for extending the level of operational flexibility of the TCV EC-system the integration of the dual-frequency gyrotrons adds a significant complexity in the evacuated 63.5mm-diameter $HE_{11}$ transmission line system connected to the various TCV low-field side and top launchers. As discussed in [1], an important part of the present TCV-upgrade consists in inserting a modular closed divertor chamber. This will have an impact on the X3 top-launcher which will have to be reduced in size. For using the new compact launcher we are considering employing a Fast Directional Switch (FADIS), combining the two 1MW/126GHz/2s rf-beams into a single 2MW rf-beam.

## 1 Introduction

TCV is presently undergoing major heating upgrades [1], with the installation of a neutral beam for direct ion heating and increasing the EC power injected in X-mode at the third harmonic (X3-upgrade). The neutral beam, with energies of 20-35keV and power up to 1MW for 2s, is operational since spring 2016 and a second neutral beam with higher energy (50-60keV, 1MW, 2s) is foreseen in 2018. The so called X3-upgrade project consists of adding two dual-frequency gyrotrons (126 GHz/84 GHz) with a total power at 126 GHz (for top-launch X3) or at 84 GHz (for low field side (LFS) injection X2) of 2MW. The three existing gyrotrons used for X3-ECH and operating at 118 GHz will be relocated to inject power from the LFS using the existing X2 transmission lines and launchers.

The heating upgrade will give access to plasma regimes and configurations with high normalized plasma pressure and wide range of temperature ratios, including $T_e/T_i$~1. As shown in Fig. 1, which is obtained from an ASTRA simulation [2] of a standard TCV configuration, to approach the ITER relevant range of normalized $\beta_N$ values ($\beta_N$~2-2.5), X3 power levels higher than the presently available 1.5MW are necessary. Considering the X3-upgrade with 2MW of additional power (total 3.5MW) and 1MW of Neutral Beam Injection NBI, the simulations show that, for the reference discharge described in Fig.1, central electron and ion temperatures as high as $T_e(0)$=4keV and $T_i(0)$=2.5keV can be reached.

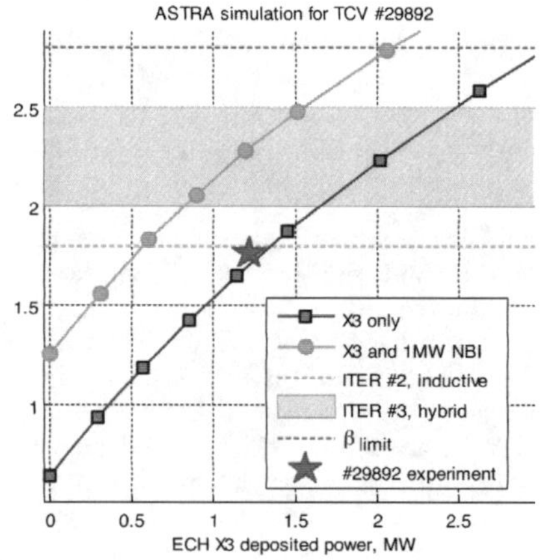


**Fig. 1.** ASTRA simulation of the TCV plasma performance as a function of X3 absorbed power based on an actual scenario (H-mode discharge 29892, with $I_p$=410kA, $B_T$=1.44T, $Z_{eff}$=2.6, $n_e(0)=6.5\times10^{19}m^{-3}$, $T_e(0)$=2.8keV, $T_i(0)$ = 0.9keV, $q_{95}$=2.5; the shaping parameters are $\kappa_{95}$=1.68 and $\delta_{95}$=0.36).

## 2 X3 Single-pass absorption properties

As shown in Fig.2, for the central electron temperature and density given in Fig.1, TORAY[3] simulations

* Corresponding author: stefano.alberti@epfl.ch

DOI: 10.1051/epjconf/201715703001

predict a single-pass absorption in excess of 80% with X3 top-launch. This absorption level has been experimentally confirmed in the past [4]. Based on these results and the ASTRA simulations, including the 1MW NBH, full single-pass absorption in H-mode plasmas will then be ensured with the 2MW/126GHz injected from the top launcher. With the efficient electron bulk heating using the top-launch, single pass absorption well exceeding 70% is to be expected when injecting the 1.5MW/118GHz rf-power for X3 heating from the low-field side. This will enable localized power deposition necessary for MHD control. The ECCD efficiency for the X3 injected power from the LFS is negligible.

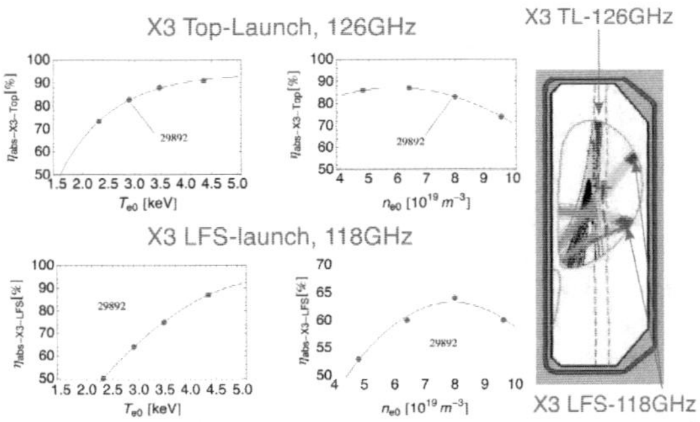


**Fig. 2.** TORAY simulations of the single-pass absorption for the X3 top-launch @126GHz (TL) and the X3 LFS-launch @118GHz versus central electron temperature and density. The decrease of the absorption at high density is due to the density dependence of the optical-depth via the index of refraction. The reference plasma parameters for discharge 29892 are given in Fig. 1.

As discussed later on (see Fig. 6), when the new dual-frequency gyrotron will be operational, the transmission lines of the three 118GHz gyrotron will be reconfigured for low-field side injection using the existing X2 launchers. This configuration has been used and validated during the first X3 heating experiments [5].

## 3 Dual-frequency gyrotron

The gyrotron design is essentially based on modifications of the W7-X gyrotron [6-8] in which some sub-components have been redesigned (see Fig. 3 [9]) using the most advanced models.

The electron beam operating parameters have been chosen to be compatible with the existing HV power supplies, and the magnetic field profile provided by the W7-X gyrotron magnets has been rescaled for an operation at 84GHz and 126GHz. The operating modes have been chosen such that their coupling to the electron beam is close to optimal, with frequencies (84GHz and 126GHz) as close as possible to those presently used on the TCV ECH system (82.7GHz and 118GHz), with the additional constraint that their wavelengths should correspond to an integer number of half-wavelengths in the gyrotron CVD diamond window. This resulted in the choice of the modes $TE_{17,5}$ at 83.91GHz and $TE_{26,7}$ at 126.16GHz repectively.

The main redesigned component is the electron gun for which, instead of a diode gun (W7-X), a triode gun will be implemented, providing flexibility in the range of achievable electron beam parameters at both frequencies.

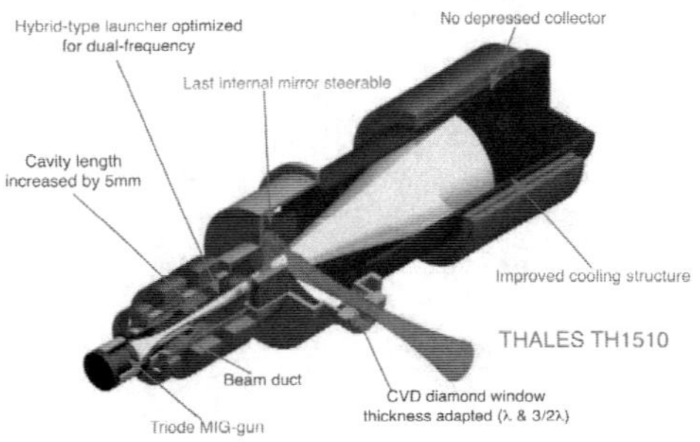


**Fig. 3.** Schematic of the dual-frequency gyrotron (TH1510) manufactured by Thales Electron Devices. Indicated are the parts which have been redesigned from the TH1507 (140GHz, W7-X) with highlighted in red the major redesigned components.

Recent design criteria such as the absence of potential wells and the avoidance of trapped electrons emitted from any part of the gun have been considered in the gun design [10].

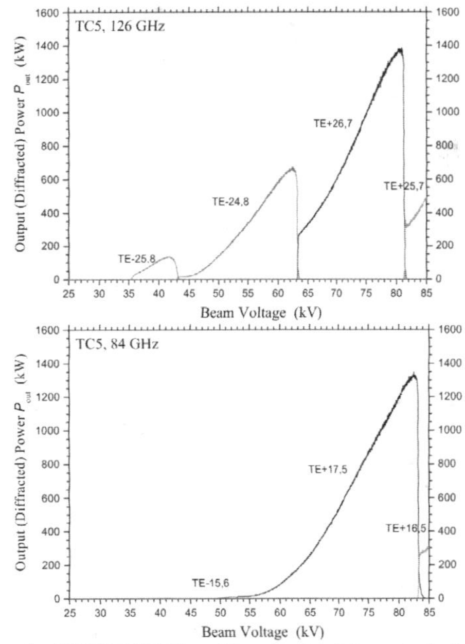


**Fig. 4.** EURIDICE[11] multimode simulations including the time dependence of the beam parameters and a 16% rms spread in the electron beam velocity ratio. The beam current is 40A and the pitch factor is 1.3 in both cases. Top: mode $TE_{26,7}$ @ 126GHz. Bottom: mode $TE_{17,5}$ @ 84GHz.

The gyrotron cavity underwent a very minor redesign essentially consisting in lengthening by 5mm the constant radius section in order to increase the quality factor and the efficiency at frequencies lower than the original one (140GHz). Monomode and multi-mode time-dependent simulations were carried out to validate the mode choices and perform a fine tuning of the magnetic field. As shown in Fig. 4, single-mode operation with a diffracted power in excess of 1MW at the operating voltage (78kV) is predicted at both frequencies, with a stability range of 2-4 kV.

In order to reach the highest Gaussian mode content, a significant effort has been placed in the redesign and optimisation of the launcher and the internal mirrors [12]. A satisfying compromise between the Gaussian content, the beam size and centering at the window and the beam waist location could be found and is illustrated on Fig. 5, where the field distribution for the mode $TE_{26,7}$ is shown with a linear scale. The beam waist is of the order of 20mm and is located approximately 100mm after the window. The results were cross-checked with SURF3D[13], which additionally estimated the stray radiation level to < 4%. The gyrotron design parameters are summarized in Table 1.

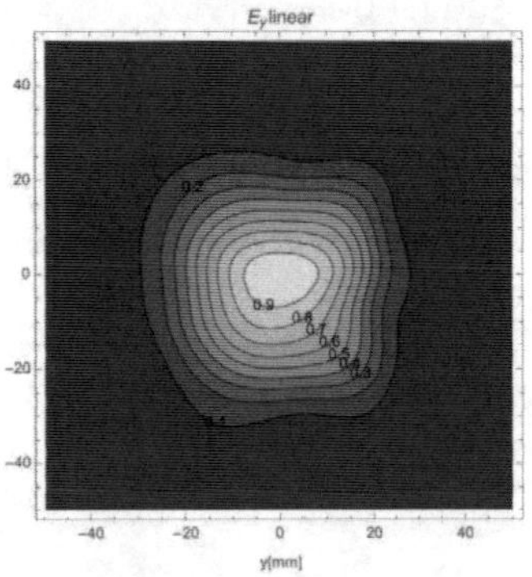


**Fig. 5.** Normalized electric field distribution at the window (linear scale), mode $TE_{26,7}$. The estimated Gaussian content is 97.6% and the stray radiation level is estimated at 2.5%.

Based on the fact that on TCV only 2s rf-pulses will be used, together with the constraint of using the existing high-voltage power supplies, it has been shown, using advanced thermo-hydraulic simulation models, that the collector loading allows the operation without a depressed collector. Compared to the W7-X gyrotron, and based on the simulations an improved cooling structure in the collector region has been implemented.

**Table 1.** Dual-frequency gyrotron design parameters[14].

| Operating cavity mode | $TE_{17,5}$ | $TE_{26,7}$ |
|---|---|---|
| Frequency [GHz] | 83.91 | 126.16 |
| RF output power [MW] | 1.05 | 1.2 |
| Cavity wall-loading [kW/cm$^2$] | 1.1 | 2.1 |
| Beam current [A] | 40 | 40 |
| Beam energy [keV] | 78 | 78 |
| Mod-anode voltage [kV] | -38 | -26 |
| Pitch angle, $\alpha$ | ~1.3 | ~1.3 |
| Cavity magnetic field, [T] | 3.31 | 4.98 |
| Electronic efficiency [%] | 35 | 41 |
| Gaussian content [%] | 97.7 | 97.6 |

The two gyrotrons (TH1510) are being manufactured by the French company Thales Electron Devices, with the first unit foreseen to be operational on TCV in the first half of 2018. In parallel to the gyrotron manufacturing, two cryogen-free superconducting magnets are being procured from the American company Cryomagnetics. Apart from the cryogen-free aspect, the magnet design is based on the superconducting magnets used for the gyrotrons at W7-X.

# 4 System Implementation

In parallel to the gyrotron development, for extending the level of operational capability of the TCV EC-system, the integration of the dual-frequency gyrotrons adds a significant complexity in the evacuated 63.5mm-diameter $HE_{11}$ transmission line transporting the rf-power from the gyrotrons to the various TCV low-field side and top launchers. Fig. 6 shows the layout of the upgraded EC-system.

The two dual-frequency gyrotrons will be connected to their respective evacuated Matching Optics Units (MOU) which will contain the quadratic surface-contour function optics for maximizing the coupling to the $HE_{11}$ waveguides (63.5mm internal diameter) for each frequency. The first mirror after the gyrotron window will be movable under vacuum for redirecting the rf-beam towards the corresponding mirrors set for each frequency, 84 or 126GHz. Each mirror set consists of two fixed mirrors. With this design the MOU also plays the role of a high-power evacuated rf-switch. The two MOU outputs will be connected to the corresponding $HE_{11}$ transmission lines manufactured by General Atomics (GA).

The rf-wave polarization control will be performed at the transmission line miter bends. For the two lines transmitting the 84GHz frequency and connected to low-field side launchers, real-time full-polarization control will be possible by using two sets of remotely controlled miter bend pairs. For the two lines transmitting the 126GHz and connected to the top-launcher only the linear polarization needs to be controlled which will be realized via a single non-motorized miter bend on each line.

With the redirection of the 118GHz power toward the low-field side launchers, these LFS launchers will be each launching either 82.7GHz(0.5MW/2s) or, alternatively, 84GHz(1MW/2s) or 118GHz(0.5MW/2s). This flexibility is realized by using rotary switches with three inputs and one output. All the transmission line components will be procured from GA and integrated into the existing system purchased from GA in the period 2000-2005.

As discussed in [1], a further upgrade of TCV consists in inserting a modular closed divertor chamber [15]. This will have an impact on the X3 top-launcher which will have to be reduced in size with the aim of rendering this system mountable and dismountable on a significantly shorter time scale than in the present situation. For using the new compact launcher we are considering to employ a Fast Directional Switch [16]

 DOI: 10.1051/epjconf/201715703001

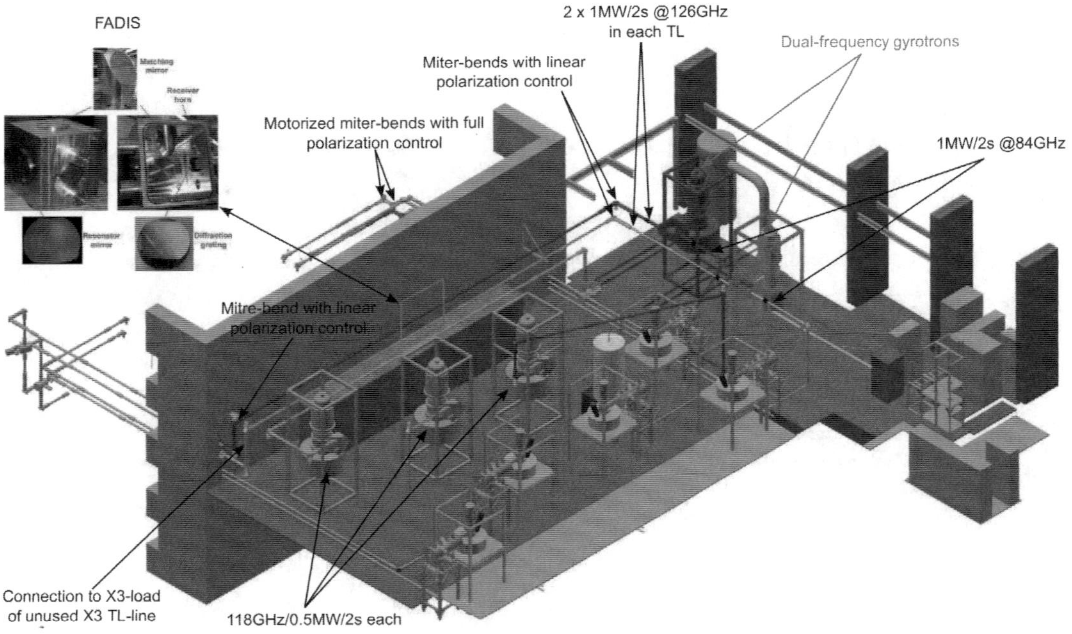


**Fig. 6.** Layout of the upgraded TCV EC-system. The envisaged position of the FADIS is indicated by the red square "traversed" by the two 126GHz transmission lines. In the inset, an example of the FADIS vacuum-vessel and some internal components are shown. The 5 gyrotrons, without labels, in the lower part of the figure are 82.7GHz gyrotrons used for X2-ECH/ECCD.

(FADIS), combining the two 1MW/126GHz/2s rf-beams into a single 2MW rf-beam. As schematically shown in Fig. 6, the FADIS will be placed in the region of the long-leg straight 126GHz transmission line section. One output of the FADIS will be connected to the compact X3 top-launcher, wheras the second output will be connected on an existing rf-power load. For the FADIS resonant-channel, a feedback-controlled mirror drive tracks the diplexer resonance [17] according to the gyrotron frequency chirp which is expected to shift by approximately 300-400MHz over typically the first 500ms of the gyrotron pulse. This frequency chirp is a consequence of the thermally induced gyrotron cavity expansion and, considering that the maximum pulse-length in TCV is of 2s, it is important that the mirror feedback-control locks to the gyrotron frequency on a time scale significantly shorter than the cavity thermal time constant. The first tests of the FADIS in ITER relevant conditions are foreseen for the end of 2018.

In conclusion, the additional power generated by the two dual-frequency gyrotrons will complete and maintain the heating capabilities on TCV in view of regimes of increased relevance for burning plasma conditions. The TCV multi-gyrotron system together with the complex transmission line system, including the FADIS, is an ideal test-bed for further validating ITER relevant EC-technology.

Work supported by the Ecole Polytechnique Fédérale de Lausanne (EPFL).

## References

1. A. Fasoli et al., Nucl. Fusion **55**(4), 043006 (2015).
2. G. Pereverzev et al., ASTRA. Automated System for Transport Analysis in a Tokamak, (Max-Planck Institute for Plasma Physics, Garching, IPP) Report 5/98 (2002).
3. K. Matsuda, IEEE Trans. Plasma Sci. **17**, 6 (1989)
4. L. Porte et al., Nucl. Fusion **47**(8), 952 (2007)
5. S. Alberti et al., Nucl. Fusion 42(1), 42 (2002)
6. S. Alberti et al., Fusion Engineering and Design **53**, 387 (2001)
7. M. Thumm et al., IEEE Trans. on Plasma Science, **35**(2), 143 (2007).
8. J. Jelonnek et al., IEEE Trans. on Plasma Science, **42**(5), 1135 (2014).
9. S. Alberti et al., *Proc. 41st IRMMW-THz Conf., Copenhagen (2016),* W2E4 - 3134041
10. I. Pagonakis et al., Phys. Plasmas **23**, 023105 (2016)
11. K.A. Avramides et al., *Proc. of the 17th Joint Workshop on Electron Cyclotron Emission and Electron Cyclotron Resonance Heating*, Deurne, The Netherlands, 7-11 May 2012, EPJ Web of Conferences 32, 04016 (2012).
12. J. Jianbo et al., IEEE Trans. on MTT , **65**(3), 699 (2017)
13. J. Neilson, *Proc. 29th IRMMW-THz Conf., Karlsruhe p. 667-668 (2004),*
14. J.-Ph. Hogge. et al., *Proc. 40th IRMMW-THz Conf., Hong-Kong (2015),* W2E4 - 3134041
15. H. Reimerdes et al., Nucl. Materials and Energy, (2017) http://dx.doi.org/10.1016/j.nme.2017.02.013
16. W. Kasparek et al., Nucl. Fusion **48**, 054010 (2008)
17. W. Kasparek et al., Nucl. Fusion **56**, 126001 (2016)

DOI: 10.1051/epjconf/201715703002

# The Role of the Plasma Current in Turbulence Decrease with Lower Hybrid Waves

*Ghassan* Antar[1,*], *Marc* Goniche[2], and *Annika* Ekedahl[2]

[1]American University of Beirut, Riad el-Solh, Beirut 1107-2020, Lebanon
[2]CEA, IRFM, F-13108 Saint Paul-lez-Durance, France

**Abstract.** We study the effect of lower hybrid waves on edge turbulence using a set of two Langmuir probes inserted in the scrape-off layer of the Tore Supra tokamak. We use the cross-correlation coefficient to assess the interplay between turbulence and the LH waves. We found that turbulence is affected at high plasma current even at relatively low LH powers. When the plasma current is below 1 MA, this effect disappears. Finally, we show that the level of fluctuations and the cross-correlation amplitude are correlated whereas the increase of the first leads to the increase of the second. The role of the plasma current could reflect that the exchange of the RF waves with edge turbulence is occurring in the closed field lines rather in the open field lines range.

## 1 Introduction

The interaction between the radio frequency (RF) waves and the edge turbulence has resurfaced recently after experimental results showing that turbulence in the SOL decreases when the ion cyclotron frequency heating (ICRH) is switched on [1,2]. Using GHz-range lower hybrid (LH) waves to drive current into tokamak plasmas, this issue presented mixed results. Some tokamaks (FTU & HT-7) have shown a net decrease of turbulence [3,4] while others (Tore Supra) did not [5,6]. In a recent paper, we showed that these apparent discrepancies among tokamaks and RF wave frequencies were removed where it is found that turbulence large-scale structures in the scrape-off layer decrease at high enough plasma currents ($I_p$) on the Tore Supra tokamak [7]. We distinguished three regimes: At low $I_p$ ($I_p$<0.7 MA) no modification is detected with statistical properties of turbulence similar to ohmic plasmas even with a lower hybrid power $P_{LH}$ reaching 4.8 MW. Moderate plasma currents (1.2>$I_p$>0.7 MA), turbulence is modified at high LH power. At high plasma currents ($I_p$>1.2 MA), a net decrease in the level of turbulence of about 30% even with moderate LH power is recorded. The level of turbulence is defined as the standard deviation of the signal normalized by its average value.

The main goal of this paper is to show that at turbulence large-scale structures decrease when both the lower hybrid power and the plasma current are moderately high. Then, we shall correlate the level of turbulence to the amplitude of the cross-correlation coefficient and deduce that at both the plasma current plays an important role in the interaction between the RF waves and edge turbulence. This leads to a change in the confinement properties as we show by comparing the dependence of the diamagnetic energy on the plasma current at low vs high RF power.

## 2 Plasma Scenarios

The experiments studied here are performed on the Tore Supra tokamak. The plasma current as well as the LH power are varied while keeping the same toroidal magnetic field, $B_\phi$=3.82 T. The major and minor radii are constant set to 2.4 and 0.7 m respectively and we verified that the distance between the last closed flux surface (LCFS) and the probes is constant as this could modify the level of fluctuations. The LH power is delivered by two multi-junction LH launchers C3 and C4 with a peak of the $N_{\|0}$ spectrum launched at 1.82. The power is distributed almost equally between the two launchers. The two probes used are in the C4 launcher and a constant bias of -100 V is applied, leading to signals reflecting the ion saturation current $I_{sat}$. They are flush-mounted at the mid-plane with a diameter of 5 mm and a distance of 1 cm between them. They are positioned at 3.158 and 3.164 m from the torus axis for the shots 47654-47666 and 47744-47757 respectively, the LCFS being at 3.09 m and the plasma decay length is 2-3 cm. The launcher front face and thus the probes are protected by side limiters protruding by 2 mm. The field lines in front of the probes are always magnetically connected to the active lower hybrid launchers but not at the same place. The field lines passing in front of the probes are always magnetically connected to the upper-half of C3. With respect to C4, they are connected to the bottom of few waveguides at low plasma currents ($I_p$ ~ 0.7 MA) whereas they are connected to most of the

* Corresponding author: ghassan.antar@aub.edu.lb

DOI: 10.1051/epjconf/201715703002

waveguides at high plasma current ( ~ 1.2 MA) (see Fig. 1). The signals studied here are sampled with DFLUC at 1 MHz, with 13,600 points per trigger per channel.

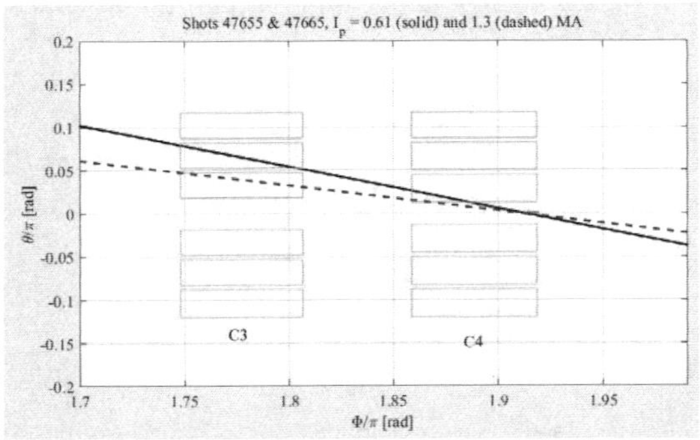


**Fig. 1.** The magnetic field lines in front of the probes. The magnetic field is 3.77 T with plasma minor and major radii equal respectively to 0.72 amd 2.38 m.

## 3 The Cross-correlation Coefficient Behavior

The effect of the plasma current and power on the fluctuations in the SOL is measured by the two Langmuir probes. We use the cross-correlation coefficient defined as

$$C_{12}(\tau) = \frac{\langle I_1(t) I_2(t+\tau) \rangle}{\langle I_1^2 \rangle^{1/2} \langle I_2^2 \rangle^{1/2}}$$

where $I_1(I_2)$ is deduced from the ion saturation current of probe 1(2) after subtracting the average value leading to $I_1 = I_{sat,1} - \langle I_{sat,1} \rangle$. We plot the cross-correlation coefficient $C_{12}$ as a function of time for five cases in Fig. 2. Data is recorded before the discharge shows that data are not correlated at all between probe 1 and 2. During ohmic discharge, the amplitude of $C_{12}$ reaches 0.8 and a width about 100 μs. The width of $C_{12}$ describes the duration of an event. It decreases from about 100 to 10 μs with increasing $I_p$ with $P_{LH}$=4.6 MW. This could be caused by, (1), an increase in the poloidal and/or toroidal convection velocity leading to structures moving faster by the probe, (2), the decrease of the large-scale structures or (3) a combination of the two. Consequently, with the width of $C_{12}$ we are unable to distinguish among the three possibilities. The amplitude of the cross-correlation coefficient (Max($C_{12}$)) on the other hand is dominated by structures with length scales greater or equal than the separation between the probes, that is, about 1 cm. It is not directly affected by the convective velocity in the poloidal or toroidal directions. For $P_{LH}$=4.6 MW one can see that the amplitude drops from 0.65 to 0.38 indicating a net decrease in the structures which have length scales greater than 1 cm; This is not the case when increasing the LHCD power from 0 to 2.2 MW. We deduce that at least part of the width decrease is caused by a net decrease in the turbulent structures' length scales. Hereafter, we shall use Max($C_{12}$) to assess the existence of large scales in the turbulent plasma since, unlike the width, it is independent of the velocity of the structures.

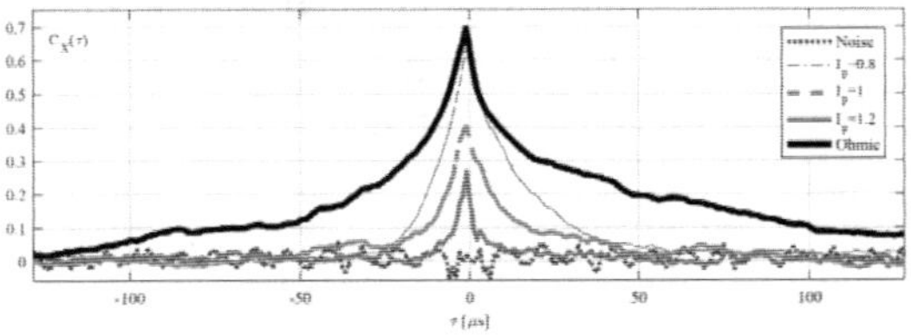


**Fig. 2.** The dependence on the time lag τ of $C_{12}$ under different conditions: no plasma reflecing instrumental noise (dotted line), ohmic plasma (thick solid), LH current drive plasmas with different plasma currents. Note that only at high plasma currents that the amplitude of the cross-correlation decreases whereas the width, which could be dominated by the convection velocity decreases as soon as $P_{LH}$ is turned on.

## 4 The Cross-Correlation Amplitude as a Function of the LH Power and Plasma Current

We determine Max($C_{12}$) for a variety of discharges and the result is shown in Fig. 3. The plasma discharges have in common most of the plasma parameters such as the main magnetic field and most importantly the distance between the probe and the last closed flux surface. On the other hand, they have different plasma currents and LH powers; The plasma density was not the same but we have shown that the turbulence parameters are not modified with increasing density [7].

As a function of the plasma current, the subplot in Fig. 3(a) shows that for $I_p$<1 MA, Max($C_{12}$) is almost unchanged and is about 0.8. However, one can note that for Ip ~1 MA, a couple of points lie around 0.6. At higher plasma currents, $I_p$>1.2, most of the recorded amplitudes are below 0.6 reaching 0.4 for $I_p$>1.4 MA. This figure clearly shows that the amplitude of the cross-correlation coefficient decreases in the presence of LH waves but this behavior appears to depend on $I_p$.

As a function of the LH power, we plot Max($C_{12}$) in subplot Fig. 3(b). For low $P_{LH}$ most of the amplitudes are about 0.8 with the exception of two data points for $P_{LH}$ equals to 2.5 and 3 MW. For high power values, the error around Max($C_{12}$) becomes large with values varying between 0.8, the low power estimate, to 0.4. When putting the two dependencies together, it appears that in order to obtain a smooth transition, one must have high current and high LH power. The same data points are now plotted as a function of $I_p P_{LH}$ in order to highlight this fact in Fig. 3(c). Clearly now when the product is below $3 \times 10^{12}$ WA, no dependence is reported whereas when it is greater a net decrease is observed and is proportional to the product. More theoretical work is needed to understand this dependence.

DOI: 10.1051/epjconf/201715703002

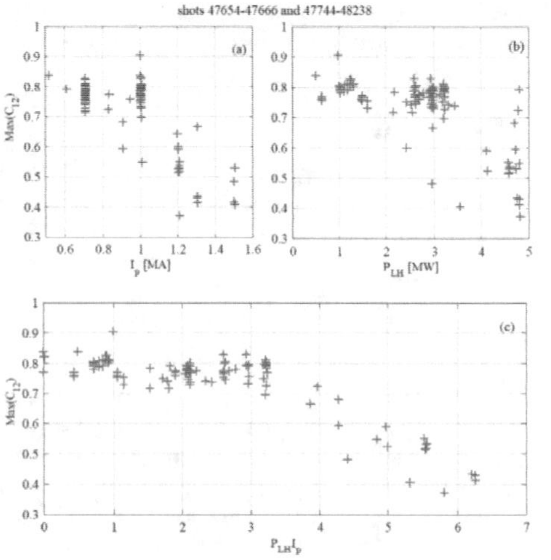


**Fig. 3.** (a) the cross-correlation amplitude as a function of the plasma current, (b) the same but with respect to the LH power, and (c) the same data points plotted against the product of the LH power by the plasma current.

## 5 Relationship between The Cross-Correlation Amplitude and the Level of Fluctuations

Turbulence transport is mainly driven by the large-scale structures which dominate its mixing properties [10]. When plasma is transported outside the last closed flux surface, it becomes connected to the solid objects, either a limiter or a divertor. Consequently, and because of the parallel motion, the small-scale structures are mostly affected leading to the decrease of their presence in the scrape-off layer. For the above results, we used the amplitude of the cross-correlation coefficient which is directly linked to the size the turbulent structures. In Fig. 4, $Max(C_{12})$ is plotted against the level of fluctuations, defined as the standard deviation normalized by the average value of the signals, denoted by $\delta I/I$. One can deduce that higher correlation amplitudes lead to a larger level of fluctuations confirming the above hypothesis and allowing us to link the effects of the plasma current and LH power to the level of fluctuations

## 6 Level of Fluctuations and confinement

At a constant LH power, Fig. 5 shows the behavior of the level of fluctuations with respect to the plasma current. It decreases with increasing $I_p$ in agreement with the decrease of $Max(C_{12})$. We approximated this dependence by a power law of the type $I_p^{-1/2}$ in reference to earlier results on the Tore Supra tokamak [8].

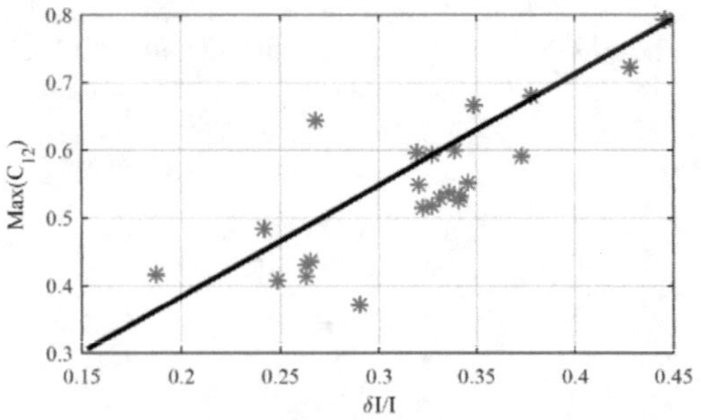


**Fig. 4.** The maximum of the cross-correlation coefficient is plotted against the level of fluctuations ($\delta I/I$) where the increase of the second is correlated to the increase of the first.

In squares the energy content of the plasma increases with $I_p$. The same behavior is obtained after the Rebut-Lallia-Watkins (RLW) scaling is applied to the discharge parameters. Meanwhile, the level of turbulence decreases with $I_p$. This could reflect the link between global confinement properties, the plasma current and SOL turtbulence.

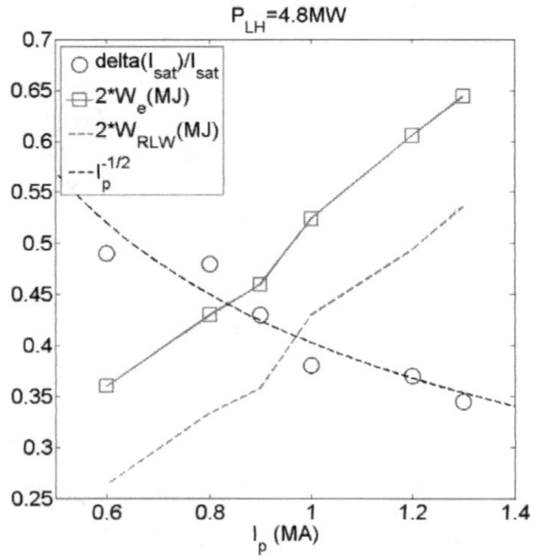


**Fig. 5.** The level of fluctuations In (o) as a function of the plasma current fitted by $I_p^{-1/2}$, in squares the total electron energy and In dashed line the RLW scaling.

Consequently, for these discharges the confinement of the electrons (no ion heating) follows the general trend observed on Tore Supra for LHCD discharges [9]. At $P_{LH}$=400 kW, the electron stored energy follows the Rebut–Lallia-Watkins scaling and at high power ($P_{LH}$~4.8MW), it is above the RLW scaling by a factor increasing when $I_p$ is decreasing. This could be the result of the internal transport barrier being built when increasing fraction of current is driven by LH waves; Further analysis of the plasma profiles is needed to support this hypothesis. So, at high power we find a good correlation between the fluctuation rate measured near the LHCD antenna and the confinement when $I_p$ is changed. At low power, we varied $I_p$ by more than 70% going from 0.8 MA to 1.3 MA and the fluctuations rate was found to be almost unchanged. The behavior of the level of fluctuations with respect to $I_p$ is depicted in Fig. 6 for low and high power RF waves. The two behaviors as a function of the current can be clearly distringuished because of the LH power.

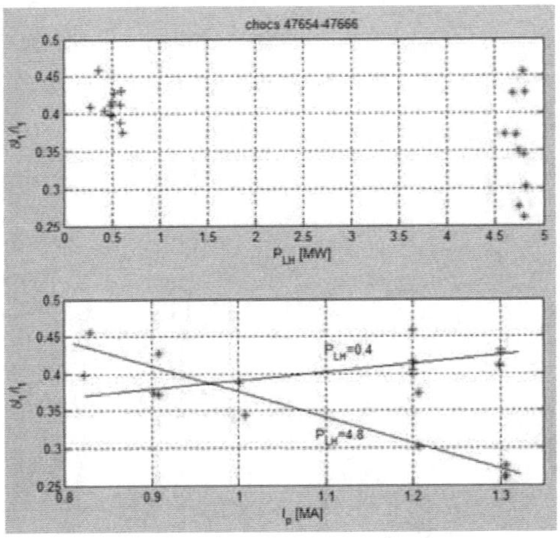


**Fig. 6.** The level of fluctuations sa a function of $I_p$ for the low RF power data (+) and for the high RF power (*).

## 7 Conclusion

In agreement with our previous publications, the present results show the important role of the plasma current in the way the LH power affects edge turbulence. We use the amplitude of the cross-correlation coefficient between two Langmuir probes positioned in the SOL and toroidally distanced by 1 cm. Several conclusions can be drawn from this work:

1- We confirm that the apparent discrepancy among tokamaks is lifted where turbulence decrease due to RF waves is now observed on all tokamaks where it has been studied.

2- We also confirm that the apparent descripency between RF waves is also lifted. RF waves around 50 MHz are used for plasma heating and those around 1 GHz for current drive. They both are capable of affecting turbulence in the SOL and in quite a similar fashion.

3- Clearly, the plasma currrent plays an important role in the way the RF power at the GHz frequencies affect turbulence. This opens up the question about the interplay between the edge where some power is deposited and the SOL where the measurements are made.

4- The effect of magnetic connection may not be important since the field lines in front of the probes are always connected to the active coupler. This is supported by the fact that turbulence decreases at a constant $I_p$ but with increasing LH power. In this case the magnetic field lines are unmodified but the LH power is.

For future work, we will continue studying the issue of the threshold and the role of the plasma current. We will be comparing the effects of the RF waves at the ICRH frequencies to those at the LH frequencies on the turbulence.

## References

[1] Antar, G., et al. "Convective transport suppression in the scrape-off layer using ion cyclotron resonance heating on the ASDEX Upgrade tokamak." *Physical review letters* 105.16 (2010).

[2] Antar, G. Y., et al. "The role of power and magnetic connection to the active antenna in the suppression of intermittent structures by ion cyclotron resonance heating." *Nuclear Fusion* 52.10 (2012): 103005.

[3] Ding, B. J., et al. "Effect of MHD on edge electrostatic fluctuation with lower hybrid current drive in the HT-7 tokamak." *Plasma physics and controlled fusion* 46.9 (2004): 1467.

[4] Ridolfini, V. Pericoli, et al. "Lower hybrid current drive efficiency in tokamaks and wave scattering by density fluctuations at the plasma edge." *Nuclear Fusion* 51.11 (2011): 113023.

[5] T Oosako et al., In *RADIO FREQUENCY POWER IN PLASMAS: Proceedings of the 19th Topical Conference*, volume **1406**, 239 (2011).

[6] Goniche, M., et al. "Lower hybrid current drive at high density on Tore Supra." *Nuclear Fusion* 53.3 (2013): 033010.

[7] Antar, G., et al. "The role of the plasma current in turbulence decrease during lower hybrid current drive." *Physics of Plasmas* 24.3 (2017): 032307.

[8] Vermare, L., et al. "Density fluctuation measurements using X-mode fast sweep reflectometry on Tore Supra." *Nuclear fusion* 46.9 (2006): S743.

[9] Hoang, G. T., et al. "Improved confinement in high li lower hybrid driven steady state plasmas in Tore Supra." *Nuclear fusion* 34.1 (1994): 75.

[10] Diamond, P. H., S.-I. Itoh, and K. Itoh. *Modern Plasma Physics: Volume 1, Physical Kinetics of Turbulent Plasmas*. Cambridge University Press, 2010.

EPJ Web of Conferences **157**, 03003 (2017) DOI: 10.1051/epjconf/201715703003
*22 Topical Conference on Radio-Frequency Power in Plasmas*

# Ray Tracing Study on Top ECCD Launch in KSTAR

*Young-soon Bae*[1,*], W. Namkung[2], and M.H. Cho[2]

[1]National Fusion Research Institute, 169-148 Gwahak-ro, Yuseong-gu, Daejeon 34133, Korea
[2]Pohang University of Science and Technology, 77 Cheongam-Ro. Nam-Gu. Pohang 37673, Korea

**Abstract.** The current drive efficiency of electron cyclotron (EC) wave is typically low compared with other RF and neutral beam heating system in tokamak. It is known that EC current drive by outboard launch suffers from low current drive efficiency due to electron trapping. However, the heating and current drive by EC wave is being regarded as a strong candidate for DEMO reactor due to the simplicity of the launcher, none of its interaction with plasma, and no coupling issue at the plasma edge. Also, off-axis heating and current drive by EC wave plays an important role of steady state operation optimization. To enhance the current drive efficiency in DEMO-relevant operation condition having high density and high temperature, the top launch of EC wave is recently proposed in FNSF design [2]. In FNSF, a top launch makes use of a large toroidal component to the launch direction adjusting the vertical launch angle so that the rays propagate nearly parallel to the resonance layer increasing of Doppler shift with higher $n_{||}$. The results shows a high dimensional efficiency for a broad ECCD profile peaked off axis. In KSTAR, the possibility of efficient off-axis ECCD using top launch is investigated using the ray tracing code, GENRAY [3] for the operating EC frequencies (105 GHz or 140 GHz, and 170 GHz). The high current drive efficiency is found by adjusting the toroidal magnetic field and the radial pivot position of the final launcher mirror for fundamental O-mode and second harmonic X-mode. A large Doppler shift is not quite sure in the typical plasma profile in KSTAR, but the simulation results show high current drive efficiency. This paper presents ray tracing results for many cases with the wave trajectories and damping of EC by scanning the launching angle for specific launcher pivot positions and toroidal magnetic field, and two equilibriums of the KSTAR.

## 1 INTRODUCTION

KSTAR aims to develop DEMO-compatible advanced tokamak operation mode with broad off axis current drive using off-axis NBI and RF heating devices [1]. In the study of FNSF (Fusion Nuclear Science Facility) [2], top launch of electron cyclotron (EC) wave shows improvement in EC current drive (ECCD) efficiency with broad ECCD profile peaked off-axis. It is understood that the top launch of EC wave enables nearly vertical trajectories on the low field side from the cold resonance, so that the resonance approaches slowly with high Doppler shift and waves interact with higher energy electrons, which are less collisional and provide larger ECCD. Eq. (1) is the Doppler-shifted relativistic EC resonance condition at $l$'th harmonic.

$$\omega = \frac{l\Omega}{\gamma} + k_{||}v_{||} \quad (1)$$

$$\frac{v_\perp^2}{c^2} = \left(1 - \frac{\omega^2}{l^2\Omega^2}\right) + 2n_{||}\frac{\omega^2}{l^2\Omega^2}\frac{v_{||}}{c} - \left(1 + n_{||}^2\frac{\omega^2}{l^2\Omega^2}\right)\frac{v_{||}^2}{c^2} \quad (2)$$

Where, $\gamma$ is the relativistic factor, $\omega$ is EC wave frequency, $\Omega$ is the cyclotron resonance frequency, $k_{||}$ ($k_\perp$) is the parallel (perpendicular) component of the wave number of EC wave to the magnetic field, $n = kc/\omega$ is the refractive index, and v is the velocity of electron. Fig. 1 shows ellipses in velocity space using the Doppler-shifted relativistic resonance condition at $l$'th harmonic. Higher is the parallel component of the refractive index ($n_{||}$), the higher is the maximum energy of the resonant electrons.

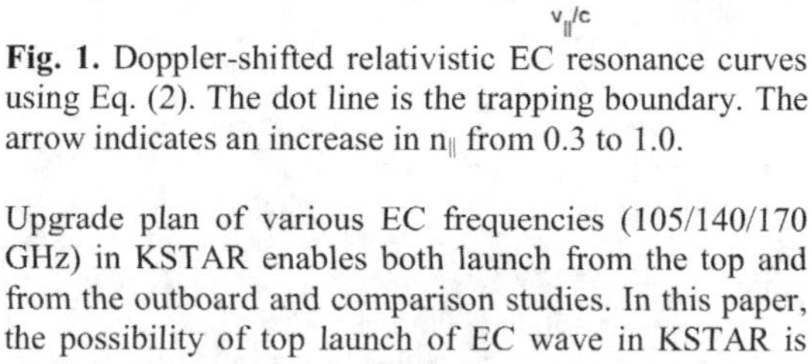


**Fig. 1.** Doppler-shifted relativistic EC resonance curves using Eq. (2). The dot line is the trapping boundary. The arrow indicates an increase in $n_{||}$ from 0.3 to 1.0.

Upgrade plan of various EC frequencies (105/140/170 GHz) in KSTAR enables both launch from the top and from the outboard and comparison studies. In this paper, the possibility of top launch of EC wave in KSTAR is studied using ray tracing simulations.

* Corresponding author: ysbae@nfri.re.kr

DOI: 10.1051/epjconf/201715703003

## 2 Ray Tracing Study

### 2.1 Simulation code

The EC wave propagation and power absorption is calculated using the ray tracing code, GENRAY and Fokker-Planck code, CQL3D [3]. In some case of the top launch conditions, a computing error in implicit solver in CQL3D code occurred. We found this error is removed using the recent fully-neoclassical finite-orbit-width version of CQL3D [4]. In the ray tracing simulation, the toray module in the genray code is used with the finite beam divergence at the launch position.

### 2.2 Top launch of 140 GHz EC wave

In this calculation, the plasma density and temperature profile is assumed by the typical H-mode profile with pedestal height and width. The central density is 5 x $10^{19}$ $m^{-3}$ and the central temperature is 7 keV. Two plasma equilibriums are compared in this calculation. One (equilium I) is basically double null shape with elongation, $\kappa = 2$ with the plasma current $I_p = 1.14$ MA and $B_t = 2.1$ T at the major radius of 1.8 m, in which $q_{95} = 4$. The other one (equilibrium II) is single null shape with lower $\kappa = 1.7$ which is typical shape in recent KSTAR experiments.

Figure 2 shows the ray tracing of 140 GHz X2 EC wave launched from the top position $R_{ant} = 1.62$ m, $Z_{ant} = 1.0$ m with specific injection angles for equilibrium I. The ray tracing shows nearly vertical trajectory parallel to cold resonance layer and broad driven current profile. The cold resonance is located at $R_{X2} = 1.512$ m which corresponds to the toroidal magnetic field 2.1 T at the major radius $R_0 = 1.8$ m.

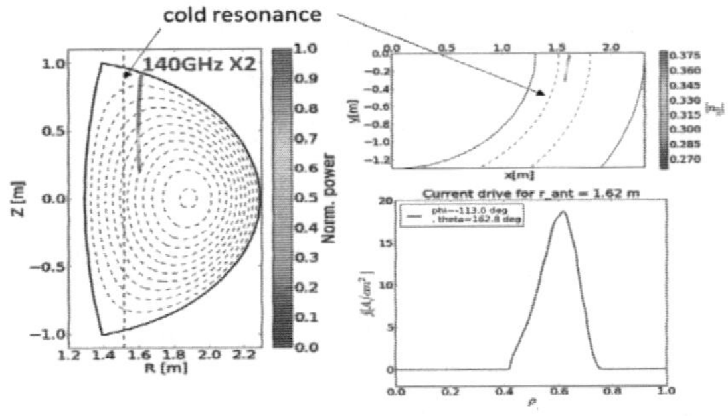


**Fig. 2.** Top launch of 140 GHz X2 EC wave for ne(0) = 5 x $10^{19}$ $m^{-3}$ and $T_e(0) = 7$ keV and for equilibrium I.

Figure 3 shows the current drive and peak position of driven current profile for launch angles ($\phi$ and $\theta$) for equilibrium I. The $\phi$ is the toroidal angle from the radial direction, and the $\theta$ is the poloial angle from the vertical axis. The current drive changes significantly with small change of the launch angles. The highest current drive is about 50 kA/MW for $n_e(0) = 5 \times 10^{19}$ $m^{-3}$ which corresponds to the current drive efficiency, $\eta = I_{CD}R\langle n_e\rangle/P_{abs} = 0.39 \times 10^{19}$ $Am^{-2}W^{-1}$, where, $I_{CD}$ is the total driven current in A, R is major radius in meter, $\langle n_e\rangle$ is the average density in $m^{-3}$, and $P_{abs}$ is the absorbed power. The current drive profile is ranged with its peak position of $\rho = 0.5 \sim 0.7$. The $\rho$ is the normalized minor radius defined by $r/a$, where $a$ is the plasma minor radius and 0.5 m in the KSTAR. For equilibrium II, top launch has the similar characteristics on the current drive and peak position as seen fig. 4. It should be noted that the antenna position of $R_{ant} = 1.62$ m for equilibrium I and 1.66 m for equilibrium II are found to be appropriate position to have high current drive with the vertically aligned ray propagation to the cold resonance layer.

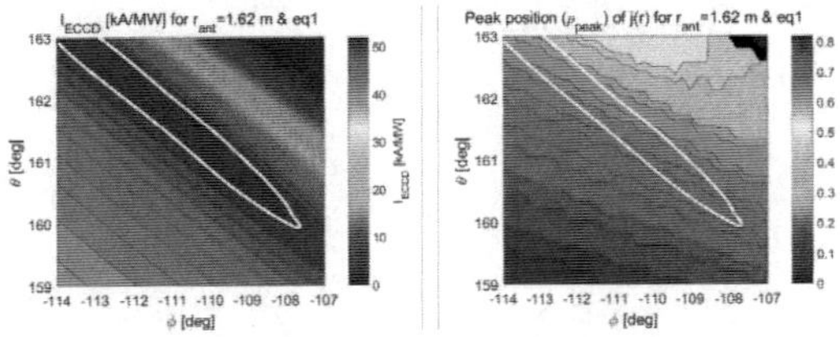


**Fig. 3**. The current drive and its peak position for various launch angles for equilibrium I. The toroidal angle is clockwise direction for the co-CD. The region enclosed by white line indicates the maximum current drive (left) obtained by ray tracing and the corresponding peak locations (right).

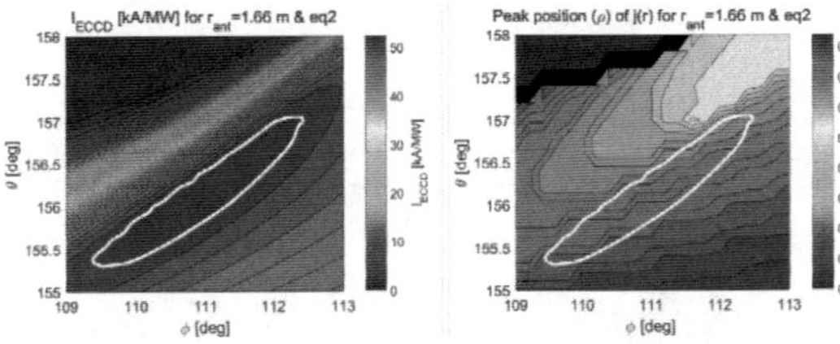


**Fig. 4.** The current drive and its peak position for various launch angles for equilibrium II. In this case, the toroidal angle is counter-clockwise for co-CD due to the opposite plasma current in equilibrium II. The region enclosed by white line indicates the maximum current drive (left) obtained by ray tracing and the corresponding peak locations (right).

Figure 5 shows the EC wave ray tracing and ECCD profile for launching angles corresponding to current drive comprised between 35 and 52 kA. Depending from the launching angles, the CD peak is located between $\rho = 0.6$ and 0.85.

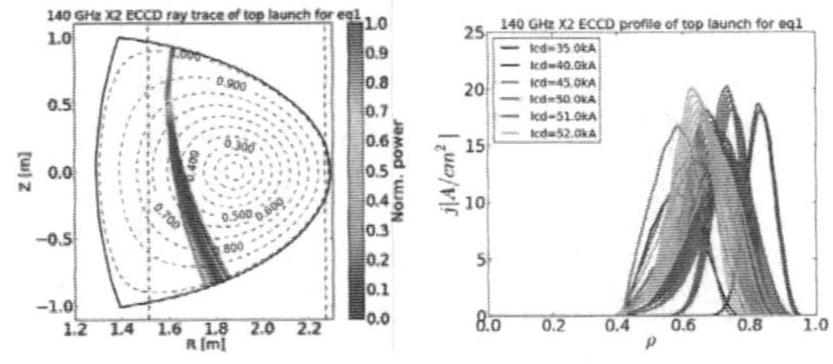


**Fig. 5.** The ray tracing and ECCD profile by 140 GHz X2 wave top launch for various launch angles.

In this scheme for X2 140 GHz top launch, the outboard launch of 140 GHz EC wave is applicable. But the 3rd harmonic parasitic power loss is observed near the plasma edge hence the lower current drive. The downward injection for outboard launch is better to reduce the 3rd harmonic parasitic power loss as seen in Fig. 6.

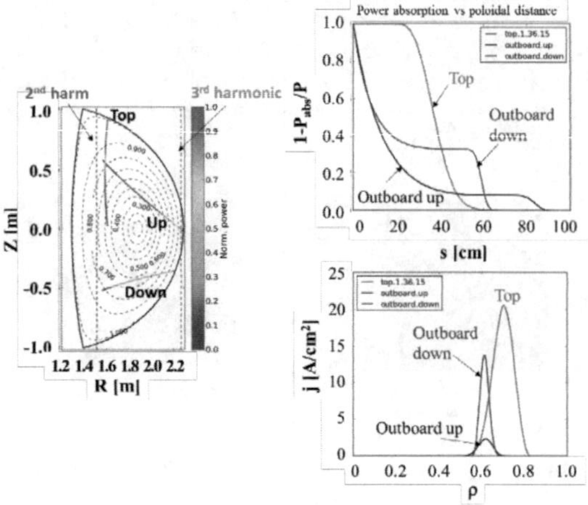


**Fig. 6.** The comparision between the top launch and two outboard launches of 140 GHz X2 EC wave.

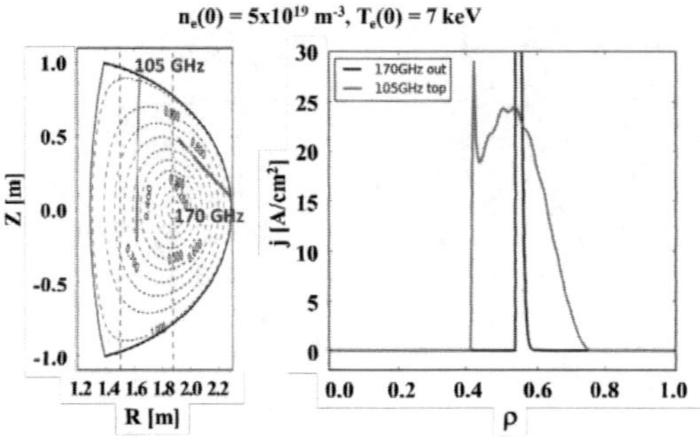


**Fig. 7.** The ray tracing of top launch of 105 GHz O1 EC wave and the outboard launch of 170 GHz X2 EC wave. Since the peak value of the driven current density is 100 A/cm$^2$ for the case of outboard launch, the scale is truncated for the comparison between two cases.

## 2.3 Fundamental Harmonic Top launch of 105 GHz EC wave

In addition to 105 GHz/140 GHz dual frequency gyrotron in KSTAR, the 170 GHz ECH system upgrade is also under consideration. Considering the possibility of higher toroidal magnetic field operation and the use of 170 GHz EC wave, the toroidal magnetic field of 3.125 T at the major radius of 1.8 m is found to provide good current drive scheme for both top launch and outboard midplane launch using 105 GHz O1 mode for top launch and 170 GHz X2 mode for outboard launch. In this case, the cold resonance magnetic field of O1 mode of 105 GHz is 3.75 T located at $R_{res}$ = 1.5 m, and the antenna pivot position is $R_{ant}$ = 1.65 m. For 170 GHz X2 cold resonance is at $R_{res}$ = 1.88 m. Fig. 7 shows two launch cases of top lauch of 105 GHz and outboard launch of 170 GHz. The very broad current density profile is obtained for the 105 GHz O1 top launch, while the peaked current dentisy profile for the 170 GHz outboard launch. The resulting current drive is 83 kA/MW ($\eta$ = 0.67 x $10^{19}$ Am$^{-1}$W$^{-1}$) for 105 GHz and 34 kA/MW for 170 GHz. For 105 GHz O1 top launch, the current drive efficiency is high even for the high density at $n_e(0)$ = 1 x $10^{20}$ m$^{-3}$ near the O1-mode cut off density (1.33 x $10^{20}$ m$^{-3}$ for 105 GHz) with $T_e(0)$ = 7 keV. Fig. 8 shows the ray tracing and current drive profile for both plasma densities of top launch of 105 GHz O1 mode.

## 2.4 3D Fokker-Planck calculation for 140 GHz ECCD top launch

As described above, the new version of CQL3D with finite orbit with is used for Fokker-Planck calculation. Fig. 9 shows the current drive profiles for different launch angles and ray tracing from GENRAY and Fokker-Planck simulation. In the Fokker-Planck calculation, the power absorption is generally calculated with non-Maxwellian distribution function. As seen in the right figure in fig. 9, the ray propagates further due to the different power absorption, hence the different driven current density profile with the peak position shifted inward. Fig. 10 shows the broad specrum of QL diffusion (DQL) for resonant electrons in velocity space for a specific launch case.

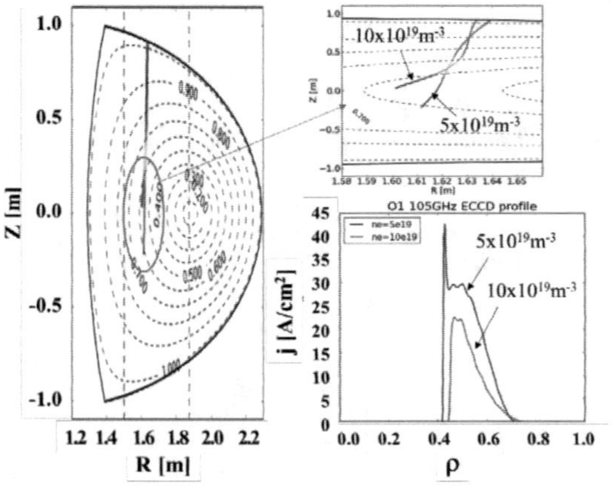


**Fig. 8**. Ray tracing of top launch of 105 GHz O1 EC wave for two plasma densities, 0.5 x $10^{20}$ m$^{-3}$ and 1 x $10^{20}$ m$^{-3}$.

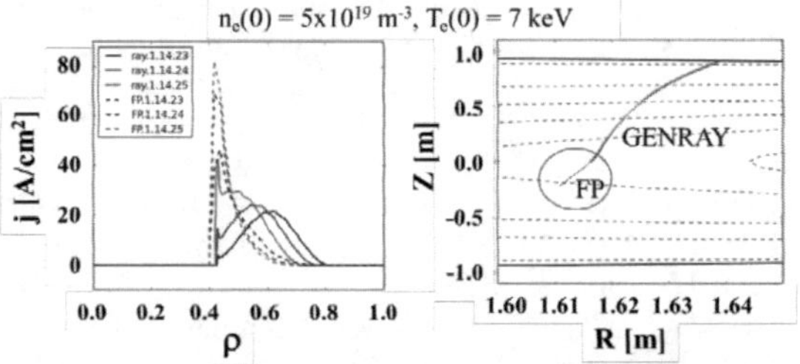


**Fig. 9.** The Fokker-Planck code CQL3D calculation of top launch of 140 GHz X2 EC wave. The solid line is for the ray tracing and the dash line is for the CQL3D calculation.

Each current drive profile is obtained for different launch angle.

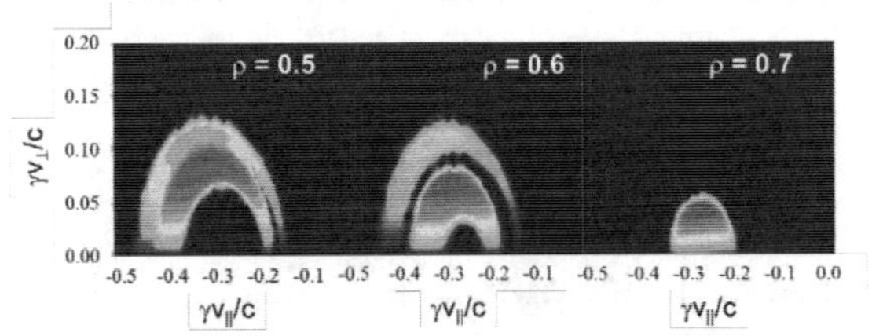


**Fig. 10**. The quasi-linear diffusion coefficient (DQL) obtained from CQL3D for the top launch of 140 GHz X2 EC wave for a specific case among of Fig. 8.

## 2.5 Conceptual design of top launcher in KSTAR ECCD

Figure 11 shows a proposed conceptual design for top launcher in KSTAR. It uses basically the vertical port and four mirrors inside vacuum vessel near the upper divertor region. It may requires additional engineering design to accommodate the four mirrors inside vacuum vessel in the aspect of interface issues of mirrors near upper divertor sectors and upper passive plates. Depending on the design of mirrors, the upper divertor may need to be removed allowing only lower single null operation in KSTAR. This is the reason that equilibrium II with single null configuration is included in the ray tracing study of top launch ECCD in KSTAR. From the ray tracing study of ECCD top launch, the best choice of last steerable mirror pivot position would be $R_{ant}$ = 1.62-1.65 m and $Z_{ant}$ = 1.0 m. Considering 170 GHz EC system in the upgrade plan, the fundamental 105 GHz top launch is very attractive to have high current drive at the off-axis with pivot position, $R_{ant}$ = 1.65 m of the last steerable mirror and Bt = 3.125 T.

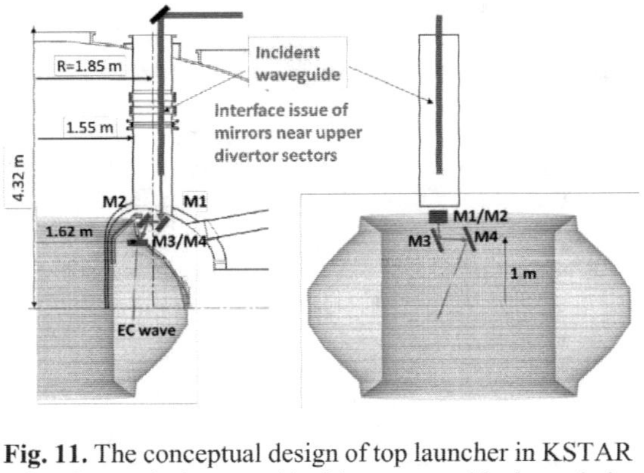


**Fig. 11.** The conceptual design of top launcher in KSTAR using the vertical port and incident waveguide through the port.

## 3. Conclusions

The ray tracing simulations for top launch ECCD shows higher current drive efficiency than low field side injection. The 105/140 GHz dual frequency gyrotron and 170 GHz gyrotron will be main power sources in KSTAR ECH system. For the 140 GHz ECCD top launch, the appropriate toroidal magnetic field and antenna pivot position are found for high current drive efficiency considering the operation window of the KSTAR toroidal magnetic field and through scanning of the antenna pivot position for various launching angles. The highest current drive of ~50 kA/MW is obtained for $n_e(0) = 5 \times 10^{19}$ m$^{-3}$ and $T_e(0)$ = 7 keV. It corresponds to the current drive efficiency $\eta = 0.39 \times 10^{19}$ Am$^{-2}$W$^{-1}$. The fundamental O-mode 105 GHz ECCD top launch is also investigated for the higher toroidal magnetic field operation than 3 T in KSTAR. We obtained very high current drive efficiency (83 kA/MW, $\eta = 0.67 \times 10^{19}$ Am$^{-2}$W$^{-1}$) for the same density as above. When the density increases to $1 \times 10^{20}$ m$^{-3}$, the driven current is similar as that of 140 GHz second harmonic ECCD top launch without the cut-off. The high toroidal magnetic field has an advantage of the current drive at the same flux surface using a second harmonic 170 GHz EC wave from the low field side injection at the same time.
The main task in this paper is to investigate the characteristics and performance of the driven current and the absorbed power for the top launch using single-ray approach. But the Fokker-Planck simulation shows the very different driven current density profile for the specific case of top launch. Therefore, the Fokker-Planck calculation might be needed for more precise top launch study. The driven current density profile also depends on the beam divergence in the single-ray approach. Depending on the launcher mirror design (focused or straight beam), the divergence need to be changed. Note that there is no significant difference in the driven current profile between single-ray tracing and and multi-ray tracing simulation. In conclusion, the top-launch approach of ECCD is very attractive and strongly recommended for the future KSTAR research mainly covering issues relevant for DEMO-compatible advanced tokamak operation scenario with limited available heating power sources.

## References

1. Y.S. Bae et al., "NUBEAM simulation for off-axis NBI design in KSTAR," American Physical Soceity (Nov. 16-20, 2015, Savannah, Georgia).
2. A.M. Garofalo et al., "Progress in the physics basis of a Fusion Nuclear Science Facility based on the Advanced Tokamak concept," Nucl. Fusion **54**, 073015 (2014).
3. Smirnov A.P. and Harvey R.W. Report CompX-2000–01 Ver 2, (www.compxco.com/genray.html).
4. Yu V. Petrov et al., Plasma Phys. Cont. Fusion **58**, 115001 (2016).

EPJ Web of Conferences **157**, 03004 (2017)
*22 Topical Conference on Radio-Frequency Power in Plasmas*
DOI: 10.1051/epjconf/201715703004

# Self-consistent calculation of the effects of RF injection in the HHFW heating regimes on the evolution of fast ions in toroidal plasmas

*Nicola* Bertelli[1,⋆], *Marina* Gorelenkova[1], *Mario* Podestà[1], *Ernest* Valeo[1], *David* Green[2], and *Eric* Fredrickson[1]

[1] *Princeton Plasma Physics Laboratory, Princeton, NJ 08543, USA*
[2] *Oak Ridge National Laboratory, PO Box 2008, Oak Ridge, TN 37831-6169, USA*

**Abstract.** A critical question for the use of ion cyclotron range of frequency (ICRF) heating in the ITER device and beyond is interaction of fast waves with energetic ion populations from neutral beam injection (NBI), fusion reactions, and minority ions accelerated by the RF waves themselves. Several experiments have demonstrated that the interaction between fast waves and fast ions can indeed be strong enough to significantly modify the NB ion population. To model the RF/fast ion interaction and the resulting fast ion distribution, a recent extension of the full wave solver `TORIC` v.5 that includes non-Maxwellian effects has been combined with the Monte Carlo `NUBEAM` code through an RF "kick" operator. In this work, we present an initial verification of the `NUBEAM` RF "kick" operator for high harmonic fast wave (HHFW) heating regime in NSTX plasma.

## 1 Introduction

Self-consistent modeling of the interactions between fast waves (FWs) and fast ions, introduced either from neutral beam injection (NBI) or from fusion-generated alpha particles, is important for both present-day experiments and for ITER. The fast-ion population changes the wave propagation and absorption, while the wave damping on fast-ions modifies their distribution. The latter implies that fast-wave heating could impact and perhaps give leverage over Alfvénic instabilities driven by beam ions. Specific to NSTX [1], FWs can modify the behavior of - and, under certain conditions, even suppress - fast ion driven instabilities such as toroidal Alfvén eigenmodes (TAEs), global Alfvén eigenmodes (GAEs) and fishbones [2, 3]. Moreover, because of the lower toroidal field of the spherical tokamak, FW heating may accelerate fast ions to loss orbits, and this power-loss mechanism must be studied and then minimized. This paper is structured as follows: the main equations of the RF-kick operator implemented in `NUBEAM` are presented in Section 2. In Section 3, a brief description of the recent extension of the full wave code `TORIC` [4, 5] is presented. In Section 4, first and preliminary tests/results are shown and discussed for NSTX plasmas.

## 2 RF-kick operator implemented in `NUBEAM`

The original implementation of the RF-kick operator in `NUBEAM` [6, 7] was presented in [8, 9]. However, a clear verification of such implementation is still missing. In this section we briefly show the main equations of the RF-kick operator implemented in `NUBEAM`. The RF-kick operator implemented in `NUBEAM` is based on the original works by C. F. Kennel, F. Engelmann [10], T. H. Stix [11], and X. Q. Xu and M. N. Rosenbluth [12]. The strong damping of FWs on fast ions occur at the ion-cyclotron resonance in the plasma and the RF resonance condition for a RF angular frequency $\omega_{\rm rf}$ is given by

$$\Lambda^{\rm rf}(B,n,k_\parallel,E,\mu) \equiv n\frac{q_s B}{m_s} + k_\parallel\sqrt{\frac{2}{m_i}(E-\mu B)} - \omega_{\rm rf} \quad (1)$$

where $B$ is the magnetic field, $n$ is the harmonic number, $k_\parallel$ is the parallel component (to the magnetic field) of the wave-vector, $E \equiv m_s v_s^2/2$ is the kinetic energy of the particle, $\mu \equiv m_s v_{\perp,s}^2/(2B)$ is the particle magnetic moment. Finally, $q_s$ and $m_s$ are the particle charge and mass, respectively. Applying a similar derivation of the Coulomb collision operator with a Monte-Carlo scheme in [12] to our problem of interest, one can evaluate the new parallel $v_\parallel'$ and perpendicular component $v_\perp'$ of particle velocity from the old velocities $(v_\parallel, v_\perp, v)$ by

$$v_\parallel' = v_\parallel(1-\nu^{\rm rf}_{\parallel s}\Delta t) + \quad (2)$$

$$+ \; 2\sqrt{3}(R_1-0.5)\sqrt{\nu^{\rm rf}_\parallel v^2 \Delta t} \quad (3)$$

$$v_\perp'^2 = v_\perp^2 - \nu^{\rm rf}_{\perp,s} v^2\Delta t + 2\sqrt{3}(R_2-0.5)\times \quad (4)$$

$$\times \; v^2\sqrt{\left[\nu^{\rm rf}_\perp - \frac{(\nu^{\rm rf}_{\parallel\perp})^2}{\nu^{\rm rf}_\parallel}\right]\Delta t} +$$

$$+ \; 2\sqrt{3}(R_1-0.5)\frac{\nu^{\rm rf}_{\parallel\perp}}{\nu^{\rm rf}_\parallel} v\sqrt{\nu^{\rm rf}_\parallel v^2\Delta t} \quad (5)$$

$$(6)$$

where $\nu^{\rm rf}_\perp, \nu^{\rm rf}_\parallel, \nu^{\rm rf}_{\parallel\perp}$ indicate the perpendicular and parallel velocity RF diffusion processes and $R_1, R_2$ are two

⋆e-mail: nbertell@pppl.gov

independent random variables that are distributed between 0 and 1. Finally, $\Delta t$ represents the integration time step (during the particle orbiting). Moreover, the leading term of the RF quasi-linear diffusion operator operator $D_{\rm rf}$ can be written as [11, 13]

$$D_{\rm rf} \propto |E_+|^2 J^2_{n-1}\left(\frac{k_\perp v_\perp}{\omega_{\rm ci}}\right) \tag{7}$$

where $E_+$ is the left-handed wave electric field and $J_n$ represents the Bessel functions with argument $k_\perp v_\perp/\omega_{\rm ci}$ ($k_\perp$ and $\omega_{\rm ci}$ are the perpendicular component of the wave vector and the ion cyclotron angular frequency, respectively). In Eq. 7 and in the TRANSP implementation the right-handed wave field, $E_-$, and the parallel wave field, $E_\parallel$ have been currently neglected.

## 3 TORIC non-Maxwellian extension

The version of TORIC implemented in TRANSP [14], which is able to solve the Maxwellian equations in the high harmonic frequency heating regimes [4], has been recently extended to include non-Maxwellian ion effects [5]. The extension of the code consists mainly in the implementation of the full hot plasma dielectric tensors for arbitrary distribution function instead of the Maxwellian distribution function as originally implemented. The HHFW version of TORIC is based on the so called "Quasi-local" approximation (the reader is referred to [4] for more details). The structure of the differential form of the wave equation implemented in TORIC in the finite-Larmor-radius (FLR) approximation is maintained and 0th-order FLR coefficients of the wave equation are replaced by the corresponding elements of the full hot-plasma dielectric tensor in which the $k_\perp$ value in the argument of the Bessel functions is obtained by solving the local dispersion relation for FWs only. This code extension can now deal with analytical functional forms of the distribution functions (such as, anisotropic Maxwellian and slowing down distributions) and numerical distributions. In fact, the main goal of this extension was to enable the code to deal with numerical distribution provided by Fokker-Planck (for instance, CQL3D code [13]) and/or Monte Carlo particle codes (for instance, NUBEAM [6, 7]). More specifically, in this work we want to use the distribution function obtained from the Monte Carlo NUBEAM code for an NSTX [1] discharge. To do that, we make use of the P2F code [15, 16], which converts a discrete particle list to a 4-D continuum distribution function. The 4 dimensions are 2 cylindrical in space $(R, z)$ and 2 cylindrical in velocity space $(v_\parallel, v_\perp)$. This conversion is necessary because the full hot dielectric dielectric tensor for arbitrary distribution functions, implemented in the code, takes velocity space derivatives of the distribution function, therefore function must be smooth enough for the derivatives to be robust. Several validations and tests with analytical and numerical distribution of these code have been presented in [5] and we are referring the reader to that reference. However, in Fig. 1 we show an example of the beam distribution function obtained from the P2F code starting with a NUBEAM particle list.

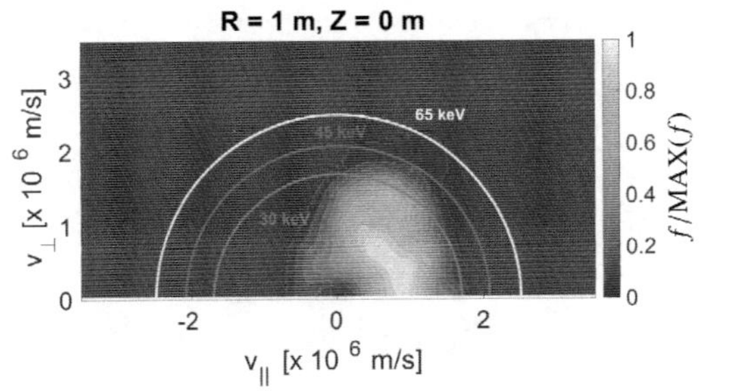


**Figure 1.** Fast ion distribution function at $R = 1$ m and $z = 0$ m (approximately corresponding to the magnetic axis) obtained from the P2F code starting from a NUBEAM particles list for an NSTX discharge 128733 at time = 0.199 s (pure NBI distribution without the RF effects).

## 4 Preliminary test of RF kick operator

In the TRANSP code, the TORIC and NUBEAM code are currently coupled in the following way: TORIC provides the wave electric field, the perpendicular wave vector and the toroidal mode number to NUBEAM in order to run the RF-kick operator and in turn, NUBEAM provides to TORIC the profiles of the flux surface average of the total energy density and the density of the beam ions. These two quantities are used to get an effective temperature given by [17]

$$T_{\rm eff} = \frac{2}{3}\frac{\mathcal{E}}{n_{\rm fi}} \tag{8}$$

where $\mathcal{E}$ and $n_{\rm fi}$ are the total energy density profile and the density of the beams ions, respectively. Furthermore, currently in TRANSP there is a re-normalization factor introduced to enforce the total power absorbed by fast ions evaluated by NUBEAM to match the absorbed power evaluated by the TORIC code in order to have consistency between the NUBEAM and TORIC calculations [9]. This normalization factor varies on a case-by-case basis. In this paper, we present an initial verification of the NUBEAM RF "kick" operator for high harmonic fast wave (HHFW) heating regime in NSTX plasma. In particular, Figure 2 shows the NSTX discharge # 128733 used in these numerical simulations. NSTX operates with a toroidal field of 0.5 T, with typical density $(3-10)\times 10^{19}$ m$^{-3}$, temperature $T_{\rm e} \approx T_{\rm i} < 1.5$ keV. For this discharge there are two main heating systems: neutral beam injection (NBI) and HHFW. During the flattop phase 1 MW of HHFW and NBI are injected. The NB injection energy is 65 keV (see also Figure 1). The antenna phase is 90 degree, which corresponds to a toroidal wave vector $k_\phi = 8$ m$^{-1}$ and the antenna frequency is 30 MHz. Full wave simulations for NSTX plasmas commonly found a significant amount of RF power absorbed by the fast ion population. In fact, several IC cyclotron resonances, where the strong interaction between FW and particles occurs, are present in the plasma, as shown in Figure 3 for NSTX discharge 128733.

The black curve in Figure 4 shows the measured value of the neutron rate for the NSTX discharge 128733,

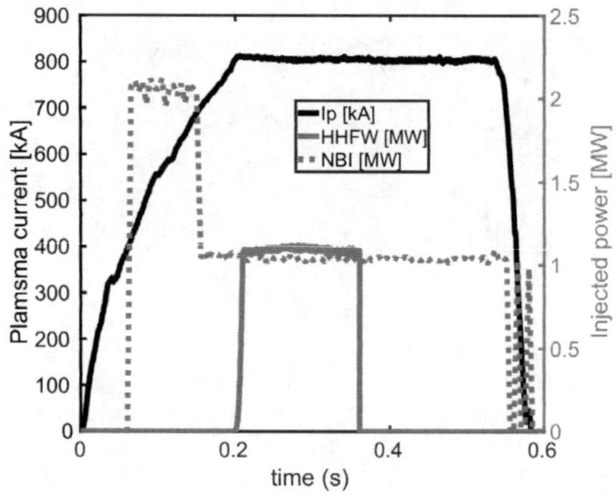


**Figure 2.** Plasma current (black curve), Injected NB (dashed red curve) and HHFW (solid red) power for NSTX discharge # 128733.

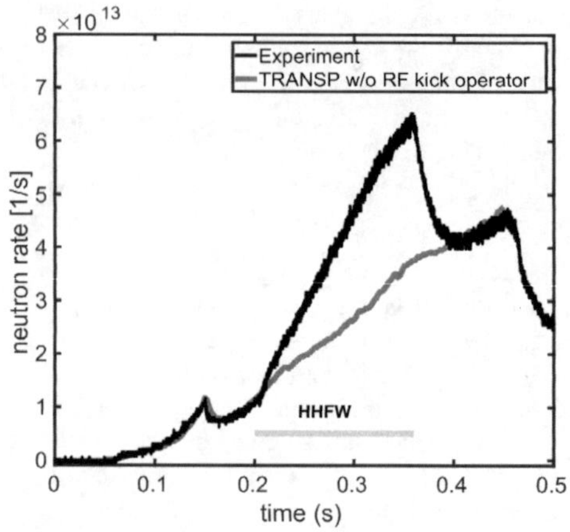


**Figure 4.** Neutron rate for NSTX discharge 128733. The black curve indicates the measured values, whereas the red curve is the predicted neutron rate obtained from TRANSP without the RF kick-operator. The green horizontal line shows for reference when HHFW is applied.

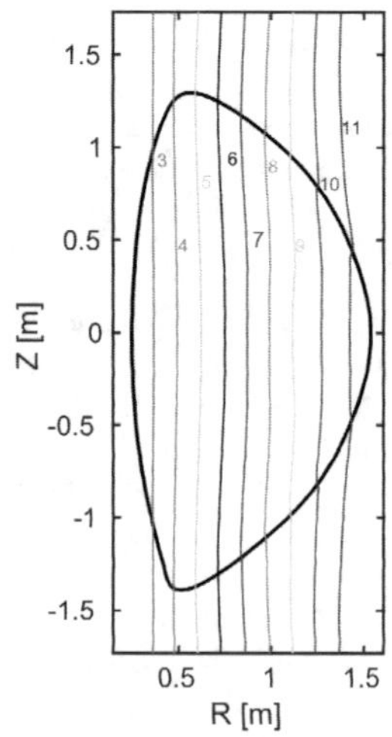


**Figure 3.** Deuterium cyclotron resonances for NSTX discharge # 128733 with $\omega/2\pi = 30$ MHz and $B \simeq 0.5$ T. The numbers represent the harmonic number.

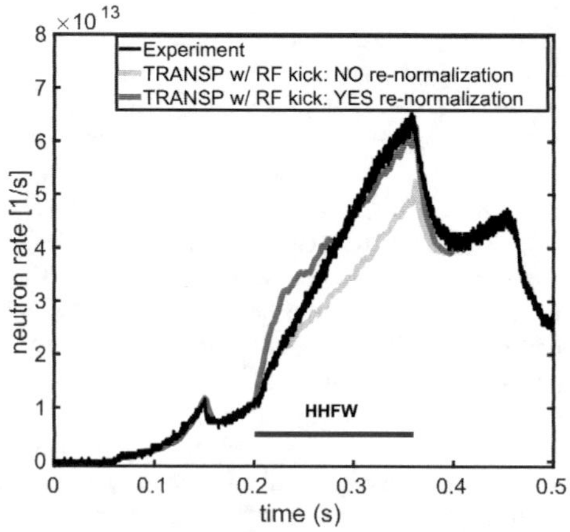


**Figure 5.** Neutron rate for NSTX discharge 128733. The black curve indicates the measured values. The green and red curves are the predicted neutron rate obtained from TRANSP including the RF kick-operator without and with the re-normalization factor, respectively . The blue horizontal line shows for reference when HHFW is applied.

whereas the red curve is the predicted neutron rate obtained from TRANSP without the application of the RF kick-operator in NUBEAM. One can clearly see that the level of neutron emission rate rises significantly during the HHFW heating (as indicated by the green horizontal line for reference). Because the neutron rate is dominated by beam-plasma reactions, this indicates that a significant fraction of high-energy fast ions are accelerated by the RF heating. In fact, this is commonly found in the RF simulations for NSTX plasmas when a NBI species is considered [17]. If this effect is neglected, TRANSP completely fails in predicting the rise in the neutron rate. As a first test, we repeat the same TRANSP simulation including the RF kick operator and the contribution from TORIC. The results are shown in Figure 5. We show two cases: green and red curves represent two TRANSP runs with and without the re-normalization factor between TORIC and NUBEAM, respectively. The predicted neutron rate without re-normalization factor tends to underestimate the measured neutron rate except for the very beginning of the HHFW application. On the other hand, the predicted neutron rate assuming a re-normalization factor tends to be closer to the measured one. However, it overestimates the neutron rate for about the first half of the HHFW application.

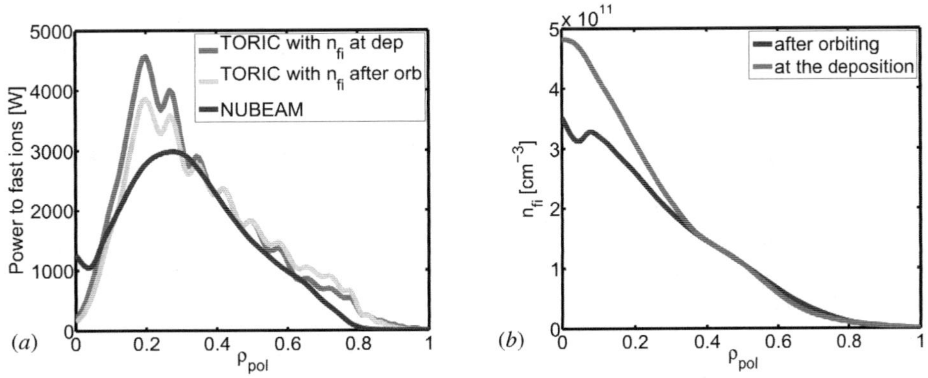


**Figure 6.** (a) ICRH power absorption profiles of the beam ions as a function of the square root of the normalized poloidal flux, $\rho_{\text{pol}}$ evaluated by NUBEAM (blue curve) with "RF-kick" operator and TORIC with the fast ion density at the deposition (red curve) and after following the particles (green curve) including fast particle losses in NUBEAM (see blue curve in Fig. b.). (b) Fast ion density at the deposition (red curve) and after following the particles including fast particle losses (blue curve) as a function of the square root of the normalized poloidal flux, $\rho_{\text{pol}}$.

In the analysis above, TORIC is using a Maxwellian distribution via the effective temperature (cf. Eq. 8) although NUBEAM provides a general fast ion distribution function. Regarding this issue, as a starting point, we have compared the power deposition profiles obtained by TORIC and NUBEAM including the RF-kick operator for an NSTX plasma described in [5]. More specifically, we have assumed a Maxwellian distribution function for the fast ion population at the deposition in NUBEAM. Figure 6(a) shows the HHFW power profiles from TORIC (red and green curves) and NUBEAM (blue curve). The green and red curves are obtained assuming the fast ion density at the deposition and the fast ion density after following the particles in a single NUBEAM time step. In both cases, an effective temperature is used as shown in Eq. 8. Figure 6(b) shows the corresponding flux surface average of the fast ion density at the deposition (red curve) and after following the particles in a single NUBEAM time step (blue curve). From 6(a), the agreement between TORIC and NUBEAM is quite acceptable. The peak locations and the profiles are similar. Moreover, TORIC's result tends to be closer to NUBEAM's result when using the fast ion density after following the particles for a first time step in NUBEAM. This preliminary exercise seems to suggest that when the fast ion distribution function between TORIC and NUBEAM are similar there is a reasonable agreement between the power profiles from the two codes. However, it is important to stress that this is still a preliminary and a single result. Several other tests are underway and they will be shown in a future work. In particular, the consideration of a realistic distribution obtained starting from NUBEAM and affected by RF, will be part of a future work together with additional tests on the "RF-kick" operator in NUBEAM.

## Acknowledgments

This work was supported by the SciDAC Center for Wave-Plasma Interactions under DE-FC02-01ER54648 and the US DOE under DE-AC02-CH0911466.

## References

[1] S.M. Kaye, Nucl. Fusion **55**, 104002 (2015)
[2] M. Podestà, et al, Nucl. Fusion **52**, 094001 (2012)
[3] E.D. Fredrickson, etal, Nucl. Fusion **55**, 013012 (2015)
[4] M. Brambilla, Plasma Phys. Control. Fusion **44**, 2423 (2002)
[5] N. Bertelli, et al, Nucl. Fusion **57**, 056035 (2017)
[6] R.J. Goldston, et al, J. Comput. Phys. **43**, 61 (1981)
[7] A. Pankin, et al, Comput. Phys. Commun. **159**, 157 (2004)
[8] J.H. Kwon, et al, in *APD-DPP conference* (2006)
[9] J.H. Kwon, et al, in *APS-DPP conference* (2007)
[10] C.F. Kennel, F. Engelmann, Phys. Fluids **9**, 2377 (1966)
[11] T.H. Stix, Nucl. Fusion **15**, 737 (1975)
[12] X.Q. Xu, M.N. Rosenbluth, Phys. Fluids B **3**, 627 (1991)
[13] Y.V. Petrov, R.W. Harvey, Plasma Phys. Control. Fusion **58**, 115001 (2016)
[14] R.J. Hawryluk (1980), Vol. 1, p. 19
[15] D.L. Green, et al, Journal of Physics: Conference Series **180**, 012058 (2009)
[16] D.L. Green, et al, AIP Conf. Proc. **1187**, 569 (2009)
[17] N. Bertelli, et al, AIP Conf. Proc. **1580**, 310 (2014)

DOI: 10.1051/epjconf/201715703005

# Characterization of 3-strap antennas in ASDEX Upgrade

*Volodymyr* Bobkov[1,*], *Roberto* Bilato[1], *Laurent* Colas[2], *Ralph* Dux[1], *Eric* Faudot[3], *Helmut* Faugel[1], *Helmut* Fünfgelder[1], *Albrecht* Herrmann[1], *Jonathan* Jacquot[1], *Arne* Kallenbach[1], *Daniele* Milanesio[4], *Riccardo* Maggiora[4], *Rudolf* Neu[1,5], *Jean-Marie* Noterdaeme[1,6], *Roman* Ochoukov[1], *Thomas* Pütterich[1], *Wouter* Tierens[1], *Wei* Zhang[1,6], the ASDEX Upgrade Team and the EUROfusion MST1 Team[**]

[1] Max-Planck-Institut für Plasmaphysik, Boltzmannstr. 2, 85748 Garching, Germany
[2] IRFM, CEA Cadarache, 13108 Saint Paul lez Durance, France
[3] IJL UMR 7198, U. de Lorraine P2M, Fac. Des Sciences, BP 70239, F-54506 Vandœuvre Cedex, France
[4] Politecnico di Torino, Torino, Italy
[5] Technische Universität München, Boltzmannstr. 15, 85748 Garching, Germany
[6] Applied Physics Department, University of Ghent, B-9000 Ghent, Belgium

**Abstract.** The response of the local RF current measured at limiters of 3-strap ICRF antenna to variations of power balance and phasing at $f_{ICRF}$=30MHz agrees qualitatively well with EM calculations by TOPICA and RAPLICASOL codes. Measurements of tungsten sputtering yield and DC current at the limiters correlate strongly with the local RF current. In contrast to findings for the 2-strap antennas, values of DC current are predominantly positive, and negative only for some locations and feeding parameters. Explanations can involve more physical mechanisms than only parallel sheath dynamics.

## Introduction

Tungsten (W) production due to Ion Cyclotron Range of Frequencies (ICRF) power was drastically reduced by installation and proper feeding of the so-called 3-strap antennas in ASDEX Upgrade (AUG) [1]. Control of the RF image current pattern at the antenna limiters, monitored by local RF probe measurements, was crucial to minimize the ICRF-specific W sources. The measurements were also found to be in a qualitative agreement with electromagnetic (EM) calculations of local near-field at the antenna limiters by TOPICA [2]. This confirmed strong evanescence of the slow wave in the typical AUG conditions of the far scrape-off layer plasma [3]. It also proved the relevance of the approach to optimize the antenna design by lowering local RF currents and RF electrical field at the antenna limiters [4]. Recent experiments in Alcator C-Mod [5] have confirmed that conditions with reduced RF image currents correspond to reduced impurity release and reduced RF-induced plasma potentials.

In this work, we extend the characterization of local quantities in the conditions close to the RF image current cancellation at the limiters of the 3-strap antennas in AUG [1,2], both experimentally (section 1) and theoretically (section 2).

## 1 Experimental characterization

In [2], experimental characteristics were studied for the case of $f_{ICRF}$=36.5MHz. Here we discuss the responses of measurements to variations of strap phasing and power balance of the 3-strap antenna for $f_{ICRF}$=30MHz. This frequency is more convenient for EM modelling, as explained in section 3.

The experimental conditions are similar to those described in [2]: ELM mitigated H-mode plasma at high density, with $P_{ICRF}$=1MW with H-minority heating and $P_{NBI}$=5MW, but at $B_t$=2.0T in discharge #33105. Figure 1 shows the 3-strap antenna 4 (*a4*) in sector 12 of AUG and the locations of the following measurements: RF current (expressed as 50 Ohm equivalent voltage $V_{RF}$), DC current $I_{DC}$ at limiter tiles numbered *01* to *12*, as well as spectroscopic measurements of W sputtering yield $Y_W$ close to limiter tiles *01* to *03* and *07* to *09*. Figures 2 and 3 presents $V_{RF}$ and $I_{DC}$ expressed as 2D functions of the strap power balance $P_{central}/P_{outer}$ (where $P_{central}$ is the power to central strap and $P_{outer}$ is the sum of the power to the outer straps) and of the deviation of phasing between the central strap and the in-phase outer straps from dipole $\Delta\Phi$, for left and right antenna sides, respectively. Figure 4 shows $Y_W$.

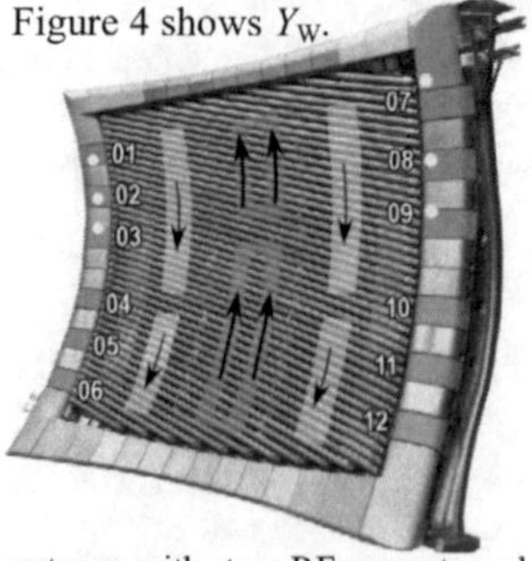

**Fig. 1.** 3-strap antenna with strap RF currents and measurement locations highlighted.

* Corresponding author: bobkov@ipp.mpg.de author
** See H. Meyer et. al. Nuclear Fusion FEC 2016 Special issue (2017)

DOI: 10.1051/epjconf/201715703005

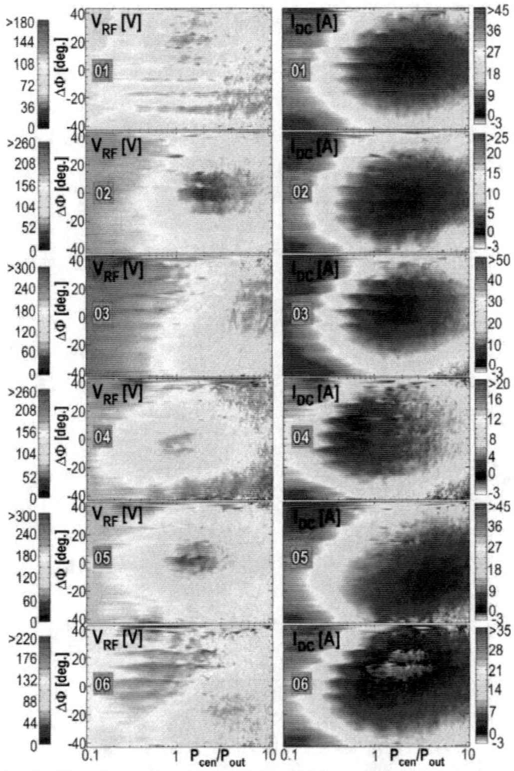


**Fig. 2.** $V_{RF}$, $I_{DC}$ as functions of $P_{central}/P_{outer}$ and $\Delta\Phi$ at various locations at the left antenna side (note different colour scales).

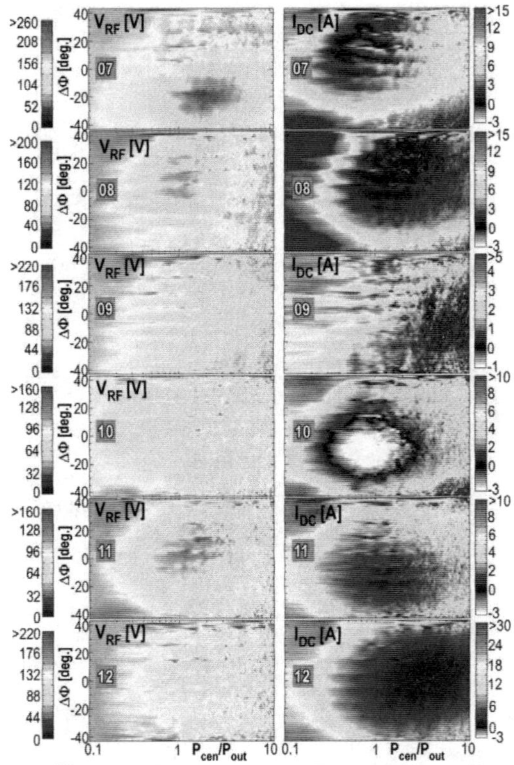


**Fig. 3.** $V_{RF}$, $I_{DC}$ as functions of $P_{central}/P_{outer}$ and $\Delta\Phi$ at various locations at the right antenna side (note different colour scales).

Although $V_{RF}$ measurements at $f_{ICRF}$=30MHz react less sensitively to changes of $P_{central}/P_{outer}$ and of $\Delta\Phi$ than those in the case of $f_{ICRF}$=36.5MHz in [2], the main characteristics of the measurements remain similar. Clear location-dependent areas with minima of $V_{RF}$ exist in the 2D strap power balance – phasing space. These areas indicate the best parameters for local RF image current cancellation. Horizontal stripes on $V_{RF}$ are visible in Fig.2 and Fig.3 and can be explained by evolving conditions during scans of $P_{central}/P_{outer}$ conducted at fixed $\Delta\Phi$, as well as by associated non-linearities. The 3D density profile in front of the antenna is likely evolving and affecting the overall balance of RF image currents at the antenna [2], as shown by calculations in [6]. Measurements of DC current $I_{DC}$ have not been reported before for the 3-strap antennas and require a particular attention.

Previously, the DC measurements were described for some of the locations of the 2-strap antennas [7] in standard dipole phasing. Values of $I_{DC}$ were predominantly negative on active antennas and positive on non-active antennas. However, these currents are predominantly positive for the active 3-strap antenna close to dipole phasing, as is shown in Fig. 2 and Fig. 3. Only several of the locations, such as *06, 07, 08* and *10* have large areas with negative $I_{DC}$. The areas of negative $I_{DC}$ are enclosed in the areas with $I_{DC}$=0, in many cases corresponding approximately to the minimum of $V_{RF}$. However, crossing the condition $I_{DC}$=0 could point to a change of dominant mechanism for collection of net DC current on the relatively large limiter tile, rather than to an optimized parameters range with a minimum of W sputtering. It is therefore questionable if the $I_{DC}$ signals can be used for optimization of the antenna feeding parameters, although these signals have the highest sensitivity to those.

The observations indicate that the DC circuit for the 3-strap antenna differs from that of the 2-strap antenna. Among the physical mechanisms deemed responsible for DC current collection, two are known to play a role: a) sheath rectification along the field lines, i.e. parallel sheath dynamics, modelled with such codes as SSWICH [8]; b) collection of ion current transverse to the magnetic field, which is observed in small RF plasma experiments with well-defined electrode geometry, such as Aline [9]. The second mechanism is not included in SSWICH and can be a key to explaining the positive $I_{DC}$. It is also enhanced by perpendicular RF electric field at the limiters. If small experiments [9] can prove the importance of the second mechanism, the transverse components of RF electric field should be considered in addition to the parallel component for sheath rectification and W sputtering. Luckily for the antenna design optimizations, the transverse field components in front of the limiters are also decreased when the RF currents at the limiters are decreased.

As for the tungsten sputtering characterized by $Y_W$ presented in Fig.4, it reacts to the feeding scans approximately in the same manner as $V_{RF}$ in the corresponding locations, i.e. showing strong correlation with $V_{RF}$ and $I_{DC}$. The overall picture is very similar to that presented in [2] where $f_{ICRF}$=36.5MHz.

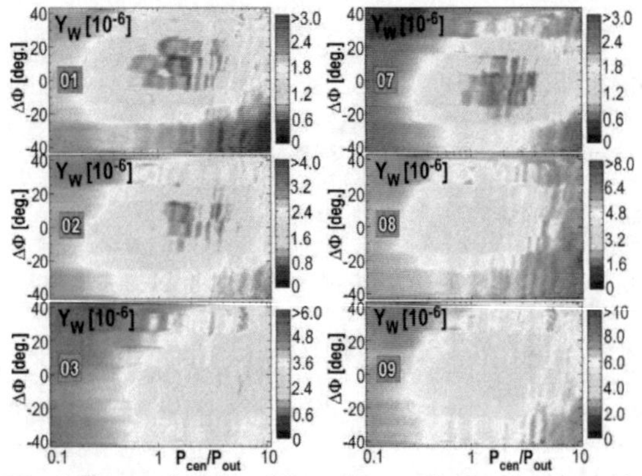


**Fig. 4.** $Y_W$, as a function of $P_{central}/P_{outer}$ and $\Delta\Phi$ at locations *01, 02, 03* at the left and *07, 08, 09* at the right antenna side (note different colour scales).

## 2 Theoretical characterization

The case with $f_{ICRF}$=30MHz is better suited for modelling with EM codes, because the frequency is significantly lower than the resonant frequency of the straps, in particular that of the outer ones. Depending on model and plasma conditions, the latter is in the range between 36.5 and 36.9MHz. Crossing this resonance frequency does not affect the near-field distribution significantly However working at $f_{ICRF}$=30 MHz helps to resolve issues in calculations with the RAPLICASOL [10] code used in this work.

We consider three cases modeled with EM codes, all using the same set of plasma profiles corresponding to the scenario of discharges #31515 and #33105. The first one uses a flat model of the 3-strap antenna in TOPICA as in [2]. The second one introduces a more realistic model of the antenna in curved geometry in TOPICA [11]. The third case applies the same flat geometry as in the first case, but in RAPLICASOL code.

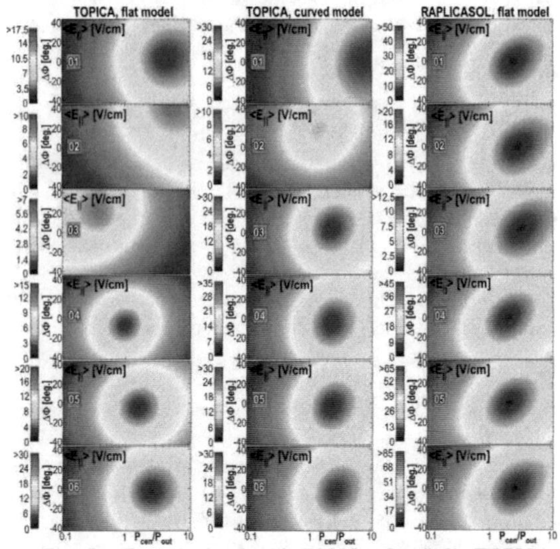


**Fig. 5.** $<E_{\|}>$ calculated with TOPICA flat (left), TOPICA curved (middle) and RAPLICASOL flat (right) models at the left antenna side locations (note different colour scales).

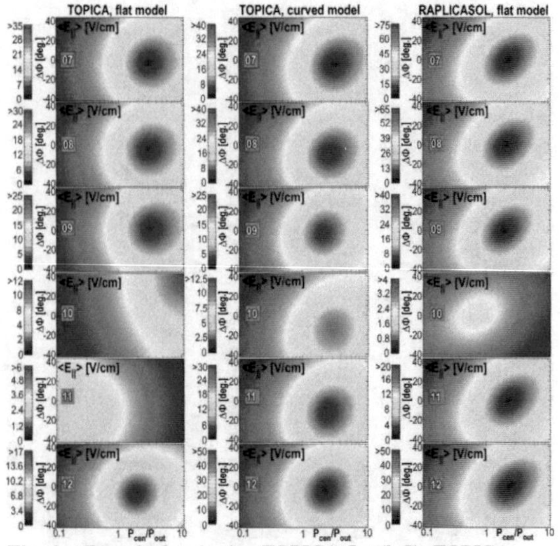


**Fig. 6.** $<E_{\|}>$ calculated using TOPICA flat (left), TOPICA curved (middle) and RAPLICASOL flat (right) models at the right antenna side locations (note different colour scales).

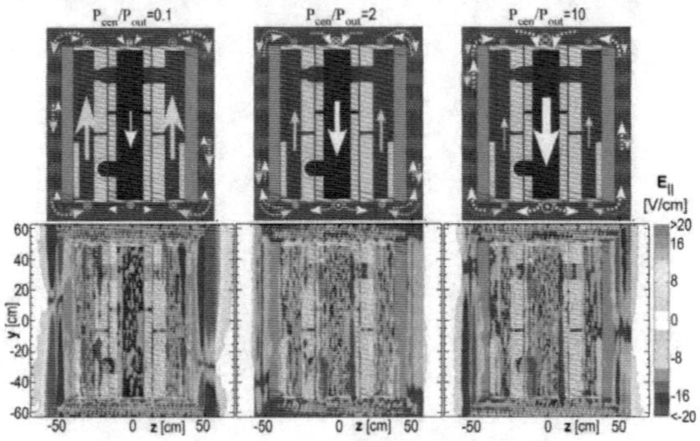


**Fig. 7.** $E_{\|}$ for flat TOPICA model for different cases of power balance in dipole for 30 MHz, with graphics showing RF current circulation at limiters (top). Purely vertical dashed arrows on the limiters indicate locations of sign reversal of parallel component of RF current.

As a theoretical "proxy", we use spatially averaged values of the $E_{\|}$ field $<E_{\|}>$ calculated just in front of the antenna limiters at locations *01-12*. Firstly, due to strong evanescence of the slow wave, $<E_{\|}>$ is the main contributor to the RF sheath driving voltage and rectified sheath voltage, as has been discussed in [2] and has been shown theoretically in [3,12]. Secondly, this quantity is primarily a linear function of local RF current (in particular, of its parallel component) at fixed geometrical and loading conditions. Thirdly, $<E_{\|}>$ is an engineering quantity which can easily be used for antenna optimization with practically any EM code. The ICRF power is scaled to 500 kW for all the simulations.

Figures 5 and 6 show the $<E_{\|}>$ calculations in the three different modelling cases (columns) for the left and the right antenna sides, respectively. Location-dependent areas with minima close to the dipole phasing and to $P_{central}/P_{outer}\approx 2$ are well reproduced in all three modelling cases compared to the experimental $V_{RF}$ diagrams, with details which differ. These minima correspond to the conditions of the local cancellation of all components of

RF current. Spatial $E_{\|}$ distribution is shown in Fig. 7 for $P_{central}/P_{outer}$ of 0.1, 2.0 and 10.0 in dipole phasing calculated for flat model in TOPICA, together with qualitative presentation of the RF image current circulation on the limiters.

According to Fig.5 and Fig.6, minima of $<E_{\|}>$ become broader somewhat above the midplane for the left side (locations *02*, *03*) and somewhat below the midplane on the right side (locations *10*, *11*). In experiment, broader regions with minima of $V_{RF}$ are also observed on the right side at about the midplane (locations *09*, *10* in Fig. 3). Close to the regions with broader minima, $<E_{\|}>$ are generally lower, according to the calculations. Here, parallel projection of the RF image current of individual straps, flowing on the limiters, changes sign (unbalanced cases in Fig.7), as well as the total RF current at the limiters is reduced. This results in minima at the $E_{\|}$ maps. However in the experiment, there is no clear tendency for significantly lower values in the locations corresponding to these.

The more detailed model with curved geometry in TOPICA produces coupling parameters which match experimental values significantly better [11]. Consequently, a more realistic relation between the power balance and the voltage balance inside the antenna (presented in [1]), is reproduced and the confidence in results using the curved antenna model is higher.

Although the general trends of the local quantities are well described by the simulations, and one can speak of at least a qualitative agreement; quantitative one-to-one correspondence between the experiments and the calculations is difficult due to existing limitations and uncertainties. As can be inferred from [6], details of 3D density profile have a significant effect on the RF image currents and $E_{\|}$ at the limiters. On top of that, the profiles are not stationary in experiments [2]. Even if a better qualitative correspondence were found by refining experimental input and by detailing the models further, the variety of the profiles in the experiment would make it very challenging to optimize antenna design further, in order to have minima at the same values of $P_{central}/P_{outer}$ and $\Delta\Phi$ at all limiter locations, and this for a broad variety of plasma conditions.

One of alternative approaches to optimize the 3-strap antenna further in experiments is increasing the number of actuators by making the antenna limiters active, fed independently by feedback-controlled RF sources. This could allow counteracting the residual RF currents at the limiters by destructive interference. However this approach needs testing on small-scale machines. Another actuator which already is available in the experiments and planned to be used in the future to tackle RF fields in the scrape-off layer is the phasing between antenna pairs.

## Summary

Experimental local values of RF current, DC current, W sputtering yield were presented for multiple locations at the 3-strap antenna as functions of variations of antenna strap power balance and strap phasing. The quantities experience a minimum close to dipole phasing when power of the central strap is close to the double of the sum of the power to the outer straps.

The DC current can however reverse sign and is predominantly positive in contrast to previous studies with 2-strap antennas [7]. This implies a complicated nature of the DC circuit, which can also involve a collection of ion current transverse to the magnetic field.

Comparison of measured RF current at the limiters with EM calculations using a flat antenna model in TOPICA and RAPLICASOL as well as a more realistic curved antenna model in TOPICA shows that qualitative features of the experimental observations are captured by the calculations. Better quantitative agreement between simulations and measurements is presenting a challenging task, given the uncertainties and variations in the experimental input [2,6]. An independent method to reduce the RF currents by actively feeding the antenna limiters and synchronizing operation of all four antennas in ASDEX Upgrade could be tested in the future.

We refer to [13] where the recent progress in solving the problems of the antenna-plasma interface in ASDEX Upgrade is summarized.

This work has been carried out within the framework of the EUROfusion Consortium and has received funding from the Euratom research and training programme 2014-2018 under grant agreement No 633053. The views and opinions expressed herein do not necessarily reflect those of the European Commission.

## References

1. V. Bobkov et al., Nucl. Fusion **56** 084001 (2016)
2. V. Bobkov et al., Plasma Phys. Control. Fusion **59** 014022 (2017)
3. L. Colas et al., Plasma Phys. Control. Fusion **59** 025014 (2017)
4. V. Bobkov et al., Nucl. Fusion **50** 035004 (2010)
5. S. Wukitch et al., "Towards ICRF Antennas Compatible with High Performance Plasmas: Characterization and Mitigation of ICRF Antenna – Plasma Edge Interaction", *this conference*, Inv-08
6. W. Tierens et al., "3-Dimensional density profiles in edge plasma simulations for ICRF heating", *this conference*, A-25
7. V. Bobkov et al., J. Nucl. Mater. **415** S1005 (2011)
8. J. Jacquot, Phys. Plasmas **21**(6) 061509 (2014)
9. E. Faudot et al., *this conference*, "DC currents collected by a RF biased electrode quasi-parallel to the magnetic field", A-33
10. J. Jacquot et al., *AIP Conf. Proc.* **1689** 050009 (2015)
11. D. Milanesio et al., "Analysis of the ASDEX Upgrade 3-strap antenna with TOPICA code: curved vs. flat 3D geometry", *this conference*, C-22.
12. W. Tierens et al., "Nonlinear plasma sheath potential in the ASDEX Upgrade 3-strap antenna: a parameter scan", *submitted to* Nucl. Fusion
13. W. Zhang, "Recent progress on improving ICRF coupling and reducing rf-specific impurities in ASDEX Upgrade", *this conference*, Inv-07

EPJ Web of Conferences **157**, 03006 (2017)
*22 Topical Conference on Radio-Frequency Power in Plasmas*
DOI: 10.1051/epjconf/201715703006

# Current Drive operation in a In-blanket ICRF Ring Array

G.Bosia[1*], R. Ragona[2]

[1]*Dept of Physics, University of Turin and INFN, Via P. Giuria 1 10125 Turin**
[2] *Ghent University, 9000 Ghent, Belgium*

**Abstract.** The currently predicted Current Drive (CD)efficiency in a tokamak reactor plasma is $\chi_{CD} \leq 50$ kA/MW for all auxiliary systems (NBCD, ICCD, ECCD). This would imply a relatively low reactor $Q = P_{electric} / P_{auxiliary}$ and an extremely high power density (>10 MW/m$^2$) at the plasma boundary, if injected through the vacuum vessel ports or by in-port launchers..
In a previous paper the concept of an Ion Cyclotron Ring Array (RA) of many elements, integrated in the reactor blanket first wall was proposed. An analysis of this RA operating in steady state Fast Wave Current Drive (FW CD) with all elements individually controlled in magnitude and phase is presented in this paper. For CD operation the RA is operated at low frequency (f ≈ 20MHz) so as to avoid the otherwise overwhelming RF power absorption by ions and α-particles. It is shown that the RA would feature a very directive radiation spectrum, a low power density operation (avoiding the drawbacks associated with high AC and DC potentials) and a potentially high flexibility of operation..

## 1. Introduction

Current drive (CD) generation in a tokamak is the result of a rather complex physics process requiring a toroidal momentum transfer (by ELD + TTMP) to the plasma electrons travelling at an average toroidal thermal speed resonant with the toroidal phase velocity of a FW injected at the plasma boundary. The competition of this process with the more efficient power absorption by ions or fusion-born α–particles significantly impairs the CD efficiency and needs to be avoided either by using specific frequency windows where ion absorption is minimized or, alternatively, by operation at a frequency sufficiently low to keep ion resonances outside the plasma cross section, as proposed here.

Steady state operation is regarded to be essential for a future commercial reactor, but the predicted CD efficiency of the "injected" auxiliary systems (NBCD and ECCD) is rather low ($\chi_{CD} \leq 0.05$ *A/W*) even in a high $T_e$ reactor [1], requiring, among other problems, an extremely high power density at the plasma edge and the use of large apertures in the vacuum vessel and blanket for direct plasma access. An ICRF launcher, necessarily located inside the vessel because of the required proximity to the plasma edge imposed by the wave evanescence, would suffer a similar power density problem, if restricted in a port. If instead a wider surface of the blanket first wall could be used, it could operate at a significantly lower AC and DC electric fields and therefore avoid the drawbacks associated with the FW evanescence (voltage breakdowns and impurity influxes due to rectified sheath potentials).

The FW CD efficiency computed for a launching structure similar to the (four straps) ITER launcher installed in DEMO (commercial fusion reactor conceptual study [2]) by using the Ehst & Karney approximate formula [3] is predicted at values close to those of the other auxiliary systems [4, 5]. It is found that the CD efficiency depends on the so called E&K parameter $w = v_{ph}/v_e$, increasing for $w > 1$ and sharply decreasing for $w < 1$ as in this range the RF power is preferentially coupled to off-axis, magnetically trapped electrons, not contributing to CD.

On the other hand, the Single Pass Absorbed (SPA) power by electrons is expected to vary as $p_e \approx w\, e^{-w^2/2}$ [6], i.e. to rapidly decrease for w > 1. All studies devoted to DEMO steady state operation use the ITER in-port array (defined by its toroidal mode number $n_\varphi \approx 45$) as reference. It is shown in [7] that the competition with ions and fusion born α-particles absorption, would restrict FWCD to relatively narrow high frequency windows (e.g. at f ≈ 51 MHz). The best predicted CD efficiency for an equatorial FW launch is $\eta_{CD} = 0.03$ *A/W*, whereas in [8] it is shown that for operation at low frequency (f < 20 MHz in DEMO, avoiding ion absorption competition), the CD efficiency of an ITER-like antenna would reduce to ≈ 0.012 A/W, due to a high off-axis magnetic trapping (against a value of ≈ 0.086.1 A/W in absence of trapping).

In [9] an all-metal Ion Cyclotron Ring Array (RA), in principle compatible with reactor operation, is proposed for heating and CD use in DEMO.

In this paper the CD performance of a toroidally resonant RA launcher radiating from the low field side are examined in a DEMO-like reactor and approximate relations useful as a guide of a realistic RA design are deduced.

## 2. Ring Array operation

The RA is a toroidal periodic structure of $N_a$ poloidal strip line elements with poloidal length $l_y \ll \lambda_y$, ($\lambda_y$ designed to be integrated in the first wall of the blanket of a tokamak fusion reactor. The strip lines currents $I_0$ (or the input voltages $V_0$) are individually controlled at an equal magnitude and toroidal phase advances $\theta_{\varphi_r}$ *(r = 1,..N$_a$)* by balanced feedback-controlled current sources (Fig.1).

The RA is designed to be integrated in the first wall of the blanket of a tokamak fusion reactor. All elements are mechanically supported at mid length to the back plate or, if present, to a slotted Faraday shield (both at ground potential). and fed at both ends The array of short strip line elements $(l_y \ll \lambda_y)$ is designed to operate at a power level of several tens of MW and features a small radial depth (15-20 cm), for minimum impact with the blanket functions. As radiation losses to

* Corresponding author: gbosia@to.infn.it

the plasma are periodically restored by the current sources, the RA may be regarded as an overall lossless toroidal periodic structure. According to the Floquet theorem, the field components of the structure can be expanded in spatial harmonics in the form:

$$\psi(R,\varphi) = \sum_{n=-\infty}^{\infty} \psi_n(R) e^{-i(\beta+n\,N_a))\varphi} \qquad (1)$$

where $\beta = \omega / v_{ph_\phi}$ is the propagation constant and $v_{ph_\phi}$ is the toroidal phase velocity. By setting the toroidal boundary conditions it can be shown that the array toroidal mode number $n_\varphi$, the toroidal phase velocity and the operating frequency are related to the number of elements $N_a$ and to phase advance $\theta_\varphi$, as listed in Table 1

Table 1.

| | |
|---|---|
| Toroida.mode number | $n_\varphi = N_a \frac{-\theta_\varphi}{2\pi}$ |
| Electrons therm. velocity | $v_{th.e} = c\sqrt{T_e/m_e c^2}$ |
| Toroidal. phase velocity | $v_{ph.\varphi} = \frac{\omega\, R_a}{n_\varphi}$ |
| E &K parameter | $w = \frac{R_a \omega}{n_\varphi c} \frac{-\theta_\varphi}{\pi} \sqrt{m_e c^2/T_e}$ |
| frequency | $f = w n_\varphi \sqrt{T_e/m_e c^2}\,(c/2\pi R_a)$ |

Due to the limitations in space of this paper the reader is referred to [11] or to the accompanying poster for more details on this subject.

## 3. Ring Array Power Spectrum

The current density per unit length (p.u.l.) $J_y(\varphi)$ (A/m) of a ring structure with elements located at angles $\varphi_n$ can be written as:

$$J_y(\varphi) = J_o \sum_{n=1}^{N_a} e^{i\varphi_n}\, \Theta_n(\varphi) \quad \text{(A/m)} \qquad (2)$$

with a rectangular function for the $\varphi$ dependance

$$\Theta_n(\varphi) = e^{-i(n-1)\theta_\varphi} \begin{cases} 1, if\ (n-1)p_\varphi < \varphi < (n-1)\ p_\varphi + w_\varphi < 0 \\ 0, otherwise \end{cases} \qquad (3)$$

where $p_\varphi = \frac{2\pi}{N_a}$ is the array period, $w_\varphi = \xi p_{\varphi.}$ is the stripline angular width such that $I_0 = \int_{-w_\varphi/2}^{w_\varphi/2} J_{0.y}(\varphi) d\varphi$ $= J_0\, l_y w_\varphi$. If a phase advance $\varphi_n = n\theta_\varphi$ is assumed

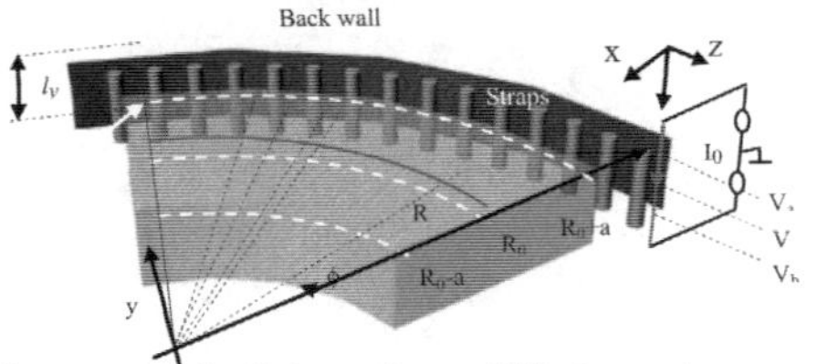


Fig 1. Geometry of a RA section of 15 elements

the Fourier transform of $J_y(\varphi)$ is:

$$J_y(k_\varphi, p_\varphi, \theta_\varphi) = \frac{J_0}{\sqrt{2\pi}} w_\varphi e^{-i\frac{k_\varphi w_\varphi}{2}} \frac{\sin\left(\frac{k_\varphi w_\varphi}{2}\right)}{\frac{k_\varphi w_\varphi}{2}} S(k_\varphi) \quad (4)$$

with $k_\varphi = \sum_{n=0}^{N_a-1} e^{-in(k_\varphi p_\varphi + \theta_\varphi)}$. The active power $P_a(k_\varphi, p_\varphi) \propto \int_{-\infty}^{+\infty} |J_y(k_\varphi,)|^2\, dk_\varphi$ coupled to the plasma is proportional to $|S(k_\varphi)|^2$ which depends on $n_\varphi$ and $\theta_\varphi$. For $N_a$>> 1 the power spectrum consists of sharply peaked distributions, with the main peak of emission at $k_{\varphi.m} = \frac{-\theta_\varphi}{p_\varphi}$ and width $\Delta k_\varphi \approx -\theta_\varphi$. In the asymptotic case $N_a \to \infty$, $|J_y(k_\varphi, \pi/2)| \to \infty$ and $\Delta k_\varphi \to 0$. Therefore, the spectral power distribution tends to $\delta(k_\varphi - n_{\varphi.m})$. Plots of $|J_y(k_\varphi, \Delta\theta)|^2$ for a RA of radius $R_a$ = 11.32 m with phase advance $\theta_\varphi = \pi/2$, f = 20 MHz and $n_\varphi =$ = 5, 10, 15, 20, 45 are shown in Fig. 2 as function of $k_\varphi$.

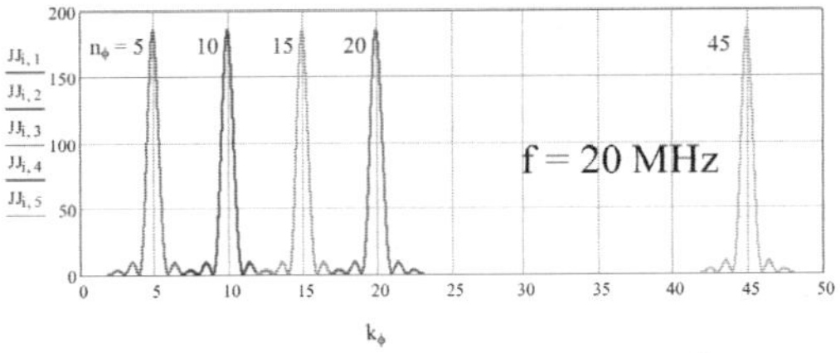


Fig 2. Power spectra vs. increasing values of $n_\varphi$

## 4. Plasma loading

The amount of power tunnelling through the evanescence layer is estimated by a 2D model of antenna coupling theory in quasi static approximation ($2\pi/k_0$ >> a, b, d, $w_z$) as function of $k_z = k_\varphi/R_a$, (the same symbols as in the original paper are used). Invariance with y ($\frac{\partial}{\partial y} = 0$) and a rectangular current density profile vs. $\varphi$ is assumed for the array elements. Therefore, all electrical quantities are defined per unit length (p.u.l). The effects of feeders and inter element coupling (which in turn depend on plasma coupling) are ignored. This implies an array geometry of loosely coupled elements. For $l_y \ll \lambda_y$, the strip lines operate close to anti-resonance. Therefore, the effects of the distributed capacitances are small and will be neglected. For the same reason, the assumption of a constant inductive strip line reactance p.u.l. is assumed. Using the definitions adopted in [10], the complex plasma loading impedance p.u.l of a single element is

$Z_y = Z_A / N_a = R_y + iX_y$, with:

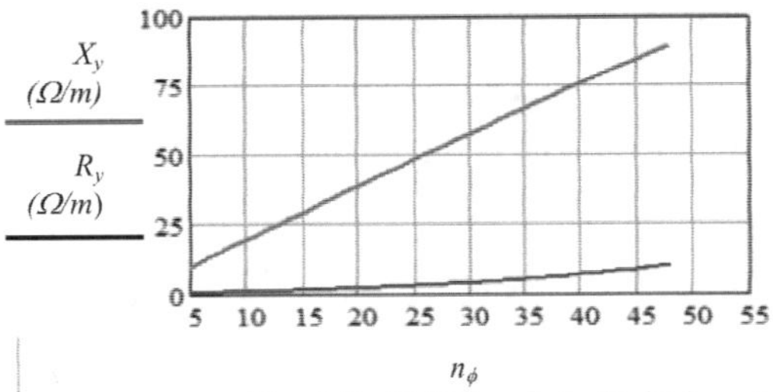


Fig. 3 $Re(Z_A/N_A)$ and $Im(Z_A/N_A)$ vs. $n_\varphi$

$$R_y = \frac{\omega\mu_0}{2\pi N_a}\int_{-\infty}^{+\infty} g(k_z)\left(\frac{|J_{a.y}(k_x)|}{I_{a.0}}\right)^2 dk_z \text{ and} \quad (5)$$

$$X_y = \frac{\omega\mu_0}{2\pi N_a}\int_{-\infty}^{+\infty} h(k_\varphi)\left(\frac{|J_{a.y}(k_z)|}{I_{a.0}}\right)^2 dk_z\ (\Omega/\text{m}).$$

$R_y$ and $X_y$ are plotted in Fig. 3 as function of (integer values) of $n_\varphi$.

## 5. FW Current Drive

As noted above, we shall assume an operation at a sufficiently low frequency $f_e$ (in DEMO $f_e \approx 25$ MHz) so as to exclude from the plasma cross section all ion and fusion-born α–particle resonances and to let electron damping emerge. The active power of a FW launched at $\theta_\varphi = \pi/2$ and $f < f_e$ is then entirely dissipated on electrons by non-collisional damping and can generate an electron convection current density due to the asymmetry of the power spectrum. The amount of power transferred depends on the E&K parameter $w(\varrho, n_\varphi,)$ which, in turn, depends on the electron temperature profile $T_e$. We shall use DEMO - like electron $T_e$ and $n_e$ profiles as [8]:

$$T_e(\rho) = (T_{e.0} - T_{e.1})(1-\rho^2)^{\alpha_T} \text{ and}$$
$$n_e(\rho) = (n_{e.0} - n_{e.1})(1-\rho^2)^{\alpha_n}$$

$\rho = r/a$, is the normalized radial coordinate such as $R(\rho) = R_0 + a\rho$ . The E&K dimensionless formula for radio frequency (RF) CD efficiency $\tilde{\eta}(\epsilon, \vartheta, w)$ in presence of magnetic trapping [3] is a function of i) the inverse aspect ratio of the flux surfaces $(\epsilon)$, ii) the poloidal location where the RF power is dissipated $(\vartheta)$ and iii) the ratio of toroidal phase velocity and electrons thermal speed $w = \frac{v_{ph.\varphi}}{v_{th.e}}$, called in the following E&K parameter.

From a power flow point of view a RA can be regarded as round conductor of radius $R_a$ of complex impedance p.u.l. $Z_A$ carrying a surface current $I_A$ . The Poynting vector $\boldsymbol{P} = (\boldsymbol{E}\times\boldsymbol{H}) \approx -E_y H_\varphi$ is almost everywhere radial and directed toward the torus axis. The power density p.u.l. through the cylindrical surface at $R \approx R_0 + a_0$ and absorbed by plasma according to the Pointing theorem is:

$$\int_V P_e dV = \int_V Re(\boldsymbol{P})dV = Re(Z_A)I_A^2 \approx \int_V f_P\, dV \quad (6)$$

with a radial distribution depending on E&K parameter

$$w(\rho, T_e) \text{ as } f_P(\rho, T_e) \approx w(\rho, T_e)e^{-\frac{w(\rho, T_e)}{2}} \quad (7)$$

which is maximum for $w = 1$ and rapidly decreases for $w > 1$. As function of $\rho = \frac{R-R_0}{a}$ and $n_\varphi$

$$w(\rho, n_\varphi) = \frac{v_{ph.\varphi}}{v_{th.e}} = \frac{1}{n_\varphi}\frac{2\pi R_a f}{c}\sqrt{\frac{m_e c^2}{T_e(\rho)}} \quad . \quad (8)$$

$f_P(\rho, T_e)$, is plotted in Fig 4 vs. $\rho$ and $n_\varphi$. As seen, it tends to increase with $n_\varphi$ because of the increasing loading R (and therefore of the power available to the electrons) . However, the power is mainly deposited in the plasma core at low $n_\varphi$ whereas it disperses to the plasma edges at high $n_\varphi$.

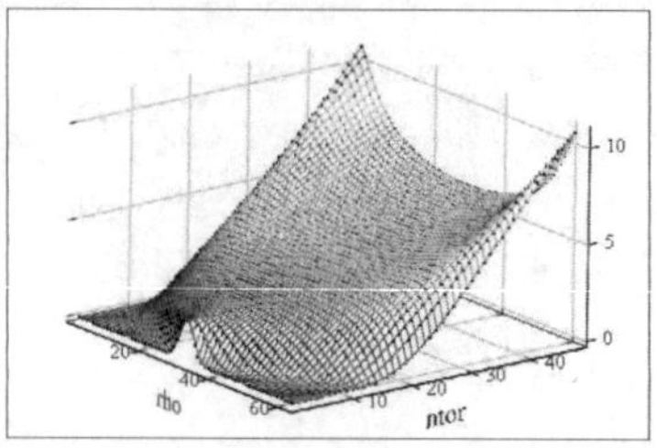


Fig 4 $f_P(\rho, n_\omega)$

Only the fraction $f_{CD}$ contributes to CD because of toroidal trapping, a condition only depenent on the ratio $v_{e\parallel}/v_{e\perp}$. The circulating electrons are those within the loss cone $\theta_S = arcsin\sqrt{\frac{1}{R_S(\rho)}}$. where the mirror ratio $R_S(\rho) = \frac{B_{\max}(\rho)}{B_{\min}(\rho)} \approx \frac{R_0+r}{R_0-r)}$ and, for a poloidal angle $(\vartheta = 0)$ launch, $r = a_0\rho$. The power fraction $f_{CD}$ dissipated on circulating electrons (and obtained by setting $P_A$ to zero for $\rho > R_S(\rho)$) is plotted in Fig. 5.

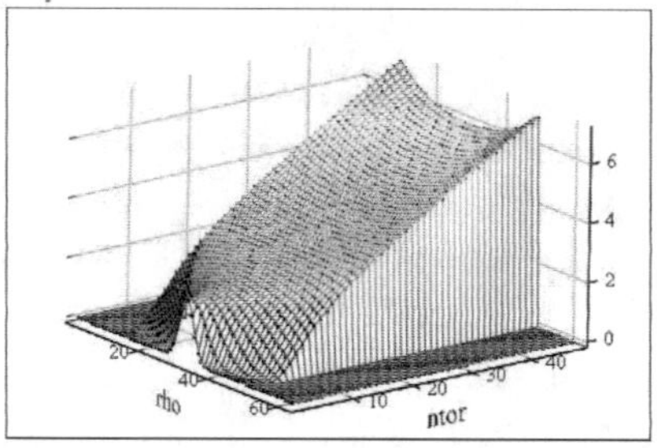


Fig. 5 $f_{CD}(\rho, n_\varphi)$

Fig 5 shows that at low frequency magnetic trapping significantly reduces (by ≈ 50%) the power available for CD at high $n_\varphi$ values, in qualitative agreement with [7]. The power density p.u.l. $P_e$ available for CD can be estimated from (6) as:

$$2\pi\int_{-\rho}^{\rho} P_{CD}(\rho, n_\varphi) f_{CD}\, \rho' d\rho' = Re(Z_A(n_\varphi)\, I_A^2 \quad (9)$$

and the CD driven in the plasma p.u.l can be estimated from [5] as:

$$J_{CD}(\rho, n_\varphi) = \frac{0.38}{ln\lambda}\frac{T_e(keV)}{n_{e.20}} P_{CD}(\rho, n_\varphi)\tilde{\eta}(\rho, n_\varphi)$$

where $\tilde{\eta}(\rho, n_\varphi)$ is the dimensionless E&K CD efficiency computed for a LF equatorial launch $(\vartheta = 0)$ which can be approximately fitted, by :

$$\tilde{\eta}(\rho, n_\varphi) = 17.22\, w(\rho, n_\varphi)^{0.22} \text{-} 13.36 \quad (10)$$

plotted in Fig. 6 for $0.3 < w < 3$, and $Z_{eff} = 1$. $J_{CD}(\rho, n_\varphi)$ (A/m$^3$) is plotted in Fig 7 for an on-axis $T_e(0) = 39.5\ keV$, a phase advance $\theta_\varphi = \frac{\pi}{2}$, $f \approx 20$ MHz and $I_A = 1$A.

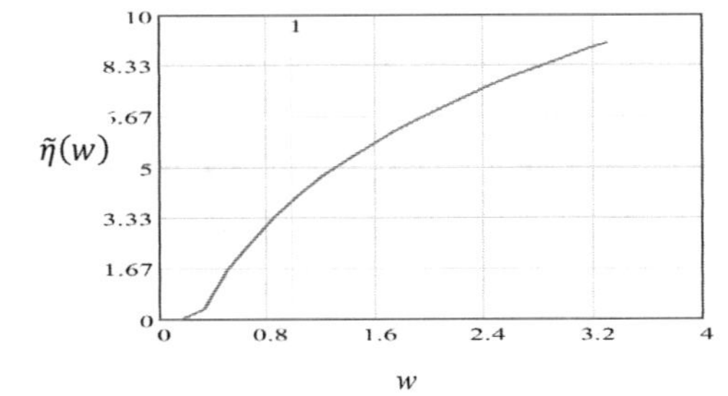


Fig. 6 $\tilde{\eta}(w, \vartheta = 0\ Z_{eff} = 1)$

The best on-axis CD efficiency is obtained for $w(0, n_\varphi) = 1$, or for an optimum $n_\varphi$:

$$\widehat{n_\varphi} = \frac{2\pi R_a f}{c}\sqrt{\frac{m_e c^2}{T_e(R)}} \approx 10 \qquad (11)$$

-

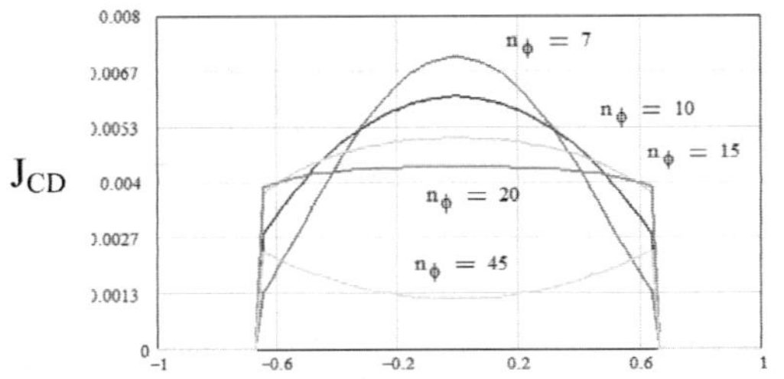


Fig 7 $J_{CD}(\rho, n_\varphi)$ (A/m$^3$)

i.e. for a number of elements $\widehat{N_a} = 4\widehat{n_\varphi} = 40$. For on-axis $T_e(0) = 39.5\ keV$, $n_{e.20} = 1$ a phase advance $\theta_\varphi = \frac{\pi}{2}$ and $f \approx 20$ MHz an integrated CD efficiency $\eta_{CD} \approx 130$ kA/MW results (Fig. 8).

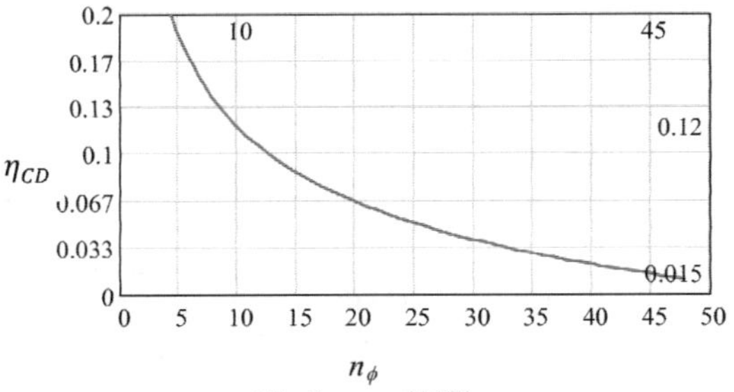


*Fig 8. $\eta_{CD}$ (A/W)*

## 6. Conclusions

In this paper, the CD operation of a Ring Array FW launcher introduced in [9] is dicussed using a simple model. The RA appears to have a number of advantages compared to existing and proposed in-port ICRF launchers operating at high $n_\phi$:

- Low frequency operation (f < 25 MHz in DEMO) avoids competition with much more efficient ion and fusion born α-particle absorption .
- High E-fields usually resulting from the FW evanescence and causing undesiderably high AC and DC potential at the plasma edge can be reduced to a safer level.
- The proposed individual and independent real time control of phase and magnitude of the RA currents (associated with an automatic matching by an overall frequency control) allows real time regulations of the power spectrum that could be used to improve stability of the drive in variable plasma conditions.
- The best CD performances of the RA low frequency operation are obtained at values of $n_\varphi(\approx 10)$. For an array phase advance $\theta_\varphi = \frac{\pi}{2}$, (maximum momentum per power expenditure) the array number of elements is $N_a \approx 40$ .
- A full metal construction of the in-vessel components and an acceptable integration in a modular blanket appears possible.
- Although not discussed here, bulk plasma heating and ion minority CD could also possible, the latter used for reactor particles exaust . H and CD mode(s) of operation could be used in conjunction in bringing the reactor to full operation.
- The present analysis shows a CD efficiency higher than that predicted by present estimates for a ITER-like types of launchers. It also shows that the results depend considerably on plasma parameters at present not easily predictable in a commercial reactor. Critical plasma parameters are low electron damping (not necessarily described by single pass absorption) and the control of the electron temperature profile, affected by the power dissipated at the edge by trapped electrons.
- In view of the significant difference between the RA and ITER like arrays so far investigated, an analysis including more refined plasma descriptions is planned

### References


[1] H. Zohm et al, Proc. EPS Conf. on Plasma Phys. ECA37 DO3, 108 (2013)

[2] G.Federici et al, Fus. Eng. and Design **89** (2014) 882–889

[3] D.Ehst, C. Karney (1991) Nucl. Fusion **31** 1933

[4] R. Bilato et.al, J.of Phys.Conf. Series **561**, (2014).

[5] E. Lerche, et al. AIP Conf. Proc. 1580, 338 (2013).

[6] M.Porkolab et al, Plasma Phys. Cont. Fus. **40** A35 (1998)

[7] Ye.Kazakov et al, Plasma Phys and Cont.Fus. **57**,(2015), 0.25014.

[8 M.Brambilla et al, Nuclear Fusion (2015) **55**(2) :023016

[9] G.Bosia, Fus.Eng. and Design.92(2015) 8-15

[10]A.M.Messiaen et al. Proc. 3rd Joint V-G Int. Symp. on Fus.Tech.**1**,675 (1984)

[11] G.Bosia "Heating and Current Drive in a Ring Array", Paper to be submitted for publication

 DOI: 10.1051/epjconf/201715703007

# Assessment of quasi-linear effect of RF power spectrum for enabling lower hybrid current drive in reactor plasmas

*Roberto* Cesario[1], *Alessandro* Cardinali[1], *Carmine* Castaldo[1], *Luca* Amicucci[2], *Silvio* Ceccuzzi[1], *Alessandro* Galli[2], *Francesco* Napoli[1], *Luigi* Panaccione[1], *Franco* Santini[1], *Giuseppe* Schettini[3], *Angelo Antonio* Tuccillo[1]

[1]ENEA Fusion and Nuclear Safety Department, C. R. Frascati, Via E. Fermi 45, 00044 Frascati (Rome) Italy
[2]Università Roma La Sapienza, Dipartimento di Ingegneria, Rome Italy
[3]Università Roma Tre, Dipartimento di Ingegneria, Rome, Italy

**Abstract.** The main research on the energy from thermonuclear fusion uses deuterium plasmas magnetically trapped in toroidal devices. To suppress the turbulent eddies that impair thermal insulation and pressure tight of the plasma, current drive (CD) is necessary, but tools envisaged so far are unable accomplishing this task while efficiently and flexibly matching the natural current profiles self-generated at large radii of the plasma column [1–5]. The lower hybrid current drive (LHCD) [6] can satisfy this important need of a reactor [1], but the LHCD system has been unexpectedly mothballed on JET. The problematic extrapolation of the LHCD tool at reactor graded high values of, respectively, density and temperatures of plasma has been now solved. The high density problem is solved by the FTU (Frascati Tokamak Upgrade) method [7], and solution of the high temperature one is presented here. Model results based on quasi-linear (QL) theory evidence the capability, w.r.t linear theory, of suitable operating parameters of reducing the wave damping in hot reactor plasmas. Namely, using higher RF power densities [8], or a narrower antenna power spectrum in refractive index [9,10], the obstacle for LHCD represented by too high temperature of reactor plasmas should be overcome. The former method cannot be used for routinely, safe antenna operations, Thus, only the latter key is really exploitable in a reactor. The proposed solutions are ultimately necessary for viability of an economic reactor.

## 1 Introduction

The main research on fusion energy utilises toroidal plasmas magnetically trapped in toroidal machines (*tokamaks*). As major scientific problem for an economic reactor, we must understand how to lead strongly heated plasma to sustain a high fusion gain, while a large fraction of current is self-produced via pressure gradient naturally present especially at large radii of the plasma column.

Strong powers of heating and current drive (CD) injected into the plasma ensure that the current induced by the transformer continues to flow, maintaining high fusion gain required for energy production [9]. The huge costs of the heating and CD systems should be reduced by exploiting, via particle transport (*bootstrap*) effect, a strong fraction of current that the plasma steadily self-produces, in the presence of pressure gradient across the plasma column mostly occurring at large radii [1]. Consequently, the pressure profile should be mandatorily controlled (for at least several hours for enabling energy production) by properly shaping the plasma current density radial profile, $j(r)$. This allows indeed suppressing turbulence that impairs thermal insulation and pressure tight of the plasma column, which in turn enables sustaining the plasma current.

Current drive (CD) is necessary for turbulence suppression [2–4], but tools envisaged so far are unable accomplishing this task, while efficiently and flexibly matching the natural self-generated current profiles [1]. The lower hybrid current drive (LHCD) [5] can satisfy this need, but, although considered long time ago as a major breakthrough for a reactor, this tool has been mothballed on JET and not considered so far a priority for ITER. This should represent a serious mistake because LHCD offers the peculiar feature of matching the self-produced current profiles by the plasma [1].

LHCD is based on the use of strong radio-frequency (RF) power coupled to lower hybrid (LH) plasma waves. They are efficiently damped via Landau wave-particle resonance (in the phase and drifting velocities, respectively) with a tail of the electron distribution function (EDF) [11]:

$$u_{\Phi//} \equiv \frac{\omega_0}{k_{//}} = \frac{c}{n_{//}} \approx 3u_{the} \qquad (1)$$

where: $u_{\Phi}$ and $k_{//}$ are, respectively, the components of wave phase velocity and wavevector parallel to $\mathbf{B}_0$, $c$ is light speed, $n_{//} \equiv \dfrac{ck_{//}}{\omega_0}$ is the parallel refractive index

* Corresponding author: roberto.cesario@enea.it

component, and $u_{the} = \sqrt{\frac{2k_B T_e}{m_e}}$ is the electron thermal velocity, where $k_B$ is the Boltzmann constant, $T_e$ is the electron temperature, and $m_e$ is the electron mass.

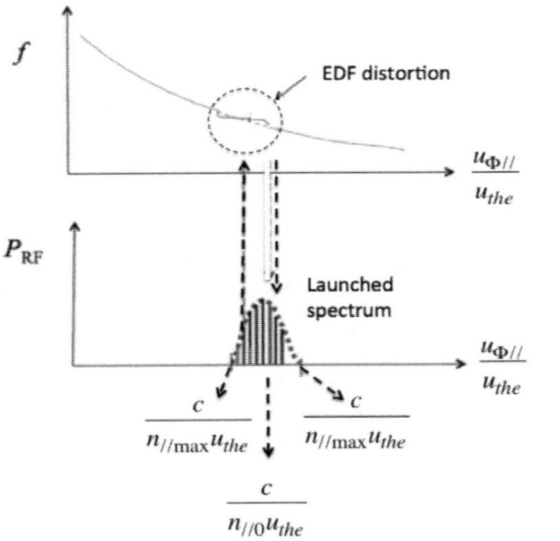


**Fig. 1.**

Top: electron distribution function (EDF). Bottom: Gaussian antenna spectrum of multi-megawatt radio-frequency power (in arbitrary units), coupled to plasma at several gigahertz, useful for producing the LHCD effect.

Accordingly to Eq. 1, plasma electrons Landau-resonate more intensely with spectral components with slower phase velocity. LH waves are electron plasma modes particularly prone to drive current, i.e, to transfer energy and momentum to plasma electrons along the direction of the confinement magnetic field, where they are free to drift. Consequent to slow-wave behaviour of LH modes (which have indeed refractive index component: $n_{//}>1$), electrons flowing with speed close to the phase velocity of the wave *see* its electric field as almost constant. The particles can thus *surf* on the wave, which results, via collisions, in an irreversible transfer of wave energy and momentum to the electrons. This produces the LHCD effect.

Consider, in the bottom box of the figure, the $n_{//}$ power spectrum of LH waves launched by the antenna, centred at $n_{//0}$, with spectral components in the interval ($n_{//max}$-$n_{//min}$). Spreading radially towards the warmer regions of the plasma, the slowest components of the spectrum (with $u_{*//}\sim u_{*//min}$ i.e., $n_{//}\sim n_{//max}$) are beginning to first to interact more intensely with the plasma, in accordance with Eq. 1. Consequent to accelerated population, some EDF flattening occurs as shown in the top box of the figure (and demonstrated by analytical model of Sec. 2).

This distortion determines the modality of the wave packet absorption and evidences the role of the spectral width[7].

Figure 1 schematizes the LHCD concept [5], proven by experiment on PLT (Princeton Large Torus) [12], and successfully extrapolated on FTU (Frascati Tokamak Upgrade) to reactor-graded, high plasma densities [7,13]. The latter method was established following the guidelines of previous theoretical prediction of reduced parasitic wave interaction at the plasma edge under higher temperature of the latter [14,15].

This can be accomplished by proper numerical tool able to display aspects of LHCD mechanism remained hidden so far, namely, in regard to the role of the width of the launched antenna power spectrum in the waveleght domain in determining the $j_{LH}$ profile in reactor relevant condition of hot plasma that deterines full RF power absorption at single half of radial pass. Too hot plasma envisaged even at large radius of pedestal represents a major problem for LHCD exploitation in a reactor [16].

Here we present model results based on quasi-linear theory, evidencing the role, w.r.t linear theory, of reducing the wave damping via high RF power densities, as previously reported [6]. The latter option cannot be exploited for removing the obstacle represented by too high temperatures because too high RF power density should be required, which exceeds the power handling capability of antenna [17]. A useful way should therefore be considered, as recently proposed [7,8]. To solve the remaining major conceptual problem of LHCD, consisting in how to enable this effect in fusion relevant condition of high plasma temperature is ultimately necessary for viability of an economic reactor.

## 2 Results

### *2.1 Analytical model*

We summarise here the wave damping analysis in 1-D quasilinear approximation. Consider the Gaussian spectrum with $n_{//0}$ and $\Delta n_{//}$ parameters ($\Delta n_{//}$ is kept at half of the peak value). Following QL theory [18], competing effects are produced: from one side, by the LH wave–particle interactions, Eq. 1, and related diffusion in the velocity space; from the other side, by collisions that tend restoring Maxwellian EDF. We analyse hereafter the condition for the EDF plateau formation, and show the link of wave packet with spectrum, schematized in Fig. 1. Assume slab plasma geometry where density and temperature gradients are directed along the $x$-direction defined above. In steady-state conditions, the spectral components $P(n_{//})$ of the LH wave power are absorbed accordingly with the equations [19]:

$$\frac{dP}{dx} = \Gamma_{QL} P \tag{2}$$

where $\Gamma_{QL}$ is the quasilinear damping rate inferred from the 1-D distribution function of plasma electrons, and depending on the local RF power spectral density.

Following Landau [20] and considering the steady-state solution of the Fokker-Planck equation [21,22] in the limit of LH waves, $|k_\perp| >> |k_{//}|$, the power damping can be approximated as [9]:

$$\Gamma_{QL} \approx \frac{\sqrt{\pi}}{k_\perp u_{the}^2} \omega_{pe}^2 \frac{u_0}{1 + u_0^3 D_{QL}(u_0)} e^{-u_1^2} \tag{3}$$

 DOI: 10.1051/epjconf/201715703007

where: $u_0 \equiv \frac{c}{n_{//0}u_{the}}$, $u_1 \approx \frac{c}{(n_{//0} + 2\Delta n_{//})u_{the}}$ and $D_{QL}(u_0)$ is given by the expression [9]:

$$D_{QL}(u_0) \approx \frac{9}{(Z_i + 2)\ln\Lambda^{ee}} \frac{\sqrt{Te[keV]}}{\left(n_e\left[10^{20}m^{-3}\right]\right)^{\frac{3}{2}}} \frac{n_{//0}^2 - 1}{\Delta n_{//}} p_{RF}\left[\frac{MW}{m^2}\right] \quad (4)$$

In Eq. 4, $Z_i$ is the effective ion charge, $\ln\Lambda^{ee}$ is the Coulombian logarithm that considers the electron-electron collision effect, and $p_{RF}$ is the total power density of the spectrum.

Eq. 4 represents the QL diffusion coefficient normalized to the collision operator, both calculated at the centre of the EDF plateau (i.e., for $u_{//}=u_{0//}$). The level of the plateau is fixed by the exponential of Eq. 3 via the value: $n_{//\max} \approx \left(n_{0//} + 2\Delta n_{//}\right)$. This is the highest value of the launched spectrum (see Fig. 1) where $D(u)$ falls below one from values typically higher than 1000 in the plateau.

Neglecting QL diffusion, i.e. assuming an EDF of Maxwellian shape, the absorption coefficient in the linear wave limit is:

$$\Gamma_L \approx \frac{\sqrt{\pi}}{k_\perp u_{the}^2} \omega_{pe}^2 u_0 e^{-u_0^3} \quad (5)$$

Hence, the ratio of the QL and linear damping coefficients is:

$$\frac{\Gamma_{QL}}{\Gamma_L} \approx \frac{e^{\left(\frac{m_e c^2}{k_B T_e}\right)\left(\frac{2\Delta n_{//}}{n_{//}^3}\right)}}{1 + u_0^3 D_{QL}(u_0)} < 1 \quad (6)$$

The inequality $\Gamma < \Gamma_M$ holds for:

$$\Delta n_{//} < \frac{k_B T_e}{m_e c^2} n_{//0}^3 \ln\left(1 + u_0^3 D_{QL}(u_0)\right) \quad (7)$$

Condition of Eq.7 is widely verified for practical cases, i.e.: $\Delta n_{//} \lesssim 0.5$, $T_e \sim 10$ keV, $n_e \sim 10^{20}$ m$^{-3}$, $p_{RF} \sim 30$ MW/m$^2$, $n_{//0} \sim 1.8$, and $Z_i \gtrsim 1$.

For given plasma parameters, Eq. 6 shows conditions that QL physics effect is useful for reducing the RF power damping w.r.t. case of linear wave approximation. In the same framework of QL limit, Eq.4 and Eq.5 indicate that, for a certain $T_e$, the wave damping should be reduced by means of higher values of $p_{RF}$, as already reported [13]. This option is however prevented by the need of operating in safe operation compatible with the routinary power handling capability of antenna [14].

Importantly, Eq. 6 shows that a smaller $\Delta n_{//}$ value pays a major role in reducing the RF power damping , owing to the exponential dependence in Eq. 5.

Numerical approach allows retaining effects of RF diffusion in 2-D dimensions in $u_{//}$ and $u_\perp$ of electron velocity, useful for taking into account important details of the quasilinear diffusion coefficient, neglected in too rough approximation of Eq. 4, but necessary for assessing the LHCD profiles in practical cases.

### *2.2 Summary of the numerical model*

The LH$^{star}$ package of numerical codes [13,7,8] has been revealed fully suitable for evidencing the QL physics shown in Sec. 2.1, namely, the effect of reduction of the RF power damping w.r.t. case of linear wave approximation, under higher values of higher $p_{RF}$ or smaller $\Delta n_{//}$, for given reactor plasma parameters [7,8].

Differently from other numerical approaches [19,23], the LH$^{star}$ tool, in order to assess the $j_{LH}$ profiles, treats the intrinsically LHCD single pass regime of reactor in simpler way because not affected by complicated architectures seeking to consider multi-pass regime, conceived with the aim of considering the case of LHCD experiments in experiments performed so far in less warm plasmas.

Consequenlty, by using antenna spectra sufficiently narrow and however compatible with constraints imposed by QL theory (trapping effect) it is possible enabling the LHCD effect and shaping the profile, spanning over the whole outer radial half of plasma of DEMO [7,8].

More specifically, assuming linear wave theory, Eq. 5, the main $j_{LH}$ peak dislocates markedly more externally (at $r/a \sim 0.9$) than in the quasi-linear case of narrow spectrum ($r/a \sim 0.5$ for $\Delta n_{//} \approx 0.083$), taken as reference. In order to assess the QL effect produced by the RF power density, the $j_{LH}$ profile has been recalculated assuming a lower launched $p_{RF}$ ($\approx 3$ MW/m$^2$ in place of 30 MW/m$^2$). Consequently, the main peak of $j_{LH}$ dislocates only slightly more externally (from $r/a \sim 0.5$ to $r/a \sim 0.6$). Therefore, the effect of a narrow spectrum plays a major role than a larger $p_{RF}$ in reducing the wave damping, i.e., the former favours more strongly the penetration of the wave packet into hot plasmas, consistently with analytical modelling of Sec. 2.1.

## 3 Discussion

Ray-tracing analysis consists in solving the equation system for the position and wave vector allowing the reconstruction of the wave-phase and the RF power damping rate along the trajectory. The LH$^{star}$ tool utilises ray-tracing approach that however respects the geometric limit approximation. Importantly, the latter holds for LH waves propagating far from the cut-off layers located at the plasma edge [10]. Consequently, owing to failure of that limit in the case of weak damping per single pass occurring in most tokamak experiments performed so far, the effect of toroidicity in up-shifting the $n_{//}$ antenna spectrum is only guessed by this approach, as well originally done for interpreting the outcomes of LHCD experiments. In important experiment of JET of reactor-graded strong absorption per single pass [24,14,15], the geometric limit approximation holds and the LH$^{star}$ tool module is able to assess the $j_{LH}$ profiles without need of any guess. In the LH$^{star}$ tool at each radial step, the quasi-linear diffusion coefficient is evaluated considering the

2-D Fokker-Planck relativistic equation solved for the electron distribution function in velocity space, and the quasilinear damping is consistently taken into account in the LH wave power equation. On this basis, the RF power density and LH-wave-driven current density radial profiles are calculated with useful robustness for reactor relevant modelling.

Assuming linear wave theory, Eq. 5, numerical results show that the main $j_{LH}$ peak dislocates markedly more externally (at $r/a\sim0.9$) than in the QL case of narrow spectrum (at $r/a\sim0.5$ for $\Delta n_{//}\cong0.083$), taken as reference. In order to assess the QL effect produced by the RF power density, the $j_{LH}$ profile has been recalculated assuming a lower launched $p_{RF}$ ($\approx3$ MW/m$^2$ in place of 30 MW/m$^2$). Consequently, the main peak of $j_{LH}$ dislocates only slightly more externally (from $r/a\sim0.5$ to $r/a\sim0.6$).

Therefore, the effect of a narrow spectrum plays a major an really exploitable role than a larger $p_{RF}$ in reducing the wave damping. The former favours more strongly than the latter the penetration of the wave packet into hot plasmas, consistently with analytical modelling of Sec. 2.1, without jeopardising routinary, safe antenna operations. Details on the method and engineering feasibility are discussed in Ref. 9.

## 4 Conclusive remarks

The natural current profiles produced by tokamak plasma are usefully matched by lower hybrid current drive. This was successfully extrapolated to reactor graded high plasma densities [12–15] and, now, also with larger flexibility, to high temperatures envisaged for a thermonuclear reactor [7,8], thus enabling his economic viability.

## References

1. J. Freidberg, J., *Plasma Physics and Fusion Energy*, (Cambridge Univ. Press, 2007).
2. G. L. Jackson, et al., Regime of very high confinement in the boronized DIII-D tokamak.
3. G. D. Conway, et al., Suppression of Plasma Turbulence During Optimized Shear Configurations in JET, *Phys. Rev. Letters* **84**, 1463 (2000).
4. A. Fasoli, Computational challenges in magnetic-confinement fusion physics, *Nature Physics*, nphys3744 (2016). G. Giruzzi, et al., Modelling of pulsed and steady-state DEMO scenarios. *Nuclear Fusion*, **55**, Number 7, 073002 (2015).
5. Giruzzi, G., et al., Modelling of pulsed and steady-state DEMO scenarios. *Nuclear Fusion*, **55**, Number 7, 073002 (2015).
6. N. Fisch, Theory of current drive in plasmas, *Rev. Modern Physics* **59**, 175-234 (1987).
7. R. Cesario, et al., Current drive at density required by thermonuclear reactors, *Nature Communications*, 1 (5) 55 (2010).
8. C. F. F. Karney and N. J. Fisch, Numerical studies of current generation by radio-frequency traveling waves, *Phys. Fluids* **22**, No. 9, 1817 (1979).
9. A. Cardinali et al., to be pub. *Scientific Reports*
10. A. Cardinali et al., to be pub. on *Plasma Phys Control. Fusion*
11. M. Brambilla, *Kinetic Theory of Plasma Waves* (Oxford Press, 1999).
12. S. Bernabei, et al., Lower-Hybrid Current Drive in the PLT Tokamak, *Phys. Rev. Letters* **49**, 1255-1258 (1982).
13. R. Cesario, et al., Spectral broadening of parametric instability in lower hybrid current drive at a high density. *Nucl. Fusion* **54**, 043002 (2014).
14. R. Cesario, et al., Modelling of Lower Hybrid Current Drive in Tokamak Plasmas by including the Spectral Broadening induced by Parametric Instability, *Phys. Rev. Letters* 92 17 175002 (2004).
15. R. Cesario, et al., Spectral broadening of lower hybrid waves produced by parametric instability in current drive experiments on tokamak plasmas, *Nucl. Fusion* **46**, 462-476 (2006).
16. H. Zohm et al., Assessment of the heating and current drive system capabilities for DEMO, *Proc. of the 40$^{th}$ European Phys. Society Conf. on Plasma Physics,* (2013).
17. C. Gormezano, et al., Lower-hybrid plasma heating via a new launcher – the multijunction grill, *Nucl. Fusion* **25,** 419 (1985).
18. Diamond, P.H., Itoh, S. and Itoh, K., *Modern Plasma Physics* (Vol.1, Cambridge University Press, 2010).
19. Bonoli, P., and Englade, R.C., "Simulation Model for Lower Hybrid Current Drive", *Phys. Fluids* **29**, 2937 (1986).
20. Landau, L., On the vibration of the electronic plasma, *Journal of Physics,* Vol X, No. 1, 25 (1946).
21. Diamond, P.H., Itoh, S. and Itoh, K., *Modern Plasma Physics* (Vol.1, Cambridge University Press, 2010).
22. Drummond W. E. and Pines D., "Non linear stability of plasma oscillations", *Nuclear Fusion Suppl.,* Part 3 1049 (1962).
23. Peysson, Y., Advances in modeling of lower hybrid current drive, *Plasma Physics and Controlled Fusion* **58** , 044008 (2016).
24. Crisanti, F., et al., JET Quasistationary Internal-Transport-Barrier Operation with Active Control of the Pressure Profile, *Phys. Rev. Lett.,* **88**, 145004 (2002).

DOI: 10.1051/epjconf/201715703008

# Experimental study of high power mm-waves scattering by plasma turbulence in TCV plasmas

*O.* Chellaï[1,*], *S.* Alberti[1], *I.* Furno[1], *T.* Goodman[1], *A.* Koehn[2], *L.* Figini[3], *D.* Ricci[3], *L.* Hizanidis[4], *P.* Papagiannis[4], *C.* Tsironis[4], and *the TCV* team[1]

[1] *Swiss Plasma Center, Ecole Polytechnique Fédérale de Lausanne, Switzerland*
[2] *Max Planck Institute for Plasma Physics, Garching, Germany*
[3] *Instituto di Fisica del Plasma, CNR, Milan, Italy*
[4] *National Center for Scientific Research Demokritos, National Technical University of Athens, Greece*

**Abstract.** Understanding the propagation of high power mm-wave in plasmas is of tremendous importance in the route to fusion considering their extensive use in magnetically confined fusion devices. Mm-beams, launched from the outside of the vessel must propagate through plasma edge-turbulence before reaching their target region. Until recently, the effect of edge-turbulence on the beam propagation was neglected, but it has been estimated for ITER that it could lead to significant differences in the time-averaged and instantaneous beam profiles, leading to a loss of efficiency in their use. In this paper, we present first direct experimental measurements of high power beam after propagation in simple magnetized toroidal plasmas in TCV.

## 1 Introduction

High power mm-waves are extensively used in tokamaks by locally driving plasma current to stabilize neoclassical tearing modes (NTMs) [1]. A narrow beam is launched from the outside of the vessel, targeting the core plasma. Along its propagation, the mm-beam traverses the highly turbulent Scrape-off Layer (SOL) region. The SOL is localized at the edge of the tokamak plasma and is characterized by large electron density fluctuations (up to 100% of the background density) associated with field-aligned filaments, commonly referred to as blobs [2]. The inhomogeneity of the electron density is responsible for scattering of the mm-wave. Unlike in present devices, in future large tokamaks, such as ITER, the path length of the beam will be long enough to enable small perturbations in the SOL to cause significant changes in both the time averaged and instantaneous beam profiles. This may lead to its broadening on average [3] and consequently to a loss of efficiency in NTM stabilization [4]. Only recently has the scattering process been investigated for beam propagation estimations in future larger devices like ITER. Analytical model for mm-wave scattering by blobs are developed [5] and state-of-the-art codes are upgraded to account for the effect of edge turbulence [3] [6]. Yet, there is still a need for experimental data to validate the simulation tools and reach predictive capabilities for ITER and beyond.

In this paper, we present first direct poloidal measurements of high power ($\approx$ 100kW) third electron cyclotron harmonic beam in the Tokamak à Configuration Variable (TCV) after propagation in a turbulent plasma in a simple magnetized torus (SMT) configuration (see section 2.1). Electron density measurements in the plasma edge are presented together with measurements of the transmitted mm-power.

## 2 Experimental setup

In this section, we present the experimental setup (Fig. 1). First, the TCV experiment is described together with plasma conditions. Then a detailed presentation of the mm-beam diagnostic is given. The main experimental challenges are mentioned at the end of this section.

| | |
|---|---|
| Major radius R | 0.88m |
| Minor radius a | 0.25m |
| Toroidal field $B_T$ | 1.3 - 1.43 T |
| Electron density $n_e$ | 0.7 - 4x$10^{18}m^{-3}$ |
| Second harmonic ECRH (X2) | 82.7 GHz / 3.6mm |
| Third harmonic ECRH (X3) | 117.8 GHz / 2.5mm |
| Total RF power | 550 kW |

**Table 1.** TCV parameters in typical simple magnetized toroidal discharges

### 2.1 Tokamak à Configuration Variable (TCV)

The TCV tokamak (see Table 1.) is characterized by its high shaping capabilities enabled by a plasma control system made of 16 independent shaping coils. TCV possesses a unique 4.5MW electron cyclotron resonance heating (ECRH) system distributed over two frequencies

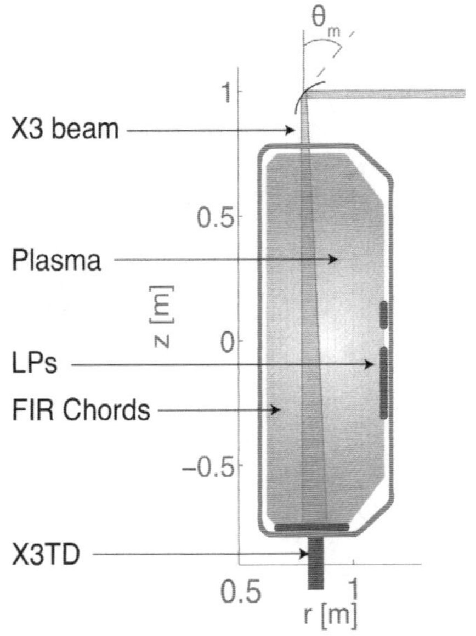


**Figure 1.** Experimental setup: a mm-beam is injected from the top of the vessel through a steerable mirror and traverses the plasma. The power reaching the floor of the vessel is detected by the X3 transmission diagnostic (X3TD). Line-integrated densities are measured along the vertical chords of the Far Infrared Interferometer (FIR). Plasma edge density fluctuations are measured by wall mounted Langmuir probes.

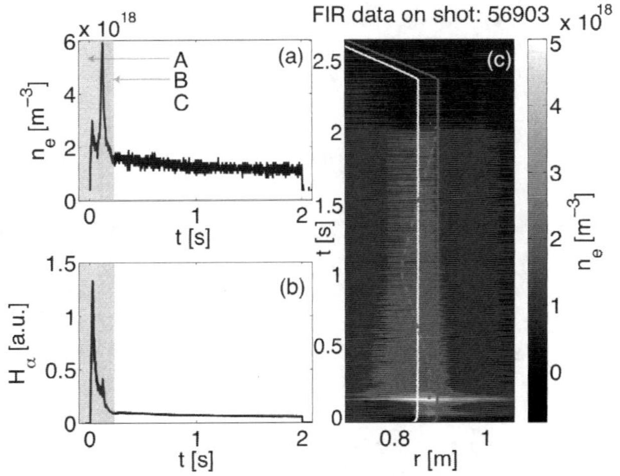


**Figure 2.** A typical SMT discharge in TCV: $D_2$ gas is ionized by the injection of mm-power at the second electron cyclotron harmonic in X-mode (phase A, (a) and (b)). During this phase, the plasma is localized around the beam. The second phase (B) is a transition phase towards a steady-state, EC-sustained, SMT (phase C). (c) : Time evolution of the line integrated electron density measured by the FIR. The X2 cold resonance is represented in white and the X3 cold resonance in magenta. The red dotted line is a visual aid : it represents the radial position of the X3 beam at the floor of the machine, with no refraction in the case of a double sweep of the launching mirror poloidal angle $\theta_m$

(second and third harmonic X-mode, respectively X2 at 82.7GHz and X3 at 117.8 GHz) [7]. The path length of the X3 beam to the X3 transmission diagnostic (see section 2.3 for details) in TCV is about 1.5m, which is comparable to the path length of the edge-to-resonance path in ITER. The versatility of TCV enabled the development of an SMT plasma scenario [8].

## 2.2 Simple magnetized toroidal plasmas in TCV

In the SMT configuration [2], a vertical magnetic field $B_v$ is imposed on a toroidal magnetic field field $B_T$ resulting in helical open field lines with $\nabla B$ and curvature, terminating on the upper and lower wall of the TCV vessel. The plasma is generated and sustained by the injection of mm-power at the second harmonic of the electron cyclotron resonance frequency (X2). In this configuration, plasma turbulence is dominated by interchange instabilities, characterized by field-aligned elongated structures - referred to as blobs - with $k_{\parallel} << k_{\perp}$ ($k_{\parallel}$ and $k_{\perp}$ being the parallel and the perpendicular wave number respectively) [2]. Blobs are formed in the crest of an interchange wave, and their internal electric field gives rise to an $\mathbf{E} \times \mathbf{B}$ radial motion. Fluctuations of electron density $n_e$ can reach up to 100% of the time-averaged density.
Line-integrated density is measured inside the vessel using the Far Infrared Interferometer [9]. Biased ($-100V$ bias) wall-mounted Langmuir Probes (LP) measure the ion saturation current, which is proportional to the electron density, at the edge of the vessel [10].
Breakdown in SMT configuration is achieved in TCV by the injection of 450kW X2 mm-power in the vessel filled with $D_2$ gas, Fig. 1. The first breakdown phase (phase A) lasts about 0.1s, see Fig. 2. This phase corresponds to the ionisation of the plasma. Images from a fast framing camera shows that during this phase, the plasma is locally formed around the X2 beam. In a second phase (phase B) (typically from 0.1 to 0.2s), the plasma is spread throughout the vessel and reaches a stationary phase (phase C). During this phase, electron densities are typically between 0.7 and $4 \times 10^{18} m^{-3}$ and electron temperatures between 30 and 50eV. The duration of the plasma was only limited by the gyrotron pulse length, and could in principle be extended to 4s using back-to-back pulses, as previously done for fully Electron Cyclotron Current Driven (ECCD) plasmas [7].

## 2.3 X3 Transmission Diagnostic (X3TD)

An X3 beam is launched from the top of the vessel using a steerable mirror. Sweeping the mirror poloidal angle $\theta_m$ in vacuum enables the reconstruction of the beam poloidal profile [11] . The power reaching the floor of the vessel is coupled to a corrugated $HE_{11}$ graphite waveguide after $-55$dB attenuation. The attenuation is achieved by an hexagonal array of 241 circular holes with diameter $d = 1.2$mm such that the cut-off frequency is $f_c = 146.4GHz$. The attenuation waveguide is placed

 DOI: 10.1051/epjconf/201715703008

inside the vacuum vessel to protect the sapphire window that provides the vacuum feedthrough. The sapphire window is matched for a minimum of reflection at $118GHz$. The transmission of the attenuation cap is independent of its orientation around the waveguide, therefore no precaution is needed during the installation. An additional $20dB$ attenuation is introduced by an absorbing foam placed within the $HE_{11}$ waveguide that is attached to the sapphire window outside of the vacuum vessel. The $HE_{11}$ mode is then coupled to a $TE_{10}$ mode in a $WR6.5$ rectangular waveguide through a special Gaussian Optical Lens-like Antenna (GOLA). The mm-power is measured by a Schottky diode and digitally acquired at 250kHz acquisition frequency.

### 2.4 Experimental challenges

The SMT experimental campain on TCV was experimentally challenging. We found that the plasma parameters are sensitive to wall conditions. Therefore operation conditions are to be predetermined for each experimental session.
Another challenge is due to the nature of the experiment itself. High power mm-power is launched in the vacuum vessel with no first-pass absorption from the plasma. Direct reflections at the floor of the vessel must be avoided. They can destabilize the operation of the gyrotron by causing arcs or mode losses. TCV is also equipped with an extensive stray radiation protection system. The signals of the stray detectors are combined with a comparison between the expected and actual electron density to determine if stray radiations are potentially hazardous to other components and if so stop the discharge. Good knowledge of beam propagation inside the vessel is therefore required for successful transmission.

## 3 Results

In this section, we present the first measurements of the beam profile through plasma density fluctuations. Previous investigations of X3 beam profiles and absorption in TCV in vacuum are presented in [11] and show that the beam profile is Gaussian as expected in the design.

### 3.1 First measurements of beam profiles through plasma

X3 power $P_{X3}$ is injected (Fig. 3) after the plasma breakdown (phase A and B, Fig. 2) through the SMT plasma with no sign of absorption. A gyrotron delivers constant power for 1.7s Fig. 3. It takes about 0.5s to 0.6s for the gyrotron frequency to stabilize Fig. 3 a). as indicated by significant fluctuations in the forward power measurement in the transmission line (due to mode beating). The nature of the absorption cap of the X3TD makes it frequency sensitive. Analysis of the mm-transmitted power can thus not be performed on the first part of the discharge in Fig. 3. Unless very short pulses are used, this imposes a minimum duration for plasma discharges.

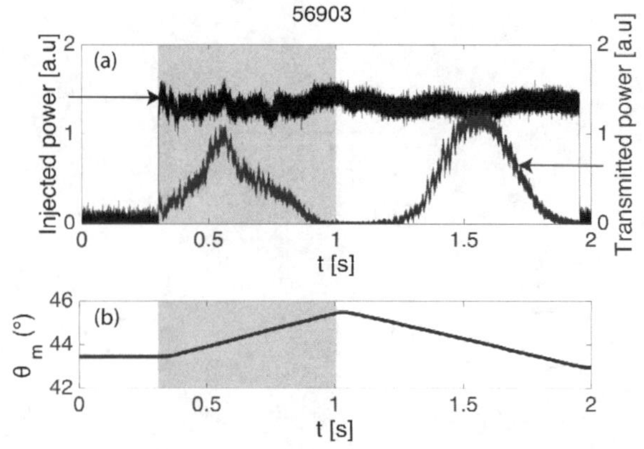


**Figure 3.** A double sweep of the X3 beam across the transmission diagnostic (a) is performed by sweeping the poloidal angle $\theta_m$ of the launching mirror (b). (a) shows an asymetry between the received powers. The stabilization time of the gyrotron frequency is identified in gray.

Fig. 4 shows the beam profile for two different plasma electron densities $n_e = 2.3\text{x}10^{18}\text{m}^{-3}$ and $n_e = 3.4\text{x}10^{18}\text{m}^{-3}$. Strong fluctuations of the transmitted power (up to 70% of the peak power) are caused by the fluctuations $\delta n_e(\mathbf{x}, t)$ of the electron density . The correlation between $\delta n_e(\mathbf{x}, t)$ and fluctuations of the transmitted power $\delta P_{T,X3}(\mathbf{x}, t)$ are investigated in the next section. Influence of the electron density on the beam propagation is under investigation. Refraction effects tied to the electron density are visible in Fig. 4 : as the launcher and receiver location are fixed in space a higher electron density requires a larger $\theta_m$ to compensate for refraction. Work is under way to estimate the contribution of the refractive effect to the measured beam profile, as well.

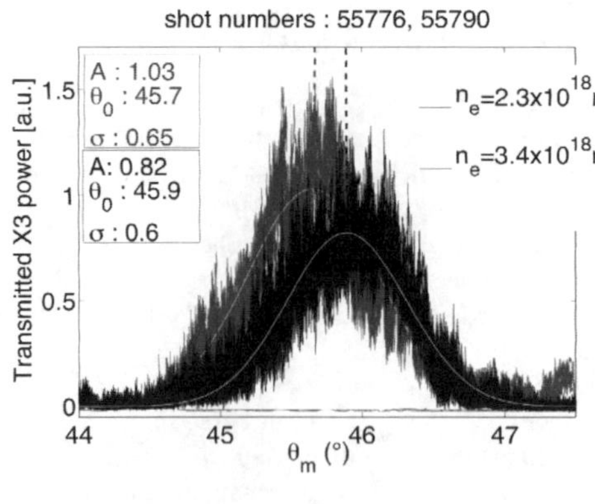


**Figure 4.** Transmitted mm-power for two different plasma electron densities. The large fluctuations of the signal are dominated by electron density fluctuations. The peak is displaced by refraction. Top-left parameters are the fitting parameters of the mm-signal to a Gaussian of the form: $A_0 e^{-(\frac{\theta_m-\theta_0}{\sigma})^2}$

### 3.2 Statistical analysis of transmitted power and electron density fluctuations

Analysis of the transmitter mm-power signal has shown a 10% correlation with the electron density fluctuations measured by Langmuir probes located at the TCV floor. Results are shown in Fig. 5.

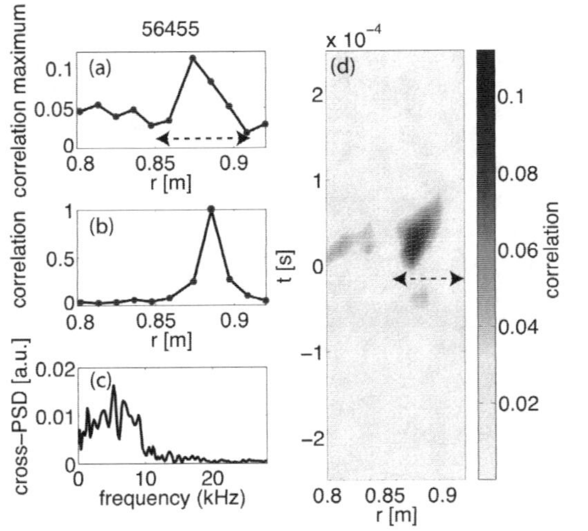


**Figure 5.** (a) and (d): correlation between the mm-power transmitted signal and the floor Langmuir probes. The black-dotted arrow shows the extent of the $HE_{11}$ waveguide of the X3TD. (b): Correlation between a LP facing the $HE_{11}$ waveguide and the neigbouring probes. (c) Cross power spectral density between a LP facing the $HE_{11}$ waveguide and the transmitted mm-power, most of the information between the two signals is exchanged between 0 and 10kHz. (d) The time scale is derived from m in equation (1).

To avoid confusion between several definitions of the correlation function found in literature, we define the one used in this paper. The computation of the cross-correlation is the following :
Let $x$ and $y$ be two discrete signals of length $N$, $f$ and $g$ are the corresponding standardized functions ($f_i = \frac{x_i - <x>}{\sigma_x}$ and $g_i = \frac{y_i - <y>}{\sigma_y}$). The cross-correlation function $\hat{R}_{xy}$ is given by:

$$\hat{R}_{xy}(m) = \begin{cases} \frac{1}{N} \sum_{n=0}^{N-m-1} f_{n+m} g_n, & m \geq 0 \\ \hat{R}_{xy}(-m), & m < 0 \end{cases} \quad (1)$$

where $m = 0, 1, ..., 2N-1$. The correlation function has a maximum value if 1 if $x = y$ and is zero if $x$ and $y$ are uncorrelated. For a signal sampled at the frequency $f_s$, we can convert $m$ in time units as follow: $t(m) = \frac{1}{f_s}(N - m)$ Fig. 5 shows that the LP signals exhibiting the best correlation with the transmitted mm-power signal are the LP probes which are located at the same radial position as the X3TD $HE_{11}$ waveguide. The information between them is mostly shared at frequencies ranging between 0 and $10KHz$ (Fig. 5 (c)). In addition, the spatial scale of the turbulent edge structures is obtained by computing the correlation between one LP located at the floor of the vessel and the neighboring probes. The correlation indicates the presence of turbulent structures with a spatial scale length typically in the order of 2cm as shown in Fig. 5b).

## 4 Summary and outlook

The effect of plasma edge turbulence on the propagation of a high-power mm-beam is investigated in TCV. A plasma scenario in the simple magnetized toroidal configuration is developed in TCV. Steady-state plasmas are obtained and sustained by the injection of mm-power at the second electron cyclotron frequency. Several electron densities are achieved. A high-power diagnostic is installed in TCV to measure the X3 beam profile after passing through the plasma turbulence. The mm-power fluctuations are shown to be correlated with edge measured electron density fluctuations. Methods to vary the fluctuation level along the beam trajectory are presently investigated; e.g. by moving the source region relative to the X3TD by changing the magnetic field. A numerical effort is also under way to compare experimental measurements with WKBeam [3], GRAY [12] and IPF-FDMC [6].

## Acknowledgements

*One of the authors, O. Chellaï, would like to gratefully thank Basil P. Duval, Antoine Merle and Miguel Silva for the interesting discussions and advices during the experimental sessions. This work has been carried out within the framework of the EUROfusion Consortium and has received funding from the Euratom research and training programme 2014-2018 under grant agreement No 633053. The views and opinions expressed herein do not necessarily reflect those of the European Commission.*

## References

[1] O. Sauter, et al., Plasma Physics and Controlled Fusion **52**, 025002 (2010)
[2] I. Furno, et al., Physical Review letters **100**, 055004 (2008)
[3] H. Weber, et al., in *EPJ Web of Conferences* (EDP Sciences, 2015), Vol. 87, p. 01002
[4] C. Tsironis, et al., Physics of Plasmas **16**, 112510 (2009)
[5] A. Rham, et al., Physics of Plasmas **20**, 056110 (2013)
[6] A. Köhn, et al., Plasma Physics and Controlled Fusion **58**, 105008 (2016)
[7] S. Coda, et al., Nuclear Fusion **51**, 094017 (2011)
[8] A. Fasoli, et al., Physics of Plasmas **13**, 055902 (2006)
[9] S. Barry, et al., in *Proceedings of the 8th international symposium on Laser-aided Plasma Diagnostics (LAPD-8)* (1997)

[10] L. Tonks, et al., Physical Review **34**, 876 (1929)
[11] T. Goodman, et al., in *EPJ Web of Conferences* (FEC, IAEA, 2016)
[12] D. Farina, et al., Fusion Science and Technology **52**, 154 (2007)

DOI: 10.1051/epjconf/201715703009

# Maintenance and preparation of the 3.7 GHz LHCD system for WEST operation

*Léna Delpech*[a,]*, *Joelle Achard*[a], *Arthur Armitano*[a], *Paul Bienvenu*[a] *and Marc Prou*[a]
[a] CEA, IRFM, F-13108 Saint Paul-lez-Durance, France.

**Abstract :** The generator for the 3.7 GHz Lower Hybrid Current Drive (LHCD) system on Tore Supra consists of 16 klystrons capable of delivering 600kW/1000s each on plasma. Such a powerful installation needs to be operated on a regular basis in order to preserve the specifications of the klystrons in terms of output power and pulse duration. This has been of particular importance during the long shutdown between the last Tore Supra campaign in 2011 [1] and the start of LHCD experiments on WEST in 2017. The TH2103C klystrons have been operated on matched load once a year during the shutdown. A reduction of 14% of the available RF power for the experimental program is found, which is partly due to the loss of one klystron. Another important aspect of the maintenance procedure is to maintain the knowledge of the operating team at a good level. This paper describes the procedure and tests performed during six years of shut-down. It also summarizes the technical problems encountered and the consequences on the test schedule, and highlights the importance of maintaining such large plants in operating condition during shutdowns.

*Keywords*: klystron, Lower Hybrid, RF heating system, generator

## 1. Introduction

The complete refurbishment of the Tore Supra tokamak into WEST tokamak equipped with a divertor dedicated to test Plasma Facing Components (PFC) for ITER, took place from 2011 to 2016 [2]. For the experimental campaigns, Lower Hybrid Current Drive (LHCD) heating system is one of the main heating to maintain long pulse operation and to ensure the power flux on the PFC targets is reached [3] [4]. The LHCD system consists of 16 klystrons TH2103C capable of delivering 600 kW /1000 s each on plasma, powering two launchers at 3.7 GHz [5][6]. To guarantee the reliability of such powerful system, the generator needs to be operated during the shutdowns. Not only the maintenance has allowed keeping the klystrons operational, but also the tests revealed that some parameters of the tubes evolved and sometimes some serious failures appeared. The traceability over the years of the state of the klystrons and the plant is essential as well as to maintain operator's skill maintenance. The tests and target to maintain the plant in good operating condition are described and explained. The paper summarizes the evolution of the 16 klystrons parameters during 6 years of shutdown in section 2 and finally, in section 3, it highlights the main failures the operators had to face.

## 2. Klystrons maintained in operational conditions

WEST LHCD generator composed of 16 powerful klystrons (700kW/1000s on matched load and 600kW/1000s on plasma) was upgraded in 2009-2010 (Fig 2). Half of the LH generator composed of 8 klystrons was commissioned both on matched load and on plasma with the Full Active Multijunction (FAM) launcher reaching routinely 3.5 MW of coupled power [7]. On the other hand, the 8 klystrons powering the Passive Active Multijunction (PAM) launcher have so far been commissioned on matched load only.

### 2.1 Klystrons maintenance

A klystron is a linear beam vacuum tube used as a RF amplifier. The klystron TH2103C in operation for WEST creates an electron beam of 21A accelerated under a voltage of -80 kV. The beam interaction with a 3W drive RF power travelling through resonant cavities amplify the drive RF power up to 700 kW in the case of the TH2103C. The vacuum in the tube is maintained by two ion pumps. The klystron and its auxiliaries are housed in a tank filled with 1000 litres of oil to avoid arcs between power cords, connections and material (Fig 1).

The shape of the electron beam travelling in the klystron, from the gun to the collector, is controlled by the means of three focusing coils. The settings of the focusing coil currents control the RF output power. Cavity temperatures and RF losses measured in the klystron as well as outgassing of the klystron can be the sign of abnormal interactions between the electron beam and the internal parts of the klystron itself. Also after a long period of shutdown, the pressure in the klystrons can increase as soon as the filament is heated or the electron beam and RF power is on. An interlock, on the ion pumps current level > 15 μA, prevents the electron beam to propagate to avoid an arc and then the destruction of the klystron.

* *Corresponding author*
*Email adress* : lena.delpech@cea.fr

DOI: 10.1051/epjconf/201715703009

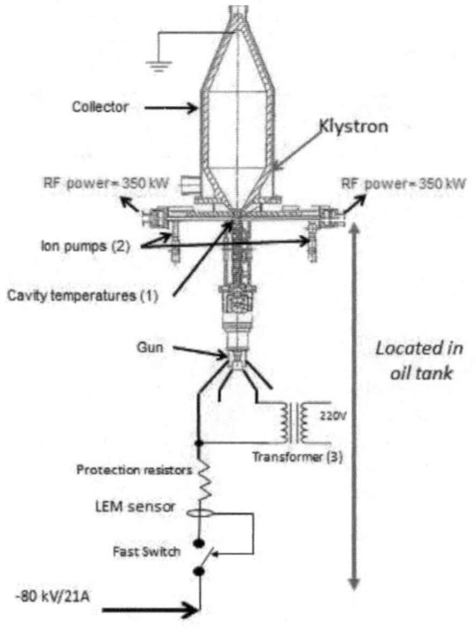


Fig 1 : One klystron and its auxiliaries installed in the LH generator (oil tank)

It is crucial to restart the klystrons quite often to check that the focusing coil currents are adapted to deliver the nominal RF power, to check cavity temperatures and RF losses and also to maintain the klystron operational and avoid over pressure. The criterion of the achievement of three shots of 550 kW/300s on matched load for each klystron was defined to provide 4.4MW of RF power for each launcher. Since 2013, the klystrons have been tested with the following steps :

- Start the klystron in diode mode (no RF power), up to the electron beam nominal value -74kV/21A. Check the vacuum conditioning and the perveance, as defined by $p = \dfrac{I_{beam}}{(V_{Cathode})^{\frac{3}{2}}}$ ($I_{beam}$= electron beam current, $V_{cathode}$ = accelerating voltage of the electron beam)
- Increase RF input power gradually while checking the cavity temperatures, RF losses, vacuum conditioning, klystron gain (dB) and efficiency (%). Comparison to the data obtained during the factory acceptance tests (Fig 3)
- Three shots at a RF power equal to 550kW/300s with no breakdown

Due to LH plant specifications, the klystrons cannot pulse simultaneously on matched load and three days for each are necessary to reach the target of 550 kW/300 s. The frequency of the klystron tests on matched load was led by the manpower available, the time necessary to operate the plant and also the shutdown for modifications or maintenance of the LH generator itself, the subsystems as HVPS or the cooling loop. Each klystron was tested at least once per year. From 2013, a specific annual report per year depicting the maintenance of the generator including: the klystron results and settings, the failures the operator has to face and the actions for the next year has been made available.

Fig 2 : Half of the generator powering the FAM launcher

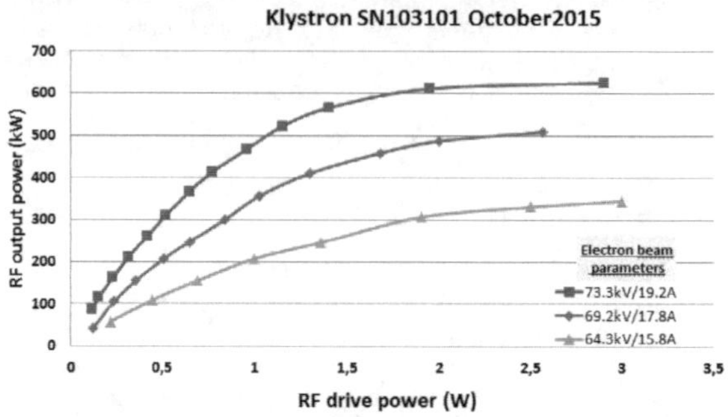


Fig 3 : Klystron Serial Number 103101, specifications that are checked during the tests. Evolution of the RF output power vs RF drive power and High Voltage power supply.

## 2.2 Evolutions of klystron parameters

Eighteen klystrons (16 for the LH generator and 2 spares) have been commissioned on matched load on the CEA test bed in 2009-2010. A first overview of the RF power available for plasma operation after the installation and commissioning of 16 klystrons in the generator (2011), shows a decrease of 20% compared with the total RF power reached on the CEA test bed. Some limits are due to the generator constraints and the fact that 4 klystrons share the same High Voltage power supply. After 6 years of shutdown some klystrons have evolved and the trend is a decreasing RF output power as shown in Fig 4 (decrease of 10% in 2017 compared with the value in 2011).

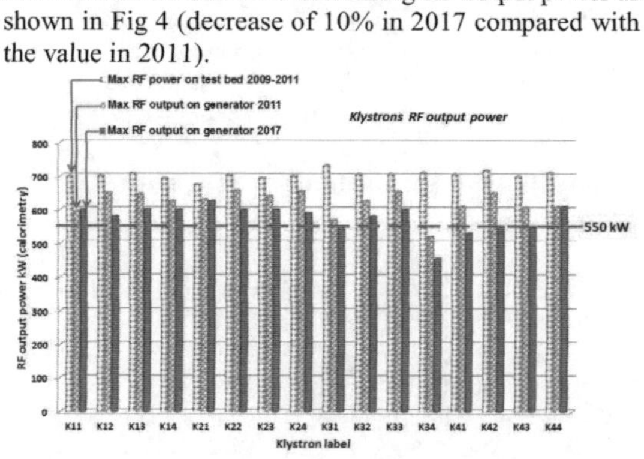


Fig 4 : Evolution of the RF output power of the 16 klystrons of the LH generator from 2010 to 2017

DOI: 10.1051/epjconf/201715703009

During the shutdown, the objective of 550 kW/300 s has been achieved for all the klystrons expect for the ones labelled K34 and K41. In particular, the efficiency of klystron K34 is very low (27%) compared to the value obtained during the factory acceptance test and on the CEA test (45%) Fig 4. Two cavity temperatures are monitored and protected with an interlock at 200°C in case of abnormal conditions (Fig 1). The values of the two temperatures are usually in the same range depending of the RF power except for the klystron K34. It has to be noted that a strong dissymmetry was found as illustrated on Fig 5a and Fig 5b for two different sets of current values in the focusing coils. The currents were optimized to decrease the temperature in the cavity 1 (Fig5 (b)) and the three shots at 450kW/300s were performed. Neither the dissymmetry nor the degradation of the RF performances of the klystron can be explained by a special event or failure and no sign of outgassing during its operation occurred. The klystron will be replaced if the efficiency drops below 25% to maintain the RF power available for the PAM launcher at the required value.

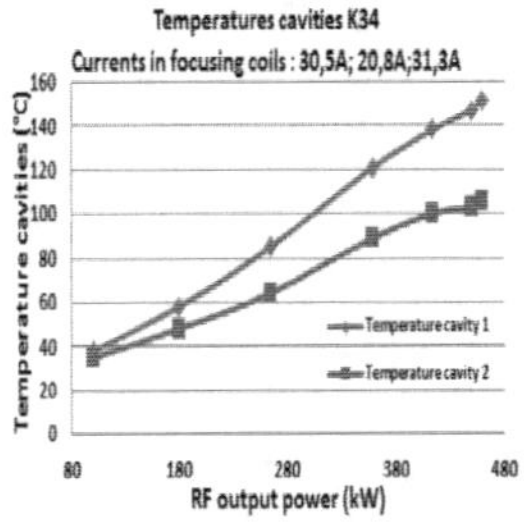


Fig 5(a)

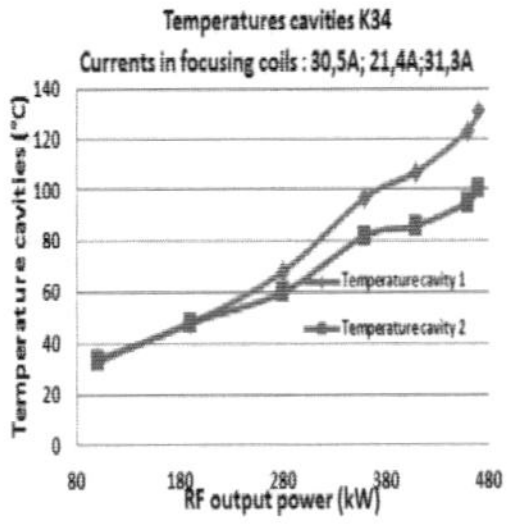


Fig 5(b)

Fig 5 Strong dissymmetry on temperature cavities in K34 measured (see Fig 1) with two different settings in the focusing coils

For the klystron K41, the focusing coils will be optimized to increase its gain in a near future. It has not been done due to lack of time.

## 3. Six years of shutdown: overview of the LH generator maintenance and failures

To increase the generator reliability, several LH subsystems have been replaced such as RF amplifiers, power supplies to heat klystron filaments and filament transformers inside oil tanks. During the period of replacement of the material or modification of the Programmable Logic Controller (PLC), the LH plant was not available for testing the klystrons on matched load.
As it is shown in Fig 1 and 2, the klystrons operating at a nominal value of -74kV/21A, are housed in an oil tank to avoid arcs between power cords, connections and material. Such voltages render crucial the choice of the material and the arrangement of the connections to avoid arcs in the tank. Also the dielectric strength of the oil is controlled every three years or in case of arc in the tank. A solid state switch to protect the klystron, a transformer to heat the filament, protection resistors and a LEM sensor to measure the electron beam current are connected with a cable compatible with oil and is 80 kV isolated. In case of a problem on the latter, to expertise the default and to repair, it is necessary to remove the oil, and in some cases to remove the klystron with a minimum of 5 days for the LH generator shutdown.

### 3.1 Replacement of eight transformers in oil tanks

The LH generator was upgraded when the klystrons TH2103A were replaced by klystrons TH2103C. This upgrade was performed in two phases. In 2009, a first phase dedicated to the replacement of the 8 klystrons, powering the FAM launcher, with the transformers to heat the klystron filament with new specification: 90kV/750VA. The second phase took place in 2010 for the 8 klystrons connected to the PAM launcher.
A new transformer (Transfo Industrie), made with resin was improved during 6 months on the CEA test bed. The choice of Transfo Industrie material was led by the compact size of the component. Eight transformers were installed in the LH generator in 2009 while the CEA test bed was still in operation. While the upgrade was ending, breakdowns occurred with the test bed transformer for HV > 5 kV due to arcs between the transformer's secondary and the screen. Following an expertise of the transformer, it was found that the fabrication process with the resin had created bubbles in the insulator. After a long period of operation, this part of the transformer being submitted to high electric fields, a slow process of degradation had created a crack as seen in Fig 6. As soon as the crack appeared the insulation was reduced from 80kV to 5 kV. The operation with the transformer was no longer possible. The problem was identified but facing the uncertainty that the transformers installed in the LH generator tanks suffer from the same defect, and because the replacement of the transformers for the klystrons powering the FAM required eight months of the transmitter shutdown, the replacement of the transformers took place in 2015 . The commissioning of the klystrons with Transfo Industrie transformers and the first experimental campaign in 2011 was performed successfully with 3.8MW/5s of LH power coupled to the plasma.

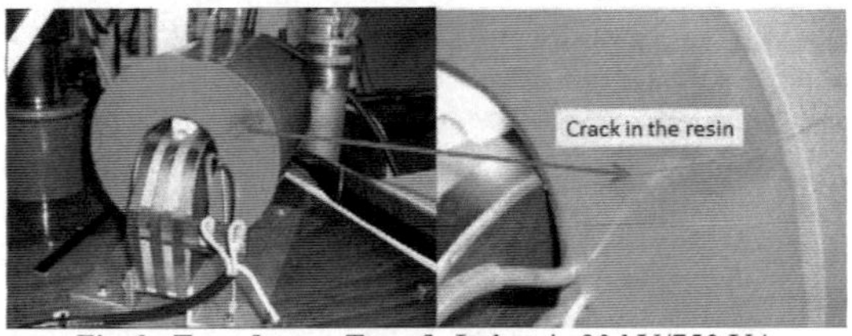


Fig 6 : Transformer Transfo Industrie 90 kV/750 VA

The replacement of the 8 transformers took place during the long shutdown in 2015. Six months were necessary to change the transformers including moving the klystrons and their auxiliaries. Following the hardware modifications, the commissionning of the 8 klystrons on

matched load with the objective of 550kW/300s, was achieved in April 2016.
For the upgrade of the klystrons powering the PAM launcher in 2010, 8 transformers fabricated by ATESYS company were choosen.

### 3.2 Arcing in oil tank

Two major failures, that needed a deep analysis and heavy interventions in the oil tank, occurred on the klystron K33 in June 2015 and klystron K13 in January 2016 after the replacement of the 8 transformers. The signs of fault were similar : arcs for HV > 60 kV. The diagnostic was difficult as the arcs took place in the oil tanks. To avoid damaging the klystrons and also for the safety of the operators, a HV power supply 100kV/4mA was used to test the HV insulation. For klystron K33, the expertise took a long time because for each modification the oil was removed from the tank, then a inspection in the oil tank with a difficult access did not permitted to locate the defect easily, the tank was refilled with oil and the HV insulation test was performed. This operation was repeated 6 times before the operator identified the fault. The defect was located on the HV cable connecting the limiting resistors the tube to the transformer as seen in Fig 7. It is the same process as the transformer : a slow degradation of the conductor insulation. The cable was changed and the klystron was commissioned on matched load.

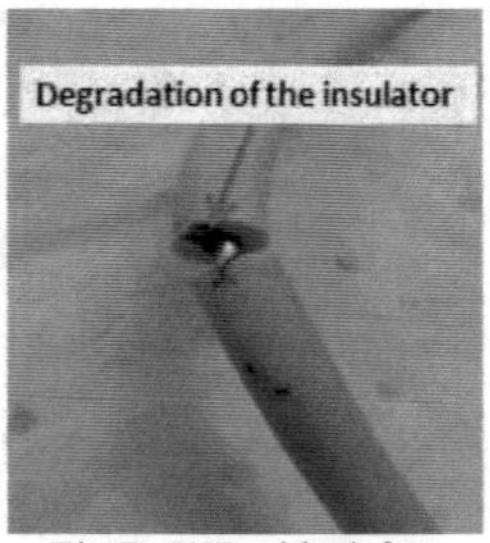


Fig 7 : HV cable defect

For klystron K13, the same defect appeared on the insulator of the cable and the arc occurred between the cable and the current transducer made by LEM to measure the electron beam current. The cable was replaced and the klystron commissioned again.
It is not foreseen for now, to replace all the HV cables of the 16 klystrons : it is not obvious all the cables have the same defects. The replacement of all the cables, the commissioning of the klystrons would take 16 months.

### 3.3 Klystron failures

The LH generator is equipped with 16 klystrons. Two spares TH2103C are stored and pumped with ion pumps and can replace a faulty klystron in the generator. These klystrons were commissioned on the CEA test bed in 2010 and are stored since then. A long period of conditioning will be needed to reach the 550kW/300s RF power with these two klystrons. As discussed in section 2.2, if the efficiency of klystron K34 decreases below 25%, this klystron will be replaced during a long shutdown. To change a klystron the estimated time is one month including the hardware and the conditioning of the tube on match load. During this period, the LH generator is shutdown.
In March 2017, while the LH generator was in shutdown (no circulation of the cooling water, the filaments were cold and no HV applied on the tubes), the pressure increased suddenly in the klystron K13. The current values measured by the ion pumps are recorded even when the generator is shutdown. The water analysis shows pH parameters in the range of klystron specifications. The klystron will be sent back in 2017 to Thales Electron Devices for expertise and will be repaired if possible. This klystron was replaced by a spare which will be commissioned after WEST experimental campaign in 2017.
These overview of a such powerful installation equipped with 16 klystrons is not exhaustive and only faults or maintenance having a major impact on the shutdown of the generator for maintenance are reported.

## Conclusion

The LHCD system is mandatory for sustaining long pulse operation in WEST and essential for testing the PFC-technology for ITER in the framework of the WEST project. The generator, with 7.8 MW of available RF power after the loss of K13 (2017), was maintained during 6 years of shutdown to be operational on request. Taking into account all the constraints, manpower, schedule and shutdown of the plant to change auxiliaries (RF amplifiers, filament power supply…), all of the 16 klystrons were tested at least once a year. It also highlights the fact that, maintaining such plant operational upon request, requires constant work. Another important aspect is to keep operator's skill maintenance at the right level to deliver a reliable plant for experimental sessions. Following a 6 year shutdown, and thanks to a regular maintenance, all klystrons pulsed with RF power from 50 kW to 200 kW on vacuum (10 ms pulses) after only six days of conditioning on vacuum in May 2017, rendering the LHCD generator ready for WEST plasma operation.

## References

[1] R. Dumont et al., *Plasma Phys. Control. Fusion* ***56*** *(2014) 075020*
[2] J.Bucalossi et al., *Fusion Eng. Des.* ***89*** *(2014) 907-912*
[3] C. Bourdelle et al., *Nucl. Fusion* ***55*** *(2015) 063017*
[4] A.Ekedahl et al, *AIP Conference Proceedings* ***1689****, (2015) 030013*
[5] L. Delpech et al., *Nucl. Fusion* ***54*** *(2014) 103004*
[6] A.Beunas et al., *2009 IEEE Trans. Electron Devices* ***56*** *864*
[7] M. Goniche et al., *Phys. Plasmas* ***21*** *(2014), 061515*

# Reinstated JET ICRF ILA: Overview and Results

*Pierre* Dumortier[1,2,*], *Frédéric* Durodié[1], *Trevor* Blackman[2], *Walid* Helou[3], *Philippe* Jacquet[2], *Ernesto* Lerche[1,2], *Igor* Monakhov[2], *Craig* Noble[2], *Volodymyr* Bobkov[4], *Richard* Goulding[5], *Michael* Kaufman[5], *Dirk* Van Eester[1], and JET Contributors[**]

EUROfusion Consortium, JET, Culham Science Centre, Abingdon, OX14 3DB, UK

[1]LPP-ERM/KMS, TEC partner, Brussels, Belgium
[2]CCFE, Culham Science Centre, Abingdon, United Kingdom
[3]IRFM CEA Cadarache 13108 Saint Paul lez Durance France
[4]Max-Planck-Institut für Plasmaphysik, D-85748 Garching, Germany
[5]Oak Ridge National Laboratory, PO Box 2008, Oak Ridge, United States

**Abstract.** The works undertaken to reinstate the JET ICRF ILA are reviewed. The vacuum matching capacitors were replaced, an extensive calibration of all the measurements in the RF circuit was carried out, new simulation tools were created and new control algorithms were implemented for the – toroidal and poloidal – phase control of the array as well as for the matching of the second stage. A review of the contribution of the reinstated ILA to the JET programme during the last campaigns is given showing namely that the new controls allowed extending the range of the operation to lower (29MHz) and higher (51MHz) frequencies than previously achieved and allowed more flexible and reliable operation. Operation with coupled power levels up to 2.8MW and voltages up to 40kV was achieved. ILA results on plasma are discussed and emphasis is given to the features of interest for ITER.

## 1 Introduction

The JET ICRF ITER-Like Antenna (ILA) is a close-packed antenna array composed of four Resonant Double Loops (RDLs) arranged in a 2 toroidal by 2 poloidal array. Each RDL consists of two poloidally adjacent short straps fed through in-vessel matching capacitors from a common vacuum transmission line. Two toroidally adjacent RDLs are fed through a 3dB combiner-splitter.

The ILA antenna has been operating essentially at 33 and 42MHz in 2008-9 until a capacitor failure prevented further operation. The assessment of the results identified that achieving routine and full array operation required a better understanding of the RF circuit, more accurate calibrations of the RF measurements and feedback control algorithms for phase control and second stage matching [1].

## 2 Pre-operation work and new features

Tests on a spare capacitor identified the cause of the capacitor failure in March 2009: a micro-leak caused by the cycle wear of the capacitor's internal bellows appears after about 4500 cycles degrading the capacitor's private vacuum. All capacitors were subsequently replaced and a new operating scheme minimizing the number of full-stroke requests was adopted. All capacitors were tested prior to installation and reconditioned if necessary [2].

The 3dB combiner-splitters feeding two toroidally adjacent RDLs were removed in 2008-9 due of phase control stability issues [1]. The new algorithms described in Section 3 allow operating with the 3dB combiner-splitters in the circuit without any phase control stability issue. They moreover provide an isolation of the generators from the rest of the circuit resulting in a much more stable operation of the generators.

The Transmission Line (TL) layout was modified in order to allow operation of the antenna in toroidal dipole (0 π) as in Current Drive (CD – 0 -π/2) [3].

A thorough calibration campaign of all RF quantities has been undertaken: all directional couplers in the TL system, all voltage probes in the antenna [4], all the RF cables and all the electronics modules and RF components have been calibrated and the necessary adaptors characterised [2,3]. This allowed achieving the level of accuracy required for the control of the antenna (see Section 3) and the operation of the S-Matrix Arc Detection (SMAD) protection system.

* Corresponding author: pierre.dumortier@rma.ac.be
** See the author list of "Overview of the JET results in support to ITER" by X. Litaudon et al. to be published in Nuclear Fusion Special issue: overview and summary reports from the 26th Fusion Energy Conference (Kyoto, Japan, 17-22 October 2016)

 DOI: 10.1051/epjconf/201715703010

Several other improvements were implemented including a full integration of the control system in Level 1 (RF pilot interface), an improved capacitor thermal protection in the RF Local Manager, a revised protection limiting the voltage on the capacitors and the re-commissioning and improvement of the SMAD [2].

A low-power commissioning of the ILA using a network analyser was performed before the high power commissioning in order to determine the initial matching settings for the different operating frequencies and chosen conjugate-T impedances ($Z_T$). This also allowed testing and optimising the control algorithms.

Finally, new numerical modelling and simulation tools were developed, which were used in the calibration of the voltage probes as well as in the implementation of the new control algorithms [4-6].

## 3 New control algorithms [6]

Fig. 1 displays a schematic of the ILA circuit illustrating the feedback control loops.

The first stage matching controls the 4 complex conjugate-T point impedances ($Z_T$) by adjusting the 8 in-vessel variable capacitors in each RDL branch (4 feedback loops; 8 controlled parameters). $Z_T$ is estimated from the Antenna Pressurised Transmission Line (APTL) forward and reflected directional couplers measurements using a CST MicroWave Studio® [7] model of the APTL with the vacuum ceramic window, the bellows and the transition between the 30Ω circular feed and the quarter-wave transformer rectangular section whereas it was previously estimated using a TL model of this section. The revised control algorithm was implemented using Field-Programmable Gate Array (FPGA) technology.

No Second-Stage Matching (SSM) control algorithm was implemented in 2008-9 leading to a very tedious procedure of manual adjustments between pulses and no real-time control to adjust to the varying plasma conditions. A new algorithm was implemented, which controls the lengths of the 4 SSM stubs and of the 4 SSM trombones (4 feedback loops; 8 controlled parameters). To optimize the matching conditions in ELMy plasmas, the algorithm has been designed to work on 2 time scales: a fast one (2μs) and a slow one (2000 points on the pulse length, i.e. 5ms for a 10s pulse). On the fast time scale, fast enough to track the coupling variations during the ELMs, the Main Transmission Line (MTL) complex reflection coefficients are processed and recorded onto a polar map divided in 8 phase bins and 15-20 amplitude bins; the points are kept in the bins for a certain time before being removed. On the slow time scale, the average reflection coefficient is estimated from the polar reflection map; the normalised admittance at the MTL side of the SSM stub is then calculated, from which error signals used to drive the SSM stub and trombone are generated. This results in an effective offset match that minimizes the positive Voltage Standing Wave Ratio (VSWR) excursion during an ELM cycle (rather than the inter-ELM VSWR value). The algorithm outputs three parameters: *move* if the element

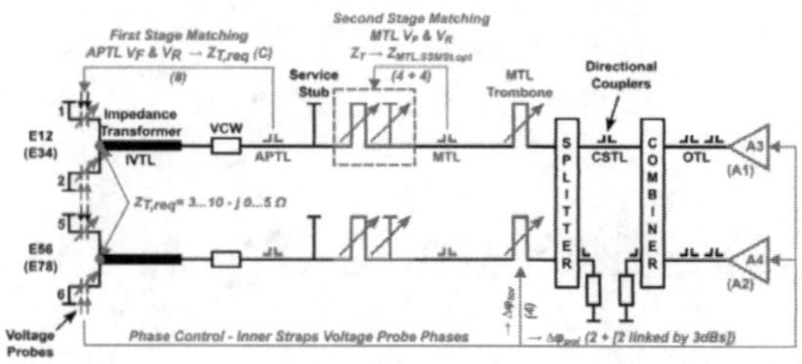


**Fig. 1.** Schematic of the ILA circuit for the ILA upper half (2 toroidal RDLs) illustrating the feedback control loops. The bottom half circuit is similar but not represented for clarity.

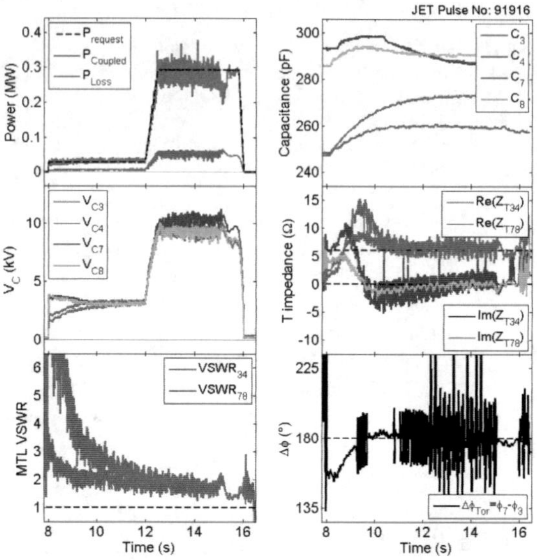


**Fig. 2.** Example of feedback control on a JET pulse (bottom half array operation). Top left: power request, coupled power and power losses. Middle left: capacitor voltages. Top right: capacitor values actuated by the first stage matching to achieve the requested $Z_T = 6 + 0j$ Ω (middle right graph). Bottom left: the SSM actuates the SSM stub and trombones to bring the MTL VSWR down towards an offset matched condition. Bottom right: the toroidal phase feedback actuates the MTL trombones to obtain the requested toroidal phasing (dipole phasing – 180° in this case).

(stub or trombone) error signal is above a threshold, *direction of movement* depending on the sign of the error signal and *fast* that will be true if the amplitude of the signal is above a certain value and will cause a fast movement of the element. The algorithm was extensively simulated [5] and proved stable with respect to the other control algorithms (first stage matching and phase control).

The phasing of the antenna array is controlled by the phase of the voltage probe signals of the four inner straps.

The phase difference between toroidally adjacent RDLs is controlled by acting on the 4 MTL trombones located at the output of the 3dB combiner-splitters (2 feedback loops; 4 controlled parameters).

The phase difference between poloidally adjacent RDLs is controlled with respect to a separate phase reference and is adjusted by acting on the 4 generators'

DOI: 10.1051/epjconf/201715703010

output phases. Two generators are feeding one 3dB combiner-splitter thereby constraining the two generators' output phases (poloidal phasing: 2 feedback loops; 2 independent parameters).

The matching and phasing of the whole array is consequently controlled by 12 feedback loops actuating 22 parameters: the 8 variable in-vessel capacitors, the 4 SSM trombones, the 4 SSM stubs, the 4 MTL trombones and the 4 generator output phases constrained by the 2 3dB combiner-splitters.

Fig. 2 illustrates the operation of the feedback controls during a pulse (bottom half array operation). The pulse starts with a low power phase to home the capacitors in and achieve a reasonable match while staying below the VSWR arc detection trip level. The power is then raised towards the requested value.

The efficiency of the SSM algorithm to optimise the match for the ELM-resilience has been discussed in [6].

## 4 Overview of operation results

The new control algorithms, in particular the SSM control algorithm, together with the improved modelling and a comprehensive calibration of all RF signals have significantly eased the operation of the ILA and allowed to extend the frequency range of the operation towards lower (29MHz) and higher (49MHz for the full array, 51MHz for the upper half) frequencies. Commissioned frequencies are: 29, 33, 37, 42, 44, 47, 49 and 51MHz. It also allowed for the first time operation in toroidal CD (0 $-\pi/2$) phasing besides the usual toroidal dipole phasing operation (0 $\pi$). CD operation however gives rise to voltage imbalance between the currents and voltages of the two toroidal RDLs due to enhanced mutual coupling [2], which eventually also impacts on the radiated power spectrum. Fig. 3 illustrates a toroidal phase scan from CD to dipole during a pulse showing the voltage imbalance during CD operation and voltage balance when reaching dipole operation (low power phase for matching from 8 to 10s). The mutual coupling also generates an imbalance in the reflection coefficients in two toroidal MTLs leading to a higher power diverted to the load of the corresponding 3dB hybrid splitter. The levels however remain small (1-3% for CD phasing).

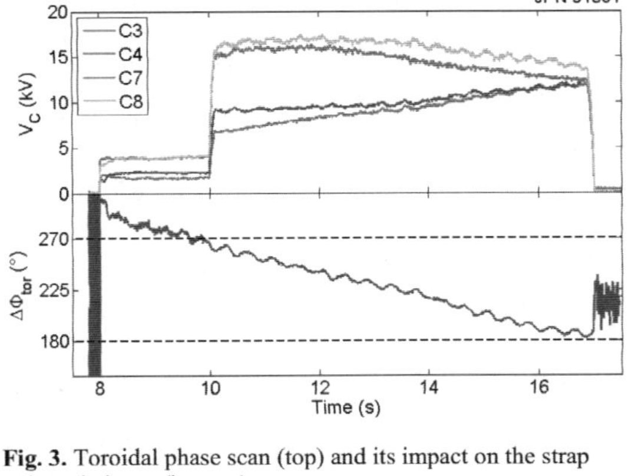


**Fig. 3.** Toroidal phase scan (top) and its impact on the strap voltage balance (bottom).

The ILA has significantly contributed to the experimental programme of JET C35-C37 campaigns up to power levels of 2.8MW coupled power (L-mode half array) and about 40kV maximum voltage. It has been particularly useful at the lower frequencies where the power delivered by conventional A2 antennas is limited by the poor coupling to the plasma and operation limited by the maximum allowed voltage [8]. At 33MHz, half ILA array is able to deliver as much power as 2 A2s, which represents a ten times higher power density.

The operation of the ILA in 2008-9 in JET with a carbon wall has demonstrated that the heating efficiency and the impurity production of the ILA were similar to the ones of the A2s despite the higher power density [1,9]. First indications from the operation of the ILA in JET with the ITER-like wall show likewise that there are no adverse effects arising from the higher power density. Fig. 4 shows an example of such a comparison, where 2MW of ICRH power was coupled from the ILA (full array), antennas A+B (3dB system) and antennas C+D (external conjugate-T) consecutively in an L-mode plasma with $B_t$=2.7T, $I_p$=1.8MA, H minority heating at 42MHz (dipole phasing), fixed antenna-plasma distance and constant top gas injection throughout the discharge. This is further supported by investigation of the RF-enhanced plasma-surface interactions due to near-antenna electric fields ([10-11]) and impurity (W and Ni) concentration measurements [12]. The impact of the impurity production on the operation of next-step devices such as ITER is further discussed in [12].

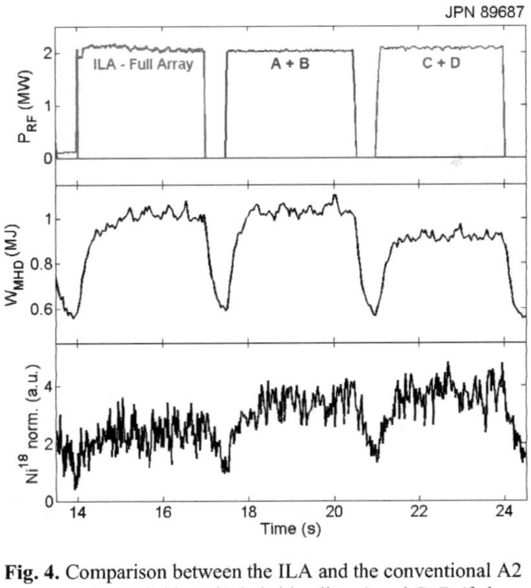


**Fig. 4.** Comparison between the ILA and the conventional A2 antennas: A+B (fed by 3dB hybrid splitters) and C+D (fed through an external conjugate-T). Top: coupled power, middle: plasma stored energy, bottom: $Ni^{18}$ line emission normalised to the plasma edge density.

The coupling resistance for the ILA is defined per RDL as follows:

$$R_{Cxy} = 2\, P_{coupled,xy} \,/\, (I_x^2 + I_y^2) \qquad (1)$$

where $x$ and $y$ are the strap numbers (1...8), $xy$ is the corresponding RDL number, $P_{Coupled,xy}$ is the power coupled to the plasma by the RDL $xy$ (calculated from the MTL power by subtracting the estimated losses) and $I_x$, $I_y$ are the strap currents estimated from the corresponding capacitor voltage probe. The voltage probes are located at the level of the variable capacitors [4] and an equivalent transmission line circuit determined by numerical modelling is used to relate the measured voltage to the strap short circuit current.

Fig. 5 shows the coupling resistance ($R_C$) variation with the average distance between the strap and the last closed magnetic surface for L-mode discharges with $I_p$=1.8-2MA, $B_t$=2.7T and $f$=42MHz. The dashed curves correspond to the fit of the coupling resistance of RDL12 for comparison. The scatter in $R_C$ is due to the different operating conditions (gas fuelling rate and location, etc.), which can have a significant impact on $R_C$. As the heat loads were below the acceptable limits the antenna was moved towards the plasma by 7mm (from pulse 89904 on) in order to improve the coupling. The closed/open symbols correspond to data from pulses respectively before/after the 7mm shift. The diamonds correspond to operation with the outer strike point of the plasma located on the bulk W divertor and featuring low pumping (Tile 5). The squares correspond to operation with the outer strike point on the W-coated divertor close to the pumping throat (Tile 6). Fig. 5 shows that all the RDLs have comparable coupling resistances for operation on Tile 6 and that the bottom-half is more sensitive to changes in the Scrape-Off Layer (SOL) conditions, as seen by the somewhat larger spread of the data. With Tile 5 operation the top-half coupling resistances are only slightly modified but the coupling of the bottom RDLs is substantially better than with Tile 6. This is attributed to the fact that the pumping efficiency is lower when the strike point is on Tile 5 and the density profiles in the far-SOL can be considerably enhanced, in particular below mid-plane.

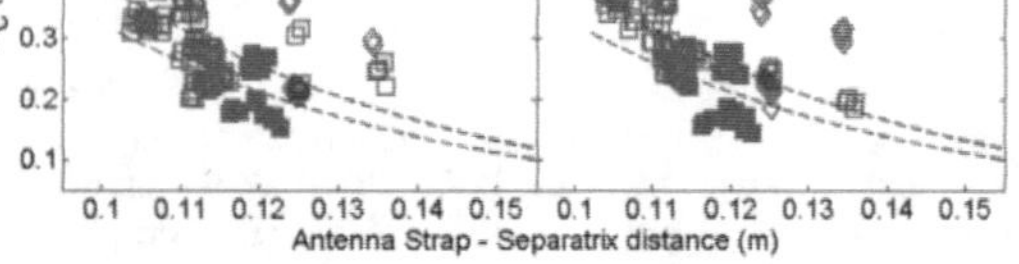


**Fig. 5.** Coupling resistance vs antenna strap-separatrix distance for the different RDLs.

Whilst the poloidal phase control was not yet fully commissioned most of the pulses during the JET campaigns were operated with half (upper or lower) array. A hydraulic fluid leak at one of the capacitor actuators unfortunately occurred in July 2016. The issue was first identified as a low voltage arc [2] but further investigation identified that the hydraulic fluid eventually flooded the pressure box of RDL56 and part of the corresponding APTL section causing arcing at a ceramic support in the APTL. It also damaged one of the position sensors. This prevented further operation of the upper half of the ILA and hence progress on the commissioning of the full array phase control. Remedial actions are currently being undertaken, which, if successful, would bring the upper half back into operation. Limited but promising results were however achieved for the full ILA array phase control.

## 5 Conclusions

The reinstated ILA has significantly contributed to the experimental programme of the JET C35-C37 campaigns (typically 20-40% of the total RF power). Improved numerical modelling, tighter calibration of all RF signals and new algorithms have greatly eased the operation, increased the flexibility and improved the reliability of the ILA. The frequency range has been extended towards lower (29MHz) and higher (51MHz) frequencies. Dipole and current drive toroidal phasing have been operated.

Feedback control of the two matching stages and of the toroidal phase was routinely used during operation on half array (5 feedback loops; 10 actuators). Limited but promising results were obtained for the control of the full array (12 feedback loops; 22 actuators).

The main features of interest of the ILA for ITER, which have already been demonstrated in JET with the carbon wall [1], have been confirmed in JET featuring the ITER-Like Wall, in particular the operation of a high power density close-packed array antenna without degradation of the performances (coupling, heating efficiency, impurity production) with voltages up to 42kV (corresponding to 2.5kV/mm electric fields).

The issues encountered by the ILA are related to the internal matching components (capacitor bellow in 2009 and hydraulic fluid leak in the capacitor actuator in 2016), which validates the choice of external matching elements for ITER.

This work has been carried out within the framework of the EUROfusion Consortium and has received funding from the Euratom research and training programme 2014-2018 under grant agreement No 633053. The views and opinions expressed herein do not necessarily reflect those of the European Commission.

## References

1. F. Durodié et al., Plasma Phys. Contr. Fus. **54**, 074012 (2012).
2. P. Dumortier et al., '*Commissioning and first results of the reinstated JET ICRF ILA*', accepted for publication in Plasma Phys. Contr. Fus.; http://dx.doi.org/10.1016/j.fusengdes.2017.05.050

3. P. Dumortier et al., AIP Conf. Proc. **1689**, 070003 (2015); http://doi.org/10.1063/1.4936510.
4. W. Helou et al., Rev. Sci. Instr. **87**, 104705 (2016).
5. F. Durodié et al., AIP Conf. Proc. **1689**, 070013 (2015); http://doi.org/10.1063/1.4936520
6. F. Durodié et al., '*ITER-Like antenna for JET first results of the advanced matching control algorithms*', accepted for publication in Plasma Phys. Contr. Fus. ; http://dx.doi.org/10.1016/j.fusengdes.2017.05.113.
7. CST GmbH, CST Microwave Studio, User Manual.
8. I. Monakhov et al., this conference.
9. A Czarnecka et al., Plasma Phys. Contr. Fus. **54**, 074013 (2012).
10. C.C. Klepper et al., this conference.
11. A. Křivská et al., this conference.
12. P. Jacquet et al., this conference.

DOI: 10.1051/epjconf/201715703011

# Lower hybrid current drive experiments in support of high confinement long pulse operation in EAST

*Annika* Ekedahl[1,*], *Bojiang* Ding[2], *Xianzu* Gong[2], *Marc* Goniche[1], *Miaohui* Li[2], *Yves* Peysson[1], *Jinping* Qian[2], *Julien* Hillairet[1], *Tuong* Hoang[1], *Fukun* Liu[2], *Chengming* Qin[2], *Yuantao* Song[2], *Xiaojie* Wang[2], *Xinjun* Zhang[2], *Yanping* Zhao[2], *Xiao-Lan* Zou[1] and the EAST Team[2]

[1]CEA, IRFM, F-13108 Saint Paul-lez-Durance, France.
[2]Institute of Plasma Physics, Chinese Academy of Sciences, Hefei 230031, P. R. China.

**Abstract.** The lower hybrid current drive (LHCD) system plays a crucial role in the mission of the Experimental Advanced Superconducting Tokamak (EAST) and is a prerequisite for reaching long pulse, high confinement plasmas on EAST [1, 2]. LHCD experiments and modelling [3] have been carried out on EAST in 2015-2016, with the aim to optimising EAST long pulse scenarios, and at the same time gain experience for the exploitation of WEST [4]. Experiments have been carried out to study the LH current drive efficiency in different plasma configurations (Upper Single Null and Lower Single Null). The effect of the gas feed location on the LH wave coupling was investigated by comparing gas fuelling from high field side, low field side and upper divertor. In view of long pulse H-mode scenarios, a series of H-mode experiments were conducted where all the heating power was provided by RF heating methods only, i.e. LHCD, ECRH and ICRH. H-modes were sustained in both Upper Single Null (W divertor) and Lower Single Null (carbon divertor) configurations, with loop voltage maintained as low as 50 mV.

## 1 Introduction

The radiofrequency (RF) heating and current drive systems play a crucial role in the mission of the Experimental Advanced Superconducting Tokamak (EAST) and are prerequisites for reaching long pulse, high-confinement (H-mode) plasmas on EAST [1, 2]. This paper reports on LHCD experiments and modelling carried out on EAST in 2015-2016, within the framework of the Associated Laboratory ASIPP-IRFM, with the aim to optimize EAST long pulse scenarios, and at the same time gain experience before the exploitation of the WEST device [4]. The unique Lower Hybrid Current Drive (LHCD) capability, i.e. two LHCD systems at different frequencies (2.45 GHz [5] and 4.6 GHz [6]), has allowed comparing the CD efficiency of the two LHCD systems [7]. Extensive modelling with the ray-tracing + Fokker-Planck codes C3PO/LUKE [8] has been performed in support to the experiments [3]. Several experiments, mainly focusing on the 4.6 GHz launcher, have been carried out to study the CD efficiency in different plasma configurations and with different gas feeding locations. In view of long pulse H-mode scenarios, a series of H-mode experiments has been conducted where all the heating power was provided by RF heating methods only, i.e. LHCD, ECRH (Electron Cyclotron Resonance Heating) and ICRH (Ion Cyclotron Resonance Heating).

In addition to the experimental results presented here, EAST experiments have revealed some operational issues important for long pulse operation. During the 2015 campaign, arcing at the 4.6 GHz launcher, accompanied by emission of flakes near the launcher and divertor regions, was observed via visible cameras [9]. Following the in-vessel inspections, several corrective actions were taken before the 2016 campaign: i) new LH launcher guard limiters were installed; ii) the LH launcher mouth was positioned radially closer to the guard limiters in order to reduce the electric field at launcher mouth; iii) all in-vessel components were cleaned from deposits that had accumulated during the previous plasma operation campaign.

## 2 LH wave coupling

EAST uses two LHCD systems operating at different frequencies. The 2.45 GHz launcher has 20 modules of eight waveguides, mounted in five rows and four columns [5]. The 4.6 GHz launcher is composed of 24 modules of 3×24 waveguides, mounted in four rows and six columns [6]. The launched wave spectra for different phasing between modules for the 4.6 GHz launcher are shown in Figure 1. The main peak is located at around $n_z$ = -2 (negative $n_z$ corresponds to the co-current direction). Approximately 20% of the power is launched in the counter-current direction, but with a higher $n_z$ thus creating less energetic electrons in the counter-current direction. The launched wave spectra for the 2.45 GHz launcher have quite similar properties as the 4.6 GHz launcher, but with a larger width of the main peak.

* Corresponding author: annika.ekedahl@cea.fr

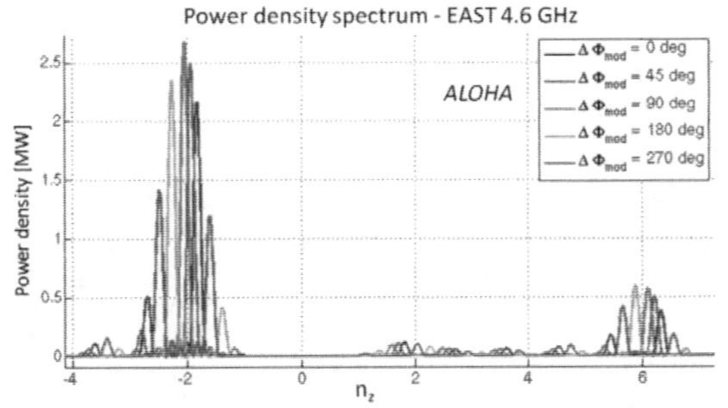


**Fig. 1.** Launched wave spectra for the EAST 4.6 GHz launcher, as computed with the ALOHA code [10].

The effect of particle fueling on the average power reflection coefficient (RC) of the 4.6 GHz launcher was studied using gas injection from different locations, combined with Supersonic Molecular Beam Injection (SMBI). These experiments were carried out in the Upper Single Null (USN) configuration ($B_T = 2.5$ T, $I_P = 0.5$ MA). Most of the particles were provided by the standard gas injection and the SMBI was used for the feedback control of the line averaged density which was ramped from $2.5\times10^{19}$ m$^{-3}$ to $4\times10^{19}$ m$^{-3}$. Three different gas injections were tested: high field side (HFS), low field side (LFS) and upper divertor). The RC is almost identical in the whole density range (Figure 2). It is noticeable that the RC is almost constant at 6% for quite a large part of the density ramp and then drops to 4% when the density exceeds $3.5\times10^{19}$ m$^{-3}$. This drop is correlated with the phase of the discharge where the largest amount of particle fueling comes from SMBI, a method that has been found to increase radial transport [11]. This effect results in an increase in density in the far scrape-off layer, which is clearly beneficial for LH coupling.

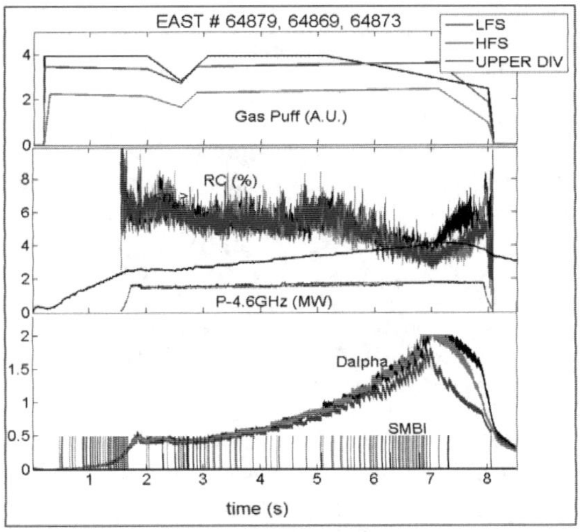


**Fig. 2.** Average reflection coefficient (RC) on the 4.6 GHz launcher for different gas fueling locations.

## 3 LHCD efficiency

### 3.1 LH wave propagation effects

To assess the effect of the magnetic configuration on the LH wave propagation, similar discharges with a density ramp were conducted in the USN and the Lower Single Null (LSN) configuration ($B_T = 2.5$ T, $I_P = 0.5$ MA) using 1.7 MW from the 4.6 GHz launcher. The electron cyclotron emission (ECE) from fast electrons driven by the LH wave, which is down-shifted by the relativistic effect and therefore measured at the very periphery of the plasma ($R = 2.30$ m, $r/a = 1$), was used as a qualitative measure of the difference in LHCD efficiency. The ECE signal was significantly larger in the USN configuration (around 30% larger) in the whole density range explored, as seen in Figure 3. The fast electrons detected via ECE are likely to be in the energy range 50 - 100 keV, as this corresponds to an emission zone of $0.2 < r/a < 0.7$, which is compatible with C3PO/LUKE modelling of the LH wave deposition [3]. These fast electrons are also detected via the Hard X-ray diagnostic in EAST, showing that the bremsstrahlung in the hard X-ray range is higher in the USN configuration. The internal inductance is reduced in USN and the total energy content is found to be slightly higher (+ 10%) for the USN configuration. In addition, the RF spectrum in the 4.58 - 4.61 GHz range was recorded by a magnetic loop outside the vacuum vessel. Figure 4 shows that when the density exceeds $3\times10^{19}$ m$^{-3}$, the width of the pump wave (at f = 4.6 GHz) is larger for the LSN configuration. This could indicate that the wave number spectrum is affected, either by scattering on density fluctuations, or by a parametric decay into an ion-sound wave (f ~ 2-3 MHz) [7].

In order to quantify the propagation effects and the resulting up-shift of the wave index ($n_{//}$) on the LHCD efficiency, two discharges were performed in the same plasma configuration (identical to discharge #64842) and the same power (0.4 MW from the 4.6 GHz launcher), launched by either the three lower rows of waveguides or by the three upper rows. The average reflection coefficient (RC) was very similar for the two discharges in the entire density range (RC ~ 3%), which indicates that the launched $n_{//}$ spectra can be considered to be the same. It has to be noted that the directivity of the wave spectrum can have a considerable effect on the CD efficiency, as found in Tore Supra [12]. A slight difference on the loop voltage and the total stored energy at the benefit of the top launch was found. When the line averaged density ($n_e$) exceeded $3.4\times10^{19}$ m$^{-3}$, a strong interaction of the launcher with the edge plasma occured, which was also related to an increase of the gas injected by SMBI, and the stored energy decreased.

 DOI: 10.1051/epjconf/201715703011

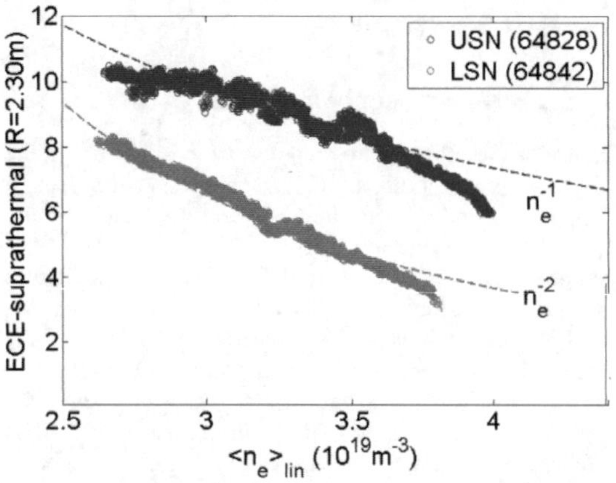


**Fig. 3.** Non-thermal Electron Cyclotron Emission for USN discharge #64828 and LSN discharge #64842 ($P_{LH}$ = 1.7 MW).

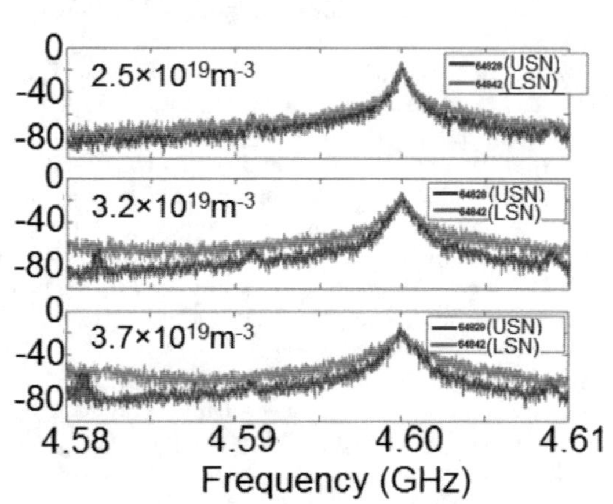


**Fig. 4.** RF spectra at three different densities for USN discharge #64828 and LSN discharge #64842 ($P_{LH}$ = 1.7 MW).

### 3.2 Effect of LH launcher frequency

In long pulse operation, both LHCD launchers are used in order to maximize the non-inductive current fraction and thus extend the pulse duration. H-mode discharges have been sustained by LHCD + ICRH ($P_{LHCD}$ = 1.8-3.2 MW; $P_{ICRH}$ = 1.2-1.5 MW), in which most of the plasma current ($I_P$ = 0.4 MA) is non-inductively driven (loop voltage < 140 mV). At low loop voltage ($V_{Loop}$), it can be shown that $V_{Loop}$ varies almost linearly with $P_{norm}$, where $P_{norm} = P_{LHCD}/(n_e.R.I_P)$, and that the CD efficiency is given by $1/P_{norm}$ at $V_{Loop} = 0$ [13]. In the density range of interest, i.e. line averaged density $n_e = 2.4 - 3.2\times10^{19}$ $m^{-3}$, the loop voltage decreases linearly with $P_{norm}$, only if the contribution of the power launched by the 2.45 GHz launcher to $P_{LHCD}$ is weighted by a factor 0.5. Similar result was inferred from $V_{Loop}$ measurements at modest LHCD power (~ 1 MW), launched by the two launchers separately in the same discharge [7]. The same weighting also leads to less spread when fitting the hard X-ray data and the total plasma energy. Neglecting the bootstrap current for these low beta discharges, the global efficiency is found to be close to $0.78\times10^{19}$ $A/(W.m^2)$.

The result above suggests that a fraction of the power launched from the 2.45 GHz launcher may be damped at the plasma edge and is not entering the plasma core. This interpretation is partially consistent with RF spectrum measurements in the scrape-off layer, which indicate broader pump width for the 2.45 GHz launcher [7]. Electron acceleration at the plasma edge due to Landau damping on high $n_{//}$-components does not seem to be the cause, since infrared measurements of the divertor do not indicate higher heat load with the 2.45 GHz launcher. In addition, calculations of the electron acceleration using the electric field from ALOHA [10] and a model for the electron dynamics in the LH electric field [14], show that the heat flux produced by edge electron acceleration is comparable for the two LH launchers [15].

## 4 H-mode experiments

A study of the H-mode access with RF heating systems was carried out on EAST in 2016, in a plasma configuration as close as possible to that foreseen in WEST [4]. The maximum available heating power in these particular experiments was limited to ~ 3.6 MW ($P_{LHCD}$ ~ 2.8 MW, $P_{ICRH}$ ~ 0.4 MW, $P_{ECRH}$ ~ 0.4 MW). The LHCD power (up to 2.8 MW) was mainly supplied by the 4.6 GHz launcher (up to 2 MW), while the 2.45 GHz launcher gave up to 0.8 MW. H-mode plasmas lasting 6 s, close to zero loop voltage ($V_{Loop}$ ~ 50mV) could be achieved, as shown in Figure 5. The plasma configuration in the discharge in Figure 5 (#64971) was LSN, i.e. with the strike point on the carbon divertor, and with $B\times\nabla B$ directed towards the target, which is expected to be favourable for the H-mode transition [16]. The $H_{98(y2)}$-factor was found to be ~ 0.76 in #64971. The L-H transition occurred at t = 2.2 s, as characterized by the onset of Type I ELMs, after the application of 2.5 MW LHCD. The ECRH power was applied slightly later, at t = 2.6 s. In these experiments, it was found that low power ECRH and/or ICRH could be beneficial for sustaining the H-mode. When changing the toroidal magnetic field direction so that $B\times\nabla B$ was directed away from the target, while keeping the plasma configuration and density the same, no H-mode transition was obtained even though the total injected power was 3.6 MW.

The same set of experiments was later repeated in the Upper Single Null (USN) configuration, i.e. with the strike points on the W-divertor. One discharge with $B\times\nabla B$ directed upwards towards the W-divertor, which again is expected to be favourable for the L-H transition, is shown in Figure 6. The H-mode is again sustained throughout the duration of the heating phase, at higher density and with similar H-factor ($H_{98(y2)}$ ~ 0.78) as in the shot in Figure 5. When reversing the $B\times\nabla B$ direction in USN configuration, and maintaining the same heating power as in #65072 from the three systems (LHCD, ICRH and ECRH), only a short lasting H-mode (~ 1 s) was obtained, even though the heating power remained constant (~ 3.6 MW).

DOI: 10.1051/epjconf/201715703011

In the discharges studied in this set of experiment ($I_P$ = 0.45 MA, $B_T$ = ±2.5 T), high density ($n_e$ ~$3\times10^{19}$ $m^{-3}$) was required in most discharges to trigger the H-mode. The additional power required to reach H-mode was around 2.5 MW with B×∇B towards the target, and around 3.6 MW with B×∇B away from the target.

It could be noted that H-modes were also obtained with LHCD alone at low power ($P_{LHCD}$ ~ 1.6MW), low density ($n_e$ = $2.5\times10^{19}$ $m^{-3}$) and relatively high H-factor ($H_{98y}$ ~ 1.1), using the USN configuration and B×∇B directed upwards. However, these discharges transited back to L-mode after approximately 2 s.

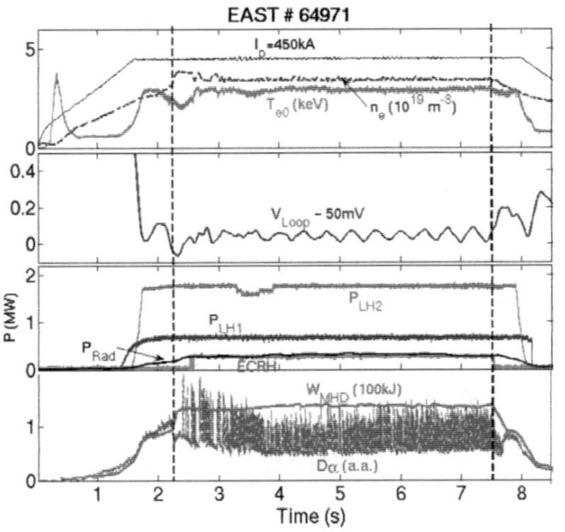


**Fig. 5.** Time traces of H-mode discharge in LSN configuration, $B_T$ = -2.5T and B×∇B directed downwards, towards the target .

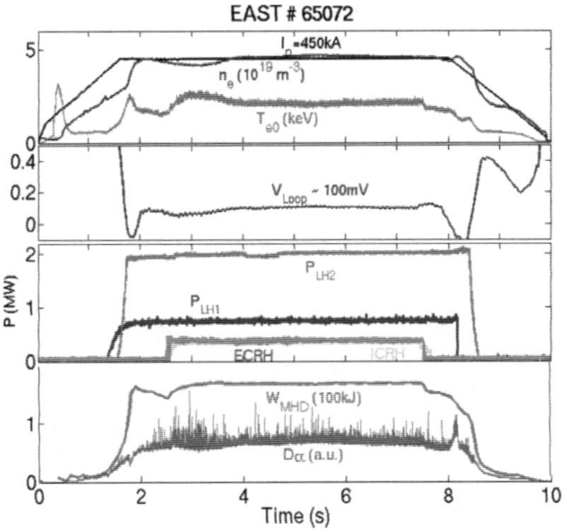


**Fig. 6.** H-mode discharge in USN configuration (W-divertor), $B_T$ = 2.5T and B×∇B directed upwards, towards the target.

## 5 Summary

Several experiments have been carried out on EAST with the goal to optimizing the efficiency of the LHCD systems in view of long pulse H-mode operation. The LHCD experiments indicate that the USN configuration could be more favourable than the LSN for the current drive efficiency. The different gas feeding location tested (high field side, low field side and upper divertor) do not affect the LH wave coupling differently. However, gas fueling with SMBI has shown to reduce the reflection coefficient, due the increase of density in the scrape-off layer, linked to increased radial transport during SMBI.

During H-mode experiments, 6 s long H-modes were obtained with loop voltage as low as 50 mV, using LHCD as main heating and current drive source, accompanied by low power ECRH and/or ICRH. Easier access to H-mode was indeed found when the B×∇B drift was directed towards the target. This was confirmed in both LSN (carbon divertor) and USN (W-divertor) configurations. In summary, all results obtained here also give valuable information in view of the exploitation of the WEST tokamak, which has similar size to EAST, and which will use RF heating methods only in W-environment.

The CEA/IRFM members warmly acknowledge the hospitality of the ASIPP Team during the visits to EAST. This work is supported by the National Magnetic Confinement Fusion Program of China (Grant No 2015GB10200) and the Associated Laboratory CEA/IRFM – CAS/ASIPP. This work has been carried out within the framework of the EUROfusion Consortium and has received funding from the European research and training programme under grant agreement N° 633053. The views and opinions expressed herein do not necessarily reflect those of the European Commission.

## References

1. B. Wan et al., Nucl. Fusion **57**, 102019 (2017)
2. F.K. Liu et al., this conference
3. Y. Peysson et al., this conference
4. C. Bourdelle et al., Nucl. Fusion **55**, 063017 (2015)
5. E.H. Kong et al., Plasma Phys. Control. Fusion **56**, 075016 (2014)
6. F.K. Liu et al., Nucl. Fusion **55**, 123022 (2015)
7. M.H. Li et al., Phys. Plasmas **23**, 102512 (2016)
8. Y. Peysson and J. Decker, Fusion Sci. Technol. **65**, 22 (2014)
9. Y.L. Li et al., this conference
10. J. Hillairet et al., Nucl. Fusion **50**, 125010 (2010)
11. X.L. Zou et al., Proc. 24th IAEA FEC (2012), San Diego, CA, paper PD/P8-08
12. E. Nilsson et al., Nucl. Fusion **53**, 083018 (2013)
13. M. Goniche et al., Phys. Plasmas **21**, 061515 (2014)
14. V. Fuchs et al., Nucl. Fusion **43**, 341 (2003)
15. L. Valade et al., AIP Conf. Proc. **1689**, 080007 (2015)
16. G.R. McKee et al., Nucl. Fusion **49**, 115016 (2009)

# Electron Cyclotron interaction in the early stage of EC-assisted plasma breakdown

*Daniela* Farina[1,⋆]

[1] *Istituto di Fisica del Plasma, Consiglio Nazionale delle Ricerche, Milano, Italy*

**Abstract.** Electron Cyclotron (EC) wave interaction with very low temperature electrons in conditions like those occurring for EC-assisted plasma breakdown is investigated via the Hamiltonian adiabatic formalism. Estimates of interest for applications to present tokamak experiments are derived. The analysis allows ultimately to derive analytically the particle energy gain per single beam crossing and to characterize the dependency of the maximum possible energy variation as a function of the wave frequency, harmonics and polarization and EC power and beam width. The investigation confirms that electrons can easily gain energies well above the ionization energy in most conditions, as observed in experiments.

## 1 Introduction

As pointed out long ago in the literature [1], wave interaction with an electron crossing the microwave beam in single transit is in general non linear in the initial stage of EC pre-ionization. In the very first phase of a plasma discharge with EC-assisted breakdown, the electron motion can be described as single particle in free space under the action of a localized e.m. beam in a static magnetic field. The EC beam is assumed to be injected into the plasma with a Gaussian-like profile such that wave electric field amplitude is $E = E_M \exp(-r^2/w^2)$, with $r$ the beam radius, $w$ the waist and $E_M$ its maximum value. For a magnetized electron at room temperature ($T_e$ = 0.03 eV) and for EC beam parameters typical of ITER ($P$ = 1 MW, width $w$ = 0.05–0.1 m, $f$ = 170 GHz) as well as of present experiments, the interaction regime is characterized by $\tau_{cycl} \ll \tau_{trap} \ll \tau_{flight} \ll \tau_{coll}$, with $\tau_{cycl} \simeq 6\times10^{-12}$ s, $\tau_{trap} \approx 10^{-8}$ s, $\tau_{flight} \approx 1-5\times10^{-6}$ s, $\tau_{coll} \approx 5\times10^{-4}-5\times10^{-3}$ s, the cyclotron period, the trapping time in the wave field, the time flight in the beam and the collision time, respectively. Being $\tau_{trap}$ much shorter than the time scale of the variation of the e.m. wave amplitude, the above conditions define an adiabatic nonlinear regime (widely investigated in the past, see e.g. [1–4]). This regime can be described theoretically via a Hamiltonian formalism, making use of the adiabatic invariants. A rigorous identification of the relevant elementary physical processes can then be performed within this approach.

## 2 Hamiltonian electron motion

The motion of an electron interacting with an EC wave with frequency $\omega$, propagating perpendicularly to the static uniform magnetic field $\mathbf{B}_0 = B_0\mathbf{e}_z$, can be described in the Coulomb gauge (with zero scalar potential) by the Hamiltonian function $H(\mathbf{x},\mathbf{P}) = mc^2\gamma = mc^2[1 + |\mathbf{P}/mc + e/c\mathbf{A}|^2]^{1/2}$, where $\mathbf{P} = \mathbf{p} - e/c\mathbf{A}$ is the canonical momentum conjugate to $\mathbf{x}$, $\mathbf{A}(\mathbf{x},t) = \mathbf{A}_0(\mathbf{x}) + \mathbf{A}_1(\mathbf{x},t)$ is the vector potential, sum of a static and e.m. part. In the following normalized variables are used, i.e., time, length, momentum, energy and Hamiltonian, and vector potential over $\Omega^{-1}$, $c\Omega^{-1}$, $mc$, $mc^2$, and $e/mc^2$, respectively being $\Omega = eB/mc$ the cyclotron frequency. The static and e.m. vector potentials can be written as $\mathbf{A}_0 = -y\hat{\mathbf{x}}$, $\mathbf{A}_1 = \epsilon(\mathbf{x})[e_x \cos\chi\hat{\mathbf{x}} + e_y \sin\chi\hat{\mathbf{y}} + e_z\cos\chi\hat{\mathbf{z}}]$, being $\epsilon = eE/mc\omega$ the normalized amplitude, $\chi = \nu(Nx - t)$ the phase, $\mathbf{e} = \mathbf{E}/E$ the wave polarization vector, $N$ the refractive index, and $\nu = \omega/\Omega$ the frequency ratio.

Following the approach described in [3], a new time-independent Hamiltonian with two degrees of freedom, valid for $\omega$ close to a cyclotron harmonic $n\Omega$, and at first order in the wave amplitude $\epsilon \ll 1$ can be obtained via canonical transformations

$$H(z,\theta,P_z,I) = \gamma - \nu_n I = H_0(P_z,I) + H_1(z,\theta,P_z,I) \quad (1)$$

with $\nu_n = \nu/n = \omega/n\Omega$, $H_0(P_z,I) = \Gamma - \nu_n I$, and $H_1(z,\theta,P_z,I) = \epsilon(z)\,\Theta_n(P_z,I)/\Gamma\cos n\theta$, where $I$ and $\theta$ are action-angle variables, $P_z$ is the parallel canonical momentum conjugate to $z$, and $\Gamma = (1 + 2I + P_z^2)^{1/2}$. For the unperturbed system ($\epsilon = 0$), $\theta$ is the phase of the gyromotion, $I = p_\perp^2/2$, and $P_z = p_z$. The polarization term $\Theta_n$ reads $\Theta_n = \sqrt{2I}[e_- J_{n-1}(b) + e_+ J_{n+1}(b)] + P_z e_z J_n(b)$, with $e_\pm = (e_x \pm e_y)/2$ and $b = \nu N\sqrt{2I}$.

In Figure 1, two examples of solution of the Hamiltonian equations of motion are shown for ITER parameters, second harmonic extraordinary mode (XM) polarization. The initial conditions refer to thermal electrons at $T_e$ = 0.03 eV, just outside the beam region, i.e., at $t$ = 0,

⋆Corresponding author: farina@ifp.cnr.it

$z_0 = -4w$ ($w = 0.1$ m), $P_{z0} = \beta_{th}$ and $I_0 = \beta_{th}^2/2$ being $\beta_{th} = v_{th}/c = (T_e/mc^2)^{1/2}$ the thermal velocity normalized to the speed of light. In the two cases, which differ in the initial phase value $\theta$ only, two quite distinct final energy values are obtained, in one case the electron recovers its initial energy value while in the other one it gains a large amount of energy as a consequence of beam crossing. While the electron crosses the EC beam, it becomes non linearly trapped in the wave field for $\epsilon = \epsilon_s \simeq 0.6\epsilon_M$ (i.e.,$|z| \simeq 0.1$ m), and then its energy experiences large variations, getting the maximum value at $z \sim 0$ (up to $\sim 180$ eV). When the e.m. field decreases (for $z > 0$) the electron becomes untrapped (for the same $\epsilon_s$ value) and exits the beam region either with final energy $\approx 0.03$ eV (blue dotted curve) or $\simeq 60$ eV (red solid curve), much larger than the initial one. For the chosen parameters, the time required to travel the distance $\Delta z = 2|z_0|$ is $\approx 8\ \mu$s. In contrast to the perpendicular energy large variations, the parallel canonical momentum $P_z$ shown in Fig. 2 is almost constant while the electron crosses the beam region, with maximum variation of the order of $\sim 3\%$ and negligible final net variation ($\sim 0.2\%$).

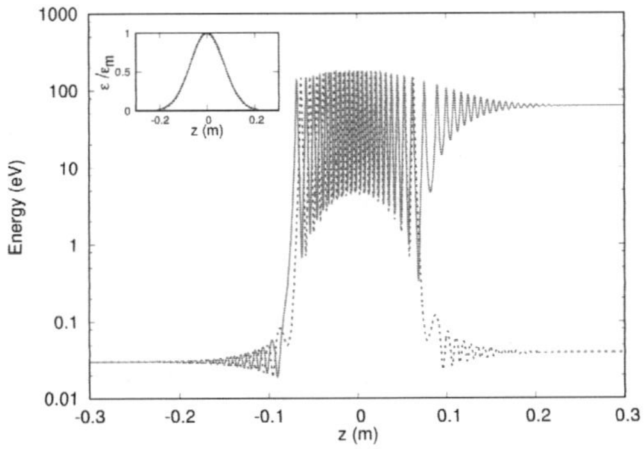


**Figure 1.** Electron energy $W = mc^2(\gamma - 1)$ (logarithmic scale) versus coordinate $z$ for two different initial phase values $\theta$. For reference, the electric field amplitude versus $z$ is shown in the inset plot on the top left. The chosen parameters are $P = 1$ MW, beam waist $w = 0.1$ m, XM second harmonic, $1 - \nu_n = 6 \times 10^{-5}$.

A peculiar behaviour of the system is reported in Figure 3, where the evolution of the action is shown for various input power keeping the other parameters constant. The same quite large net action variation is found independently of the e.m. input power, provided that the electron becomes trapped in the wave field. The threshold power for trapping is $\sim 250$ kW for the considered beam parameters and initial particle energy, and electrons recover their initial energy for all phase values in the lowest power case for which trapping does not occur. Quite similar results are found at first harmonic, but with a much larger energy variation (see below).

## 3 Adiabatic nonlinear theory

The electron motion is governed by the Hamiltonian function (1), and is characterized by a couple of fast action-

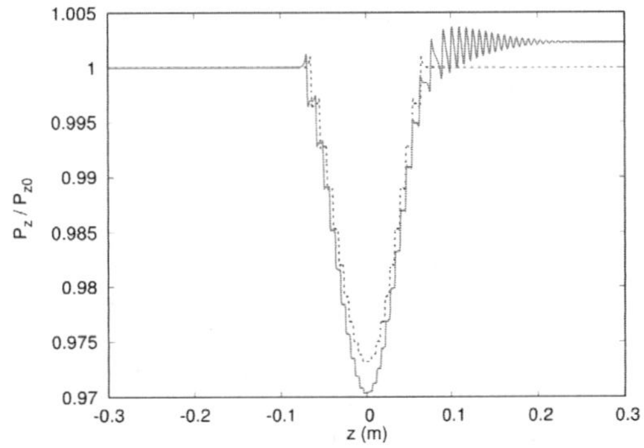


**Figure 2.** Parallel canonical momentum versus coordinate $z$ for the same cases of Fig. 1.

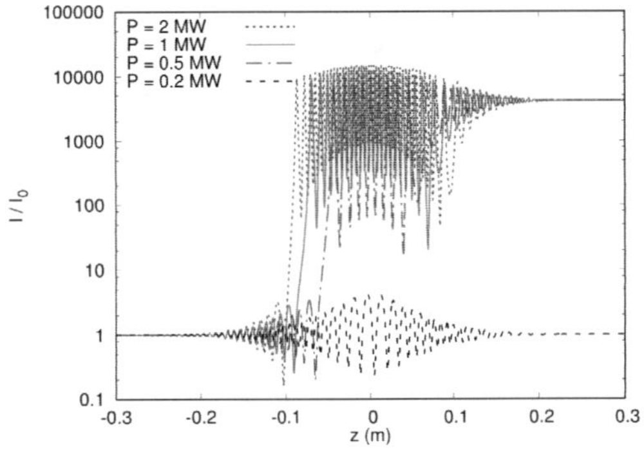


**Figure 3.** Evolution of canonical action $I$ normalized to initial values as a function of coordinate $z$ for various input EC power, $P = 0.2, 0.5, 1, 2$ MW. Other parameters as in Fig. 1.

angle variables $(\theta, I)$ and of slow canonically conjugate variables $(z, P_z)$ in the considered regime of variation of the e.m. field. Whenever the e.m. field variation experienced by the electron while crossing the beam is sufficiently slow, the action integral $J(z, P_z) \equiv n/2\pi \oint I\, d\theta$ is an adiabatic invariant of the motion, that breaks only at separatrix crossing between trapped and untrapped orbits. Breaking of the invariant via trapping/untrapping allows to connect different regions in phase space and provides the mechanism for net energy variation. This physics process for EC nonlinear interaction was investigated in the past in [1–4].

The basic process can be described recalling that the initial and final states are determined by the unperturbed Hamiltonian $H_0 = const$ (valid for particles outside the e.m. illumination region), with initial and final states related by $\Delta\gamma = \nu_n \Delta I$, and $\Delta P_z^2 = \Delta\gamma[\Delta\gamma - 2(1/\nu_n - \gamma_0)]$. In addition, $\Delta P_z \simeq 0$ is satisfied under adiabatic conditions in perpendicular propagation. A particle with initial $I_0$ and $P_{z0}$ belonging initially to a given branch of the Hamiltonian will enter the beam region and eventually cross the separatrix for $\epsilon = \epsilon_s$ for increasing $\epsilon$ (if $\epsilon_s \leq \epsilon_M$), being trapped and then untrapped for decreasing $\epsilon$ for $\epsilon = \epsilon_s$, and

 DOI: 10.1051/epjconf/201715703012

will exit the beam region with final values $I_f$ and $P_{zf}$, with $P_{zf} \simeq P_{z0}$, within the present approximation. A net energy variation can occur only if $H_0$ is double valued in action $I$, condition satisfied if $\omega < n\Omega$ and the resonant condition $\partial H_0/\partial I = 0$ is met for an action value $I_r$ corresponding to the unperturbed resonant action at constant $P_z$.

In Figure 4, the interaction process is sketched in the phase space of the fast variables for increasing $\epsilon$ values: motion across the EC beam corresponds to snapshots from top left to low right and viceversa, i.e., to first increasing and decreasing $\epsilon$ values. The particle will lie initially on either the lower or the upper branch depending if $I_r$ is larger or smaller than $I_0$ (see upper left snapshot). At an intermediate $\epsilon$ value not shown here, the two branches merge, and the particle is nonlinearly trapped (snapshots in lower row). After beam crossing both branches can be populated. Note that $\Delta\gamma$ will be different from zero only if the particle after the last separatrix crossing will not lie on the same branch of the Hamiltonian as initially.

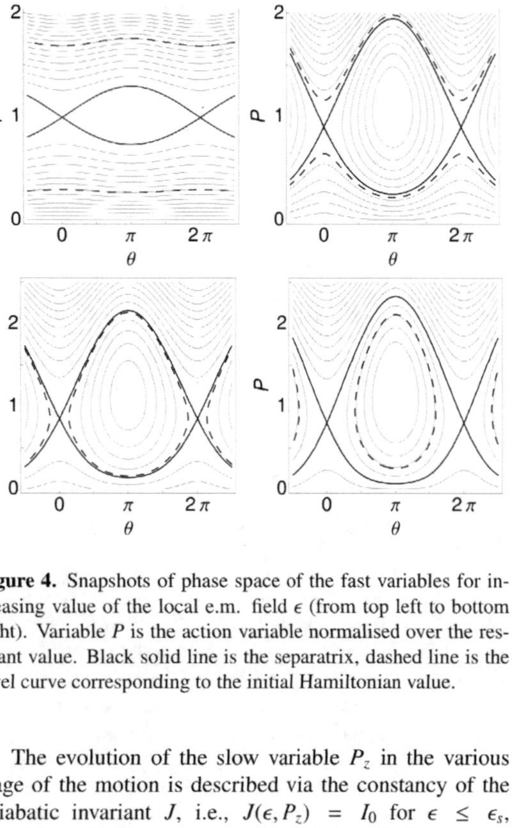


**Figure 4.** Snapshots of phase space of the fast variables for increasing value of the local e.m. field $\epsilon$ (from top left to bottom right). Variable $P$ is the action variable normalised over the resonant value. Black solid line is the separatrix, dashed line is the level curve corresponding to the initial Hamiltonian value.

The evolution of the slow variable $P_z$ in the various stage of the motion is described via the constancy of the adiabatic invariant $J$, i.e., $J(\epsilon, P_z) = I_0$ for $\epsilon \le \epsilon_s$, $J(\epsilon, P_z) = J_s$ for $\epsilon_s \le \epsilon$, and $J(\epsilon, P_z) = I_f$ for $\epsilon \le \epsilon_s$, being $J_s = A_s/2\pi$ with $A_s$ the area of the phase space region inside the separatrix at its crossing. Being the motion adiabatic in each stage, the global variation of the action variable $I$ is equal to the jump of the adiabatic integral at the separatrix

$$\Delta\gamma = \nu_n(I_f - I_0) = \nu_n(|J_f(\epsilon_s, P_{zs})| - |J_0(\epsilon_s, P_{zs})|), \quad (2)$$

where $P_{zs}$ is the $P_z$ value at the separatrix, and $J_{0,f}(\epsilon, P_z)$ is the action integral in the initial and final stage of the motion, i.e., before and after the separatrix crossings. The $\epsilon$ and $P_z$ values at the separatrix are related to the initial (or final) momentum values by means of the following system of equations

$$I_{0,f} = |J_{0,f}(\epsilon_s, P_{zs})|, \quad (3)$$
$$H_0(I_{0,f}, P_{z0,f}) = H_s(\epsilon_s, P_{zs}), \quad (4)$$

where $H_s$ is the Hamiltonian computed at the separatrix.

As shown in Fig. 3, the energy variation is independent on the peak value of the e.m. field $\epsilon_M$, provided trapping occurs. The maximum field amplitude determines the set of initial momenta for which an energy transition can occur: $\epsilon_s(I_0, P_{z0}) \le \epsilon_M$. The above expressions for the momentum variation in case of trapped motion can be computed explicitly in the weakly-relativistic approximation, that is well justified for the electron energies in the physical regimes of interest here.

In the following, the analysis of the first two harmonics $n = 1, 2$ is briefly summarized (see, e.g., [1, 3, 4] for detailed derivations). Expanding Eq. (1) for small energy values up to second order in the action in $H_0$, and to lowest order in the polarization term, one obtains

$$H_0 = P_z^2/2 + I_r(P_z)I - I^2/2, \quad (5)$$
$$H_1 = \epsilon(z)\, T_n(P_z)\, [2I]^{n/2} \cos n\theta$$

with $I_r(P_z) = 1 - \nu_n - P_z^2/2$ the resonant action, and $T_n(P_z) = (N\nu)^{n-1}[(e_x - e_y)^q + (e_z N \nu_n P_z)^{1-q}]/[2^n(n-1)!]$, with $q = 0, 1$ for OM, XM.

To determine the separatrix crossing and compute the adiabatic invariants it is convenient to investigate the Hamiltonian (5) in the phase space of the fast variables for fixed $\epsilon$ and $P_z$. The singular points $(\theta_p, I_p)$ of the Hamiltonian $H = H_0 + H_1$ are the solutions of the set of equations $\partial H/\partial I = 0$, $\partial H/\partial\theta = 0$. In the resonant case ($I_r > 0$, i.e., $P_z^2/2 < 1 - \nu_n$ and $\nu_n < 1$), a separatrix is found for $\epsilon \le \epsilon_{cn} = \beta_{cn} I_r^{2-n/2}/T_n$, with $\beta_{c1} = (2/3)^{3/2}$ and $\beta_{c2} = 1/2$. In the resonant case, the phase space topology is pendulum-like for low $\epsilon$ values, while it shows quite different features for $\epsilon \lesssim \epsilon_{cn}$. The explicit expressions of the adiabatic integrals $J_{nl}$ and $J_{nu}$ computed on the lower and upper branch of the separatrix are reported in [3], the relation $J_{nl} + J_{nu} = 2I_r(P_{zs})$ being valid for $n = 1, 2$.

Let us consider a very low energy particle with initial momentum $I_0$, $P_{z0}$ outside the EC beam region, flying towards it, with initial action $I_0$ lower than the resonant action $I_r$. The separatrix crossings are determined by Eqs.(3,4), in the first crossing one has $J_{nl} = I_0$, while in the last one either $J_{nl} = I_f$ or $J_{nu} = I_f$, depending on the final branch of the Hamiltonian, lower or upper. In the first case $I_f = I_0$, so the electron recover its initial energy ($\Delta\gamma = 0$), while in the last case it experiences a net energy variation

$$\Delta\gamma = \nu_n(J_{nu}(\epsilon_s, P_{zs}) - J_{nl}(\epsilon_s, P_{zs})) = 2\nu_n[I_r(P_{zs}) - I_0]. \quad (6)$$

Equation (6) clearly shows that the nonlinear energy variation due to trapping in the wave field depends on the initial energy and frequency mismatch only. Moreover, for a given set of launching parameters the maximum energy gain occurs for particle with the lowest possible energy. On

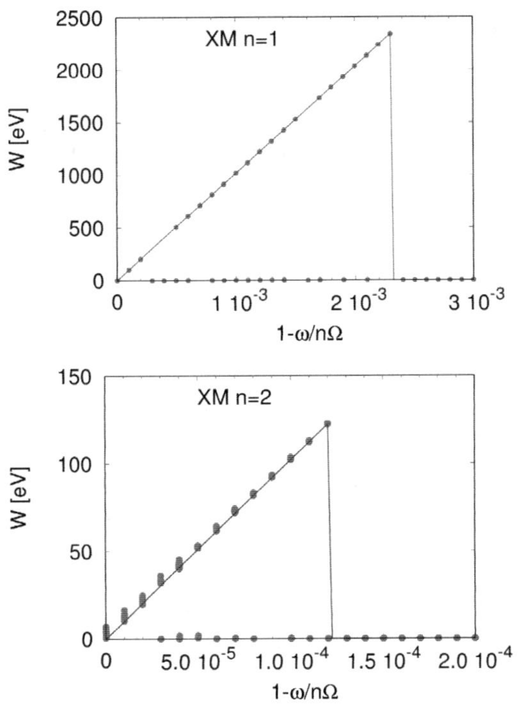


**Figure 5.** Electron final energy $W$ as a function of the EC frequency ratio for $P = 1$ MW, $w = 10$ cm, $f = 170$ Ghz and XM1, XM2 perpendicular propagation. Solid blue curve correspond to the analytical estimate (7) of the maximum energy, red points to the numerical solution of Hamiltonian equations (50 uniformly distributed $\theta_0$ initial values with $I_0 = \beta_{th}^2/2$, $P_{z0} = \beta_{th}$ are evaluated for each frequency ratio). Note the different scales for the two harmonics.

the contrary, the maximum field amplitude $E_M$ determines under which conditions trapping occurs.

Simple analytical estimates for the energy variation can be derived for electrons with very low initial energy. The analysis demonstrates that $P_{zs} \simeq P_{z0}$ and $I_r(P_{zs}) \simeq I_r(P_{z0}) = 1 - \nu_n - {P_{z0}}^2/2$, so that

$$\Delta\gamma \simeq 2\nu_n(1 - \nu_n - I_0 - {P_{z0}}^2/2). \quad (7)$$

Figure 5 shows the very good agreement between the analytical estimate (7) and the results obtained from the solution of the Hamiltonian equations of motion (as in Sec. 2) for various initial phase values and same thermal initial action $I_0$ and $P_{z0}$. The energy increases linearly with $1 - \nu_n$, i.e. with the frequency mismatch from the resonance, up to a maximum value which depends on the maximum field amplitude $\epsilon_M$, and drops to zero afterwards. Since very low energy particles cross the separatrix for $\epsilon \simeq \epsilon_{cn}$, a simple estimate of the maximum energy $W = mc^2(\gamma - 1)$ for increasing $1 - \nu_n$ is obtained by imposing $\epsilon_M = \epsilon_{cn}$

$$W_{max} \simeq 2mc^2 \left[\epsilon_M P_z^{1-q}/(2\beta_{cn})\right]^{2/(4-n)} \quad (8)$$

The scaling of the maximum energy $W_{max}$ with the injected EC beam parameters is obtained recalling that $\epsilon_M \propto P^{1/2}/(fw)$. In case of the X-mode one has

$$W_{max,XM1} \simeq 15.6\, P^{1/3}/(f^{2/3}w^{2/3}), \quad (9)$$

$$W_{max,XM2} \simeq 2.1\, P^{1/2}/(fw), \quad (10)$$

where $W, P, f, w$ are measured in keV, MW, GHz and meters, respectively. The above estimates are valid in the considered low density cold plasma. Note that the EC wave energy transfer to cold electrons is much less efficient in case of the OM, due to the different scaling of the polarization term

## 4 Conclusions

The analysis performed here provides a theoretical description of the EC wave–particle non linear interaction in the pre-ionization phase (much before collisions can play a role). Simple analytical estimates, in very good agreement with numerical particle simulations, are derived for the maximum energy gain as a function of the peak electric field amplitude, and expressed in terms of the EC power, frequency and beam spot size. The results confirm that cold electrons can easily gain energies well above the ionization energy in most conditions, as observed in many experimental conditions [5, 6].

The work leading to this publication has been funded partially by Fusion for Energy under Grant F4E-615. This publication reflects the views only of the author, and Fusion for Energy cannot be held responsible for any use which may be made of the information contained therein.

## References

[1] E. V. Suvorov, M. D. Tokman, Sov. J. Plasma Phys., **14**, 557 (1988)
[2] W. M. Nevins, T. D. Rognlien, and B. I. Cohen, Phys. Rev. Lett., **59**, 60 (1987)
[3] D. Farina, R. Pozzoli, Phys. Fluids B, **3**, 1570 (1991)
[4] I. A. Kotelnikov, G. V. Stupakov, Phys. Fluids B **2**, 881 (1990); J. Plasma Phys., **45**, 19 (1991).
[5] G. L. Jackson, *et al.*, Nuclear Fusion, **47**, 257 (2007)
[6] G.Granucci, *et al.*, Nuclear Fusion, **55**, 093025 (2015)

EPJ Web of Conferences **157**, 03013 (2017) DOI: 10.1051/epjconf/201715703013
*22 Topical Conference on Radio-Frequency Power in Plasmas*

# DC currents collected by a RF biased electrode quasi-parallel to the magnetic field

E. Faudot[1,*], S. Devaux[1], J. Moritz[1], V. Bobkov[2], and S. Heuraux[1]

[1]Institut Jean Lamour, Facult des sciences, 54500 Vandoeuvre-les-Nancy, France
[2]Max-Planck-Institut fr Plasmaphysik, Boltzmannstrasse 2, D-85748 Garching, Germany

## Abstract

Local plasma biasings due to RF sheaths close to ICRF antennas result mainly in a negative DC current collection on the antenna structure. In some specific cases, we may observe positive currents when the ion mobility (seen from the collecting surface) overcomes the electron one or/and when the collecting surface on the antenna side becomes larger than the other end of the flux tube connected to the wall. The typical configuration is when the antenna surface is almost parallel to the magnetic field lines and the other side perpendicular. To test the optimal case where the magnetic field is quasi-parallel to the electrode surface, one needs a linear magnetic configuration as our magnetized RF discharge experiment called Aline. The magnetic field angle is in our case lower than 1 relative to the RF biased surface. The DC current flowing through the discharge has been measured as a function of the magnetic field strength, neutral gas (He) pressure and RF power. The main result is the reversal of the DC current depending on the magnetic field, collision frequency and RF power level.

## 1 Introduction

The negative DC currents measured on ICRF antenna structure [1] are driven by the self biasing of the plasma due to near RF electric fields and RF sheath. These DC currents can also be positive in some specific circumstances. The tilting angle (angle between magnetic lines and the wall or electrode surface) plays a role in the current collection as well as the RF biased surface compared to the connected wall surface on the other side of the flux tube. This study aims at understanding the main mechanisms responsible for the reversal of the current collected by a RF biased surface. To investigate the most favorable case to get positive current, we have biased a RF cathode with a surface parallel to the magnetic field. This configuration can be found in our experimental setup Aline.

*Corresponding author: *eric.faudot@univ-lorraine.fr*

Table 1: Comparison of Tokamak and Aline parameters.

| | Tokamak (SOL) | Aline |
|---|---|---|
| Density (m-3) | $10^{17} - 10^{19}$ | $10^{14} - 10^{17}$ |
| Ti (eV) | 20-40 | 0.026 |
| Te (eV) | 10-20 | 1-10 |
| gas | H, D | D, He, Ar |
| B (T) | 1 - 5 | 0 - 0.11 |
| $\omega_{ci}/\omega_{pi}$ | 0.02 0.8 | 0 0.4 |
| $\omega_{ce}/\omega_{pe}$ | 1 49 | 0 34 (for He) |
| $\omega_{rf}/\omega_{ci}$ | 2 | 10-100 |
| $V_{rf}/Te$ | 1 - 100 | 1 - 200 |
| $\omega_{ci}/\nu_{iN}$ | $\gg 1$ | 0 - 10 |

## 2 Experimental setup : Aline

Aline [2] is a linear magnetized plasma device dedicated to RF sheath study and plasma-wall interaction in RF magnetized discharges. The main advantages of this reactor is to be equipped with a broadband RF amplifier usually dedicated to RF communication antenna. This amplifier can sustain up to 400 W of reflected RF power, which allows to make it work without matching box to achieve a direct coupling as it is the case for RF biased structure close to ICRF antenna in Tokamak SOL. The second asset is the RF compensated probe (planar or cylindrical) mounted on a 3 axis movable arm, able to draw 3D map automatically in a volume (50x10x10 cm) surrounding the RF cathode. This allows to study the flux tubes connected to the cathode and the wall and investigate every topics linked to RF biased magnetized plasma. The typical plasma parameters in Aline can be compared to Tokamak edge plasma in table 1.

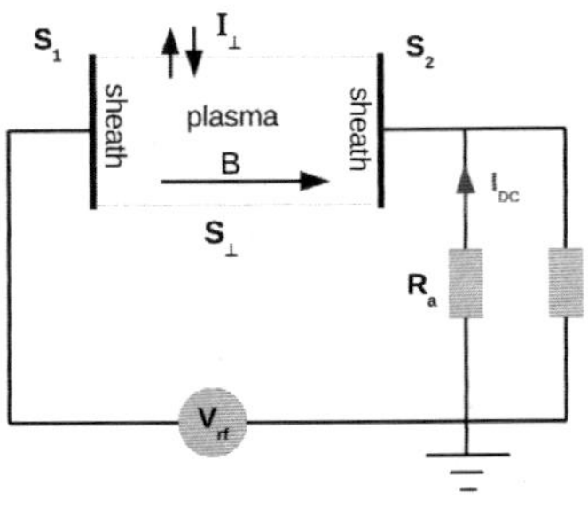


Figure 1: Electric schematic of the RF magnetized plasma discharge with direct coupling.

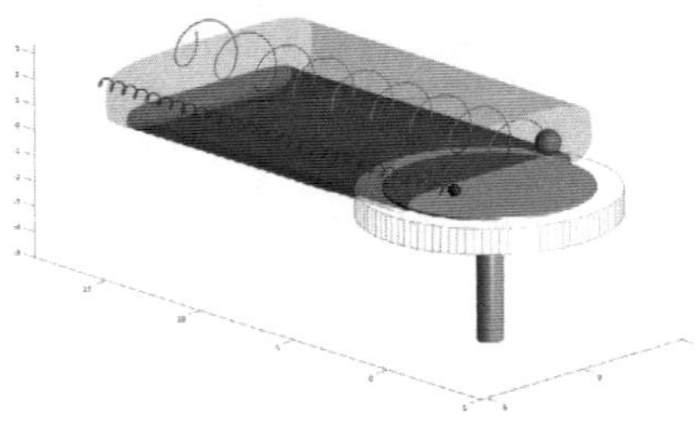

Figure 2: Ion (red) and electron (blue) magnetic flux tubes connected to the RF cathode.

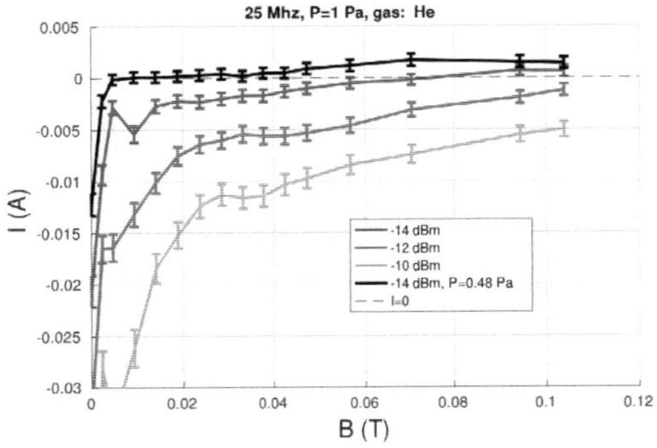


Figure 3: DC current with respect to B for 3 values of RF power and 2 pressures (both at -14 dBm).

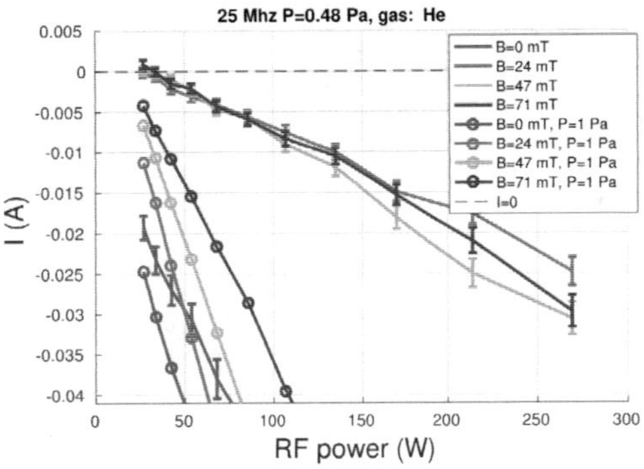


Figure 4: DC current with respect to RF power for 4 values of the magnetic field and 2 pressures.

## 3 DC current measurement

The DC current flowing in the electric circuit is measured according to schematic in figure 1. An ammeter is connected between the ground side of the amplifier and the wall chamber. Which means the ammeter measures a fraction of the total current flowing to the ground. To get the total current, one makes 2 measurements with 2 different ranges with their own internal resistance to deduce the resistance of the other branch. This measurement can not be done on the RF power side because the RF disturbs the measurement.

The DC current flows between the upper RF cathode surface (R=4 cm) parallel to the magnetic field (tilting angle lower than 1) and the wall chamber. When the magnetic field is high enough the collecting surfaces are modified and can be sketched as in figure 2.

3 parametric sweeps have been achieved.

The first is a magnetic field sweep as depicted in figure 3 for 3 values of the RF power (-14 dBm=27 W, -12 dBm=43 W, -10 dBm=68 W). One can see that the current without magnetic field B is negative (electron current received by the cathode) but rises up very fast at low pressure to become positive for magnetic field on the order of 50mT at 0.48 Pa for He gas, and 100 mT at 1 Pa.

The second sweep (figure 4) is a RF power sweep for 4 values of the magnetic field. which shows that the higher the RF power and the higher is the negative current. The current is positive only for low RF power level (-14 dBm, or 27 W), low pressure (only 0.48 Pa) and high magnetic field (over 70 mT).

The third sweep (figure 5) is a pressure sweep for 4 values of the magnetic field. Again the current magnitude increases with the pressure and the current is negative except at low pressure (lower than 1 Pa). At higher pressure, ions can be considered as unmagnetized because $\omega_{ci} < \nu_{iN}$. For example at 1 Pa, the ion-neutral collision frequency is on the order of 200 kHz (Ti=2eV), which corresponds to a magnetic field of 0.05 T. But it is not easy to know the ion temperature at the edge of the flux tube where the RF electric field is strong.

DOI: 10.1051/epjconf/201715703013

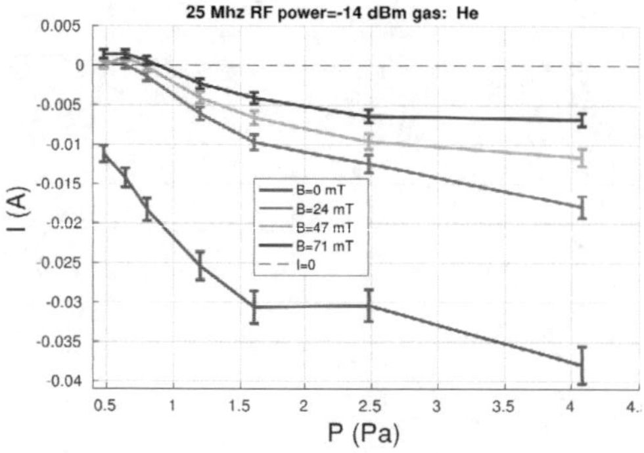


Figure 5: DC current with respect to pressure for 4 values of the magnetic field.

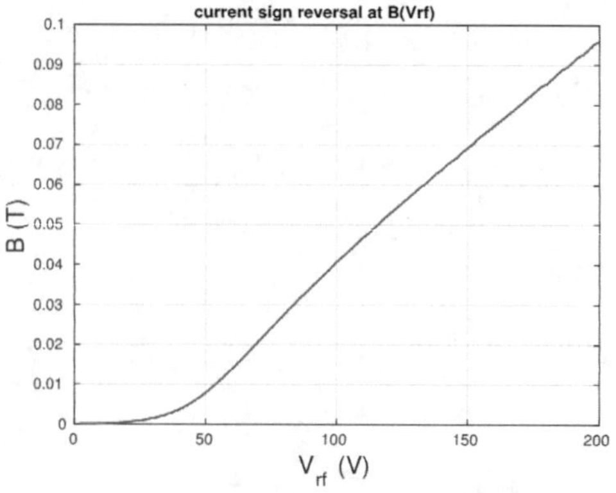


Figure 6: B field needed to reverse the current as a function of the RF potential according to equation 2.

# 4 Modeling

This is a double asymmetric probe model with Boltzmann electrons. The model is based upon a previous model [3] assuming the surfaces of the cathode $S_1$ and anode $S_2$ are the same for both electrons and ions. While the classical symmetric double probe model can be written with a *tanh* function, the asymmetric model with perpendicular current depends on 4 collecting surfaces (figure 2) and one perpendicular current term whose nature is chosen as a function of the plasma type (discharge, fusion plasma [3]). The current conservation at both ends of the flux tube is mainly driven by classical RF sheath theory (unmagnetized) because the plasma potential is still higher than the average potential at the cathode (0V), and then electrons are repelled by a negative potential well around the cathode. The saturation ion and electron currents are then deduced from the collecting surfaces. One can distinguish 4 collecting surfaces, with $D$ the cathode diameter, $r_i$ the ion Larmor radius and $r_e$ the electron Larmor radius :

$$S_1^i = r_i(D + 2r_i) \qquad S_1^e = r_e(D + 2r_e)$$
$$S_2^i = S_1^i \qquad S_2^e = S_2^i$$

with $S_1^i$, $S_1^e$, $S_2^i$ and $S_2^e$ the collecting surface for ions and electrons at the cathode and at the wall respectively.

The DC current is then given by :

$$I = j_i S_1^i \left( \frac{1 - \frac{S_1^e}{S_2^e}\left(\frac{S_2^i}{S_1^i} + \frac{I_\perp}{j_i S_1^i}\right)\exp(\phi_{rf}(t))}{1 + \frac{S_1^e}{S_2^e}\exp(\phi_{rf}(t))} \right) \tag{1}$$

with $j_i$ the ion saturation current and $\phi_{rf} = e.V_{rf}/T_e.sin(\omega t)$. This time averaged current over one RF period must be computed (integration not possible analytically). Assuming $S_2^e = S_2^i$ on the wall side because the surface is perpendicular to the magnetic field, this formula can be written as follows :

$$I = j_i S_1^i \left( \frac{1 - \exp(\phi_{rf}(t) + \delta\phi_1)}{1 + \exp(\phi_{rf}(t) + \delta\phi_2)} \right) \tag{2}$$

$$\delta\phi_1 = \ln\left( \frac{S_1^e}{S_1^i} + \frac{I_\perp}{j_i S_1^i} \right) \tag{3}$$

$$\delta\phi_1 = \ln\left( \frac{S_1^e}{S_2^e} \right) \tag{4}$$

From equation 2, we can deduce for which magnetic field the current is reversed at a given RF potential (collisionless model). Figure 6 reveals that at low RF potential only a small magnetic field is required while at higher RF potential the dependence is quasi linear. Above the curve the current sign is positive and below it is negative. From equation 3, one can see that the effect of the surface ratio and $I_\perp$ are similar to $\phi_{rf}$ effect.

The main effect of the perpendicular current is to shift the current (figure 7) so that it can sometimes never been reversed (at $-I_{sat}$ and $-2I_{sat}$), with $I_{sat}$ the ion satutation current at the cathode ($I_{sat} = en_0 C_s S_1^i$). The sign of the perpendicular current is very important. In our device Aline, this current is supposed to be low and depends mainly on ion-neutral collision [4]. This current sign is the same as the sign of the perpendicular electric field. If the perpendicular ion current enters into the flux tube, $I_\perp$ is negative as defined in assumptions and contributes to reverse the collected current at low B, while a positive perpendicular current (higher than $I_{sat}$ here)

DOI: 10.1051/epjconf/201715703013

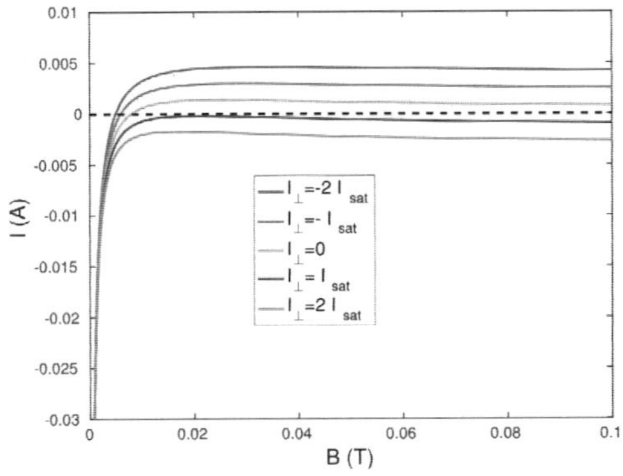


Figure 7: Effect of the perpendicular current on the I(B) characteristics.

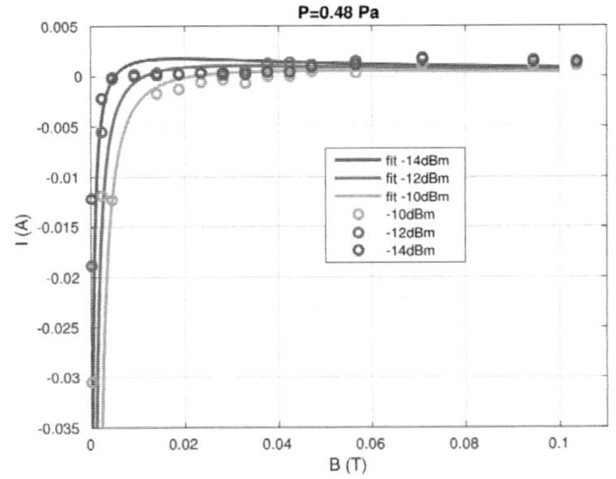


Figure 8: Fit of the I(B) characteristics according to equation 6.

means that no positive current can be collected by the cathode.

Finally, in figure 8, the I(B) measured at P=0.48 Pa and for 3 values of the rf power, can be fitted by equation 1 with the typical parameters measured by the probe and the rf potential measured by the coupler at the cathode. For the -14 dBm rf power, the fit does not match very well the measured data because the positive current first increases and next decreases unexpectedly according to measured values. The model still needs to be improved but gives the main physics involved in current reversal (RF potential, magnetic field). The pressure and thus the collisionnality should be added to the model to explain the pressure dependence.

# 5 Conclusion

The DC current collected on a RF biased surface is not necessarily negative as usually measured on the ICRF antenna structure in tokamaks or in RF capacitive discharges. When the magnetic field is high enough, the RF power at low level and at low pressure (low collisionnality), the DC current can be positive. This has been measured in the linear plasma device Aline in which the magnetic field, the gas pressure, the RF level and frequency can be swept as needed. These measurements have been compared to an asymmetric double probe model including perpendicular current and collecting surface for ions and electrons. The current reversal comes mainly from the collecting surface ratio on the RF and wall side for electrons and ions. If the surface ratio tends to one (symmetric case) and due to the higher mobility of ions perpendicular to the magnetic field, then the ion flux is higher on the RF biased structure. The optimal case is with a biased surface parallel to the magnetic field. The amount of positive current should again depend on the tilting angle, which could be investigated in a future work.

# References

[1] R. Van Nieuwenhove and G. Van Oost, Plasma Phys. Controlled Fusion 34, 525 (1992)

[2] E. Faudot et al, Review of scientific instruments 86, 063502 (2015)

[3] E. Faudot, Physics of plasmas 22, 083506 (2015)

[4] V A Rozhansky et al, Nuclear Fusion, Vol. 39, No. 5 (1999) 613-628

EPJ Web of Conferences **157**, 03014 (2017) DOI: 10.1051/epjconf/201715703014
*22 Topical Conference on Radio-Frequency Power in Plasmas*

# Helicon wave-generated plasmas for negative ion beams for fusion

*Ivo Furno*[1,*], *Riccardo Agnello*[1], *U. Fantz*[4], *Alan Howling*[1], *Remy Jacquier*[1], *Claudio Marini*[1], *Gennady Plyushchev*[1], *Philippe Guittienne*[2], *and Alain Simonin*[3]

[1]Ecole Polytechnique Fédérale de Lausanne (EPFL), Swiss Plasma Center, CH-1015 Lausanne, Switzerland
[2]Helyssen, Route de la Louche 31, CH-1092 Belmont-sur-Lausanne, Switzerland
[3]CEA, IRFM, F-13108 St Paul lez Durance, France
[4]Max-Planck-Institut für Plasmaphysik, Boltzmannstr. 2, 85748 Garching, Germany

**Abstract.** In the next generation of fusion reactors, such as DEMO, neutral beam injectors (NBIs) of high energy (0.8-1 MeV) deuterium atoms with high wall-plug efficiency (>50%) will be required to reach burning plasma conditions and to provide a significant amount of current drive. The present NBI system for DEMO assumes that 50 MW is delivered to the plasma by 3 NBIs. In the Siphore NBI concept, negative deuterium ions are extracted from a long, thin ion source 3 m high and 15 cm wide, accelerated and subsequently photo-neutralized. This requires the development of a new generation of negative ion sources. At the Swiss Plasma Center, a novel radio frequency helicon plasma source, based on a resonant network antenna source delivering up to 10 kW at 13.56 MHz, has been developed and is presently under study on the Resonant Antenna Ion Device (RAID). RAID is a linear device (1.9 m total length, 0.4 m diameter) and is equipped with an extensive set of diagnostics for full plasma characterization. In this work, the principles of operation of resonant antennas as helicon sources are introduced. We present absolute spectroscopy, Langmuir probe, and interferometry measurements on helicon plasmas. We characterize the performance of the source in terms of hydrogen/deuterium dissociation and negative ion production as a function of the input power. Furthermore, first results with the helicon birdcage antenna installed on the Cybele negative ion source at CEA-IRFM are presented, as a first step towards the validation of the Siphore concept.

## 1 Introduction

The next generation of fusion reactors, such as DEMO, will require neutral beam injectors (NBIs) of high energy (0.8-1 MeV) deuterium atoms with high wall-plug efficiency (>50%) to reach burning plasma conditions and provide a significant amount of current drive. The present NBI system for DEMO assumes that 50 MW is delivered to the plasma by 3 NBIs, which can only be obtained using a negative ion-based NBI system. This has stimulated a R&D program on negative ion production from caesiated surfaces. Recently, the use of helicon sources has been suggested as a possible candidate for plasma generation in negative hydrogen ion sources for NBIs for fusion applications [1]. For example in the Siphore concept [2], negative deuterium ions are extracted perpendicularly in a sheet-like geometry (3 m high, 15 cm wide) from a plasma column produced by helicon sources, accelerated and subsequently photo-neutralized.

Since the first experiments performed by Boswell [3], helicon sources have been extensively studied and they are proven to be very efficient for high-density $10^{18}$-$10^{19}$ $m^{-3}$ plasma production (in noble gases, typically argon) with moderate injection power at low pressure. Historically, they have found applications in semiconductor manufacturing and ion thruster systems. For fusion applications, helicon sources may have the following advantages over traditional inductively-coupled plasma (ICP) generators: 1) reduced required radio-frequency (RF) power, leading to wider operational domain; 2) stable operation at low pressure (< 0.3 Pa), reducing negative ion losses by electron stripping; 3) lower electron temperatures in the peripheral plasma region, resulting in higher efficiency of negative ion production in volume production mode; 4) high degree of molecular dissociation in hydrogen/deuterium plasmas, which would be favourable in a caesiated source, although disadvantageous for volume production. However, high power helicon sources are today unexplored both from the physics and technological point of view. While an extensive literature can be found on sources at low to moderate RF powers, ~100 W to 2 kW, there are only a few measurements at larger RF powers [4], and almost none in the 10-50 kW range operating in steady-state (>0.1 s), as required for example for the DEMO NBI system [5].

A new helicon plasma source has been recently developed at the Swiss Plasma Center of the EPFL, the

* Corresponding author: ivo.furno@epfl.ch

DOI: 10.1051/epjconf/201715703014

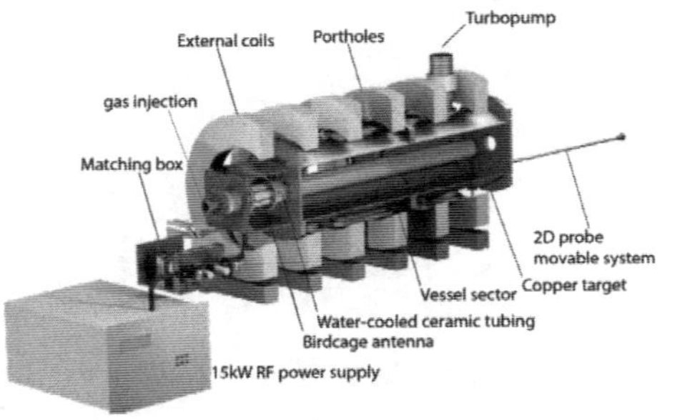


**Fig. 1.** Schematic cross-section of the Resonant Antenna Ion Device (RAID) with the main elements. See Fig. 2 for details of the birdcage antenna.

Resonant Antenna Ion Device (RAID). This has a planned steady-state operating power of 10 kW in $H_2/D_2$ and is currently operating in steady-state at 5 kW. Although a single 10 kW helicon generator will probably not achieve the relevant plasma density required for a fusion-grade NBI source, the 10 kW helicon source is an intermediate step towards larger helicon powers, which will provide an opportunity to explore the main technology and physics issues related to high power, steady state helicons.

This article is organized as follows. In Sec. 2, we describe the main elements of RAID. Section 3 details the birdcage resonant antenna and its working principles. In Sec. 4, we review the first results in hydrogen and deuterium plasmas, showing that a helicon antenna-based system may provide the negative ion injectors considered necessary for fusion reactor development. Future opportunities for research are outlined in Sec. 5.

## 2 The Resonant Antenna Ion Device – RAID

The Resonant Antenna Ion Device (RAID) is a flexible device allowing for full diagnostic access to study helicon wave-produced plasmas and their potential for negative ion production. Plasmas are produced and sustained using a 10 kW birdcage resonant network at 13.56 MHz using different gases with typical electron densities $n_e \sim 10^{18}$ $m^{-3}$ and temperatures $T_e \sim 5$ eV in hydrogen/deuterium.

RAID has a linear vacuum chamber with radius R=0.2 m and length 1.5 m. A drawing of the device with the main elements is shown in Fig. 1. The circular cross-section vacuum chamber is made of stainless-steel and is divided into two sectors. Each sector is water-cooled and possesses many port-holes of different shapes and sizes for diagnostics access. The linear sectors are installed on rails and are individually extractable, so that each sector can be easily removed to facilitate the installation of diagnostics. Spare sectors with different port-hole geometries are available and can be modified for new experiments, without stopping the experimental activity. One turbo-pump (nominal pumping speed of 345 l/s for

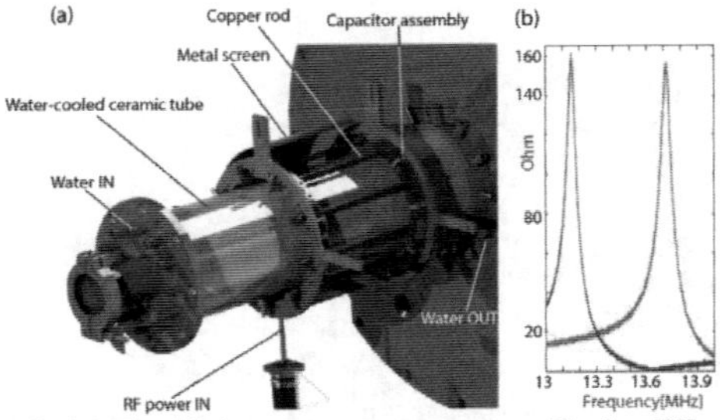


**Fig. 2.** (a) The birdcage resonant antenna with the different components. (b) Magnitude of the antenna impedance in vacuum for two positions of the metal screen.

nitrogen), backed by a primary pump, maintains a base pressure < $10^{-6}$ mbar inside the vessel. Gas, or a mixture of different gases (Ar, He, $H_2$, $D_2$, $N_2$, Ne) is injected either at one axial position or from the back of the helicon antenna, as shown in Fig. 1, via calibrated mass flow controller. Typical gas flow rates are 3 ml/min and 10 ml/min in $H_2$ and Ar gas, respectively, at 0.3 Pa base pressure. The birdcage resonant antenna is installed on the end flange of the vacuum vessel around a dielectric tube, as shown in Fig. 1. This tube consists of an assembly of two coaxial tubes (internal diameter: 9.5 cm, external diameter: 11.5 cm, length: 38 cm) made of high thermal conductivity alumina ($Al_2O_3$). The external ceramic cylinder has 8 channels for cooling the assembly by laminar water flow permitting continuous operation at high RF power. The plasma generated by the antenna forms a column of approximately constant radius terminating on a water-cooled copper plate (target), placed at the end of the chamber. The target can be grounded, floated or biased to test several types of boundary conditions.

Six water-cooled copper coils for the generation of the magnetic field surround the vacuum vessel. The spacing between adjacent coils is 30 cm and each coil can be independently powered up to a maximum coil current of 600 A. This corresponds to a maximum magnetic field on-axis of ~800 Gauss when the coil currents are all in the same direction. By independently powering the magnetic coils, a variety of magnetic field configurations can be produced, including quasi-constant magnetic field, gradients and expanding magnetic field configurations.

## 3 The birdcage resonant antenna

RAID is equipped with a new type of helicon source based on the concept of a birdcage resonant antenna [6], which is shown in Fig. 2(a), using a CAD drawing. The antenna is mounted axially on one end of the vacuum chamber and is made of N=9 conducting parallel legs distributed around a dielectric tube in a cylindrical configuration. The antenna diameter is 13 cm and the 15 cm-long conducting parts are made of copper tubing for water cooling of the system. Each leg is connected at both ends to its closest neighbours by capacitor assemblies. These are made of parallel arrangements of 6 high quality factor mica capacitors. The total individual

 DOI: 10.1051/epjconf/201715703014

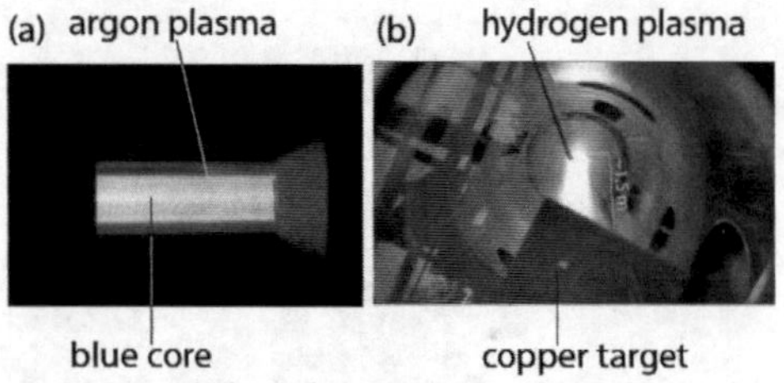


**Fig. 3.** (a) Lateral view of argon plasma with a clearly visible blue core indicating helicon mode (~800 Gauss, 0.3 Pa, 4 kW). (b) End on view of visible light emission from hydrogen plasma (150 gauss, 0.3 Pa, 3.5 kW).

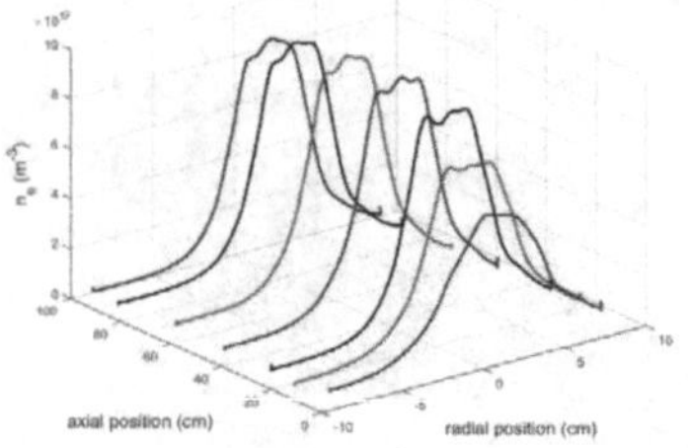


**Fig. 4.** Electron density profile for $H_2$ plasma at 2kW. The centre of birdcage antenna is located at the axial position z = -28

capacitance of each arrangement is $C$ = 3840 pF. A review of the theory of birdcage antennas, extensively used for nuclear magnetic resonance applications, can be found in Ref. [7]. An open birdcage antenna exhibits N-1 resonant frequencies corresponding to the normal modes of the structure. When excited at one of its resonant frequencies a strong, azimuthal sinusoidal, distribution of current amplitude is generated in the antenna legs, all these currents being temporally in phase. In the neighbourhood of each resonant frequency, the antenna behaviour can be modelled in a first approximation as an L, R, C parallel resonance circuit. In the present design on RAID, a resonance of the birdcage at 13.56 MHz is used to generate a homogenous distribution of the RF field in the centre of the structure, corresponding to a sinusoidal current distribution in the antenna legs. The choice of this frequency is dictated by the ISM standard in the industrial plasma field and the availability of RF components.

In previous studies [6], the birdcage helicon antenna was shown to be very efficient in terms of helicon wave excitation in noble gases (typically Ar, Xe, and Ne). Using interferometry measurements of the electron density, a transition from inductive to helicon mode was observed in argon at relatively low injected RF power (~200W). Electron densities of the order of $5x10^{18}$ $m^{-3}$ were obtained in a 10 cm discharge tube, with argon at 1 Pa and for 300 W of power injection. It was also shown that the birdcage resonator allowed many limitations inherent to conventional antenna designs to be overcome (pressure range for which helicon regimes can be sustained, high gas flow rate of electronegative gases, non-uniform static magnetic field). An example of helicon plasma in argon in RAID is shown in Fig. 3(a). A visible light image is taken from a lateral port (perpendicular to the plasma column) and shows an intense blue core on axis (blue being the dominant ion Ar II emission), typically used as an identification of the helicon mode.

Beside its ability to efficiently excite helicon modes, the resonant network has very interesting properties in terms of its input impedance. In fact, these antennas present a large, quasi-real input impedance at the resonance, even when coupled to plasmas. This results in small electrostatic potentials of the antenna with respect to the ground thus mitigating sputtering problems and avoiding the use of Faraday shield at relatively high powers. In terms of impedance matching between the RF generator and its load it also constitutes a major advantage with respect to conventional antennas, which have almost purely reactive impedances and are then strongly mismatched with the standard real 50 Ohm output impedance of the RF generators. This is important to minimize power losses due to dissipation in the matching network. The electrical characterization of the RAID birdcage antenna was performed by using a commercial RF network analyzer to measure the impedance in the absence of a plasma. The magnitude of the impedance is shown in Fig. 2(b), which exhibits a clear resonance around 13.56 MHz with a real value of approximately 160 Ohms. Similarly to flat antennas [8], during plasma operation, the position of the resonance shifts towards higher frequencies and the real impedance decreases, indicating that power is dissipated in the plasma. This frequency shift is a disadvantage when working with fixed-frequency RF power generators, since it results in an impedance mismatch condition and possibly high reflected power. Although the birdcage antenna would work even when operated slightly off its resonant frequency, the 50 Ohm impedance matching condition at the resonance is a great advantage since it minimizes the power dissipated in the matching network. At high power, this will become of crucial importance. In RAID, the frequency shift is compensated by radially moving two semi-cylindrical metallic screens, disposed around the coil on the air side. Adjusting the screen position allows tuning the antenna resonant frequency with 2 MHz range. The effect of the screen on the birdcage impedance is also shown in Fig. 2(b) in vacuum. Based on the measured shift in the resonant antenna frequency and the measured real impedance of the antenna with and without plasma, the power transfer efficiency from the RF antenna to the plasma is approximately 90% [8].

## 4 First characterization of hydrogen and deuterium plasmas

Here, we present first results obtained with the birdcage antenna in hydrogen and deuterium gases, which are required for neutral beam applications. In particular, we focus on conditions satisfying Siphore requirements [9] for the plasma source, namely a low base neutral pressure, < 0.3 Pa and magnetic field ~ 150 G, which will provide the first step towards the validation of the Siphore concept. The discharges are steady-state with the plasma lasting for minutes at all levels of power. As an

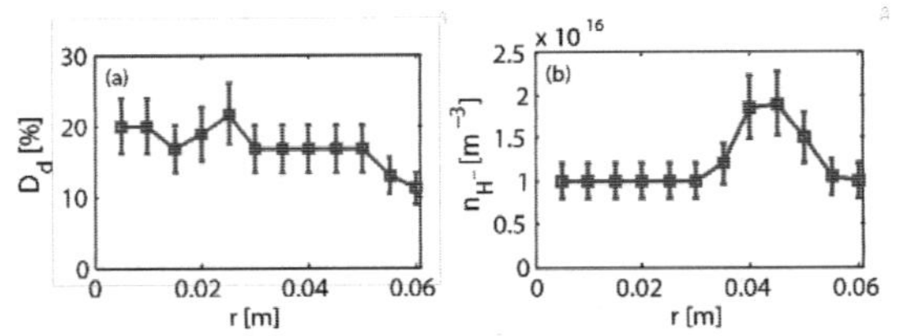


**Fig. 5.** Profiles of (a) the degree of dissociation $D_d$, and (b) the $H^-$ density estimated by YACORA for an input power of 3 kW in hydrogen. The presence of negative hydrogen in a shell around $r$=4.5 cm is clearly visible in (b).

example, Fig. 3(b) shows a visible light image of a hydrogen plasma with 3.5 kW of injected RF power. The camera was positioned end on to provide a full axial view of the plasma column. The pressure of 0.3 Pa, also required for operation in Siphore, is one of the source requirements for ITER and DEMO NBIs to minimize the negative ion losses due to collisions with molecules in the accelerator. As clearly seen, although the birdcage antenna is positioned behind the end flange in Fig. 3(b), the plasma column extends in the axial direction over the entire length of the vacuum vessel, consistent with a volumetric plasma production from helicon waves. The advantage of the birdcage helicon antenna, in particular its ability to produce dense and long plasma columns, with respect to the ICP driver for the Siphore ion source has already been discussed in Ref. [9]. Using a 2D system for a Langmuir probe (LP) in ion saturation mode and a 100 GHz interferometer, we measured the profile of the electron density $n_e$ in a $H_2$ plasma, shown in Fig. 4 for 2 kW of injected RF power. The electron density increases from ~$4x10^{17}$ $m^{-3}$ near the birdcage antenna to ~$1.1x10^{18}$ $m^{-3}$ at 30 cm away from it and remains approximately constant up to the end flange. This feature is particularly promising for the Siphore concept, which will require a homogenous plasma column over a ~3 m length, to extract constant negative ion current over the entire length of the device. LP measurements (in swept mode) of the electron temperature profile are presently ongoing.

Optical emission spectroscopy (OES) measurements, detailed in Ref. [10], were interpreted using the collisional-radiative code YACORA [11] to measure the density of the six species H, $H^-$, $H_2$, $H^+$, $H_2^+$ and $H_3^+$, together with the electron density $n_e$ and temperature $T_e$. Figure 5 shows the radial profiles of (a) the dissociation degree, and of (b) the $H^-$ density estimated by YACORA 68 cm away from the birdcage antenna in a $H_2$ plasma with 3.5 kW of RF power. The dissociation degree $D_d=f / (f+2)$ ($f$ being the ratio of the atomic to molecular hydrogen is almost flat across the entire radius of the plasma column. The most striking feature is the presence of a radial shell between $r=3$ and $r=5$ cm, where the negative ion density reaches values of the order of $2x10^{16}$ $m^{-3}$. Similar OES measurements were performed in $H_2/D_2$ plasmas at different powers revealing that For $H_2$, the negative ion density increases linearly with power, reaching ~$3x10^{16}$ $m^{-3}$ at 5 kW, while for $D_2$ it saturates at ~$4.5x10^{16}$ $m^{-3}$. These values are of interest for a negative source development operating in volume production mode.

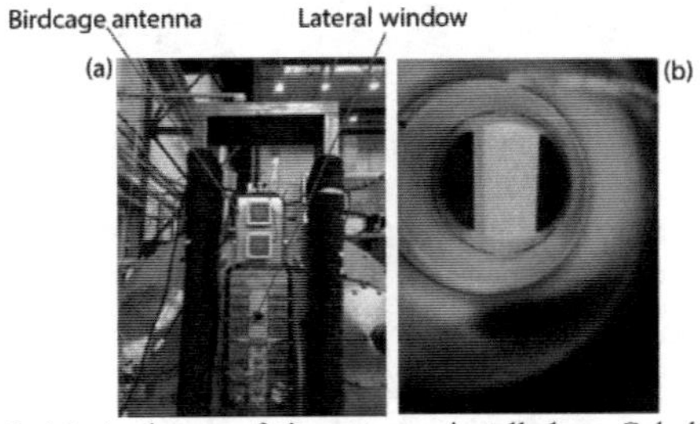


**Fig. 6.** (a) A picture of the antenna installed on Cybele. (b) First helicon plasma in argon in Cybele seen from a lateral window.

## 5 Summary and Outlook

The RAID device recently came online to investigate the physics and technology of high power helicon sources for NBI systems for fusion. The novel 10 kW birdcage resonant antenna has shown great potential for producing highly dissociated and negative ion-rich hydrogen/deuterium plasmas, therefore opening perspectives for NBI fusion applications. A twin prototype of the helicon antenna was recently installed on the Cybele ion source [12] at the IRFM of CEA, Cadarache. Cybele is a tall and narrow ion source with a rectangular aspect ratio, which is particularly relevant to Siphore. Cybele is also equipped with a 30 keV pre-accelerator with slotted apertures allowing to extract a thin blade-like shaped negative ion $H^-$ beam [2]. A picture of the antenna mounted on top of Cybele is shown in Fig. 6(a). Preliminary successful tests of plasma production in argon were performed up to a power of 2 kW. An example of Ar plasma is shown in Fig. 6(b) through a lateral porthole of Cybele. Detailed optimization studies of the magnetic field are ongoing to maximize the homogeneity of the plasma along its axis. The second phase of the developments both on the RAID test bed and on Cybele will investigate the ability of the helicon source at the nominal power of 10 kW to produce a high negative ion density in pure volume production, and in a second step with caesium seeding on Cybele.

This work has been carried out within the framework of the EUROfusion Consortium and has received funding from the Euratom research and training programme 20142018 under grant agreement no. 633053. The views and opinions expressed herein do not necessarily reflect those of the European Commission.

## References

1. U. Fantz and W. Bohm, RF workshop at 14th ICIS (2011).
2. A. Simonin et al., New J. Phys. **18**, 125005 (2016).
3. R.W. Boswell, Plasma Phys. Control. Fusion **26**, 1147 (1984).
4. J. B. O. Caughman et al., Journal of Vacuum Science & Technology A: Vacuum, Surfaces, and Films **35**, 03E114 (2017); B. D. Blackwell et al., Plasma Sources Sci. Technol. **21**, 55033 (2012); J. F. Caneses and B. D. Blackwell, Plasma Sources Sci. Technol. **25**, 55027

DOI: 10.1051/epjconf/201715703014

(2016).
5. P. Sonato et al., Nuc. Fusion **57**, 056026 (2017).
6. P. Guittienne, E. Chevalier and Ch. Hollenstein, Journal of Applied Physics **98**, 083304 (2005).
7. J. Jin, *Electromagnetic analysis and designs in magnetic resonant imaging*, (Boca, Raton, FL: CRC Press), 1988.
8. Ch. Hollenstein et al., Plasma Sources Science and Technology **22**, 055021 (2013).
9. A. Simonin et al., Nuc. Fusion **55**, 123020 (2015).
10. C. Marini et al., Nuc. Fusion **57**, 036024 (2017).
11. D. Wunderlich, S. Dietrich, and U. Fantz, J. Quant. Spectrosc. Radiat. Transfer **110,** 62 (2009).
12. A. Simonin and P. Garibaldi, Rev. Sci. Instrum. **12** 77 03A525 (2006).

# Modelling of combined ICRF and NBI heating in JET hybrid plasmas

*Dani* Gallart[1,*], *Mervi* Mantsinen[1,2], *Clive* Challis[3], *Domenico* Frigione[4], *Jonathan* Graves[5], *Joerg* Hobirk[6], *Eva* Belonohy[6], *Agata* Czarnecka[7], *Jacob* Eriksson[8], *Marc* Goniche[9], *Carl* Hellesen[8], *Philippe* Jacquet[3], *Emmanuel* Joffrin[9], *Natalia* Krawczyk[7], *Damian* King[3], *Morten* Lennholm[10], *Ernesto* Lerche[11], *Ewa* Pawelec[12], *George* Sips[6], *Emilia* Solano[13], *Maximos* Tsalas[14], *Marco* Valisa[15], and JET Contributors[**]

[1]Barcelona Supercomputing Center (BSC), Barcelona, Spain
[2]ICREA, Pg. Lluís Companys 23, 08010 Barcelona, Spain
[3]CCFE, Culham Science Centre, Abingdon, Oxon, OX14 3DB, UK
[4]ENEA, Fusion and Nuclear Safety Dep., C.R. Frascati, Frascati (Roma), Italy
[5]Ecole Polytechnique Fédérale de Lausanne (EPFL), Swiss Plasma Center (SPC), CH-1015 Lausanne, Switzerland
[6]Max-Planck-Institut für Plasmaphysik, D-85748 Garching, Germany
[7]Institute of Plasma Physics and Laser Microfusion, Hery 23, 01-497 Warsaw, Poland
[8]Uppsala University, SE-75120 Uppsala, Sweden
[9]CEA, IRFM, F-13108 Saint Paul Lez Durance, France
[10]European Commission, B-1049 Brussels, Belgium
[11]Laboratory for Plasma Physics, LPP-ERM/KMS, Renaissancelaan 30 Avenue de la Renaissance B-1000, Brussels Belgium
[12]Institute of Physics, Opole University, ul. Oleska 48, Opole 45-052, Poland
[13]Laboratorio Nacional de Fusión, CIEMAT, Madrid, Spain
[14]ITER Organization, Route de Vinon, CS 90 046, 13067 Saint Paul Lez Durance, France
[15]Consorzio RFX, corso Stati Uniti 4, 35127 Padova, Italy

**Abstract.** During the 2015-2016 JET campaigns many efforts have been devoted to the exploration of high performance plasma scenarios envisaged for ITER operation. In this paper we model the combined ICRF+NBI heating in selected key hybrid discharges using PION. The antenna frequency was tuned to match the cyclotron frequency of minority hydrogen (H) at the center of the tokamak coinciding with the second harmonic cyclotron resonance of deuterium. The modelling takes into account the synergy between ICRF and NBI heating through the second harmonic cyclotron resonance of deuterium beam ions which allows us to assess its impact on the neutron rate $R_{NT}$. We evaluate the influence of H concentration which was varied in different discharges in order to test their role in the heating performance. According to our modelling, the ICRF enhancement of $R_{NT}$ increases by decreasing the H concentration which increases the ICRF power absorbed by deuterons. We find that in the recent hybrid discharges this ICRF enhancement was in the range of 10-25%. Finally, we extrapolate the results to D-T and find that the best performing hybrid discharges correspond to an equivalent fusion power of ~7.0 MW in D-T.

## 1 Introduction

During the 2015-2016 JET campaigns many efforts have been devoted to the exploration of high performance plasma scenarios envisaged for ITER operation [1]. The inductive (baseline) scenario and the hybrid scenario have shown major improvements during these campaigns surpassing the previous ITER-like-wall fusion record of 2.3x1016 n/s, thus showing good progress towards demonstrating the fusion rate needed for DT (milestone for DT-ready plasma is $5\times10^{16}$ n/s). The hybrid scenario reached with 33 MW of combined ICRF and NBI power a record neutron rate $R_{NT}$ of $2.9\times10^{16}$ n/s. Experimental results related to optimizing the use of ICRF waves in the hybrid scenario including the discharge time-behaviour are given in [2,3].

We report on simulations aimed at modelling of combined NBI and ICRF heating for high fusion yield in these recent hybrid discharges. We evaluate the performance of H minority heating in the presence of D beam ions using the ICRF modelling code PION [4] coupled to the beam deposition code PENCIL [5]. Our modelling takes into account the synergy between ICRF and NBI heating through the second harmonic cyclotron resonance of deuterium beam ions, which allows us to assess its impact on the fusion performance. We model a number of key discharges with H minority that were designed in order to evaluate key ICRF aspects as the

* Corresponding author: daniel.gallart@bsc.es
** See the author list of X. Litaudon et al. Nucl. Fusion 57 (2017) 102001.

avoidance of impurity accumulation with ICRF waves and its dependence on the ICRF resonance position, the ICRF fusion performance enhancement and the impact of H concentration on the ICRF damping mechanisms, i.e, the fundamental H resonance, the $2^{nd}$ D hearmonic resonance and the direct electron damping. Finally, we model and extrapolate a record discharge to a 50:50 DT fuel mixture.

## 2 Resonance position scan for ICRF impurity control

Dedicated discharges were carried out to assess the dependence of ICRF impurity control on the ICRF resonance position in hybrid plasmas. In these five discharges up to 5 MW of ICRF power at a frequency of 42 MHz and up to 25 MW of NBI power were used. The magnetic field $B_T$ was modified for each discharge and, therefore, the ICRF resonance was located at different major radii ranging from the high-field-side to the low-field-side. The main plasma parameters were the same for each discharge; see Ref. 2 for more details on the experimental setup and results.

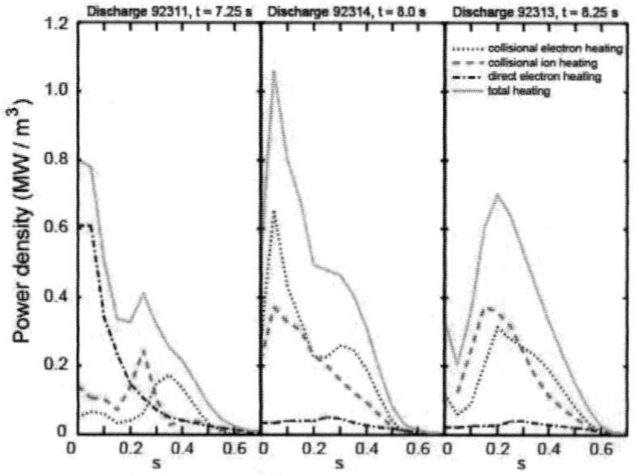


**Fig 1.** The radial profiles of the collisional ion heating, collisional electron heating and direct electron heating due to ICRF waves as given by PION for discharges 92311, 92314 and 92313 with a HFS, central and LFS ICRF resonance, respectively. Here, s is the square-root of the normalised poloidal flux and the profiles are shown after 1.75 s from the start of the main heating.

Figure 1 shows the radial profiles of the collisional ion heating, collisional electron heating and direct electron heating due to ICRF waves as given by PION for discharges 92311, 92314 and 92313 with a HFS, central and LFS ICRF resonance, respectively. The profiles are shown after 1.75 s from the start of the main heating.The maximum bulk ion heating is located at $s \approx 0.25$ for the discharges with an off-axis resonance, where s is the square-root of the normalised poloidal flux. According to PION, direct electron damping becomes dominant in the plasma core for HFS heating (92311) while is almost negligible for LFS heating (92313) and central heating (92314). Nevertheless, this discharge as the other discharge with an off-axis resonance $|R_{res} - R_0| > 15$cm suffered impurity accumulation unlike the discharge with a central resonance, giving rise to overall better central heating [2].

## 3 Hydrogen concentration scan

A set of discharges were carried out to assess the impact of the H concentration on the ICRF heating and fusion performance. These discharges were prepared in the same way except for the different H concentration, see Table 1 [3]. Here, we quote the H concentration $n_H/(n_H+n_D)$ as deduced from the ratio of the $D_\alpha$ and $H_\alpha$ light collected along lines of sight through the plasma. Penning gauge spectroscopy in the divertor gave somewhat higher $n_H/(n_H+n_D)$ of 3-4%. According to our modelling, however, the experimental results are more consistent with the values deduced from the $D_\alpha$ and $H_\alpha$ light. Note that only a few percent of hydrogen is used and there is only a small difference between the hydrogen concentration between the discharges but it is large enough to have an impact in the way plasma damps the ICRF wave energy and consequently, the plasma performance.

**Table 1**. The H concentration for discharges in the H concentration scan.

| | 92321 | 92322 | 92323 |
|---|---|---|---|
| $n_H/(n_H+n_D)$ (%) | ~2.0 | ~1.5 | ~3.0 |

As in the previous case, ICRF power was applied with a central $\omega = \omega_{cH} = 2\omega_{cD}$ resonance. Figure 2 shows the impact of H concentration on the power partition between the H and D ions at t = 11.5 s during the main heating phase. In first order, the ratio of H to D damping scales roughly as $n_H/(n_H+n_D)$, as expected. Typically, for this ICRF scenario, the hydrogen minority is the main absorber at low plasma densities and temperatures that take place during the ramp up. Once the deuterium beams are injected and the plasma gets hotter, damping by resonant D ions becomes the main damping mechanism in the plasma core. As the H concentration decreased from ~3% to ~1.5%, D damping increased from 35 % to 50% and H damping decreased from 55% to 40%, while direct electron damping stayed roughly the same, i.e. 10%, of the total ICRF power.

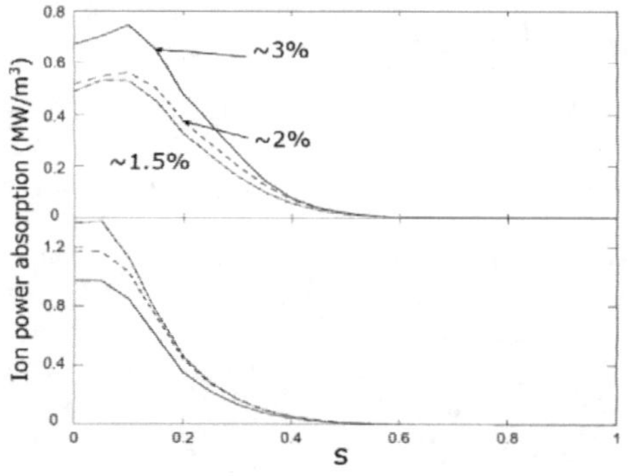


**Fig 2.** ICRF ion power absorption profile for fundamental H (top) and $2^{nd}$ D harmonic resonance (bottom) at t = 11.5 s.

 DOI: 10.1051/epjconf/201715703015

The second harmonic D damping has an advantageous effect on the fusion yield. This is due to the fact that an energetic tail develops in the D distribution function, which enhances the fusion yield. Figure 3 shows the simulated neutron rates for the three discharges which are consistent with the measured neutron rates [2]. The difference between the discharge with the lowest H concentration and the discharge with the highest concentration is about 10-25% which is in good agreement with the experimental results.

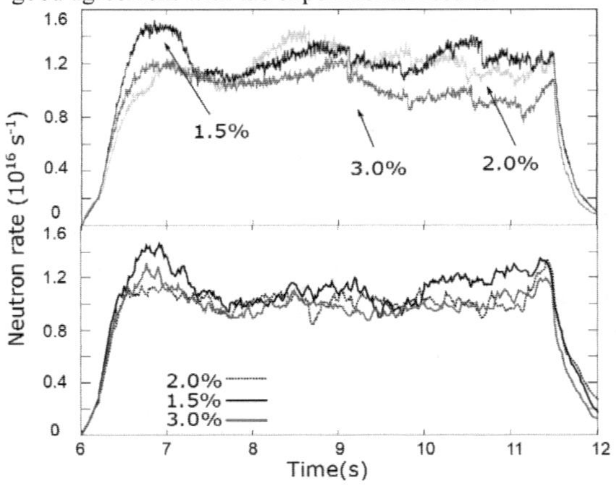


**Fig 3.** Comparison of experimental neutron emission rate (top) and modelled neutron emission rate (bottom) of the H scan discharges.

## 4 High-performance hybrid discharge

One of the main goals of the recent experiments with hybrid plasmas was to improve the fusion performance with respect to the previous neutron rate record of $2.3\times10^{16}$ n/s. Here, we analyse one of the best discharges, i.e. 2.8 T/2.2 MA discharge 92398. In this discharge the ICRF antenna frequency was 42.5 MHz and 5 MW of ICRF power was applied tuned to a central fundamental H and second D harmonic resonance. Together with deuterium NBI power of 26 MW, the total external heating power was 31 MW. For more details on the experimental scenario see Ref. [2].

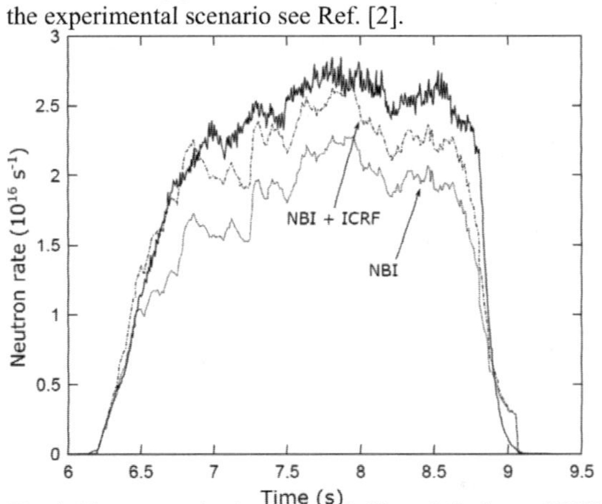


**Fig. 4.** Neutron production rate modelling of discharge 92398 with ICRF and without ICRF.

The experimental and modelled neutron rates are shown in Fig. 4. In these simulations, we have used a Ti/Te ratio deduced from the X-ray crystal spectroscopy data. This data suggests that the Ti/Te ratio is around 1.25-1.6. Here, we apply the H concentration (~2.0%) $n_H/(n_H+n_D)$ as deduced from the ratio of the $D_\alpha$ and $H_\alpha$ light collected along lines of sight through the plasma. Figure 4 shows good agreement between the modelled and the experimental neutron yield. However, there are uncertainties in the prediction of the neutron yield due to uncertainties in some of the measured quantities such as the impurity content, with impact on dilution and uncertainties in the Ti measurement. Figure 4 also shows the modelled neutron rate evaluated with no ICRF heating while keeping all the other parameters exactly the same. The difference between the modelled neutron rates with and without ICRF power allows us to estimate the ICRF enhancement of the neutron rate (Fig.5).

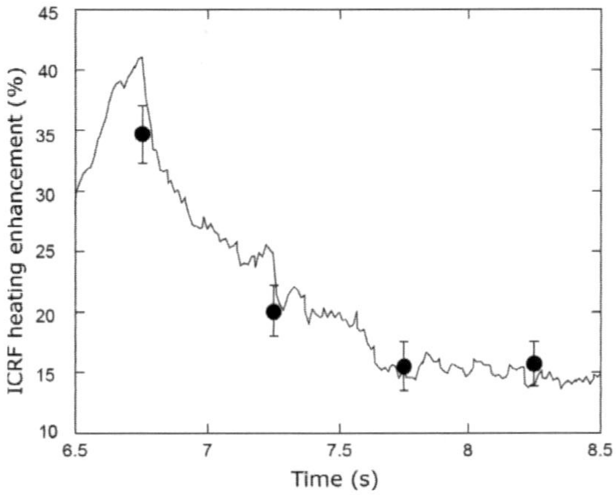


**Fig. 5.** ICRF enhancement of neutron yield for discharge 92398. Comparison of PION (solid line) and experimental results based on TOFOR (black dots) measurements.

As we can see, this enhancement is of the order of 30-15% and decreases in time. The reason of this decrease in time is due to the increasing temperature and density. The neutron emission from NBI increases with these parameters reducing the impact of ICRF in the neutron yield. These results are in good agreement with the ICRF enhancement estimated from a spectroscopic analysis of data collected with the neutron time-of-flight spectrometer TOFOR.

Figure 6 shows ICRF power partitioning between the three ICRF wave power absorption mechanisms. The same ICRF physics prevail in this discharge as in the H concentration scan explained above. Fundamental H damping dominates during the ramp up and the 2$^{nd}$ D harmonic damping during the main heating phase.

The simulated collisional ion and electron heating as well as direct electon damping are shown in Fig. 7. As we can see, collisional ion heating is dominant through the discharge after the transient ramp-up phase where collisional electron heating dominates.

 DOI: 10.1051/epjconf/201715703015

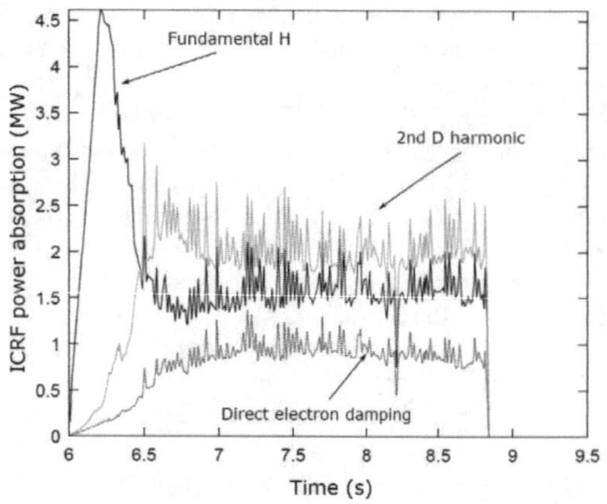


**Fig. 6.** ICRF power absorption mechanisms, fundamental H, 2nd D harmonic and direct electron damping for discharge 92398.

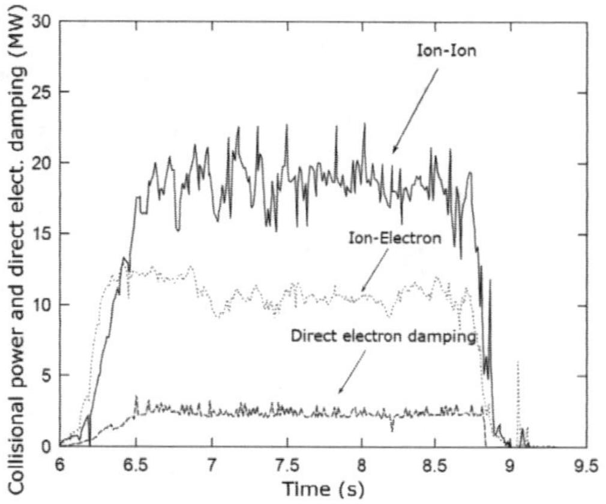


**Fig. 7.** Direct electron damping and collisional power transferred to ions and electrons from ion-ion collisions and io-electron collisions, respectively.

### 4.1 Predictions for D-T

For the prediction of the high-performance hybrid discharge 92398 to D-T, we have kept all the plasma parameters identical to those in the experiment. The only modification consisted of replacing half of the deuterium ions in the main plasma and in NBI by tritium, hence modelling a 50%:50% DT plasma.

In this scenario, the third T harmonic resonance is present in addition to the ICRF damping mechanisms discussed previously. The dominant mechanism is the second D harmonic once the beams are injected. However, as opposed to the D-D fusion reaction, the D-T reaction has an optimal energy at ~120 keV where its cross section peaks. Beyond this energy the cross section value drops drastically, and, therefore, if ICRF heating induces a high energy tail in the distribution function of the resonant particles it could eventually decrease the neutron production rate. In this analysis we took into account the same plasma density and ICRF power output as in the experiment and, thus, we have not assessed how the ICRF performance could be optimised, for more details see [6].

The resulting ICRF enhancement of the neutron rate is about ~5% in D-T and thus lower than that in the D-D. This is because the D beams have an energy around 100 keV which is close to the optimal energy for D-T fusion reactions to occur. Therefore, almost all fusion reactions are produced through NBI. According to our prediction, the equivalent D-T fusion power is about 7 MW.

## 5 Conclusions and further steps

The development of the hybrid plasma scenario has progressed significantly in the recent JET campaign. A higher neutron emission rate and extended duration has been achieved with regards to the previous ILW neutron record.

A number of key topics have been modelled and analysed in this paper, such as the impurity control using ICRF waves, the impact of H minority concentration on the ICRF performance and the ICRF properties and neutron yield enhancement of a high performance hybrid discharge in addition to its D-T prediction. The final conclusions are that heating with ICRF waves centrally has shown to be beneficial in order to avoid impurity accumulation while heating further away from the centre can cause impurity accumulation in some cases. Low H minority concentration tends to make 2nd D resonance more dominant with an advantageous effect in the fusion yield. Finally, the ICRF enhancement in the neutron yield of the high performance discharge has demonstrated to be relevant while in the D-T prediction its importance is diminished.

From the experimental point of view, using $^3$He minority scheme instead of H minority may be beneficial, as modelling predicts improved bulk ion heating and improved fusion performance with $^3$He minority heating as compared to H minority heating [1,6].

*This work has been carried out within the framework of the EUROfusion Consortium and has received funding from the Euratom research and training programme 2014-2018 under grant agreement No 633053. The views and opinions expressed herein do not necessarily reflect those of the European Commission. Dani Gallart would like to thank "La Caixa" for support of his PhD studies.*

## References

1. J. Garcia, et al., Plasma Physics and Controlled Fusion **59**, 014023 (2017).
2. M. Mantsinen, et al., this conference.
3. M. Mantsinen, et al., Europhysics Conference Abstracts vol. 41F, O3.110 2017.
4. L.-G. Eriksson, et al., Nuclear Fusion **33**, 1037 (1993).
5. P.M. Stubberfield and M.L. Watkins, Multiple Pencil Beam, JET-DPA(06)/87,1987.
6. D. Gallart, et al., Europhysics Conference Abstracts, vol. 40A, P2.003 2016.

# Experimental Results of the EU ITER Prototype Gyrotrons

G. Gantenbein[a,*], F. Albajar[b], S. Alberti[c], K. Avramidis[a], W. Bin[d], T. Bonicelli[b], A. Bruschi[d], J. Chelis[e], F. Fanale[d], F. Legrand[f], V. Hermann[f], J.-P. Hogge[c], S. Illy[a], Z. C. Ioannidis[a], J. Jin[a], J. Jelonnek[a], W. Kasparek[g], G. P. Latsas[e], C. Lechte[g], M. Lontano[c], I. G. Pagonakis[a], T. Rzesnicki[a], C. Schlatter[c], M. Schmid[a], I. G. Tigelis[e], M. Thumm[a], M. Q. Tran[c], J. L. Vomvoridis[h], A. Zein[a], A. Zisis[e]

[a]Karlsruhe Institute of Technology (KIT), IHM, Kaiserstr. 12, 76131 Karlsruhe, Germany
[b]European Joint Undertaking for ITER and the Development of Fusion Energy (F4E), Barcelona, E-08019, Spain
[c]Swiss Plasma Center (SPC), Ecole Polytechnique Fédérale de Lausanne, CH-1015 Lausanne, Switzerland
[d]Istituto di Fisica del Plasma, Consiglio Nazionale delle Ricerche, Via R.Cozzi 53, 20125 Milano, Italy
[e]Faculty of Physics, National and Kapodistrian University of Athens, Zografou, GR-157 84, Athens, Greece
[f]Thales Electron Devices, 2 rue Marcel Dassault, Vélizy-Villacoublay, F-78141, France
[g]University of Stuttgart, IGVP, Pfaffenwaldring 31, 70569 Stuttgart, Germany
[h]National Technical University of Athens, Greece

**Abstract.** The European 1 MW, 170 GHz CW industrial prototype gyrotron for ECRH&CD on ITER was under test at the KIT test facility during 2016. In order to optimize the gyrotron operation, the tube was thoroughly tested in the short-pulse regime, with pulse lengths below 10 ms, for a wide range of operational parameters. The operation was extended to longer pulses with a duration of up to 180 s. In this work we present in detail the achievements and the challenges that were faced during the long-pulse experimental campaign.

## 1 Introduction

The European 1 MW, 170 GHz industrial prototype CW gyrotron for ECRH&CD on ITER is a conventional (hollow-cavity) gyrotron, which has been developed by the European GYrotron Consortium (EGYC) in cooperation with the industrial partner Thales Electron Devices (TED), under the coordination of the European Joint Undertaking for ITER and the Development of Fusion Energy (F4E).

According to the EU development strategy, the design of the CW gyrotron is completely based on a corresponding modular short-pulse (SP) prototype and the 1 MW, 140 GHz gyrotron development for W7-X [1]. The SP prototype was developed in order to validate the scientific design in terms of RF output power, total efficiency and quality of RF output beam.

The SP prototype was extensively tested at the KIT test facility in 2015 and the scientific design of the individual tube components has been verified, with the gyrotron delivering an output power of higher than 1.2 MW with 40 % total efficiency (in single-stage depressed collector operation) and with a RF output beam Gaussian mode content of about ~98 % [2].

The CW industrial prototype was delivered to the KIT test facility in early 2016 and the experiments started immediately in the short-pulse regime (with pulse length below 10 ms). The main goal of this first validation phase was to optimize the gyrotron alignment with respect to the magnetic field axis, as well as to determine the best operational parameters, namely accelerating voltage, beam current and magnetic field profile, in terms of generated RF power. A summary of the first short-pulse experiments at KIT with the CW prototype is presented in [3, 4].

The optimal operational parameters at short-pulses were used as a starting set for the conditioning of the CW tube with RF and the progressive increase of the pulse length. After some weeks of conditioning, it was possible to operate the gyrotron with pulse lengths of up to 180 s, which is the limit at the High-Voltage (HV) power supply currently available at KIT.

In this work we focus on the performance of the tube during the second validation phase, i.e. long-pulse operation at the KIT test facility, details regarding the operating points are presented, the performance of the tube during the short-pulse phase of the experiments is summarized and the long-pulse phase of the experiments with pulses up to 180 s is analyzed.

* Corresponding author: gerd.gantenbein@kit.edu

## 2 Operating Points for the CW Prototype

In order to have higher flexibility in the operation of the CW gyrotron, two operating points in terms of the accelerating voltage have been defined. The nominal one is the High Voltage Operating Point (HVOP), which corresponds to an accelerating voltage of 79.5 kV and a beam current of 40 A, the alternative one is the Low Voltage Operating Point (LVOP), which corresponds to an accelerating voltage of 71 kV and a beam current of 45 A. Further details regarding the operational parameters of both operating points, as well as the corresponding simulated performance of the gyrotron, are given in Table 1.

The optimization of the gyrotron operation in the short-pulse regime started and focused on the LVOP, whereas only brief experiments with the HVOP were performed. Due to the limited time of the experimental campaign the experiments in the long-pulse regime were continued in the LVOP, as the tube was specifically conditioned for such operating parameters. In the following paragraphs the presented results refer always to the LVOP, except if it is stated differently.

**Table 1.** Operating points for CW operation

| Parameter | HVOP | LVOP |
|---|---|---|
| Operating mode | $TE_{32,9}$ | $TE_{32,9}$ |
| Cavity magnetic field | 6.78 T | 6.69 T |
| Accelerating voltage | 79.5 kV | 71.0 kV |
| Depression voltage | 35 kV | 30 kV |
| Beam current $I_b$ | 40 A | 45.3 A |
| Cavity beam radius $R_b$ | 9.44 mm | 9.44 mm |
| Cavity pitch factor $\alpha$ | 1.29 | 1.22 |
| Output power at window | 1 MW | 1 MW |
| Frequency | 170.23 GHz | 170.23 GHz |
| Interaction efficiency | 35% | 35% |
| Total efficiency w/o depressed collector | 32% | 32% |
| Total efficiency w/ depressed collector | >50% | >50% |
| Peak cavity wall loading | 2.1 kW/cm$^2$ | 2.1 kW/cm$^2$ |

## 3 Short-Pulse Operation

A basic description of the performance of the tube in SP operation is presented in [3-4]. Herein, we report on further details for the operation of the CW gyrotron in the SP regime. It should be noted that the gyrotron performance described in this section corresponds to pulses not longer than 5 ms and always to non-depressed collector operation of the tube.

The alignment verification of the gyrotron cavity with respect to the magnetic field has been performed in the short-pulse regime using a method developed at KIT, which is based on the study of symmetry in the excitation regions of the main mode [5]. The concept is to introduce a radial displacement of the electron beam in the cavity area using a set of super-conducting (SC) transverse field dipole coils integrated inside the KIT magnet and monitor the influence of this displacement on the excitation of a cavity mode. By changing the current of the dipole coils, the switching point between the main mode and its neighboring competitors is recorded. The graphical representation of the switching points forms a circle, around the dipole currents values, where the main mode is properly excited. The distance from the center of this circle to the origin of the dipole coils coordinate system (zero current in both directions) provides an indication regarding the misalignment of the magnetic field axis with respect to the gyrotron cavity axis.

During the alignment phase it was found that these current values correspond roughly to a (0.56 mm, 0.41 mm) shift of the electron beam. By operating the gyrotron at the LVOP with the above mentioned coil current values, the maximum RF output power obtained with near nominal voltage/current parameters was 930 kW with 26 % total efficiency (in non-depressed collector operation). More details about the gyrotron performance under these conditions are given in [4].

The thermal footprint of the gyrotron output RF beam has been imaged on two different target plates of 200 mm diameter, one made from PVC and a second one from regular paper. The plates are placed along the RF beam path at various distances between 180 mm and 1300 mm from the gyrotron output window. The distance was progressively increased using a motorized movement stage. According to the post processing of the measurements with a phase retrieval algorithm, the Gaussian mode content of the beam was found to be at least 97 %, which fulfills the ITER specifications.

## 4 Long-Pulse Operation

In order to proceed from the short-pulse operation to 180 s long-pulse operation, several weeks were needed including the optimization of the test-stand auxiliaries as well as of the control system. Recall that due to limitations of the HV power supply at KIT, the maximum pulse length that can be achieved with beam current up to 50 A is 180 s with a 1/10 duty cycle. Due to these time limitations the experimental campaign was focused on the LVOP (beam current ~45 A and accelerating voltage ~71 kV).

Fig. 1 presents the control system recordings for one of the first 180 s pulses that were achieved. In this figure the black curve corresponds to the accelerating voltage, which is the sum of the cathode voltage (green curve) and the depression voltage (red curve), whereas the blue curve is the emitted electron beam current. In the current curve it is obvious that during the first seconds of the pulse the emitted current is slightly reduced because of the cathode cooling effect and then recovers towards the targeted value. This recovery is achieved by employing a current boosting scheme, which controls the filament current before and during the pulse.

In the same figure, the olive and yellow curves correspond to the vacuum pressure signal of the ion getter pumps. During most of the pulses including this specific one, the vacuum level remained better than $10^{-8}$ mbar. The purple curve represents the output voltage of the mode-loss RF diode, which deactivates the gyrotron operation as soon as the signal drops below a predefined level. This could happen for example in case that the accelerating voltage was increased above the value which the nominal operating mode could tolerate. In general, no unexpected mode excitation took place during the experimental campaign.

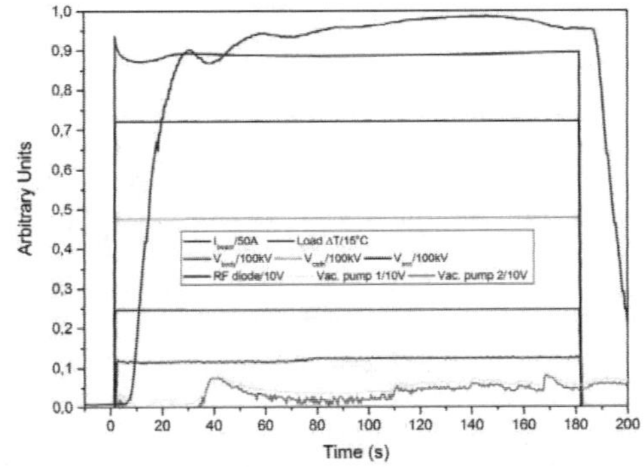


**Fig. 1.** Typical 180 s pulse achieved during the experiments, the temperature measurement in the load (proportional to RF power) is delayed and shows oscillations at the beginning of the pulse due to the KIT cooling system only (all measurements normalised to the indicated values).

The generated power and the dissipated losses on the internal components of the gyrotron as well as of the auxiliaries in the microwave box are monitored through the various cooling circuits that are connected to a calorimetry system.

Table 2 presents an example of the power that was measured in the different circuits of the calorimetry system during a typical 0.8 MW pulse. It should be noted that the CW gyrotron prototype is equipped with three cooling circuits for the beam tunnel, the cavity and the quasi-optical launcher. These circuits also support the cooling of internal mirror 1 (M1) and mirror 2 (M2) of the quasi-optical system of the gyrotron, which are connected in series. The last mirror of the system, namely mirror 3 (M3) has a completely individual cooling circuit.

Considering that the losses on M1 and M2 should be comparable to those on M3, it is evident from Table 2 that the measured losses in the beam tunnel were negligible (the accuracy of the calorimetric measurements is in the order of 1%).

In the last column of the table the power measurements are expressed with respect to the total input power of the electron beam, that is calculated by the cathode voltage and the beam current, which for this pulse were 47.4 kV (depressed collector operation of the gyrotron) and 47.5 A, respectively. The internal losses (internal components of the tube) are in the level of 5 % with respect to the RF power at the synthetic diamond window, whereas the external losses (inside the microwave box) are in the range of 2-3 % with respect to the power at the window.

**Table 2.** Calorimetry measurements for a typical pulse

| Measurement | Power (kW) | % of input power |
|---|---|---|
| Cavity (M1-M2) | 28 | 1.24 |
| Beam tunnel (M1-M2) | 1.26 | 0.06 |
| Launcher (M1-M2) | 11 | 0.49 |
| Mirror 3 (M3) | 1 | 0.04 |
| Window | 1 | 0.04 |
| Mirror box | 3 | 0.13 |
| Relief load | 2 | 0.09 |
| Collector | 1365 | 60.67 |
| Water load | 792 | 35.20 |
| Microwave box | 21 | 0.93 |
| Total | 2225 | 99 |

Setting properly the three independent currents of the gyrotron magnet (Oxford Instruments), a quite wide area of magnetic field angles $\varphi_B$ at the cathode emitter and beam radii $R_b$ in the cavity was examined. The magnetic field angle $\varphi_B$ is an important operating parameter that is strongly related to the pitch factor of the electrons [6].

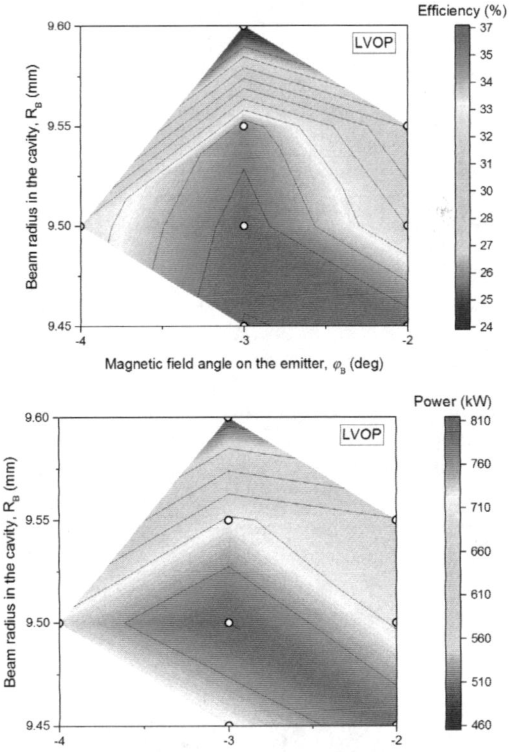


**Fig. 2.** Operational map – RF power (bottom) and total efficiency (top) with respect to the magnetic field angle at the cathode emitter and the radius of the electron beam in the cavity.

DOI: 10.1051/epjconf/201715703016

Fig. 2 presents the operational maps of the gyrotron in long-pulse operation in dependence of $\varphi_B$ and $R_b$. In these figures the white circles correspond to the position where the actual measurements took place. It should be noted that the performance reported for each ($\varphi_B$, $R_b$) combination corresponds to the optimal voltage and beam current operating parameters found at this point, whereas the collector depression voltage is usually set in the range 20 - 25 kV without special optimization.

The optimal performance of the gyrotron in terms of output power was found at the operating point ($\varphi_B = -3°$, $R_b = 9.50$ mm), where 800 kW were generated in most of the pulses. The maximum RF power value recorded is 811 kW with a total efficiency of 36 % (in depressed collector operation).

Fig. 3 shows the RF power and the corresponding total efficiency versus the collector depression voltage for the operating point ($\varphi_B = -3°$, $R_b = 9.50$ mm). The measurements were recorded at 60 s long pulses. By increasing the depression voltage, the generated RF power remains almost constant, while the efficiency practically increases linearly, reaching a maximum value of 38 %.

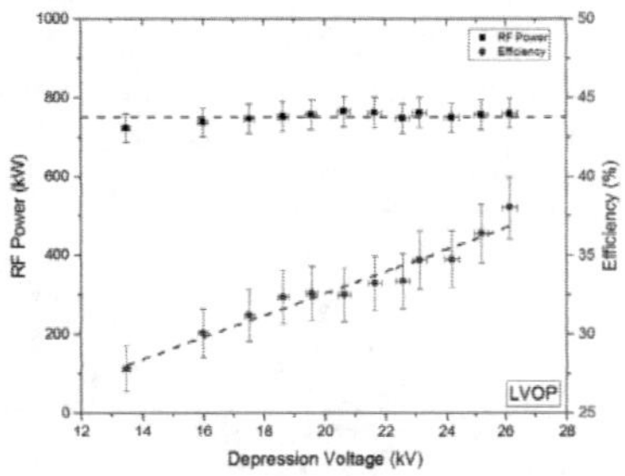


**Fig. 3.** RF power and corresponding total efficiency versus the depression voltage for the LVOP at the operating point ($\varphi_B = -3°$, $R_b = 9.50$ mm).

The maximum value of the applied depression voltage, where stable operation of the gyrotron was achieved is 26 kV. The operation with higher values of the depression voltage was quite challenging, since a body current in the mA range and often instabilities during the first seconds of the pulses were observed. Such instabilities, which include arcing and beam current fluctuations, could be an indication of an increased number of reflected electrons [7].

The time dependency of the frequency of the nominal mode $TE_{32,9}$ was recorded using the KIT pulse shape analysis (PSA) measurement system [8].

Starting from approximately 170.300 GHz the operating frequency drops to 169.950 GHz ($\Delta f$= 350 MHz) within 500 ms at an output power of 650 - 700 kW. The frequency and frequency drop meet very well the ITER requirement of 170 GHz $\pm$ 0.3 GHz.

The above described frequency drop during a pulse is, in general, attributed to the neutralization of the electron beam as well as to the thermal expansion of the cavity that strongly depends on the generated RF power.

A very brief part of the experimental campaign was focused on long-pulse operation of the tube at the HVOP with pulses of 30 s and 60 s duration. A RF power above 800 kW was generated with the corresponding total efficiency of approximately 35 % (in depressed collector operation).

## 5 Conclusion

The first phase of the experiments with the 1 MW, 170 GHz CW prototype EU gyrotron for the ITER ECRH&CD system have been successfully completed at KIT. With these experiments most of the ITER specifications have been verified including the HV properties of the gyrotron, the frequency of the nominal mode, the stability of the vacuum, as well as the $TEM_{00}$ mode purity (Gaussian content higher than 97%) and the alignment of the RF beam at the diamond window. The maximum RF power that was achieved in 180 s pulses is 0.8 MW both for the LVOP and HVOP with total efficiency 38 % and 35 %, respectively. Considering that this is the first industrial CW prototype of the tube, the achieved performance in the first phase of the experiments is very good.

The experiments will be continued at SPC, Lausanne. The gyrotron was transported to SPC in March 2017 and installed in the specified (Cryogenic) magnet. Taking advantage of the flexibility of the High Voltage Power Supply (HVPS) at SPC, the goal is to further increase the pulse length. Furthermore, reliability and power modulation tests will be performed. The goal is to achieve at least 95 % success rate out of 50 subsequent 180 s pulses. Regarding the modulation tests (up to 5 kHz), 100 % power depth modulation using both the main and the body HVPS will be attempted during 180 s pulses.

**Acknowledgement**

This work is partly supported by Fusion for Energy under Grant F4E-GRT-553 to the European Gyrotron Consortium (EGYC). EGYC is a collaboration among SPC, Switzerland; KIT, Germany; HELLAS, Greece; IFP-CNR, Italy. The views expressed in this publication are the sole responsibility of the authors and do not necessarily reflect the views of F4E and the European Commission.

## References

[1] G. Gantenbein, et al., J. Infrared Milli Terahertz Waves, **32**, no. 3, pp. 320-328, Dec. 2011.
[2] T. Rzesnicki, et al., IRMMW-THz 2015, proc. 3134554, Hong Kong, Aug. 23-28, 2015.
[3] J.-P. Hogge, et al., IRMMW-THz 2016, Copenhagen, Denmark, Sep 25-30, 2016.
[4] T. Rzesnicki, et al., Fusion Eng. Des., published on-line, Feb. 2017.
[5] T. Rzesnicki, et al., 37th International Conference on Infrared, Millimeter, and Terahertz Waves, NSW, Australia, 2012.
[6] I. Gr. Pagonakis, et al., Phys. Plasmas, **24**, no. 3, 2017.
[7] I. Gr. Pagonakis, et al., 9th Int. Workshop "*Strong Microwaves and Terahertz Waves: Sources and Applications*", Nizhny Novgorod, Russia, 24-30 July 2014.
[8] A. Schlaich, et al., IEEE Trans. Microwave Theory Tech., **61**, no. 12, pp. 4660–4666, Dec 2013.

EPJ Web of Conferences **157**, 03017 (2017) DOI: 10.1051/epjconf/201715703017
*22 Topical Conference on Radio-Frequency Power in Plasmas*

# Recent Trends in Fusion Gyrotron Development at KIT

G. Gantenbein*, K. Avramidis, J. Franck, S. Illy, Z. C. Ioannidis, J. Jin, J. Jelonnek, P. Kalaria, I. Gr. Pagonakis, S. Ruess, T. Rzesnicki, M. Thumm, C. Wu

Karlsruhe Institute of Technology (KIT), IHM, Kaiserstr. 12, 76131 Karlsruhe, Germany

**Abstract.** ECRH&CD is one of the favorite heating system for magnetically confined nuclear fusion plasmas. KIT is strongly involved in the development of high power gyrotrons for use in ECRH systems for nuclear fusion. KIT is upgrading the sub-components of the existing 2 MW, 170 GHz coaxial-cavity short-pulse gyrotron to support long-pulse operation up to 1 s, all components will be equipped with a specific active cooling system. Two important developments for future high power, highly efficient gyrotrons will be discussed: design of gyrotrons with high operating frequency (~ 240 GHz) and efficiency enhancement by using advanced collector designs with multi-staged voltage depression.

## 1 Introduction

This contribution reports on activities in the field of research and development of RF sources (gyrotrons) and related components for electron cyclotron resonance heating and current drive (ECRH&CD) of magnetically confined nuclear fusion plasmas at KIT. In particular activities towards a 170 GHz, 2 MW gyrotron for a later stage of ITER and developments towards a future DEMO reactor will be addressed.

KIT is pushing forward the experimental verification of the multi-megawatt coaxial-cavity technology, by upgrading the 170 GHz 2 MW coaxial-cavity short-pulse prototype to longer pulses (up to 1 s). In order to increase the pulse length of the 2 MW coaxial cavity gyrotron all major subcomponents have been equipped with active cooling systems. An improved design of a new Magnetron Injection Gun (MIG) which respects fundamental gun design criteria for long-pulse operation and minimises beam instabilities has been performed and ordered. An inverse MIG (IMIG) has been designed and is currently under manufacturing. Several advantages of this type of gun compared to a conventional MIG will be verified experimentally.

In the frame of EUROfusion the investigation of advanced 240 GHz gyrotrons for future DEMO is ongoing, focusing on two options: conventional-cavity gyrotron and coaxial-cavity gyrotron. Conventional-cavity gyrotrons offer a robust design with, at present, 1 MW RF output power. Our investigations on limits of that design show that 1.5 MW could be achieved with this technology. The coaxial cavity technology is more challenging, but offering higher output power. A focus area in the development is the investigation of the impact of tolerances and misalignment of the electron beam and coaxial insert on stable operation of the design mode.

High efficiency operation of gyrotrons is a key issue for future application in ECRH systems of fusion reactor relevant devices. To operate such a system economically an overall efficiency of the gyrotron significantly above 50 % is required. KIT is focussing on the developments of advanced collectors which recover efficiently the non-spent energy of the electron beam. Several principles of multistage depressed collector systems (MSDC) have been studied intensively.

## 2 MW Coaxial Cavity Gyrotron Development

In order to increase the gyrotron pulse length of the existing 2 MW coaxial-cavity gyrotron [1], all subcomponents such as beam tunnel, cavity, launcher, mirrors, synthetic diamond window, and the collector will be equipped with an active cooling system. In order to take benefit of the modularity of the gyrotron independent cooling systems have been designed. This allows monitoring of the internal losses of the individual components and an energy balance of the tube during long-pulse operation.
The design of the existing conventional magnetron injection gun (MIG) has been improved in order to support the design criteria for long-pulse operation [2]. The suppression of the electron beam halo caused by magnetically trapped electrons in the region between the cathode and the cavity has been optimized by modification of the anode and cathode contour. The

* Corresponding author: gerd.gantenbein@kit.edu

 DOI: 10.1051/epjconf/201715703017

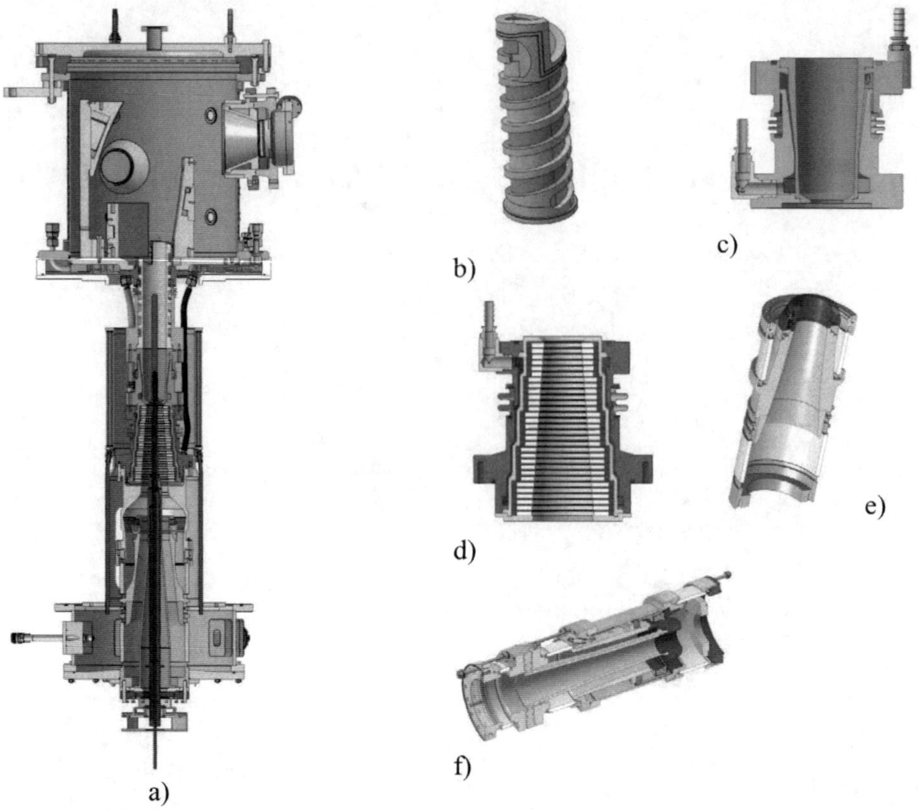


Fig. 1:Schematic view of coaxial cavity gyrotron (a) with long-pulse components, launcher (b), cavity (c), beam tunnel (d), MIG (e) and IMIG (f).

generation of secondary electrons in non-emitter regions from bombardment with trapped particles has been minimised. Additionally, an advanced type of emitter ring [3] is used in order to minimize the sensitivity of the manufacturing and alignement tolerances of the emitter and its neighbouring parts on the gyrotron performance

In parallel, an innovative Inverse Magnetron Injection Gun (IMIG) has been designed at KIT for the modular 2 MW, 170 GHz coaxial-cavity pre-prototype gyrotron [4]. The design allows the implementation of a significant larger emitter ring compared to the conventional MIG, considering the same radial tube dimensions. Due to the direct cooling of the thermally loaded components close to the emitter a better temperature profile for those components and a smaller thermal expansion of the beam forming components is achieved which reduces the possibility for unwanted emission.

A traditional beam tunnel with stacked corrugated copper and damping ceramic rings and active cooling has been designed. The development and manufacturing of a full metallic beam tunnel is also in process.

RF power generation from electron-wave interaction takes place in the cavity and is necessarily associated with ohmic surface losses. In this case the power loss at the surface of the cavity will be 50 kW with a maximum power density ~2 kW/cm² at the centre of the cavity. The approach of the active cooling system of the cavity is shown in Fig. 1. The outer water jacket is movable along the axis in order to respect the different temperature and thermal expansion coefficient of the components.

The launcher and the mirror system are responsible for the conversion of the main $TE_{34,19}$ cavity mode into a linearly polarized fundamental Gaussian beam mode. In high power operation the loading of the launcher surface, especially at the top part of the launcher, is very high. A

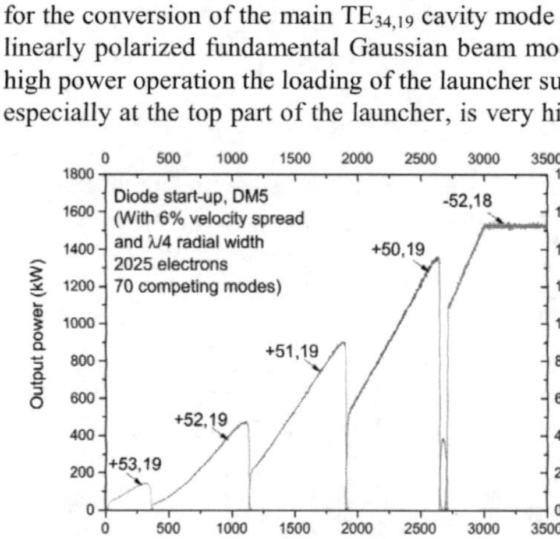


Fig. 2: Multi-mode, self-consistent, diode start-up analysis for $TE_{52,18}$ mode.

helical cooling structure has been designed in order to guarantee stable operation.

## 3 240 GHz Gyrotron Development for DEMO

Based on the (EU-) DEMO baseline of 2012 design studies on gyrotrons and key components have been performed in order to identify and overcome possible design and simulation obstacles and limitations. The two major concepts – conventional, hollow-cavity (or: cylindrical-cavity) and coaxial-cavity gyrotrons – are investigated in parallel. While conventional, hollow-cavity gyrotrons have a robust design and are expected to deliver around 1 MW output power per unit [5], coaxial-cavity gyrotrons can produce up to 2 MW (or even higher) at the cost of a more difficult design [6]. In the present study, a 236 GHz hollow-cavity design and a 237.5 GHz coaxial-cavity design are considered, with operating parameters as given in Table 1 [7].

In the case of the conventional-cavity concept, a thorough investigation on the highest mode eigenvalue $\chi_{max}$ at which stable operation is still possible at 236 GHz (assuming a diode-type start-up) was performed in order to estimate the limitations in power. This led to $\chi_{max} \sim 125$ [8]. The RF behaviour of the very high-order mode $TE_{52,18}$ (eigenvalue ~ 125) and its competing modes was studied by multi-mode, time-dependent, self-consistent simulations using EURIDICE [9] and is shown in Fig. 2. Stable operation of the $TE_{52,18}$ mode producing ~1.5 MW of power was achieved. Similar simulations for even higher-order modes verified the mode eigenvalue limit of around 125 for a 236 GHz hollow-cavity gyrotron. The effects of an off-axis electron beam on the mode excitation and on the behaviour of neighbouring modes in the hollow-cavity design have also been studied. The simulation results suggest stable output up to a beam misalignment $D = 0.30$ mm ($D/\lambda = 0.24$).

Although the coaxial concept is capable of higher output power, the tolerance on the misalignment of the coaxial insert should be estimated prior to fabrication, in order to achieve robust performance. In coaxial-cavity gyrotrons, shapes and eigenvalues of modes are altered by a misalignment of the insert versus the cavity wall, a feature not present in hollow cavities. This leads to different mode eigenvalues, to changes in the coupling to the electron beam, and to increased mode conversion and thus to increased stray radiation after the end of the insert. Extensive theoretical investigations and development of numerical tools were performed in view of the study of the insert misalignment [10]. The basic idea is to express the field of a mode in a cavity with misaligned insert as a superposition of modes of a cavity with aligned insert, as shown in Fig. 3. For the coaxial design of Table 1, a maximum acceptable insert misalignment of 0.15 mm was estimated, since, according to calculations, such a value would lead to an additional unwanted stray radiation of 3.5 %, compared to about 4 % stray radiation experimentally verified in a 170 GHz coaxial cavity gyrotron.

**Table 1.** Key parameters of the two gyrotron designs.

| Parameters (ideal beams) | Coaxial Concept | Conventional Concept |
|---|---|---|
| Frequency (GHz) | 237.5 | 236.0 |
| Interaction mode | $TE_{49,29}$ (Eigenvalue ≈ 158) | $TE_{43,15}$ (Eigenvalue ≈ 103) |
| Beam Current (A) | 69.3 | 43 |
| Beam Voltage (kV) | 85.6 | 61 |
| Magnetic Field (T) | 9.58 | 9.175 |
| Output Power (kW) | 1900 | 920 |
| Interaction efficiency $\eta_{elec}$ (%) | 33 | 36 |

## 4 Advanced Concepts for Efficiency Enhancement

The key to increase the gyrotron efficiency to a level significantly higher than 50% is the development of a highly efficient collector system for the recovery of the spent beam energy. In this context, a significant effort has been dedicated at KIT during the last years in order to design a reliable, robust and efficient multistage depressed collector system. There are two concepts for the efficient sorting of the spent beam electrons, which are confined by a magnetic field, on the collector stages. The first concept is based on unwinding of the spent electron beam cyclotron motion utilizing non-adiabatic transitions of magnetic fields [11], while the second one is based on **ExB** drift [12].

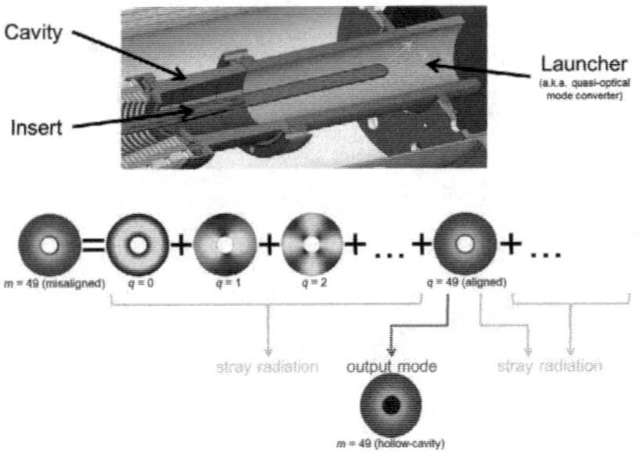


Fig. 3: Top: Cavity, insert and launcher of a coaxial gyrotron. The origin of the insert-induced stray radiation is indicated by orange arrows. Bottom: Schematic field decomposition of the misaligned-cavity mode $TE_{49,29}$ into aligned-cavity modes with azimuthal indices $q = 0, 1, 2\ldots$ After the end of the insert, all components except for $q = 49$ become stray radiation.

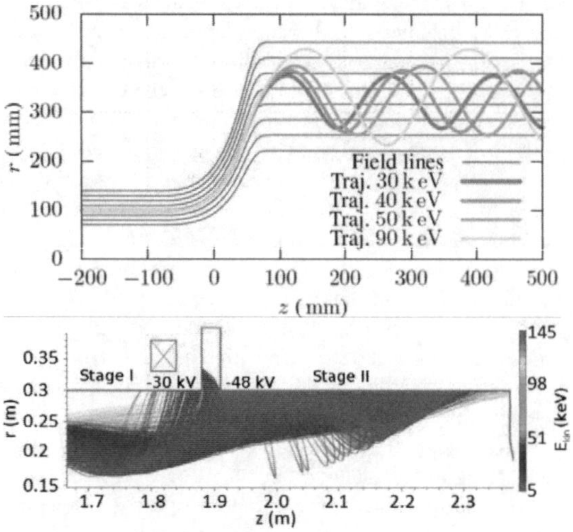


Fig. 4: Spatial modulation of electron trajectories through a non-adiabatic transition (top) and electron trajectories in the two-stage collector (bottom).

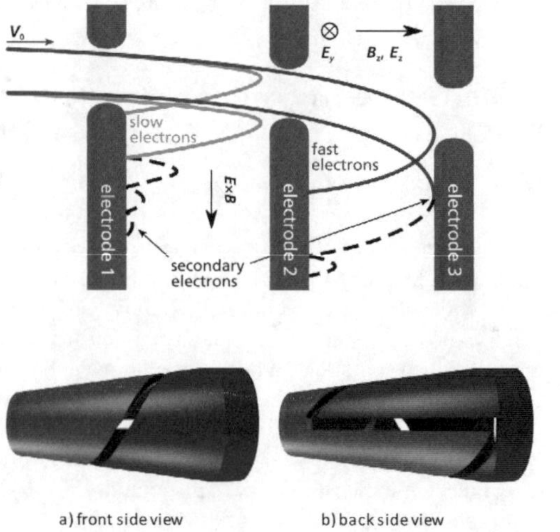


Fig. 5: **E×B** drift concept for the sorting of the electrons according to their initial energy (top) and two stages separated by a conical helix slot for the **E×B** drift (bottom).

A MSDC design based on the first concept for the EU 1 MW 170 GHz gyrotron for ITER has been designed. The spent beam passes a quickly-decreasing magnetic field. After the non-adiabatic transition, two modulations appear, as shown in Fig. 4: a) the maximal reachable radii, b) the axial periodic length; both depend on the kinetic energy of electrons. Based on this modulation, we achieved 73 % collector efficiency (without considering secondary electrons).

On the other hand, several design approaches have been developed based on the **ExB** drift concept [13, 14]. In the simplest one [15] an azimuthal electric field and the longitudinal magnetic field are used for the **E×B** drift as is schematically shown in Fig. 5. It consists of a conical helix-type slot which separates the different collector stages and induces the $\varphi$-component of the electric field. The helix slot has only one azimuthal period. The pitch of the helix defines the angle between the electric and magnetic field, and thus it is related to the strength of the **E×B** drift. Taking secondary electrons into account, PIC simulations predict an average collector efficiency of 77 %. Secondary electrons only reduce the efficiency by 1 %. There could be less than 1.5 % of the total beam current returned beyond the mirror box (including secondary electrons).

## 5 Conclusions

The upgrade of the 2 MW coaxial cavity gyrotron towards longer pulses is ongoing, the design of the cooling system of the major components has been finalised, manufacturuing is ongoing. Design studies towards high frequency operation (~ 240 GHz) of high power gyrotrons and high efficiency operation by using multi-staged depressed collector systems were discussed.

**Acknowledgement**

This work has been partly carried out within the framework of the EUROfusion Consortium and has received funding from the EURATOM research and training programme 2014-2018 under grant agreement No 633053. The views and opinions expressed herein do not necessarily reflect those of the European Commission.

## References

[1] T. Rzesnicki, et al., IEEE Transactions on Plasma Science, **38**, 1141, 2010

[2] I. Gr. Pagonakis et al., Physics of Plasmas **23**, 23105, 2016

[3] I. Gr. Pagonakis et al., Physics of Plasmas **23**, 83103, 2016

[4] S. Ruess, et al., IEEE Transaction of Electron Devices, **63**, 2104, 2016

[5] P. C. Kalaria, et al., Physics of Plasmas **23**, 092503, 2016

[6] J. Franck et al., 40th International Conference on Infrared Millimeter and Terahertz waves (IRMMW-THz 2015), 23-28 Aug. 2015, Hong Kong, Conference proceedings.

[7] M. Thumm, et al., Terahertz Science and Technology, **8**, no. 3, pp. 85-100, September 2015.

[8] P. C. Kalaria, et al., 17th IEEE International Vacuum Electronics Conference (IVEC 2016), 19-21 April 2016, Monterey, CA, USA, Conference proceedings.

[9] K. A. Avramides, et al, EPJ Web of Conferences 32, 04016 (2012). DOI: http://dx.doi.org/10.1051/epjconf/20123204016

[10] J. Franck, Ph.D. Thesis, Karlsruher Institut für Technologie (KIT), KIT Scientific Publishing, 2017, DOI: 10.5445/KSP/1000068000.

[11] R. L. Singh, et al., United States Patent, Patent Number 5,780,970, July 1998.

[12] I. Gr. Pagonakis, et al., IEEE Trans. on Plasma Sc., **36**, 469, 2008.

[13] I. Gr. Pagonakis, et al., Phys. Of Plasmas, **23**, 43114, 2016.

[14] C. Wu, et al., Phys. of Plasmas, .24, pp. 43102, 2017.

[15] C. Wu, et al., 18th International Vacuum Electronics Conference (IVEC), 2017.

# Influence of Li conditioning on Lower Hybrid Current Drive efficiency in H-mode and L- mode plasmas on EAST

*Marc* Goniche[a,*], *Miaohui* Li[b], *Yves* Peysson[a], *Yingjie* Chen[b], *Bojiang* Ding[b], *Annika* Ekedahl[a], *Haiqing* Liu[b], *Yong* Liu[b], *Jinping* Qian[b], *Xiuda* Yang[b], *Qing* Zang[b], *Tao* Zhang[b], *Sun* Zhen[b], *Xiao-Lan* Zou[a] and the EAST Team

[a]*CEA, IRFM, F-13108 Saint Paul-lez-Durance, France*
[b]*Institute of Plasma Physics, Chinese Academy of Sciences, Hefei 230031, People's Republic of China*

**Abstract.** The lower hybrid current drive efficiency on the EAST tokamak is estimated on a large database of low loop voltage discharges ($V_L$<120mV) covering the 2016 campaign starting with lithium-free plasma facing components and ending with strong cumulated deposition of lithium. The efficiency is found to vary in a wide range from 0.6 to $1.2\times10^{19}$A.W$^{-1}$.m$^{-2}$. No deleterious effect of the density on the efficiency is found between 2.3 and $3.2\times10^{19}$m$^{-3}$. The high efficiency occurs after strong lithium evaporation. The low effective charge $Z_{eff}$ and the higher temperature $\langle T_e\rangle$ of these discharges, can account for the high efficiency according to the expected scaling with $Z_{eff}$ and $\langle T_e\rangle$. Modelling with a ray-tracing code coupled to a Fokker-Planck solver supports this result, assuming that the fast electron transport is reduced in the zero loop voltage discharge with high efficiency.

## 1. Introduction

Lower Hybrid waves are recognized to be an efficient tool to drive current with high efficiency. Fully or partially ($V_{loop}$<0.1V) non-inductive long discharges have been obtained with lower hybrid current drive (LHCD) on several superconducting tokamaks: Tore Supra [1,2,3], TRIAM [4], EAST [5]. A one-minute H-mode discharge has been recently obtained on EAST [6]. This discharge, achieved with LHCD combined with other heating methods (ICRH, ECRH and NBI), has low and steady plasma radiation thanks to the lithium coating of the plasma facing components (PFC) and in particular of the divertor tungsten tiles. Lithium coverage reduces the effective charge $Z_{eff}$ of the plasma by trapping the oxygen atoms [7,8]. It also reduces the particle recycling leading to a change of the $n_e$ and $T_e$ profiles at the plasma edge, including the scrape-off layer [4]. Both effects can be beneficial to the penetration of the wave and the current drive efficiency $\eta=nRI_{LH}/P_{LH}$.

## 2. The experimental database

LHCD efficiency was estimated from a data base covering 160 L-mode and H-mode discharges with low loop voltage (-0.02V<$V_L$<0.12V), a line-averaged density ($\langle n_e\rangle_{lin}$ in the 2.0-3.5$\times10^{19}$m$^{-3}$ range. The plasma is in the upper single-null (USN) configuration with a current $I_p$ in the 0.4-0.5MA range and a toroidal field of 2.25T or 2.45T. The database, covering the 2016 experimental campaign from March to October, includes discharges with lithium-free plasma facing components (5-10 March), discharges with lithium dropping during the plasma ramp-up phase (10 March) and discharges with lithium coating before the experimental day (11 March – 11 October). Data are averaged on a time slice of 3 seconds (respectively from 1 to 3 s) for 83% (respectively 100%) of the discharges.

LHCD is provided by the 2.45GHz (0-1.5MW) and 4.6GHz (1-2.7MW) multijunction antennas [9]. The 2.45GHz (resp. 4.6GHz) antenna is an array of 5 (resp. 12) rows of 32 (resp. 48) narrow waveguides In addition, ICRH (0-0.8MW), ECRH (0-0.5MW) and NBI (0-2.7MW) provide plasma heating. Total power is in the 2-5MW range. The parallel wave index $N_{//}$ of the 2.45GHz (resp. 4.6GHz) antenna can be varied between 1.85 and 2.6 (resp. 1.6 and 2.25), but most of the discharges were performed with $N_{//}$= 1.8 or 2.1 on both antennas.

The mean power reflection coefficient RC is in the 2-7% (resp. 1-6%) range for the 2.45GHz (resp.4.6GHz) antenna. This range is consistent with the RF modelling, provided by the ALOHA code [10], indicating an equivalent range of electron density in front of the launcher of 2-10 (resp. 1.3-5) time the cut-off density at 2.45GHz (resp.4.6GHz).

The current drive efficiency for low loop voltage ($V_L$) discharges can be estimated by plotting $V_L$ as a function of the normalized LHCD power $P_{norm}=P_{LH}/(\overline{n_e}\,RI_p)$. Keeping the first-order term of the dielectric tensor, namely the hot conductivity $\sigma_{hot}$, the relative loop voltage drop $\Delta V_L/V_L = (V_0-V_L)/V_L$ with respect of the ohmic heating loop voltage $V_0$ can be expressed as

$$\rho\left[\left(-\frac{\Delta V_L}{V_L}\right)-\left(1-\frac{1}{\rho}\right)\right]=\frac{(\eta+\eta_h)P_{norm}}{1+\eta_h P_{norm}} \qquad (1)$$

With the following notations: $\rho = (R_{sp})_{LH}/(R_{sp})_{OH} = (Z_{LH}/Z_{OH})\,(T_{e,LH}/T_{e,OH})^{-3/2}$, $(R_{sp})_{LH}$ and $(R_{sp})_{OH}$ are the plasma resistivity in the LHCD and ohmic phase, respectively, $\eta$ is the LHCD efficiency, $\eta_h = \sigma_{hot}/(P_{norm}\,\sigma_{sp})$, $\sigma_{sp}$ being the Spitzer conductivity. For low loop voltage discharges $dP_{norm} = 1/\eta - P_{norm} \ll 1$ and equation (1) can be linearized as follows

*Corresponding author: marc.goniche@cea.fr

$$V_L = \frac{V_0}{\rho}\frac{\eta}{1+\eta_h/\eta}\left(\frac{1}{\eta} - P_{norm}\right) \quad (2)$$

We therefore expect the slope of the curve $V_L$ vs $P_{norm}$ to be weakly varying as long as $P_{norm}$ does not vary too much ($P_{norm}$=0.8-1.3 for these discharges). This quasi-linear relationship, verified on Tore Supra [2], allows an estimate of the efficiency with an accuracy of ±10% for $V_L$<100mV. This efficiency includes the bootstrap current in the total non-inductive current but in these low beta discharges, the fraction of bootstrap current does not exceed 10-15% of the plasma current.

On EAST, the loop voltage with 1MW launched by the 2.45GHz antenna ($V_L$=0.27V) is much higher than that with the same power launched by the 4.6GHz antenna ($V_L$=0.15V) suggesting a lower efficiency for the 2.45GHz antenna [9, 11]. On a large database where the two current drive systems were combined, the loop voltage data (but also the stored energy data) are found to be more consistent assuming the efficiency of the 2.45GHz antenna is half of that of the 4.6GHz. Even with a matched wave index the power spectra $P(N_{//})$ are slightly different and the efficiency, estimated from a $1/N_{//}^2$-weighted directivity, indicates a higher value by ~20% for the 4.6Ghz antenna. However, modelling with a ray-tracing code coupled to a Fokker-Planck solver (C3PO/LUKE) [12] does not indicate a higher LHCD efficiency of the 4.6GHz antenna. Parasitic interaction of the wave with the plasma edge such as wave scattering or parametric decay has been evoked as a possible explanation of this reduced efficiency [9,11] We therefore weighted the power launched by the 2.45GHz ($P_{2.45}$) antenna by a factor 0.5 to obtain a '4.6GHz-equivalent' efficiency. In this database, the ratio $P_{2.45}/P_{4.6}$ does not exceed 0.5 in most cases and this correction affects the efficiency by no more than ~20%.

For the range of magnetic field and plasma density of these discharges, the accessibility of the wave to the plasma core is restricted to the outer half plasma. The Stix-Golant condition $N_{//}>[1-\omega_{pi}/\omega)^2+(\omega_{pe}/\omega_{ce})^2]^{1/2}+(\omega_{pe}/\omega_{ce})$, where $\omega$, $\omega_{pi}$, $\omega_{pe}$, $\omega_{ce}$ are respectively the wave, ion plasma, electron plasma and electron cyclotron frequency, indicates a penetration of the wave up to a normalized radius r/a~0.8 ($B_t$=2.25T) - ~0.7 ($B_t$=2.45T) for a line-average density ($\langle n_e\rangle_{lin}$ $=3\times10^{19}m^{-3}$ and $N_{//}$=2. It should be kept in mind that this condition is an approximation which holds for cold plasma with no toroidal effect and further details on power absorption from C3P0/LUKE modelling will be given in section 4.

## 3. Experimental LHCD efficiency

Figure 1 shows the loop voltage as a function of the normalized power $P_{norm}$ for the $B_t/I_p$=2.25T/0.4MA case. Most of the points lie between $\eta=0.7\times10^{19}A.W^{-1}.m^{-2}$ and $\eta=0.85\times10^{19}A.W^{-1}.m^{-2}$ (broken lines) with no ordering of the efficiency with the density indicated by the colour code.

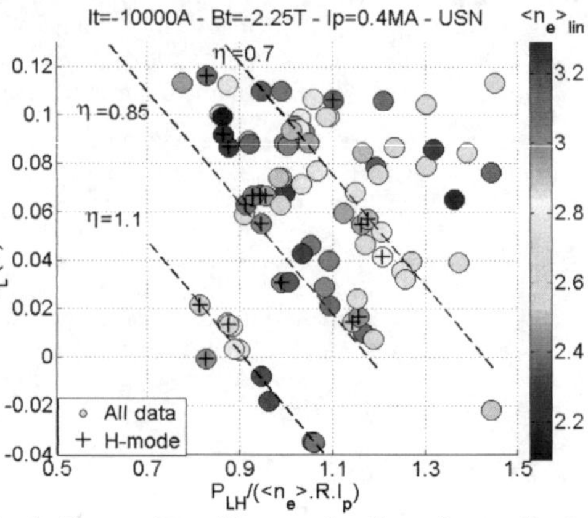


**Fig. 1.** Loop voltage $V_L$ as a function of normalized power $P_{norm}$ ($B_t/I_p$=2.25T/0.4MA, USN).

However, there are a significant number of discharges which have a higher efficiency ($\eta\sim1.1\times10^{19}A.W^{-1}.m^{-2}$). When the extrapolated efficiency is plotted as a function of the discharge number, these high efficiency discharges are all obtained late in the campaign when the cumulated amount of the lithium evaporated exceeds 150-200g (Figure 2). Lithium aerosols during the discharge, between the two broken lines of figure 2, do not seem to improve the efficiency. Most of the H-mode discharges have a high efficiency, between 0.8 and $1.15\times10^{19}A.W^{-1}.m^{-2}$.

Late in the campaign, the plasma current was raised to 0.5MA ($B_t$=2.45T). The efficiency is in the 0.85-$1.1\times10^{19}A.W^{-1}.m^{-2}$ range and, on a statistical basis, no improvement with respect of the 2.25T/0.4MA case (same period of the campaign) is observed. On other tokamaks, a beneficial effect of $I_p$ (scaling as $I_p^{1/2}$) [13,14] has been inferred and a slight increase of ~12% could be expected when the current is increased from 0.4MA to 0.5MA. The internal inductance $l_i$, from the equilibrium code EFIT, decreases with plasma density from 1.2±0.1 ($\langle n_e\rangle_{lin}=2.3\times10^{19}m^{-3}$) to 0.95±0.1 ($\langle n_e\rangle_{lin}$ = $3.0\times10^{19}m^{-3}$) with no significant effect of the LH power.

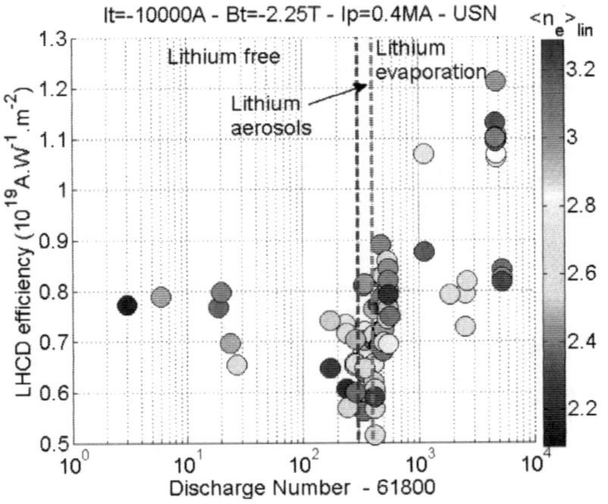


**Fig. 2.** Time history of the LHCD efficiency ($B_t/I_p$=2.25T/0.4MA, USN).

In addition to the effect of lithium coverage of the PFCs, the efficiency is closely related to the LH coupled power (figure 3-a) and η values above 1 are only obtained at moderate power launched by the 4.6GHz, between 1.4 and 1.8MW. Moreover these high efficiency discharges have all radiation from the bulk plasma, between 0.3 and 0.5MW.

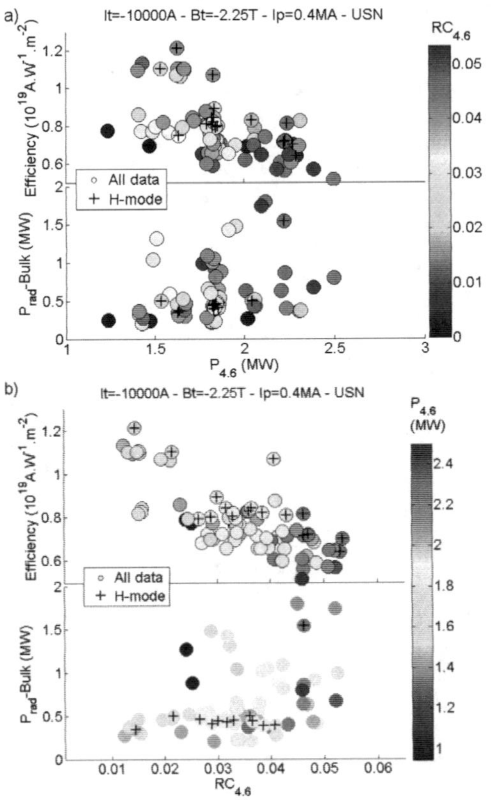


**Fig.3.** LHCD efficiency and radiated power in the main plasma as a function of a) power b) reflection coefficient of the 4.6GHz antenna.

The correlation between the efficiency and the power reflection coefficient of the 4.6GHz launcher $RC_{4.6}$ is even better (figure 3-b). The increase of the RC with the launched power indicates non-linear interaction of the RF electric field with the plasma facing the antenna, namely ponderomotive forces as observed and modelled on Tore Supra [15]. Although a change of RC from 1-2% to 4-5% cannot explain a modification of the wave spectrum such as the CD efficiency is reduced by a factor 2, it suggests that the spectrum could be further modified when the RF electric is high due to wave scattering or parametric decay [9, 11].

## 4. LHCD Modelling

From this database, 5 discharges with same line-average density ($<n_e>_{lin}=2.95\times10^{19}m^{-3}$), plasma current ($I_p$=0.4MA) and magnetic field ($B_t$=2.25T) were selected: 61824 (before any lithium evaporation), 62209-62296-62349 (after weak lithium evaporation, 12-25-37g) and 66526 (after strong lithium evaporation, >150g). The 3 weak lithium evaporation discharges were performed with different values of $N_{//}$ (1.82-2.04-1.60). The density profiles, from the Thomson scattering system, were re-scaled to match the line-average density provided by the far infra-red interferometer (with a scaling factor varying between 0.7 and 1.0). The temperature profiles are also provided by the Thomson scattering diagnostic. These profiles, along a vertical chord, were mapped to the mid-plane using the EFIT equilibrium code (figure 4). For these discharges the profiles are very similar except #66526 (high lithium case) which has a less peaked density and a more peaked temperature. The total LH power varies between 1.9MW and 2.5MW, the mean effective charge $Z_{eff}$ between 2.7 and 4.9 according to the bremsstrahlung diagnostic ($Z_{eff}$-Brem). $Z_{eff}$ was also estimated from the radiated power with the Matthews' law ($Z_{eff}$-Mat) [16] (Table1).

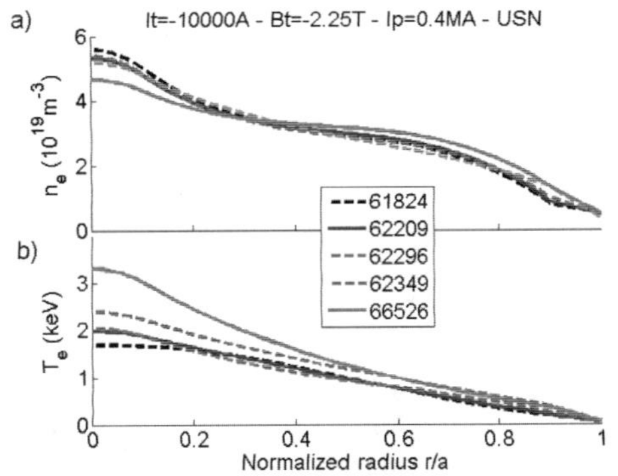


**Fig. 4.** a) density b) electron temperature profiles. The 5 pulses have the same line-average density ($<n_e>_{lin}=3.0\times10^{19}m^{-3}$).

DOI: 10.1051/epjconf/201715703018

The loop voltage varies between 59mV and 88mV for the no or weak lithium cases and is -1mV for the strong lithium case. The resulting efficiency is found to be between 0.70 and $0.80\times10^{19}$A.W$^{-1}$.m$^{-2}$ for the no or weak lithium cases but jumps to $1.2\times10^{19}$A.W$^{-1}$.m$^{-2}$ for the strong lithium case. Following the temperature scaling ($\sim T_e^{0.5}$) found on Tore Supra [13] and FTU [14] and the $Z_{eff}$ ($\sim(5+Z_{eff})^{-1}$) scaling from the 1-D Fisch's theory[17], the efficiency normalized to the volume-averaged temperature $<T_e>$=1keV and $Z_{eff}$=2 lies between 0.96 and 1.25 except for the no lithium case for which the $Z_{eff}$ could be over-estimated and the low $N_{//}$ discharge for which the wave accessibility is poor.

| Pulse # | Li (g) | $P_{LH}$ (MW) | $N_{//}$ 4.6GHz | $V_l$ (mV) | $Z_{eff}$ Brem | $Z_{eff}$ Mat | $\eta$ ($10^{19}$A. W$^{-1}$.m$^{-2}$) | Norm.$\eta$ ( $10^{19}$ A.W$^{-1}$.m$^{-2}$) |
|---|---|---|---|---|---|---|---|---|
| 61824 | 0 | 2.37 | 1.82 | 88 | 4.9 | 5.0 | 0.76 | 1.36 |
| 62209 | 12.5 | 2.03 | 1.82 | 89 | 2.7 | 2.6 | 0.72 | 0.96 |
| 62296 | 25 | 2.52 | 2.04 | 59 | 2.7 | 2.4 | 0.80 | 1.03 |
| 62349 | 37.5 | 2.16 | 1.60 | 66 | 3.5 | 2.2 | 0.70 | 0.78 |
| 66526 | >150 | 1.88 | 2.04 | -1 | 4.1 | 1.9 | 1.21 | 1.25 |

**Table 1.** Main parameters of the modelled discharges with same density ($\langle n_e\rangle_{lin}=2.95\times10^{19}$m$^{-3}$).

In order to validate further the effect of lithium, the C3PO/LUKE code was run for these 5 discharges. Most of the power is absorbed at r/a=0.5-0.7 with, for some of them, a significant part of the power absorbed in the very core (r/a<0.3). It was found that for the 4 no/weak lithium discharges, with residual loop voltage, the efficiency is correctly modelled when a diffusion coefficient of the fast electrons is included in the model. For the strong lithium discharge, the modelled and experimental efficiencies match when no transport for the fast electrons is assumed (Figure 5-a). This could be the result of an internal transport barrier for this fully non-inductive discharge which has a much higher central electron temperature. The modelled internal inductance is also consistent with the values obtained from the EFIT equilibrium code within 10-15% in most cases (Figure 5-b).

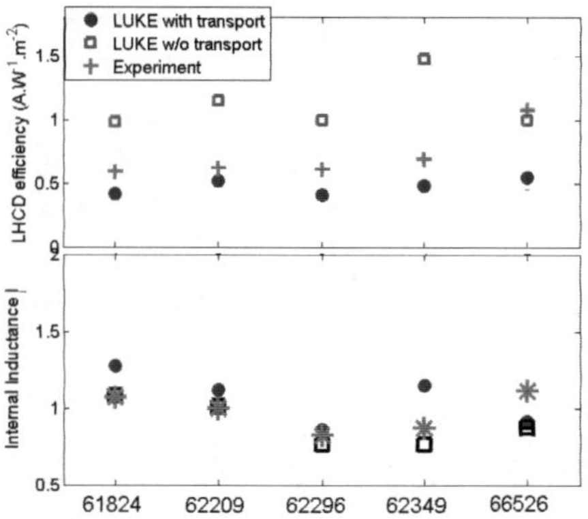


**Fig. 5** a) Efficiency b) Internal inductance frm experiment and modelling for the 5 discharges of the table 1.

## 5. Conclusions

The analysis of LHCD efficiency for discharges with more than 70% of the current is driven by the 4.6GHz antenna shows a beneficial effect of wall coverage with lithium. This was also observed with the 2.45GHz antenna [18]. This effect is partly due to the lower radiation and consequently lower $Z_{eff}$ and higher temperature of the plasma. However, the normalized efficiency of the discharge performed after strong lithium evaporation indicates a higher efficiency by about 25% and edge plasma–wave interaction affecting the wave spectrum cannot be ruled out. This assumption is also supported by the correlation between the efficiency and the RF electric field which is a function of launched power and reflection coefficient. However, no deleterious effect of the density between 2.3 and $3.2\times10^{19}$m$^{-3}$ on the efficiency is found from this large database.

The CEA/IRFM members warmly acknowledge the hospitality of the ASIPP Team during the visits to EAST. This work is supported by the National Magnetic Confinement Fusion Program of China (Grant No 2015GB10200) and the Associated Laboratory CEA/IRFM – CAS/ASIPP. This work has been carried out within the framework of the EUROfusion Consortium and has received funding from the European research and training programme under grant agreement N° 633053. The views and opinions expressed herein do not necessarily reflect those of the European Commission.

## References

[1] D.Van Houtte et al. Nucl. Fusion **44** (2004) L11–L15
[2] M.Goniche et al., Physics of Plasmas **21** 061515 (2014)
[3] R.J.Dumont et al., Plasma Phys. Control. Fusion 56 (2014) 075020 (11pp)
[4] H.Zushi et al., Nuclear Fusion 45 (2005) S142-S156.

[5] H.Y.Guo et al., Phys. Plasmas 19, 056124 (2012)
[6] B.N.Wan et al., 26th IAEA FEC, Kyoto, 17-22 October 2016
[7] G.S.Xu, Nucl. Fusion 51 (2011) 072001 (6pp)
[8] L.Chen et al., Nucl. Fusion 56 (2016) 056013
[9] M. H.Li et al. Phys. Plasmas 23, 102512 (2016)
[10] J.Hillairet et al., Nucl. Fusion 50, 125010 (2010)
[11] B.J.Ding et al., 26th IAEA FEC, Kyoto, 17-22 October 2016
[12] Y.Peysson, Y. and J.Decker, FST, 65 (2014) 22
[13] M.Goniche et al., 16th Topical Conference on Radio Frequency Power in Plasmas (2005), AIP Conf. Proc. 787, pp. 307-310; doi: http://dx.doi.org/10.1063/1.2098245
[14] V.Pericoli Ridolfini et al., Nucl. Fusion 45 (2005) 1386–1395
[15] M.Preynas et al., Nucl. Fusion 53 (2013) 013012,
[16] G.F.Matthews et al., J. Nucl. Mat. 241-243 (1997) 450-455
[17] N.J.Fisch 1987 Rev. Mod. Phys. **59** 175–234
[18] B.J.Ding et al., Nucl. Fusion 55 (2015) 093030

# Anomalous backscattering and absorption in second harmonic ECRH experiments due to low-threshold two UH plasmon decay instability in blobs

*Evgeniy* Gusakov*, *Aleksei* Popov, *and Ekatherine* Sysoeva

Ioffe Institute, 194021 St-Petersburg, Russia

**Abstract.** An universal mechanism responsible for excitation of the low-threshold two upper-hybrid (UH) plasmon absolute parametric decay instability of a pump extraordinary microwave is proposed and analysed in detail. A corner-stone of it is the 2D localization of the daughter UH waves in an intensive magnetic field aligned blob/filament and the presence of a finite-width mm-wave pump beam

## 1 Introduction

In the recent decade a substantial number of observations in the electron cyclotron resonance heating (ECRH) experiments in tokamaks and stellarators had been made which do not fit into a simple linear picture. The most eloquent among them is the anomalous backscattering phenomena studied in detail in the second harmonic X-mode heating experiments on TEXTOR [1,2]. The theoretical model [3] proposed recently explains this anomalous backscattering as a result of the two upper-hybrid (UH) plasmon parametric decay instability (PDI) possessing very low threshold due to trapping of nonlinearly excited plasmons in the vicinity of the density maximum that accompanies the magnetic island [4]. The theoretical model [3] also predicts substantial (up to 25%) anomalous absorption due to this process. This makes important further detailed analysis of the mechanisms and scenarios responsible for excitation of the low-power-threshold PDI of an extraordinary wave.

The density fluctuations can also lead to anomalous absorption of mm-wave [5]. In particular, the large amplitude density fluctuations routinely observed in experiments at the plasma edge [6-8] in the form of blobs or filaments, aligned with the magnetic field line and possessing the finite size perpendicular to it can ensure the 2D localization of the daughter UH waves and, thus, serve as potential centers of low-threshold two-plasmon parametric decay and sources of anomalous scattering of mm-waves.

In the present paper we analyze the scenario of the low-threshold PDI associated with excitation of UH waves 2D trapped in plasma blobs which is applicable to both the stellarator and tokamak second harmonic ECRH experiments. The numerical estimations for this scenario are given in the paper for conditions of the second harmonic ECRH experiment in ASDEX Upgrade [9].

## 2 Parametric excitation of the 2D trapped UH waves by the pump extraordinary wave

In order to describe nonlinear excitation of the UH waves trapped in a blob aligned with the magnetic field line, we introduce the local cylindrical and Cartesian coordinate systems $(r,\theta,z)$ and $(x,y,z)$ with their common origin located in the centre of a blob and the coordinate $z$ directed along the magnetic field line. The coordinate $x$ is related to the flux surface label, and $y$ is the coordinate perpendicular to the magnetic field line on the magnetic surface. We focus on the ECRH experiments being carried out in ASDEX Upgrade [9] and model the density profile as a sum of the background smoothly inhomogeneous term $n_0$ and a sharp perturbation arising due to the presence of a blob generated by a low-frequency instability of MHD or drift type and given in the local cylindrical coordinate systems $(r,\theta,z)$ by expression

$$\delta n = \delta_n n_0 \exp\left(-r^2 / l_b^2\right) \quad (1)$$

In the above formula $\delta_n \ll 1$ is the relative amplitude of a density perturbation and $l_b$ is the width of it. Thus, the density perturbation (1) models a structure being aligned with the magnetic field line and possessing the finite size perpendicular to it. For our numerical analysis we choose its width in the millimeter scale range $l_b = 0.3$ cm corresponding to the trapped electron mode typical for the ECRH experiments. The perturbation relative amplitude used in the analysis $\delta_n = 0.25$ is consistent with the values usually observed at the edge plasma [8]. We assume temporal variation of the density fluctuation negligible during the period or inverse growth rate of the decay waves under consideration in this paper.

* Corresponding author: evgeniy.gusakov@mail.ioffe.ru

Then, we analyze the three-wave interaction model relevant to the experiment. We assume that the pump extraordinary wave propagates almost perpendicular to the magnetic field in the density inhomogeneity direction with its dominant polarization vector component being mostly directed along the coordinate $y$. We represent the X-mode pump propagating through the decay layer as

$$\mathbf{E} = \mathbf{e}_y \frac{E_0}{2} \exp\left(ik_x x - i\omega_0 t\right) + c.c. \qquad (2)$$

where $E_0(y,z) \equiv \sqrt{8P_0/(cw^2)}\exp\left(-(y^2+z^2)/2w^2\right)$ is the wave amplitude with $P_0$ and $w$ representing the pump power and the width of a beam, $k_x(\omega_0) \approx \omega_0/c$ being a component of the wavenumber is nearly constant for the typical conditions of the 2[nd] harmonic ECRH experiments where the inequalities $\omega_0^2 >> \omega_{ce}^2, \omega_{pe}^2$ usually hold, and *c.c.* is the term derived from the first one by complex conjugation. In the cylindrical coordinate system (2) reads

$$\mathbf{E} = \frac{E_0(z)}{2}\left[\mathbf{e}_r \sin\theta + \mathbf{e}_\theta \cos\theta\right] \times \sum_{m=-\infty}^{\infty} J_m(k_x r)\exp\left(im\theta - i\omega_0 t\right) + c.c. \qquad (3)$$

where we have used the inequality $l_b << w$ and introduced Bessel function of the first kind $J_m$.

We consider the pump wave decay into a couple of the electrostatic UH waves propagating in opposite directions and represented by the potentials $\varphi_{1,2}(\mathbf{r},t) = \phi_{1,2}(\mathbf{r})\exp\left(\mp i\omega_{1,2}t\right)/2 + c.c.$, $\omega_2 = \omega_0 - \omega_1$. n weakly inhomogeneous plasma the set of equations describing this nonlinear coupling reads

$$\hat{D}_1\phi_1(\mathbf{r}) = 4\pi\rho_1, \qquad \hat{D}_2\phi_2(\mathbf{r}) = 4\pi\rho_2 \qquad (4)$$

where

$$\hat{D}_{1,2} \equiv \hat{D}(\omega_{1,2}) = l_{T1,2}^2\left(\frac{1}{r}\frac{\partial}{\partial r} r\Delta_\perp \frac{\partial}{\partial r} + \frac{1}{r}\Delta_\perp \frac{1}{r}\frac{\partial^2}{\partial\theta^2}\right) - \varepsilon\Delta_\perp^2 - \frac{\partial\varepsilon}{\partial r}\frac{\partial}{\partial r} - \eta\frac{\partial^2}{\partial z^2} - \frac{\omega_{1,2}^2}{c^2}\left(\varepsilon^2 - g^2\right),\ \Delta_\perp = \frac{1}{r}\frac{\partial}{\partial r} r\frac{\partial}{\partial r} + \frac{1}{r^2}\frac{\partial^2}{\partial\theta^2},$$

$$l_{T1,2}^2 \equiv l_T^2(\omega_{1,2}) = \frac{3}{2}\frac{\upsilon_{te}^2}{4\omega_{ce}^2 - \omega_{1,2}^2},$$

$\varepsilon$, $g$ and $\eta$ are the components of the "cold" dielectric tensor. The nonlinear induced charge densities $\rho_{1,2}$ arising on the RHS of equations (4) describe coupling of the daughter UH waves in the presence of the pump wave (2). Their explicit representation is given below.

Then, we assume the weak nonlinear coupling of UH waves and utilize the perturbation theory approach [10]. In the first step of the perturbation procedure we put $\rho_{1,2} = 0$ and analyze the homogeneous differential equations $\hat{D}_1\phi_1(\mathbf{r}) = 0$ and $\hat{D}_2\phi_2(\mathbf{r}) = 0$. We start with consideration of the 1D problem and show in figure 1 the 1D dispersion curves of the UH wave ($f_1 = 70\text{GHz}$) in the presence of the blob (1) for conditions of the second harmonic ECRH experiments in ASDEX-Upgrade

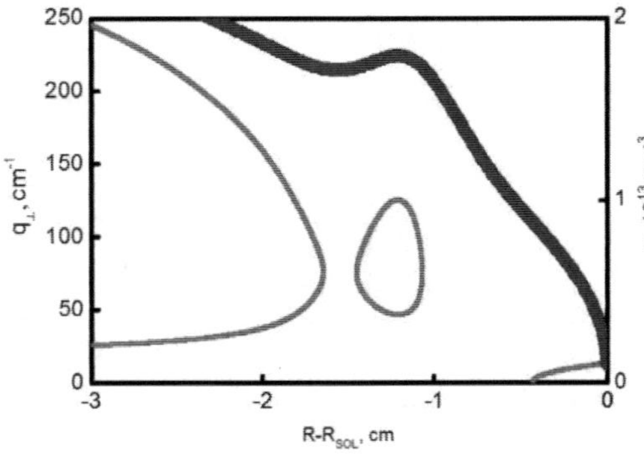


Fig. 1. (Left and bottom axes) - The 1D dispersion curves $q_\perp$ of the UH wave ($f_0/2 = 70\text{GHz}$ - dash-dotted curve. (Right and bottom axes) - the density profile (thick solid line).

($B_0 = 3.4$ T, $<n> = 9\times10^{13}\text{cm}^{-3}$, $T_{e,edge} = 150 \div 200\text{eV}$, the system operates at 140 GHz with X2 scheme) [9]. The chosen edge density profile is also depicted in the figure 1 in thick solid curve. As we can see, the UH wave can be localized in a vicinity of the local maximum of the density profile. Then, we get back to the realistic 2D problem and demonstrate the possibility of 2D localization of both the UH waves in the blob. Seeking solutions of the homogeneous equations by means of the WKB approximation and assuming the high azimuthal modes $n_{1,2} >> 1$ of the UH waves we get

$$\phi_{1,2}(\mathbf{r}) = C_{1,2}F_{1,2},$$

$$F_{1,2} = \frac{1}{2\sqrt{q_{1,2r}(r)}}\exp\left(i\int^r q_{1,2r}(\xi)d\xi + in_{1,2}\theta + iq_z z\right) + c.c.$$

where

$$q_{1,2r} = \frac{1}{l_{T1,2}}\sqrt{-\frac{\varepsilon}{2l_{T1,2}^2} - \sqrt{\frac{\varepsilon^2}{4l_{T1,2}^4} + \left(\frac{\omega_{1,2}^2 g^2}{c^2 l_{T1,2}^2} + \frac{q_z^2\eta}{l_{T1,2}^2}\right)} - \frac{n_{1,2}^2}{r^2}}.$$

In the absence of the nonlinear pump the amplitudes of the UH waves are constant, $C_{1,2} = \text{const}$. The frequencies $\omega_1^{k,l}$, $\omega_2^{k,l} = \omega_0 - \omega_1^{k,l}$ and wave-number $q_z^{k,l}$ of them obey simultaneously the Bohr-Zommerfeld quantization conditions

$$\int_{r_{1l}^*}^{r_{1r}^*} q_{r1}\left(\xi, \omega_1^{k,l}, q_z^{k,l}\right)d\xi = \pi(k + 1/2),$$
$$\int_{r_{2l}^*}^{r_{2r}^*} q_{r2}\left(\xi, \omega_0 - \omega_1^{k,l}, q_z^{k,l}\right)d\xi = \pi(l + 1/2). \qquad (5)$$

The two pairs of the coordinates $(r_{1l}^*, r_{1r}^*)$ and $(r_{2l}^*, r_{2r}^*)$ in (5) are the solutions of two equations $q_{1r}(r_{1l,r}^*) = 0$ and $q_{2r}(r_{2l,r}^*) = 0$, corresponding to the UH wave turning points in the radial direction. For the conditions used the decay of the azimuthal modes $m = \pm1$ (see equation (3)) forming the symmetric pump

$$\mathbf{E}^{m=1} = \mathbf{e}_r \frac{E_0(z)}{2} J_1(k_x r)\exp\left(i\frac{\pi}{2} - i\omega_0 t\right) + c.c. \qquad (6)$$

into the UH waves with the equal frequency $f_1 = f_2 = f_0/2 = 70\text{GHz}$ and the poloidial mode numbers $n_1 = n_2 = n = 13$ ($q_z = 0.37\text{cm}^{-1}$) seems to be the most dangerous because the decay condition for radial wavenumbers in this case is satisfied all over the

 DOI: 10.1051/epjconf/201715703019

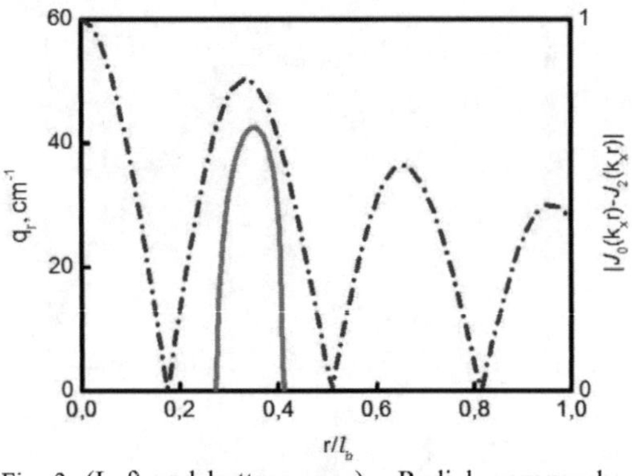


Fig. 2. (Left and bottom axes) - Radial wavenumber $q_r$ of the UH waves ( $f_1 = 70\text{GHz}$ , $n_1 = 13$ ) versus the dimensionless blob radius. (Right and bottom axes) – the radial profile of $|J_0(k_x r) - J_2(k_x r)|$ (dash-dotted line).

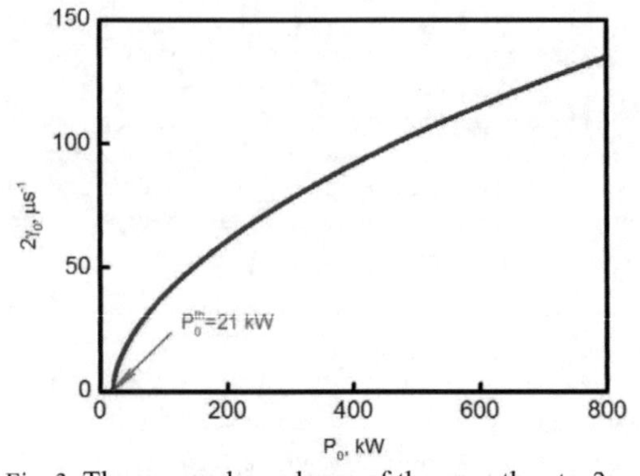


Fig. 3. The power dependence of the growth-rate $2\gamma_p$ (9); $p = 0$ .

plasmon localisation region. In (6) we have omitted the terms $\propto \exp(\pm i2\theta)$ which are unable to afford resonant nonlinear coupling with the selected daughter UH waves.

In order to describe the nonlinear coupling we introduce the charge densities $\rho_{1,2}$ which are given at $q_r << m/r$ by the following equations

$$\begin{bmatrix} \rho_1 \\ \rho_2 \end{bmatrix} = \frac{|e|}{4\pi m_e} \frac{n^2}{r^2} \frac{\omega_{pe}^2 \omega_1 \omega_2}{\Delta_0 \Delta_1 \Delta_2} \frac{\partial J_1(k_x r)}{\partial r} \begin{bmatrix} E_0(z)\phi_2 \\ E_0^*(z)\phi_1 \end{bmatrix}.$$

with $\Delta_j = \omega_j^2 - \omega_{ce}^2$ , $j = 0,1,2$ . In figure 2 we demonstrated the dispersion curves of the daughter UH waves corresponding to fundamental radial modes being merged along with the radial distribution of the spatial form-factor determining the efficiency of the nonlinear interaction of pump and daughter waves, $2|\partial J_1(k_x r)/\partial(k_x r)| = |J_0(k_x r) - J_2(k_x r)|$ . As it is seen there, the daughter UH waves possessing large azimuthal number are localised in a narrow ring within the second maximum of the form-factor $2|\partial J_1(k_x r)/\partial(k_x r)|$ . The latter makes strong nonlinear coupling possible.

In the second step of the perturbation procedure we take into account the nonlinear coupling and the daughter waves energy losses along the axis $\mathbf{e}_z$ . The amplitudes of UH waves in this case are no longer constant, *i.e.* $C_{1,2} \to C_{1,2}(z,t)$ . We substitute (2) into (4), multiply the first line by $F_1^*$ , the second one by $F_2^*$ and integrate them over the coordinate $r$ . Eventually, we arrive at

$$\begin{cases} \dfrac{\partial C_1}{\partial t} + u_z \dfrac{\partial C_1}{\partial z} = \nu_0 \dfrac{E_0(z)}{H} C_2 \\ \dfrac{\partial C_2}{\partial t} - u_z \dfrac{\partial C_2}{\partial z} = \nu_0^* \dfrac{E_0^*(z)}{H} C_1 \end{cases} \qquad (7)$$

where $u_z$ is the group velocity of both the UH plasmons averaged over their radial localization area and describing their convective energy losses along the axis $\mathbf{e}_z$ . The term describing nonlinear coupling of the UH plasmons ( $\nu_0$ ) is defined as follows

$$\nu_0 = \left. \frac{\omega_0 |\omega_{ce}| k_x c}{16\Delta_0} \right|_{x=x_b} \frac{\int_{r_{1l}}^{r_{1r}} \frac{dr}{|q_{r1}|} \left(\frac{n}{r}\right)^2 \left(J_0(k_x r) - J_2(k_x r)\right)}{\int_{r_{1l}}^{r_{1r}} \frac{dr}{|q_{r1}|} \left(\frac{n}{r}\right)^2},$$

with $x_b$ indicating a position of the blob. For the parameters used the coefficients in (7) are $u_z = 2.76\text{cm}/\mu\text{s}$ and $|\nu_0| E_0(0)/H = 11\mu\text{s}^{-1}$ . It should be stressed that returning back to the decay region of even a part of energy, lost from it convectively along the magnetic field is leading to the feed-back effect and as a result to a possibility of the temporal exponential growth of the daughter waves. Such a feed-back is created when in the spot of a finite-width pump beam the daughter waves propagate in opposite directions. Following [2,3] we seek the temporally growing solution of equations (7) by means of the WKB approximation, i.e. assuming

$$C_{1,2} \propto q_z^{-1/2}(z) \exp\left(\gamma t + i\int^z q_z(z')dz'\right).$$

This leads to the dispersion relation

$$u_z^2 q_z^2 + \gamma^2 = |\nu_0|^2 \frac{E_0(z)^2}{H^2} \qquad (8)$$

If the pump power is large enough, we use the Taylor expansion of the pump beam distribution along the magnetic field $E_0^2(z) \approx E_0^2(0)\left(1 - z^2/w^2\right)$ and get from (8) the expression for the wavenumber of the UH wave

$$q_z = \frac{|\nu_0| E_0(0)}{u_z H} \sqrt{\left(1 - \frac{\gamma^2}{|\nu_0|^2} \frac{H^2}{E_0^2(0)}\right) - \frac{z^2}{w^2}}.$$

valid at $\gamma \approx |\nu_0| E_0(0)/H$ . Then, we utilize the quantization procedure $\oint q_z dz = \pi(2p+1)$ and get explicitly the growth rate of the eigen mode of the daughter UH plasmon

$$\gamma_p = |\nu_0| \frac{E_0(0)}{H} \sqrt{1 - (2p+1) \frac{u_z}{|\nu_0| w} \frac{H}{E_0(0)}} \qquad (9)$$

The power dependence of the growth rate is shown in figure 3 where a surprisingly low power threshold of the decay of only $P_0^{th} = 21$ kW is seen. At the higher heating power $P_0 = 500\text{kW}$ typical for the ECRH experiments in ASDEX Upgrade [9] and exceeding drastically the PDI power threshold the growth rate for the fundamental mode is extremely large $2\gamma_0 \approx 10^8 s^{-1}$ similar to the case of two-plasmon decay in the magnetic island [3]. Based on this similarity one can expect that the decay instability will affect the ECRH power budget in the case of blob, as in [3]. This makes actual correct estimation of the amount of pump power gained anomalously by the daughter UH waves. To this end it is necessary to clarify a saturation mechanism of the decay and to evaluate its saturation level. Most likely, the PDI leading to amplification of the UH waves from the thermal noise level in a local area within the pump beam, the length of which is $\delta z_p = 2(2p+1)u_z wH / (|v_0| E_0(0))$, is saturated due to the nonlinear effects. The most low-threshold among them is the secondary PDI driven by the daughter UH wave which gives rise to the secondary UH wave trapped in the blob and ion Bernstein wave. The nonlinear coupling of different UH plasmons, as it is shown in [3], can lead to generation of the microwaves frequency downshifted in respect to the pump. Observation of this waves by the collective Thomson Scattering (CTS) diagnostics is interpreted as strong anomalous scattering of the powerful pump wave [9].The detailed analysis of the saturation mechanism of the primary PDI of the pump wave and of the accompanying electromagnetic radiation will be presented elsewhere.

## 3 Conclusions

In the paper we analysed the possibility of the low-power-threshold two-UH-plasmon parametric decay of the pump extraordinary wave in the second harmonic ECRH experiments. The corner-stone of the proposed model is the presence of a helical coherent structures, *i.e.* blobs or filaments, arising in the plasma turbulence, aligned with the magnetic field line and leading to the 2D nonmonotonous plasma density profile. The UH waves are shown to be easily captured in a plane perpendicular to the magnetic field in these 2D structures. The presence of the finite-width pump beam leads to a feed-back effect for the daughter UH waves propagating along the magnetic field line in opposite directions and to their temporal exponential growth, which can be interpreted as an excitation of the absolute two-UH-plasmon PDI. We derive expressions for the instability power-threshold and its growth rate for the actual experimental plasma parameters of the second harmonic ECRH experiments in ASDEX Upgrade and the realistic filament parameters. The analyzed parametric decay possesses a very low threshold power of several tens kW and the growth rate being extremely large $2\gamma_0 = 10^8 s^{-1}$, which is similar to [3]. The saturation of this strong instability could occur at the level leading to substantial power transfere into UH waves affecting the power budget of the ECRH as it according to [3] was the case in ECRH experiments in TEXTOR tokamak [1, 2]. The nonlinear coupling of the UH plasmons is a possible candidate for explanation of the anomalous scattering effect observed in the ECRH experiments in ASDEX Upgrade tokamak [9].

*The financial support of the RSF grant 16-12-10043 and of the Ioffe Institute is acknowledged.*

## References

[1] E. Westerhof, S.K. Nielsen, J.W. Oosterbeek et al. *Phys. Rev. Lett.* **103**, 125001 (2009).
[2] S.K. Nielsen, M. Salewski, et al. Plasma Phys. Control. Fusion **55**, 115003 (2013)
[3] E.Z. Gusakov and A.Yu. Popov Physics of Plasmas **23**, 082503 (2016)
[4] A.Yu. Popov, E.Z. Gusakov Plasma Phys. Control. Fusion **57**, 025022 (2015)
[5] E. Gusakov, A. Popov, A., Saveliev Plasma Phys. Control. Fusion **56**, 015010 (2014)
[6] S.I. Krasheninnikov *Physics Letters* A **283**, 368-370 (2001)
[7] S.J. Zweben, R.J. Maqueda, D.P. Stotler et al. Nuclear Fusion **44**, 134-153 (2004)
[8] B. Vanovac, I.G.J. Classen, S.S. Denk et al. Proc. 43rd EPS Conference on Plasma Physics **40A** P1.014
[9] S.K. Nielsen, W. Bongers, S. Fietz et al. Proceedings of the 9th International Workshop "Strong Microwaves and Terahertz Waves: Sources and Applications" (2014)
[10] E.Z. Gusakov, V.I. Fedorov Sov. J. Plasma Phys. **5**, 463 (1979)

EPJ Web of Conferences **157**, 03020 (2017) DOI: 10.1051/epjconf/201715703020
*22 Topical Conference on Radio-Frequency Power in Plasmas*

# Sequential modelling of ICRF wave near RF fields and asymptotic RF sheaths description for AUG ICRF antennas

*Jonathan* Jacquot[1], *Wouter* Tierens[1,*], *Wei* Zhang[1,2], *Volodymyr* Bobkov[1], *Laurent* Colas[3], *Jean-Marie* Noterdaeme[1,2], and The ASDEX Upgrade Team[1]

[1]*Max-Planck-Institut für Plasmaphysik, Boltzmannstrasse 2, D-85748 Garching, Germany*
[2]*Ghent University, Sint-Pietersnieuwstraat 41, Gent, Belgium*
[3]*CEA, IRFM, F-13108 Saint Paul-lez-Durance, France*

**Abstract.** A sequence of simulations is performed with RAPLICASOL and SSWICH to compare two AUG ICRF antennas. RAPLICASOL outputs have been used as input to SSWICH-SW for the AUG ICRF antennas. Using parallel electric field maps and the scattering matrix produced by RAPLICASOL, SSWICH-SW, reduced to its asymptotic part, is able to produce a 2D radial/poloidal map of the DC plasma potential accounting for the antenna input settings (total power, power balance, phasing). Two models of antennas are compared: 2-strap antenna vs 3-strap antenna. The 2D DC potential structures are correlated to structures of the parallel electric field map for different phasing and power balance. The overall DC plasma potential on the 3-strap antenna is lower due to better global RF currents compensation. Spatial proximity between regions of high RF electric field and regions where high DC plasma potentials are observed is an important factor for sheath rectification.

## 1 Introduction

Plasma heating with waves in the Ion Cyclotron Range of Frequencies (ICRF) is a convenient method to heat plasma and achieve the high temperatures required for a fusion reactor. However, in addition to the desired heating in the core plasma, spurious non-linear interactions with the plasma edge and with plasma limiting boundaries have been observed in magnetic fusion devices equipped with ICRF antennas [1], such as JET, ASDEX Upgrade (AUG), Tore Supra, Alcator C-Mod. These interactions often determine the operational limits of the radio-frequency systems. Their most likely origin is RF-sheath rectification [2], which is suspected to produce enhanced sputtering (which causes impurity release [3–5]), increased heat fluxes [6] on the antennas as well as objects magnetically connected to the launcher, and Scrape Off Layer (SOL) density re-distribution [7] modifying the coupling properties of the antenna. Active strap RF currents and the induced image currents on the surrounding passive structures are a priori influencing the RF-sheaths' spatial topology and magnitude. For reliable ICRF operation at high power over long pulses in metallic environment, the physical mechanisms and their interplay that underlie these spurious interactions must be understood and minimized. The present paper presents simulations results for two ASDEX Upgrade antennas (so called 2-strap and 3-straps antenna, see figure 1) obtained with two codes:

- RAPLICASOL (Radiofrequency wAve couPLing for Ion Cyclotron Antenna in Scrape-Off-Layer) to solve Maxwell's equations in the cold plasma approximation without sheaths or any non-linear phenomenon.
- SSWICH (Self-consistent Sheaths and Waves for Ion Cyclotron Heating) to self-consistently determine the non linear response of the interaction of RF waves with the edge plasma (SOL).

## 2 RAPLICASOL

RAPLICASOL [8] is a 3D finite-element model implemented in COMSOL[9], which solves Maxwell's equations in the cold plasma approximation in the neighbourhood of a realistic 3D antenna geometry as shown in figure 1. Being restricted to cold plasma, RAPLICASOL cannot model the core plasma. Instead an absorbing boundary layer is used at the core plasma (a Perfectly Matched Layer [10] or PML). On all other boundaries, metallic boundary conditions are enforced. Waves are first excited from port 1, and then from port 2, which gives us all information required to determine the scattering parameters ($S$-matrix) and related quantities. To further decrease

*e-mail: wouter.tierens@ipp.mpg.de

DOI: 10.1051/epjconf/201715703020

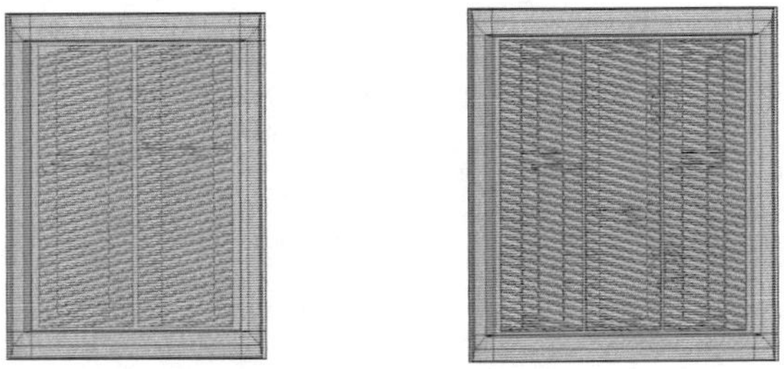

**Figure 1.** 3D model of the 2-strap antenna (left) and the 3-strap antenna (right). The 2 side straps of the 3-strap antennas are fed from the same port.

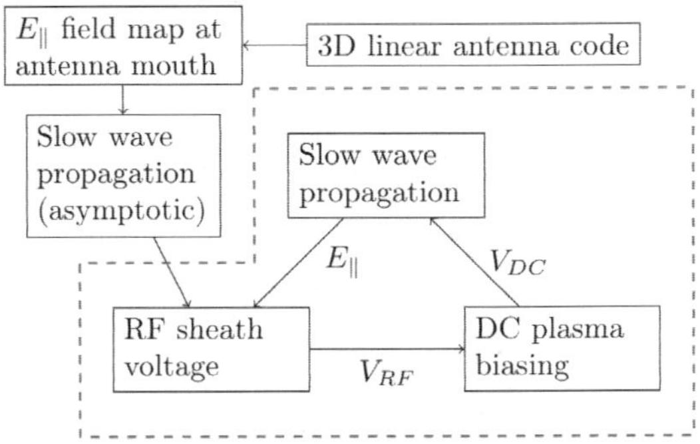


**Figure 2.** Schematic representation of the algorithm used in the SSWICH code. The three fluids are solved in their own module and are coupled together through RF sheath properties. An asymptotic version of the SW module initializes the loop to ensure convergence.

the memory requirements, an iterative solver is used [11].

A fairly detailled 3D model of the antenna geometry, as shown in figure 1, is imported into the model and placed in vacuum. To avoid numerical resolution issues in the private region of the antenna and in particular near the LH resonance [12], the density is assumed to be 0 up to the leading edge of the antenna.

## 3 SSWICH-SW asymptotic

The SSWICH [13–15] code follows a three-field approach to self-consistently model the interplay between RF waves and edge plasma DC biasing. In the SSWICH-SW version which is used in this study, the RF waves are reduced to the slow magnetosonic wave (SW) only. The simulation domain is a set of straight open magnetic field lines in the SOL of magnetized plasma in the vicinity of an IC antenna. This domain contains protruding material objects able to intersect magnetic field lines. The inclusion of the limiters allows to distinguish long field lines passing in front of the antennas from short field lines connected to the limiters or in between the limiters. The magnetic field lines are either perpendicular or parallel to the boundaries in order to be consistent with the assumption of the slow wave only[16]. Inside this domain and as shown on figure 2, three fields are solved for: the RF parallel electric field $E_{\|}$, an oscillating sheath voltage $V_{RF}$ (both varying as as $\exp(-i\omega_0 t)$ at the wave pulsation $\omega_0$) and the DC plasma potential $V_{DC}$. Sheaths at both ends of open flux tubes are described by RF and DC sheaths boundary conditions [17, 18]. The non-linear I-V characteristic of the sheaths couples the RF and DC quantities. More details can be found in [13]. The excitation field map is computed in the present paper with the antenna code RAPLICASOL. The physical model of SSWICH-SW is implemented using the COMSOL Finite Element solver. In the present study, only the asymptotic part of SSWICH is considered in order to reduce the simulation time of an extensive number of simulations (the loop in figure 2 only runs once). Due to limited computational resources, an automated multi-2D approach was used, whereby several radial/parallel planes at consecutive poloidal positions are solved independently, assuming the poloidal derivatives can be neglected with respect to radial gradient in all the equations.

## 4 Simulations results interfacing SSWICH-SW with RAPLICASOL

Numerical simulations were performed to compare both models of AUG antennas at 30MHz and $B = 1.917$T. This frequency is the standard operating frequency for H minority. The imported density profile corresponds to the scenario of #31278. Although SSWICH and RAPLICASOL can handle 3D density profiles [19], only a 1D radial density profile is considered in this paper. All calculations are scaled to 1MW coupled power to the plasma. Results from RAPLICASOL simulations are shown in figure 3 displaying the RF parallel electric field $E_{\|}$ in front of the Faraday screen in between the antenna limiters. The RF electric field pattern outside the limiters is neglected. The 2-strap antenna is configured in a standard dipole phasing (i.e. 0° phase difference between both ports) and a balanced power distribution, whereas the three straps antenna is configured with twice more power on the central strap compared to the side straps in dipole phasing. In both cases, the fieldmap is a linear superposition of the simulated fieldmap for each port one at a time normalized using the scattering matrix. It clearly shows that the $E_{\|}$ field, generated by induced RF currents, near the limiters is mitigated for the 3-strap antenna. The presence of the Faraday screen bars is creating a small scale field structure in the capacitive E fields generated in the strap boxes. The modulation of this pattern is following the number of bars in the geometry. These fieldmaps were then the source files for SSWICH simulations. All parameters are identical for both antennas. The vertical step size between two SSWICH simulations is fixed to 1cm, which is approximately to the size of a Faraday screen bar, and is the smallest length scale on which we should expect $V_{DC}$ variations.

 DOI: 10.1051/epjconf/201715703020

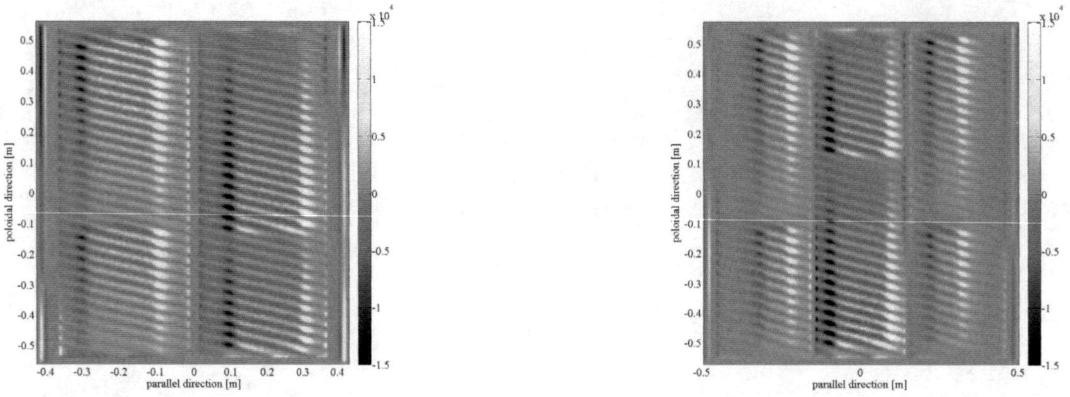


**Figure 3.** 2D map (poloidal/toroidal) in front of the Faraday screen of the real part of the parallel RF electric field for the 2-strap antenna (left) and 3-strap antenna (right). Coupled power=1MW, dipole phasing, power balance 0.5:0.5 (left) and 2/3:1/3 (right).

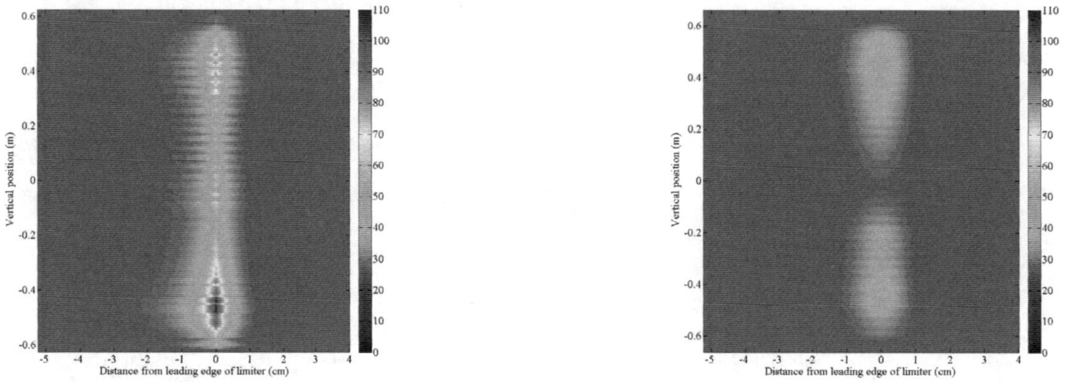


**Figure 4.** 2D map (poloidal/radial) of the DC plasma potential on the outer side of the right limiter simulated by the sequence RAPLICASOL/SSWICH-SW computed with the fieldmaps in figure 3. The left figure is for the 2-strap antenna while the figure for the 3-strap antenna is on the right. Positive distance from the leading edge denotes the direction towards the center of the torus.

The finite DC conductivity tensor establishes a circulation of DC currents from the private region of the antenna to the rest of the SOL. The compensation of RF induced currents on the 3-Strap antenna indeed translates into a significant decrease of the DC plasma potential $V_{DC}$ in SSWICH-SW on the outer side of the antenna limiter, as seen on figure 4 (60V compared to 110V). The absolute value of $V_{DC}$ is severely affected by the local density [19], but also by poorly constrained parameters in SSWICH [13]. More interesting is the relative comparison. A double-hump structure can be seen with maxima near the top and bottom of the antenna for both antennas where the highest parallel electric field are present. It results that toroidal and poloidal proximity to regions of high $E_{\parallel}$ is indeed dominant [20]. The topology of this structure has been observed on different ICRF antennas [13, 20].

Both figures contain vertically small scale structures that are attributed to the Faraday screen bars. These modulations of $V_{DC}$ can occur due to electric fields further toroidally from the limiters away and follow the modulation in the electric field map. The relative visibility of this modulation is also related to the multi-2D procedure in SSWICH. In the case of the 3D version of SSWICH, poloidal transport properties (i.e. DC transverse conductivity that is not only radial but also poloidal) would likely smooth the modulation.

A scan of the maximum value of $V_{DC}$ vs. the phasing and power balance leads to figure 5. The power balance, contrary to the convention in [4], is here defined as $p_b = \frac{P_{outer}}{P_{outer}+P_{center}}$ for the 3-strap antenna.

 DOI: 10.1051/epjconf/201715703020

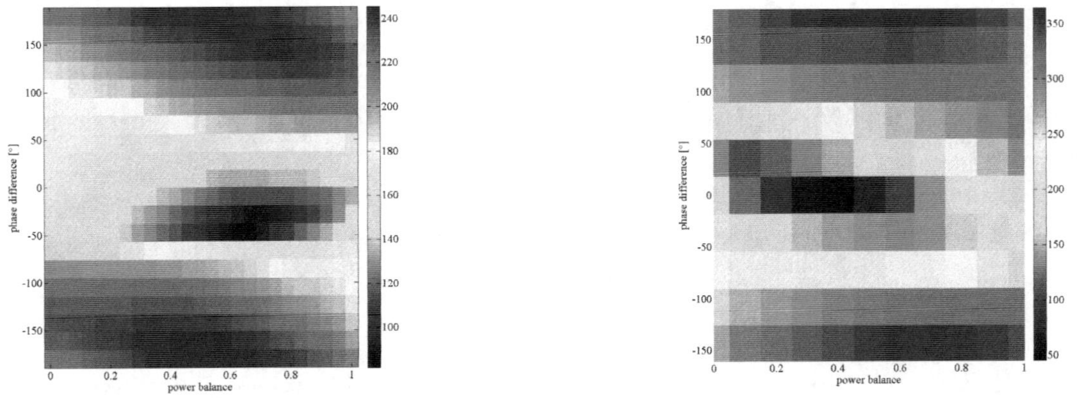


**Figure 5.** Maximum DC plasma potential $V_{DC}$ of the outer side of the right limiter vs. power balance and phasing for the 2-strap antenna (left) and the 3-strap antenna (right). Dipole phasing corresponds to a zero phase difference.

A similar definition is used for the 2-strap antenna: $p_b = \frac{P_{right}}{P_{right}+P_{left}}$. For the 3-strap antenna, the maximal $V_{DC}$ is minimized close to dipole phasing and $p_b = \frac{1}{3}$. It does also corresponds to the minimum for $E_{\|}$ as well as experimental values [4]. The situation seems to get worse than for the 2-strap antenna in case of bad power balance and phasing settings.

## 5 Conclusions and discussion

Simulations were performed with RAPLICASOL and SSWICH-SW to characterize two AUG ICRF antennas. Qualitative features of the experimental observation are well captured by the calculations. Progress was also made for quantitative predictions. We believe the accuracy is constrained by uncertainties in the experimental input and loosely constrained numerical parameters in SSWICH. Repeating the scan in power balance and phasing with a 3D density [19] could help to close the discrepancy. SSWICH-SW only accounts for the slow wave and its mechanism for generating transverse DC current is likely oversimplified. Furthermore, the RF electric field outside the antenna limiters are presently neglected. Replacing it with SSWICH-FW (Full Wave) [16] would bring the fast wave into the mix and open the way to a 3D SSWICH-FW.

**Acknowledgments** This work has been carried out within the framework of the EUROfusion Consortium and has received funding from the European research and training programme under grant agreement N°633053. The views and opinions expressed herein do not necessarily reflect those of the European Commission.

## References

[1] J.M. Noterdaeme, G.V. Oost, Plasma Physics and Controlled Fusion **35**, 1481 (1993)

[2] F. Perkins, Nuclear Fusion **29**, 583 (1989)

[3] D'Ippolito et al., Plasma Physics and Controlled Fusion **33**, 607 (1991)

[4] V. Bobkov et al., Plasma Physics and Controlled Fusion **59**, 014022 (2017)

[5] V. Bobkov et al., Nuclear Fusion **53**, 093018 (2013)

[6] L. Colas et al., Fusion Science and Technology **56**, 1173 (2009)

[7] W. Zhang et al., Nuclear Fusion **57**, 116048 (2017)

[8] J. Jacquot et al., Proc. 21th RF Top. Conf **1689** (2015)

[9] https://www.comsol.com/

[10] J. Berenger, Journal of computational physics **114**, 185 (1994)

[11] Y. Saad, *Iterative methods for sparse linear systems* (SIAM, 2003)

[12] L. Lu et al., Plasma Physics and Controlled Fusion **58**, 055001 (2016)

[13] J. Jacquot et al., Physics of Plasmas **21**, 061509 (2014)

[14] L. Colas et al., Physics of Plasmas **19**, 092505 (2012)

[15] J. Jacquot, Ph.D. thesis, Université de Lorraine (2013)

[16] L. Lu, ECA (submitted to Plasma Physics and Controlled Fusion) **40A**, 2.068 (2016)

[17] D. D'Ippolito, J. Myra, Physics of plasmas **13**, 102508 (2006)

[18] M.A. Lieberman, IEEE Transactions on Plasma Science **16**, 638 (1988)

[19] W. Tierens et al., This conference (2017)

[20] L. Colas et al., Plasma Physics and Controlled Fusion **59**, 025014 (2017)

DOI: 10.1051/epjconf/201715703021

# Time-Domain Modeling of RF Antennas and Plasma-Surface Interactions

*Thomas G.* Jenkins[1,*], and *David N.* Smithe[1]

[1]Tech-X Corporation, 5621 Arapahoe Avenue Suite A, Boulder, Colorado 80303, USA

**Abstract.** Recent advances in finite-difference time-domain (FDTD) modeling techniques allow plasma-surface interactions such as sheath formation and sputtering to be modeled concurrently with the physics of antenna near- and far-field behavior and ICRF power flow. Although typical sheath length scales (micrometers) are much smaller than the wavelengths of fast (tens of cm) and slow (millimeter) waves excited by the antenna, sheath behavior near plasma-facing antenna components can be represented by a sub-grid kinetic sheath boundary condition, from which RF-rectified sheath potential variation over the surface is computed as a function of current flow and local plasma parameters near the wall. These local time-varying sheath potentials can then be used, in tandem with particle-in-cell (PIC) models of the edge plasma, to study sputtering effects. Particle strike energies at the wall can be computed more accurately, consistent with their passage through the known potential of the sheath, such that correspondingly increased accuracy of sputtering yields and heat/particle fluxes to antenna surfaces is obtained. The new simulation capabilities enable time-domain modeling of plasma-surface interactions and ICRF physics in realistic experimental configurations at unprecedented spatial resolution. We will present results/animations from high-performance (10k-100k core) FDTD/PIC simulations of Alcator C-Mod antenna operation.

## 1 Introduction

Physical processes occurring in the edge and scrape-off layer (SOL) of magnetically confined fusion plasma can dramatically reduce the effectiveness of the RF sources used for heating or current drive within the core plasma. Sheath formation on plasma-facing antenna components, for instance, may induce sputtering of high-Z ions from these surfaces; such impurity components induce radiative cooling as transport processes carry them toward the reactor core, and may quench the fusion reaction altogether. In addition, parametric decay instabilities or other resonant phenomena may be excited in the plasma edge, reducing RF antenna efficiency as injected RF power is partially diverted away from the core into undesired edge oscillations. In these interactions, the plasma edge density plays a significant role. Higher densities near plasma-facing antenna components are associated with higher material sputtering rates and sheath potential drops, and although reduced densities and sheath potential values can be achieved by widening the gap between the antenna and the plasma's last closed flux surface, such a change may also open additional, undesired channels into which RF wave power may flow.

In this paper, numerical techniques for modeling parasitic wave excitation and the physics of plasma-material interactions in the tokamak edge and scrape-off layer will be described. We will present a self-consistent finite-difference time-domain (FDTD) model describing the concurrent evolution of RF wave fields and sheath potentials, and will describe its implementation in the VSim FDTD code [1]. Half-torus simulations of ICRF heating in C-Mod, which make use of realistic experimental configurations and plasma profiles from experimental data, will be described; these simulations run for hundreds of ICRF periods with millimeter-scale spatial resolution. We will also discuss a process for modeling sputtering impurity generation in these discharges using VSim's particle-in-cell (PIC) capabilities. These complex numerical simulations are possible only due to the increasing capabilities afforded by massively parallel supercomputing platforms, and the computations we describe here exercise these capabilities at large scales (tens of thousands of processor cores).

## 2 FDTD methods for EM waves in plasma

VSim uses electromagnetic FDTD methods [2] which are modified to account for the presence of plasma. Electric and magnetic field vector components associated with an individual grid cell are not co-located at the cell nodes, but are evaluated (respectively) on edges and faces of the grid cell in the conventional manner. Leap-frog time-staggering of the electric and magnetic fields relative to one another preserves proper time-centering; the path and surface integrals of Maxwell's equations are evaluated respectively around cell edges and over cell faces to center these equations properly in space. Reference [3]

* Corresponding author: tgjenkins@txcorp.com

describes how cold-plasma currents in the Maxwell equations can also be included in the FDTD formalism. Components of plasma current associated with an individual grid cell are co-located at cell nodes [permitting evaluation of the cold-plasma current evolution equation (3); see below]. Interpolation methods which carry edge fields to the grid nodes ("edge-to-node") and nodal fields to edges ("node-to-edge") thus become necessary when such sources are used. A locally implicit time-advance algorithm enables the numerical restrictions imposed by plasma cutoffs (when wave phase velocities diverge) to be circumvented, and preserves the physics of linear plasma dispersion in the presence of a background magnetic field.

Cut-cell techniques [4] are used to model geometrically complex boundaries. Cut edge lengths, associated with portions of individual grid cells intercepted by and lying partially outside conducting boundaries, are used in the integral Maxwell equations. Edge-to-node and node-to-edge operations for the corresponding fields must be appropriately modified to reflect the presence of the conducting boundary. At this boundary, the physics of plasma sheath formation can also be modeled.

The extreme length scale disparity between the width of such sheaths (generally a few Debye lengths) and the wavelengths of the ICRF wave phenomena of interest here precludes direct spatial resolution of the sheath. However, salient sheath effects can be represented by a "sub-grid sheath boundary condition" [5], in which local capacitive and resistive circuit elements associated with the presence of the sheath are defined at the nodes of grid cells intersecting the material boundary. This formulation enables a local sheath potential to be defined at the boundary. Reference [6] generalizes the methods of Ref. [3], such that these time-varying sheath potentials at the conducting boundary can be modeled within the locally implicit time advance. Fields at the *k*-th simulation timestep $\Delta t$ are evolved to the $(k+1)$-st timestep by the equations

$$\frac{\mathbf{B}^{k+1/2}-\mathbf{B}^{k}}{\Delta t/2} = -\nabla\times(\mathbf{M}^T\cdot\mathbf{e}^k) \quad (1)$$

$$\frac{\mathbf{e}^{k+1}-\mathbf{e}^{k}}{\Delta t} = -\frac{1}{\epsilon_0}\sum_\alpha\left(\frac{\mathbf{j}_\alpha^{k+1}+\mathbf{j}_\alpha^{k}}{2}\right) + c^2\mathbf{M}\cdot(\nabla\times\mathbf{B}^{k+1/2}) \quad (2)$$

$$\frac{\mathbf{j}_\alpha^{k+1}-\mathbf{j}_\alpha^{k}}{\Delta t} + \Omega_\alpha\times\left(\frac{\mathbf{j}_\alpha^{k+1}+\mathbf{j}_\alpha^{k}}{2}\right) = \epsilon_0\omega_{p\alpha}^2\left[\left(\frac{\mathbf{e}^{k+1}+\mathbf{e}^{k}}{2}\right)+\frac{\mathbf{n}}{\Delta_s}\left(\frac{\phi^{k+1}+\phi^{k}}{2}\right)\right] \quad (3)$$

$$\frac{\phi^{k+1}-\phi^{k}}{\Delta t} = -\frac{\Delta_s\mathbf{n}}{\epsilon_0}\cdot\sum_\alpha\left(\frac{\mathbf{j}_\alpha^{k+1}+\mathbf{j}_\alpha^{k}}{2}\right) - \nu_s\left(\frac{\phi^{k+1}+\phi^{k}}{2}\right) \quad (4)$$

$$\frac{\mathbf{B}^{k+1}-\mathbf{B}^{k+1/2}}{\Delta t/2} = -\nabla\times(\mathbf{M}^T\cdot\mathbf{e}^{k+1}) \quad (5)$$

wherein $\mathbf{e}$ is the node-centered electric field obtained by applying the non-square edge-to-node matrix operator $\mathbf{M}$ to the edge electric field $\mathbf{E}$ (with the transposed matrix $\mathbf{M}^T$ forming the node-to-edge operator), $\mathbf{B}$ is the face-centered magnetic field, $\mathbf{j}_\alpha$ is the plasma current of species $\alpha$, $\phi$ is the local sheath potential, $\mathbf{\Omega}_\alpha$ is the signed species gyrofrequency vector in the background magnetic field, and $\omega_{p\alpha}$ is the species plasma frequency; the speed of light $c$ and vacuum permittivity $\varepsilon_0$ also appear. Sheath potentials are defined only at grid nodes with one or more adjoining edges cut by the conducting boundary; vectors normal to this boundary are interpolated to the neighboring nodes outside the boundary to form the unit normal $\mathbf{n}$ associated with each such node. The sheath width $\Delta_s$ and the sheath dissipation parameter $\nu_s$ are, most generally, spatially and temporally varying numerical parameters defined on these boundary nodes. Equation (3), which computes the cold-plasma response of the species plasma currents, removes the sheath electric field (subtracting the approximate negative gradient of $\phi$ across the thin layer $\Delta_s$) to compute this response only as a function of the electromagnetic fields in the bulk plasma. Equation (4) is a Kirchoff relation for the parallel RC circuit whose behavior approximates the physics of the sheath near the wall. Equations (2) - (4) can be solved implicitly.

Explicit FDTD PIC simulations were used in Ref. [6] to quantify the circuit-like behaviors of the sheath. These simulations model the sheath width with the expression

$$\Delta_s = \lambda_{De}\left[1+\left(\frac{5q_i\langle\phi\rangle}{3T_e}\right)^2\right]^{3/8} \quad (6)$$

which captures DC Debye shielding and Child-Langmuir effects in the limit of small or large $\langle\phi\rangle$ (compare with Ref. [5]). In Eq. (6), $T_e$ and $\lambda_{De}$ are the electron temperature and Debye length, $q_i$ is the elementary charge, and $\langle\phi\rangle$ is the DC-averaged sheath potential over the RF wave period. More detailed expressions for the sheath capacitance and resistance, such as those of Ref. [7], can also be incorporated into the model, and future code development efforts to incorporate such models in VSim are envisioned.

## 3 Alcator C-Mod Simulation Geometry

Simulations are carried out on a rectangular mesh encompassing half of the Alcator C-Mod torus. Beginning with an initially metal-filled domain; toroidally symmetric boundary wall data is read from an equilibrium EQDSK file and rotated to hollow out the 3D vacuum vessel interior. Limiter boundaries corresponding to antenna surfaces are ignored in this data; the approximate shape of the interior vessel with and without antenna limiters is shown in the solid line curves of Figs. 1a and 1b. The CAD file for C-Mod's field-aligned ICRF antenna is imported into the simulation and properly aligned with the formerly axisymmetric wall. Current driven within the coaxial antenna feeds powers the antenna. Open boundary conditions are used at the rear of the coaxial feeds, while a matched absorbing layer boundary condition is used on the non-metallic portions of the plane bisecting the torus. All other boundaries in the simulation are metallic. Magnetic fields are generated from the EQDSK Grad-Shafranov data and are

DOI: 10.1051/epjconf/201715703021

extrapolated onto the rectangular VSim grid via bicubic spline operations. Density profiles from C-Mod shot

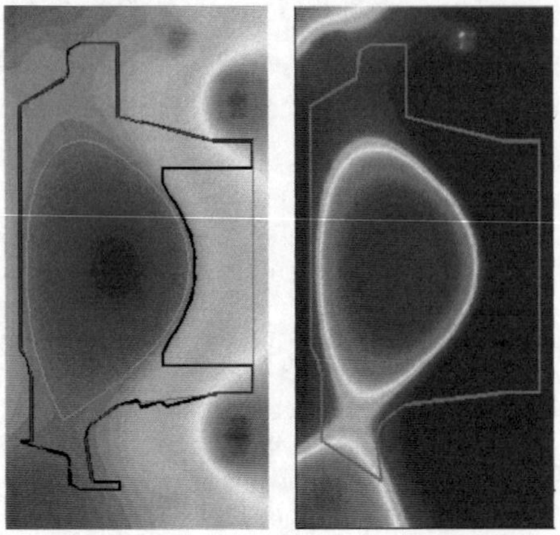

**Fig. 1.** (left) Poloidal magnetic flux from the C-Mod EQDSK equilibrium, together with non-axisymmetric limiter boundaries (dark curve) and approximate axisymmetric boundary (lighter curve also appearing at right). The plasma separatrix is also shown. (right) Plasma density profile from C-Mod shot 1050426022 extrapolated onto this 2D domain. Plasma density outside the closed boundary curve is set to zero.

1050426022 are mapped onto the simulation domain to construct the plasma profile (see Figure 1b). The entire half-torus simulation domain including the plasma (cut away on the left side to reveal the antenna placement) is shown in Figure 2.

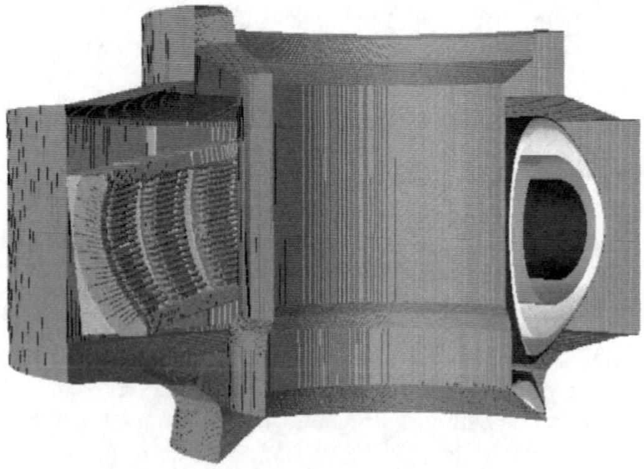

**Fig. 2.** The 3D simulation domain in its entirety, including the antenna, vacuum vessel, and contours of plasma density interpolated from the 2D values of Figure 1. Plasma contours are cut away at the left for ease of visualization.

## 4 Fast and Slow Waves

In Figure 3 we plot a typical wave field generated by the model described above. In this discharge, adjacent antenna straps (of which there are 4 in total) are driven 180 degrees out of phase from one another, yielding complex wave patterns as fast waves are excited in the edge plasma and propagate toward the plasma core with some degree of focusing. Contours in the figure show oppositely signed values of the vertical electric field. In addition to the fast waves, millimeter-scale slow wave excitations in the plasma edge and SOL are visible in the simulations; the slow wave is evanescent outside of a narrow layer near the plasma edge. One focus of present modeling efforts is on determining the fate of the power which flows from the antenna and is diverted into these slow waves, since quantifying power losses associated with parasitic slow wave excitations as plasma edge parameters are varied will enable more efficient antenna operation scenarios to be developed.

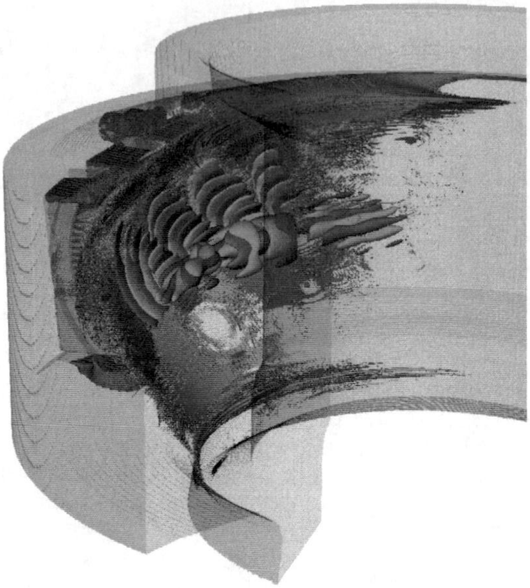

**Fig. 3.** VSim modeling of ICRF antenna operation in Alcator C-Mod; contours of vertical electric field (dark/light contours have equal magnitudes and opposite signs) are shown. Fast waves induced by the antenna form periodic wave structures of long wavelength (tens of cm) near the tokamak core; some power is lost in the SOL as millimeter-scale slow waves are excited in the thin layer between the density cutoff and the lower-hybrid resonance. This simulation used 60,000 cores on the Titan Cray XK7, simulating 250 RF antenna periods in 12 wallclock hours.

## 5 Sputtering and Impurity Modeling

To model sputtering and impurity physics, VSim uses material composition data for the various antenna surfaces in tandem with its PIC modeling capabilities. Sputtering yields vary depending on the material surface composition, so the model must represent the different plasma-facing surfaces comprising the antenna and vacuum vessel as distinct material objects with correct physical properties. Protective tiling on C-Mod's interior vessel surfaces is made of TZM (a molybdenum alloy with traces of titanium, zirconium, and carbon). Faraday shields (the many nearly-horizontal rods in Figure 4) are also made of TZM. The antenna straps (large metallic strips beneath and perpendicular to the Faraday shields in Figure 4) and limiter structures (grey notched and solid structures surrounding the straps) are high-purity oxygen-free copper plated respectively over electroless nickel flash on Inconel 625 and 316 stainless steel. VSim imports these different components from within the antenna CAD file as separate physical objects, and associates these with relevant material composition data to model sputtering effects as particles impinge on these distinct surfaces.

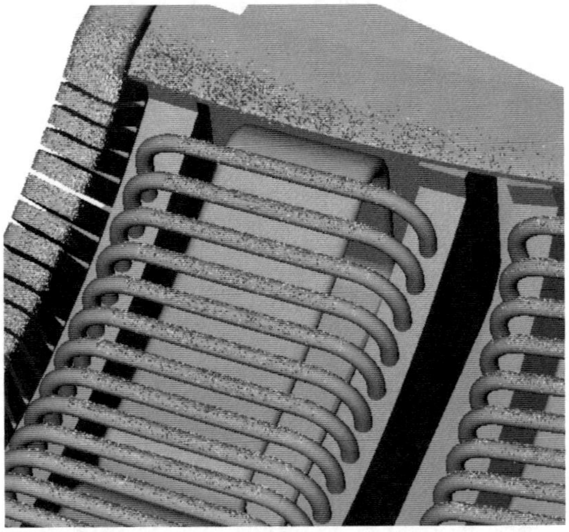

**Fig. 4.** Close-up of the C-Mod ICRF antenna showing locations of various sputtering sites. Faraday shields (the many near-horizontal rods), antenna straps (beneath and perpendicular to Faraday shields), and limiters (slotted and solid structures at left and above) are visible. Sputtering events ensue from test-particle collisions with antenna surfaces. Collision data is shown over many RF cycles.

The particles which induce such sputtering events are treated in the simulations as test particles; they respond to the electromagnetic fields of the background and RF waves but do not contribute to these fields. Particles are probabilistically loaded into the domain such that their density profiles match the experimental edge and SOL profiles. Temperature profiles for these particle distributions are also obtained from C-Mod data. As the simulation proceeds, particles which pass through the sheath to strike the material wall have their strike energy incremented in accordance with the local sheath potential at the strikepoint, enabling the sputtering computations (as well as local measurements of particle and heat flux) to accurately take into account the energetics of sheath transit. We continue to develop the model's sputtering capabilities; current areas of focus include improving ionization models in the SOL for the sputtered neutral wall atoms, as well as improving surface models for the material alloys comprising the wall and antenna surfaces. Figure 4 shows initial data from the sputtering modeling; here, the ICRF antenna was run for many RF periods to determine the locations on the antenna where ion strikepoints were most likely to occur. The possibility that such time-integrated data can be compared with wear/erosion patterns on existing antenna structures in experiments is of interest. In addition, we hope to apply the model to more generally explore sputtered impurity production, impurity transport, and the interaction of RF waves with impurity populations.

## 6 Ongoing and Future Modeling Efforts

We have begun efforts to generalize the modeling techniques used here so that they can be applied to other experiments; currently our efforts focus on ICRF modeling for ITER and EAST. The non-trivial computational resource requirements for this modeling continue to pose challenges; the development of algorithms which make use of modern GPU capabilities to increase speed and efficiency is being explored. Efforts to develop improved sheath boundary conditions, diagnostics, and impurity generation models are also ongoing. In addition to the continuing study of parasitic power losses due to slow wave excitation, we have also begun to explore the physics of density pump-out arising from ponderomotive force effects [8]. The inclusion of yet more detailed sputtering models via coupling to F-TRIDYN [9], a Monte-Carlo Binary-Collision-Approximation code which simulates the physics of plasma-wall interactions at the atomic level, is a topic of interest for integrated modeling efforts. Other potential integrated modeling opportunities include couplings to impurity transport, edge turbulence, or extended MHD codes.

Supported by U. S. Department of Energy, grants DE-FC02-08ER54953 and DE-SC0009501. This research used resources of the Oak Ridge Leadership Computing Facility at the Oak Ridge National Laboratory, which is supported by the Office of Science of the U.S. Department of Energy under Contract No. DE-AC05-00OR22725. We thank Steve Wukitch, Paul Bonoli, and Yijun Lin (MIT-PSFC) for providing data and useful details about the C-Mod experiments. In addition, we thank David Green (ORNL) and other members of the RF-SciDAC group for providing instructive feedback and discussion of this work. Animations of the simulations discussed here are available at https://nucleus.txcorp.com/~tgjenkins/movies.html.

## References

1. C. Nieter and J. R. Cary, J. Comp. Phys. **196**, 448 (2004).
2. K. S. Yee, IEEE Trans. Antennas Propag. **14**, 302 (1966).
3. D. N. Smithe, Phys. Plasmas **14**, 056104 (2007).
4. S. Dey and R. Mittra, IEEE Microw. Guided Wave Lett. **7**, 273 (1997).
5. D. A. D'Ippolito and J. R. Myra, Phys. Plasmas **13**, 102508 (2006).
6. T. G. Jenkins and D. N. Smithe, Plasma Sources Sci. Technol. **24**, 015020 (2015).
7. J. R. Myra and D. A. D'Ippolito, Phys. Plasmas **22**, 062507 (2015).
8. D. N. Smithe, T. G. Jenkins, and J. R. King, AIP Conf. Proc. **1689**, 005004 (2015).
9. J. Drobny *et al.*, https://arxiv.org/abs/1703.00315.

 DOI: 10.1051/epjconf/201715703022

# Design of a steerable launcher for the ECH system on KSTAR

*S. G. Kim*[1,*], *M.* Jeong[1], *J. W. Han*[1], S. J. Wang[1],
*D. S. Kim*[2], *M. S.* Choi[2]

[1]Heating current drive research team, KSTAR research center, National Fusion Research Institute, Daejeon, Korea
[2]Physics department, Ulsan National Institute Science Technology, Ulasn, Korea

**Abstract.** 105/140Ghz dual frequency gyrotron is used in KSTAR tokamak for an electron cyclotron resonance heating (ECRH). The ECH launcher consists of a fixed mirror, a steering mirror, cooling systems and some mechanical component for a steering mirror. We have designed the fixed ellipsoidal focusing mirror for an efficient heating and localized control. The beam radius using the existing ECH launcher system is about 45 mm at the resonance layer. On the other hand, the modified launcher mirror reduces the beam size at the resonance layer for incident RF beam to 38 mm. The design results showed good agreement compared with the computation and simulation results. Since the distance between the corrugated waveguide and the fixed focusing mirror is fixed, it is difficult to more reduce the beam size. However, reduced beam size is small enough to localize MHD control and heating.

## 1 Introduction

A gyrotron is a high powered linear beam vacuum tube and has emerged as an excellent radio frequency (RF) source, which is commonly used for plasma heating, current drive of fusion devices, dynamic nuclear polarization (DNP), nuclear magnetic resonance (NMR), and so on [21-25]. Electron cyclotron heating (ECH) and electron cyclotron current drive (ECCD) launcher system has installed on the KSTAR tokamak. An ECH/ECCD system is required for localized plasma heating, plasma startup for large superconducting tokamak device, pre-ionization for low voltage startup, non-inductive current drive, and discharge cleaning of a vacuum vessel [1-3]. Recently, ECH/ECCD system is used in MHD instability modes control such as suppression of neoclassical tearing mode (NTM), control of sawtooth and internal transport barriers for high-performance and stability of the plasma [4-6]. One of advantage of ECH/ECCD system is localizing the deposited power. In order to increase efficient plasma heating and localized absorption of the RF power, we modified a flat fixed mirror to an elliptical mirror and designed the elliptical mirror.

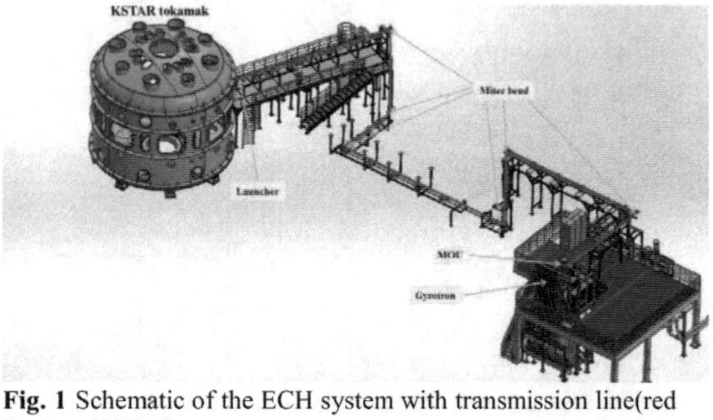

**Fig. 1** Schematic of the ECH system with transmission line(red line) on KSTAR

**Fig. 2** Fixed mirror and steering mirror of ECH launcher for KSTAR

## 2 Launcher for KSTAR ECH system

Electron cyclotron heating (ECH) and electron cyclotron current drive (ECCD) launcher system has installed on the KSTAR tokamak to inject the high power, high frequency RF beam from transmission line system. 1MW/300s/105/140Ghz dual frequency gyrotron is used in KSTAR tokamak for an electron cyclotron resonance heating (ECRH).

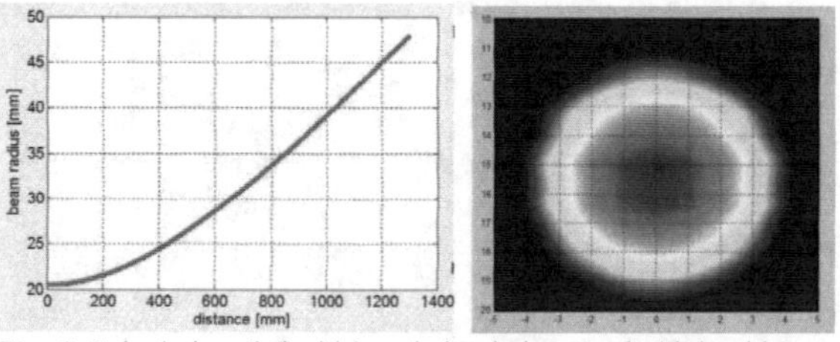

**Fig. 3** Calculation (left side) and simulation result (right side) of ECH launcher using both of flat mirror

* Corresponding author: chan0912@nfri.re.kr

DOI: 10.1051/epjconf/201715703022

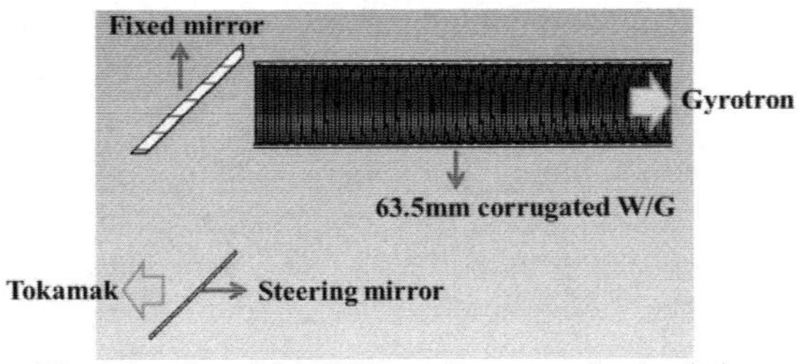


**Fig. 4** Drawing of the ECH launcher system for simulation

The gyrotron system is connected to transmission line system. A transmission line system of ECH system of KSTAR tokamak consist of several miter bend (M/B), a matching optic unit (MOU) and a hundred meters of evacuated corrugated waveguide (W/G) as shown in Fig.1. The high power, high frequency RF beam generated by the Gyrotron is transmitted by the HE11 corrugated waveguide to the launcher. The corrugated waveguide radiates the Gaussian-like RF beam with its beam waist situated at the waveguide output. Radiated RF beam from corrugated waveguide spreads quickly in free space and the beam is reflected by two mirrors. Fig.2 shows the ECH launcher of KSTAR. Both of two mirrors of the launcher has flat surface. KSTAR is a medium size tokamak with a major radius of 1.8 m and a minor radius of 0.5 m. The distance of the resonance layer from the launcher is about 1 m. Fig.3 shows the calculation result using the quasi-optical Gaussian theory and simulation result, respecrively. The beam radius is about 45 mm at the resonance layer because of two flat mirrors. As shown fig. 4, the distance is about 47mm from the corrugated waveguide to the fixed mirror and the distance is about 156 mm from the fixed mirror to the steering mirror. The traveling distance of the reflected beam by the steering mirror is about 1000 mm. In order to increase efficient plasma heating and localized absorption of the RF power, we modified a flat fixed mirror to an elliptical.

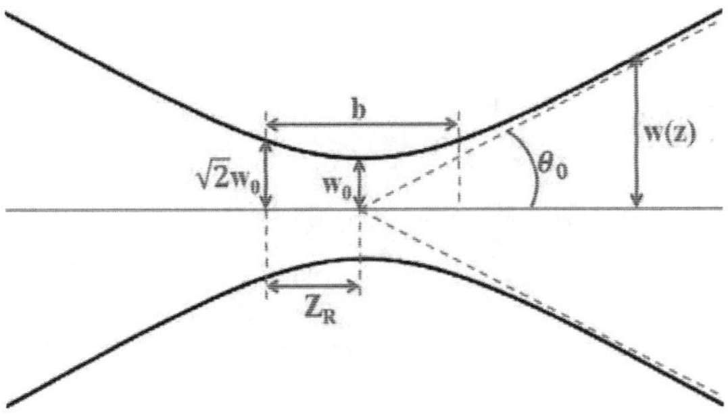


**Fig. 5** Gaussian beam width $w(z)$ as a function of the distance z from the axis of propagation with beam waist( $w_0$ ), Rayleigh range( $z_r$ ), depth of focus ( $b$ ) and far-field divergence angle ( $\theta_0$ ).

## 3 Basic principle of mirror design

As mentioned above, output beam of gyroton is transmitted in HE11 mode. Gaussian beams diverge in the axis of propagation direction as shown in Fig. 5, and in the paraxial limit the beam radius as a function of position along the direction is defined as [8, 9] .

$$w(z) = w_0\sqrt{\left[1+\left(\frac{\lambda z}{\pi w_0^2}\right)^2\right]} \tag{1}$$

where, $w_0$ is the beam waist radius. In Eq. (1), the quantity $\frac{\pi w_0^2}{\lambda}$ is called the confocal distance ( $z_c$ ) or Rayleigh range(distance from the beam waist on the axis of the propagation where the beam radius is increased by a factor of $\sqrt{2}$ ). From the axis of propagation, the power density $P(r,z)$ of the Gaussian beam is given by [10].

$$P(r,z) = P_0(z)\exp\left[-2\left(\frac{r}{w(z)}\right)^2\right]. \tag{2}$$

Here, $P_0$ is the maximum power density at $r = 0$ and $w(z)$ is the beam radius. The total power is proportional to the power integrated over the area of the beam passing any perpendicular plane on the axis of the propagation.

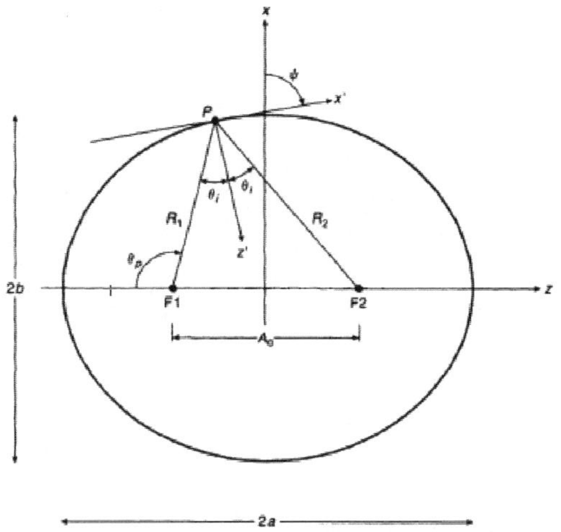


**Fig. 6** Geometry of ellipsoidal reflector. F1 and F2 are the focal points of the ellipse, and R1 and R2 are the distances from the point P on the surface to the focal points [7].

The total power is irrelevant to the axis of the propagation in lossless space and expressed as

$$P_{rad} = \frac{1}{2} P_0(z) \pi w^2(z)$$
$$P_0(z) = \frac{2P_{rad}}{\pi w^2(z)}. \quad (3)$$

Substituting Eq. (3) into Eq. (2), the power density of the Gaussian beam can be written as

$$P(r,z) = \frac{2P_{rad}}{\pi w^2(z)} \exp\left[-2\left(\frac{r}{w(z)}\right)^2\right]. \quad (4)$$

The Gaussian-like beam radiated by corrugated waveguide is converged by the focusing mirror. In order to focus the incident beam with some angle, the focusing mirror should be ellipsoidal surface. The ellipsoidal model is rotationally symmetric about its major axis, which as shown in Fig.6. The equation of the ellipsoid surface is :

$$\frac{y^2}{b^2} + \frac{\left[x^2 + z^2 + 2z(R - h_0)\right]}{a^2} = 1$$
$$b = \frac{a}{\cos\theta}, h_0 = R - \sqrt{R^2 - a^2}, r = 2F\cos\theta \quad (5)$$

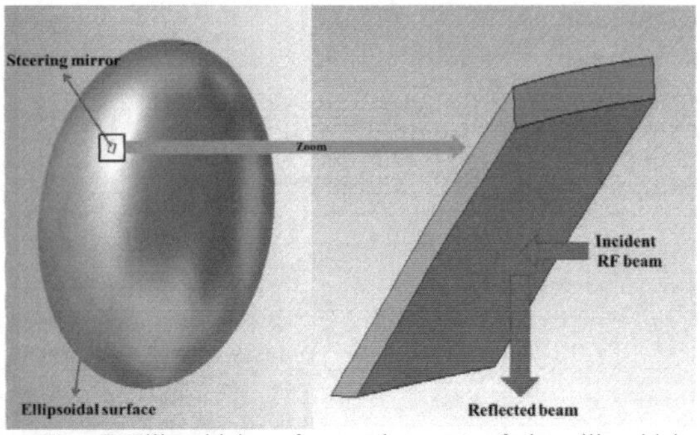


**Fig. 7** Ellipsoidal surface and a part of the ellipsoidal surface (focusing mirror)

## 4 Result and conclusion

A new quasi-optical launcher has been modified and designed for KSTAR ECH system.

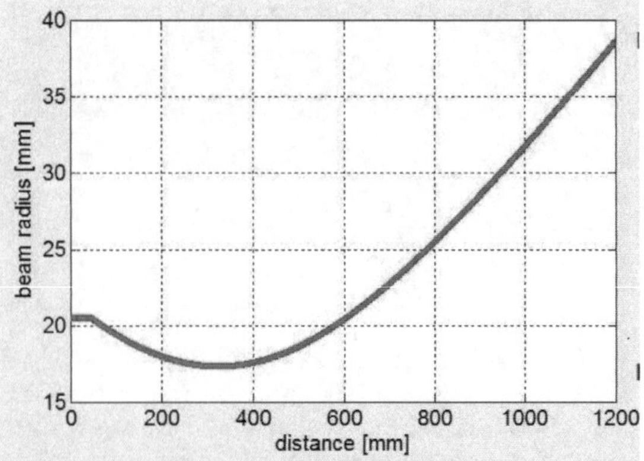


**Fig. 8** Beam radius along the propagation trajectory

Fig.7 shows the new designed fixed mirror for ECH launcher system. The beam radius along the propagation path is shown in Fig. 8. As shown in this result, the focusing mirror is located at 47 mm away from the corrugated waveguide output and the beam radius is about 38 mm at the resonance layer (about 1200mm). Fig. 8 shows that the beam size is smaller at the resonant layer compared to the reults in Fig. 3. Fig. 9 is the Surf 3D code result which is commercial code and shows the beam path trajectory result from the waveguide output to the resonance layer. In this result, the beam radius is also about 39 mm at the resonance layer. This result also shows that the beam size is reduced at the resonant layer compared to the reults in Fig. 3. It is small enough to localize MHD control and heating. Actually, the beam size can be further reduced by using a focusing mirror and it is one of ways to use ECH more effectively. However, since the distance between the corrugated waveguide and the fixed focusing mirror is fixed, it is difficult to more reduce the beam size. In order to more reduce the beam size at the resonant layer, the structure of the launcher have to changed. Since changing the structure of the launcher is out of scope of this research, it will be deal with in the future research.

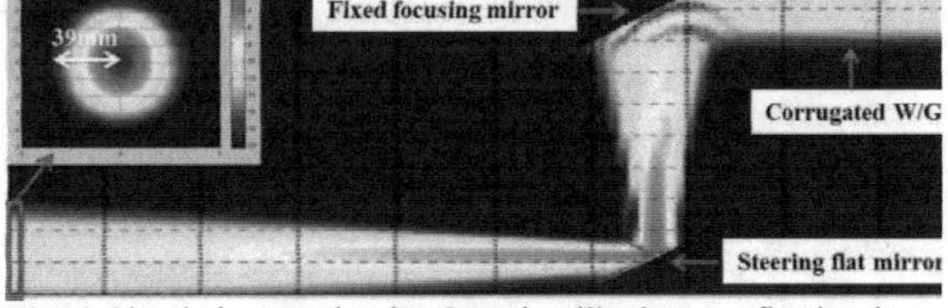


**Fig. 9** Simulation result of a Gaussian-like beam reflection in an ellipsoidal fixed mirror and flat steering mirror

## References

1. C. H. Liu, D.H. Xia, D.L. Sun, Z.J. Wang, J.X. Xiao, Z. Zeng, F.T. Cui, Z.X. Yu, G. Zhuang,

"Design of a polarization-controlled launcher for the ECH system onJ-TEXTC," Fusion Eng.Des. 112 (2016) 404–409.
2. R. Prater, "Heating and current drive by electron cyclotron waves," Phys.Plasmas 11 (2004) 2349–2376.
3. S. Gnesin, J. Decker, S. Coda, T.P. Goodman, Y. Peysson, D. Mazon "3rd harmonic electron cyclotron resonant heating absorption enhancement by2nd harmonic heating at the same frequency in a tokamak," Plasma Phys.Control. Fusion 54 (2012) 035002
4. N. Kobayashi, G. Bosia and B. Petzold "Design status of EC system in ITER — design of reliable RF beam launching system," Fusion Eng.Des. 53 (2001) 475–484
5. T. Shimozuma, H. Igami, S. Kubo, Y. Yoshimura, H. Takahashi, M. Osakabe,et al., "Optimization of the high harmonic ECRH scenario to extend a heatingplasma parameter range in LHD," Nucl. Fusion 55 (2015) 063035
6. T. Omoria, M.A. Henderson, F. Albajar, S. Albertic, U. Baruahd, T.S. Bigelowe,et al., "Overview of the ITER EC H&CD system and its capabilities," Fusion Eng.Des. 86 (2011) 951–954
7. P. F. Goldsmith, "Quasioptical system," Chap.5
8. S. G. Kim, A. Sawant, I. Lee, D. Kim, M. S. Choe, J. H. Won, J. H. Kim, J. So, W. J, and E. M. Choi, "System development and performance testing of a W-band Gyrotron," J. Infrared Milli. Terahz. Waves, vol. 14, pp. 1-14, 2015
9. S. G. Kim, J. H. Kim, J. H. Won, W. Lee, J. Y, and E. M. Choi, "Direct real-time power measurement of a high-power electron cyclotron maser by a simple one-point Schottky detector signal," IEEE Trans. THz Sci. Technol., vol. 5, pp. 779-785, 2015
10. M. E. Read, G. S. Nusinovich, O. Dumbrajs, H. Q. Dinh, D. Opie, G. Bird, K. Kreischer and M. Blank, "Design of a 3 MW, 140 GHz, gyrotron based on a $TE_{21,13}$ coaxial cavity," *in Proc. 18th Int. Cont. Infr. Millim. Waves,* vol. SPIE 2 104, p. 521, 1993

DOI: 10.1051/epjconf/201715703023

# A current drive by using the fast wave in frequency range higher than two timeslower hybrid resonance frequency on tokamaks

*Sun Ho* Kim[1*], *Yong Seok* Hwang[2], *Seung Ho* Jeong[1], *Son Jong* Wang[3], and *Jong Gu* Kwak[3]

[1]Korea Atomic Energy Research Institute, Nuclear Fusion Technology Development Division, 34057Daejeon, Korea
[2]Seoul National Univeersity, Department of Nuclear Engineering, 08826Seoul, Korea
[3]National Fursion Research Institute, Diagnostic and Heating DeviceGroup, 34133Daejeon, Korea

**Abstract.**An efficient current drive scheme in central or off-axis region is required for the steady state operation of tokamak fusion reactors. The current drive by using the fast wave in frequency range higher than two times lower hybrid resonance ($w>2w_{lh}$) could be such a scheme in high density, high temperature reactor-grade tokamak plasmas. First, it has relatively higher parallel electric field to the magnetic field favorable to the current generation, compared to fast waves in other frequency range. Second, it can deeply penetrate into high density plasmas compared to the slow wave in the same frequency range. Third, parasitic coupling to the slow wave can contribute also to the current drive avoiding parametric instability, thermal mode conversion and ion heating occured in the frequency range $w<2w_{lh}$. In this study, the propagation boundary, accessibility, and the energy flow of the fast wave are given via cold dispersion relation and group velocity. The power absorption and current drive efficiency are discussed qualitatively through the hot dispersion relation and the polarization. Finally, those characteristics are confirmed with ray tracing code GENRAY for the KSTAR plasmas.

## 1 Introduction

Continuous current drive is one of the key issues to be resolved fortokamaks to progress toward fusion reactor. Though it is expected to be achieved by self bootstrap current in advanced operation regime, the external non-inductive current drive scheme is still required to control the pressure gradient profile and fill the gap between target and bootstrap current.[1] The fast waves in various frequency ranges has been suggested as a current drive scheme for such a reactor grade tokamak. High Harrmonic Fast Wave(HHFW) current drive in frequency range of ($\omega_{ci} \ll \omega \ll \omega_{lh}$) was suggested in 1995[2] and it was observed that the current drive is in well agreement with the theory and simulation via MSE measurement.[3] The fast wave close to LHR(Lower Hybrid Resonance) with harmonic number about 20-40 ($\omega_{ci} \ll \omega < \omega_{lh}$) has beenrecently suggested for off-axis current drive. And it is planned to confirm the theory in KSTAR and DIIID.[4-6] The current drive using fast wave more close to LHR in frequency range $\omega \sim \omega_{lh}$ was noted in 1980 and the fast wave coupling is analyzed, where it is insisted that the fast wave can be operated in more lower frequency than LHCD since it does not suffer thermal mode conversion, and accessible parallel refractive index $N_{||}$ is broadened with more viable RF system requirement.[7] The fast wave experiment in this frequency range had been tried in JIPPT-IIU, JFT-2M, and PLT.[8-10] It was found that the efficient current drive is impossible above density limit as LHCD, which was explained by mode conversion of fast wave to slow wave in confluence layer. The fast waves are depicted witha CMA diagram for hydrogen plasma to compare in various frequency range in Fig.1.

In this study, the current drive by the fast wave in more high frequency range than previously explored, i.e., $2\omega_{lh} \leq \omega \ll \omega_{ce}$ is suggested and the characteristics are analyzed theoretically and numerically. Hereinafter it is called Lower Hybrid Fast Wave(LHFW). The suggestion is based on three advantages as following. First, LHFW has higher electric field parallel to the magnetic field which is favorable to efficient current drive compared to the fast waves in other frequency range. Second, it can deeply penetrate into high density plasmas compared to the slow wave in the same frequency range. Third, though the antenna RF power with $E_y$ polarization for LHFW launching is partly coupled to Lower Hybrid Slow Wave(LHSW) due to the high density gradient or other mechanisms, the parasitic coupling to slow wave can work as a LHCD without parametric instability, thermal mode conversion and ion heating occured in the frequency range $\omega < 2\omega_{lh}$ [11]. The boundary and accessibility of LHFW propagation are given via cold dispersion relation in section 2.1. The energy flow and density limit are analyzed through the group velocity comparing with LHSW in section 2.2. The energy absorption and current drive are discussed qualitatively through the hot dispersion relation and the polarizationin

*Corresponding author: shkim95@kaeri.re.kr

section 2.3. Finally, the analyzed characteristics of LHFW is verified with ray tracing code GENRAY[12] for the KSTAR plamsas in section 3.

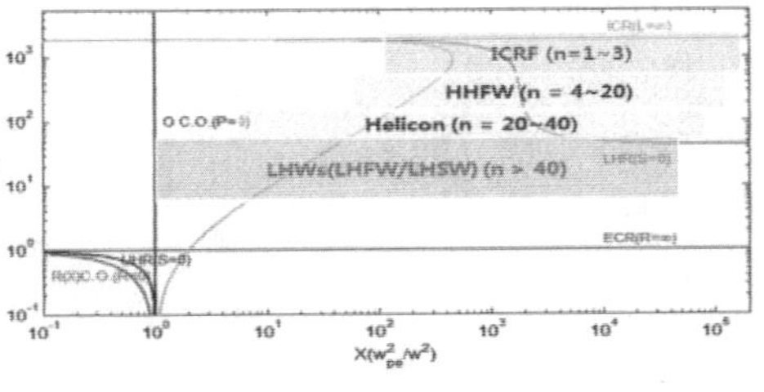


**Fig. 1.** Fast waves in various frequency range for hydrogen plasmas

## 2 Propagation, absorption and current drive of LHFW

### 2.1. Dispersion relation and propagation boundary

The dispersion relation of fast and slow wave in the frequency range of $\omega_{ci} \ll \omega \ll \omega_{ce}$ can be represented as follows.

$$N_\perp^2 = \begin{cases} -\frac{P(N_\parallel^2-S)}{S} & \text{for SW} \\ -\frac{(N_\parallel^2-R)(N_\parallel^2-L)}{(N_\parallel^2-S)} & \text{for FW} \end{cases} \quad (1)$$

,where S, P, R, and L are Stix paramters of cold palsams, and $N_\parallel$ is a refractive index parallel to the magnetic field. The launching density of LHSW and LHFW are obtained from P=0 and $N_\parallel^2 = R$ of Eq.(1) as follows.

$$n_{launch} \cong \begin{cases} \frac{m_e\epsilon_0}{e^2}\omega^2 & \text{for SW} \\ \frac{m_e\epsilon_0}{e^2}\left(N_\parallel^2 - 1\right)\omega\omega_{ce} & \text{for FW} \end{cases} \quad (2)$$

The launching density of LHFW is much greater than that of LHSW. It makes the evanescent layer in front of antenna thicker and it gives rise to difficulty in coupling of LHFW. The coupling concerning this high launching density must be resolved to apply the LHFW current drive concept to reality.

Meanwhile, the uppder density of LHFW is limited by the confluence of two wave branches. It is obtained from the more general cold plasmas dispersion relation as Eq.(3).

$$n_{confluence} \cong \frac{1}{4}\frac{m_e\epsilon_0}{e^2}N_\parallel^2\omega_{ce}^2 \quad (3)$$

It is a more generalized accessibility condition since the mode conversion takes place whenever Eq.(3) is satisfied by the change of the parallel refractive index and magnetic field during the propagation of LHWs.The confluence density in Eq.(3) is proportional to the square of the magnetic field for given $N_\parallel$. It means that the mode conversion will not easily appear in high magnetic field device. From Eq.(1)~Eq.(3), the propagative region of the fast and slow wave can be depicted on CMA diagram as shown in Fig.2.

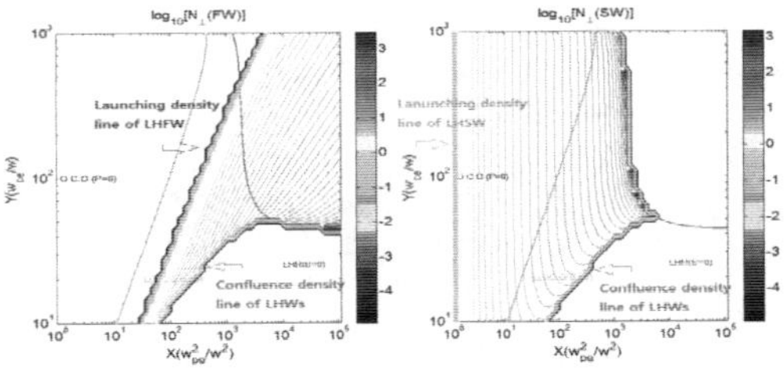


**Fig. 2.** The propagative region of FW and SW on CMA diagram($N_\parallel = 2.0$ is assumed.)

Considering that the parallel refractive index and magnetic field change as LHFW propagates in tokamak geometry, a kind of window for plasma density proiflecan be figured with Eq.(2) and (3) as shown in Fig.3. The plasma density profile must be sustained within the window for LHFW to propagate.

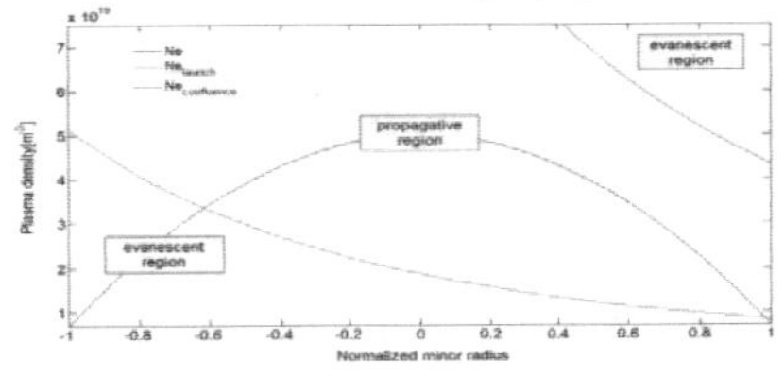


**Fig. 3.** Density window for LHFW propagation

### 2.2 Group velocity and penetration

Though the propagative region of LHFW is within the density boundaries, the real wave propagation and energy flow of LHFW can be figured with the direction of the group velocity to the magnetic field. The angle of group velocity of cold plasma waves is represented as Eq.(4).

$$\theta_g = \arctan\left\{\frac{N_\perp}{N_\parallel}\frac{2SN_\perp^2-[RL+SP-N_\parallel^2(S+P)]}{2PN_\parallel^2-[2SP-N_\perp^2(S+P)]}\right\} \quad (4)$$

It is depicted for$N_\parallel$= 2.0 in Fig 4.

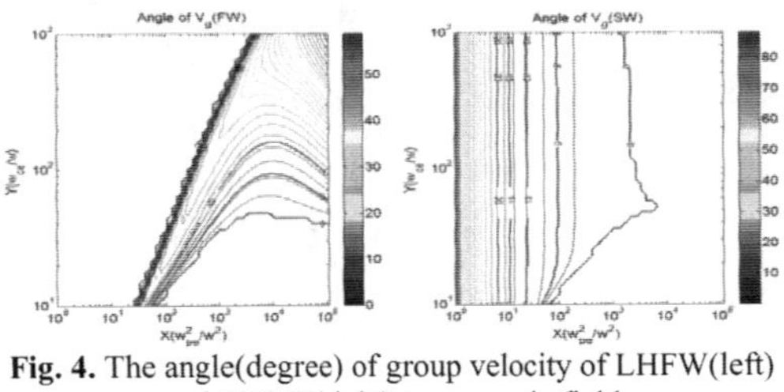

**Fig. 4.** The angle(degree) of group velocity of LHFW(left) and LHSW(right) to magnetic field

The LHSW group velocity is almost perpendicular to the magnetic field near the launching density, but it decreases rapidly with increase of density and gets to be less than that of LHFW near the launching density of LHFW. It means that the LHSW ray is easily aligned to the magnetic field during propagation into high density plasmas so it is not easy to use LHSW to drive thecurrent in high density plasmas. It might explain the LHCD density limit observed in experiments that the current drive efficiency decreases with increase of density. Meanwhile, the angle of LHFW is relatively greater than that of LHSW above LHFW launching density. So LHFW can be used to drive current inner high density plasma region than LHSW if once LHFW is launched and mode conversion is avoided. It will be shown more clearly at ray tracing simulation in section 3.

### 2.3 Absorption and current drive

The power absorption of LHFW can be obtained from the imaginary part of refractive index as given in Eq.(5) which is obtained from a hot plasma dispersion relation.

$$N_{\perp i} = \begin{cases} N_{\perp r,S}\pi^{1/2}\eta^3 e^{-\eta^2} & \text{for LHSW} \\ N_{\perp r,F}\frac{\pi^{1/2}\eta^3 e^{-\eta^2}}{\omega_{ce}^2 N_\parallel^2/\omega_{pe}^2 - 1} & \text{for LHFW} \end{cases} \quad (5)$$

Since the slow wave has very high real refractive index, the imaginary refractive index of LHSW is very large. It means that it can absorbed very quickly as it propagates into plasmas once the Landau damping condition is satisfied. Meanwhile, the absorption of fast wave increases only if the plasma density increases enough and the Landau damping condition is satisfied. It means that LHFW is suitable for bulk plasma heating and current drive since the absorption is weak in the edge low density region and becomes stronger as it propagates into high density bulk plasma region.

The current drive depends on the polarization of wave electric field and resultant power absorption. The main mechanism of electron heating and current drive of fast waves are Transit Time Magnetic Pumping(TTMP) by $B_z(E_y)$ field and Landau Damping(LD) by $E_z$ field. Generally the TTMP can be as efficient as LD. But it tends to be concentrated in central region due to the dependence on plasma betaand the trapped particle effect. In addition, TTMP is vulnerable to competitive absorption of alpha particle in reactor grade plasmas. On the contrary, the LD is very effective only if the resonance condition is satisfiedand $E_z$ is high. Therefore, higher $E_z$ compared to $E_y$ is favorable to current drive. The electric field polarization of fast waves can be represented as follows.

$$\begin{cases} \left|\frac{E_y}{E_z}\right| \simeq i\frac{N_\parallel D}{N_\perp(N_\parallel^2-S)} \sim \frac{\omega\omega_{ce}}{\omega_{pe}^2}\left(\frac{m_e}{m_i}\frac{\omega_{ce}^2}{\omega^2}-1\right)^{-1} & \text{for SW} \\ \left|\frac{E_z}{E_y}\right| \simeq -N_\parallel N_\perp\frac{D}{P(N_\parallel^2-S)} \sim \left(\frac{m_e}{m_i}\frac{\omega_{ce}^2}{\omega^2}-1\right)^{-1} & \text{for FW} \end{cases} \quad (6)$$

The field ratio of $E_z$ to $E_y$ for fast wave increases as the driving frequency increases.Therefore, it is advantageous to use the fast waves in high frequency range as possible. The electric field polarization is depicted with that of slow wave in Fig.5. The $E_z$ field of fast waves is usually lower than that of slow wave, but it increases with the driving frequency increase and it becomes comparable to that of slow wave in the frequency range of $\omega \geq 2\omega_{lh}$.

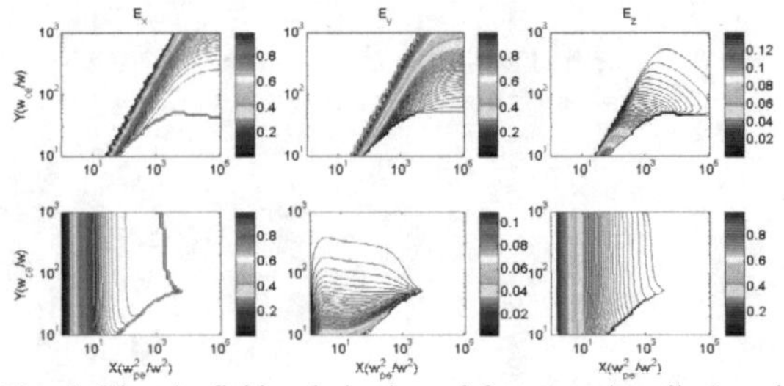

**Fig. 5.** Electric field polarization of fast wave(top line) and slow wave(botton line). They are normalized to total electric field of each wave branch.

Concering the difference of absorption and current drive between LHSW and LHFW, it was already simulated for ITER LHCD study[13] and for the identification of the role of mode converted FW in LHCD[14]. And it was found that LHFW current drive can be as efficient as LHSW.

## 3 LHFW simulation by using GENRAY for KSTAR plasmas

### 3.1. Simulation condition

The ray tracing simulation of LHWs is carried out for KSTAR plasmas to confirm the analysis. The major radius is 1.8m, the minor radius is 0.5m, and the $B_0$ is 2T. The density profile is parabolic with central and edge density of $5\text{x}10^{19}\text{m}^{-3}$ and $1\text{x}10^{19}\text{m}^{-3}$, respectively. The temperature profile is linear. The central and edge temperaturesareare 3.5 keV and 1.0keV. The RF frequency is 2.45 GHz and the $N_\parallel$ of the antenna are set to 2.3 and 2.7 to evaluate the LHFW propagations in relation to accessibility condition. The dispersion relation for two cases are shown in Fig. 6. In case of $N_\parallel$= 2.3 the accessibility condition is not satisfield. The confluence point exists and the evanescent layer develops interior of bulk plasmas. As a result, the launched fast and slow wave cannot penetrate into bulk plasmas. In the case of of $N_\parallel$= 2.7, on the contray, the confluence point does not exist so the evanescent layer disappear and the fast wave and slow wave can penetrate into central region. It is needed to note that it just give a rough probable propagation of LHWs for prediction becausethe parallel refractive index continuously changes by refraction during propagation in toroidal geometry. In somes cases the accesibility can be satisfied although it is not satisfied during intial propagation in edge region.

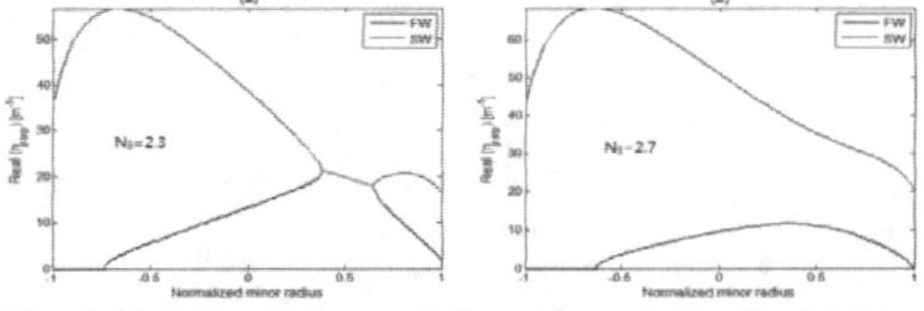

**Fig. 6.** Density window and dispersion relation for (a) $N_\parallel$ = 2.3 and (b) $N_\parallel$ = 2.7

### 3.2. Results of simulation

The ray tracing result with GENRAY is shown in case of $N_{||}$=2.3 in Fig.7. One wave branch is mode converted to the other branch in outer region, and vice-versa, several times. This is because the accessibility condition is not satisfied. As a result, wave field is intensively localized in the outer region and the wave can penertrate into central region only after several multi-pass and severe change of refractive index by the edge reflection. In this case, considerable collsional absorption is expected in the edge region though it is not included here. Meanwhile, both the two wave branches can propagate into bulk plasma region with $N_{||}$= 2.7 as shown in Fig.8 since accessibilty condition is satisfied. As expected, the LHFW penetrates into more central region than the LHSW and driven current of 147 kA for 1MW injection which are comparable to 121 kA of LHSW. The driven current density profiles for LHFW and LHSW are shown in Fig.9. The current figure of merit of LHFW is about 0.1 A/m$^2$/W which is less than the best result ~0.3A/m$^2$/W in previous LHCD experiments using LHSW. It is related with higher parallel refractive index which cuases kinetic resonance to invovle more collisioanl electrons compared to lower $N_{||}$. It is expected to increase in reactor grade tokamaks with lower $N_{||}$ and higher temperature.

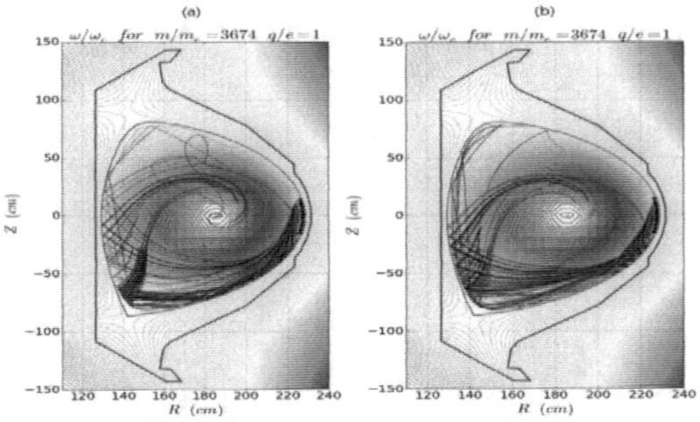


**Fig. 7.** Ray tracing result for (a) LHFW launching case and (b) LHSW launching case when $N_{||}$ = 2.3. Wave is launched at equatorial port.

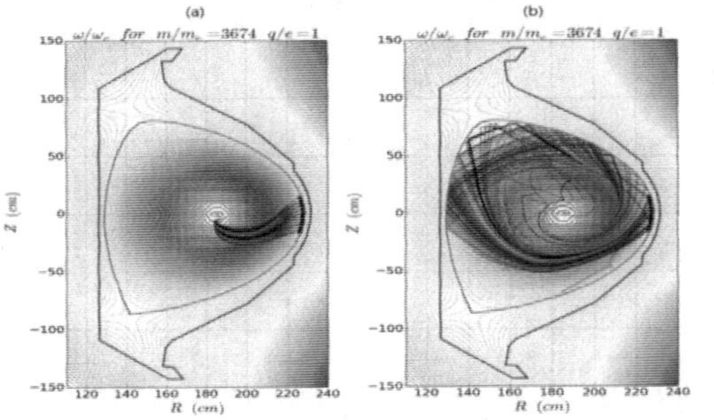


**Fig. 8.** Ray tracing result for (a) LHFW launching case and (b) LHSW launching case when $N_{||}$ = 2.7. Wave is launced at equatorial port.

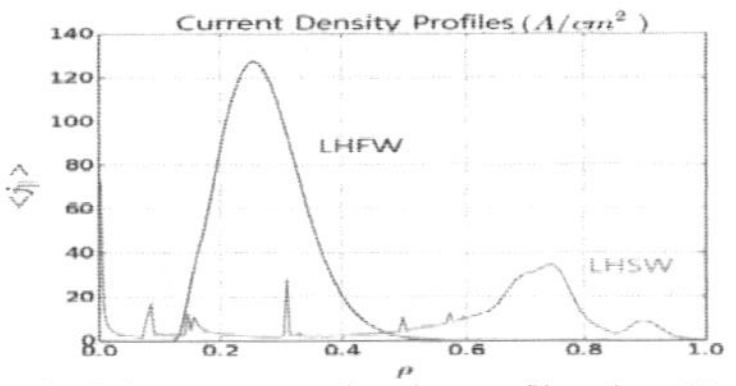


**Fig. 9.** Driven current density profile when $N_{||}$ = 2.7

## References

1. A C C Sips, "Advanced Scenarios for ITER operation", Plasma Phys. Control. Fusion, **47**, A19 (2005)
2. M. Ono, "High harmonic fast waves in high beta plasmas", Phys. Plasmas, **2(11)**, 4075 (1995)
3. C. K. Phillips et al., "Spectral effects on fast wave core heating and current drive", Nuclear Fusion, **49**, 075015 (2009)
4. V. L. Vdovin, "Current generation by helicons and lower hybrid waves in modern tokamaks and reactors ITER and DEMO. Scenrarios, Modeling and Antenna", Plasma Physics Reports, **39(2)**, 95 (2013)
5. R. Prater et al., "Application of very high harmonic fast waves for off-axis current drive in the DIII-D and FNSF-AT tokamaks", Nuclear Fusion, **54**, 083024 (2014)
6. S. J. Wang et al., "Helicon wave coupling in KSTAR plasmas for off-axis current drive in high electron pressure plasmas", Nuclear Fusion, **57**, 046010 (2017)
7. K. Theilhaber, A. Bers, "Coupling to the fast wave at lower hybrid frequencies", Nuclear Fusion, **20**, 547(1980)
8. K. Ohkubo et al., "Current drive by fast magnetosonic waves near the lower hybrid frequency in the JIPP T-IIU Tokamak", Physical Review Letters, **56(19)**, 2040 (1986)
9. T. Yamamoto et al., "Observation of direct electron heating by the fast waves in the lower hybrid range of frequencies in the JFT-2M tokamak", Physical Review Letter, **63(11)**, 1148 (1989)
10. R. I. Pinsker et al., "800 MHz fast wave current drive experiments in PLT", Proceeding of 7th topical conference on Applications of radio-frequency power to plasmas(AIP), 175 (1987)
11. M. Porkolab et al., "Observation of parametric decay instabilities in lower hybrid radio frequency heating of tokamaks", Physical Review Letters, **38(5)**, 230 (1977)
12. R. W. Harvey et al., "The GENRAY ray tracing code", CompX report CompX-20000-01.
13. J. Decker et al., "Calculations of lower hybrid current drive in ITER", Nuclear Fusion, **51**, 073025 (2011)
14. J. A. Heikkinen et al., "Role of fast waves in the central deposition of lower hybrid power", Plasma Phys. Control. Fusion, **41**, 1231-1249 (1999)

# RF Sheath-Enhanced Plasma Surface Interaction Studies using Beryllium Optical Emission Spectroscopy in JET ITER-Like Wall

*C.Christopher* Klepper[a,]*, *Laurent* Colas[b], *Alena* Křivská[c], *Volodymyr* Bobkov[d], *Philippe* Jacquet[e], *Ephrem* Delabie[a], *Carine* Giroud[e], *Krassimir* Kirov[e], *Ane* Lasa[a], *Ernesto* Lerche[c,e], *Pierre* Dumortier[c], *Frederic* Durodié[c], *Mehdi* Firdaouss[b] *and JET* contributors[f]

[a] Oak Ridge National Laboratory, Oak Ridge, Tennessee 37831-6169, USA
[b] IRFM CEA Cadarache 13108 Saint Paul lez Durance France
[c] LPP-ERM/KMS, Royal Military Academy, 30 Avenue de la Renaissance B-1000, Brussels, Belgium
[d] Max-Planck-Institut für Plasmaphysik, D-85748 Garching, Germany
[e] CCFE, Culham Science Centre, Abingdon, Oxfordshire OX14 3DB, UK
[f] See the author list of "Overview of the JET results in support to ITER" by X. Litaudon et al. to be published in NF Special issue: overview and summary reports from the 26th Fusion Energy Conf. (Kyoto, Japan, 17-22 Oct. 2016)

**Abstract.** A dedicated study on JET-ILW, deploying two types of ICRH antennas and spectroscopic observation spots at two outboard, beryllium limiters, has provided insight on long-range (up to 6m) RF-enhanced plasma-surface interactions (RF-PSI) due to near-antenna electric fields. To aid in the interpretation of optical emission measurements of these effects, the antenna near-fields are computed using the TOPICA code, specifically run for the ITER-like antenna (ILA); similar modelling already existed for the standard JET antennas (A2). In the experiment, both antennas were operated in current drive mode, as RF-PSI tends to be higher in this phasing and at similar power (~0.5 MW). When sweeping the edge magnetic field pitch angle, peaked RF-PSI effects, in the form of 2-4 fold increase in the local Be source,are consistently measured with the observation spots magnetically connect to regions of TOPICAL-calculated high near-fields, particularly at the near-antenna limiters. It is also found that similar RF-PSI effects are produced by the two types of antenna on similarly distant limiters. Although this mapping of calculated near-fields to enhanced RF-PSI gives only qualitative interpretion of the data, the present dataset is expected to provide a sound experimental basis for emerging RF sheath simulation model validation.

## 1 Introduction

Optical emission spectroscopy (OES) is used on JET ITER-Like Wall (JET-ILW) to non-intrusively study RF sheath-enhanced plasma-surface interactions (RF-PSI) in ICRF heated discharges. Such effects are observed on beryllium (Be) outboard limiters, in the form of poloidally-localized regions of enhanced (by ~2-4x) Be I and Be II light emission. RF-PSIs are also of interest for the study of RF-sheath formation and for the potential impact of localized erosion on ILW component lifetimes. In the early stages of JET-ILW (2011-12) dedicated JET pulses, combining sequential toggling of all available ICRH antennas with concurrent magnetic pitch-angle sweeping, indicated that these RF-PSIs can be traced, via magnetic field-line connection, to regions of anticipated high $E_{//}$ near-field at an active antenna [*1* - *3*]. Such effects are understood as arising from local rectification of these fluctuating near-fields by the plasma near the local limiters, with the resulting potential acting then on the observed (remote) limiter via magnetic connection. Erosion modeling showed that the enhanced optical emission was consistent with increased local erosion of the Be limiter surface [*3*, *4*].

In addition to the standard type-"A2" antennas [*5*], available in the earlier study, the ITER-Like Antenna (ILA) [*6*, *7*] has been recently recommissioned [*8*] and included in a set of new dedicated RF-PSI experiments in the JET 2015-16 Campaign. The aim of this paper is to provide first analyses of the data obtained from the new experimental sessions, with emphasis on the correlation between the optical emission amplification at the observation points, which now include points on a limiter near the ILA, to regions of high parallel components of the near-field ($\mathbf{E}_{//}$) at the active, magnetically connected antenna. Furthemore, we seek to determine whether the RF-PSI induced by the ILA is qualitatively comparable to that induced by the type-"A2" antennas for similar phasing and connection geometries. For the type "A2" antennas, the observed, enhanced, long-range RF-PSI effect was shown to be consistent with $E_{//}$ near-field A2-specific antenna modeling (*TOPICA* [*9*] maps of the active antennas as in Ref. [*3*]). For the present work, ILA-specific Topica modeling is carried out to see if these correlations also extend to the ILA.

This paper is arranged as follows: In Section 2 we briefly describe the experimental arrangement and cast it

* Corresponding author: kleppercc@ornl.gov

in the framework of improved experimental capability over the earlier study; in the same section some of the observed responses of the spectroscopic signals to the antenna toggling and simultaneous of the edge safety factor ($q_{95}$) scanning (to change the magnetic connection geometry) are presented. In Section 3, we discuss the magnetic field mapping and then show graphically some of the obtained correlations between the response of the spectroscopic signals and the mapping of these effects to the TOPICA $\mathbf{E}_{//}$ maps. In Section 4, we summarize and further discuss the findings of these first analyses and the envisioned next steps.

### 1.1 Experimental arrangement and approach

The overall approach to this experiment is similar to that described in Refs [*1, 3*]. The principal main differences were in (a) the presence of the ILA and (b) the optical access to an additional poloidal limiter. Figure 1 recalls the toroidal arrangement of the JET standard type-"A2" antennas, as well as the location of the ILA. Furthermore, the Figure shows (in cartoon form – not in exact scale) the location of the observation spots available for these newer measurements. Of most interest is the new access to observation spot "K14", which is nearly centered on a limiter near the ILA.

Observation spot "D14", which was also used in the earlier published studies, was available again with an improvement: As a result of periscope realignment, this spot was nearly centered onto the limiter to the last of Ant-D, thus sampling the leading edge, where PSI is expected to be peaked. It is recalled that, for the earlier study, this same sightline was shifted to the right, in the Figure, such that its footprint was cast onto the left-most region Ant-D's the recessed Faraday screen. As such, this sightline (coming tangentially fro the left side) would glance by the antenna side of the limiter. This made the interpretation of the measurement challenging in the limiter surface erosion effort of Refs [*3, 4*]. The shift of this spot back onto the limiter will also be of great value to the erosion modeling of this newer set of data.

As in the earlier study, all the available antennas were toggled repeatedly, while sweeping the edge safety factor, $q_{95}$, to vary the magnetic connection between the active antennas and the observation spots. In the present study, the active antennas were Ant-D, Ant-A&B (i.e. Ant-A and Ant-B, cobined as one) and the ILA.

Unfortunately, Ant-C, whose high $\mathbf{E}_{//}$ upper corner near-field cast the most notable bright emission spots onto the central region of Ant-D's left limiter in the earlier study, was not available for the present study. Making up for this, the present access to the "K14" spot provided a measurement point that is nearly equidistant from Ant-D as is "D14" from the ILA, allowing for a quantitative comparison of near-field effect on a distant limiter from each these two very different antenna technologies.

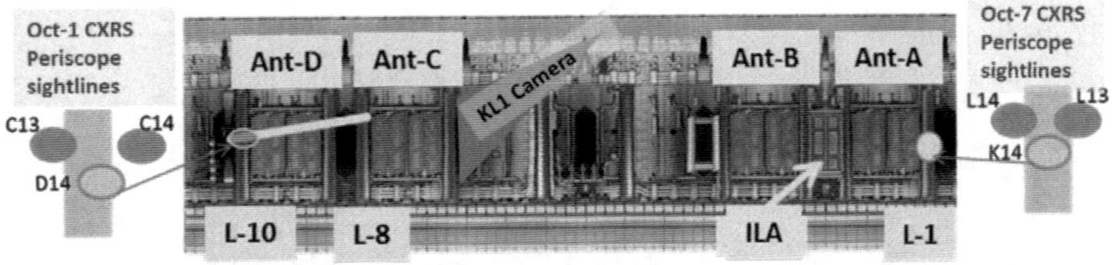


**Fig. 1.** Position of the new sightline footprints (or spots) on the plasma-facing components of interest, with respect to the position of all JET ICRH antennas in the machine at the time of the Experiment. The spots are named using conventions followed by the JET CXRS diagnostic, whose spectrometers and sightlines are used here. The outboard limiters are arbitrarily labelled for this narrative from L-1 to L-12 (not all seen here) starting with the one on the left of Ant-A, which is in JET octant 1 and going in the direction of increasing octant number.

## 2 Spectroscopic detection on antenna-specific PSI

Figure 2 shows the effect of the toggling of antennas on the local Be spectral line emission at the observation spots D14 and K14 (roughly limiter centered and mid-planar) as a function of $q_{95}$. It is important to note that, although for the pulse of this figure the measurement for both these spots was set for singly ionized beryllium (Be II), they were not using the same spectral line. Thus, the difference in absolute intensities is not of interest in this discussion. (It will become of interest in the upcoming surface erosion modeling effort, in which atomic physics models of the specific spectral lines are also included).

Of more interest in the present discussion is that the response to the antenna toggling is quite different at the two observation spots, as well as strongly dependent on which antenna is ON.

It is further noted that the local line brightness response to $q_{95}$ can be substantially different than that seen in the equivalent graph of Ref. [*1*]. In the latter, the effect of switching on Ant-C on the brightness at the observed (left-side) limiter of Ant-D had a minimum in the middle of the $q_{95}$ scan. In contrast, for the present data, the impact of switching on the ILA on (essentially the same) spot "D14" exhibits continuous monotonic and substantial increase as a function of $q_{95}$. In the same sweep, the effect of Ant-D on this same spot shows much less variation and, if anything, it has maximum rather than a minimum in the middle of the sweep.

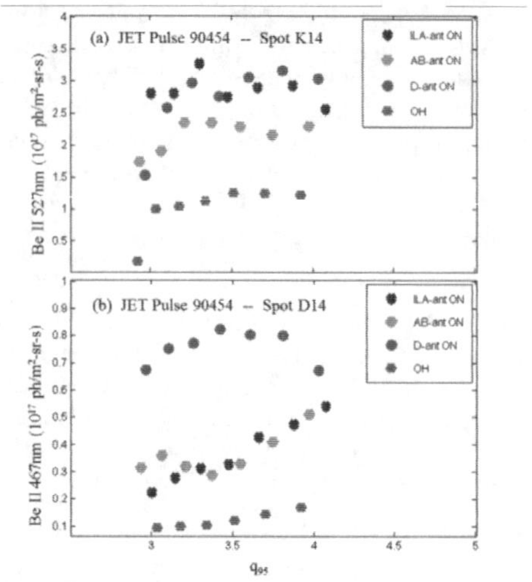


**Fig. 2.** Plateau-averages of the response of the spectral emission at the two principal spot location from Fig. 2, now plotted separately by active antenna and as a function of $q_{95}$.

As with the earlier data set, camera imaging could be used to confirm that the spot measurements indeed correspond to poloidally-localized phenomena. In this new experiment, one of the Be II-filtered, wide angle cameras provided discernable data, albeit for a different limiter than for the spot measurements (Fig. 1).

An attempt to understand some of the patterns of Fig. 2 will be made in the next section with magnetic field and $\mathbf{E}_{//}$ mapping. Meanwhile, it is important also to emphasize that almost all these data sets show a significant enhancement in the local Be emission. To see this, one must compare each point to the corresponding Ohmic phase level (the red points in both graphs). Even in the absence of absolute Be influx values, the relative change in brightness, from Ohmic to Ant-(A&B or D or ILA) ON is valid for quantitative comparison between the two observation spots.

## 3 Qualitative interpretation via magnetic field tracing to active antenna $\mathbf{E}_{//}$ near-field mapping with TOPICA

Topica $\mathbf{E}_{//}$ maps of the "A2" antennas, for conditions similar to the ones used in this new experiment, have been previously presented [*3*, *10*]. For the interpretation of the present experiment, Topica modeling was carried out for the ILA in the three configurations in which it was possible to operate it for this study: Bottom half only, top half only, full antenna, and also for both dipole and current drive phasings. A detailed description of this modeling will be available in Ref [*11*].

For analysis intended in this paper, the TOPICA $\mathbf{E}_{//}$ mappings were calculated for a plane located 5mm beyond the near-ILA poloidal limiters. This was done because it is desired here to make correlations of the measured line emission brightness to $\mathbf{E}_{//}$ variations at the limiters near the active antenna, rather than at the recessed antenna surface. Additional $\mathbf{E}_{//}$ mappings were calculated Ref [*11*] for a plane located close to the recessed surface (and thus in the antenna private region) because the aim is ultimately to model the propagation of the RF sheath effects around the limiters (and out of the private region) using a simulation of RF sheath formation and propagation in the SOL for a refined (and potentially extrapolable) interpretation of the experimental measurements [*12*].

For mapping the magnetic field connections, a code, *LHconlen*, initially developed to study the impact of the ICRH antennae on LHCD coupling [*13*] was deployed for the present study, with small modifications. In addition, a custom version of the PFCFlux code [*14*] was also deployed for this study, partly to provide a check of the 2D maps from the first approach, by providing a special output containing the coordinated at which the fieldlines, emanating from the observation spots, pass over the leading edge of the active antenna's limiters.

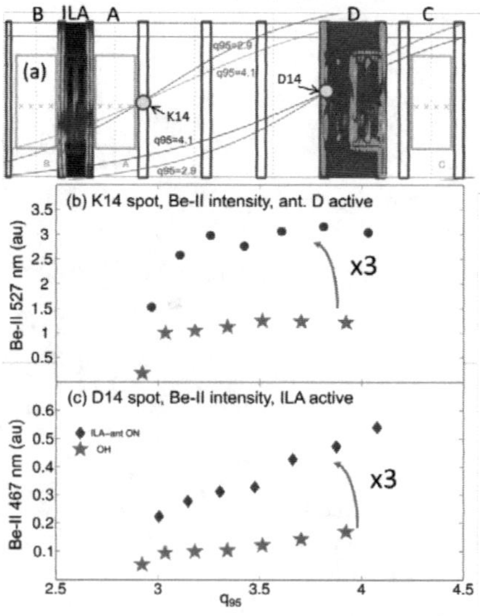


**Fig. 3.** (a) Magnetic connections for spots K14, D14 to similarly distant antennas: Type-"A2" and ILA, correspondingly. (b, c) Enhanced Be flux comparison, Type-"A2" Ant-D to ILA, at the corresponding, distant antenna points.

Figure 3 shows the magnetic connection the distant spot D14 to the ILA and the, quite similar, connection from the distant spot K14 to the Type "A2" Ant-D. It is noted that, in both cases, the measurement spot samples (via the $q_{95}$ sweep of the magnetic connection) a limited range in the very edge of the corresponding antenna limiter, as well as outside the poloidal range of the main antenna structure. However, looking also at the superimposed and corresponding Topica $\mathbf{E}_{//}$ maps, it is seen that significant $\mathbf{E}_{//}$ nearfields are indeed exhibited also at these extreme edges.

DOI: 10.1051/epjconf/201715703024

In the same Figure, the response of local Be emission for both cases are shown for all $q_{95}$ values of the antenna toggling for the same pulse as in Fig. 2. The two important features of these responses are: (1) they both decrease to nearly the Ohmic level when the magnetic fieldline connection gets beyond the poloidal range of the limiter(s) and (2) the relative increase in the emission (over Ohmic) is similar for the two different antenna technologies. Thus, with this observation, this new dataset confirms the previously published interpretation of the 2011-2012 campaign effect, in which rectification of near-fields at the distant antenna leads to the observed Be flux amplification factor (in principle corresponding to a similar local surface erosion enhancement factor) via the magnetic connection. This same effect is now observed at roughly twice the distance along the fieldlines than there was in the earlier study (i.e. the effect of Ant-C on the Ant-D's left limiter, in the connection indicated by the short diagnonal line in Fig. 1). It is now observed arising also from ILA, an antenna technology that is substrantially different than that of JET's standard "A2" antennas [*15*].

The vanishing of the amplification factor when the magnetic connection goes off the limiter helps further buttress the interpretation that these RF-PSI phenomena being studied with this RF-specific Be spectroscopy method are indeed resulting for these long-range effects (via magnetic connection) from near-field rectification at the antenna and its near-limiters. If they were resulting from far fields forming local sheaths at the observation spots, it would be difficult to explain this transition of the measured effect at the end of the poloidal range of the antenna limiter.

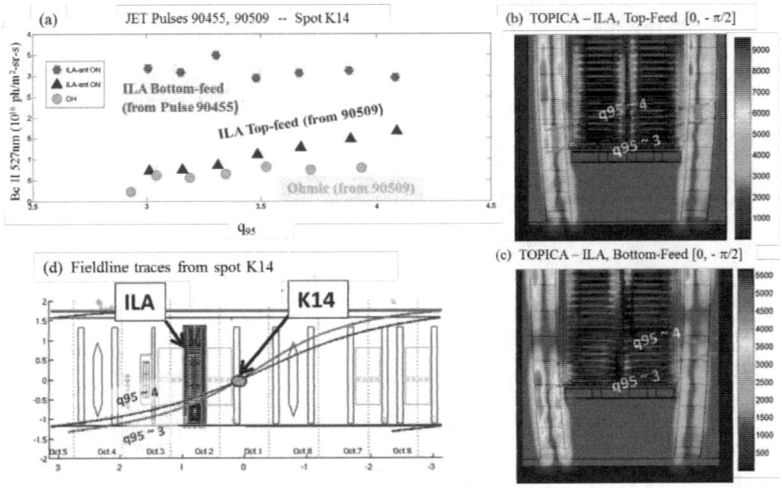


**Fig. 4.** Response of local Be emission at spot K14 to the ILA, including a comparison between two ILA operation configurations in similar pulses. (a) Plateau-averages of emission at the near ILA K14 spot location for the two configurations and as a function of $q_{95}$. (b, c) Upper and lower, correspondingly, feed configuration $\mathbf{E}_{//}$ maps obtained with TOPICA for the current drive [0, -π/2] phsing used in the pulse sequence. (d) Corresponding fieldline tracing, showing only extremes q=~3 and q=~4. The red "*" marks on top of the $\mathbf{E}_{//}$ maps are the locations where seven fieldlines for the $q_{95}$ values within this range cross the leading edge of each of the two limiters of the ILA. These at the same $q_{95}$ values as in the graphs of Be II versus $q_{95}$.

Figure 4 pertains to the response to the ILA of the local Be emission at spot K14 (relatively close to the ILA) and including a comparison between two ILA operational configurations: only top and only bottom half of the antenna being fed by the generator. It can be seen that, in the bottom-feed case (lower $\mathbf{E}_{//}$ image) the K14-emanating fieldlines "sample" a roughly uniform, strong $\mathbf{E}_{//}$ region on the left-side limiter of the ILA, while missing any significant $\mathbf{E}_{//}$ on the right-side. This is consistent with a nearly flat Be II response, which is nevertheless substantially amplified over Ohmic (also plotted in the Figure).

On the other hand, in the upper-antenna-feed case (upper $\mathbf{E}_{//}$ image), the K14-emanating fieldlines now sample regions of lower $\mathbf{E}_{//}$, with respect to the maximum for this calculation (i.e. in the images shown the values are autoscaled to the maximum value for each case). However, although operated at about the same input power (~0.5MW) the TOPICA calculation yields a much higher (~2-fold) maximum $\mathbf{E}_{//}$ for the top-feed case and this leads to non-proportional scaling between the Be emission values and the $\mathbf{E}_{//}$ values that the nea-antenna limiter regions to which they connect. Curiously, TOPICA calculations in dipole phasing (not shown here) show a substantially decrease in $\mathbf{E}_{//}$ in the lower part of the antenna for the top-feed case, in better agreement with the experiment (which was, however, carried out in current-drive phasing). This might be considered as an indication of the limits to using antenna near-field maps to interprete the observed Be emission $q_{95}$ profiles and additional motivation to pursue a full RF sheath simulation as discussed in [*11, 12*].

## 4 Summary and Conclusions

This newer, JET 2015-16 campaign experiment, follow-up to the 2011-12 RF-PWI study in an ILW environment, further confirms to potential of strong $\mathbf{E}_{//}$ near-fields to induce localized impurity sources at distant plasma-facing components, up to several meters from the active antenna, via magnetic connection. This newer dataset also included the ILA, additional observation spots, as well as larger distances on magnetic connection between active antenna and observed RF-PSI location. Local Be flux enhancement factors induced by the ILA were found to be similar to those of an "A2" antenna (at similar antenna to observed PFC distance).

A more detailed interpretation of this new dataset was sought using new antenna modeling results produced specifically for the ILA using TOPICA. Although the range of poloidal extent sampled along the antenna limiter over the available range of the $q_{95}$ scan was quite small, with respect to the significan spatial features of the modeled $q_{95}$, it was still possible to qualitatively correlate local maxima in the Be II emission to magnetically connected maxima in the $\mathbf{E}_{//}$ poloidal distributions on the active antenna's near-limiters. However, quantitative agreement is not reached, further motivating the planned pursuit of a full, RF sheath rectification model, including transport of rectified potentials and currents in the scrape-off later.

In additional to the RF sheath simulation, the surface erosion side of the study is also under way. The availability of additional spectral lines (some simultaneously for different observation spots, e.g. spots L13, L14, on the same limiter as K14, and some on similar pulses, by scanning the tunable spectrometer associated with D14) will be exploited to improve on long pulse-relevant Be component erosion estimates.

*This work was supported, in part, by the US DOE under Contract No. DE-AC05-00OR22725 with UT-Battelle, LLC. This work has been carried out within the framework of the EUROfusion Consortium and has received funding from the Euratom research and training programme 2014-2018 under grant agreement No 633053. The views and opinions expressed herein do not necessarily reflect those of the European Commission.*

## References

1. C.C. Klepper et al., J. Nucl. Mater. 438, S594–S598 (2013)
2. P. Jacquet et al., Physics of Plasmas (1994-present) 21, 061510 (2014); doi: 10.1063/1.4884354
3. C.C. Klepper et al., Phys. Scripta, vol. 2016, no. T167 (2016) 014035
4. A. Lasa et al., "ERO Modeling and Sensitivity Analysis of Erosion Enhanced by Magnetically Connected Antennas," Submitted to NF (Apr. 2017)
5. A. Kaye, T. Brown, V. Bhatnagar, P. Crawley, et al., Fus. Engin. Design 24, 1 (1994)
6. F. Durodié, at al., "Physics and engineering results obtained with the ion cyclotron range of frequencies ITER-like antenna on JET," Plasma Phys. Control. Fusion 54 (2012) 074012
7. A. Czarnecka at al., Plasma Phys. Control. Fusion 54 (2012) 074013
8. P. Dumortier et al., this conference.
9. D. Milanesio, O. Meneghini, V. Lancellotti, R. Maggiora, and G. Vecchi, Nucl. Fusion 49, 115019 (2009).
10. Vl. Bobkov et al., Journal of Nuclear Materials 438 (2013) S160–S165
11. A. Křivska et al., This Conference
12. L. Colas, et al., Overview Talk in this Conference.
13. K. Kirov et al. PPCF 2009 044003
14. M. Firdaouss et al., Journal of Nuclear Materials 438 (2013) S536–S539
15. F. Durodie et al., *Operational Experience with the ICRF ITER Like Antenna on JET*, EFDA-JET-PR(11)48 (EUROfusion Report).

EPJ Web of Conferences **157**, 03025 (2017)
*22 Topical Conference on Radio-Frequency Power in Plasmas*
DOI: 10.1051/epjconf/201715703025

# Feasibility study of Passive Optical Emission Spectroscopy for the electric field measurements in IShTAR

*Ana* Kostic[1,2,★], *Kristel* Crombé[1,3], *Rodolphe* D'Inca[2], *Jonathan* Jacquot[2], *Roman* Ochoukov[2], *Anton* Nikiforov[1], *Mari* Usoltceva[1,2,4], *Elijah* H. Martin[5], *Jean-Marie* Noterdaeme[1,2], and the IShTAR team

[1] *Ghent University, Department of Applied Physics, Ghent, Belgium*
[2] *Max-Planck-Institut für Plasmaphysik, Garching, Germany*
[3] *LPP-ERM-KMS, TEC partner, Brussels, Belgium*
[4] *Université de Lorraine, Nancy, France*
[5] *Oak Ridge National Laboratory, Oak Ridge, Tennessee, USA*

**Abstract.** Direct, non-intrusive, measurements of electric fields are essential for understanding the RF-sheath physics. This is especially true in the case of the ICRF antenna - plasma edge interaction in fusion devices. The rectification of the RF-fields near the plasma-facing components of an antenna leads to the development of DC electric fields that accelerate the ions from the plasma towards the antennas' plasma-facing components, enhancing physical sputtering and release of impurities. IShTAR is a device dedicated to the investigation of the plasma - antenna interactions in tokamak edge-like conditions. It has a simplified geometry and enables an easy access and fast modifications, which makes it a suitable environment to develop diagnostics for electric field measurements.

This paper presents the observed Stark effect on He I spectral line profile, with passive optical emission spectroscopy. To be able to fully control the operating parameters, at this initial stage, the measurements are conducted on a simple DC-biased electrode rather than the ICRF antenna. Measured line profiles are compared with the analytical models of the Stark effect in magnetised helium plasma that, as a result of the good fit, provide the electric field strength.

## 1 Introduction

Auxiliary plasma heating by waves in the ion cyclotron frequency range (ICRF) is an important heating option in magnetic confinement fusion devices. The great advantage is its ability to directly heat the ions in the plasma core in various scenarios with high efficiency. The present-day concern regarding the ICRF heating, however, is its contribution to the heavy-impurity influx originating from the ICRF antenna structure. This has been experimentally observed in various devices that employ ICRF antennas for heating [1–3].

The underlying cause for this spurious phenomenon is not yet fully understood. It is theorised that the formation of the hot spots, erosion and increased physical sputtering from the antenna's plasma-facing components are enhanced by the rectified potential that occurs in the RF sheath established around these areas [4, 5]. Due to this mechanism, also called the RF-sheath rectification, the ions are accelerated out of the plasma with greater velocities than defined by the Bohm sheath criterion, leading to the higher sputtering yields.

To address this issue, and either suppress or completely eliminate these harmful effects, two parallel efforts are ongoing. One approach is to change the design of the antenna itself [6]. At the same time, an active field of research is the understanding of RF sheath physics for which several theoretical models have been proposed. The development of a numerical description is ongoing. However, the theories are still missing a quantitative experimental verification.

The main goal of our work is to develop a diagnostics method that will provide direct measurements of electric fields in the vicinity of an ICRF antenna, in a simplified geometry compared to the one found in a tokamak, yet in representative environmental plasma-edge conditions. The obtained data will benchmark or even guide the models of the RF sheath physics. The results presented here are the initial spectroscopic measurements of Stark-effect-sensitive helium lines performed on a DC-biased electrode immersed in the helicon plasma source of IShTAR.

## 2 IShTAR

IShTAR is a test stand dedicated to the RF sheath studies, located at the Max-Planck Institute for Plasma Physics in Garching, Germany. This device is designed to mimic the tokamak edge environment (in terms of plasma density and temperature) especially for studying ICRF antenna - plasma interactions, however in rather simplified

★e-mail: ana.kostic@ipp.mpg.de

DOI: 10.1051/epjconf/201715703025

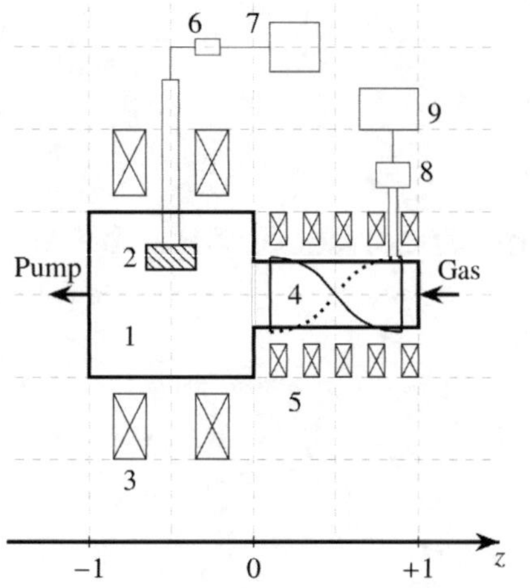


Figure 1: Schematic view of IShTAR: 1 - Main chamber; 2 - single strap ICRF antenna; 3 - Magnetic field coils (Big coils); 4 - Helicon plasma source; 5 - Magnetic field coils (Small coils); 6 - Autotuner; 7 - ICRF generator; 8 - Matching system; 9 - RF generator.

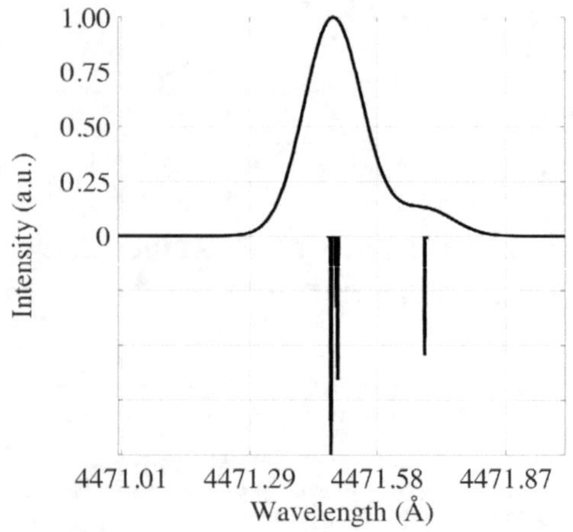


Figure 2: Triplet He-I $4^3D{-}2^3P$ line profile simulated with the EZSSS code with no external electric field (reference scenario) shows discrete and continuous results, with the main component at 4471.49 Å and the second component at 4471.6 Å

| Parameter | Achievable Value |
|---|---|
| $p$ [mbar] | $10^{-6} - 10^{-2}$ |
| Gas | Ar/He/H |
| $B_{BigCoils}$ [T] | 0.27 (10 s) or 0.4 (pulsed) |
| $B_{SmallCoils}$ [T] | 0.1 (10 s) or 1.0 (coils' limit) |
| $P_{RFHelicon}$ [MW] | 1 (presently 0.003) |
| $f_{Helicon}$ [MHz] | 2-30 |
| ICRF antenna | 1 MW (30 MHz), AUG generator<br>1 kW (5 or 15 MHz), IShTAR gen. |

Table 1: Achievable values of the base operational parameters of the IShTAR experiment

geometry required by the numerical models concentrating on this area of research. The design characteristics, choice and range of operational and plasma parameters of IShTAR have been elaborated in [7–9]. The schematics of the machine and its main components is shown in Figure 1. IShTAR consists of a helicon plasma source connected to a larger experimental chamber equipped with an ICRF antenna, referred to as the main chamber. Around both the plasma source and the main chamber magnetic field coils are wound in a Helmholtz-like configuration, providing a parallel magnetic field along the $z$−axes.

The achievable limits for various parameters are listed in Table 1.

## 3 Time-averaged passive optical emission spectroscopy

Passive optical spectroscopy is selected as the first approach to directly measure electric fields in the vicinity of an ICRF antenna, without disturbing the plasma environment. This technique enables studying the perturbation of the electronic structure of an atom caused by an external electric field, the Stark effect. These perturbations are detectable as a shift of the central wavelength of a spectral line, and the occurrence of forbidden components of the fine structure in a spectral line profile. We have focused our research on the Stark effect on $4^3D - 2^3P$ transitions in helium.

Helium is, beside argon, one of the gases available in IShTAR. However, studying the Stark effect on helium lines is beneficial for two reasons - *i*) the changes on the spectral line profiles caused by an electric field are asymmetrical since both the shift and the forbidden components occur; *ii*) due to the fact that in the so-called orthohelium one electron is always present in the ground state and the energy difference between the ground and the first excited state is large, the wave-functions of the selected transition can be accurately represented by the wave-function of hydrogen, which simplifies the modelling of the spectral lines with the Schrödinger solver. While the sensitivity of an electron in an excited state to an external electric field is increasing with an increasing prime quantum number of the state, $n$, at the same time the emission intensity is decreasing. For that reason, the $n = 4 - 2^3P$ transition has been selected for this study, since the signal-to-noise ratio of the recorded spectra was acceptable.

Once the time-averaged spectra are recorded, the spectral line profiles are compared to the simulated ones and the electric field amplitude is extracted with the method of least squares. To simulate the spectra perturbed by an electric field, in the presence of a background magnetic field, the Explicit Zeeman Stark Spectral Simulator (EZSSS) [10] was used. This code generates the discrete spectrum by solving the Schrödinger equation in electric dipole approximation, with external electric and magnetic fields as perturbations. In the second step, by convoluting the discrete spectra with Gaussian and/or Lorenzian profiles to mimmic the broadening mechanisms, the continuous spectrum is obtained.

Figure 2 depicts the modelled triplet line profile corresponding to the $4^3D - 2^3P$ He-I transition with no electric field externally imposed on the system. The discrete spectra, calculated with the EZSSS code is shown as a set of lines in the mirror image of the intensity scale. The continuous spectra, presented in the positive part of the Intensity axes is convoluted with a Gaussian distribution corresponding to the Doppler broadening with a temperature of the radiator of 0.7 eV. The distinct feature of these spectra is the occurrence of the second spectral line red-shifted from the main component, corresponding to the fine structure of the triplet transition of helium. This line has intensity of about 10% compared to the main line. Therefore, the experimental data has to be recorded with high signal-to-noise ratio in order to distinguish this component from the noise. Moreover, the spectra have to be recorded with superior spectral resolution to be able to resolve those two components.

## 4 Experimental setup

To develop a reliable method for measuring the electric fields in the vicinity of an ICRF antenna, the first step is to observe the Stark effect due to a known DC field, in the same plasma that can be found in front of the antenna. Once the concept is confirmed, it will still be used for the calibration of the measurements conducted in the sheath in front of the antenna. Thus a controlled experiment has been constructed that consisted of a planar electrode immersed in the helicon plasma source of IShTAR through the gas-feeding port, as depicted in the Figure 3.

The optical emission spectra was recorded by an Andor™Shamrock750 high resolution spectrometer (focal length 750 mm), equipped with an intensifying CCD detector. The spectrometer is based on a Czerny-Turner optical design and is equipped with 3 gratings of 600, 2400 and 3600 lines/mm. The light from the indicated line of sight was coupled to a 10 m long, 400 $\mu m$ broadband fibre via an aspherical 6 mm collimating lens with a confocal length of 8.7 nm and an adjustable focal point.

### 4.1 Operational and acquisition parameters

To detect the expected Stark-effect-induced changes on the spectral line profile, we have used a high-resolution grating with 3600 lines per mm, that provides a spectral resolution of ≈ 0.02 nm.

The discharge parameters for the dedicated set of experiments are listed in Table 2.

Two experimental scenarios have been performed to obtain the results presented here. The reference data were recorded without the voltage applied to the electrode, while the second set was obtained for the set of experiments with the electrode biased to a positive DC-voltage

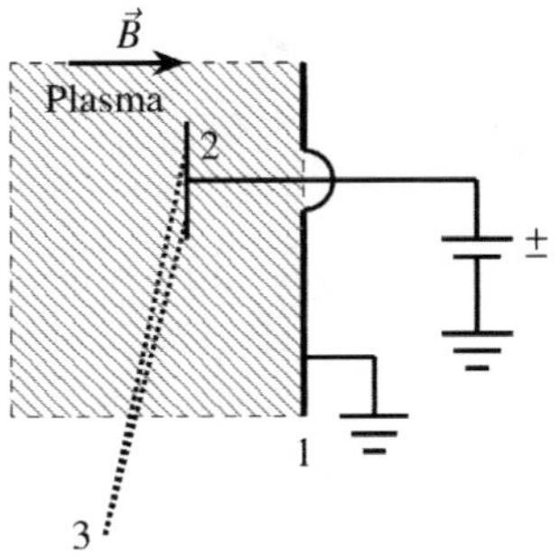


Figure 3: Schematic view of the experimental setup: 1 - Back flange of the plasma source, 2 - Electrode, 3 - Collimating lens and the line of sight indicated in dashed lines.

| Parameter | Value |
|---|---|
| $p$ | $7.8 \times 10^{-4}$ mbar |
| Gas | He |
| Gas flow rate | 60 sccm |
| $B_{BigCoils}$ | 0 T |
| $B_{SmallCoils}$ | 0.01 T |
| $P_{RFHelicon}$ | 700 W |
| $f_{Helicon}$ | 11.76 MHz |
| Discharge duration | 12 s |

Table 2: Discharge parameters for the experiments dedicated to this study

of $U_{el}$ = 1 kV, which was the maximum voltage provided by the power supply.

The obtained line profile corresponding to the $4^3D - 2^3P$ transition in He-I were recorded over the whole duration of a discharge, with an exposure time of 0.5 s, in a kinetic series of 24 scans per discharge. The intensifier in front of a CCD sensor of the spectroscopic system provided a relative gain of 8 counts per photoelectron per scan, contributing to the acceptable signal-to-noise ratio. To compare and time-average the recorded data, the scans corresponding to the stable plasma phase were selected, which was an interval of 5 s, thus counting 10 scans per discharge. With the good reproducibility of the line being confirmed, the accumulated spectra were averaged over 4 discharges per scenario (with and without applied voltage on the electrode). This way each of the experimentally recorder line profiles presented here (Figure 4) are the result of 40 averaged spectra.

## 5 Results and Discussion

The processed data shows a reproducible shift of the He-I $4^3D - 2^3P$ line (Figure 4) when the external electric field is present in the thermal sheath in front of the DC-biased electrode.

For a rough estimate of the electric field expected in a thermal sheath in front of the electrode we can assume that it originates from the difference in potentials between the

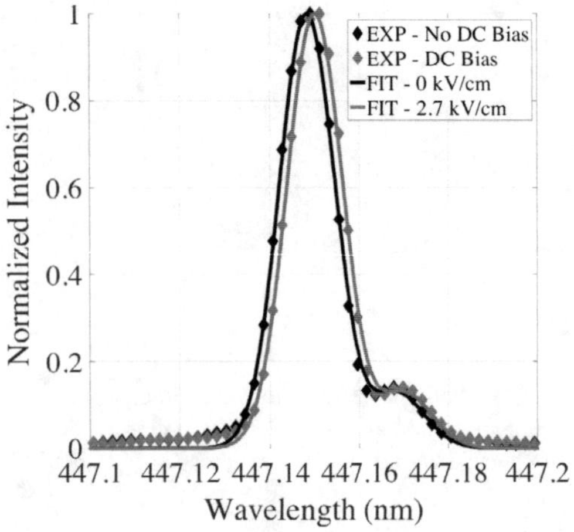


Figure 4: Experimental $4^3D - 2^3P$ He I line profile for $E_{DC} = 0\,\text{kV/cm}$ and $E_{DC} = 2.7\,\text{kV/cm}$ (black and red diamonds, respectively) fitted with the simulated theoretical spectra calculated with the EZSSS code (solid lines)

electrode and the plasma, over the distance corresponding to the sheath thickness. Compared to the potential applied on the electrode,$U_{el} = 1\,\text{kV}$, the plasma potential is $U_{pl} \approx 0\,\text{kV}$. The sheath thickness corresponding to the density of helium plasma in IShTAR of $n_e = 10^{16}\,\text{m}^{-3}$, and the electron temperature of $T_e = 5\,\text{eV}$ is proportional to the Debye length of $\lambda_D = 5 \times 10^{-4}\,\text{m}$. Therefore the electric field in the vicinity of the electrode biased to $U_{el} = 1\,\text{kV}$, can be estimated to be the applied voltage drop over the sheath thickness - $E \approx 20\,\text{kV/cm}$. However, the density in this particular plasma volume can not be measured with the diagnostics currently available in IShTAR, hence the plasma properties listed here could be greatly exaggerated, as are the estimates on the order of magnitude of the sheath thickness since they carry density dependence.

To complete the study, the He-I $4^3D - 2^3P$ transition is simulated with the EZSSS code for several values of the DC electric field, and the best match between the measured and simulated lines corresponds to a simulated spectral line exposed to the external electrical field of $E = 2.7\,\text{kV/cm}$, as depicted in Figure 4 with the solid lines.

However, it has to be taken into account that in the present setup the emission is collected not only from the sheath region, but from the plasma volume as well. Therefore, the Stark effect induced by the electric fields in the sheath region is shielded.

Even though we were able to detect the electric-field-induced changes on the spectral line profiles, the presently available equipment for these experiments and the described set-up is not suitable to provide fully trustworthy results. The results presented here are a successful conclusion of proof-of-concept experiments. It is important to note that the further improvements of the set-up are necessary, and ongoing, as it will serve as a routine calibration for the electric field measurements in the vicinity of an ICRF antenna in IShTAR.

This work has been carried out within the framework of the EUROfusion Consortium and has received funding from the Euratom research and training programme 2014-2018 under grant agreement No 633053. The views and opinions expressed herein do not necessarily reflect those of the European Commission.

## References

[1] M. Bures, J. Jacquinot, K. Lawson, M. Stamp, H.P. Summers, D.A. D'Ippolito, J.R. Myra, Plasma Physics and Controlled Fusion **33**, 937 (1991)

[2] S. Wukitch, B. Lipschultz, E. Marmar, Y. Lin, A. Parisot, M. Reinke, J. Rice, J. Terry, Journal of Nuclear Materials **363-365**, 491 (2007)

[3] V.V. Bobkov, F. Braun, R. Dux, A. Herrmann, L. Giannone, A. Kallenbach, A. Krivska, H.W. Muller, Nuclear Fusion **50** (2010)

[4] J. Myra, D. D'Ippolito, M. Gerver, Nuclear Fusion **30**, 845 (1990)

[5] F. Perkins, Nuclear Fusion **29**, 583 (1989)

[6] V. Bobkov, D. Aguiam, R. Bilato, S. Brezinsek, L. Colas, H. Faugel, H. Fünfgelder, A. Herrmann, J. Jacquot, A. Kallenbach et al., Plasma Physics and Controlled Fusion **59**, 014022 (2017)

[7] R.D. Inca, K. Crombe, F. Fischer, H. Fnfgelder, H. Faugel, R. Ochoukov, J. Jacquot, G. Siegl, S. Devaux, J. Moritz et al. (2015)

[8] K. Crombé, S. Devaux, R. D'Inca, E. Faudot, H. Faugel, H. Fünfgelder, S. Heuraux, J. Jacquot, F. Louche, J. Moritz et al., *Studies of RF sheaths and diagnostics on IShTAR*, in *AIP Conference Proceedings* (2015), Vol. 1689, p. 030006, ISBN 9780735413368, ISSN 15517616, `http://aip.scitation.org/doi/abs/10.1063/1.4936471`

[9] R. D'Inca, J. Jacquot, R. Ochoukov, I. Morgal, K. Crombe, F. Louche, D. Van Eester, S. Heuraux, S. Devaux, J. Moritz et al., *First experimental results on the IShTAR testbed*, in *AIP Conference Proceedings* (2015), Vol. 1689, p. 050010, ISBN 9780735413368, ISSN 15517616, `http://aip.scitation.org/doi/abs/10.1063/1.4936498`

[10] E.H. Martin, Ph.D. thesis, North Carolina State University (2014)

EPJ Web of Conferences **157**, 03026 (2017) DOI: 10.1051/epjconf/201715703026
*22 Topical Conference on Radio-Frequency Power in Plasmas*

# Electromagnetic simulations of JET ICRF ITER-like antenna with TOPICA and SSWICH asymptotic codes

*Alena* Křivská[1,*], *Volodymyr* Bobkov[2], *Laurent* Colas[3], *Pierre* Dumortier[1,4], *Frédéric* Durodié[1], *Ernesto* Lerche[1,4], *Philippe* Jacquet[4], *Jonathan* Jacquot[2], *Christopher* Klepper[5], *Daniele* Milanesio[6] and JET contributors[7]

[1]LPP-ERM/KMS, Royal Military Academy, 30 Avenue de la Renaissance B-1000, Brussels, Belgium
[2]Max-Planck-Institut für Plasmaphysik, D-85748 Garching, Germany
[3]IRFM CEA, Cadarache F-13108 Saint Paul lez Durance, France
[4]CCFE, Culham Science Centre, Abingdon, Oxfordshire OX14 3DB, UK
[5]Oak Ridge National Laboratory, Oak Ridge, Tennessee 37831-6169, USA
[6]Politecnico di Torino, Corso Duca degli Abruzzi 24, I-10129 Torino, Italy
[7]See the author list of "Overview of the JET results in support to ITER" by X. Litaudon et al. to be published in Nuclear Fusion Special issue: overview and summary reports from the 26th Fusion Energy Conference (Kyoto, Japan, 17-22 October 2016)

**Abstract.** Multi-megawatt Ion Cyclotron Range of Frequencies (ICRF) heating is routinely used in the JET tokamak. To increase the ICRF heating power available from the A2 antennas, the ICRF ITER-Like Antenna (ILA) was reinstalled for the 2015 JET ITER-like wall experimental campaign. The application of high levels of ICRF power typically results in increased plasma wall interaction which leads to the observation of enhanced influx of metallic impurities in the plasma edge. It is assumed that the impurity production is mainly driven by the parallel component of the Radio-Frequency (RF) antenna electric near-field, $E_{//}$ (parallel to the confinement magnetic field of the tokamak), that is rectified in a thin boundary layer (RF sheath). Torino Polytechnic Ion Cyclotron Antenna (TOPICA) code was used to compute $E_{//}$ field maps in front of the ILA and between its poloidal limiters in the presence of plasma using measured density profiles and various antenna feedings. $E_{//}$ field maps calculated between the poloidal limiters were used to estimate the poloidal distribution of RF-sheath Direct Current (DC) potential in this private region of the ILA and make relative comparison of various antenna electrical settings. For this purpose we used the asymptotic version of the Self-consistent Sheaths and Waves for Ion Cyclotron Heating Slow-Wave (SSWICH-SW) code. These estimations can help to study the formation of RF sheaths around the antenna and even at distant locations (~3m away).

## 1 Introduction

Multi-megawatt Ion Cyclotron Range of Frequencies (ICRF) heating with four A2 antennae is routinely used in the JET tokamak. To increase the ICRF heating power available from the A2 antennas, the ICRF ITER-Like Antenna (ILA) was reinstalled for the 2015 JET experimental campaign [1] with ITER-like wall (ILW). It uses compact strap array which has successfully been operated with power densities up to 6.2 $MW/m^2$ on L-mode and 4.1 $MW/m^2$ on H-mode and voltages up to 42 kV [2]. A detailed description of the ILA can be found in [3].

Application of high levels of ICRF power typically results in increased plasma wall interaction which leads to observation of enhanced influx of metallic impurities in the plasma edge [4]. Nevertheless the amount of the metallic impurities in JET plasmas is small and within tolerable limits without significantly affecting the plasma performance [5]. Previous studies (with the carbon fibre reinforced carbon wall) indicate that the RF-induced nickel influx per MW of ICRF power applied are comparable for the A2 and the ILA antennas [5], despite the higher power density (smaller radiating surface) of the latter. With the ILW, the ILA operation with only bottom half array powered leads to improvement of around 20% in terms of lower radiated power from the bulk plasma and the RF-induced tungsten content compared to A2 antennas operations. It is not yet clear from where in the machine the dominant tungsten contribution comes from [4]. Enhanced beryllium optical emission was also measured at some A2 antenna limiters that are magnetically connected to ILA [6, 7].

It is assumed that the impurity production is mainly driven by parallel component of Radio-Frequency (RF) antenna electric near-field $E_{//}$ (parallel to confinement magnetic field of the tokamak $\mathbf{B}_0$) that is rectified in thin boundary layers (RF sheaths). $E_{//}$ fields generate oscillating voltages, $V_{RF}$, across the rectifying RF-sheaths on $\mathbf{B}_0$ lines intercepting conducting obstacles. In these sheaths, the oscillating voltages $V_{RF}$ cause enhanced Direct Current (DC) plasma potentials $V_{DC}$. RF sheaths can develop over distances ranging from a few centimetres up to several

* Corresponding author: alena.krivska@rma.ac.be

DOI: 10.1051/epjconf/201715703026

meters away from the ICRF antennas, and cause plasma ion acceleration. The ions are accelerated along the confinement magnetic field lines, hit the obstacle and cause enhanced sputtering and localized heat flux. Advances in the field show that the induced currents on the passive antenna structures are the main source of $E_{//}$ fields on the lateral faces of antenna side limiters [8].

In this contribution, Torino Polytechnic Ion Cyclotron Antenna (TOPICA) code [9] was used to compute $E_{//}$ field maps in front of the ILA and between its poloidal limiters for operation of only top, only bottom and full antenna array in the presence of plasma using experimentally measured plasma density profiles. $E_{//}$ field maps calculated between the poloidal limiters were used to estimate poloidal distribution of RF-sheath DC potential $V_{DC}$ in this private region of the ILA. For this purpose we used the asymptotic version of the Self-consistent Sheaths and Waves for Ion Cyclotron Heating Slow-Wave (SSWICH-SW) code with radiating boundary conditions [10]. These estimations can help to study the formation of RF sheaths around the antenna and even at distant locations (~3m away) and make association to some variations of the Be line emission that were measured spectroscopically on some of the A2 antenna limiters when the magnetic connection from observation point to the active ILA was moved via a $q_{95}$ scan [7].

## 2 Description of the simulations

$E_{//}$ field maps were computed with TOPICA code. To perform simulation with the TOPICA code, the 3D curved model of ILA with limiters was placed into a recess and limited by an aperture. The aperture follows the curvature of the antenna structure and represents the interface between the antenna region, which is in vacuum, and the plasma region. The vacuum-plasma interface was placed 1 cm in front of the leading edge of the poloidal limiters as shown in Figure 1. First surface where we calculated the $E_{//}$ field was defined in the vacuum region 5 mm behind the vacuum-plasma interface. It follows curvature of the antenna structure too. Second surface for the $E_{//}$ field calculations was placed between the poloidal limiters, 4 mm behind their leading edge as shown in Figure 1. $E_{//}$ maps from the second surface were used as input for estimation of $V_{DC}$ between the poloidal limiters.

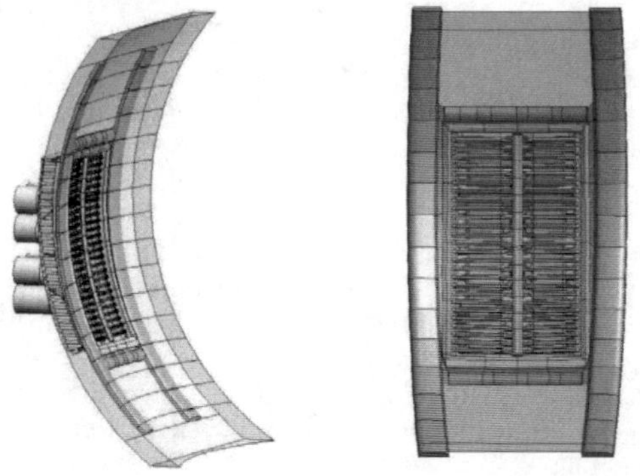

**Fig. 1.** ILA model with the vacuum plasma interface (on the left) and surface for $E_{//}$ field calculations between the poloidal limiters (on the right).

For the $E_{//}$ field maps computations we used as input plasma density and temperature profiles from L-mode shot #90454 that were measured with lithium beam emission spectroscopy system. During the shot $q_{95}$ (safety factor at 95% of the plasma minor radius) varied from 3.04 to 3.92 (Figure 2), Radial Outer Gap (ROG) was equal to 6 cm, central plasma density was around $3 \cdot e^{19}$ $m^{-3}$ and central plasma temperature was equal to 1.85 keV. Central static magnetic field was 2.375 T, 10° tilted with respect to the toroidal direction. Plasma geometry was assumed slab with temperature and density gradients along minor radius of the tokamak chamber. Simulations were done for frequency 42 MHz and minority heating scheme with 3.5% of hydrogen and 96.5% of deuterium. Different antenna feeding was adopted: a) full array ($0\ \pi$), b) full array ($0\ -\pi/2$), c) only top half array ($0\ \pi$) and d) only bottom half array ($0\ \pi$).

$E_{//}$ field maps in front of the antenna were also used for work that studies the RF sheath enhanced plasma surface interactions using optical emission spectroscopy on JET with ILW [7]. For that investigation we used the same antenna model and plasma settings as mentioned in the previous text. $E_{//}$ field maps were calculated on the first surface that was located 5 mm in front of the leading edge of the poloidal limiters. In that case simulations were performed for only top half array ($0\ -\pi/2$) and only bottom half array ($0\ -\pi/2$) antenna feedings.

For estimation of the poloidal distribution of the rectified DC plasma potential $V_{DC}$ between the poloidal limiters of the ILA we used the asymptotic version of the SSWICH-SW code with radiating boundary conditions. It was chosen because it is rapid, robust, always provides systematically one solution for each set of input parameters and convergence is guaranteed. It is suitable for relative comparisons of antenna designs and electrical settings. Principal disadvantages of the code are related to number of approximations made on the original physical model. Parallel DC plasma conductivity is taken as infinite while transverse plasma conductivity is neglected in the simplified model of DC plasma biasing. The potential $V_{DC}$ is homogeneous all along the open field lines and depends only on the voltages $V_{RF}$ at both ends of these field lines. Main input to the SSWICH-SW code is the $E_{//}$ field map evaluated between the poloidal limiters. Simple 3-dimensional parallelepiped simulation domain with walls normal to the straight magnetic field lines and only slow wave, responsible for excitation of the RF sheaths when the side walls are normal to $\mathbf{B}_0$, were only adopted. Sketch of the simulation domain can be found in [11].

DOI: 10.1051/epjconf/201715703026

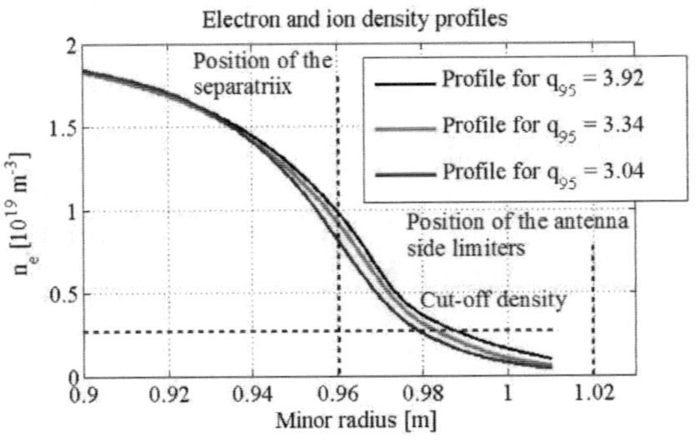


**Fig. 2.** Electron and ion density profiles with varying $q_{95}$ that were measured during shot #90454 with Li beam emission spectroscopy. Cut-off density was evaluated for main $k_{//}$ radiated by ILA in dipole phasing.

## 3 Simulation results

### 3.1 Estimation of spatial distribution of $E_{//}$ near field

$E_{//}$ field maps calculated 5 mm in front of the leading edge of the poloidal limiters for various antenna feedings are shown in Figures 3-5. These maps were obtained for the profile with $q_{95}$ = 3.34. Results were normalized to 450 kW of coupled power to the target plasma.

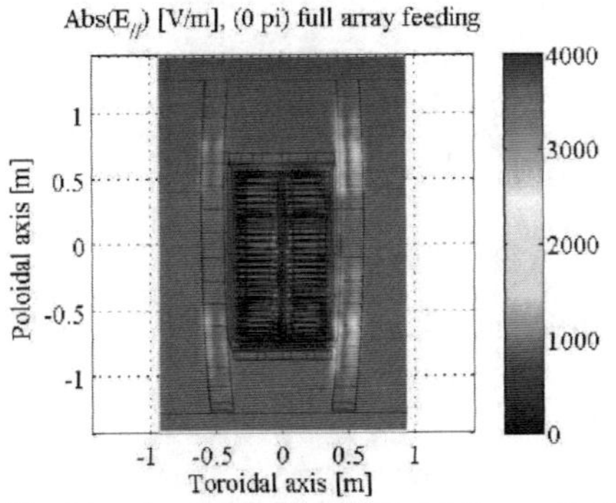


**Fig. 3.** Absolute values of $E_{//}$ field for (0 $\pi$) full array feeding.

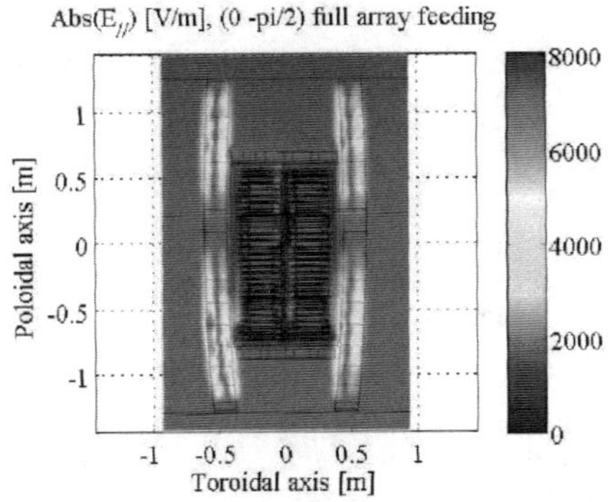


**Fig. 4.** Absolute values of $E_{//}$ field for (0 -$\pi$/2) full array feeding. Used scale is not the same as in case of the other $E_{//}$ field maps.

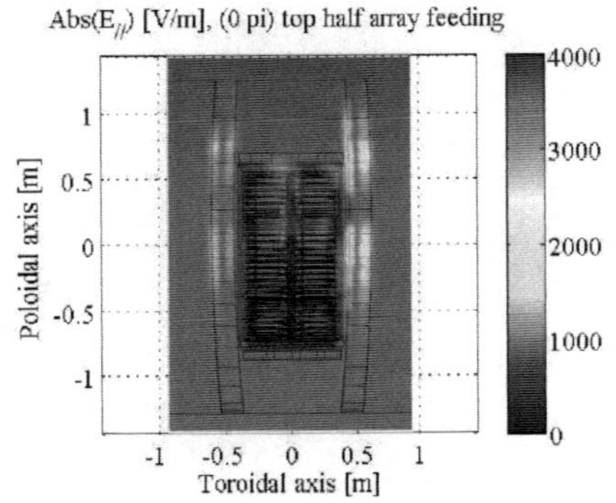


**Fig. 5.** Absolute values of $E_{//}$ field for (0 $\pi$) only top half array feeding.

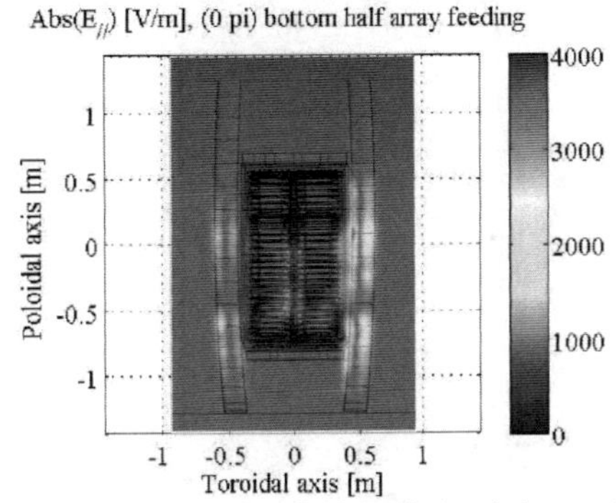


**Fig. 6.** Absolute values of $E_{//}$ field for (0 $\pi$) only bottom half array feeding.

Areas with peak values of the $E_{//}$ field are in the corners of the maps along the poloidal limiters. For full array (0 $\pi$) feeding the increased field values are in poloidal direction from y = -0.8 m to y = -0.6 m and from y = 0.4 m to y = 0.8 m. For full array (0 -$\pi$/2) feeding it is from y = -1.2 m to y = -0.2 m and from y = 0.3 m to y = 1.2 m. In case of only top half array (0 $\pi$) feeding the increased field values are from from y = -0.3 m to y = 0.15 m and from y = 0.4 m to y = 1 m. In case of only bottom half array (0 $\pi$) feeding it is from y = -0.9 m to y = -0.6 m and from y = -0.35 m to y = 0.45 m.

Highest $E_{//}$ field values were obtained for (0 -$\pi$/2) full array antenna feeding. For this setting there is low wave number parallel to the magnetic field $k_{//}$ dominant in the antenna spectrum. This ensures good coupling to plasma but absorption of the waves in the plasma is smaller and larger fraction of the RF power can excite coaxial and surface modes. Peak $E_{//}$ field values obtained for half-array feedings are higher than for (0 $\pi$) full array feeding. This is due to higher power density for half-array feedings compared to full array feeding for constant coupled power. Main source of the increased values of $E_{//}$ fields are image current densities that circulate on the lateral faces of the antenna side limiters.

### 3.2 Estimation of distribution of RF-sheath DC potential

$E_{//}$ field maps calculated in absence of sheaths on surface between the poloidal limiters, 4 mm behind their leading edge, were used as main input to the SSWICH-SW code. This location was chosen to include properly influence of the current density circulation on the poloidal limiters. Calculated $V_{DC}$ potentials were averaged in radial direction in area between the leading edge of the polodal limiters and position 4 mm behind their leading edge. Resulting distributions in poloidal direction for various antenna feedings are shown in Figures 7-10. They were normalized to 450 kW of coupled power to the target plasma. Poloidal distributions of peak values of $V_{DC}$ potential correspond to the poloidal distributions of increased values of $E_{//}$ field. It is due correlation of $V_{DC}$ potentials with amplitudes of $E_{//}$ fields at the lateral sides of the input field maps. More investigations about this effect were performed in [11, 12].

For full array (0 $\pi$) feeding the peak values of $V_{DC}$ potentials are in poloidal direction around y = - 0.7 m and y = 0.65 m. For full array (0 -$\pi$/2) feeding it is around y = - 0.7 m, y = 0.8 m and y = 0.95 m. In case of only top half array (0 $\pi$) feeding the peak values of $V_{DC}$ potentials are around y = - 0.05 m and y = 0.65 m. In case of only bottom half array (0 $\pi$) feeding it is around y = - 0.7 m and y = 0.07 m. Highest $V_{DC}$ potentials values were obtained for (0 -$\pi$/2) full array antenna feeding. Higher $V_{DC}$ potentials values were otained for profiles with lower $q_{95}$.

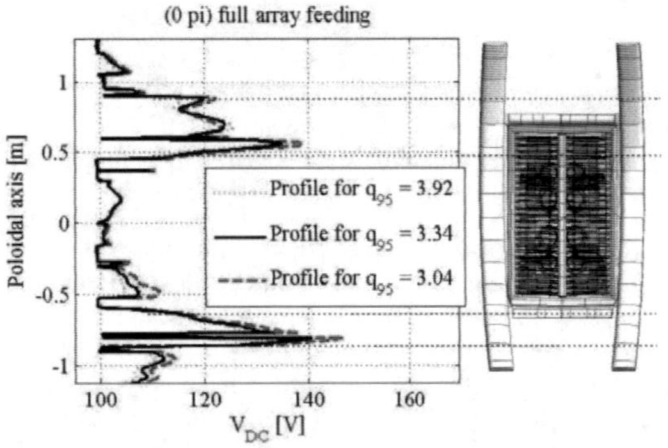


**Fig. 7.** Distribution of $V_{DC}$ potential for (0 $\pi$) full array feeding.

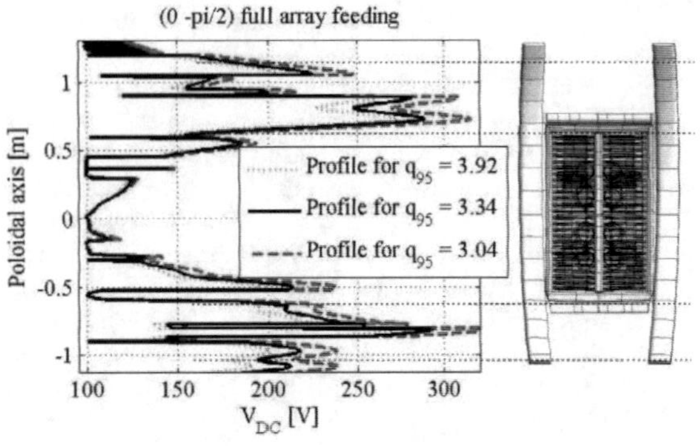


**Fig. 8.** Distribution of $V_{DC}$ potential for (0 -$\pi$/2) full array feeding. Used scale is not the same as in case of the other potential plots.

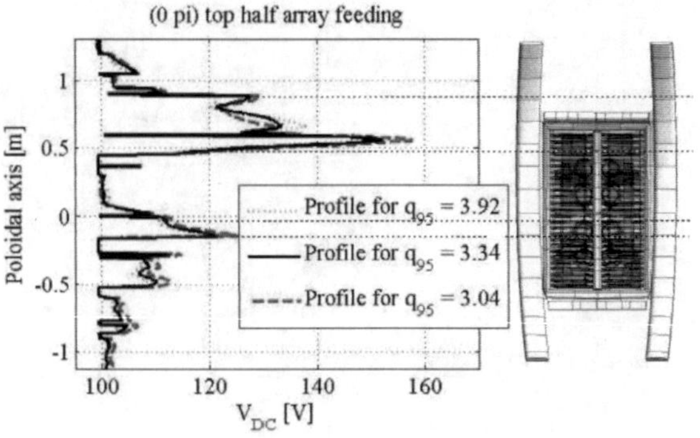


**Fig. 9.** Distribution of $V_{DC}$ potential for (0 $\pi$) top half feeding.

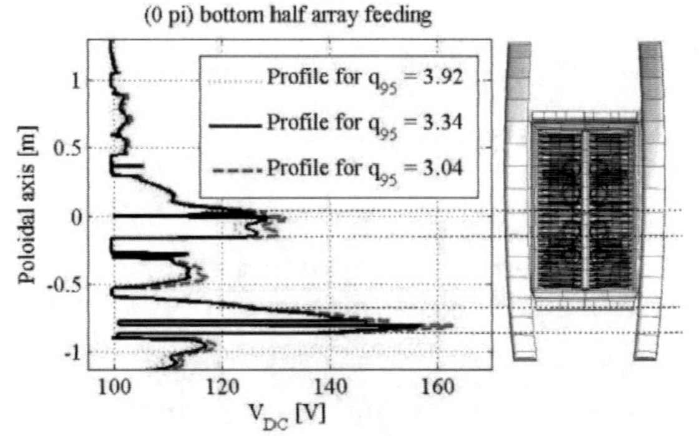


**Fig. 10.** Distribution of $V_{DC}$ potential for (0 $\pi$) bottom half feeding.

## 4 Conclusions and prospects

Distributions of $E_{//}$ field and $V_{DC}$ potential were calculated and areas with increased values were identified. Relative comparison for various antenna feedings was performed. Main source of the increased values of $E_{//}$ field and $V_{DC}$ potential was identified as image current densities that circulate on the poloidal antenna limiters. Higher $V_{DC}$ potentials values were computed for (0 -$\pi$/2) phasing and for profiles with lower $q_{95}$.

In the next step, we will investigate how distribution of $E_{//}$ field and $V_{DC}$ potential changes with changing frequency, ROG and plasma parameters. Further, heat loads will be estimated using the $V_{DC}$ potential values. The near-field and RF sheath models should be validated using experimenal data. Comparison of calculated near fields and $V_{DC}$ potentials will be performed for the A2 antenna and the ILA.

Acknowledgements. This work has been carried out within the framework of the EUROfusion Consortium and has received funding from the European research and training programme under grant agreement N° 633053. The views and opinions expressed herein do not necessarily reflect those of the European Commission. This work has been carried out using Cineca's supercomputer system Marconi.

## References

1. P. Dumortier et al., submitted to Fusion Eng. Des.
2. F. Durodié et al., Plasma Phys. Control. Fusion **54**, 7 (2012)
3. F. Durodié el al., Fusion Eng. Des. **74**, 223–228 (2005)
4. V. Bobkov et al., Nuclear Materials and Energy, in press (2016)
5. A. Czarnecka et al., Plasma Phys. Control. Fusion **54**, (2012) 074013 (13pp)
6. P. Jacquet et al., submitted to Eur. Phys. J.
7. C. C. Klepper et al., submitted to Eur. Phys. J.
8. L. Colas et al., Nucl. Fus. **45**, 767 (2005)
9. V. Lancellotti et al., Nucl. Fus. **46**, S476 (2006)
10. L. Colas et al., Phys. Plasmas **19,** 092505 (2012)
11. L. Colas et al., Plasma Phys. Control. Fusion 59, (2012)
12. A. Křivská et al., AIP Conf. Proc. **1689**, 050002 (2015)

EPJ Web of Conferences **157**, 03027 (2017) DOI: 10.1051/epjconf/201715703027
*22 Topical Conference on Radio-Frequency Power in Plasmas*

# Forecasting electron cyclotron current drive stabilization of neoclassical tearing modes in ITER

*Robert J.* La Haye[1] for the International Tokamak Physics Activity (ITPA) MHD, Disruptions and Control Topical Group Joint Experiment on Current Drive Prevention/Stabilization of Neoclassical Tearing Modes (MDC-8)

[1] *General Atomics San Diego 92186 California USA*

**Abstract.** The ITER baseline design relies on ECCD to stabilize confinement-degrading disruptive NTMs [1]. However, the EC power required will take a toll on the fusion gain Q. The MDC-8 group (in existence since 2005) has the goal to provide a range of data to benchmark the Rutherford tearing stability equation for NTM evolution, allowing predictions for ITER ECCD requirements to be validated. Experimental contributors have included ASDEX UPGRADE, DIII-D, EAST, FTU, HL-2A, JT-60U, KSTAR, TEXTOR and TCV. While any m/n tearing mode island can reduce confinement, the m=2, n=1 mode at q=2 is particularly damaging. This mode is at a relatively large minor radius in the low q95~3 safety factor of ITER and thus close to the resistive wall; with the relatively low rotation in ITER (large inertia, small torque), an uncontrolled mode will rapidly lock at low tearing mode amplitude with subsequent disruption [2]. While progress is being made in modeling of the stability space and control [3] and experiments are promising, implementation still needs to be successfully demonstrated experimentally.

The ITPA consensus is that ITER's 24 1-MW gyrotrons will provide more than sufficient EC power from the upper launch mirrors to drive narrow (but not too narrow) ECCD at q=2 for stabilization, with good alignment. Broadening of the ECCD, by edge turbulence for example, is a concern that would demand more EC power but also make alignment easier. Pre-emption at lower CW power or active stabilization by early mode onset detection and higher peak (possibly lower average) pulsed power are issues still under continuing investigation.

Most EC-NTM experimental studies so far are at relatively high q95 with smaller radius at q=2 and thus higher Te for better current drive efficiency, higher rotation and weaker wall coupling. DIII-D, for example, is now well poised to pursue ECCD NTM stabilization at both low q95 and at low rotation in the 2017 campaign. The MDC-8 as a whole is proceeding to narrow the experimental focus for a comparison of observations from different devices. This will establish the physics basis for successful stabilization in ITER.

## 1 Neoclassical tearing modes and their stabilization by ECCD

NTMs are classically stable and neoclassically metastable requiring a seeding event to destabilize the mode [4]. The destabilizing helically perturbed neoclassical bootstrap current is counteracted at small island widths less than that of the ion banana width $w_{ib}$; everything else equal the NTM scaling is unfavorable for ITER as with about the same $w_{ib}$, the ratio of $w_{ib}$ to the minor radius $r_s$ at a rational surface is much smaller. This makes for a relatively easier to seed mode with a more virulent growth rate than in existing tokamaks. Well-aligned ECCD in the direction co to the plasma current direction can replace the missing bootstrap current stabilizing the mode, i.e. removing the metastable space. A simplified equation for an NTM is given as Eqn (1) of the modified Rutherford equation for normalized growth versus normalized island size. Key terms in Eqn (1) are given in [4]. Here $K_1$ is an effectiveness parameter of order 1 for well-aligned ECCD that has width ("$w_{cd}$" for

**MRE @ q=m/n, neglecting curvature**

$$1.22^{-1}\frac{\tau_R}{r_s}\frac{dw}{dt} = \Delta_0' r_s + 3\frac{j_{boot}}{\langle j_{||}\rangle}\frac{L_q}{r_s}\left[\frac{1}{(w/r_s)} - \frac{(3\,\varepsilon^{1/2}\rho_{\theta i}/r_s)^2}{3\,(w/r_s)^3} - K_1\frac{j_{eccd}}{j_{boot}}\frac{1}{(w/r_s)}\right] \qquad (1)$$

Small island stabilization (second term in brackets); Assumed <0 ($\Delta_0' r_s$); NTM term >0 (first term in brackets); ECCD stabilization (term with $K_1$)

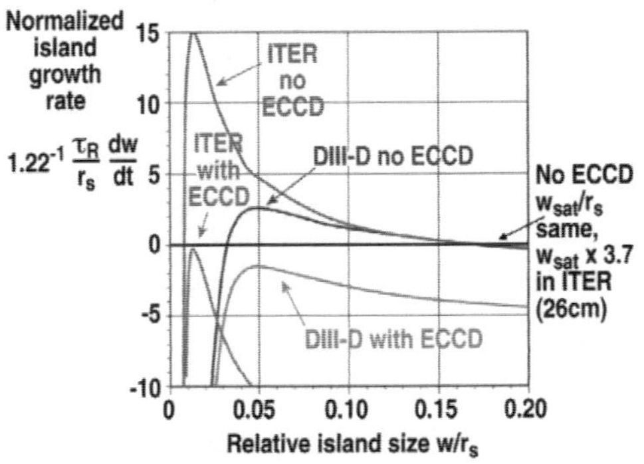


**Fig 1**. Normalized growth rates for DIII-D and ITER w/o ECCD and with well-aligned and matched by width ECCD of $j_{eccd}/j_{boot}$=1.2 for $K_1$=0.5. ITER is 3.7X larger than DIII-D and all other quantities are same.

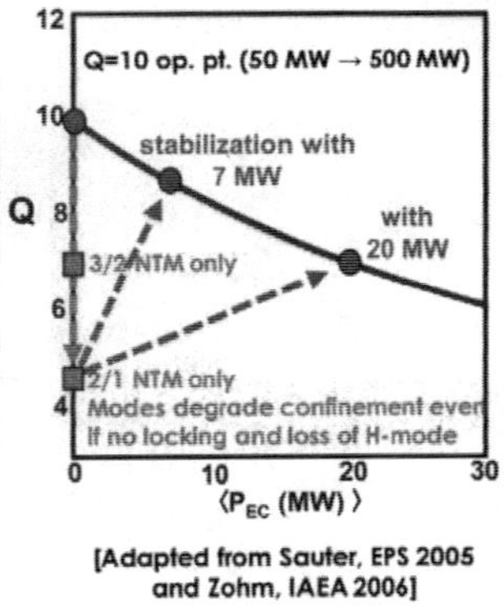


**Fig.2.** Operating point Q for fusion gain in ITER with 3/2 or 2/1 islands (no EC) and islands stabilized (with EC).

±1/e) comparable to the most virulent largest growth rate island width ("$w_{marg}$") w/o ECCD as shown in Figure 1. We will focus on the m/n=2/1 mode closest to the vessel wall that is the most damaging. The ITER upper launcher has good capability for NTM stabilization [5]. About 6 MW can drive the figure of merit $j_{eccd}/j_{boot}$=1.2 with a good match of $w_{cd}$ to the expected $w_{marg}$ of 3 times the ion banana width for q=2. Modeling is discussed more in 2 and experiments in 3.

## 2 Power requirements for controlling NTMs by ECCD in ITER

The fusion gain (Q) is improved by controlling NTMs by ECCD, and large islands that slow down, lock to the first resistive wall and induce disruption, are avoided, but the EC power required takes a toll on Q. This is predicted in Figure 2 [6]. A belt model is used to estimate how the islands degrade the confinement for fixed heating power as compared to stabilization with ECCD power added. Modeling of co-ECCD stabilizes both by replacing the missing bootstrap current and/or by making classical tearing more stable. Studies of the minimum EC power requirement for the 2/1 tearing mode in ITER make different physics assumptions but are consistent in the power requirements for ITER [7-12, 3]. This is shown in Table 1. Control should minimize average EC power for successful stabilization subject to available peak EC power in ITER. Accurate real-time location of both q=2 (MHD reconstruction EFIT with motional Stark effect pitch angle profile for example) and ECCD location (by TORBEAM) [13] could keep radial misalignment to no more than ±1 cm comparable to DIII-D but relatively 4X more accurate in ITER. Control strategies to keep the EC power on at a minimum or off until needed are given in Figure 3. Pre-emption may require less peak power by avoiding a mode being destabilized, while standby has to be turned on when a mode is detectable and may already be larger width than that of the ECCD so that $K_1$ is reduced [3]. However, catching a mode at low amplitude may be possible in ITER [14]. NTMs need seeding to be destabilized and seeding comes from predictably periodic MHD events such as q=1 sawteeth crashes, 1/1 fast

**Table 1.** Minimum CW EC power modeled needed to stabilize the 2/1 NTM in ITER with different assumptions for $w_{cd}$ and $w_{marg}$ as well as classical tearing (not shown). Based on the MRE as given in Eqn (1).

| Reference | $w_{cd}$ (cm) | $w_{marg}$ (cm) | $P_{EC}$ (MW) |
|---|---|---|---|
| Sauter 2010 | 2.5 → 5 | 4 of 2, 4, 6 scan | 7 → 9 |
| Urso 2010 | 3.0 | 3.7 | 10 starting from full $w_{sat}$ |
| Bertelli 2011 | 5.0 | 4 of 2 → 6 scan | 9 |
| van den Brand 2012 | 2.4 | 2 or 6 | 13 fixed for "latency" at 10 cm |
| E. Poli 2015 | 2.8-3.7 | 4 of 1 → 6 scan | 6 |
| La Haye 2015 | 3.1 | 3.5 | 5~6 |
| F. Poli 2016 | 4 | 4.3 | 5~7 |

 DOI: 10.1051/epjconf/201715703027

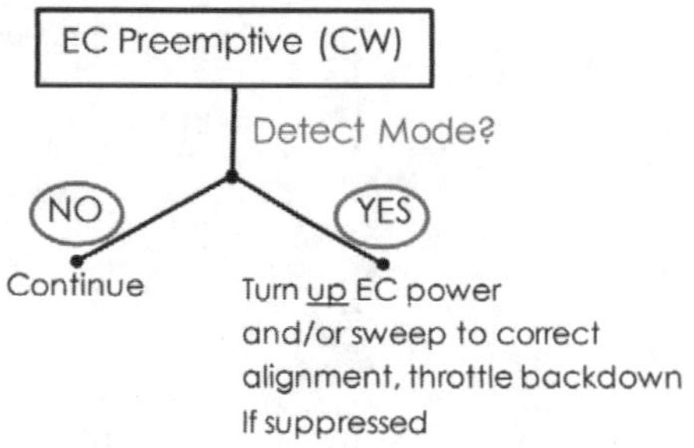


**Fig. 3**. Pre-emptive ECCD is CW with a minimum EC power with mirrors moving to stalk the q=2 surface. Standby is mirrors moving to stalk q=2 without EC power on until a mode detected and EC then turned on.

particle events known as fishbones or broad intermediate n edge localized modes (ELMs). A standby control that is off most of the time with a higher peak power when on might have lower average EC power than that of a pre-emptive control that is always on. This is being tested experimentally as described next.

### 3 Ongoing experimental studies of ECCD NTM stabilization and means of control

Much experimental progress has been made since the MDC-8 group was founded in 2005. More gyrotrons for more power and more sophisticated means of controlling the EC power have been successfully implemented. Experiments are just getting underway on EAST and KSTAR and will be a key feature of JT-60SA and WEST. Other experiments that have established and are active in establishing the physics basis were performed on ASDEX Upgrade, DIII-D, FTU, HL-2A, JT-60U, TCV and TEXTOR. Only a very few highlights can be noted here. Pre-emption requires good alignment of the ECCD on q=2 rather than on an observable island. On TCV the mirrors were commanded to make multiple ECCD sweeps back and forth across the radial position where q=2 is expected to be; the mode did not appear until the EC power was turned off and when reapplied the mode was influenced but not stabilized [15]. This is shown in Figure 4 noting that EC is also used for core heating.

In the presence of a mode, here m/n=3/2, ASDEX Upgrade has used mirror sweeping over the island location which with enough EC power can be transiently good enough for stabilization [16]. This is shown in Figure 5. The mode amplitude is periodically reduced as the alignment is good but is only stabilized when the EC power is turned up.

On DIII-D pre-emption of the 2/1 tearing mode in an ITER baseline scenario at normalized beta of $\beta_N$=1.8, safety factor at the 95% flux surface q95=3.1 and an equivalent low torque for low rotation and thus little if any tearing stabilization from flow shear is being studied with "Active Tracking". Real-time Thomson scattering for $n_e$ and $T_e$ profiles is used to aim the mirrors for ECCD at a q=2 surface determined by real-time equilibrium reconstructions including the motional Stark effect magnetic pitch angle profile diagnostic. An offset can be programmed

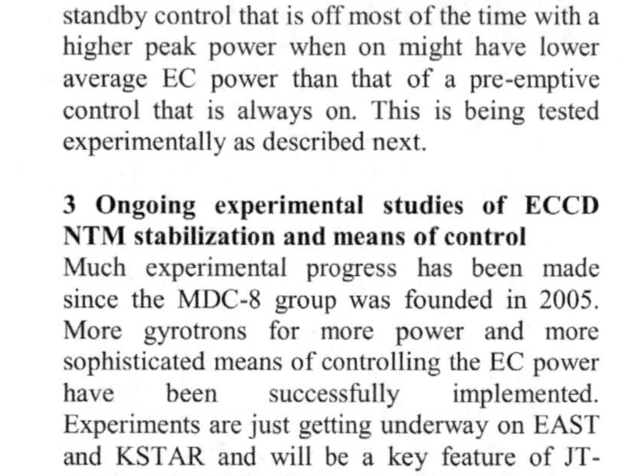


**Fig. 4.** Pre-emptive ECCD stabilization of the 2/1 tearing mode by mirror sweeping in TCV followed by the same applied to the pre-existing mode [15]. Same sweeping deposition whenever EC the power is on.

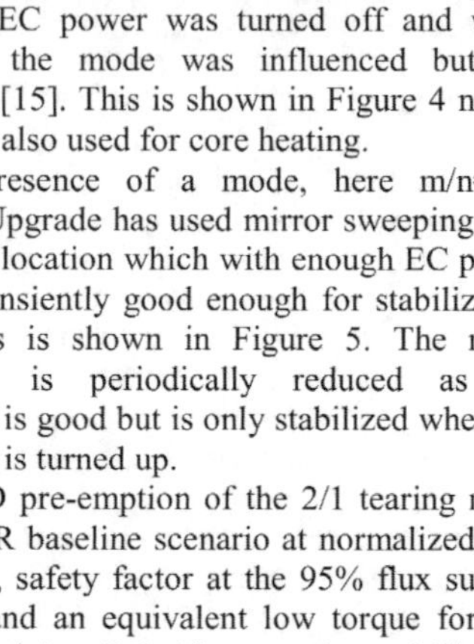


**Fig. 5.** ECCD stabilization of a 3/2 tearing mode in Asdex Upgrade with mirror sweeping so that the alignment is transiently good enough for complete suppression with sufficient EC power [16]. Lower pane black crosses island location from ECE, dashes ECCD locations from each gyrotron when powered.

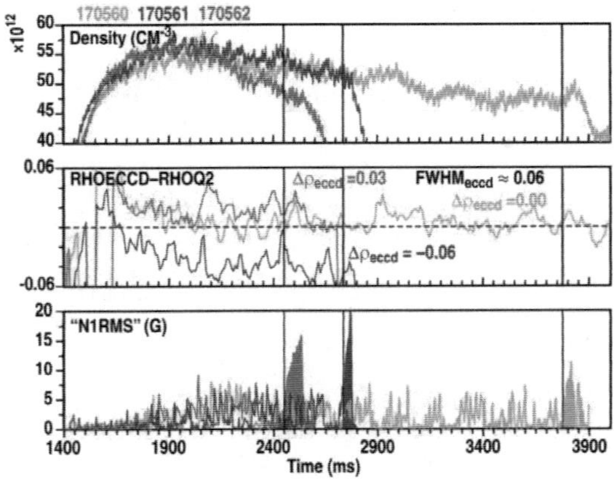


**Fig. 6.** Active tracking of ECCD on q=2 in DIII-D with PCS keeping $\Delta\rho_{eccd}$=0.00, 0.03 or -0.06 alignment.

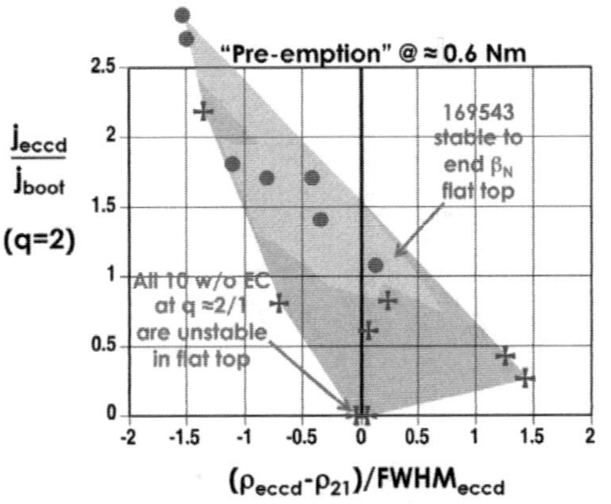


**Fig. 7.** Pre-emption by ECCD of 2/1 tearing mode in DIII-D low torque IBS. Red crosses are at onset of mode. Green dots are no mode. Contours are pink mode unstable and green mode stable

into the plasma control system (PCS) for physics study of alignment or to correct a suspected systematic error. This is shown in Figure 6 for a very recent experiment. EC from 4 gyrotrons is applied at t=1400 ms just before the L to H transition that brings up the beta and density to flat top at ~2000 ms; this is followed by a natural very slow density decay during which the ECCD refraction is reduced while q=2 remains the same, thus necessitating mirrors moving. While all 3 discharges eventually go unstable with the mode promptly locking, the longest delay is for best alignment; here power from 4 gyrotrons directed at q=2 (2.4 MW, $j_{eccd}/j_{boot}$ ≈0.8), is not enough while 5 (2.8 MW, $j_{eccd}/j_{boot}$ ≈1.0) can keep the 2/1 mode from occurring as shown in Figure 7. Here 0, 4 or 5 gyrotrons are applied with different preset offsets using either "Active Tracking" of mirrors or "Point & Shoot" fixed mirrors. The soon to be added new higher power gyrotron will allow better resolution of the stable/unstable space for x > 0 and y > 1, which is clearly not now warranted in Figure 7. Taken together the minimum in the figure of merit for successful pre-emption is about $j_{eccd}/j_{boot}$~1. This requires the EC power to be about 50% of the total input power including that of neutral beam injection.

*This work is supported in part by the US DOE under DE-FC02-04ER54698 and the DIII-D 3D Stability Physics and EC Operations Groups are acknowledged. The ITPA MDC-8 participants include M. Reich and E. Westerhof for ASDEX Upgrade, R. La Haye for DIII-D, L. Hu for EAST, S. Novak for FTU, Y. Liu for HL-2A, P. Buratti for JET, A. Isayama for JT-60SA, M. Joung for KSTAR, S. Sabbagh for NSTX-U, M. Reich and O. Sauter for TCV and P. Maget for WEST.

## References

[1] T.C. Hender, et al., Nuclear Fusion 47, S128 (2007)
[2] R.J. La Haye, C. Paz-Soldan and Y.Q. Liu, Nuclear Fusion 57, 014004 (2017)
[3] F. M. Poli et al., 26th IAEA Fusion Energy Conf, Kyoto, P4-6 (2016)
[4] R.J. La Haye, Phys. Plasmas 13, 055501 (2006)
[5] L. Figini et al., Plasma Phys. Control. Fusion 57, 054015 (2015)
[6] O. Sauter, H. Zohm, 32nd EPS Conference on Plasma Phys. Tarragona, 27 June - 1 July 2005 ECA Vol.29C, P-2.059 (2005)
[7] O. Sauter, et al., Plasma Phys. Control Fusion 52, 025002 (2010)
[8] L. Urso et al., Nucl. Fusion 50, 025010 (2010)
[9] N. Bertelli, et al., Nucl. Fusion 51, 103007 (2011)
[10] H. van den Brand et al., Plasma Phys. Control. Fusion 54, 094003 (2012)
[11] E. Poli, et al, Nuclear Fusion 55, 013023(2015)
[12] R. La Haye for the DIII-D Electron Cyclotron Operations and Dynamics & Control Groups, 21st rfPP Conference, Lake Arrowhead, 27-29 April, Volume 1689, 030018 (2015)
[13] E. Kolemen, et al., Nucl. Fusion 54, 073020 (2014)
[14] K E J Olofsson, et al., Plasma Phys. Control. Fusion 56, 095012 (2014)
[15] D. Kim, PhD thesis 2015 EPFL https://infoscience.epfl.ch/record/205040/files/EPFL_TH6539.pdf
[16] M. Reich, et al., Fusion Science and Technology 61, 309 (2012)

EPJ Web of Conferences **157**, 03028 (2017) DOI: 10.1051/epjconf/201715703028
*22 Topical Conference on Radio-Frequency Power in Plasmas*

# A new quasilinear formulation for ICRF plasmas in a toroidal geometry

*Jungpyo Lee[1,*], John Wright[1], Nicola Bertelli[2], David Smithe[3], Ernest Valeo[2], Yuri Petrov[4], Erwin F. Jaeger[5], Lee Berry[5], Robert Harvey[4], and Paul Bonoli[1]*

[1]MIT Plasma Science and Fusion Center, Cambridge, MA, 02139, USA
[2]Princeton Plasma Physics Laboratory, Princeton, NJ, 08540, USA
[3]Tech-X, Boulder, CO, 80303, USA
[4]CompX, Del Mar, CA, 92014, USA
[5]XCEL Engineering, Oak Ridge, TN, 37830, USA

**Abstract.** We present a new formulation for quasilinear velocity space diffusion for ICRF plasmas that considers two different aspects: (1) finite Larmor radius approximation and (2) includes the effect of toroidal geometry and constructs a positive definite form. In the first aspect, the Kennel-Engelmann (K-E) quasilinear diffusion coefficients are successfully approximated in a small Larmor radius limit and implemented for the numerical codes (TORIC-CQL3D). In the second aspect, the quasilinear diffusion is reformulated in a toroidal geometry in order to include the parallel dynamics in the inhomogeneous plasmas and magnetic fields. We use these two quasilinear formulations to simulate ITER plasmas with ICRF injection for minority fundamental heating and Tritium second harmonic cyclotron heating.

## 1 Introduction

The ion cyclotron range of frequency (ICRF) waves transfer their energy and momentum to plasmas, and as a result a significant amount of fast ions are produced. The kinetic description of the non-Maxwellian distribution plasmas is obtained using quasilinear diffusion coefficients due to the RF waves in a Fokker-Planck equation. The quasilinear diffusion coefficients are derived by Kennel and Engelmann (K-E) [1] with the assumptions of the homogenous plasmas and magnetic fields along the particle trajectory. In this proceeding, we present the theoretical modifications of the K-E diffusion coefficients in two different aspects.

In the first aspect, the coefficients are modified to be consistent with the dielectric tensor of the reduced model in the small Larmor radius approximation. In the reduced model, the dielectric tensor for the plasma current is expanded by a small parameter $k_\perp \rho_i$, in which $k_\perp$ can be replaced by the differential operator [2]. Accordingly, the computation time of the reduced model can be significantly reduced by a factor of $O(n_r^2)$ compared to the full model [3] that uses $k_\perp$ explicitly without the small Larmor radius approximation. Here, $n_r$ is the number of radial coordinate grid. In Section 2, we summarize the derivation of the equivalent quasilinear diffusion to the reduced model as in our recent publication [4], and present an additional example using the implementation in TORIC-CQL3D.

TORIC is already coupled with a sophisticated Fokker-Planck module, SSFPQL [5, 6]. It is worth noting the different features between our approach in TORIC-CQL3D [4] and that in TORIC-SSFPQL [5, 6]. CQL3D is the Fokker-Planck code, which has been validated against demonstrated by many RF wave experiments. It has fully nonlinear and relativistic collision operators and it can calculate the time evolution of distribution function for the tail formation and relaxation by transient RF waves. For the quasilinear diffusion, our formulation in TORIC-CQL3D guarantees the consistency with the dielectric tensor, while it cannot include the contribution from the higher order in $k_\perp \rho_i$ that is retained in the K-E coefficients used in TORIC-SSFPQL. As the result of the self-consistency, the power absorptions by dielectric tensor and quasilinear diffusion theoretically match each other and the numerical iterations between the Maxwell's equation solver and the Fokker-Planck code likely converge without any normalization factor to the diffusion coefficients [4].

In the second aspect, we modify the formulation of K-E coefficients to be positive-definite in the toroidal geometry. The positive definite form can include some important features of the diffusion in the toroidal geometry which cannot be captured in the K-E coefficients, as will be explained in Section 3. Using this new positive-definite from, we can eliminate the unphysical growing mode violating H-theorem and the related numerical errors in the coupled code TORIC-CQL3D. To best of our knowledge, it is the first implementation of the positive-definite form for the quasilinear diffusion coefficient in a continuum Fokker-Planck solver for a toroidal geometry.

## 2 FLR approximations

### 2.1. Kinetic energy change ($\dot{W}$)

The rate of the kinetic energy change due to the RF waves ($\dot{W}$) is used to define the quasiliner diffusion coefficients in a specific numerical formulation. For example, in the full model using a full Fourier spectral $\boldsymbol{k}$

* Corresponding author: Jungpyo@psfc.mit.edu

 DOI: 10.1051/epjconf/201715703028

space, the K-E quasilinear diffusion coefficients can be defined by

$$\dot{W} = \frac{1}{2} Re \left[ \sum_{k_1} \sum_{k_2} e^{i(k_1 - k_2)\cdot r} (\boldsymbol{E}_{k_1} \cdot \mathbf{W}_l \cdot \boldsymbol{E}_{k_2}) \right]$$

$$= \int \boldsymbol{dv} \frac{mv^2}{2} Q(f) \,, \quad (1)$$

where $\mathbf{W}_l$ is the resonance Kernel (e.g. Eq. (11) of [3]). For energy transfer, the quasilinear diffusion coefficients are gyroaveraged, and they can be described in the velocity space of speed-pitchangle coordinate $(v, \vartheta)$ by

$$Q(f) = \frac{1}{v^2} \frac{\partial}{\partial v} \left\{ B \frac{\partial}{\partial v} + C \frac{\partial}{\partial \vartheta} \right\} f$$

$$+ \frac{1}{v^2 \, \mathrm{SIN}\, \vartheta} \frac{\partial}{\partial \vartheta} \left\{ E \frac{\partial}{\partial v} + F \frac{\partial}{\partial \vartheta} \right\} f. \quad (2)$$

The relation between the coefficients $B, C, E$, and $F$ are given to preserve the diffusion directions in velocity space [6].

In the reduced model, the kinetic energy can be defined to be consistent with the approximation of plasma currents using the expansion of the gyromotion around the gyrocenter. For example, when evaluating $\dot{W}$, the electric field at the gyrocenter is approximated by

$$\boldsymbol{E}(\boldsymbol{r}_g) = \left[ \mathbf{1} - (\boldsymbol{r} - \boldsymbol{r}_g) \cdot \boldsymbol{\nabla} + \frac{1}{2} (\boldsymbol{r} - \boldsymbol{r}_g)(\boldsymbol{r} - \boldsymbol{r}_g) : \boldsymbol{\nabla\nabla} \right] \boldsymbol{E}(\boldsymbol{r}), \quad (3)$$

and it results in the expansion of $\dot{W}$ in terms of the small parameters $k_\perp \rho_i$,

$$\dot{W} = \dot{W}^{(0)} + \dot{W}^{(1)} + \dot{W}^{(2)} + \cdots, \quad (4)$$

where the superscript indicates the order of $k_\perp \rho_i$. The detailed derivations for Eq. (4) are in Section 2 of [4].

## 2.2. Quasilinear diffusion coefficient

The K-E quasilinear diffusion coefficient corresponding to $\dot{W}$ in the full model of Eq. (1) is

$$B = \frac{\pi \epsilon w_p}{2m\, n_s} \sum_n Re[\sum_{k_2} v_\perp \chi^*_{k_2,n} e^{-i k_2 \cdot r}$$

$$\sum_{k_1} v_\perp \chi_{k_1,n} e^{i k_1 \cdot r} \delta(\omega - n\Omega - k_{\parallel,1} v_\parallel)] \,, \quad (5)$$

where $n$ is the integer to determine the harmonic number of ion cyclotron resonance and $\chi_k$ is the effective potential [3].

To derive the quasilinear diffusion coefficients corresponding to $\dot{W}$ in the reduced model of Eq. (4), three important properties of the diffusion need to be considered : positive-definiteness, wave polarization, and, diffusion direction. We found a way to formulate the coefficients to preserve these three properties [4]. The coefficient $B$ is formulated from the lowest order of $\dot{W}$, and other coefficients are calculated from the original relations with $B$.

Accordingly, for the fundamental damping (**n**=1), the coefficient is defined by $\dot{W}^{(0)}$, giving

$$B^{(n=1)} = \frac{\pi \epsilon w_p}{2m\, n_s} Re[\sum_{k_2} v_\perp E^*_{+k_2,n} e^{-i k_2 \cdot r}$$

$$\sum_{k_1} v_\perp E_{+k_1,n} e^{i k_1 \cdot r} \delta(\omega - n\Omega - k_{\parallel,1} v_\parallel)]. \quad (6)$$

For the second harmonic damping ($n$=2), the coefficient is defined by $\dot{W}^{(2)}$, giving

$$B^{(n=2)} = \frac{\pi \epsilon 2 w_p^2}{8m\, n_s} Re[\frac{v_\perp^2}{\Omega^2} \sum_{k_2} v_\perp E^*_{+k_2,n} e^{-i k_2 \cdot r}$$

$$\partial_- (\sum_{k_1} v_\perp \partial_+ E_{+k_1,n} e^{i k_1 \cdot r} \delta(\omega - n\Omega - k_{\parallel,1} v_\parallel))], \quad (7)$$

where the detail derivation can be found in [4]

## 2.3 Examples

The quasilinear diffusion coeffcients for the reduced model are evaluated in the wave code by the reduced model, TORIC [2]. Using the diffusion coefficients, TORIC is coupled with a Fokker-Planck code CQL3D [6], to have the self-consistent solutions of distribution and wave fields. In the recent paper [4], we simulate the scenario for ITER 10MW He3 minority species damping ($n$=1) by the reduced model TORIC-CQL3D, and compared the results with the full model by AORSA-CQL3D [3]. The comparisons show that the reduced model is acceptably valid when the considerable amount of fast ions satisfies $k_\perp \rho_i \lesssim 1.0$ and the wave power density is less than 1MW/m$^3$ in the scenario.

In this proceeding, we introduce the results of the second harmonic damping ($n$=2) scenarios of ITER 5.3T. We simulate two ion species with the ratio of (D, T)= (50,50)%, and the ICRF wave frequency is 52.5MHz. The dominant wave power is absorbed by the second harmonic damping of the major species Tritium around the magnetic axis because $\omega = 2\Omega$ at R = $R_0$+0.17m. The power decompositions are reasonably similar between TORIC and AORSA: for Maxwellian plasmas, 49% of T second harmonic damping and 48% electron damping in TORIC, while 48% of T second harmonic damping, and 51% electron damping in AORSA. For self-consistent non-Maxwellian plasmas, 48% of T second harmonic damping, and 46% electron damping in TORIC, while 61% of T second harmonic damping, and 39% electron damping in AORSA.

Figure 1 and 2 show the distribution functions of TORIC-CQL3D and AORSA-CQL3D, which show the similar patterns but some difference in the tails around $v_\parallel \simeq 0$. The main reason of difference could be the error of the approximation on the Bessel function factor $J_1(k_\perp \rho_i) \simeq k_\perp \rho_i / 2$ for the high energetic Tritium of several MeV. The error of the small Larmor radius approximation can be larger for n=2 than that of n=1 in [4].

DOI: 10.1051/epjconf/201715703028

(a)

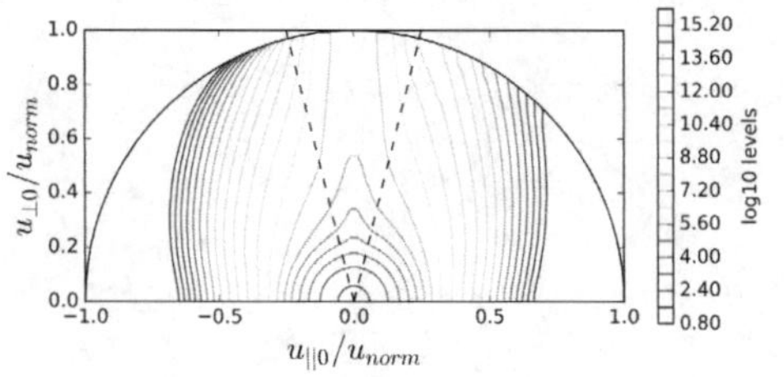


(b)

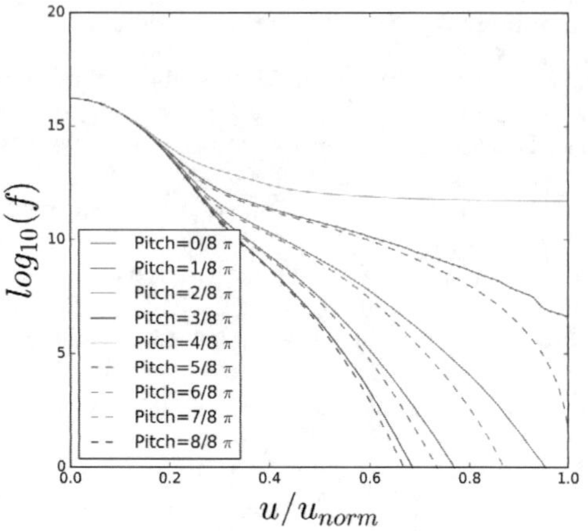


**Fig. 1.** Reduced model results by TORIC-CQL3D : (a) 2-D contour plots of the distribution function in velocity space at r/a=0.1 (b) 1-D distribution functions in terms of speed for several pitch-angles. Here, $u_{norm}$ is the momentum corresponding to the energy of Tritium 4 MeV.

(a)

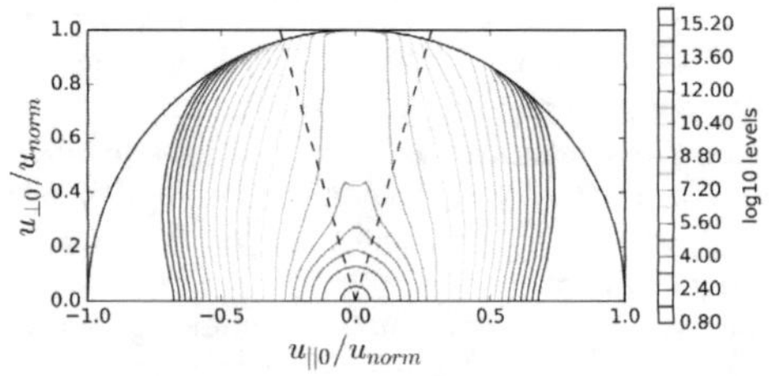


(b)

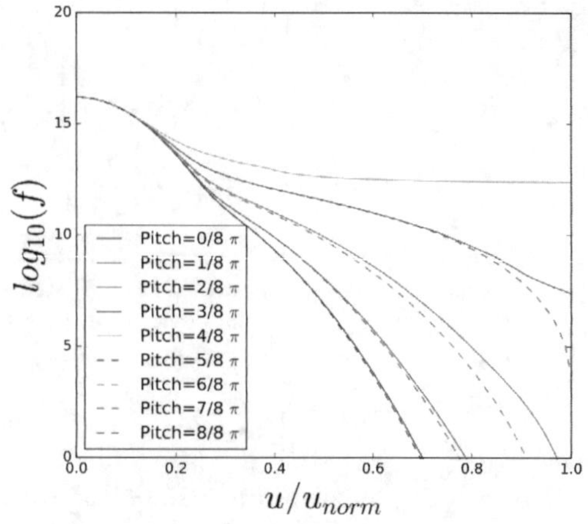


**Fig. 2.** Full model results by AORSA-CQL3D : (a) 2-D contour plots of the distribution function in velocity space at r/a=0.1 (b) 1-D distribution functions in terms of speed for several pitch-angles.

## 3 Toroidal geometry effect

### 3.1. Positive definite form

In a toroidal geometry, the magnetic field is inhomogeneous along the particle trajectory and the assumption of K-E coefficients is violated. The violation is typically ignorable because of the small correlation length of the wave-plasma interaction compared to the variation of the magnetic fields. However, we found that it does not result in the positive-definiteness when the quasilinear diffusion coefficients are bounce-averaged for the bounce-averaged Fokker-Planck code, CQL3D [6]. As a consequence, the negative values in the bounce-averaged diffusion coefficient $\langle B\rangle_b = \int dl\, B/v_{\parallel}$ cause unphysical growing modes and numerical errors when enforcing to eliminate them.

The positive-definiteness of the bounce-averaged coefficients are recovered when treating the trajectory integral and the bounce integral as the same way in the toroidal geometry [7]. For example, the bounce-averaged coefficient is reformulated by

$$\langle B\rangle_b = \frac{\pi\epsilon w_p}{2m\, n_s}\int_{-\infty}^{0} dt\, Re[\textstyle\sum_{\boldsymbol{k_2}}(\ldots \boldsymbol{E}_{\boldsymbol{k_2}})e^{-i\boldsymbol{k_2}\cdot\boldsymbol{r}(t)}$$
$$\langle\sum_{\boldsymbol{k_1}}\int_0^{\infty} d\tau e^{i2\gamma t}\, e^{i\omega\tau}(\ldots \boldsymbol{E}_{\boldsymbol{k_1}})e^{-i\boldsymbol{k_1}\cdot\boldsymbol{r}(t-\tau)}\rangle_{\boldsymbol{g}}]$$
$$= \frac{\pi\epsilon w_p}{2m\, n_s} Re\left[\textstyle\sum_{\boldsymbol{n}}\sum_{\boldsymbol{k_2}}\sum_{\boldsymbol{k_1}} E^{*}_{+\boldsymbol{k_2},\boldsymbol{n}} A(\boldsymbol{k_1},\boldsymbol{k_2}) E_{+\boldsymbol{k_1},\boldsymbol{n}}\right], \quad (8)$$

where $\langle\ldots\rangle_{\boldsymbol{g}}$ is the gyroaverage and the symmetric operator $A(\boldsymbol{k_1},\boldsymbol{k_2}) = A^{*}(\boldsymbol{k_2},\boldsymbol{k_1})$ is obtained by the change of variables $t_1 = t-\tau$ and $t_2 = t$, giving

$$A(\boldsymbol{k_1},\boldsymbol{k_2}) = \frac{1}{2}\int_{-\infty}^{0} dt_1 \int_{-\infty}^{0} dt_2 v_{\perp}(t_1)$$
$$v_{\perp}(t_2)e^{i\Phi_n(t_2,k_{\parallel,2})-i\Phi_n(t_1,k_{\parallel,1})}, \quad (9)$$

and it guarantees the positive-definiteness in the toroidal geometry. In Eq. (5), K-E coefficients have the Dirac-delta function depending on only $k_{\parallel,1}$ because of the

 DOI: 10.1051/epjconf/201715703028

uniform phase only in the trajectory integral, and it breaks the symmetry shown in Eq. (9).

### 3.2 Trajectory integral

The evaluation of the trajectory integral in Eq. (9) is not trivial because of the phase variation in the toroidal geometry,

$$\frac{\partial \Phi_n}{\partial t} = \omega - n\Omega(t) - k_{\parallel}(t)v_{\parallel}(t) \quad (10)$$

The stationary approximation around the resonance point $\partial \Phi_n/\partial t = 0$ is useful for the short correlation length. By expanding $\Phi_n$ in terms of the small temporal distance from the resonance $t - t_r$ up to third order, it results in the Airy functions ($Ai$), as many previous studies investigated [8,9],

$$\int_{-\infty}^{0} dt v_{\perp}(t) e^{i\Phi_n(t,k_{\parallel})} \simeq v_{\perp}(t_r) e^{i\Phi_n(t_r,k_{\parallel})} R(t_r) \quad (11)$$

$$R(t_r) = \frac{\pi}{\alpha^{\frac{1}{3}}} e^{-i\alpha\left(x_1 t_+ + \frac{t_+^3}{3}\right)} Ai(\alpha^{\frac{3}{2}} x_1), \quad (12)$$

where $\alpha = |0.5\partial^3\Phi_n/\partial t^3|$ , $t_+ = |\partial^2\Phi_n/\partial t^2/\partial^3\Phi_n/\partial t^3|$, and $x_1 = -t_+^2$ at $t = t_r$, result in the negative argument of the Airy function.

When $\partial^2\Phi_n/\partial t^2 = 0$ is satisfied, the correlation length becomes large so that the variation of the phase could be important even for the non-resonance locations with $\partial\Phi_n/\partial t \neq 0$. This condition typically occurs at the inner-midplane and outer–midplane for passing particles and trapping tips and outer-midplane for trapped particles [9]. We can include this non-resonant contribution to the diffusion by the positive arguments of the Airy function,

$$\int_{-\infty}^{\infty} dt v_{\perp}(t) e^{i\Phi_n(t,k_{\parallel})} \simeq v_{\perp}(t_{nr}) e^{i\Phi_n(t_{nr},k_{\parallel})} N(t_{nr}) (13)$$

$$N(t_{nr}) = \frac{\pi}{\alpha^{\frac{1}{3}}} \left(Ai\left(\alpha^{\frac{3}{2}} x_2\right) + Ai\left(\alpha^{\frac{3}{2}} x_2\right)^*\right) \quad (14)$$

where $x_2 = 2|\partial\Phi_n/\partial t/\partial^3\Phi_n/\partial t^3|$ at the specific conditions $t = t_{nr}$.

The resonant contribution in Eq. (11) and the non-resonant contribution in Eq. (13) are smoothly connected by the Airy function, and it results in the continuous diffusion coefficients in velocity space.

### 3.3 Implementation

This positive definite form of the bounce-averaged coefficients in Eq. (8) is implemented in TORIC. The resonant contribution is evaluated when the spectral mode has at least one resonance location on a flux surface, and the non-resonant contribution is included when the spectral mode has no resonant location.

The phase integral $\Phi_n$ is not poloidally periodic on a general flux surface with the parallel spectral mode number (m+nq) not being an integer. Thus, the evaluation of the diffusion coefficient depends on the initial position of the integral range, and the average diffusion changes depending on the number of poloidal periods in the evaluations. We will show the importance of the initial position and evaluation range in a future publication.

Additionally, the positive definite form can show the correlation between the consecutive resonance crossings, while the kicks by the Dirac-delta function of the K-E diffusion coefficients are likely randomized. This correlation between resonances are important in the two consecutive resonance around the location of $t = t_{nr}$ (e.g. outer-midplane), where the toroidal geometry effect needs to be considered within the relatively long correlation length.

This new quasilinear diffusion is expected to be beneficial for ICRF modelling in a low aspect ratio tokamak that has the significantly varying magnetic field along particle trajectories. In spite of the good features to capture the toroidal effects in the quasilinear diffusion, this positive definite form is computationally expensive because it requires about $n_v n_m n_{ch} n_{int} {\sim} 10^6$ times more number of floating-point operations than evaluating K-E quasilinear diffusion in Eq. (5). Here, $n_v{\sim}100$, $n_m{\sim}100$, $n_{ch}{\sim}10$, and $n_{int}{\sim}10$ are the number of velocity space grid in each dimension, the number of polodal spectral grid, the order of interpolation, the required number of interpolation, respectively.

This work was supported by US DoE Contract No. DE-FC02-01ER54648 under a Scientific Discovery through Advanced Computing Initiative.

## References

1. C.F. Kennel, and F. Engelmann Phys. Fluids **9**, 2377 (1966)
2. M. Brambilla, Plasma Physics and Controlled Fusion **41**, 1 (1999).
3. E. F. Jaeger, L. A. Berry, E. DAzevedo, D. B. Batchelor, and M. D. Carter, Physics of Plasmas **8**, 1573 (2001).
4. J. P. Lee, J. C. Wright, N. Berterlli, E. F. Jaeger, E. Valeo, R. W. Harvey, and P. T. Bonoli, Phys. Plasmas **24**, 052502 (2017)
5. M. Brambilla and R. Bilato, Nuclear Fusion **46**, S387-S396 (2006).
6. M. Brambilla and R. Bilato, Nuclear Fusion **49**, 085004 (2009).
7. R. W. Harvey, and M. G. McCoy, in Proc. IAEATCM on Advances in Sim. and Modeling of Thermonuclear Plasmas, available through USDOC/NTIS No. DE93002962 (1992) , pp. 489–526
8. D. Smithe, Plasma Physics and Controlled Fusion **31**, 1105 (1989)
9. I.B. Bernstein and D.C. Baxter, Phys. Fluids **24**, 108 (1981)
10. P.J. Catto and J.R. Myra, Phys. Fluids B **4**,187 (1992)

DOI: 10.1051/epjconf/201715703029

# Sawtooth pacing with on-axis ICRH modulation in JET-ILW

E. Lerche[1,2], M. Lennholm[1], I. S. Carvalho[1,3], P. Dumortier[1,2], F. Durodie[2], D. Van Eester[2], J. Graves[4], Ph. Jacquet[1], A. Murari[5] and JET contributors[1]

EUROfusion Consortium, JET, Culham Science Centre, Abingdon, OX14 3DB, UK

[1] *CCFE, Culham Science Centre, Abingdon, United Kingdom*
[2] *LPP-ERM/KMS, Association EUROFUSION-Belgian State, TEC partner, Brussels, Belgium*
[3] *Instituto de Plasmas e Fusão Nuclear, EUROFUSION-IST, University of Lisbon, Portugal*
[4] *CRPP-EPFL, Association EUROFUSION - Confédération Suisse, Lausanne, Switzerland*
[5] *Consorzio RFX, Association EUROFUSION – CNR/ENEA, Padova, Italy*

**Abstract.** A novel technique for sawteeth control in tokamak plasmas using ion-cyclotron resonance heating (ICRH) has been developed in the JET-ILW tokamak [2]. Unlike previous ICRH methods, which explored the destabilization of the internal kink mode when the radio-frequency (RF) wave absorption was placed near the $q$=1 surface, the technique presented here consists of stabilizing the sawteeth as fast as possible by applying the ICRH power centrally and inducing a sawtooth crash by switching it off at the appropriate instant. Since it employs on-axis ICRH, this technique is aligned with other ICRH applications that require central plasma heating such as high-Z impurity control and fusion power enhancement. The validation of this method in JET-ILW L-mode discharges is presented.

## 1 Introduction

In tokamak plasmas with a central safety factor lower than unity ($q_0$<1), the central plasma parameters commonly exhibit a non-linear oscillation, known as the sawtooth instability [3]. The sawtooth cycle manifests itself through a slow increase in the plasma temperature (and other parameters) in the central part of the plasma, followed by an abrupt crash, usually resetting the plasma profiles approximately to those at the start of the cycle. Large sawtooth crashes can, however, trigger more deleterious instabilities, such as neo-classical tearing modes (NTMs) [4,5]. Such modes will reduce the plasma confinement significantly and, if the instability grows sufficiently large, they can lead to a plasma disruption. Large sawtooth crashes following long quiescent periods without a crash have been seen to be much more likely to trigger NTMs than crashes associated with a shorter inter-crash interval. For this reason it is important to assure that the sawtooth period ($\tau_{SWT}$) remains shorter than the threshold for NTM triggering. Nevertheless, excessive shortening of the sawtooth period does result in a moderate confinement reduction and for this reason it is desirable to maintain it as long as possible without inducing NTMs. It is therefore useful to develop methods to control the sawtooth period to a preset value for high performance scenario development.

A number of different methods have been developed to control sawteeth. The traditional actuators for implementing such control are Electron Cyclotron Resonance Heating and Current Drive (ECRH) and Ion Cyclotron Resonance Heating and Current Drive (ICRH).

ECRH can affect the sawtooth period by driving current near the $q$=1 surface, modifying the local current profile in this region [5-12]. In continuous control schemes, the ECRH power is kept constant while the ECRH deposition location is controlled in real-time to adjust the sawtooth period [8]. Alternatively the ECRH power can be modulated with a modulation period equal to the desired sawtooth period. Such control scheme is referred to in the literature as sawtooth pacing or sawtooth locking [10-12].

In addition to affecting the local current profile near $q$=1, ICRH influences the sawtooth period by modifying the pressure and the distribution function of the RF-heated ions in the plasma [13-14]. Recent work [15] has also highlighted the influence of the fast ion orbit topology near the $q$=1 surface on the sawtooth dynamics. Sawtooth control based on the modification of the fast ion distribution function near $q$=1 with ICRH has been demonstrated in JET [16-21], including real time control of the deposition location through variation of the RF frequency.

The current paper describes the first demonstration of sawtooth control through modulation of the ICRH power. The concept is similar to the ECRH pacing and locking experiments [10,11], the difference being that the main effect of the ICRH modulation is to change the fast ion population properties (pressure, anisotropy, etc.) in the plasma centre rather than perturb the $q$-profile near the $q$=1 surface. Sawtooth pacing acting on the fast ion stabilization properties (local heating) is inherently slower than pacing based on $q$-

profile perturbations (local current-drive) since the slowing-down dynamics of the fast ions is part of the pacing process. The main advantage of the former technique is that it relies on central ICRH deposition rather than on off-axis power deposition ($q$~1) and therefore the location of the ICRH power is not critical as long as a sufficiently large fraction is deposited inside the $q$=1 surface. As a consequence the ICRH can be deposited at the optimal location required by other considerations, such as for avoiding central accumulation of heavy impurities in metallic machines [22-31] or for enhancing fusion power [32].

## 2 Experimental results

The first experiments for sawtooth control with on-axis ICRH modulation in JET-ILW were done in L-mode plasmas with $B_0$=2.7T, $I_P$=2MA, $n_{e0}$=(5±0.5)x10$^{19}$/m$^3$. Central hydrogen minority ICRH at $f$=42.5MHz (with dipole phasing) was used with $P_{icrh}$=3-4MW. Figure 1 shows the comparison between a discharge with constant ICRF power (left) and two pulses with ICRH power modulation with $f_{mod}$=3Hz and 5Hz (duty-cycle 75%).

The non-modulated discharge is characterized by a sequence of long sawtooth build-ups followed by rapid crashes (b). The long sawteeth (with periods exceeding 0.5s) always trigger considerable magnet-hydro-dynamic (MHD) activity, though in this pulse not causing lasting NTMs (c). It is interesting to note that, while this mode is present, the sawteeth are shorter and the plasma performance is degraded (d). When the ICRH power is modulated in otherwise identical plasma conditions, the sawteeth are efficiently paced at the RF modulation frequency and the $N$=2 MHD activity is fully suppressed. Despite the 25% lower averaged ICRF power applied, the peak temperatures are comparable to the constant ICRH power case. The total plasma stored energy is also similar in the two discharges, probably due to the absence of $N$=2 MHD activity in the modulated case which partially compensates for the lower time averaged ICRH power applied. The insert in Fig.1b-right shows the details of the sawtooth pacing dynamics: The sawtooth crash is delayed with respect to the RF notch because it takes some time for the change in the fast particle population to affect the core electron dynamics as described in section 3.

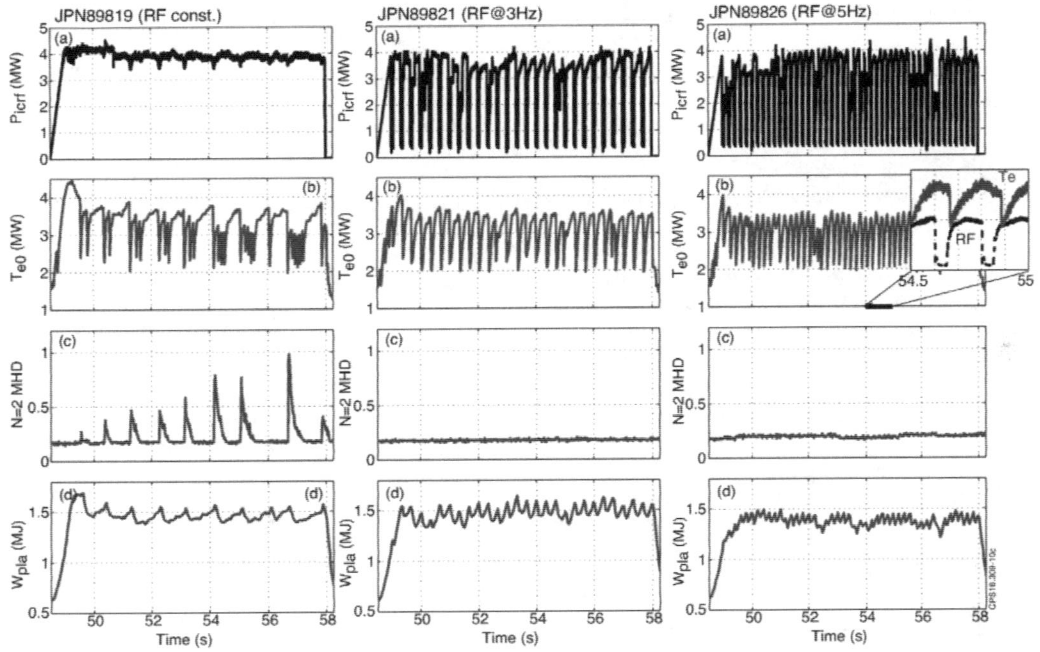


**Fig.1:** Comparison of three L-mode discharges with constant (89819), 3Hz (89281) and 5Hz (89826) modulated ICRF heating, illustrating the successful sawtooth pacing and total suppression of the N=2 MHD activity in the latter: (a) ICRF power; (b) central electron temperature; (c) N=2 MHD mode amplitude; (d) plasma stored energy.

One of the main objectives of these experiments was to determine the range of modulation frequencies that can be used to reliably pace sawteeth in JET plasmas. The main results are summarized in Fig.2, which shows the experimental sawtooth periods as function of time obtained in similar discharges with different RF modulation frequencies, $f_{mod}$ = 0 (RF const.), 2Hz, 3Hz and 5Hz, all with duty-cycle=75%. The maximum RF power was roughly constant ($P_{icrf}$~4MW) and the H concentration was similar ($n_H/n_e$=2-3%) in all discharges.

In the non-modulated case one recognizes the composite sawteeth dynamics with long (0.5-0.7s) sawteeth interleaved with very short ones. When the ICRF power is modulated at 2Hz, the maximum sawtooth period is, as expected, reduced to $\tau_{SWT}$=0.5s but the shorter sawteeth still exist as the $N$=2 modes are not fully suppressed. In the 3Hz and 5Hz RF modulation cases, the $N$=2 activity is absent and the large majority of the sawteeth follow the RF modulation frequency, illustrating the efficient sawteeth control obtained in these conditions. The relative number of RF-triggered sawteeth (normalized to the total number of sawteeth detected) increases with modulation frequency from 45% at 2Hz to 70% and 85% in the 3Hz and 5Hz cases, respectively, since less spontaneous sawteeth occur in the latter. The triggering efficiency, defined as the number of RF-triggered sawteeth divided by the number of RF notches is high in all cases, 80% at 2Hz, 83% at 3Hz and 88% at 5Hz, and does not depend strongly on the RF modulation frequency in the studied conditions.

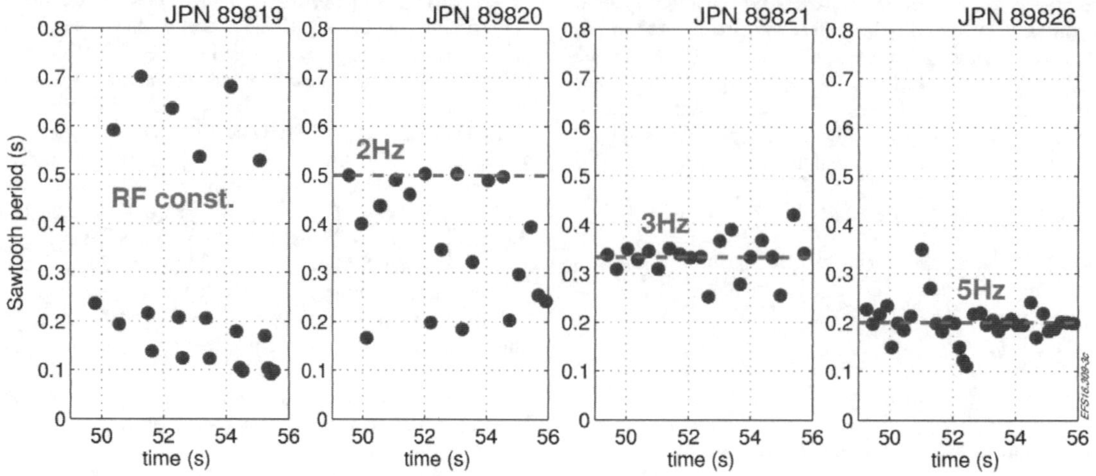


**Fig. 2:** Sawtooth period vs. time for 4 similar L-mode discharges with constant and modulated ICRF power.

## 3 Time delays (slowing-down)

Although the actual physics of sawtooth stabilisation by fast particles is quite complex and depends on details of the fast particle distribution function in specific regions of the plasma [16,17], the basic characteristics of the described method can be illustrated via first principle plasma parameters such as the slowing down time of the RF-induced fast particles. The slowing-down time of the fast ions onto the electrons does not depend on the fast particle energy and is given by $\tau_{SD} = C.(k_B T_e)^{3/2} / n_e$ [33], where $n_e$ and $T_e$ are the local electron density and temperature, respectively, $k_B$ is the Boltzman's constant and $C$ is a constant that depends on the fast ion mass (protons in this case) and the Coulomb logarithm. Figure 3 shows the time intervals between the RF switch-off instants and the respective sawtooth crashes in an L-mode discharge with an RF modulation frequency of 3Hz. This pulse was run at $B_0$=2.7T / $I_P$=2MA with approximately 4MW of on-axis modulated H minority ICRH at $f$=42.5MHz (dipole phasing). The peak core electron temperature was $T_{e0}$=3.5keV and the central plasma density was $n_{e0}$=5x10$^{19}$/m$^3$. The dotted traces represent the electron slowing-down time $\tau_{SD}$ computed from the experimental $n_{e0}$ and $T_{e0}$ signals. The fluctuation in this trace is due to the variation of the central electron temperature during the sawtooth cycles. As expected from first order theory, the time delays between the RF switch-off instants and the sawtooth crashes are comparable to the slowing down time of the fast ions throughout the discharge.

The correlation between the slowing-down dynamics of the RF accelerated ions and the sawteeth triggering times is further illustrated in Fig.4, where the results of a statistical analysis of a series of L-mode discharges are shown. The curves represent the experimental distribution function of the waiting time from the start of the ICRH notch to the next sawtooth crash. This distribution time represents the relative number of triggering events detected in a narrow time window as function of the time delay between the ICRH power notches and the subsequent sawtooth crashes. The data are taken from a series of similar discharges (with different RF modulation frequencies). The statistical results confirm that the highest sawtooth triggering probability is indeed centred around the lower boundary of the average electron slowing-down time of the fast ions in these pulses, $\tau_{SD}$=40-70ms. More sophisticated statistical analysis methods have also been employed and confirmed these observations [34].

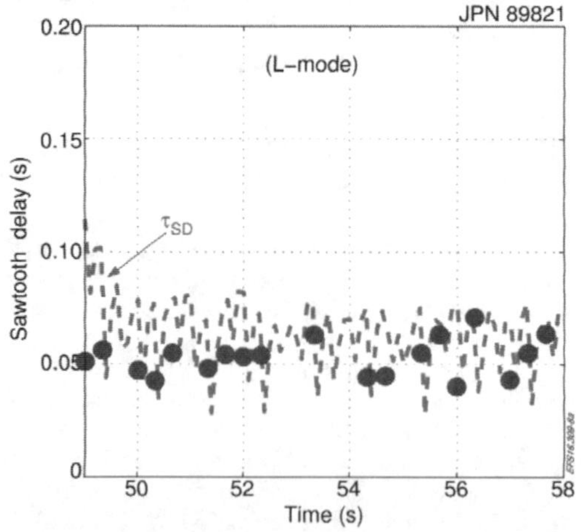


**Fig. 3:** Time delays between the RF switch-off instants and the sawtooth crashes for an L-mode discharge as function of time. $P_{icrh}$=4MW of central ICRH modulated at 3Hz was applied. The dashed curve represents the fast ions slowing-down time.

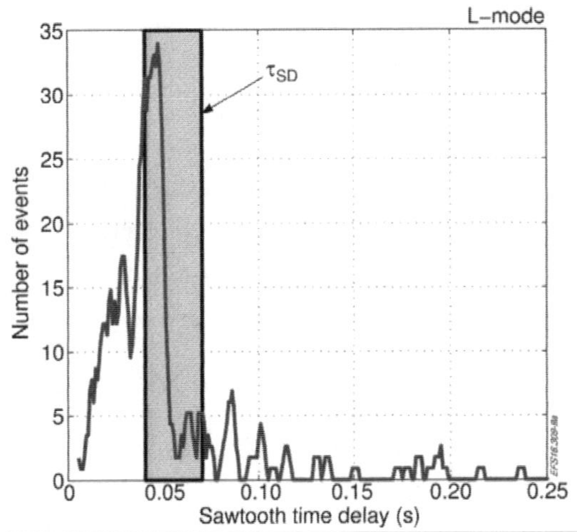


**Fig. 4:** Statistical analysis of the time delays between the RF switch-off instants and the sawtooth crashes for a series of L-mode discharges. The grey area represents the range of the slowing-down times of the fast ions in these pulses.

## 4 Summary & Discussion

Reliable sawtooth control was achieved for the first time with modulated on-axis ICRF heating in JET-ILW L-mode discharges with $B_0$=2.7T, $I_P$=2MA. Central H minority ICRH at $f$=42.5MHz (dipole phasing) with $P_{icrh}$=3-4MW was used with modulation frequencies ranging from 2Hz to 5Hz. While the non-modulated reference plasmas were characterized by long sawteeth ($\tau_{SWT}$>0.5s) and considerable $N$=2 MHD activity following the sawtooth crashes, the discharges in which the sawteeth were paced with $f_{mod}\geq$3Hz showed complete suppression of the $N$=2 modes. The triggering efficiency (number of RF-triggered sawteeth over number of RF notches) is typically above 80% in the cases studied. The lag time between the RF switch-off instant and the sawtooth crash is well correlated to the electron slowing-down time of the fast ions and does not depend on the RF modulation frequency in the studied range. A real-time control algorithm that imposes the RF power notch after a pre-programmed time interval has elapsed from the last detected sawtooth crash was also tested [35]. It showed remarkable pacing efficiency (close to 100%) and it minimizes the number of notches and their duration thus maximizing the averaged ICRF power applied.

Preliminary studies (not described in this paper) indicate that this technique can also be applied to moderate NBI power H-modes ($B_0$=2.7T, $I_P$=2MA, $P_{NBI}$~10MW) at modest ICRH power levels (2-4MW), despite the stabilizing contribution of the NBI ions [2]. In this case, reliable sawtooth pacing requires more careful tuning of the ICRF heating parameters (RF power, H conc.) according to the discharge properties, but once the optimal conditions are reached, the triggering efficiency is comparable to L-mode (≥80%). The delays between the RF switch-off instants and the sawtooth crashes also correlate with the slowing-down time of the fast ions, but there are indications that other effects – such as the NBI stabilization – contribute to the effective time delay for a sawtooth to be triggered. One important question is whether this technique can also be used in DT fusion plasmas, where the α-particles (in addition to the NBI ions) also contribute to the background sawteeth stabilization. Numerical tools that properly describe the competition between the different fast ion sources taking into account their different particle distributions and the synergy between them are still under development. Another question is whether pacing the sawteeth at an optimal frequency is beneficial for minimizing high-Z core impurity accumulation and how it compares with the impurity screening obtained by applying constant RF power on-axis (and not controlling the sawteeth).

## Acknowledgements

This work has been carried out within the framework of the EUROfusion Consortium and has received funding from the EURATOM research and training programme 2014-2018 under grant agreement No 633053. The views and opinions expressed herein do not necessarily reflect those of the European Commission.

## References

[1] "*Overview of the JET results in support to ITER*" by X. Litaudon *et al.*, Proc of 16th IAEA FEC (Kyoto, 2016)
[2] E. Lerche *et al.*, *Nucl. Fusion* 2017 **57** 036027
[3] S. von Goeler *et al.*, *Phys. Rev. Lett.* 1974 **33** p.1201
[4] O. Sauter *et al.*, *Phys. Rev. Lett.* 2002 **88** 105001
[5] I. Chapman *et al.*, *Nucl. Fusion* 2010 **50** 102001
[6] C. Angioni *et al.*, *Nucl. Fusion* 2003 **43** 455
[7] A. Mück *et al.*, *Plasma Phys. Contr. Fus.* 2005 **47** 1633
[8] M. Lennholm *et al.*, *Phys. Rev. Lett.* 2009 **102** 115004
[9] G. Witvoet *et al.*, *Nucl. Fusion* 2012 **52** 074005
[10] T. Goodman *et al.*, *Phys. Rev. Lett.* 2011 **106** 245002
[11] M. Lauret *et al.*, *Nucl. Fusion* 2012 **52** 062002
[12] S. Nowak *et al.*, *Nucl. Fusion* 2014 **54** 033003
[13] R. White *et al.*, *Phys. Rev. Lett.* 1989 **62** 539
[14] F. Porcelli *et al.*, *Plasma Phys. Contr. Fus.* 1996 **38** 2163
[15] J. Graves *et al.*, *Phys. Rev. Lett.* 2009 **102** 065005
[16] J. Graves *et al.*, *Plasma Phys. Contr. Fus.* 2015 **57** 014033
[17] J. Graves *et al.*, *Nat. Commun.* 2012 **3** 624
[18] I. Chapman *et al.*, *Nucl. Fusion* 2012 **52** 063006
[19] M. Lennholm et al., Nucl. Fusion, 2011 **51** 073032
[20] L.-G. Eriksson *et al* 2004 *Phys. Rev. Lett.* **92** 235004
[21] L.-G. Eriksson *et al* 2006 *Nucl. Fusion* **46** S951
[22] F. Romanelli *et al.*, *Nucl. Fusion* 2013 **53** 104002
[23] R. Neu *et al.*, 2013 *Phys. Plasmas* **20** 056111
[24] G. Matthews *et al* 2013 *J. Nucl. Mater.* **438** S2–10
[25] E. Lerche *et al.*, *Nucl. Fusion* 2016 **56** 036022
[26] M. Goniche *et al.*, *Plasma Phys. Contr. Fus.* 2017 **59** 055001
[27] E. Joffrin *et al* 2014 *Nuclear Fusion* **54** 013011
[28] M. Beurskens *et al.*, *Plasma Phys. Contr. Fus.* 2013 **55** 124043
[29] R. Neu *et al.*, *Plasma Phys. Contr. Fus.* 2007 **49** B59-B70
[30] T. Pütterich *et al.*, *Plasma Phys. Contr. Fus.* 2013 **55** 124036
[31] R. Dux *et al.*, *Journal Nucl. Mat.* 2009 **390–391** 858
[32] M. Mantsinen *et al.*, these proceedings
[33] T.H. Stix, "Waves in Plasmas" (1992) AIP, New York
[34] A. Murari et al., *Nucl. Fusion* 2016 **56** 076008
[35] M. Lennholm *et al.* accepted in *Fus. Eng. & Design* (2017).

EPJ Web of Conferences **157**, 03030 (2017)
*22 Topical Conference on Radio-Frequency Power in Plasmas*
DOI: 10.1051/epjconf/201715703030

# ICRF mode conversion in three-ion species heating experiment and in flow drive experiment on the Alcator C-Mod tokamak

*Y.* Lin[*,1], *S.J.* Wukitch[1], *E.* Edlund[1], *P.* Ennever[1], *A. E.* Hubbard[1], *M.* Porkolab[1], *J.* Rice[1] and *J.* Wright[1]

[1]MIT Plasma Science and Fusion Center, Cambridge, MA, USA

**Abstract.** In recent three-ion species (majority D and H plus a trace level of $^3$He) ICRF heating experiments on Alcator C-Mod, double mode conversion on both sides of the $^3$He cyclotron resonance has been observed using the phase contrast imaging (PCI) system. The MC locations are used to estimate the species concentrations in the plasma. Simulation using TORIC shows that with the $^3$He level <1%, most RF power is absorbed by the $^3$He ions and the process can generate energetic $^3$He ions. In mode conversion (MC) flow drive experiment in D($^3$He) plasma at 8 T, MC waves were also monitored by PCI. The MC ion cyclotron wave (ICW) amplitude and wavenumber $k_R$ have been found to correlate with the flow drive force. The MC efficiency, wave-number $k$ of the MC ICW and their dependence on plasma parameters like $T_{e0}$ have been studied. Based on the experimental observation and numerical study of the dispersion solutions, a hypothesis of the flow drive mechanism has been proposed.

## 1 Introduction

Alcator C-Mod is a compact ($a$ = 0.22 m and $R$ = 0.68 m) tokamak running at high field ($B_{t0}$ < 8.1 T) and high current ($I_p$ < 1.6 MA). The ICRF system on Alcator C-Mod is able to provide reliable heating power up to 6 MW from three antennas: D antenna (80 MHz), E antenna (80.5 MHz) and J antenna (78 MHz).

Recently we have carried out the 3-ion species ICRF heating experiment (Section 2) and further investigated the mode conversion (MC) flow drive experiment in D-$^3$He plasmas (Section 3). In both experiments, we used the phase contrast imaging (PCI) diagnostic to detect the RF waves. In the 3-ion species heating, the PCI observation of double mode conversion has been used to estimate the species concentration, thus providing the input parameters for TORIC wave simulation. In the flow drive experiment, PCI measured wave amplitude and $k_R$ together with the numerical solutions of hot-plasma full-wave dispersion equations and TORIC simulation have been used to form a hypothesis of flow drive mechanism.

## 2 Double mode conversion in 3-ion species ICRF heating experiments

Three-ion species ICRF heating was discovered inadvertently on JET and the first controlled experiment was performed on Alcator C-Mod [1]. On both tokamaks, the enhanced heating effectiveness was confirmed and energetic ions were observed. In this paper, we discuss the feature of double mode conversion in the Alcator C-Mod experiment.

Effective ion heating happens when the $^3$He level is so small that the $^3$He ion cyclotron layer is located closely in the vicinity of two hybrid layers on both the high-field-side (HFS) and the low-field-side (LFS). Near a hybrid layer $n_{\parallel}{}^2$ = $S$, where $S$ is the first component of the dialectic tensor in Stix's notation and $n_{\parallel}$ is the component of the index of refraction parallel to the magnetic field, mode conversion from the fast wave (FW) to the ion Bernstein wave (IBW) and ion cyclotron wave (ICW) can occur. Moreover, the left-polarization component of the fast wave is greatly enhanced in the region. At a low $^3$He level, the distance between the IC layer and the two hybrid layers are both sufficiently small so that the Doppler broadened IC layer can interact with the fast wave near the hybrid layers and result in strong power absorption to $^3$He ions.

To carry out the experiment, we need to know the precise value of the species concentrations: $X[^3\mathrm{He}]$ = $n_{\mathrm{He3}}/n_e$ and $X$[H]. Fortunately, we can estimate the species concentrations between plasma shots from the PCI observation of the RF waves. PCI has been used to study different mode conversion scenarios on Alcator C-Mod, including $^3$He-H [2], D(H), D($^3$He) [3]. A detailed numerical study has been performed to understand the measured PCI wave signals. [4] In this 3-ion ICRF heating experiment, we successfully used PCI to monitor the species concentration based on the observation of double mode conversion.

In Fig. 1-(b), the amplitude of the PCI signal at the RF frequency is plotted vs. the major radius R and time t. Two MC regions are clearly visible on the LFS and HFS of the $^3$He cyclotron resonance, respectively: One at $R \approx$ 0.65 m, the other at 0.71 m, on HFS and LFS sides of the the $^3$He IC layer at $R \approx$ 0.69 m ($B_{t0}$ = 7.83 T and RF frequency 78 MHz).

* Corresponding author: ylin@psfc.mit.edu

Increasing $X$[H] would move both hybrid layers toward the HFS, and increasing $X[^3\text{He}]$ would separate the two layers farther apart. As a result, both $X$[H] and $X[^3\text{He}]$ can be uniquely determined from the two MC locations. For this particular plasma shot, we estimate that $X[^3\text{He}] \approx 0.9\%$, $X[\text{H}] \approx 65\%$, assuming that all the remaining ions (mostly D plus small amount of impurities) have $A/Z = 2$.

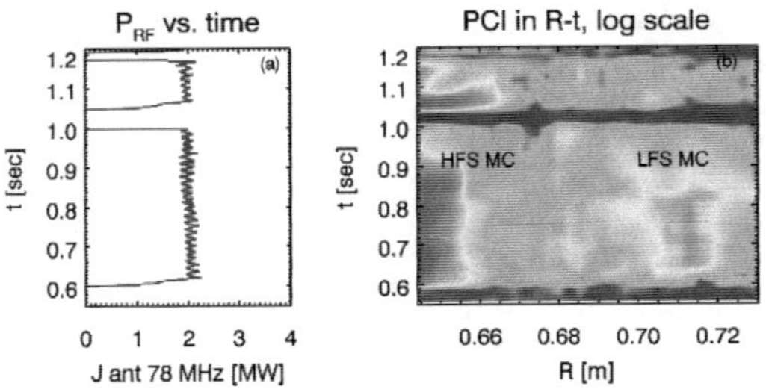


**Fig. 1.** (a) RF power; (b) Contour plot of PCI RF signal vs. $R$-$t$.

In Fig. 2, we show the numerical solution of the full dispersion equation for the waves in the region. [2, 5]. The HFS and LFS MC locations and cut-offs are also labelled. In this 3-ion heating scenario, while the MC waves provide direct electron heating, the important and unique feature is that the fast wave is strongly absorbed by the $^3$He ions near the $^3$He IC layer. Note that $Im(k_\perp)$ of the fast wave in this region ($0.67\text{ m} < R < 0.72\text{ m}$) is more than one order of magnitude larger than the FW branches at $R < 0.65$ m and $R > 0.74$ m.

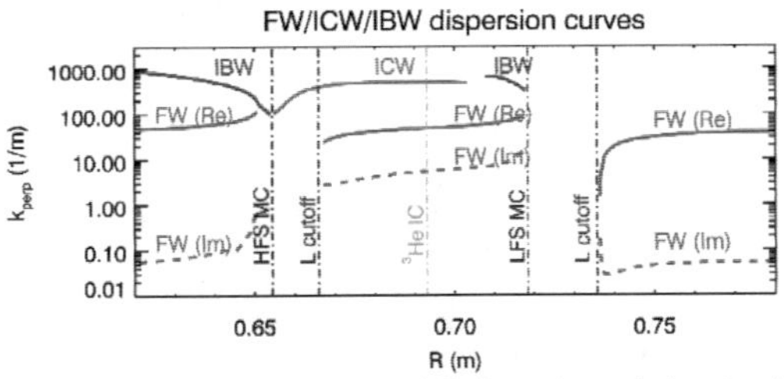


**Fig. 2.** Numerical solution of the full dispersion solution for the waves. For the fast wave, both $Im(k_\perp)$ and $Re(k_\perp)$ are shown.

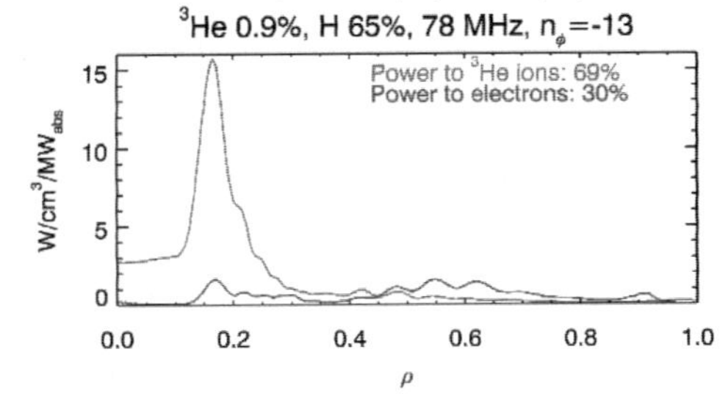


**Fig. 3.** Power deposition profiles from TORIC simulation.

The strong FW ion heating has been confirmed by 2-D full wave simulation using TORIC. As shown in Fig. 3, about 70% of the RF power is absorbed by $^3$He ions via localized strong ion absorption near axis. In contrast, the electron absorption via the fast wave and MC wave is broad and weak. For some other plasmas in the experiment with higher $X[^3\text{He}]$, e.g., ~3%, the simulation shows that most RF power is to the electron heating via mode conversion.

## 3 ICRF wave measurements in D($^3$He) flow drive experiments

### 3.1 Experimental observation

Mode conversion flow drive has been studied extensively on Alcator C-Mod [6, 7]. A detailed scaling law has been obtained vs. main plasma parameters and RF parameters. However, in all the previous flow drive experiments, PCI was unavailable for direct RF wave measurement. In lack of directly supporting experimental evidence, we established a plausible correlation between the greatly enhanced rotation with RF power in D($^3$He) plasmas with $X[^3\text{He}]$ from 10%-15% and the MC process by the comparison with the intrinsic rotation in D(H) minority heating and by TORIC simulation. Lately, we have further investigated the MC flow drive experiment with an improved PCI system for RF wave measurement. This measurement has provided a more direct comparison of the wave physics and the driven rotation.

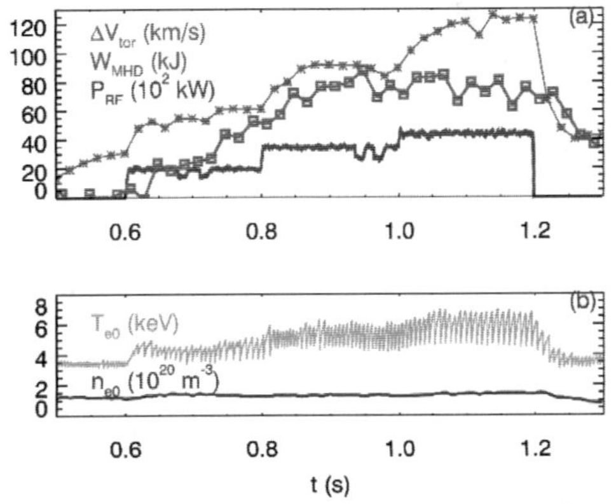


**Fig. 4.** Traces of parameters for a plasma in MC flow drive experiment: (a) $W_{\text{MHD}}$, $\Delta V_{\text{tor}}$, and $P_{\text{RF}}$; (b) $T_{e0}$ and $n_{e0}$.

In Fig. 4, we compare the time traces of major plasma parameters from one of the 6 plasmas in the experiment. $I_p = 1.2$ MA, $X[^3\text{He}] \sim 14\%$, and $B_{t0} = 7.99$ T. The direction of B field is in the direction that is unfavourable for H-mode on C-Mod, i.e., $B \times \nabla B$ pointing away from the divertor. The plasma starts in L-mode confinement and enters I-mode at t ~ 0.9 sec according to the signature of a weakly coherent mode in the plasma edge (not shown here) [8, 9]. In Fig. 4-(a), the change of the central toroidal rotation after the RF power is applied, $\Delta V_{\text{tor}}(t) = V_{\text{tor}}(t) - V_{\text{tor}}(t = 0.6\text{s})$, is shown together with the traces of the stored energy $W_{\text{MHD}}$ and total RF power $P_{\text{RF}}$. The time traces of $T_{e0}$ and $n_{e0}$ are shown in Fig. 4-(b). In this plasma shot, $P_{\text{RF}}$ has three steps at 2 MW, 3.5 MW and 4.5 MW, each lasting for 0.2 second. It is evident that along with the stepping up of $P_{\text{RF}}$, $W_{\text{MHD}}$ and $T_{e0}$ both show definite increases in steps, even for the highest RF power step. In contrast, $\Delta V_{\text{tor}}$ rises rapidly at the first two $P_{\text{RF}}$ steps but stops increasing at the last power step even if the plasma is in I-mode, in which the rotation is expected to rise from the intrinsic rotation drive from the pedestal region. [10] Such rapidly rising and saturating behaviour in $\Delta V_{tor}$ is similar for all other plasmas in this MC flow drive experiment. It has also been observed in some previous MC flow drive experiments in I-mode at high $P_{RF}$ aiming at, but unsuccessfully, synergetic linear supposition of the I-mode intrinsic rotation and MC driven rotation. Note that the intrinsic rotation and the MC driven flow cannot be easily separated, especially in the I-mode phase. To some extent, the root cause of the intrinsic rotation in I-mode can also be traced back to the ICRF heating. In this preliminary study, we only study the change of the total rotation vs. the RF.

As shown in Fig. 4, at the low to middle levels of $P_{RF}$, $\Delta V_{tor}$ is strongly correlated with $W_{mhd}$ and $T_{e0}$, probably due to the common effect of the RF heating. However, the saturation of $\Delta V_{tor}$ at high $P_{RF}$ (albeit with better plasma confinement and higher $T_{e0}$) indicates something more subtle is at work and PCI observation provides a clue.

In Fig. 5, PCI measured RF signal for this plasma is shown vs. $R$-$t$ and vs. $k_R$-$t$. Strong MC waves are detected for $t \approx 0.6$-$1.0$ sec, but the MC signal becomes weaker in time and becomes nearly undetectable after $t > 1.0$ sec although the RF power is higher. The MC wave is mostly at $k_R \sim 3$-$7$ cm$^{-1}$ and the fast wave peak is at $k_R < 1$ cm$^{-1}$.

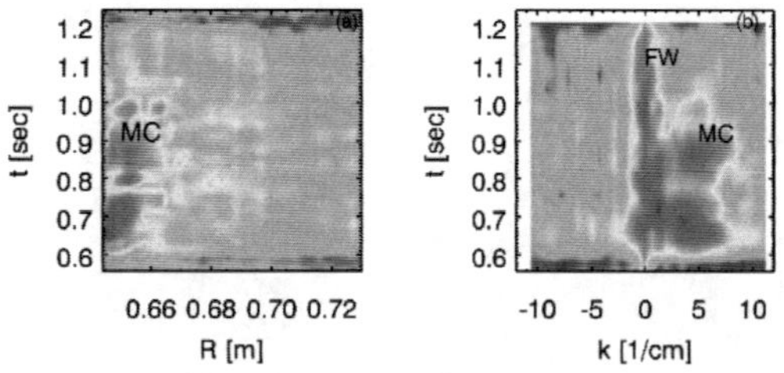


**Fig. 5.** Contour plot of the PCI RF signal (a) vs. $R$-$t$; (b) vs. $k_R$-$t$. The RF power trace (J antenna at 78 MHz) is in Fig.6-(b).

To better show the time evolution of the PCI observation, in Fig. 6 we compare the PCI observed FW amplitude and MC wave amplitude (mostly from the MC ICW). The MC ICW is from the J port antenna (also shown in Fig. 5), and the FW is from E port antenna at (80 MHz). In Fig.6-(a), the PCI measured FW signal appears to broadly follow the applied RF power level. This behaviour is expected because the PCI signal $\propto |E|^2$. On the other hand, in Fig. 6-(b), the PCI MC ICW signal is very different. It starts with a quite high level at the beginning of the RF pulse ($P_{RF} \sim 1$ MW for J antenna) at a moderate $T_{e0} \sim 3$ keV, and with the stepping up of $P_{RF}$ and the rise of $T_{e0}$, the MC wave amplitude level actually drops precipitously. By the time at $t > 1.0$ sec when the RF power reached the 2 MW level and the peak electron temperature reaches $T_{e0} \sim 6$ keV, the MC amplitude in PCI has become less than 5% of its peak level at $t = 0.65$ sec. In Fig. 4-(a), it is shown that the rotation increases rapidly when the RF power is at the low to middle level but the rotation speed levels off at the highest RF power level.

As shown in Ref. [4], the level of the PCI signal has been found to be very difficult to calculate from given plasma parameters and RF parameters. Many factors prevent us from obtaining quantitative interpretation of the PCI RF measurement. For instance, PCI takes a snapshot of a cross-section far away from the antenna but the MC process is essentially a 3-D problem. In this study, we limit our scope only to the trend of the PCI signal (but not the absolute level) vs. the observed rotation. In Fig. 7, we show the time traces of $n_{e0} \times dV_{tor}/dt$, i.e., a proxy of the driving force, vs. log$_{10}$(PCI MC amplitude) for all 6 shots in the experiment. For the entire data set, the correlation between these two quantities is about 0.39, while the correlation between the drive force and other parameters (e.g., $T_{e0}$ and PCI FW amplitude) are close to zero (<0.1). Fig.7 shows that the correlation possibly exists and persists for each and all plasma shots.

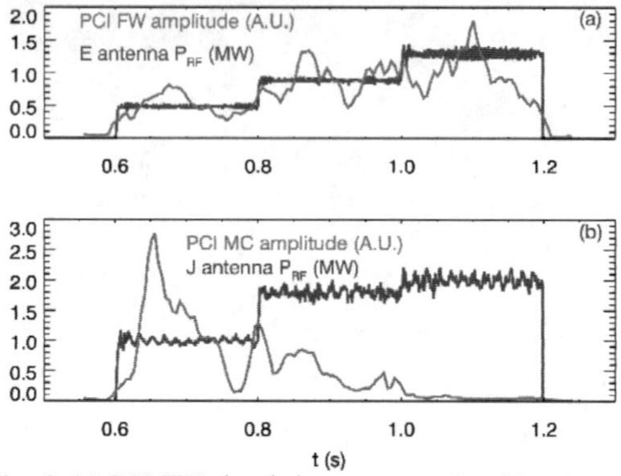


**Fig. 6.** (a) PCI FW signal time trace vs. $P_{RF}$ (E antenna, 80 MHz); (b) PCI MC wave signal trace vs. $P_{RF}$ (J antenna, 78MHz).

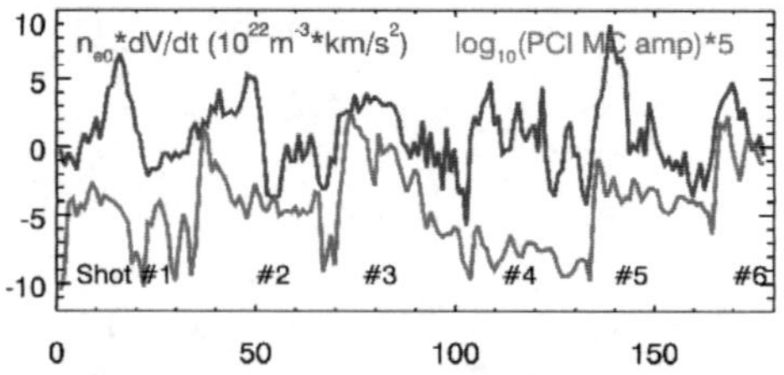


**Fig. 7.** Scaled log$_{10}$(PCI MC amplitude) vs. the proxy rotation drive force. The time traces are from consecutive 6 plasma shots in the experiment.

### 3.2 Dispersion equation solution and $T_{e0}$ dependence

In this exploratory experiment, we were aiming at finding a correlation with the PCI measurement although it was not specific what to expect. Can the correlation shown in Fig. 7 indicate possible causality? In order to further understand whether the correlation would indicate possible causality, we have studied the dispersion solution of the MC process for these plasmas, especially that of the MC ICW.

As shown in Ref. [2, 3, 5], the MC ICW solution can be obtained from the full dispersion equation combined with the poloidal field induced wave-vector $k$ transformation equation. This approach, albeit imprecise, is better in pointing out the possible physics mechanisms than full-wave simulations using TORIC or AORSA. In Fig. 8-(a), the dispersion solution of all three waves: the

DOI: 10.1051/epjconf/201715703030

fast wave, the MC IBW and ICW are plotted vs. major radius for the case where $n_{e0} = 1.5\times10^{20}$ m$^{-3}$ and $T_{e0} = 3$ keV. For the MC ICW, both real and imaginary components of the complex $k_\perp$ are shown. Note both $Re(k_\perp)$ and $Im(k_\perp)$ of the MC ICW increases along with the propagation path toward the LFS. Its $k_\|$ also increases as constrained by the $k$-vector transformation for a given $k_\phi$. A normalized power deposition of the MC ICW, calculated from the ratio of $Re(k_\perp)$ and $Im(k_\perp)$, is in Fig. 8-(b), where the peak deposition is at about 0.8 cm from the MC layer and nearly no more power is left after propagating for 2 cm in the direction perpendicular to B field (equivalently ~25 cm in the direction parallel to B). This is consistent with the PCI observation in Fig. 5-(b).

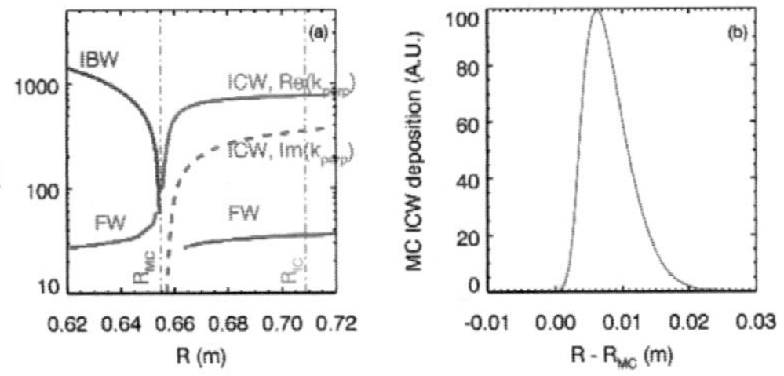


**Fig. 8.** (a) Wave solution from hot-plasma full-wave dispersion equation; (b) Deposition profile of the MC ICW.

In Fig. 9-(a) and 9-(b), we plot the result of a $T_{e0}$ scan of the dispersion solution. Both $Re(k_\perp)$ and $k_\|$ of ICW decrease quickly vs. $T_{e0}$. Since geometrically PCI $k_R \sim k_\perp$, the numerical result agrees with the PCI observation in Fig. 5-(c), where $k_R$ of the PCI MC signal starts at $k_R \sim 7$ cm$^{-1}$ and then decreases vs. time to $k_R \sim 3$ cm$^{-1}$, along with the rise of plasma $T_{e0}$. In other words, the MC ICW is absorbed at larger $k_\perp$ and $k_\|$ at lower $T_{e0}$. In addition, Fig. 9-(c) shows that there is also more RF power absorption via the MC ICW (plus a small portion from the FW) according to the TORIC calculation ($n_\phi = 13$).

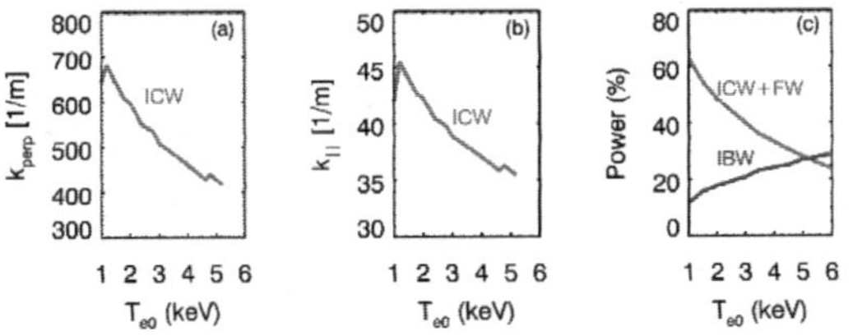


**Fig. 9.** Dispersion solution and MC power vs. $T_{e0}$. (a). $Re(k_\perp)$; (b) $k_\|$; (c) The fraction of total RF power to the MC ICW and the MC IBW from TORIC calculation.

In other words, the experimental result, dispersion solution and TORIC simulation together suggest that at lower $T_{e0}$, there is more RF power converted to the MC ICW and the MC ICW then deposits power at larger $k_\|$ and $k_\perp$ than at higher $T_{e0}$.

### 3.3 Hypothesis on the flow drive mechanism

Experimentally, we have observed that the PCI MC ICW amplitude is much larger at lower $T_{e0}$, so is the force driving the rotation. The correlation between these two observations appears meaningful. From the analysis in Section 3.2, we can explain the behaviour of the PCI signal from the mode conversion physics, but we still need to find a physical mechanism that may connect the MC waves and the force driving the rotation, i.e., $n_{e0}\times dV_{tor}/dt$.

Note that a wave at larger wavenumber $k$ (i.e., slower velocity) would generally have large momentum per unit of power. A MC ICW with larger amplitude and at larger $k_\|$ at the absorption would in theory mean a larger momentum transfer to the plasma (both ions and electrons) after its power has fully deposited. In addition, the MC process to ICW is up-down asymmetric due to poloidal field effect. [3] As a result, the MC ICWs above the mid-plane and below the mid-plane are at different $r/a$ locations in terms of origin, propagation and absorption. The MC process might work like a powerful "engine" that churns the core plasma and re-distributes momentum and then the plasma ultimately gains net momentum from the friction with the wall. This "engine" would be more powerful at lower $T_{e0}$ where more RF power is converted to the MC ICW and at higher $k$, consistent with the observed dependence of the driving force.

We may find more supportive evidence from data mining and simulation, but the hypothesis needs to be tested in future experiments. For example, would the rotation continue rising if we can maintain a high level of PCI MC amplitude using moderate RF power and at moderate $T_{e0}$? Would the plasma rotate faster if not entering I-mode? After the closure of Alcator C-Mod in 2016, these questions can only be answered by future experiments on other devices.

## 4 Summary

Double mode conversion has been observed in the 3-ion ICRF heating scenario by PCI. Plasma composition can be accurately determined and used as input for full-wave simulation. In the MC flow drive experiment, the fast wave and MC waves have been detected directly for the first time and interesting correlation has been found between the PCI MC amplitude vs. the rotation drive.

This work was supported by US DoE Cooperative agreement DE-FC02-99ER54512 at MIT using the Alcator C-Mod tokamak, a DOE Office of Science user facility.

## References

1. Y.O. Kazakov et al, Nature Phys. (19 June 2017) doi: 10.1039/nphys4167.
2. E. Nelson-Melby et al, Phys. Rev. Lett. **90**, 155004 (2003).
3. Y. Lin et al, Plasma Phys. Controlled Fusion **47**, 1207 (2005).
4. N. Tsujii et al, Phys. Plasma **22**, 082502 (2015).
5. E. F. Jaeger et al, Phys. Rev. Lett. **90**, 195001 (2003).
6. Y. Lin et al, Phys. Rev. Lett. **101**, 235002 (2008).
7. Y. Lin et al, Nucl. Fusion. **51**, 063002 (2011).
8. D. G. Whyte et al, Nucl. Fusion **50**, 105005 (2010).
9. A. E. Hubbard et al, Nucl. Fusion **52**, 114009 (2012).
10. J. Rice et al, Phys. Rev. Lett. **106**, 215001 (2011).

DOI: 10.1051/epjconf/201715703031

# Simulation of cold magnetized plasmas with the 3D electromagnetic software CST Microwave Studio®

*Fabrice* Louche[1,*], *Alena* Křivská[1], *André* Messiaen[1], and *Tom* Wauters[1]

[1] *Laboratory for Plasma Physics, ERM/KMS, EUROfusion Consortium Member, Brussels, Belgium, TEC Partner*

**Abstract.** Detailed designs of ICRF antennas were made possible by the development of sophisticated commercial 3D codes like CST Microwave Studio® (MWS). This program allows for very detailed geometries of the radiating structures, but was only considering simple materials like equivalent isotropic dielectrics to simulate the reflection and the refraction of RF waves at the vacuum/plasma interface. The code was nevertheless used intensively, notably for computing the coupling properties of the ITER ICRF antenna. Until recently it was not possible to simulate gyrotropic medias like magnetized plasmas, but recent improvements have allowed programming any material described by a general dielectric or/and diamagnetic tensor. A Visual Basic macro was developed to exploit this feature and was tested for the specific case of a monochromatic plane wave propagating longitudinally with respect to the magnetic field direction. For specific cases the exact solution can be expressed in 1D as the sum of two circularly polarized waves connected by a reflection coefficient that can be analytically computed. Solutions for stratified media can also be derived. This allows for a direct comparison with MWS results. The agreement is excellent but accurate simulations for realistic geometries require large memory resources that could significantly restrict the possibility of simulating cold plasmas to small-scale machines.

## 1 Introduction

During the two last decades a significant number of Ion Cyclotron Resonance Frequency (ICRF) systems for heating, current drive and wall conditioning have been developed. The accurate design and the geometry optimisation of these systems were made possible with the development of commercial softwares like CST Microwave Studio®[1] (MWS), or codes developed in the community like TOPICA[2] and ANTITER II[3]. MWS is based on the "Finite Integration Technique" (FIT) and uses special broadband frequency sweep techniques in order to derive the full spectrum from a limited number of frequency samples. Solids are meshed with a tetrahedral meshing algorithm with iterative refinement techniques. One of the main advantages of MWS is the user-friendly graphical interface that allows to develop very accurate 3D geometries. It has been used for the design and the optimisation of the ICRF antennas for devices like JET [4], ITER[5], W7-X[6], TOMAS[7] and IShTAR[8] notably. The input impedance matrices computed by the program were used to develop matching systems, while the RF fields and currents maps were integrated to evaluate thermal properties and plasma production possibilities. Unfortunately all these evaluations did not incorporate plasma loading. Nevertheless it was proven numerically and experimentally [9] that a dielectric load can fairly simulate plasma loading, with a proper choice of dielectric permittivity for a given set of plasma parameters ($K_D \sim c^2/{v_A}^2$, where $c$ is the light velocity and $v_A$ is the Alfven velocity).

General tensor materials can now be incorporated into MWS and we have written a Visual Basic macro-command that allows to create a cold magnetized plasma. As long as thermal effects can be neglected, i.e. far from resonances and confluence regions, the propagation of waves in plasmas can be described by the cold plasma model. In this model we consider a plasma immersed in a static magnetic induction field $\vec{B_0}$, aligned with the $z$-axis of the Cartesian coordinate system. The components of the cold-plasma dielectric tensor $\boldsymbol{\varepsilon}$ were given by Stix [10]. In the ICRF, when $\omega \sim \omega_{ci}$ and $\omega_{ce} \gg \omega_{ci}$, the perpendicular term can be approximated by $\varepsilon_\perp \approx -\sum_i \omega_{pi}^2/\left(\omega^2-\omega_{ci}^2\right)$ where the sum is limited to the ion contributions, the off-diagonal term becomes $\varepsilon_{xy} \approx \sum_s \omega_{ps}^2 \Omega_{cs}/\left((\Omega_{cs}^2-\omega^2)\omega\right)$, while the parallel term includes only the electron contribution.

## 2 Definition of a test case

The Stix's tensor for a cold plasma was implemented into MWS via a macro-command with an intuitive user interface. The macro must first be tested in cases for which an exact or analytical solution can be derived. The routine must reproduce the classical features of cold plasma waves, like the co-existence of two modes (bi-refringence) with wavelengths given by the solutions of the dispersion equation. But we also want to compare the field maps computed by MWS with analytical solutions.

*e-mail: fabrice.louche@rma.ac.be

DOI: 10.1051/epjconf/201715703031

We therefore consider a linearly polarized monochromatic plane wave $E_y$ excited in vacuum and incident to a given plasma medium. The plasma is bounded by a metallic wall or by an open boundary that imposes a zero reflection condition (see figure 1). For this latter case a perfectly matched layer (PML) must be defined (see Sect. 3.2). The wave propagation is purely longitudinal with respect to the static magnetic field $\vec{B_0} = B_0 \vec{1_z}$. Analytical solutions can be obtained for homogeneous plasmas, stratified plasmas, and for various types of continuous plasma profiles. This allows to directly compare MWS results with analytical solutions and to identify discrepancies coming from the formulation of the numerical problem within MWS.

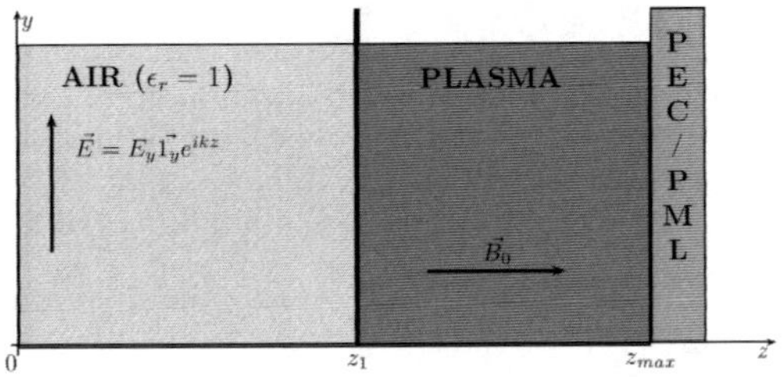


**Figure 1.** 1D model of a linearly polarized monochromatic plane wave $E_y$ with longitudinal propagation w.r.t. the magnetic field direction

When the propagation is parallel to the magnetic field, the two solutions to the dispersion equation reduce to $k_\pm = k_0 \sqrt{\epsilon_\perp \pm \epsilon_{xy}} = k_0 \sqrt{\epsilon_\pm}$, with $k_0 = \omega/c$ and the components of the dielectric tensor being given in Sec. 1. Both modes are characterized by the absence of electric and magnetic fields component in the direction of propagation ($E_z, H_z = 0$). They are circularly polarized: $(E_x/E_y)_\pm = \mp i$ for respectively Left Handed Circular Polarization (LHCP) and Right Handed Circular Polarization (RHCP), the latter being mostly evanescent for typical tokamak plasma parameters in the ICRF. The problem of the longitudinal propagation of a plasma wave in a magnetized plasma can be reduced to the propagation of two circularly polarized plane waves in isotropic dielectrics with permittivities $\epsilon_+$ and $\epsilon_-$ respectively. The methodology for the classical 1D transmission/reflection problem of a plane wave incident to layered or continuous dielectric media can be exploited for treating the case of a longitudinal plasma wave with an appropriate change of frame from Cartesian coordinates to circular coordinates.

All the results presented in the following sections will be for a Deuterium-Tritium plasma with equal ion concentrations. Two sets of parameters will be considered: "plasma#1" with $B_0$=2.5 T and $n_e$=5 $10^{13}$ $cm^{-3}$ and "plasma#2" with $B_0$=5 T and $n_e$=9 $10^{13}$ $cm^{-3}$. In these cases the RHCP mode is mostly evanescent, it only becomes propagative at low frequency for the high density case "plasma#2".

## 3 MWS results for 1D longitudinal propagation

We define the $z$-axis as the direction of propagation and inhomogeneity (see figure 1). The linearly polarized $E_y$ is excited at $z = 0$. The boundary at $z = z_{max}$ will be discussed in the following sections. Periodic boundary conditions are imposed at the 4 other boundaries: this guarantees that the numerical problem is strictly equivalent to a 1D propagation case. An important parameter for MWS models is the mesh density. The accuracy of the solution of the numerical problem as stated in MWS is strongly dependent on the meshing of the spatial domain. A fine mesh is needed to accurately describe the spatial variations of the electromagnetic fields. Each medium (plasma, air, ...) will be defined by its specific sub-grid tuned by a parameter $\delta_z$ defined as the number of cells needed to discretize a characteristic reference wavelength in the medium: the side of a cell will roughly be $\lambda_{ref}/\delta_z$.

### 3.1 Subcase #1: Homogeneous plasma bounded by a metallic wall

We begin with the case of a 1D plasma with length $l_2$ bounded by a perfect metallic boundary at $z = z_{max}$. In each medium (air, plasma) the solution is a superposition of a forward and a backward wave. At the air/plasma interface the reflection coefficient is given by $\Gamma_1 = (Z_{pl,\pm} - \eta_0)/(Z_{pl,\pm} + \eta_0)$, with $\eta_0$ is the intrinsic impedance of vacuum, and $Z_{pl,\pm} = -i\eta_\pm \tan k_\pm l_2$ is the impedance viewed towards the plasma [11].

For the specific problem of a plane wave reflected at a metallic wall, two successive voltage nodes are separated by a distance equal to $\lambda_{ref} = \lambda/2$ in each medium. This is the minimum length that must be accurately resolved by the parameter $\delta_z$. Figure 2 displays the amplitude of some components of the electric field as a function of $\delta_z$. We see that an excellent agreement between the MWS results and the analytical solution is obtained, but it requires a very fine mesh defined by $\delta_z > 300$ to reduce discrepancies. For the set of parameters discussed here this leads to a sub-millimetric size of a typical elementary mesh. Also it should be noted that for the cases where both modes are propagative the bi-refringence is correctly described by MWS (see figure 2-(c)).

### 3.2 Subcase #2: Infinite homogeneous plasma

In usual antenna modeling with MWS and dielectric loading the single pass absorption in a plasma bulk was simulated by a purely outgoing wave condition at boundaries of the simulation domain. In our 1D test case this is an elementary problem to solve analytically, but it cannot be straightforwardly implemented into MWS for plasmas. Indeed the so-called "open boundary conditions" emulating radiating boundaries in MWS cannot be used in contact with anisotropic materials. The idea of a "Perfectly Matched Layer" (PML) was proposed in [12] and extended by many authors in specific situations. In particular, the use of PMLs for magnetized plasmas has been studied in

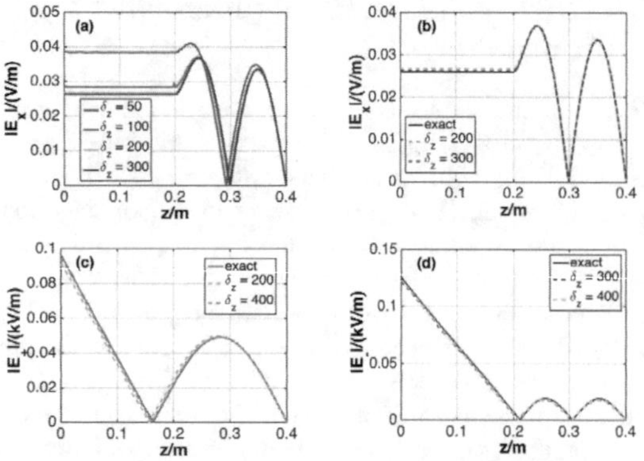


**Figure 2.** Amplitude of the electric field for various levels of mesh accuracy: (a) $E_X$ component at 50 MHz for plasma#1 at low mesh density; (b) $E_X$ component at 50 MHz for plasma#1 at high mesh density; (c) $E_+$ at 20 MHz component for plasma#2; (d) $E_-$ component at 20 MHz for plasma#2.

[13] and for ICRF plasmas in [14]. A PML is an additional boundary layer put at the edge of the computation domain where perfect absorption is desired. This layer is made of an artificial lossy material whose properties are tailored in such a way to absorb and exponentially attenuate any incoming wave to guarantee zero-reflection at the boundary with the plasma medium, while the PML is followed by a perfect metallic boundary.

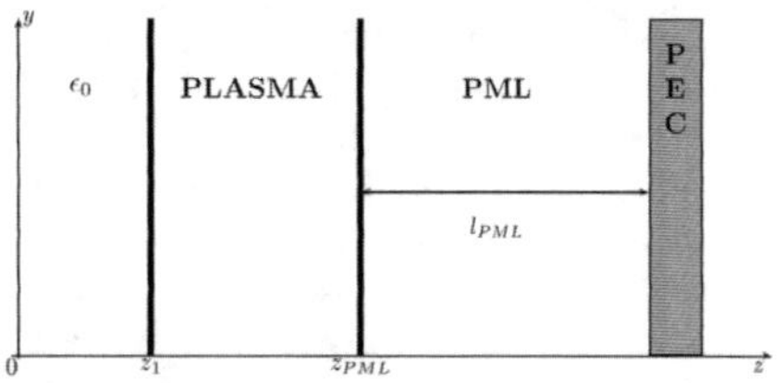


**Figure 3.** Definition of 1D PML

For our simplified 1D case the PML is characterized by a transformed dielectric tensor given by

$$\boldsymbol{\varepsilon_{PML}} = \begin{pmatrix} \varepsilon_{\perp} S_z & -i\varepsilon_{xy} S_z & 0 \\ i\varepsilon_{xy} S_z & \varepsilon_{\perp} S_z & 0 \\ 0 & 0 & \varepsilon_{\parallel}/S_z \end{pmatrix},$$

and a similar formula for the (relative) diamagnetic tensor. A common choice for the stretching function is to use a polynomial form as:

$$S_z(z) = 1 + (S'_z + iS''_z)\left(\frac{z - z_{PML}}{l_{PML}}\right)^{p_z}$$

where $z_{PML}$, $l_{PML}$, $p_z$, $S'$ and $S''$ are respectively the location of the plasma/PML interface, the PML depth (see fig. 3), the order of the stretching function and the stretch parameter (real and imaginary part). The absorption of the LHCP wave in the PML will be optimized if $p_z = 1$ and the imaginary part of the stretching is sufficiently large, and for sufficiently long PML to allow damping.

Before we implement the PML medium into CST MWS, we must discuss an additional approximation. The PML is a medium where the dielectric properties vary *continuously* in space. As such media cannot be modelled in MWS, the continuous PML will be approximated by a stratified medium with each layer having the same thickness and being characterized by constant PML electric and magnetic permittivities. The Helmholtz equation for a longitudinal plasma wave in a stratified medium can be solved by an iterative method which involves to compute recursively the reflection coefficients at each interface [11]. We can then obtain the exact reflection coefficient at the plasma/PML boundary for each mode and compare to the one of a perfect PML. This gives a precise way to assess the minimum number of layers needed to satisfactorily discretize the PML. Typical results can be seen in figure 4-(a): we note that a large number of layers is obviously needed for reproducing the variations and amplitude of the field magnitude. If this number is low an artificial high amplitude standing wave pattern is visible. The important parameter here is the local wavelength in the PML given by $\lambda_{PML}(z) = 2\pi k_{PML} = \lambda_{\pm}/S_z(z)$. If we plot the reflection coefficient as a function of the number of layers *per wavelength*, where the reference value is the minimum $\lambda_{PML}$ at the back of the PML, we observe satisfactory convergence for all stretching functions when the medium is approximated by at least 100 layers (less that 5% error): figure 4-(c). In practical situations where MWS is used the choice of the number of layers will be a compromise between accuracy of the absorption by the PML and reasonable thickness for the layers. Indeed for a case with very good absorption like $S_z = 1 + 3i$ and $p_z = 1$, the minimum wavelength is 3 cm only which would lead to layers of sub-millimeter thickness. If we look at the value of the reflection coefficient for 10 layers we obtain a reflection of 3 $10^{-6}$, which is sufficient to guarantee an effective PML. This is a pragmatic choice which will make the MWS models much more reasonable.

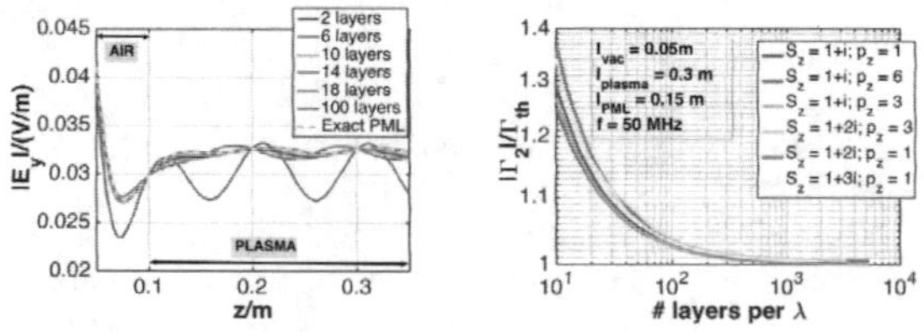


**Figure 4.** Study of stratified PML; (a) electric field profile for $S_z = 1 + i$, $p_z = 1$ and $l_{PML} = 0.15$m,compared with the solution for the continuous PML for various number of layers in the PML; (b) ratio between the reflection coefficient at the plasma/PML interface for stratified PML and continuous PML .

Let us now come to the MWS model. Due to the reduction of the wavelength along the PML we have incorporated a specific sub-meshing for this medium, with an increased meshing density. While we use the same $\delta_z = 300$ as defined previously to tune the plasma meshing (for $\lambda_{ref} = \lambda \simeq 0.2$m), we now use for the PML a reduced reference wavelength. Varying this quantity gives results such as figure 5. We obtain an excellent agreement with the exact solution for a stratified PML. But clearly this comes at the price of a substantially finer mesh than in the plasma. It should also be noted that an almost perfect agreement also requires a very detailed mesh in the vacuum/air region. Reducing the length of the elementary vacuum mesh $z_{vac}$ from 10mm to 5mm was needed to avoid spurious oscillations in this region.

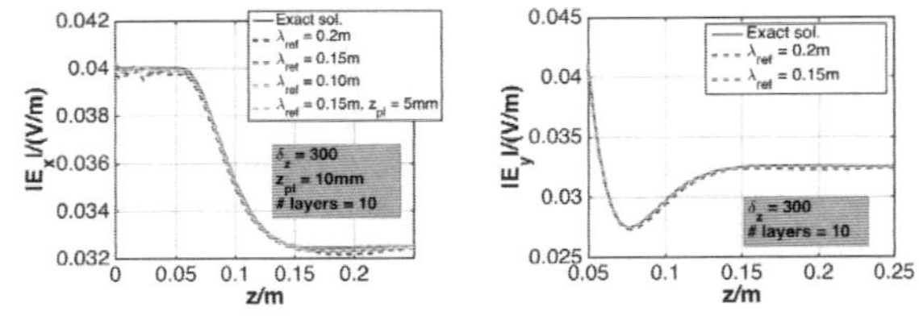


**Figure 5.** Electric field components $E_x$ and $E_y$ computed by MWS for a given PML ($S_z = 1 + 3i$, $p_z = 1$ and $l_{PML}$=150mm) for various reference wavelengths for the meshing in the PML. "Plasma #1" set of parameters. Frequency is 50 MHz.

### 3.3 Subcase #3: Linear density profile

We finally consider the case of a linearly increasing density profile followed by a plasma bulk at constant density: see figure 6. For this case an exact analytical solution for the fields in each medium can be derived. The field in the linearly varying density region is given by the sum of two Airy functions [15]: $E_2(z) = C_1 Ai(\xi) + C_2 Bi(\xi)$ with $\xi = -\left(k_0{}^2\alpha\right)^{1/3}\left(z + \alpha^{-1}\right)$ and $\alpha$ is the slope of the density profile. The two constants $C_1$ and $C_2$, the reflection coefficient in the air region $\Gamma_1$, and the transmitted amplitude $E_0'$ can be exactly computed by imposing the continuity of both the electric and the magnetic field at each interface.

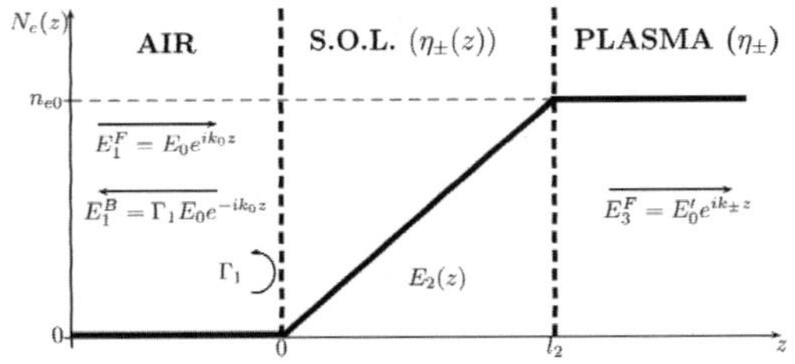


**Figure 6.** Definition of the parameters for linear density profile bounded by a homogeneous plasma.

We have run MWS simulations for various types of profiles, characterized by different slopes, at 50 MHz. The mesh parameter is $\delta_z = 300$, while the usual PML stretching parameter $1+3i$ is used, with 10 layers in the PML. We compare the solution for an increasing number of layers in the profile region with the exact analytical solution in figure 7. A minimum number of 50 layers is needed to obtain the best agreement. It should be noted that the profile discussed here does not lead to any resonance or cut-off at 50 MHz and therefore a homogeneous stratification of the profile is possible. In a situation where the fast wave encounters a cut-off, an adaptive algorithm of stratification will be necessary. This is still under study.

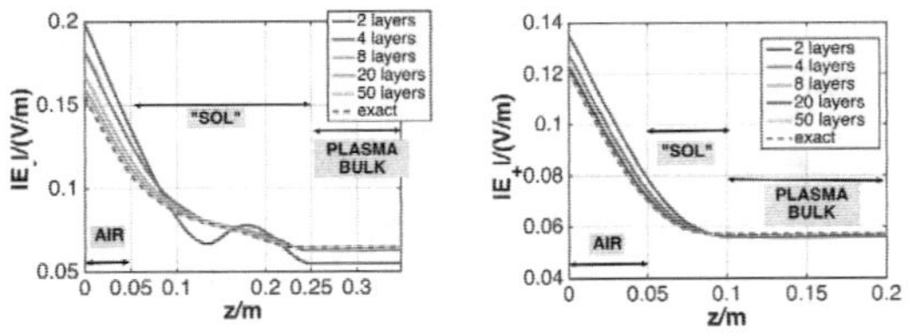


**Figure 7.** LHCP component of the electric field for various number of layers in the density profile region, compared with the analytical solution. Frequency is 50 MHz. Left: plasma #1; right: plasma#2 (higher density).

## 4 Conclusions

The work presented here is a first step in the direction of more realism in the predictions obtained from MWS. The plasma module developed for this study has been benchmarked for a 1D model of plasma and the results obtained were in perfect agreement with the exact analytical solutions. Nevertheless the accuracy of the spatial discretization of the plasma material needed to reach this agreement is extremely high and the modeling of realistic 3D antennas facing magnetized plasmas should require huge hardware resources, in particular for cases with radiative boundaries where a PML must be added to the simulation. The possibility to use a parallelised version of MWS will be exploited in a near future.

## References

[1] *CST STUDIO SUITE®*, CST AG, Germany
[2] V. Lancellotti et al., Nucl. Fusion **46**, S476 (2006)
[3] A. Messiaen et al., Nucl. Fusion **50**, 025026 (2010)
[4] P. Lamalle et al., Nucl. Fusion **46**, 432 (2006)
[5] F. Louche et al., Nucl. Fusion **51**, 103002 (2011)
[6] F. Louche et al., Fusion Eng. Des. **96-97**, 508 (2015)
[7] F. Louche et al., Fusion Eng. Des. (2017) **http://dx.doi.org/10.1016/j.fusengdes.2017.04.123.**
[8] F. Louche et al. *AIP Conf. Proc.*, **1689** 070016, 2015.
[9] A. Messiaen et al. *Fusion Eng. Des.*, **86** 978, 2011.
[10] T.H. Stix, *Waves In Plasmas* (New-York: AIP, 1992)
[11] S. Orfanidis, *Electromagnetic Waves and Antennas* (ECE Department Rutgers University, www.ece.rutgers.edu/orfanidi/ewa/, 2014)
[12] J. Berenger, J. Comput. Phys. **114**, 185 (1994)
[13] N. Gondarenko, J. Comput. Phys. **194**, 481 (2004)
[14] J. Jacquot et al., Plasma Phys. Control. Fusion **55**, 115004 (2013)
[15] K. Yeh, C.H. Liu, *Theory of Ionospheric Waves* (Academic Press, 1972)

DOI: 10.1051/epjconf/201715703032

# The Role of Combined ICRF and NBI Heating in JET Hybrid Plasmas in Quest for High D-T Fusion Yield

*Mervi* Mantsinen[1,2,*], *Clive* Challis[3], *Domenico* Frigione[4], *Jonathan* Graves[5], *Joerg* Hobirk[6], *Eva* Belonohy[6], *Agata* Czarnecka[7], *Jacob* Eriksson[8], *Dani* Gallart[1], *Marc* Goniche[9], *Carl* Hellesen[8], *Philippe* Jacquet[3], *Emmanuel* Joffrin[9], *Damian* King[3], *Natalia* Krawczyk[7], Morten Lennholm[10], *Ernesto* Lerche[11], *Ewa* Pawelec[12], *George* Sips[10], *Emilia R.* Solano[13], *Maximos* Tsalas[14], *Marco* Valisa[15] and JET Contributors[**]

[1]Barcelona Supercomputing Center (BSC), Barcelona, Spain
[2]ICREA, Pg. Lluís Companys 23, 08010 Barcelona, Spain
[3]CCFE, Culham Science Centre, Abingdon, Oxon, OX14 3DB, UK
[4]ENEA, Fusion and Nuclear Safety Dep., C.R. Frascati, Frascati (Roma), Italy
[5]Ecole Polytechnique Fédérale de Lausanne (EPFL), Swiss Plasma Center (SPC), CH-1015 Lausanne, Switzerland
[6]Max-Planck-Institut für Plasmaphysik, D-85748 Garching, Germany
[7]Institute of Plasma Physics and Laser Microfusion, Hery 23, 01-497 Warsaw, Poland
[8]Uppsala University, SE-75120 Uppsala, Sweden
[9]CEA, IRFM, F-13108 Saint Paul Lez Durance, France
[10]European Commission, B-1049 Brussels, Belgium
[11]Laboratory for Plasma Physics, LPP-ERM/KMS, Renaissancelaan 30 Avenue de la Renaissance B-1000, Brussels Belgium
[12]Institute of Physics, Opole University, ul. Oleska 48, Opole 45-052, Poland
[13]Laboratorio Nacional de Fusión, CIEMAT, Madrid, Spain
[14]ITER Organization, Route de Vinon, CS 90 046, 13067 Saint Paul Lez Durance, France
[15]Consorzio RFX, corso Stati Uniti 4, 35127 Padova, Italy

**Abstract.** Combined ICRF and NBI heating played a key role in achieving the world-record fusion yield in the first deuterium-tritium campaign at the JET tokamak in 1997. The current plans for JET include new experiments with deuterium-tritium (D-T) plasmas with more ITER-like conditions given the recently installed ITER-like wall (ILW). In the 2015-2016 campaigns, significant efforts have been devoted to the development of high-performance plasma scenarios compatible with ILW in preparation of the forthcoming D-T campaign. Good progress was made in both the inductive (baseline) and the hybrid scenario: a new record JET ILW fusion yield with a significantly extended duration of the high-performance phase was achieved. This paper reports on the progress with the hybrid scenario which is a candidate for ITER long-pulse operation (~1000 s) thanks to its improved normalized confinement, reduced plasma current and higher plasma beta with respect to the ITER reference baseline scenario. The combined NBI+ICRF power in the hybrid scenario was increased to 33 MW and the record fusion yield, averaged over 100 ms, to $2.9 \times 10^{16}$ neutrons/s from the 2014 ILW fusion record of $2.3 \times 10^{16}$ neutrons/s. Impurity control with ICRF waves was one of the key means for extending the duration of the high-performance phase. The main results are reviewed covering both key core and edge plasma issues.

## 1 Introduction

The application of ICRF waves in fusion plasmas is known to have a number of advantageous effects [1]. In addition to providing a fairly localised power deposition, the ICRF-driven fast ions arising through the resonant interactions between the ICRF waves and plasma ions can play a number of important roles for example by suppressing turbulence, enhancing the fusion reactivity and affecting MHD stability. In this paper, we report on the experiments carried out on the JET tokamak to optimise the use of ICRF waves in high-performance hybrid plasmas in preparation of the forthcoming D-T campaign. The optimization involves not only the selection of the ICRF scheme, the magnetic field and resonance location given the available ICRF frequencies but also the optimization of local gas puffing close to ICRF antennas for good ICRF coupling, optimization of the fuel ion mixture for optimal ICRF power partitioning and the choice of ICRF antenna phasing. Given that all of them affect the plasma performance in multiple ways, the optimization of the use of ICRF waves is an integral part of plasma scenario optimization and development. Its overall aim was to improve the fusion performance

* Corresponding author: *mervi.mantsinen@bsc.es*
** See the author list of X. Litaudon et al. Nucl. Fusion 57 (2017) 102001.

DOI: 10.1051/epjconf/201715703032

and extend the duration of high-performance phase with respect to the 2014 record 2.5MA/2.9T hybrid discharge 86614 [2]. This discharge reached a normalised beta $\beta_N$ of 2.1, a neutron rate $R_{NT}$ of $2.3x10^{16}$ neutrons/s and a confinement factor H98(y,2) of 1.1. The goal for D-D fusion is $R_{NT}\approx 5x10^{16}$ neutrons/s, to ensure a D-T fusion power of ~15 MW. Combined ICRF and NBI heating in discharge 86614 was analysed in detail in [3]. In the present paper, experimental results on ICRF optimization in hybrid plasmas are presented while the results of ICRF+NBI modelling, including details on ICRF power partitioning and prediction for a 50%:50% D-T plasma, are reported in an accompanying paper [4].

Given the available ICRF frequencies and range of the toroidal magnetic field in JET, the ICRF schemes relevant for high-performance JET D-T plasmas are hydrogen (H) and $^3$He minority heating which coincide with second harmonic deuterium and tritium heating, respectively. The latter is the main ITER ICRF scheme for D-T plasmas and thus is of particular importance. However, all the experiments carried out in high-performance hybrid plasmas in the 2015-2016 JET experimental campaign were carried out with H minority heating. This was due to limited experimental time at high power together with operational restrictions on the ICRF use after $^3$He puffing that are currently in force in JET.

Nevertheless, extensive computer modelling involving several ICRF codes were undertaken to prepare experiments with $^3$He minority heating. The modelling results showed improved bulk ion heating and improved fusion performance with $^3$He minority heating as compared to H minority heating [5, 6]. Further beneficial effects on fusion performance are expected to arise from the stabilization of plasma microturbulence by ICRF-accelerated $^3$He minority ions which was not included in the modelling.

In this paper we discuss the results obtained with H minority heating. They include the avoidance of impurity accumulation with ICRF waves and its dependence on the ICRF resonance location, the dependence of the fusion performance on the H minority concentration, and the achieved record fusion performance.

## 2 ICRF waves for the avoidance of impurity accumulation

Impurity control with ICRF waves was demonstrated as one of the key means for extending the duration of the high-performance phase in hybrid pulses. Figure 1 shows the main plasma parameters for two 1.5MA/2.1T hybrid discharges prepared in the same way except for the heating power waveforms. In discharge 90350 only NBI heating was applied while in discharge 90348 NBI power was reduced in the early part of the discharge and 2 MW of ICRF heating was added at a frequency of 32 MHz for a central resonance.

As we can see from Fig 1, in the early phase of the discharges, up to 6 s, discharges perform in the same way despite the fact that the total input power was about 10-20% lower in discharge 90348 with ICRF heating. However, from 6 s onwards, discharge 90350 with NBI heating only suffers impurity accumulation which leads to a deterioration of the fusion performance. In comparison, the performance is maintained throughout the heating phase in discharge 90348 with ICRF heating. The reason for this difference is believed to be the central heating provided by ICRF waves. It resulted in a more peaked electron temperature profile and a flatter density profile in the plasma core which were favourable for the avoidance of impurity accumulation. Our results are in line with previous results on ICRF heating for core impurity control, see .e.g. [7, 8, 9] and references therein, extending them to high-performance hybrid plasmas of long duration for D-T with JET-ILW.

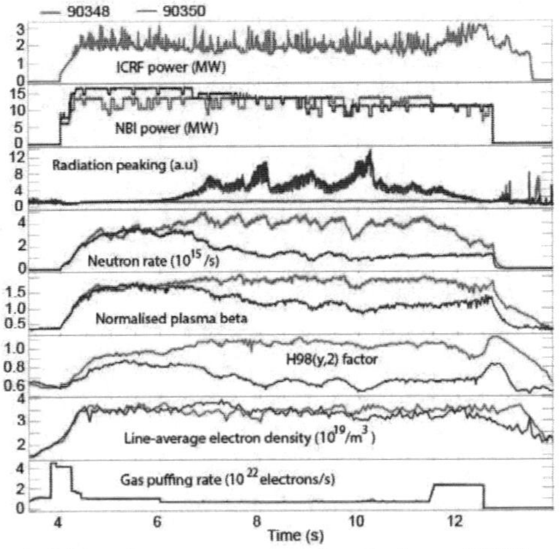


**Fig. 1.** Main plasma parameters for discharge 90348 with combined NBI and ICRF heating and for discharge 90350 with NBI heating only. The radiation peaking factor is the ratio of three most central bolometer channels to three channels off-axis and is an indication of the impurity content in the plasma center.

## 3 Dependence of ICRF impurity control on the ICRF resonance location

The dependence of impurity control on the ICRF resonance location was studied in plasmas with a normalised plasma beta $\beta_N$ of about 2 and a safety factor profile optimized for MHD control. Figure 2 shows five hybrid discharges with 5 MW of ICRF power at a frequency of 42 MHz and up to 25 MW of NBI power which were prepared in the same way except for the magnetic field $B_T$ and the start time of ICRF and NBI heating. $B_T$ was varied to change the ICRF resonance location from a high-field-side resonance at $R_{res}\approx 3.2$ m to a low-field-side resonance at $R_{res}\approx$ 2.75 m. The magnetic axis was located at $R_0\approx$ 3.0 m. Furthemore, the start time of ICRF and NBI heating was shifted by 0.25 s for each $B_T$ change of 0.1 T to keep the central safety factor as similar as possible at the heating onset. The H concentration $n_H/(n_H+n_D)$, as deduced from the ratio of the $D_\alpha$ and $H_\alpha$ light collected along lines of sight through the plasma, was 1-2%. Penning gauge spectroscopy in the divertor gave somewhat higher $n_H/(n_H+n_D)$ of 3-4 %.

The results shown in Fig. 2 suggest that a central resonance with $|R_{res}-R_0|<15$cm is favourable in order to prevent impurity accumulation and deleterious MHD in the present plasma conditions. It should be noted, however, that before the appearance of MHD and impurity accumulation discharge 92311 with a resonance furthest away on the high-field side produced similar or even better performance in terms of plasma energy content, $\beta_N$, neutron rate $R_{NT}$ and electron and ion temperature as compared to discharges with a more central resonance. For this discharge, our modelling [4] predicts stronger direct fast wave electron damping until the onset of MHD activity. Unlike ion cyclotron damping, electron damping increases with the electron density and temperature and thus typically peaks in the plasma center. It could have played a role in the early phase of this discharge by increasing the central electron heating power deposition. This could be of interest if further ICRF optimization will be pursued e.g. with polychromatic operation.

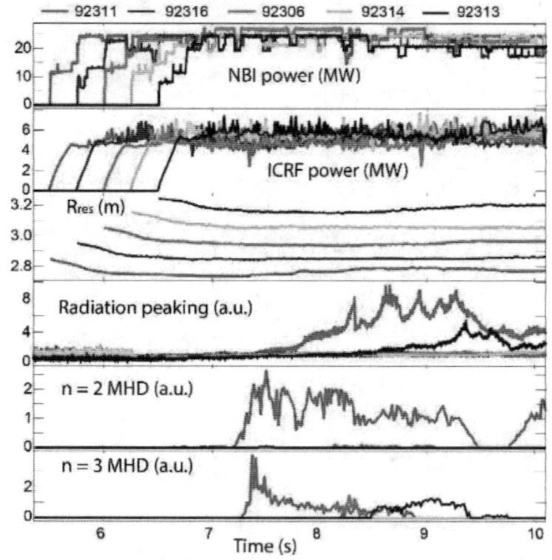


**Fig. 2.** Overview of five hybrid discharges with combined NBI+ICRF heating at different toroidal magnetic fields to vary the ICRF resonance location: 2.6, 2.7, 2.8, 2.9 and 3.0 T in discharge 92311, 92316, 92306, 92314 and 92313, respectively. The start time of external heating was modified to match the central safety factor at the heating onset.

# 4 Dependence of fusion performance on H minority concentration

Another aim of the ICRF optimization in the recent 2015-2016 experiments was to clarify the role of the H concentration for fusion performance. The H concentration is known to play a key role for ICRF power partitioning between the H minority ions and the majority and beam-injected deuterium ions and thereby can affect the bulk plasma heating and the ICRF enhancement of $R_{NT}$ due to ICRF-accelerated deuterons [10].

Figure 3 shows the main plasma parameters for three 2.8T/2.4MA hybrid discharges with 5 MW of ICRF power and 20 MW of NBI power which were prepared in a same way except for H gas puffing. The resulting $n_H/(n_H+n_D)$ as deduced from the Penning gauge spectroscopy is approx. 3.5, 3 ad 6.5 % in discharges 92321, 92322 and 92323, respectively. The ratio of the $D_\alpha$ and $H_\alpha$ light spectroscopy indicated similar relative changes but lower absolute $n_H/(n_H+n_D)$ values of 1.5-3 %.

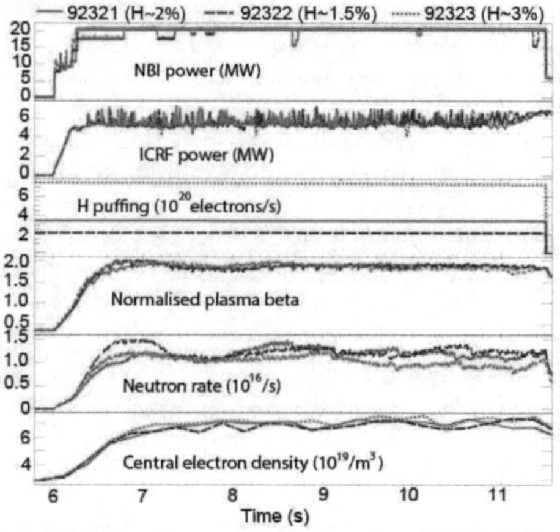


**Fig. 3.** Main plasma parameters for 2.8T/2.4MA hybrid discharges 92321, 92322 and 92323 with different H gas puffing rates to vary the H concentration.

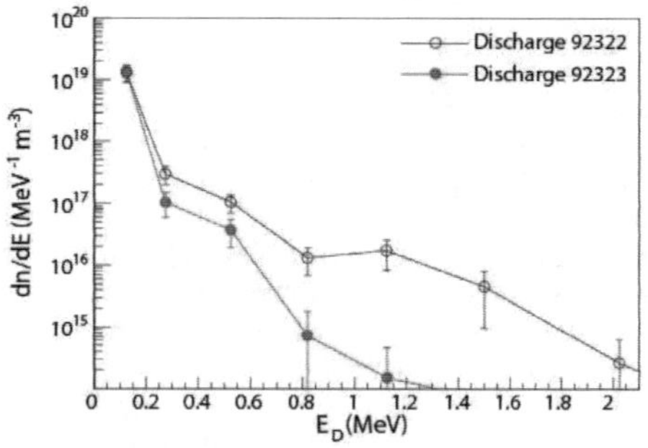


**Fig. 4.** D distribution functions as deduced from measurements with the neutron time-of-flight spectrometer TOFOR [11] for discharges 92322 and 92323 with a H concentration of ~1.5% and 3%, respectively.

The main difference between the discharges in the early part of the heating phase up to t= 8 s is the fast initial rise of $R_{NT}$ in the discharge with lowest H puffing. A lower H concentration in this discharge leads to a larger D damping and acceleration of D ions to higher energies than in other discharges. The difference in $R_{NT}$ reduces as the plasma density increases in time. Nevertheless, in the later steady phase of discharges from t = 8 s onwards, $R_{NT}$ is 10-25% lower in discharge 92323 with the highest H concentration than in the discharges with lower H concentrations. This difference is consistent with a stronger ICRF-accelerated fast deuterium tail in discharges with lower H concentrations as deduced from the measurements with the neutron time-of-flight spectrometer TOFOR [11] (cf Fig. 4). It is also consistent with modelling which shows that central deuterium damping increases and hydrogen damping decreases when the H concentration is decreased while direct fast wave electron damping remains virtually the

DOI: 10.1051/epjconf/201715703032

same [4]. We conclude that ICRF heating with a low H concentration is advantageous for maximising the D-D fusion yield through ICRF-acceleration of D ions while preventing impurity accumulation. The ICRF enhancement of $R_{NT}$ was typically 10-20% in the recent hybrid discharges [4]. Due to the different dependence of the D-T and D-D fusion cross section on the deuterium energy, the ICRF-enhancement of equivalent D-T fusion yield is smaller, of the order of 5 % [4].

## 5 Record fusion performance

Figure 5 shows the main plasma parameters of the hybrid discharge with the highest fusion yield in the 2015-2016 experimental campaign. It was carried out at a plasma current of 2.2 MA and a magnetic field of 2.8 T using 27 MW of NBI heating and 4.5 MW of ICRF heating. The ICRF H minority resonance was in the plasma center and the H concentration was 1-2% as deduced from the ratio of the $D_\alpha$ and $H_\alpha$ light. $R_{NT}$, $\beta_N$ and H98(y,2) steadily increased to their steady state values of $2.9\times10^{16}$ $s^{-1}$, 2.9 and 1.2, respectively. The radiated power increased slowly throughout the discharge and reached 12 MW towards the end of the heating phase. The dominant impurities were mid-Z impurities Ni, Fe, Cu and Mo. At t = 7.9 s, the discharge was terminated by the machine protection system that detected a main chamber hot spot and a concomitant raise of the temperature of one of the poloidal limiters above the permitted limit of 940 °C. The reason for the hot spot and the correct actions for its avoidance are still under investigation. In subsequent shots, the hot spot was mitigated with local gas puffing close to the limiter but it was not possible to progress further due to limited experimental time at the end of the campaign.

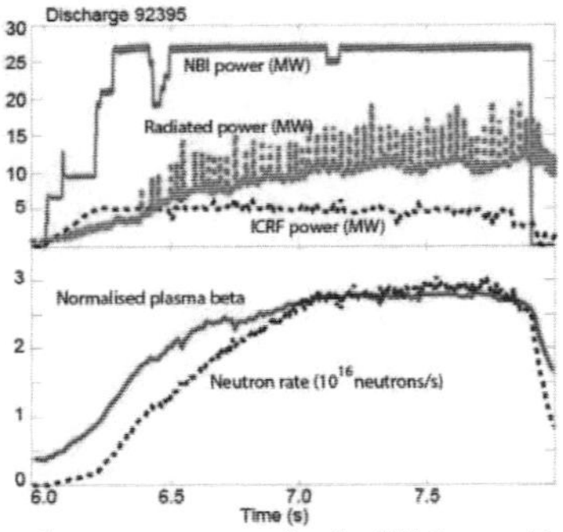


**Fig. 5.** Main plasma parameters for the 2016 record hybrid discharge 92395 with combined NBI + ICRF heating.

## 6 Conclusions and future prospects

Significant progress has been made in the development of the JET high-performance hybrid scenario compatible with the ITER-like wall in preparation of the future D-T campaign. A new record JET ILW fusion yield and an extended duration of the high-performance phase have been achieved. H minority heating with a central resonance and a low H minority concentration played key roles in avoiding central impurity accumulation and maximising the fusion performance and its duration. Future plans include making use of a higher total ICRF+NBI power of ~40 MW.

Three main topics remain for further optimization of combined ICRF and NBI heating in JET hybrid plasmas. First, $^3$He minority heating needs to be experimentally demonstrated in high-performance D-T plasmas in preparation of the D-T campaign as discussed above. Second, the use of +90° degree ICRF phasing requires further experimental exploration. While all the experiments reported above used the dipole phasing of the JET A2 ICRF antennas, three hybrid discharges with +90° phasing of the A2 antennas were carried out. Figure 6 compares one of them with a similar discharge with dipole phasing. In broad terms similar plasma performance was obtained. However, one important difference was observed: the ICRF coupling resistance was larger by a factor of about two for all three discharges with +90° degree ICRF phasing as compared to all nine discharges with dipole phasing carried out in the same experimental session. This implies that it may be possible to increase ICRF power from the A2 antennas using +90° degree phasing, which would give important new prospects of increased input heating power and better fusion performance in JET D-T. Finally, the use of polychromatic ICRF operation may become relevant with high ICRF power at high $\beta_N$.

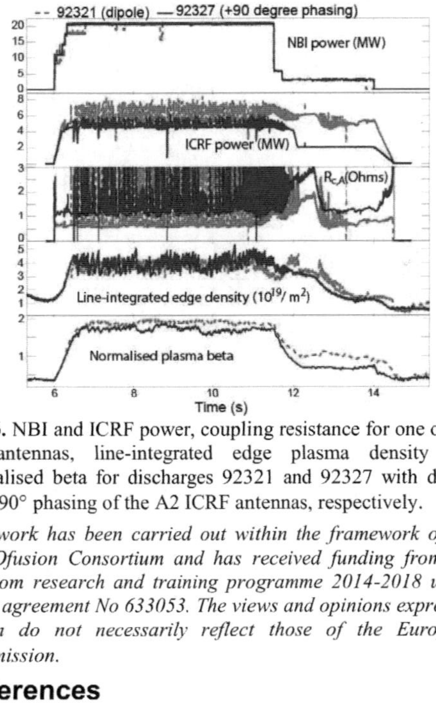


**Fig. 6.** NBI and ICRF power, coupling resistance for one of the A2 antennas, line-integrated edge plasma density and normalised beta for discharges 92321 and 92327 with dipole and +90° phasing of the A2 ICRF antennas, respectively.

*This work has been carried out within the framework of the EUROfusion Consortium and has received funding from the Euratom research and training programme 2014-2018 under grant agreement No 633053. The views and opinions expressed herein do not necessarily reflect those of the European Commission.*

## References

1. M.J. Mantsinen et al., PPCF 45 (2003) A445 -A456.
2. Frigione et al., Europhysics Conference Abstracts 2015, vol. 39E, pp. P2.116.
3. M.J. Mantsinen et al., Europhysics Conference Abstracts 2015, vol. 39E, pp. P2.171.
4. D. Gallart et al., this conference.
5. D. Gallart et al., Europhysics Conference Abstracts 2016, Vol. 40A, P2.003.

6. J. Garcia et al., PPCF 59 (2017) 014023.
7. C. Angioni et al., Nuclear Fusion 57 (2017) 056015.
8. M. Goniche et al., PPCF 59 (2017) 055001.
9. E. Lerche et al., Nuclear Fusion 56 (2016) 036022.
10. M.J. Mantsinen et al., PPCF 41 (1999) 843-865.
11. M. Gatu Johnson et al., Nucl. Instrum. Methods A, 591 (2008) 417.

EPJ Web of Conferences **157**, 03033 (2017)
*22 Topical Conference on Radio-Frequency Power in Plasmas*
DOI: 10.1051/epjconf/201715703033

# Modeling of the Traveling Wave Antenna in view of the ICRF heating of DEMO

*Andre* Messiaen[1,*] and *Riccardo* Ragona[1,2]

[1]Laboratory for Plasma Physics, Royal Military Academy, (LPP-ERM/KMS), BE-1000, Brussels, Belgium

[2]Ghent University, Department of Applied Physics, 9000 Ghent, Belgium

**Abstract.** An upgraded version of the fast semi-analytical code ANTITER is used to model the refueled TWA sections of any arbitrary number of straps facing a low coupling plasma profile. The code computes the **Y**, **Z** and **S** matrices of the antenna array from which the performances of the TWA sections and of their feeding system are deduced as a function of the geometry of the TWA. The model incorporates the feeding of the section by a resonant ring circuit recirculating the output power of each section. Approximate scaling laws guiding the design are derived for the antenna main parameters. A first optimization of the design for the DEMO antenna section is made and the resulting power capability of a DEMO system evaluated.

## 1 Modeling of a TWA section

### 1.1 Introduction

A Traveling Wave Antenna (TWA) system is presently considered for a reliable ICRF heating of DEMO in low coupling conditions [1]. The TWA is integrated in the first wall. It can be continuous all around the machine but has to be periodically refueled by power sources. We prefer to divide it in individually fed TWA sections to improve the reliability of the entire system and avoid large cross-coupling effects between the generators [2]. The paper studies the optimization of the TWA parameters and proposes a first conceptual electromagnetic design for DEMO.

### 1.2 Basic relations and iterative impedance

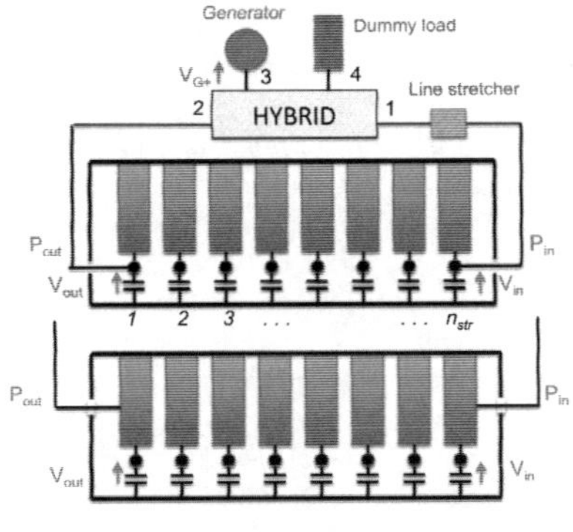


**Fig. 1.** TWA section with its feeding network (top without tap: $Z_0$=$Z_{it}$, bottom with tap: $Z_0$ = 50Ω).

Figure 1 shows a TWA section of $n_{str}$ straps and its feeding circuit. Each strap is tuned by its own capacitor. Only the first and last strap-capacitor connection points are connected to an external feeder. The admittance matrix $\mathbf{Y}_{oc}$ seen towards the $n_{str}$ connection points is the sum of the strap admittance matrix $\mathbf{Y}_S$ and the capacitor one $\mathbf{Y}_C$: i.e. $\mathbf{Y}_{oc}=\mathbf{Y}_S+\mathbf{Y}_C$. The array impedance matrix $\mathbf{Z}_a$, between the connection points to the two feeders, is immediately obtained from the terms of $\mathbf{Z}_{oc}=\mathbf{Y}_{oc}^{-1}$ by the relation:

$$\mathbf{Z}_a = [\mathbf{Z}_{oc}(1,1)\ \ \mathbf{Z}_{oc}(1,n_{str})\ ;\ \mathbf{Z}_{oc}(n_{str},1)\ \ \mathbf{Z}_{oc}(n_{str},n_{str})] \quad (1)$$

The iterative impedance of the TWA section is such that $Z_{in}$=$Z_{it}$ if the section is terminated on the same impedance $Z_{out}$=$Z_{it}$. If the section is symmetric (i.e. $\mathbf{Z}_{oc}(1,1)$ = $\mathbf{Z}_{oc}(n_{str},n_{str})$ and $\mathbf{Z}_{oc}(1,n_{str})$ = $\mathbf{Z}_{oc}(n_{str},1)$) it is given by the relation:

$$Z_{it} = (\mathbf{Z}_{oc}(1,1)^2 - \mathbf{Z}_{oc}(1,n_{str})^2)^{1/2} \quad (2)$$

From $\mathbf{Z}_a$, the corresponding scattering matrix $\mathbf{S}_a$ is derived for a characteristic impedance $Z_0$ = $Z_{it}$ or by approximation $Z_0$ = Re($Z_{it}$). A tap on the first and last strap can be used for the connection to the feeders to bring $Z_0$ close to a standard $Z_0$ value (as 50Ω: see the bottom of figure 1). We recall that the TWA section behaves as a lossy transmission line section in its passband (see ref. [1]).

### 1.3 Computation of the strap matrix Z

ANTITER II [3] computes the field distribution by a Fourier integral expansion in the medium in front to the antenna box and an orthogonal function expansion inside

* Corresponding author: a.messiaen@fz-juelich.de

the antenna box. The two expansions are matched at the box aperture to insure tangential fields continuity. First the contribution of each strap of the box to the field distribution is computed separately for a unit current flowing in the strap. As the field equations are linear the total field excited by the strap array is obtained by summing on the product of each contribution by its current complex amplitude. The active radiated power is obtained either by the flux of the Poynting's vector at the antenna aperture or by the e.m.f. method applied on the strap surface in the antenna box; The two computations are giving exactly the same result.

The impedance matrix $\mathbf{Z}=\mathbf{Y}^{-1}$ is computed by associating in pair the single strap contributions. The $(i,j)$ term of the $\mathbf{Z}$ matrix is obtained from the strap pair $(i,j)$ with even or odd excitation ($I_i = +I_j$ or $-I_j$). Applying the e.m.f. method to compute the corresponding complex power for the two excitations we have $2\mathbf{P}_{even}(i,j) = \int(E_i + E_j).(J_i + J_j)^*dS$ and $2\mathbf{P}_{odd}(i,j) = \int(E_i - E_j).(J_i - J_j)^*dS$. For $|I_i|$=1A we have also $2\mathbf{P}_{even}(i,j) = \mathbf{R}_{even}(i,j) + i\mathbf{X}_{even}(i,j) = \mathbf{Z}_{even}(i,j)$ and $2\mathbf{P}_{odd}(i,j) = \mathbf{R}_{odd}(i,j) + i\mathbf{X}_{odd}(i,j) = \mathbf{Z}_{odd}(i,j)$. Having computed the $\mathbf{Z}_{even}$ and $\mathbf{Z}_{odd}$ matrices then the impedance matrix $\mathbf{Z} = \mathbf{Y}^{-1}$ is obtained by the relation $\mathbf{Z} = (\mathbf{Z}_{even} - \mathbf{Z}_{odd})/2$. The figure below shows the normalized values $\mathrm{Re}(\mathbf{Z}_n(i,j))$ and $\mathrm{Im}(\mathbf{Z}_n(i,j))$ of $\mathbf{Z}$ for each value of $i$ as a function of $j$ for an $n_{str}$ = 8 TWA section. The impedance $\mathbf{Z}_n$ is expressed in Ω/m and has to be multiplied by the strap length $l_{str}$ to find $\mathbf{Z}$ (for $l_{str} << \lambda$).

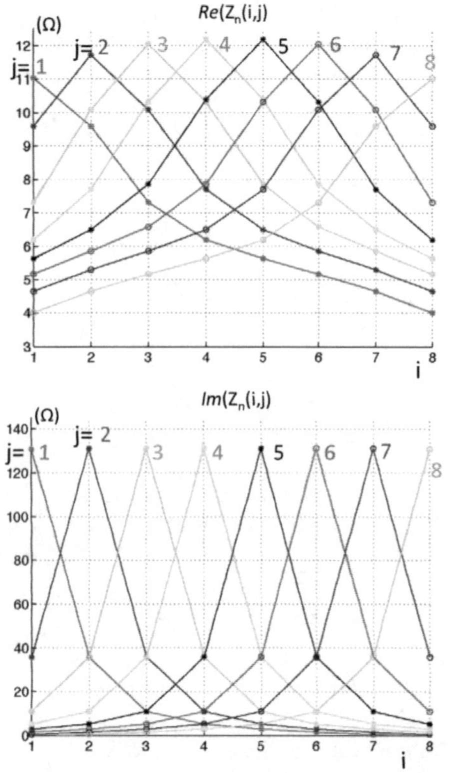


**Fig. 2.** Resistive and reactive parts of $\mathbf{Z_n}$ versus strap number for the conditions given in section 2.1 and the reference plasma profile.

### 1.4 Approximate relations and optimization of TWA

*Frequency band and iterative impedance*. The frequency band is mainly determined by the choice of the tuning capacitance $C$ and the mean values of strap self-inductance $L = \mathrm{Im}(<\mathbf{Z}(i,i)/\omega_o>)$ and mutual inductance to the adjacent straps $M_1 = \mathrm{Im}(<\mathbf{Z}(i,i\pm1)/\omega_o>)$. We take the same value of the capacitance $C = (L\omega_0^2)^{-1}$ for all the straps. Simple formulae are derived from the $n_{str} = 2$ case that remain approximately valid for any strap number:
(i) The midband frequency of the TWA is $f_0 = \omega_0/2\pi$ with $\omega_o = (LC)^{-0.5}$ and the useful frequency band ranges between $f_1 = (2\pi)^{-1}(L+M_1)^{-1/2}$ and $f_2 = (2\pi)^{-1}(L-M_1)^{-1/2}$.
(ii) The iterative impedance is given by:

$$Z_{it} = \omega_0(L/M_1)(L^2-M_1^2)^{0.5} \quad (3)$$

A large frequency band and a low $Z_{it}$ value requires a large ratio $M_1/L$ i.e. tightly magnetically coupled straps.
*Decay of strap voltage amplitude in the TWA section due to radiation*. A TWA section can sustain traveling wave propagation in both directions [2]. We consider a traveling wave propagating in one direction from the input to the output of the TWA section. The radiated power by the strap i is given by: $P_{str,i} = G_{aeff}|V_{str,i}|^2/2$ where $G_{aeff}$ is the array effective conductance. The incoming power along the TWA to the strap $i$ is given by: $P_{in,i} = |V_{str,i}|^2/(2Z_{it})$ while the outgoing power from strap $i$ to the next strap by: $P_{out,i} = P_{in,i+1} = P_{in,i} - P_{str,i}$. Therefore the radiated power decay along the TWA section can be computed as a function of the parameters $G_{aeff}$ and $Z_{it}$. We have $|V_{i+1}|^2 = |V_i|^2(1-G_{aeff}Z_{it}/2)^2$ and the total radiated power by the section is given by the following geometric progression:

$$P_{TWA}=(1/2)\Sigma_1^N G_{aeff}|V_{in}|^2(1-G_{aeff}Z_{it}/2)^{2(N-1)} \quad (4)$$

where $N = n_{str}$.This sum is equal to $P_{TWA,max} = (n_{str}/2) G_{aeff}|V_{in}|^2$ if all straps have same voltage amplitude $|V_{str,i}| = |V_{in}|$. This relation shows that the product $G_{aeff}Z_{it}$ has to be minimized to avoid a too large decay of the radiated power along the TWA section. The output power of the section is given by:

$$P_{out} = 0.5\ G_{aeff}|V_{out}|^2 = 0.5\ G_{aeff}|V_{in}|^2(1-G_{aeff}Z_{it}/2)^{2(N-1)} \quad (5)$$

This is the power to be recirculated by the feeding system.
*Resulting discussion for the optimization of the design Iof the TWA section*. $G_{aeff}$ has to be maximized while $Z_{it}$ minimized in order to maximize the radiated power of a TWA section for a given value of $|V_{in}|$.
$G_{aeff}$ strongly depends on the plasma profile in front of the TWA. Moreover it also depends on the TWA geometry mainly due to its dependency on the selected $k_{//MAX} = \Delta\Phi/S_z$ for which the radiated power spectrum $P_{rad}(k_{//})$ is maximum ([5], see also Fig. 4). $\Delta\Phi$ is the mean phase difference between adjacent straps and it is directly related to the choice of the frequency in the passband of the TWA section. At midband frequency $f_0$ we have $\Delta\Phi \cong 75°$.

The $Z_{it}$ value determines the power $P_{in}$ fed to the TWA section for an input strap voltage amplitude $|V_{in}|$ and acts on the voltage decay along the TWA section by the product $Z_{it}G_{aeff}$. $Z_{it}$ (see Eq.3) has to be chosen as low as possible to reduce the voltage amplitude and increase the efficiency of the TWA section. This means a high mutual coupling ratio $M_1/L$ and a small strap length $l_{str}$. Low $L$ requires straps with large width $2w_z$ and high $M_1$ requires small values of mid-strap inter-distance $S_z$ ($S_z$ and $w_z$ are defined in Fig. 3). A compromise has to be found between the contradictory request of small $S_z$ with: (i) good voltage standoff between adjacent straps, (ii) large strap width to decrease $L$ and (iii) sufficiently small $|k_{//MAX}| = \Delta\Phi/S_z$ value to maximize the coupling. A short strap length is also incompatible with good coupling. This is the reason why we propose, for increasing the effective strap length seen by the plasma, to cut a long strap in several grounded short straps having their currents in phase (see Fig. 6).

# 2 Electromagnetic design for DEMO

## 2.1 Modeling conditions

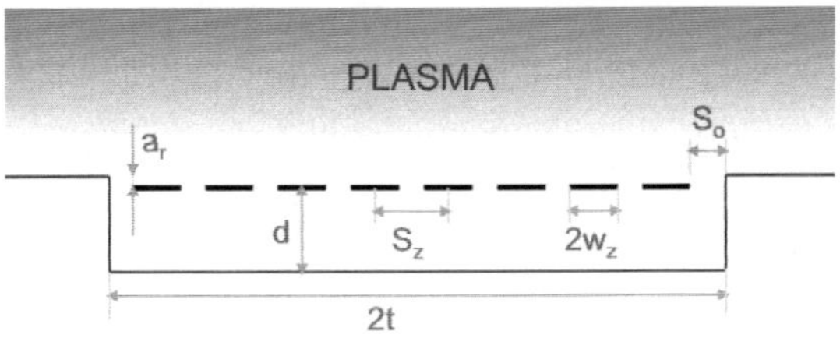


**Fig. 3.** Crossection of the TWA section in front of plasma.

*Geometrical parameters*. The geometrical parameters indicated on Figure 3 have been chosen as a compromise taking into account the preceding discussion ($a_r$ = 1.5cm, $d$ = 20cm, $S_z$ = 25cm, $S_o$ = 42.5cm, $w_z$ =10cm, strap length $l_{str}$ = 20cm, number of strap per section $n_{str}$ = 8).

*Plasma profile*. The profiles choosen are the ITER reference profile for low coupling condition "2010 Low" (considered as the most pessimistic one [4] (see profiles in [5] ) and the same profile shifted by 5 or 10 cm away from the antenna in order to decrease further the coupling. The distances antenna-LCMFS (last closed magnetic flux surface) are then respectively 23, 28 and 33cm. We choose a DT($He^3$) plasma with $B_{edge}$≈3.5T at $f$=50MHz. $He^3$ minority heating is one good single pass absorption scenario, characteristic that is assumed for the boundary condition in the plasma bulk.

*Resonant ring feeding network*. The system is shown in Fig. 1 and it is used to recirculate the output power of the TWA section. It has many advantages which are discussed in reference [1]. The modeling is made by connecting the $\mathbf{S}_a$ matrix of the TWA section to the 4X4 matrix of the hybrid junction $\mathbf{S}_{HYB}$. The characteristic impedance of all the system is taken equal to $Z_{it}$ (its value can be adjusted to e.g. 50Ω by a tap connection to the first and last straps as shown in Fig. 1). The complete network equations are solved for a given excitation by the generator. The line stretcher position is adjusted to obtain the ring resonance [1] (i.e. the phase difference of the traveling wave after one complete turn in the ring equal to $p2\pi$ with $p$ = 1, 2, 3...).

## 2.2 Main modeling results

Figure 4 shows (i) the mean phase difference <ΔΦ> of the current of adjacent straps versus frequency of the TWA section tuned at 50MHz and (ii) some corresponding radiated power $P_{rad}(k_{//})$ spectra in the bandpass for 42<$f$<56MHz and the reference plasma profile.

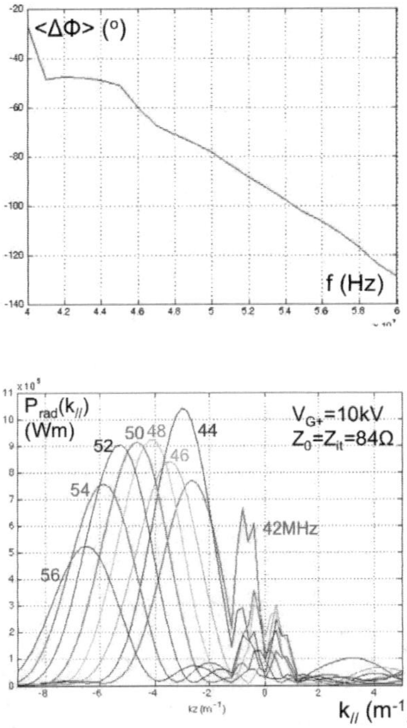


**Fig. 4.** <ΔΦ> and $P_{rad}(k_{//})$ versus frequency of the considered TWA section.

The most excited $k_{//}$ is given by $k_{//MAX}$ = $\Delta\Phi/S_z$ if the strap current amplitude $|I_{str}|$ and the inter-strap current phasing ΔΦ are the same [5]. This relation is still valid for the TWA with decaying $|I_{str}|$ but with broader peak. When $f$ varies from 42 to 56MHz the $|k_{//MAX}|$ value increases from 2.6 to 6.5m$^{-1}$. Note the larger excitation of coaxial modes (occurring for $|k_{//}|<k_0$) at low $|k_{//MAX}|$. The decay of strap current and voltage amplitude along the TWA section is shown in Figure 5 at midband and for the reference and both removed plasma profiles.

The performances of the TWA section at midband are summarized in Table 1 for the 3 plasma profiles and for a maximum strap voltage amplitude of 15kV. The TWA section couples to the plasma 930kW but this value is reduced to 430kW when the TWA distance to the plasma is increased by 10cm. Note that the interstrap voltage is larger and amount 18kV. Indeed this voltage $\Delta V=|V_{str,i}$ - $V_{str,i+1}|$ exceeds $|V_{str,i}|$ when 60°<ΔΦ<300° if $|V_{str,i}|$ ≅ $|V_{str,i+1}|$ and at 50MHz we have ΔΦ≈80°

 DOI: 10.1051/epjconf/201715703033

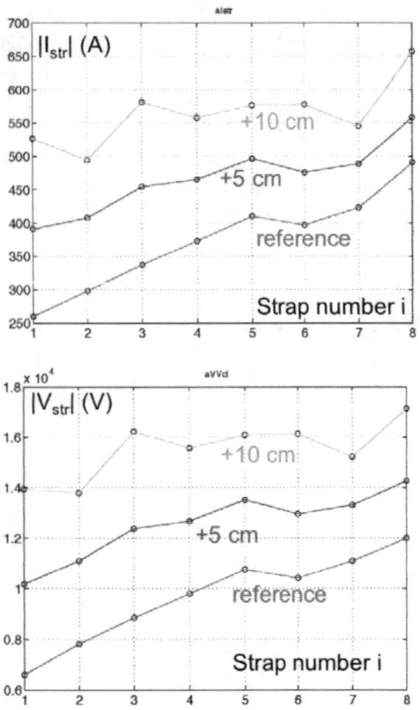


**Fig. 5.** Current and voltage amplitude of the straps for the three considered plasma profiles and a generator forward voltage $V_{G+}$=10kV.

**Table 1.** Voltages and power of the TWA section (as defined in Fig. 1) for the three considered plasma profiles.

| Antenna plasma distance | $P_G$=$P_{rad}$ (MW) | $P_{in}$ (MW) | $P_{out}$ (MW) | $V_{in}$ (kV) | $V_{out}$ (kV) | $V_{G+}$ (kV) | $V_{G+}$ 50Ω (kV) |
|---|---|---|---|---|---|---|---|
| +10cm | 0.428 | 1.18 | 0.75 | 15 | 12.2 | 8.75 | 6.58 |
| +5cm | 0.636 | 1.25 | 0.62 | 15 | 10.7 | 10.52 | 7.98 |
| reference | 0.930 | 1.30 | 0.37 | 15 | 8.24 | 12.5 | 9.65 |

## 2.3 First electromagnetic design for DEMO

18 sections of quadruple TWA arrays are considered. Each section is constituted of two superposed and centrally grounded double tuned straps (see figure 6). The current are flowing in phase in each row of the four grounded straps: the effective strap length is then four times the one of the single strap $l_{str}$ considered in section 2.1. Therefore the total radiated power is 18x4 the one computed for the single array. The expected total power capability for a maximum strap voltage of 15kV is summarized in Table 2. We show for comparison (i) the larger expected radiated power with all individually fed straps and (ii) the power reduction due to vertical septa between the straps.

The main conclusions are:

• A set of 18 sections has a power capability of 67MW at 50MHz and exceeding 60MW in the frequency band 44-52MHz for a maximum strap voltage of 15kV and a maximum inter-strap voltage of 18kV and for the reference low coupling plasma profile of ITER. The power capability of 67MW is reduced to 45 or 31MW if the distance antenna-plasma is increased by 5 or 10cm, i.e. distance antenna-LCMFS of 28 and 33cm.

**Table 2.**Total rf power coupled to DEMO by the proposed TWA system for the three considered plasma profiles.

| Antenna plasma distance | Resonant ring feed: $|I_{str}|$ decay | Individual strap feed: no $|I_{str}|$ decay | Individual strap feed: septa between straps |
|---|---|---|---|
| +10cm | 30.8 MW | 53.2 MW | 27.7 MW |
| +5cm | 45.4 MW | 77.5 MW | 43.1 MW |
| reference | 66.4 MW | 127.1 MW | 73.3 MW |

• The peak of maximum radiation varies from $|k_{//MAX}|$ = 2.6 to 6.5m$^{-1}$ when changing the frequency from 42 to 56MHz. For a maximum strap voltage of 15kV the radiated power exceeds 39MW in this frequency band and exceeds the value of 59MW from 42 to 52 MHz. This power capability could be further improved via edge gas puffing [6]. As a comparison, ANTITER predicts for the tight 24 straps array of the present ITER design facing the same reference plasma profile at 53MHz a power capability ranging from 9MW (for $k_{//MAX}$ = 7.5m$^{-1}$) to 15MW (for $k_{//MAX}$ = 3.1m$^{-1}$) for a strap voltage amplitude of 41kV.

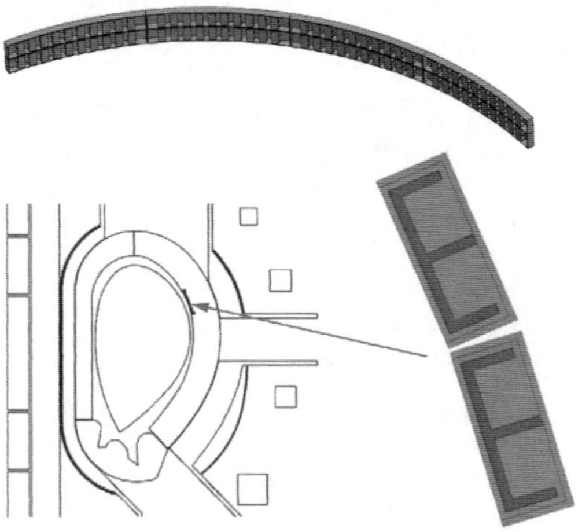

**Fig. 6.** Cross-section of the proposed TWA antenna system for DEMO. The sides of the centrally grounded double straps are the tuning capacitances. On top an overview of four segments.

This work has been carried out within the framework of the EUROfusion Consortium and has received funding from the Euratom research and training programme 2014-2018 under grant agreement No 633053. The views and opinions expressed herein do not necessarily reflect those of the European Commission.

## References

1. R. Ragona and A. Messiaen, Nucl. Fusion **56** (2016) 076009.
2. R. Ragona and A. Messiaen, this conference.
3. A. Messiaen et al, Nucl. Fusion **50** (2010) 025026.
4. S. Carpentier and R. A. Pitts Report ITER_D_33y59M_v2,3 (April 2010).
5. A. Messiaen and R. Ragona, European Conference Abstracts (ECA) vol. **40A** (2016), Leuven, Belgium.
6. P. Jacquet et al, Nucl. Fusion **56** (2016) 046001

DOI: 10.1051/epjconf/201715703034

# Analysis of the ASDEX Upgrade 3-strap antenna with TOPICA code: Curved vs. flat 3D geometry

*Daniele* Milanesio[1,*], *Riccardo* Maggiora[1], *Volodymyr* Bobkov[2], ASDEX Upgrade Team, and EUROfusion MST1 Team

[1] Politecnico di Torino, Dipartimento di Elettronica e Telecomunicazioni (DET), Torino, Italy
[2] Max-Planck-Institut für Plasmaphysik, Boltzmannstr. 2, 85748 Garching, Germany

**Abstract.** Before the 2015/2016 experimental campaign, the ASDEX Upgrade (AUG) 2-strap ICRF antennas with tungsten-coated limiters were replaced by 3-strap antennas. The main goal of the 3-strap launcher was to reduce the release of tungsten (W) in order to improve the ICRF operation, which appeared to be troublesome after the all-W wall installation [1]. In this paper, we analyse the behaviour of the 3-strap antenna with the help of TOPICA code [2], a numerical tool able to take into account a realistic antenna geometry and an accurate plasma description. By loading an experimental plasma profile from the AUG campaign, we characterize the antenna both in terms of input parameters and of radiated fields. In particular, we compare TOPICA results obtained with a simplified 3D flat model adopted during the design phase with the exact 3D curved geometry installed on the AUG experiment. In particular, the curved model predicts a lower coupling to plasma and higher RF electric fields with slightly different distribution in front of the launcher. The capability to include a fully 3D curved model is of great importance to correctly account for all geometrical effects on the antenna performances. The advantages and disadvantages of both geometrical representations are eventually outlined, trying to estimate how the curvature of the antenna can affect code predictions. Comparisons between measured experimental results and simulated ones are presented in [8].

## 1 Introduction and objectives

TOPICA [2] is a numerical tool realized for the 3D/1D simulation of Ion Cyclotron (IC) antennas, i.e. accounting for realistic 3D geometry with an accurate 1D plasma model. While referring the interested reader to [2] for a more detailed analysis of TOPICA formulation, we would like to enlighten here only the features that have been exploited in this paper. In particular, the code permits to compute both the input parameters of an IC launcher, and hence the power coupled to plasma, and the electric field distribution everywhere inside the antenna enclosure and in the plasma column. Moreover, with the help of post-processing tools it is possible to determine the local electric field value thought to be the most important element for driving Radio Frequency (RF) sheaths, and to directly verify the impact of the geometrical accuracy of the antenna on the field values themselves. From the plasma side, 1D FELICE code [3] has been adopted, affording density and temperature profiles, and finite Larmor radius effects.

Provided the mentioned tool, the goal of this paper is to compare TOPICA results obtained with a simplified 3D flat model of the new 3-strap AUG antenna (adopted during the design phase) with the real 3D curved geometry installed on the experiment. Both the power transferred to plasma and the local electric fields will be analyzed in order to show the differences between the two simulated versions of the same launcher. A set of nine artificial plasma profiles will be adopted for this task, together with an experimentally measured one from 2015 campaign.

## 2 Antenna models and plasma profiles

As already mentioned, before the 2015/2016 experimental campaign, the AUG 2-strap ICRF antennas with tungsten-coated limiters were replaced by 3-strap antennas, whose main goal was to reduce the release of tungsten in order to improve the ICRF operation [1]. Most of the 3-strap antenna design was actually performed with a simplified 3D flat model with the help of HFSS suite [4] and TOPICA; Figure 1 reports the TOPICA version of the 3D flat launcher.

The real 3D curved model has been tailored for an electromagnetic simulation code only quite recently; it is worth mentioning that this model was directly imported from the technical drawings and, therefore, all the geometrical details of the antenna are consistent with the launcher actually installed on the machine. Figure 2 reports the 3D curved version of the launcher simulated with TOPICA.

For sake of comparison, Figure 3 also reports the real antenna model.

* Corresponding author: daniele.milanesio@polito.it

 DOI: 10.1051/epjconf/201715703034

**Fig. 1.** 3D flat model of the 3-strap AUG antenna.

**Fig. 2.** 3D curved model of the 3-strap AUG antenna.

**Fig. 3.** Real antenna model installed on AUG

The reader may immediately notice that not only the introduced curvature represents quite a difference between the two models, but also several other geometrical details appear as simplified in the flat antenna geometry, in particular in the plasma exposed antenna region. Despite this, we tried to reduce any source of difference due to TOPICA between the two groups of simulations: we imposed a similar mesh density on both antennas (about 4cm resolution), we loaded the same density profiles at the same distance from the radiating elements (about 4.7cm), i.e. keeping constant the strap-cutoff distance in both cases, we computed the electric field distribution on the same surface (located about 3mm in front of the limiters). The different geometrical accuracy is clearly witnessed by the mesh requirements: the flat antenna model is discretized by approximately 65k unknowns while the curved one requires more than 250k unknowns to account for all geometrical details. This translates in heavier computational requirements too, namely from 576 CPUs to 900 CPUs for about 5 hours, on the new Marconi High Performance Computing (HPC) system at CINECA [5].

Nine "artificial" density profiles were loaded for this comparison, as documented in Figure 4. These profiles sets correspond approximately to the cases for various gas injection sources described in [6]. The same temperature profile was loaded for both ions and electrons in all cases. In addition to them, a measured profile, namely shot 31515, was also included into the simulation database.

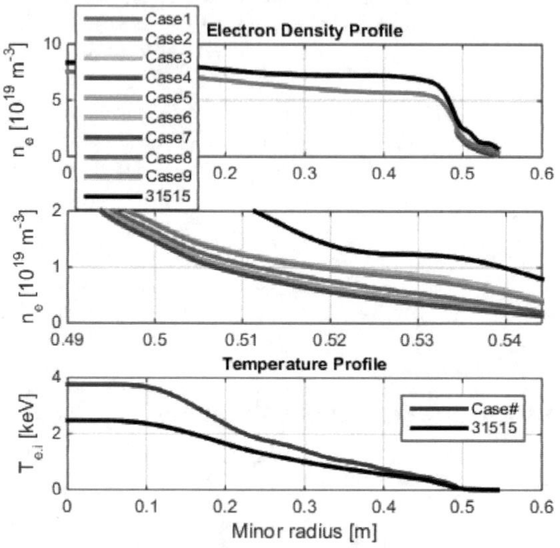


**Fig. 4.** Loaded plasma profiles: electron density (top) with zoomed view of the edge density profile (middle) and ion/electron temperature profile (bottom).

## 3 Analysis of the results

Provided the aforementioned loadings, the two geometries are first compared in Figure 5 in terms of power transferred to plasma, assuming infinite coaxial lines connected to the two antenna ports withstanding a maximum voltage of 30kV and imposing $0\pi0$ phasing at the straps. The working frequency is 30MHz. Figure 5 reports the same relative behavior for both geometries but the difference in loading is approximately of a factor of 2. For both geometries the measured plasma profile, i.e. shot number 31515, generates a higher coupled power which is basically due to the slightly higher edge density with respect to the other "artificial" cases.

The reader should remind that, while TOPICA can handle a fully curved geometry, the coupled plasma code, namely FELICE, is a 1D model. The two codes are joined together with the help of a procedure named

DOI: 10.1051/epjconf/201715703034

stretching in [7, App. A], which allows to flatten the curved interface between the antenna and the plasma. Thanks to this procedure, the simulated results obtained with the curved antenna geometry better match the experimental evidence. A further improvement could be to directly couple a 3D plasma model.

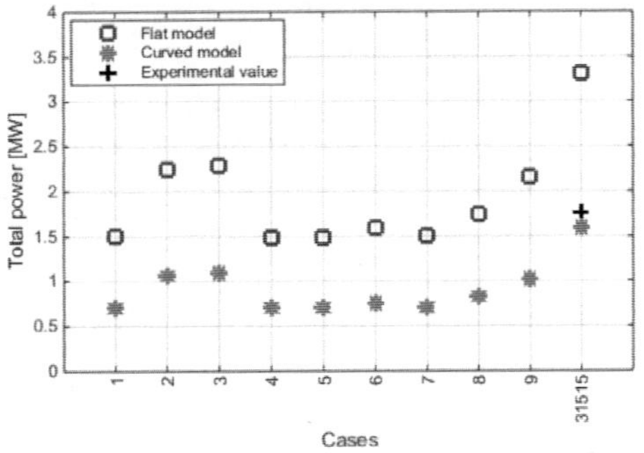


**Fig. 5.** Power transferred to plasma for different plasma cases with flat (blue squares) and curved (red stars) geometries. The black cross reports the experimental measure.

To conclude the comparison between the flat and curved launchers, the parallel electric field is also calculated. $0\pi0$ phasing is imposed at the straps with 1MW of total power transferred to plasma; besides, to minimize the parallel electric field in front of the limiters, a ratio of two is imposed between the power radiated by the central strap and the one radiated by the sum of the two outer ones. Figure 6 and Figure 7 reports the real part of the electric field parallel to the magnetic field lines (11° tilted respect to the toroidal axis, shown on the picture with a dashed line) for the flat and curved models respectively, loading Case3 plasma profile.

The overall pattern of the parallel electric field distribution is not that different for both geometries, even though one should notice that the curved model shows a more pronounced field distribution on the central part of the antenna.

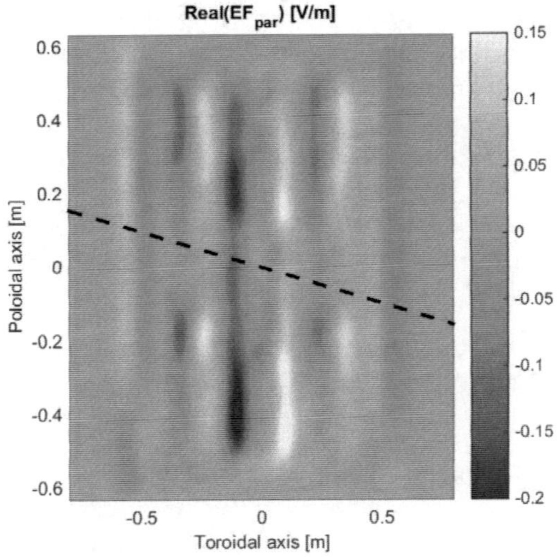


**Fig. 6.** Parallel electric field computed with $0\pi0$ input phasing for the flat antenna model, assuming 1MW of delivered power.

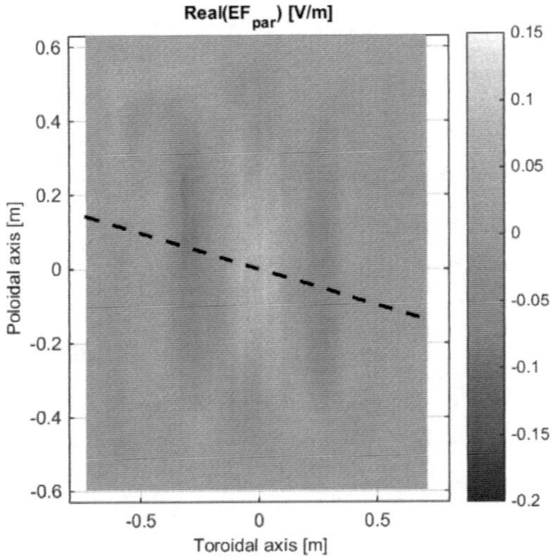


**Fig. 7.** Parallel electric field computed with $0\pi0$ input phasing for the curved antenna model, assuming 1MW of delivered power.

As fully detailed in [9], recent experiments proved that the quantities relevant for antenna-plasma interactions cannot be described by a single value for a given magnetic field line. Therefore, instead of computing the integral of the parallel electric field along the magnetic field lines crossing the antenna, we preferred to follow the approach outlined in [9], namely to determine the average electric field in front of the limiters, i.e. the most protruding conducting structures. The parallel component of the electric field is still taken into account, but for each poloidal coordinate in front of both limiters, its value is averaged along the toroidal extension of the limiter itself. Figure 8 and Figure 9 show the so obtained curves for the antenna left and right limiters, respectively, loading three plasma profiles of the database.

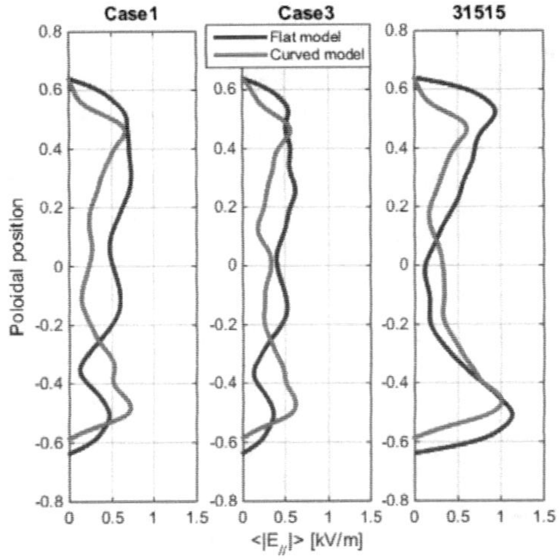

**Fig. 8.** Averaged electric field (parallel component) computed in front of the left limiter (seen from plasma) with 0π0 input phasing for both geometries, assuming 1MW of delivered power.

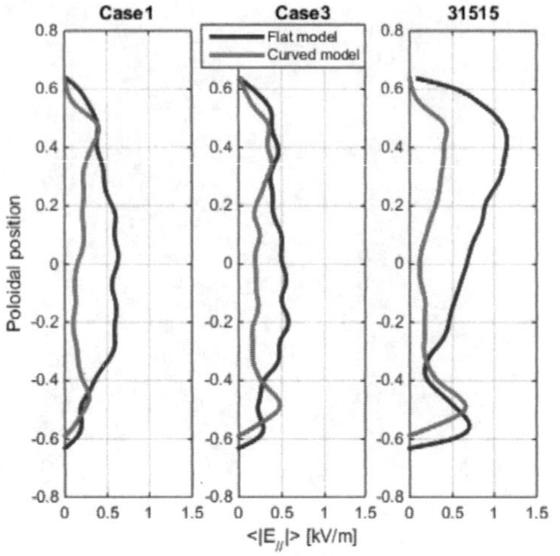


**Fig. 9.** Averaged electric field (parallel component) computed in front of the right limiter (seen from plasma) with 0π0 input phasing for both geometries, assuming 1MW of delivered power.

Both the value and the position of the top and bottom peaks and, more generally, the poloidal behavior are rather different for the two geometries. At least for the maxima poloidal position, one should take into account that the flat model can be seen as the results of a stretching of the curved one; as a consequence, it is not surprising to observe a shift of the maxima towards the poloidal extremities of the plot. The reader should notice that the averaged parallel field shares the same qualitative pattern on both limiters in case of the curved geometry, while a strong asymmetry can be observed when looking at the flat launcher. This may be partially explained by the fact that, even though the total coupled power is identical, the ratio between the central and the outer straps is slightly different for the two geometries. Eventually, one should also observe that, for the setup adopted for this last analysis (the electric fields are computed few millimeters from the limiter front surface), the behavior of the electric fields reported in Figure 8 and Figure 9 follows (and it is indeed determined) the electric current distribution flowing on the side limiters.

We refer the interested reader to [8] for the analysis of the antenna behavior during the experiment and the comparison with numerical predictions.

## 4 Conclusions

The detailed analysis carried out with the help of TOPICA code indicates that a flat model can be a reasonably good approximation of a real curved launcher even though some differences still remain. It is unfortunately impossible for the time being to identify general rules to quantify how the adoption of a curved model instead of a flat one influences the simulated results except for a general reduction of the loading and a significant modification in the poloidal behavior of the RF sheaths. It is also clear that the differences between the simulated results obtained with a curved model instead of a flat one strongly depends on the specific antenna geometry and must be evaluated case by case.

This said, we firmly believe that a realistic curved antenna geometry should always be analyzed during any design process, especially if one is interested in more localized phenomena, such as RF sheaths, where even the smallest details can have an influence on the final result.

This work has been carried out within the framework of the EUROfusion Consortium and has received funding from the Euratom research and training programme 2014-2018 under grant agreement No 633053. The views and opinions expressed herein do not necessarily reflect those of the European Commission. The authors gratefully acknowledge the contribution of W. Zhang concerning the profile generation. The complete list of all MST contributors can be find in "Overview of progress in European Medium Sized Tokamaks towards an integrated plasma-edge/wall solution" by H. Meyer et al., to be published in Nuclear Fusion Special issue: Overview and Summary Reports from the 26th Fusion Energy Conference (Kyoto, Japan, 17-22 October 2016).

## References

1. V. Bobkov et al., Nucl. Fus. **56**, 084001 (2016)
2. D. Milanesio et al., Nucl. Fus. **49**, 115019 (2009)
3. M. Brambilla, Plasma Phys. Control. Fusion **35**, 41-62 (1993)
4. V. Bobkov et al., Nucl. Fus. **53**, 093018 (2013)
5. www.hoc.cineca.it/hardware/marconi
6. W. Zhang, Nucl. Fusion **56**, 036007 (2016)
7. V. Lancellotti et al., Nucl. Fus. **46**, S476-S499 (2006)
8. V. Bobkov et al., Characterization of 3-strap antennas in ASDEX Upgrade, this conference
9. V. Bobkov et al., Plasma Phys. Control. Fusion **59**, 014022 (2017)

# ICRH system performance during ITER-Like Wall operations at JET and the outlook for DT campaign

*Igor* Monakhov[1,*], *Trevor* Blackman[1], *Pierre* Dumortier[1,2], *Frederic* Durodié[2], *Philippe* Jacquet[1], *Ernesto* Lerche[1,2], *Craig* Noble[1], and JET Contributors[3]

[1]CCFE, Culham Science Centre, Abingdon, OX14 3DB, United Kingdom
[2]LPP-ERM/KMS, TEC partner, Brussels, Belgium
[3]See the author list of "Overview of the JET results in support to ITER" by X. Litaudon et al. to be published in Nuclear Fusion Special issue: overview and summary reports from the 26th Fusion Energy Conference (Kyoto, Japan, 17-22 October 2016)

**Abstract.** Performance of JET ICRH system since installation of the metal ITER-Like Wall (ILW) has been assessed statistically. The data demonstrate steady increase of the RF power coupled to plasmas over recent years with the maximum pulse-average and peak values exceeding respectively 6MW and 8MW in 2016. Analysis and extrapolation of power capabilities of conventional JET ICRH antennas is provided and key performance-limiting factors are discussed. The RF plant operational frequency options are presented highlighting the issues of efficient ICRH application within a foreseeable range of DT plasma scenarios.

## 1 Introduction

In 2011, the carbon first wall at JET was replaced with a metal ITER-Like Wall (ILW). Reliable injection of high RF power into ELMy plasmas is essential for control of heavy impurities in high-performance JET discharge scenarios [1]. This requirement fully applies to planned DT experiments which, in addition, will benefit from high-power ICRH for boosting the fusion reaction [2].

## 2 Status of the RF plant at JET

The RF plant at JET (Fig. 1) comprises four conventional 4-strap antennas energised in pairs by two different power-splitting systems [3,4] and one 8-strap main-port ITER-Like Antenna (ILA) [5]. The ELM-tolerant 3dB hybrid system (antennas A&B) is operational since 2006. The External Conjugate-T (ECT) system (antennas C&D) was installed in 2009 to provide ELM-tolerant operations at four selected frequencies [4]; in 2016, the number of commissioned ECT frequencies was increased to eleven covering the 25-57MHz band. It has to be noted that the existing ECT transmission line configuration does not allow achieving the ELM-tolerance at frequencies close to 25MHz, 29MHz and 37MHz. The ILA was first introduced in 2008; however it remained inactive during initial period of ILW operations due to a capacitor failure. In 2015, the system was restored and successfully recommissioned at several frequencies within the 29-51MHz band demonstrating adequate tolerance to moderate ELMs. Presently, one half of the ILA is under investigation for emerged capacitor control problems which may affect its availability in future.

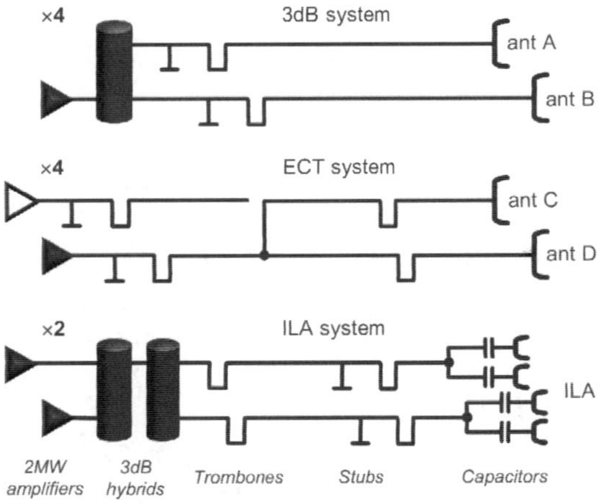


**Fig. 1.** Schematic representation of three independent RF systems used during ILW ICRH operations at JET.

## 3 Total RF power coupled to ILW plasmas

Statistical data (Fig. 2) demonstrate the steady increase of maximum and average total RF power coupled to plasma during JET pulses since the ILW installation. This encouraging trend can be attributed to several factors including the re-introduction of the ILA in 2015, development of gas injection scenarios for coupling improvement [6], achieving statistically higher maximum antenna voltages (see Sect. 3.1), growing operator experience and ongoing efforts to improve diagnostics, tuning and maintenance of the RF plant sub-systems. In 2016, the peak and pulse-average total

* Corresponding author: igor.monakhov@ukaea.uk

coupled power reached respectively 8MW and 6MW; notably, comparable results were achieved in low-, medium- and high-frequency bands. Due to poor coupling, high-power operations at low frequencies remain challenging for the conventional JET antennas (see Sect. 3.2) and availability of the ILA is essential for delivering high RF power to plasmas at f<35MHz.

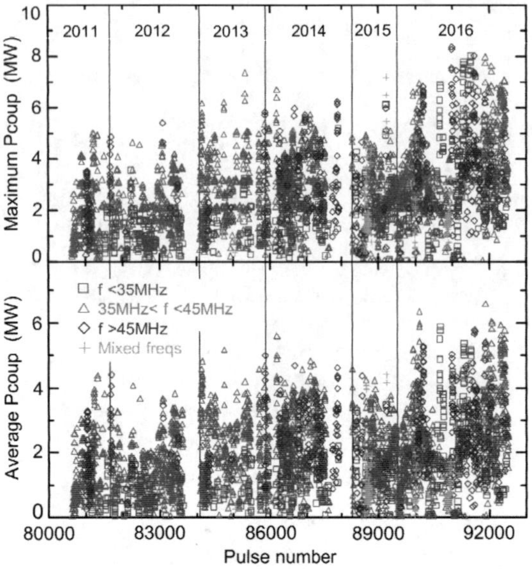


**Fig. 2.** The maximum (top) and average (bottom) total RF power coupled to plasma during JET ILW pulses in different frequency bands.

## 4 Capabilities of conventional antennas

### 4.1 Statistics for performance-limiting factors

Technically, power limitations of conventional ICRF antennas at JET can be attributed to two principal factors: antenna & transmission line arcing at high voltages and performance issues and fault-free power capabilities of generators feeding the antenna straps.

Figure 3 provides statistics for the maximum voltages $V_{strap}^{max}$ reached on the antenna straps during JET pulses. The data indicates that higher voltages have been routinely achieved since the ILW installation. Although open to other interpretations, the increase can be related to improved electrical strength of the RF system due to ongoing transmission line refurbishments and less dusty carbon-free ILW vacuum environment. The operational $V_{strap}^{max}$ limit of 30-35kV is rather fuzzy and it is subject not only to variable physical thresholds of arcing but also to a number of practical compromises (conditioning/deconditioning, probability/confidence, risk/gain etc.).

Statistical assessment of the RF generator performance is given on Fig. 4. It clearly shows that even in presence of the power-splitting 3dB & ECT systems, the 2MW generators typically remain loaded only partially, i.e. hardware upgrades for higher nominal maximum power output are not needed. At the same time, improving reliability and fault-free performance of the existing generators, especially at the output power above 1MW, is necessary for enhancing coupled power capabilities of the conventional antennas. In preparation for the DT campaign, a modern FPGA-based data collection system is being installed at JET which will allow in-depth diagnostics of the RF amplifier stages and facilitate their tuning, optimisation and fault-finding.

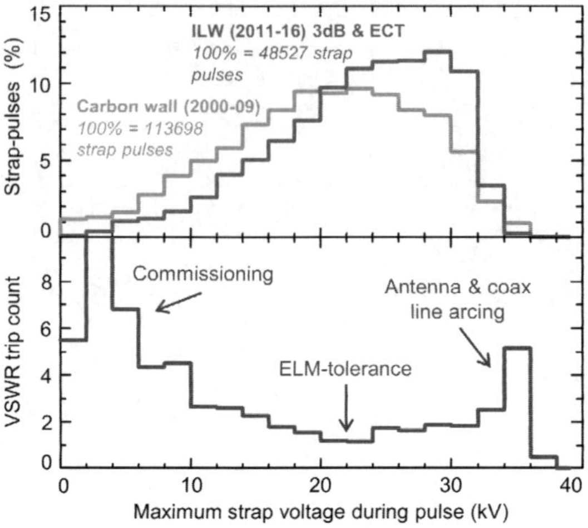


**Fig. 3.** Percentage of antenna strap pulses reaching given maximum voltage during the carbon-wall and ILW operations (top) and the average number of high-VSWR protection trips during a ILW pulse per the RF generator feeding the strap at given maximum voltage (bottom).

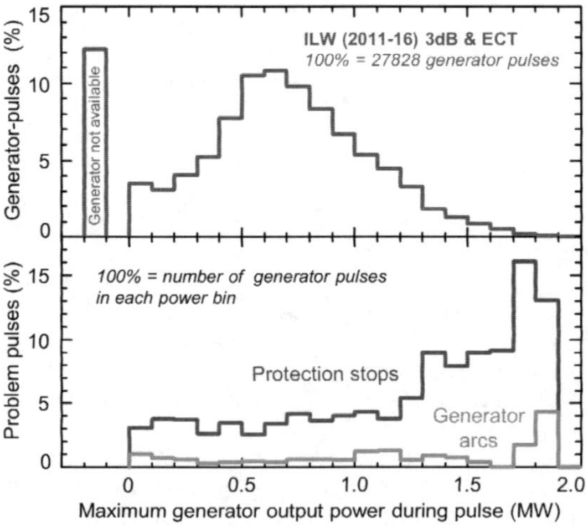


**Fig. 4.** Percentage of generator pulses reaching given output power during the ILW operations (top) and percentage of problematic generator pulses in each power bracket (bottom).

### 4.2 Actual & feasible maximum coupled power

The power characteristics of conventional ICRH antennas energised by the 3dB & ECT systems during the ILW pulses are illustrated by Fig. 5. An assessment of feasible maximum power during same pulses is also given in assumption of perfect RF plant control and

DOI: 10.1051/epjconf/201715703035

performance within realistic hardware limits. Note that, in practice, the maximum voltages $V_{strap}^{max}$ and coupling resistances $R_{strap}$ of individual antenna straps are different and it is more convenient to use global parameters of strap-equalised maximum voltage $V_{eq}^{max}$ and strap-averaged coupling resistance $R_{aver}$ such that:

$$P_{coup} = \sum P_{strap} = \frac{1}{2Z_0^2}\sum \left(V_{strap}^{max}\right)^2 R_{strap} \equiv$$

$$\equiv \frac{\left(V_{eq}^{max}\right)^2}{2Z_0^2}\sum R_{strap} = \frac{n\left(V_{eq}^{max}\right)^2}{2Z_0^2} R_{aver} \qquad (1)$$

where $n$=16 is the total number of straps in the conventional antennas and $Z_0$=30 Ohm is the characteristic impedance of transmission lines at JET.

Figure 5 demonstrates that both the actual (<7MW) and feasible (4-10MW) maximum power coupled by the antennas strongly depends on specific discharge conditions which define an order of magnitude spread of $R_{aver}$ during the ILW operations. For this reason, giving universal predictions of antenna power capabilities for the DT campaign is extremely challenging and optimisation of discharge scenarios with respect to antenna-plasma coupling will remain as important as the RF plant optimisation.

The nature of antenna power limitations largely depends on operational frequency: at f<35MHz, the coupling resistance is statistically noticeably lower than in the mid- and high-frequency bands. Thus, at low frequencies, the maximum voltage constraints typically dominate over generator performance issues. Antenna coupling and power capabilities in the mid- and high-frequency bands are similar while the generator issues play more significant role than the $V_{strap}^{max}$ limitations.

The gap between data points representing the actual and feasible maximum power during the ILW pulses (Fig. 5) indicates that the conventional ICRH antennas at JET have not achieved their full potential yet. Together with hardware performance and reliability improvements, intelligent real-time power balance control between the individual antenna straps and generators is required for bridging this gap. The necessity for such a control can be explained as follows: the limits of fault-free operations and protection thresholds are individual to the 16 antenna straps and 8 generators; the existing real-time power control software responds to changing plasma conditions by modifying the generator global output power request, however it is not capable of automatically adjusting the power balance between the generators during a pulse such that the straps voltages remain optimally distributed within the antennas while the generators stay within their individual operational limits. Consequently, attempts of the ICRH performance maximisation often result in the power being clamped by protection systems or, worse, with the RF pulse abnormally terminated due to issues with a single weakest component while the rest of the RF plant has not reached its limits. Inadequate balance optimisation also explains the significant difference between the maximum voltages achieved on individual antenna straps (Fig. 3) and the strap-equalised maximum voltages (Fig. 5) characterising the plant as a whole. The existing situation requires from the RF plant operators to make several optimisation pulses and to leave substantial fail-safe confidence margins while reaching the maximum coupled power. A major upgrade of real-time RF power control software is presently underway at JET which will address the aforementioned issues.

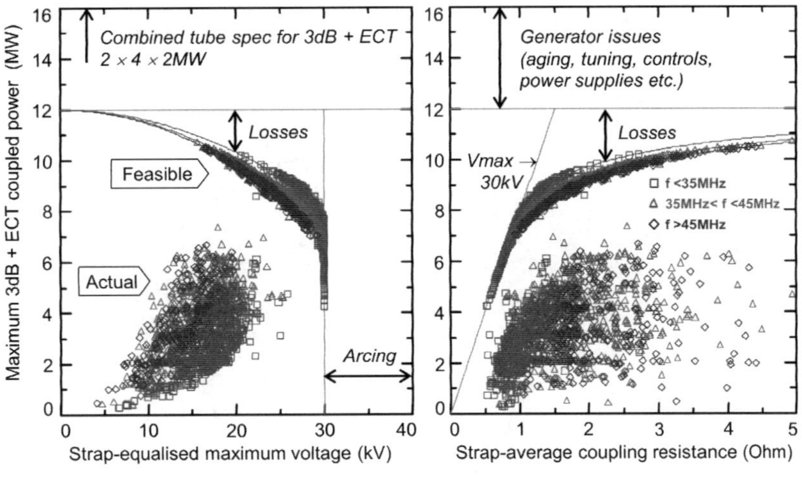


**Fig. 5.** Dependencies of actual and feasible maximum RF powers coupled to the ILW plasmas during combined operations of the 3dB and ECT systems on the strap-equalised maximum voltage (left) and on the strap-average coupling resistance (right). Each symbol represents a JET pulse with all the 16 straps of conventional antennas operational; the values of the maximum voltage and coupling resistance correspond to the moment of time during a pulse when the total power coupled by the 3dB&ECT systems reaches the maximum. Feasible power is assessed on the basis of coupling resistances of individual straps measured during the actual pulses and assuming perfect RF plant performance within 30kV maximum strap voltage limit and 75% power capability of each generator.

## 5 Operational frequency options and implications for DT ICRH scenarios

Efficient ICRH application in a range of possible DT scenarios with central toroidal magnetic field within $B_T$=2.7-4.0T requires careful consideration of the RF plant operational frequency options. The frequencies and corresponding coupled power capabilities of individual RF systems are subject to a number of constraints related to transmission line configuration, matching element variation and antenna RF characteristics. The ICRH-assisted tungsten control and enhancement of DT fusion reaction rely on different ICRH mechanisms which may require operations at different frequencies. Additional difficulties arise from undesirable presence of ion cyclotron resonances on the inner or outer vessel walls at certain frequencies; such potentially harmful conditions are avoided thanks to operational instructions presently adopted at JET.

Figure 6 summarises the available frequency options for the 3dB, ECT and ILA systems and shows corresponding $B_T$ compatibility windows for the RF power deposition in central plasma regions (specifically, in the regions where relevant ion cyclotron resonances are within $-5cm<R-R_0<+15cm$). Note that the system maximum powers refer to all kind of ILW plasmas in 2011-16, i.e. the data are not directly relevant to ELMy high-performance DT discharges. At the same time, no changes to the commissioned frequencies of the 3dB & ECT systems are expected unless the transmission and matching hardware is modified.

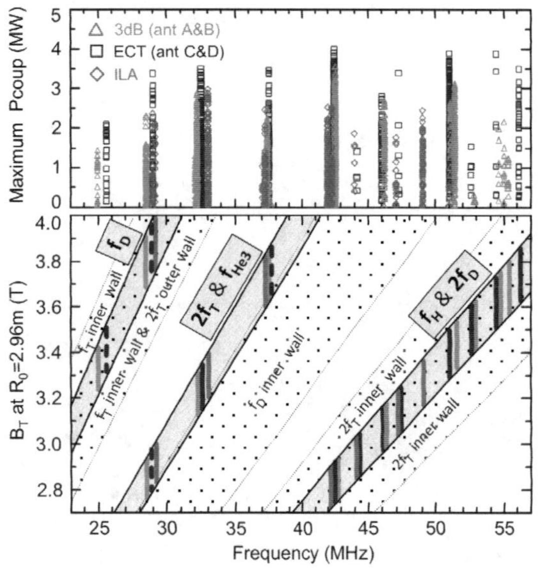


**Fig. 6.** Maximum power coupled to ILW plasmas by individual RF systems at the commissioned frequencies (top) and toroidal magnetic field induction at the JET vessel centre required for efficient ICRH application during DT experiments depending on the operational frequency (bottom). Shaded areas indicate undesirable conditions when parasitic ion cyclotron resonances will be present on the inner or outer vessel walls.

On the basis of currently available information, the outlook for applicability of main ICRH scenarios during the DT experiments at JET is the following:

- $f=f_H=2f_D$ – good coverage of the $B_T$ range; high power expectations regardless of the ILA availability; unavoidable inner wall $f=2f_T$ resonance
- $f=2f_T=f_{He3}$ – reduced coverage of the $B_T$ range; high power operations require the ILA; no major issues with the wall resonances
- $f=f_D$ – low power expectations except for the ILA-assisted high-$B_T$ option; unavoidable inner wall $f=f_T$ and marginally avoidable outer wall $f=2f_T$ resonances

Combining the ICRH heating schemes by simultaneously operating different RF systems in different frequency bands might be advantageous both for maximising the total RF coupled power and the DT physics program. Dedicated assessment of implications of the vessel wall $f=f_T$ & $f=2f_T$ parasitic resonances in presence of tritium neutral beam injection needs to be done in preparation to the DT campaign.

## 6 Conclusions

Steady progress has been made at JET in increasing the total RF power coupled to plasma since installation of ILW; the maximum pulse-average and peak values reached respectively 6MW and 8MW in 2016. Further efforts are required for improving reliability & performance of the RF generators and implementation of advanced real-time power balance control which will allow approaching the full potential of the RF plant.

At the same time, the maximum power coupled to plasma at present and the expectations for future DT campaign strongly depend on antenna-plasma coupling conditions specific to particular discharge scenarios and on the $B_T$-dependent selection of the RF plant operational frequencies.

The main concerns for high-power ICRH operations during the DT campaign are related to uncertainties of the ILA availability in view of recently emerged capacitor control issues, insufficient ELM-tolerance of the ECT system at 25MHz, 29MHz & 37MHz and the presence of parasitic tritium cyclotron resonances on the vessel inner wall.

This work has been carried out within the framework of the EUROfusion Consortium and has received funding from the Euratom research and training programme 2014-2018 under grant agreement No 633053. The views and opinions expressed herein do not necessarily reflect those of the European Commission.

## References

1. E. Lerche, et al., Nucl. Fusion **56** 036022(2016)
2. M. J. Mantsinen, et al., this conference
3. M. Graham, et al., Plasma Phys. Control. Fusion **54**, 074011 (2012)
4. I. Monakhov, et al., Nucl. Fusion **53**, 083013 (2013)
5. F. Durodié, et al., Plasma Phys. Control. Fusion **54**, 074012 (2012)
6. P. Jacquet, et al., Nucl. Fusion **56**, 046001 (2016)

EPJ Web of Conferences **157**, 03036 (2017) DOI: 10.1051/epjconf/201715703036
*22 Topical Conference on Radio-Frequency Power in Plasmas*

# Conceptual design studies of the Electron Cyclotron launcher for DEMO reactor

*Alessandro* Moro[1,*], *Alex* Bruschi[1], *Thomas* Franke[2,3], *Saul* Garavaglia[1], *Gustavo* Granucci[1], *Giovanni* Grossetti[4], *Kyriakos* Hizanidis[5], *Ioannis* Tigelis[6], *Minh-Quang* Tran[7], *Christos* Tsironis[5]

[1]Istituto di Fisica del Plasma, Consiglio Nazionale delle Ricerche, Milano, Italy
[2]EUROfusion Consortium, Boltzmannstr. 2, D-85748 Garching, Germany
[3]Max-Planck-Institut für Plasmaphysik, Boltzmannstr. 2, D-85748 Garching, Germany
[4]IAM-AWP, Karlsruhe Institute of Technology (KIT), D-76131 Karlsruhe, Germany
[5]School of Electrical and Computer Engineering, NTUA, 157 73 Athens, Greece
[6]Faculty of Physics, NKUA, 157 84 Athens, Greece
[7]Swiss Plasma Center, EPFL, CH-1015 Lausanne, Switzerland

**Abstract.** A demonstration fusion power plant (DEMO) producing electricity for the grid at the level of a few hundred megawatts is included in the European Roadmap [1]. The engineering design and R&D for the electron cyclotron (EC), ion cyclotron and neutral beam systems for the DEMO reactor is being performed by Work Package Heating and Current Drive (WPHCD) in the framework of EUROfusion Consortium activities. The EC target power to the plasma is about 50 MW, in which the required power for NTM control and burn control is included. EC launcher conceptual design studies are here presented, showing how the main design drivers of the system have been taken into account (physics requirements, reactor relevant operations, issues related to its integration as in-vessel components). Different options for the antenna are studied in a parameters space including a selection of frequencies, injection angles and launch points to get the best performances for the antenna configuration, using beam tracing calculations to evaluate plasma accessibility and deposited power. This conceptual design studies comes up with the identification of possible limits, constraints and critical issues, essential in the selection process of launcher setup solution.

## 1 Introduction

The design of a DEMOnstration Fusion Power Plant includes heating systems as mandatory to achieve controlled burning plasma and reactor relevant conditions. In the framework of EUROfusion Consortium activities, the Work Package Heating and Current Drive (WPHCD) is performing the engineering design and R&D for the Electron Cyclotron (EC) [2, 3], Ion Cyclotron (IC) and Neutral Beam (NB) systems for the DEMO reactor. In the present stage of DEMO design, the EC target power to the plasma is 50 MW. This power, when confirmed by detailed scenarios simulation, must be assured during the entire DEMO pulse length. The different EC tasks foreseen in DEMO [4] (listed in Table 1) require different deposition locations in the plasma (from $\rho < 0.3$ up to $\rho = 0.86$, being $\rho$ the normalized plasma radius) and a defined amount of power. A launcher with a good degree of flexibility and compatible with the main technical constraints is also required. In order to cover all the EC applications a candidate option for DEMO launcher could be based on Remote Steering Antennas (RSA), able to grant a continuous (but limited) steering range with no movable parts or mirrors in plasma proximity. As an alternative, simpler truncated-waveguide antennas can also be considered, in combination with step-tunable gyrotrons presently under development [5].

The status of a conceptual design study for an EC launcher is here presented, with evaluation of possible antenna options and potentials for multi frequency gyrotrons, launching performance, plasma accessibility, possible integration into port plug and preliminary evaluation of required apertures for antenna assembly.

**Table 1.** DEMO EC tasks and operation modes with corresponding required power and deposition location in terms of normalized radius.

| EC Task | Mode | Power [MW] | Deposition [$\rho$] |
|---|---|---|---|
| Assisted Breakdown | Heating | 6-10 | <0.3 |
| Ramp up and L-H transition | Heating/CD | 50 | <0.3 |
| Heating | Heating/CD | 50 | <0.3 |
| Sawtooth control | CD | 2-10 | <0.3 |
| NTM control (q=2;q=3/2) | CD | 10-15 | 0.86; 0.76 |
| Ramp down | Heating | 40 | [0.3-0.5] |

## 2 EC Launcher configuration definition

The conceptual design study presented here relies on the EU DEMO1 2015 baseline for a pulsed machine with aspect ratio AR = 3.1 and $B_T$ = 5.7 T [4]. The EC system is configured with high modularity and organized in clusters (with 8x2 MW gyrotron per cluster) [6]. The 8 EC beams of a cluster delivers to a single plug-in launcher composed by 8 independent antennas per port. An exception under study is considered for the gyrotrons dedicated to NTM control that would use two different

* Corresponding author: moro@ifp.cnr.it

launchers (4 beams each) located in two different ports. The total number of cluster considered is 4+1 for main EC tasks and one for NTM stabilization.

A selection of candidates fixed launching points, corresponding to waveguide termination locations have been identified (see points coordinates in Table 2), taking into account locations distributed in the equatorial port plug and in the larger major radius region of the vertical port plug (that could also be considered as hypothetical terminations for an upper port at ITER). To simulate EC injection in a wide range of possible configurations the parameters for beam tracing calculations have been chosen variable in steps, in angular and frequency ranges (square brackets) listed in Table 2. Wave frequency range includes 170 GHz, seeking compatibility with the EC system of ITER, for heating, 204 GHz for current drive tasks and discrete frequencies for heating and current drive operations in line with multi-frequency (e.g. 136/170/204/238GHz) or step-tunable gyrotron sources presently under development [5]. Even if the final choice will be done when the EU DEMO conceptual design phase will be completed, the selected range for this study focuses on 170 GHz and 200 GHz.

**Table 2.** Selected parameters space for beam tracing calculations.

| Parameter | | Value/Range |
|---|---|---|
| EC wave frequency | $f$ | [140 GHz -240 GHz] |
| Beam waist | $w_0$ | 20.43 mm |
| Equatorial launch points {X,Y,Z} [m] | $EPP_i$ | $EPP_3$={13.5, 0, 0}, $EPP_{1,2}$={13.5, 0, ±0.67} |
| Vertical/Upper launch points {X,Y,Z} [m] | $UPP_i$ | $UPP_1$={12.2, 0, 4.2}, $UPP_2$={10.7, 0, 3.9} |
| Toroidal angle | $\beta$ | [0°,40°] for $EPP_i$ |
| Poloidal angle | $\alpha$ | [-30°, 30°] for $EPP_i$ [0°, 50°] for $UPP_i$ |

The analysis was done with the beam tracing code TORBEAM [7] and a self consistent plasma scenario for DEMO1 obtained with ASTRA [8] code. TORBEAM runs provide the resulting beam trajectories, from which numerical parameters including the deposition location $\rho$ of EC power absorption, the current drive efficiency $\eta$, the absorbed power $P_{ABS}$, the total driven current $I_{CD}$ and the deposition profile width $\Delta\rho$ as function of frequency $f$ and injection angles $\alpha$ and $\beta$, organised in a multi-dimensional matrix.

## 3 Remote Steering Antenna (RSA) option

The selection of favourable launching configurations requires the identification of an antenna axis and a steering plane capable to cover the widest range of deposition locations, corresponding to the required tasks, and, for NTM stabilization, with a sufficient amount of driven current up to $\rho$ = 0.86. A candidate option for DEMO EC launcher conceptual design is the RSA. A maximum steering range $\Delta\gamma$= ±15° is studied, centered at a nominal injection direction given by the angles $\beta_0$, $\alpha_0$ and the steering plane is defined by the angle $\theta$ of its normal versor with respect to a horizontal direction (as shown in Fig. 1). The pair of poloidal and toroidal are expressed in terms of the cylindrical components of the wave vector **k** as follows: $k_R$=cos$\alpha$cos$\beta$, $k_\varphi$=sin$\beta$, $k_z$=−cos$\alpha$sin$\beta$ to give tan$\alpha$=$k_z/k_R$ and sin$\beta$=$k_\varphi$. Results in terms of deposition location accessibility $\rho$ (normalized plasma radius) and driven current $I_{CD}$ are mapped in Fig. 2 as a function of the two angles $\alpha$ and $\beta$, for a beam at two frequencies (170 GHz top, 200 GHz bottom) and waist at the launching position $w_0$=20.4 mm) launched from an equatorial port plug point ($EPP_3$). The toroidal and poloidal injection angles in the maps vary in the range $\beta$=[0°, 30°] and $\alpha$=[-30°, 30°], respectively.

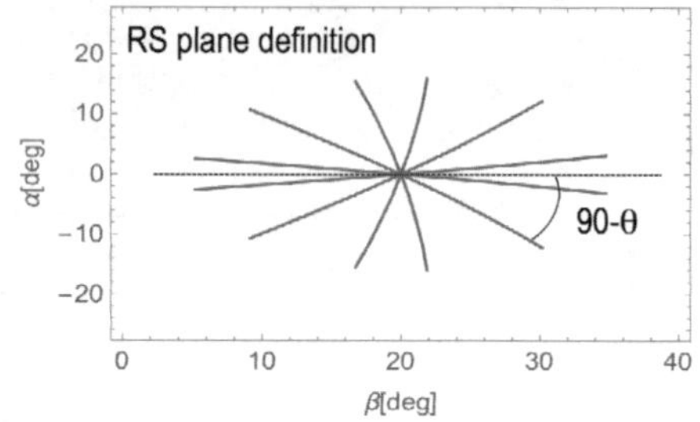


**Fig. 1.** Examples of possible launching angles pairs $\alpha$ and $\beta$, when steering is performed ($\Delta\gamma$ = ±15°), centered on pair of angles $\beta_0$=20°, $\alpha_0$=0°. Each curve corresponds to a given steering plane orientation.

Widest coverage of plasma regions is found for steering planes orthogonal to the iso-radius curves, although this is not straightforwardly accompanied with high values of total driven current, which are strongly frequency-dependent.

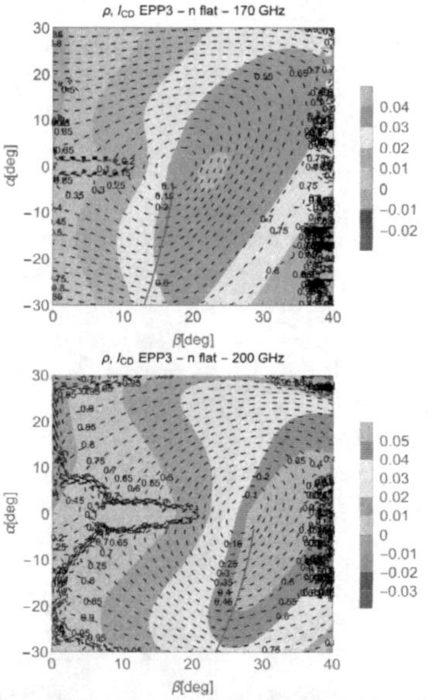


**Fig. 2.** Contour plots for normalized deposition location $\rho$ (black dashed curves) and total driven current $I_{CD}$ (color code, MA/MW) as a function of the injection angles ($\alpha$, $\beta$) with different possible steering planes (red lines) in the parameters

 DOI: 10.1051/epjconf/201715703036

space. The cases of 170 GHz frequency (Top) and 200 GHz (bottom) are shown.

A given steering plane, proved to be good for one frequency, is not suitable for higher (or lower) frequencies, as shown in Fig 3, where the total amount of driven current $I_{CD}$ is plotted as a function of $\rho$ (along one specific steering plane) in the case of 170 GHz and 200 GHz. A steering plane with parameters $\beta_0$=15°, $\alpha_0$ =-20°, $\theta$=-5° ensures that only 170 GHz EC beams covers large plasma regions (red curve in Fig. 3 left) when compared to the 200 GHz case. The opposite situation is found with steering plane parameters $\beta_0$=26°, $\alpha_0$=-20°, $\theta$=-5° promising at the higher frequency better plasma coverage and total driven current compared to the 170 GHz case with same orientation (Fig. 3 right).

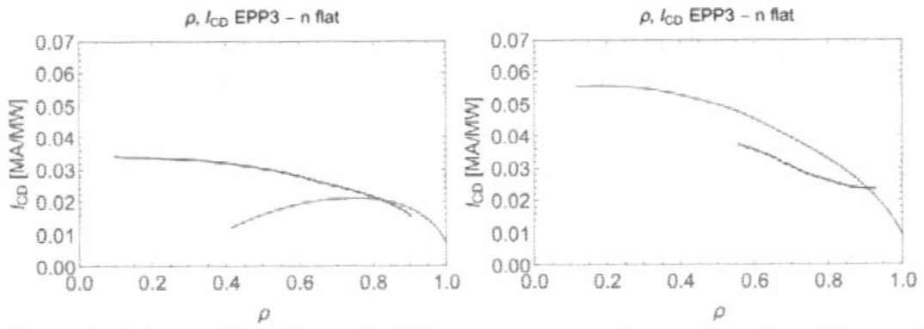


**Fig. 3.** Normalized total driven current $I_{CD}$ as a function of accessible deposition location ρ with the beam launched from EPP3 with steering plane parameters $\beta_0$=15°, $\alpha_0$=-20°, $\theta$=-5° (left) for 170 GHz (red) and $\beta_0$=26°, $\alpha_0$ =-20°, $\theta$=-5° (right) for 200 GHz (green).

A comparison of the current drive efficiency $\eta$ for the two optimal cases is shown in Fig. 4, where it appears that highest frequencies are more efficient for inner deposition, while at outer locations the difference is not so appreciable.

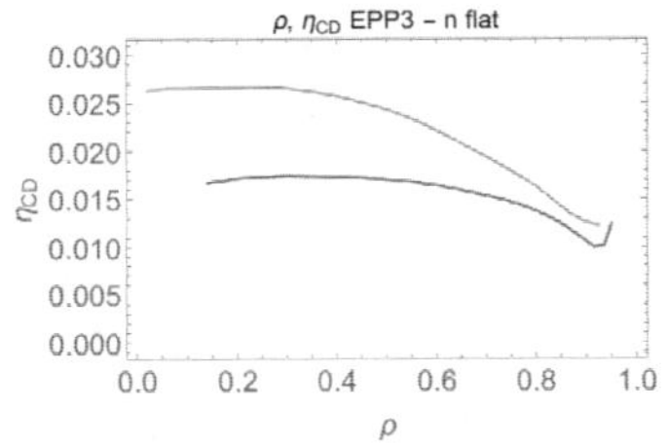


**Fig. 4.** Calculated current drive efficiency along the steering planes as a function of deposition location. Higher efficiency along the range is found for 200 GHz (green curve) with respect to 170 GHz (red curve).

The study of the overall performance of a given configuration cannot neglect to consider also the deposition profile width that characterises the EC absorption. It is modelled assuming that it is close to a gaussian, with most of the power released at a deposition location $\rho \pm \Delta\rho/2$, being $\Delta\rho$ defined as the full current density profile width at *1/e* of the peak value. Smaller deposition profile widths are beneficial of course, in particular for EC applications requiring current density with narrow profile as in the case of NTM control, where deposition within the magnetic island is mandatory for mode stabilization. In the cases considered, the total driven current around the $q$=2 location ($\rho$ = 0.86) is $I_{CD}$ = 25.3 kA/MW with $\Delta\rho$ = 0.06 in the case of 170 GHz, $I_{CD}$ = 27.6 kA/MW, $\Delta\rho$ = 0.14 in the case of 200 GHz. The expected lower deposition profile width with reduced length of the beam path from launcher to EC absorption region, motivated the analysis of a launching point from a higher position with respect to the equatorial port. The outcome is reported in Fig. 5, which shows the contour map of the total driven current obtained with $UPP_1$ launch, considering promising steering planes for 170 GHz and 200 GHz. In these cases smallest $\Delta\rho$ are found at 200 GHz, with $\Delta\rho$<0.03. It has to be pointed out that even if larger amount of driven current are achievable at larger toroidal angles ($\beta$>30° in 200 GHz case), these angles have the drawback to give even larger deposition profile width $\Delta\rho$, so toroidal injection in the range 15°<$\beta$<30° seems a better trade-off.

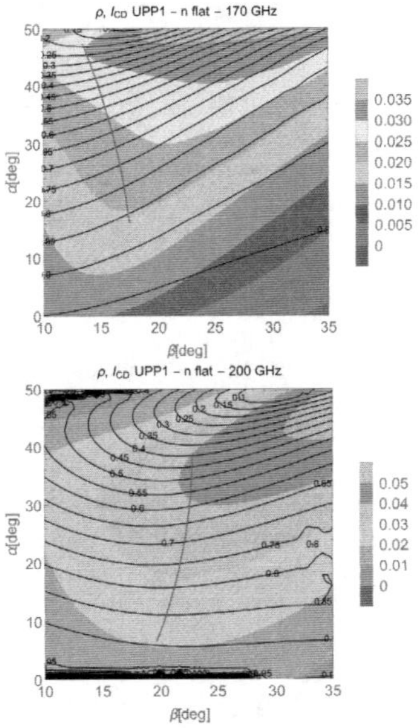


**Fig. 5.** Contour plots for normalized deposition location $\rho$ (black dashed curves) and total driven current $I_{CD}$ (color code, MA/MW) as a function of the injection angles ($\alpha$, $\beta$) with different possible steering planes (red lines) in the parameters space. The cases of 170 GHz frequency (top) and 200 GHz (bottom) are shown.

A comparison similar to the $EPP_3$ shows that 200 GHz ensures higher values of $I_{CD}$ with reduced plasma region accessibility (Fig 6). The lower frequency has a slightly wider range of accessible locations with lower efficiency.

As a general result of the conceptual studies done so far on the RSA it could be a valid option for inner EC tasks in terms of deposition location in the plasma, where large steering required to cover regions ranging from $\rho <$ 0.2 up to $\rho \sim$ 0.5, and operating at one frequency (~200 GHz). For NTMs control the simpler truncated waveguide will be investigated as an alternative, to be used at fixed orientation and exploiting the multi-frequency and step tunability of the sources under development for the fine-tuning of the deposition.

DOI: 10.1051/epjconf/201715703036

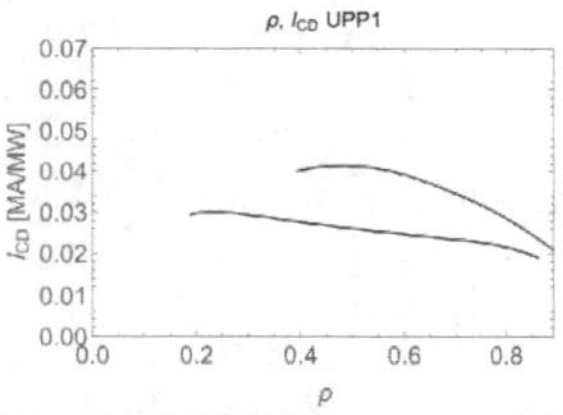


**Fig. 6.** Normalized total driven current $I_{CD}$ as a function of accessible deposition location ρ with the beam launched from $UPP_1$. The steering plane parameters are $\beta_0$=16°, $\alpha_0$=32°, $\theta$=-2° for 170 GHz (purple) and $\beta_0$=22°, $\alpha_0$=22.5°, $\theta$=-15° for 200 GHz (blue).

## 4 Port integration studies

The RSA option has limits that must be taken into account for launcher integration since the early stage of a conceptual design [9]. In particular, good beam characteristics can be obtained for a limited angular range (10°-15°) [10] centered on the waveguide axis and this range affects the width of the required apertures on the blanket modules. Preliminary estimation of a group of 8 waveguides packed in a row and providing a steering in the poloidal direction requires an aperture at blanket level $A$=0.19 m$^2$ (~0.5 m$^2$ for the whole pack and ~2.5m$^2$ considering five ports). The waveguide length in a RS arrangement is of the order of several meters and increases with frequency. In the case of a waveguide aperture $a$=75 mm and $f$=204 GHz the required length is $L_{RS}$≅15 m. Moreover, the waveguide routing within the plug towards ex-vessel DEMO environment must be compliant with the constraint that mitre bends can be inserted only in positions not too close to the waveguide (WG) termination, with bends and doglegs allowed only in the plane perpendicular to the chosen steering plane. An example of a possible setup showing blanket apertures and waveguide routing is sketched in Fig. 7, where a top view of the arrangement (left) and a front view of the waveguide apertures to allow beam steering (right) are presented, in case of a single row arrangement for the 8 waveguides with central injection at port aperture.

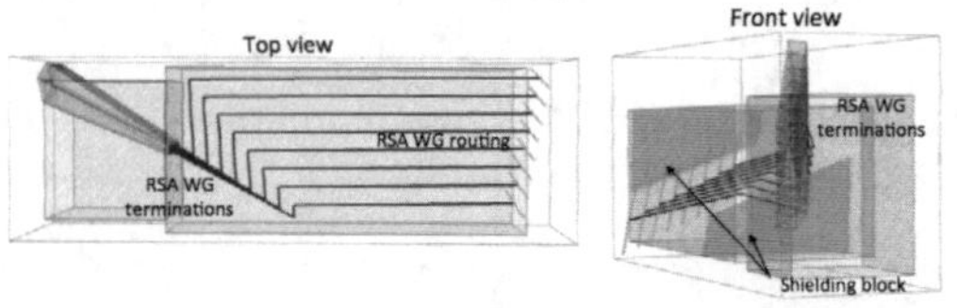


**Fig. 7.** Left: Sketched top view of a possible RSA pack assembly with a single vertical row of eight antennas toroidally inclined in an equatorial DEMO port. Right: RSA pack as viewed from the plasma, with gray regions representing volumes available for neutron shielding blocks.

Alternative configurations could also be considered, with different waveguide arrangements (for example using 2 rows per port, and side injection at port aperture, as shown in Fig. 8). Any configuration has to be evaluated in terms of neutronics issues, mechanical impact on breeding blanket, tritium breeding ratio, bioshield interactions and interfaces. An evaluation of the impact of apertures of the EC launchers in terms of tritium breeding ratio degradation has been performed starting from similar antenna design [11].

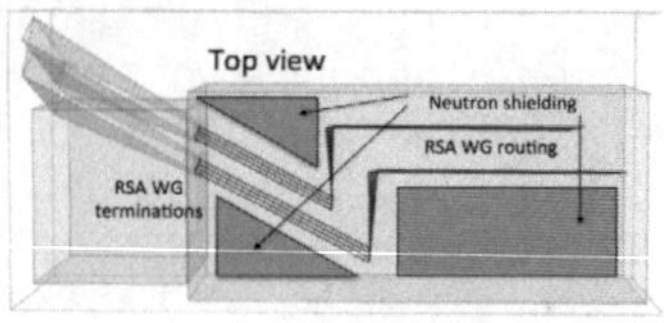


**Fig. 8.** Sketched top view of a possible RSA pack assembly with 2 vertical rows of four antennas toroidally inclined in an equatorial DEMO port. Gray regions represent volumes available for neutron shielding blocks.

## 5 Conclusions

Guidelines for a conceptual design of an EC launcher for DEMO reactor were presented in this paper with preliminary illustration of an antenna design, that aims at fulfilling the requirements with efficient and reliable EC deposition. The choice of the preferred antenna solution (steering plane orientation in the case of RSA or combined RSA and truncated waveguides properly oriented) should be validated through more detailed beam tracing calculations presently ongoing (in order to evaluate in a quantitatively way the performances in terms of EC injection and deposition) and through port plug integration feasibility study with an iterative approach to adapt the solution to engineering and physics requirements.

*This work has been carried out within the framework of the EUROfusion Consortium and has received funding from the Euratom research and training programme 2014-2018 under grant agreement No 633053. The views and opinions expressed herein do not necessarily reflect those of the European Commission.*

## References

1. F. Romanelli et al., Fusion Electricity, A roadmap to the realization of fusion energy, EFDA – Nov. 2012 - ISBN 978-3-00-040720-
2. S. Garavaglia et al., AIP Conference Proceedings **1689**, 090009 (2015)
3. G. Federici et al., Fusion Eng, and Design **109-111**, Pages 1464-1474 (2016)
4. R. Wenninger et al., Nucl. Fusion **57** (2017) 016011
5. J. Franck et al, IEEE International Vacuum Electronics Conference (2014) 31-32R
6. S. Garavaglia et al., EPJ Web of Conf. 147 (2017) 04002
7. E. Poli et al., Comput. Phys. Comm. 136, 90 (2001)
8. G.V. Pereverzev, P.N. Yushmanov, IPP-Report 5/98, (2002)
9. G. Grossetti et al. Proc. of 26th FEC – Kyoto (2016) FIP/P7-4
10. W. Kasparek et al., EPJ Web of Conferences **87**, 04005 (2015)
11. G. Granucci et al., Nucl. Fusion **57** (2017) 116009

EPJ Web of Conferences **157**, 03037 (2017) DOI: 10.1051/epjconf/201715703037
*22 Topical Conference on Radio-Frequency Power in Plasmas*

# A post-processing method to simulate the generalized RF sheath boundary condition

*James R.* Myra[1,*] and *Haruhiko* Kohno[2]

[1] Lodestar Research Corporation, Boulder, Colorado, 80301, USA
[2] Kyushu Institute of Technology, Iizuka, Fukuoka, 820-8502, Japan

**Abstract.** For applications of ICRF power in fusion devices, control of RF sheath interactions is of great importance. A sheath boundary condition (SBC) was previously developed to provide an effective surface impedance for the interaction of the RF sheath with the waves. The SBC enables the surface power flux and rectified potential energy available for sputtering to be calculated. For legacy codes which cannot easily implement the SBC, or to speed convergence in codes which do implement it, we consider here an approximate method to simulate SBCs by post-processing results obtained using other, e.g. conducting wall, boundary conditions. The basic approximation is that the modifications resulting from the generalized SBC are driven by a fixed incoming wave which could be either a fast wave or a slow wave. The method is illustrated in slab geometry and compared with exact numerical solutions; it is shown to work very well.

## 1 Introduction

For applications of ion cyclotron range of frequencies (ICRF) power in fusion devices, control of RF specific interactions with the scrape-off layer plasma and material surfaces is of great importance. Near these surfaces, sheaths on the scale of the Debye length form. For typical parameters of interest, i.e. RF sheath voltages much larger than the local temperature, these sheaths are strongly dominated by the presence of RF waves. The sheaths impact the wave fields and the local wave fields impact the sheaths. The rectification of rf voltages across the sheath is believed to contribute to power losses [1] and enhanced sputtering, the latter motivating the investigation of new antenna designs [2, 3].

Because of the scale separation between the sheath and global wave physics, and strong sheath nonlinearities, it is not practical to directly simulate RF sheaths in global ICRF waves codes. Instead, sheath boundary conditions (SBCs) and related methods have been developed [4 - 7] and implemented in global codes [8]. A recent general formulation [9] of the SBC provides an effective surface impedance for the interaction of the RF sheath with the waves. Furthermore, it enables quantities of interest for material interactions, such as the surface power flux and rectified potential energy available for sputtering, to be calculated.

Unfortunately, many legacy ICRF codes do not implement the SBC. For them it would be desirable to have an approximate method to simulate sheath boundary conditions by post-processing results obtained using other, e.g. conducting wall, boundary conditions. This paper describes such a method, expected to be useful when the conducting wall code can accommodate boundaries which are not restricted to coincide with a flux surface. Furthermore, in codes with an SBC, iteration is required for a converged self-consistent solution since the SBC is nonlinearly dependent on the RF wave amplitude. The method may also be useful in speeding up convergence in these cases.

The basic approximation enabling the method is that the modifications resulting from the generalized SBC are driven by a fixed incoming wave which could be either a fast wave or a slow wave. Code data is first post-processed to obtain the amplitude of the incoming waves. Then holding this amplitude fixed, the wave equations are solved again, approximately, in the vicinity of the wall sheath, using the SBC. The resulting solution contains modified RF fields in the vicinity of the wall as well as a modification of any reflected (outgoing) waves.

A general description of the method and algorithm is presented in Sec. 2. In Sec. 3 the post-processing method is applied to a sample problem in slab geometry, and the results are compared with a numerical solution using the rfSOL code [8, 10], a finite element RF wave code which now implements the full generalized SBC. Finally a discussion and conclusions are given in Sec. 4

## 2 Description of the method

The generalized sheath BC may be written in the form.

$$\mathbf{E}_t = \nabla_t (J_n z_{sh}) \tag{1}$$

* Corresponding author: jrmyra@lodestar.com

where $J_n$ is the rf current normal to the wall, $\mathbf{E}_t$ is the projection of $\mathbf{E}$ tangential to the wall surface, $\nabla_t$ is the tangential gradient operator, and $z_{sh}$ is the sheath impedance. The rf sheath voltage is

$$\Phi_{sh} = -J_n z_{sh} \tag{2}$$

and the sheath BC simply matches the $\mathbf{E}_t$ of the wave at the entrance to the sheath to that of the sheath itself. Inside the sheath $\Phi$ hence $\mathbf{E}_t$ decays to zero on the surface of the conducting wall.

Although the method is quite general, for the sake of clarity we will describe the method for the case where the original code result was obtained using a conducting wall BC, $\mathbf{E}_t = 0$. This corresponds to an original solution using $z_{sh} = 0$ in Eq. (1). For our post-processing method, the data required from the original code result consists of $J_n$ and the tangential RF magnetic field $\mathbf{B}_t$ for all points on the boundary.

In the vicinity of the wall four scalar equations describe the four possible modes in a cold plasma, two each describing different propagation directions for the fast wave (FW) and slow wave (SW). We will label these modes as m = 1, 2, 3, 4. For each patch of wall region, taking the local conductivity tensor to be constant in this near-wall region, using the local eikonal approximation, and choosing a local coordinate system with the x-direction normal to the wall, we have

$$\sum_m \mathbf{e}_x \cdot \overline{\overline{\sigma}} \cdot \mathbf{E}^{(m)} = J_x \tag{3}$$

$$\sum_m E_y^{(m)} = ik_y J_x z_{sh} \tag{4}$$

$$\sum_m E_z^{(m)} = ik_z J_x z_{sh} \tag{5}$$

$$\sum_m [n_z E_x^{(m)} - n_x E_z^{(m)}] = B_y \tag{6}$$

Eq. (3) is the definition of the normal current in terms of the conductivity tensor while Eqs. (4) and (5) are the two components of the SBC. Eq. (6) is one component of the Maxwell equation $\mathbf{n}\times\mathbf{E} = \mathbf{B}$ where $\mathbf{n} = \mathbf{k}c/\omega$. Instead of $B_y$ the equation for $B_z$ could be employed. ($B_y$, $B_z$ and $J_x$ are not linearly independent.)

The procedure is as follows:

(i) Post-process the original code result (with conducting wall BC) to obtain $J_x$ and $B_y$ on the surface

(ii) Solve Eqs. (3) – (6) with $z_{sh} = 0$ using $J_x$ and $B_y$ as inputs to obtain the amplitudes of the four modes, $E^{(m)}$, m = 1, 2, 3, 4. Note that the polarization vectors of the modes are known from the eikonal solution so only the amplitudes are unknown and there are four equations for these four unknowns.

(iii) Solve Eqs. (3) – (6) with $z_{sh}$ specified as the desired sheath impedance, using as inputs the amplitudes of the two incoming modes from step (ii) (in general, one FW and one SW). Solve for new values of $J_x$ and $B_y$ as well as new amplitudes of the two outgoing modes.

Finally for a self-consistent sheath impedance, the dependence $z_{sh}(|\Phi_{sh}|)$ on $\Phi_{sh}$ must be taken into account, i.e. the equation

$$\Phi_{sh} = -J_n z_{sh}(|\Phi_{sh}|) \tag{7}$$

must also be satisfied at each point on the surface. A root-finder for Eq. (7) may additionally be imposed on the procedure. In some cases, iteration may be sufficient, in which case we may add a final step

(iv) Calculate the new sheath potential from Eq. (7), and using the updated value of $z_{sh}$ for this sheath potential, repeat steps (iii) and (iv) until convergence.

Note that the $k_y$ in Eq. (4) assumes that the spatial variation of the envelope $|\Phi_{sh}|$ in Eq. (7) is slow compared with that of the phase; this could be improved by employing Fourier transforms in y on the whole term as discussed subsequently. Note that in some cases Eq. (7) [e.g. coupled to Eq. (18)] may have multiple roots so that a single solution with metallic BCs may result in more than one solution with sheath BCs. Further discussion of multiple roots is beyond the scope of this paper; the issues are independent of the post-processing method.

## 3 Sample application

As an example application, intended mainly as a first proof-of-principle test of the procedure, we now consider the case of a perpendicular sheath (i.e. one for which the magnetic field is perpendicular to the surface) in a slab geometry with constant plasma parameters. The geometry is illustrated in Fig. 1 and is the same as described in Ref. [10] for the 2D slab version of the rfSOL code.

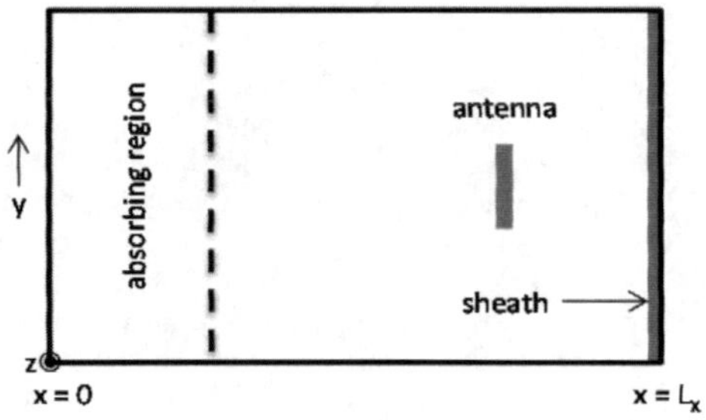


**Fig. 1.** Geometry of the sample application.

A sheet current antenna launches waves that propagate (or are evanescent) both to the left and right of it. The right propagating branch interacts with the sheath

DOI: 10.1051/epjconf/201715703037

boundary condition of surface impedance $z_{sh}$ at $x = L_x$. In general there is reflection from the surface; this left going branch together with the one launched directly by the antenna are absorbed before they reach the left domain boundary at $x = 0$ creating an outgoing/evanescent wave condition in this direction. The simulations are periodic in y and for the perpendicular incidence problem considered here the background magnetic field is $\mathbf{B}_0 = \mathbf{e}_x B_{0x}$. With walls normal to $\mathbf{B}_0$ the FW does not contribute to RF sheath excitation and a SW analysis will suffice.

The rfSOL code was first run with conducting wall boundary conditions at $x = L_x$ equivalent to $z_{sh} = 0$. From this solution we extract the incoming wave amplitude (i.e. the wave incident on the $L_x$ boundary) as follows. The general solution for the RF wave in the electrostatic limit (to be generalized subsequently) is

$$\mathbf{E} = -i\mathbf{k}_{sw}\Phi^{(1)}e^{ik_{sw,x}(x-L_x)} + i\mathbf{k}_{sw}\Phi^{(2)}e^{-ik_{sw,x}(x-L_x)} \quad (8)$$

where $\Phi^{(1)}$ and $\Phi^{(2)}$ are the amplitudes of the electrostatic potential for the incoming and outgoing waves respectively. By convention $k_{sw,x}$ is taken as the root with $k_{sw,x} > 0$. The total potential at the wall is

$$\Phi = \Phi^{(1)} + \Phi^{(2)} \quad (9)$$

which must be zero for the conducting wall (cw) case. Therefore at the wall

$$E_x = -2ik_{sw,x}\Phi^{(1)} \quad (10)$$

The total $E_x$ and $D_x \equiv iJ_x/\omega$ at the wall are related by

$$\varepsilon_0\varepsilon_{||}E_x = D_x \quad (11)$$

where $\varepsilon_0$ is the vacuum dielectric constant and $\varepsilon_{||}$ is the parallel (diagonal) component of the cold plasma dielectric tensor. Thus we have

$$\Phi^{(1)} = \frac{iD_{x,cw}}{2\varepsilon_0\varepsilon_{||}k_{sw,x}} \quad (12)$$

Although it is not necessary for the method, here for simplicity we make the assumption that the waves near the boundary can be treated in the local approximation $k_z > k_y$ where $k_y \approx k_\perp$ is a typical wavenumber in the y-direction. (In general for $k_y \sim k_z$ each Fourier mode in y would be treated independently and summed up). The SW dispersion relation gives

$$k_{sw,x}^2 = k_{||}^2 = \frac{-k_\perp^2\varepsilon_\perp}{\varepsilon_{||}} \quad (13)$$

completing the determination of $\Phi^{(1)}$.

Next we turn to the solution for a specified value of $z_{sh}$. The basic assumption is that $\Phi^{(1)}$ remains unchanged from the value in Eq. (12). Thus we solve the system

$$\varepsilon_0\varepsilon_{||}E_x = D_x \quad (14)$$

$$\Phi^{(1)} + \Phi^{(2)} \equiv \Phi_{sh} = -i\omega D_x z_{sh} \quad (15)$$

now regarding $\Phi^{(1)}$ as known and $\Phi^{(2)}$ and $D_x$ as the unknowns. $E_x$ in Eq. (14) is expressed in terms of $\Phi^{(1)}$ and $\Phi^{(2)}$ from the expansion given in Eq. (8). After a small amount of algebra, the result is

$$D_x = \frac{D_{x,cw}}{1-\omega\varepsilon_0\varepsilon_{||}k_{sw,x}z_{sh}} \quad (16)$$

$$\Phi^{(2)} = -\frac{iD_{x,cw}}{2\varepsilon_0\varepsilon_{||}k_{sw,x}}\left(\frac{1+\omega\varepsilon_0\varepsilon_{||}k_{sw,x}z_{sh}}{1-\omega\varepsilon_0\varepsilon_{||}k_{sw,x}z_{sh}}\right) \quad (17)$$

The RF sheath voltage $\Phi_{sh} = \Phi^{(1)} + \Phi^{(2)}$ is therefore

$$\Phi_{sh} = \frac{-i\omega D_{x,cw}z_{sh}}{1-\omega\varepsilon_0\varepsilon_{||}k_{sw,x}z_{sh}} \quad (18)$$

If $z_{sh}$ is very small then it is clear that the conducting wall solution is recovered. In the opposite limit of large $z_{sh}$ (the quasi-insulating limit) the current $J_x$ into the wall, proportional to $D_x$, goes to zero and the sheath voltage is $\Phi_{sh} = iD_{x,cw}/\varepsilon_0\varepsilon_{||}k_{sw,x}$. In intermediate cases, if $z_{sh}$ and $k_{sw,x}$ are nearly pure imaginary (possible in the capacitive sheath limit with evanescent SWs), sheath-plasma resonance occurs and can result in large values of $\Phi_{sh}$.

It is not difficult to show that in the electromagnetic (EM) case the SW dispersion relation, Eq. (13) is replaced by

$$n_\perp^2\varepsilon_\perp + n_{sw,x}^2\varepsilon_{||} = \varepsilon_\perp\varepsilon_{||} \quad (13')$$

and the final results in Eqs. (16) – (18) only require the replacement of $k_{sw,x}$ with $q_{sw,x}$ given by

$$q_{sw,x} = k_{sw,x}\frac{n_\perp^2}{n_\perp^2 - \varepsilon_{||}} \quad (19)$$

where $\mathbf{n} = \mathbf{k}c/\omega$.

For a test to compare with rfSOL we considered an evanescent EM SW case. The excitation of the FW can be neglected. The main parameters of interest here are $B_{0x} = 1$ T, density $n_e = 2\times10^{18}$ m$^{-3}$, $T_e = 15$ eV, RF wave frequency $f = \omega/(2\pi) = 80$ MHz, antenna-sheath distance = 0.2 m, and the specified wavenumber $k_z = 160$ m$^{-1}$. The dominant mode launched by the antenna is $k_y = 120$ m$^{-1}$ and is taken as an input parameter here. The antenna current is 1.1 kA/m which results in a maximum (in y) sheath voltage of about 130 V. Other parameters, of less interest for present purposes, are given in Ref. [10]; see Fig. 11(a).

For these parameters, $\omega/\omega_{pi} = 0.38$ resulting in a complex impedance with significant real part, and $|n_\perp^2/\varepsilon_{||}| = 0.57$ implying important EM effects. Furthermore, it turns out that $z_{sh}$ is not so large as to put the final result in the quasi-insulating limit; thus, the results are still sensitive to the value of $z_{sh}$. This case therefore provides a robust test of the method. Iteration, updating the nonlinear dependence [9] of $z_{sh}(|\Phi_{sh}|)$, was carried out using the same fits as in rfSOL [10] to obtain a self-consistent solution at each value of y. Starting from an initial constant value of either essentially infinite $z_{sh}$ or from $z_{sh} = 0$ the solution converges rapidly.

As can be seen from Fig. 2, the comparison between the post-processing solution and the rfSOL result is quite good. The main differences are likely the result of employing local theory for the post-processing, which begins to lose accuracy when only the dominant harmonic in $k_y$ is retained, as done here. As explained previously, local theory is not essential to the method, but is convenient for the present test.

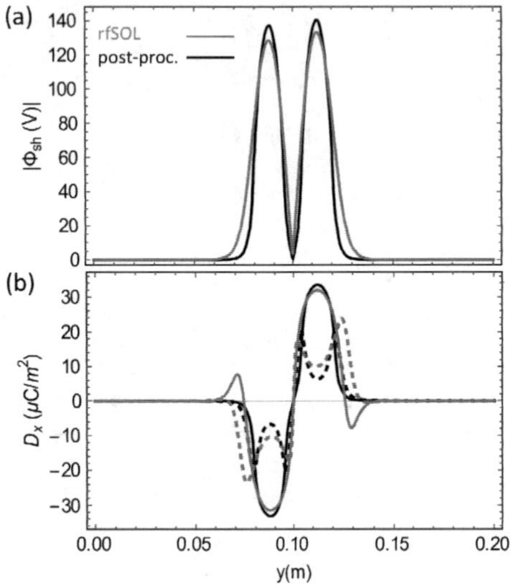


**Fig. 2.** Comparison of the post-processing solution (black) with the numerical solution obtained from rfSOL (red) for the electromagnetic test case: (a) the magnitude of the rf sheath voltage; (b) real (solid) and imaginary (dashed) parts of $D_x$.

## 4 Discussion and conclusions

The main assumption of the method is that the amplitude of the incoming waves is not changed by the BC. In the example this is assured by the fact that the SW is evanescent on a scale shorter than the distance between the walls and (in general if necessary) the absorbing layer. In addition, the present analysis has also assumed locally constant plasma parameters which enable a local solution for $k_x$ given specified values of $k_y$ and $k_z$. Here $k_y$ and $k_z$ may be obtained from the Fourier transform of the normal current ~ $D_x$ on the surface. For 2D codes $k_z$ may be calculated from the mode number in the toroidal direction. If there is no dominant $k_y$, then the procedure may be applied to each $k_y$ in the spectrum. The value of $k_x$ for each mode i.e. propagation direction of SW and FW (m = 1,2,3,4) is given from the local dispersion relation for the chosen $k_y$ and $k_z$. In the illustration of the method in Eqs. (3) – (6), the tangential **k** is assumed to be equal for FW and SW, but this could be generalized. The use of local wavenumbers $k_x$, $k_y$ and $k_z$ to describe the fields near the surface is expected to be rigorous when the surface is locally flat on the scale of the waves and the plasma parameters are constant on that scale. Note that in the case of FW→SW conversion due to wall shape (e.g. a wall "bump") [8] the conducting wall solution for $D_x$ will contain $k_y$ values characteristic of the bump scale.

We have shown that a sheath post-processing method can be used to transform a solution obtained with a conducting wall BC to one with a general impedance sheath BC. Used in this way, the method allows legacy rf codes (with conducting wall BCs) to obtain approximate rf sheath properties by post-processing. In fact, the same procedure can be used to transform between any two different cases of an impedance boundary condition (for example from an insulating sheath solution where $z_{sh} \rightarrow \infty$ to a finite $z_{sh}$ solution).

In the case of a locally curved surface, or if the density or magnetic field vary significantly with respect to the SW wavelength near the wall, then the eikonal ansatz could introduce significant approximations. Further work will be required to assess the accuracy of the method for practical applications in such cases.

This material is based upon work supported by the U.S. Department of Energy Office of Science, Office of Fusion Energy Sciences under Award Numbers DE-FG02-97ER54392 and DE-FC02-05ER54823. The digital data for this paper can be found at https://doi.org/10.5281/zenodo.579548.

## References

1. R. J. Perkins, J. C. Hosea, M. A. Jaworski, J.-W. Ahn et al., Phys. Plasmas **22**, 042506 (2015)
2. V. Bobkov, D. Aguiam, R. Bilato, S. Brezinsek et al., Plasma Phys. Control. Fusion **59**, 014022 (2017)
3. S. J. Wukitch, M. L. Garrett, R. Ochoukov, J. L. Terry et al. Phys. Plasmas **20**, 056117 (2013)
4. E. F. Jaeger, L. A. Berry, J. S. Tolliver, D.B. Batchelor, Phys. Plasmas **2**, 2597 (1995)
5. D. A. D'Ippolito and J. R. Myra, Phys. Plasmas **13**, 102508 (2006)
6. L. Colas, J. Jacquot, S. Heuraux, E. Faudot, K. Crombé et al., Phys. Plasmas **19**, 092505 (2012)
7. D. Van Eester, K. Crombé and V. Kyrytsya, Plasma Phys. Control. Fusion **55**, 055001 (2013)
8. H. Kohno, J. R. Myra, and D. A. D'Ippolito, Phys. Plasmas **22**, 072504 (2015); and Phys. Plasmas **23**, 089901 (2016)
9. J. R. Myra and D. A. D'Ippolito, Phys. Plasmas **22**, 062507 (2015)
10. H. Kohno and J. R. Myra, in press, Comp. Phys. Comm. (2017)

EPJ Web of Conferences **157**, 03038 (2017) DOI: 10.1051/epjconf/201715703038
*22 Topical Conference on Radio-Frequency Power in Plasmas*

# Major upgrades of the high frequency B-dot probe diagnostic suite on ASDEX Upgrade

*Roman* Ochoukov[1,*], *Volodymyr* Bobkov[1], *Cecilia* Bruhn[1,2], *Rodolphe* D'Inca[1], *Helmut* Faugel[1], *Helmut* Fuenfgelder[1], *Rachael* McDermott[1], *Philip* Schneider[1], *Guillermo* Suarez-Lopez[1,3], *Jean-Marie* Noterdaeme[1,4], *ASDEX Upgrade* Team[**] and *MST1* Team[***]

[1]Max Planck Institute for Plasma Physics – Garching, Germany
[2]Technical University of Munich – Garching, Germany
[3]Ludwig Maximilian University – Munich, Germany
[4]Ghent University – Ghent, Belgium

**Abstract.** The high frequency B-dot (HFB) probe diagnostic on the ASDEX Upgrade tokamak has undergone a considerable upgrade during the 2016 opening of the torus. The probe coverage is now greatly expanded toroidally, as well as radially with the addition of probes on the high field side and the removable manipulator head. A new 2-channel fast digitizer now allows to examine and record radio frequency (RF) wave emissions emanating from the plasma in the ion cyclotron range of frequencies (ICRF). Possible studies that can be achieved now include: a study of core ICRF power absorption efficiency; a study of ion cyclotron emissions from the plasma generated by energetic ions; and study of ICRF wave/plasma turbulence interactions in the scrape-off layer region.

## 1 Introduction

Radio frequency waves in the ion cyclotron range of frequencies (ICRF) are successfully used to heat magnetized plasmas to fusion relevant temperatures. Examples of tokamaks using ICRF systems are ASDEX Upgrade [1], Alcator C-Mod [2], Tore Supra/WEST [3], and JET [4]. ITER, the next generation tokamak [5], and DEMO, the first generation fusion power plant [6], are also going to rely on ICRF power for their operation. The ICRF system on ASDEX Upgrade is tasked with studying ICRF power coupling and mitigating unwanted plasma-material interactions, while maintaining high performance core plasma [1, 7]. In order to assist with this task, a number of probes have been installed that directly detect the ICRF waves' magnetic field components and their properties. This set of high frequency B-dot (HFB) probes contains an array of inductors capable of measuring the amplitude, the phase, and the polarization of the ICRF wave [8]. It recently received an upgrade with a number of individual probes that greatly expanded the diagnostic coverage toroidally and radially with a number of probes placed on the high field side of the ASDEX Upgrade tokamak. Additionally, a high frequency 2-channel digitizer has been added to the diagnostic rack, in order to resolve the ion cyclotron emission (ICE) frequency spectra emanating from the plasma. Finally, a pair of HFB probes, together with a pair of Langmuir probes, are put together for a new diagnostic head of the ASDEX Upgrade low field side midplane manipulator. This additional diagnostic aims to measure the radial profile of the ICRF wave field amplitude in the scrape off layer (SOL), as well as to study the interactions between turbulent SOL plasmas and ICRF wave fields. The overall outline of the paper is to describe the details of the newly expanded HFB diagnostic and to demonstrate the capabilities of these new additions to measure ICRF wave properties in various plasma conditions. Possible future studies will also be suggested.

## 2 Diagnostic and Experimental Procedure Description

The first set of HFB probes was installed before the 2015 ASDEX Upgrade experimental campaign and included three probe pairs and one field-aligned probe array of six probe pairs, all located on the low field side near the midplane [8]. Each pair consists of two inductors, orthogonally aligned to each other: one probe oriented to measure the fast wave polarization and the other for the slow wave. An example of a single probe, with the inductor exposed is shown in Fig. 1 (a). The latest addition for the 2017 campaign now includes 8 new probe pairs: 4 on the low field and 4 on the high field side. Another probe pair is installed, together with a pair of Langmuir probes, on a removable diagnostic head for the ASDEX Upgrade midplane manipulator (MEM), Fig. 1 (b). The purpose of the additional probes on the high field side is to help quantify the efficiency of various ICRF absorption schemes: hydrogen (H-) or helium-3 ($^{3}$He-) minority

*Corresponding author: rochouko@ipp.mpg.de
**For authors list, see H. Zohm et al., Nucl. Fusion 55, 104010 (2015).
***For authors list, see H. Meyer et al., Nucl. Fusion Special issue, Nucl. Fusion 57 (2017) 102014.

DOI: 10.1051/epjconf/201715703038

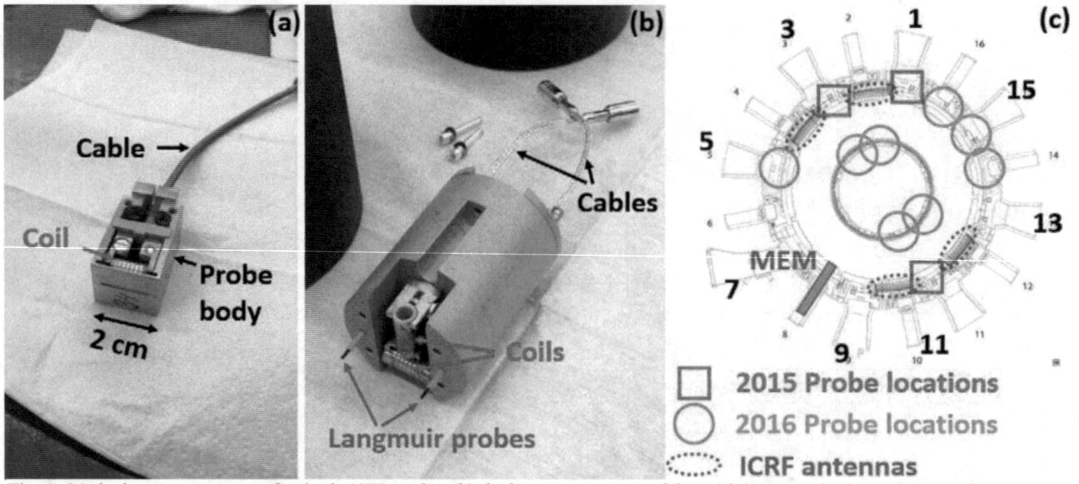


**Fig. 1.** (a) the key components of a single HFB probe; (b) the key components of the midplane manipulator (MEM) ICRF head; (c) the locations of the previous and latest installations of the HFB probes. The ICRF antennas and the MEM toroidal positions are also indicated.

heating, 3-ion heating, or higher harmonic D heating. The new MEM probe head will be used to study interactions between ICRF waves and SOL turbulence. The ICRF wave amplitude signal is generally sampled with a logarithmic RF detector that cannot distinguish frequencies in the range of 10-50 MHz [8]. However, the addition of a 2-channel ICE detector/digitizer with a sampling frequency of 125 MHz allows us to measure emitted ICE frequency spectra up to 67.5 MHz from two different probes simultaneously. The overall layout of the probes inside the ASDEX Upgrade torus is shown in Figure 1 (c) .

Both L- and H-mode discharges are used for ICRF studies with the HFB probes. The H-mode is typically achieved on ASDEX Upgrade when the total auxiliary heating power is in excess of 2 MW, which can be reached with a combination of ICRF, electron cyclotron waves, and neutral beam injection. The ICRF power is typically injected at either 30 MHz (for the on-axis toroidal field value $B_T = +/-2$ T) or 36.5 MHz (for $B_T = +/- 2.5$ T), with hydrogen minority being the dominant heating scheme in these discharges. Fig. 1 (c) shows the locations of the ICRF antennas inside the ASDEX Upgrade torus.

To determine how efficiently ICRF power is absorbed in the plasma core, we use the break-in-slope method [9], while modulating ICRF power on/off with the frequency of 8-10 Hz. Following this method the ICRF power absorption coefficient ($\eta_{ICRF}$) is estimated from the rate of change of the stored plasma energy ($dW_{MHD}/dt$) before and after the ICRF pulse is turned off: $\eta_{ICRF} \cong \Delta dW_{MHD}/dt|_{t = t\ ICRF\ off}$. The value of $\eta_{ICRF}$ is then averaged over 5-10 ICRF pulses to smoothen the effect of edge localized modes (ELMs) on the estimate. The goal is then to determine if the expected trend (weaker RF field amplitude ($V_{RF}$) on the high field side for stronger ICRF absorption coefficient) is observed in ICRF-heated discharges on ASDEX Upgrade. It is worth noting that the break-in-slope method underestimates $\eta_{ICRF}$. The method assumes a prompt plasma response, which is not entirely true for the H-minority heating scheme, where a large power fraction is absorbed on the ions and only then is transferred to the electrons through collisions [9].

## 3 Experimental Results and Discussion

The break-in-slope technique is commonly used to determine how efficiently ICRF power is absorbed in the plasma core [9]. In our case, the ICRF power was modulated at either 8.333 or 10.0 Hz (Fig. 2 (a)). The corresponding changes in the plasma stored energy are approximated with linear curves (Fig. 2 (b)), the slopes of which are then used to estimate $\eta_{ICRF}$. The core plasma density remains constant during these modulations (Fig. 2 (c)), excluding a possible effect on plasma fueling by the active antennas. The modulation of the core soft x-ray (SXR) emissions and the neutron rate (dN/dt) are in line with ICRF power being deposited in the plasma core and heating the plasma species (Figs. 2 (d) and (e)). The SXR signal is additionally modulated by sawteeth at nearly the same frequency as the ICRF power modulation (Fig. 2 (d)). The strength of the ICRF fast wave amplitude on the high field side is then recorded during these modulations (Fig. 2 (f)) to compare with the ICRF power absorption coefficient.To smoothen the effect of ELMs, which are clearly visible on the time traces of $W_{MHD}$ (Fig. 2 (b)) and $V_{RF}$ (Fig. 2 (f)), we use 5-10 ICRF pulses to obtain a single average $\eta_{ICRF}$ value. The expected trend is to have a lower ICRF fast wave amplitude on the high field side, as the ICRF power absorption coefficient improves. Such behavior is observed (Fig. 3). In total, six discharges (five H- and one L-mode) were used, where each point represents a single on/off ICRF cycle. Note that the H-minority concentration is ~4-6% in these discharges, suggesting that the reason for the higher $V_{RF}$ and lower $\eta_{ICRF}$ in H-mode discharges with ICRF power at 30.0 MHz is due to a different wavenumber spectrum of the launched ICRF wave.

The ability to study ICRF power absorption is not the only capability of the newly expanded HFB diagnostic suite. With the addition of a fast 2-channel digitizer

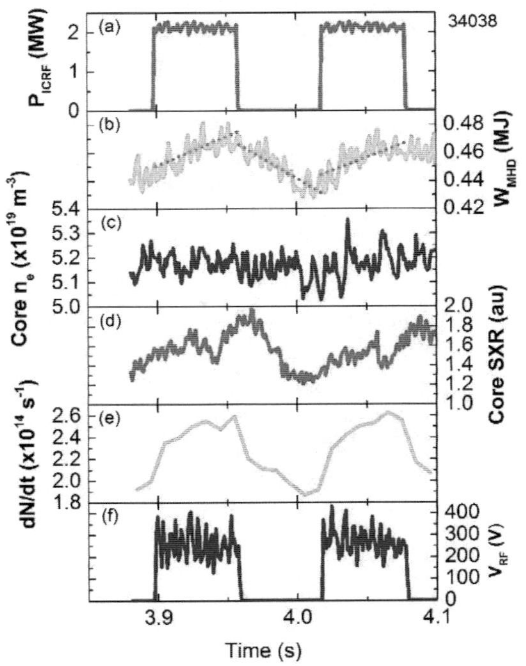


**Fig. 2.** The key plasma and machine parameters of a discharge used to examine core ICRF wave absorption. (a) the ICRF power $P_{ICRF}$; (b) the stored plasma energy $W_{MHD}$; (c) the core plasma density ne; (d) the soft x ray (SXR) signal from a chord that passes through the plasma core; (e) the neutron rate dN/dt; and (f) the fast ICRF wave amplitude $V_{RF}$ from the high field side HFB probe in Sector 2. The dashed lines in (b) are linear fits.

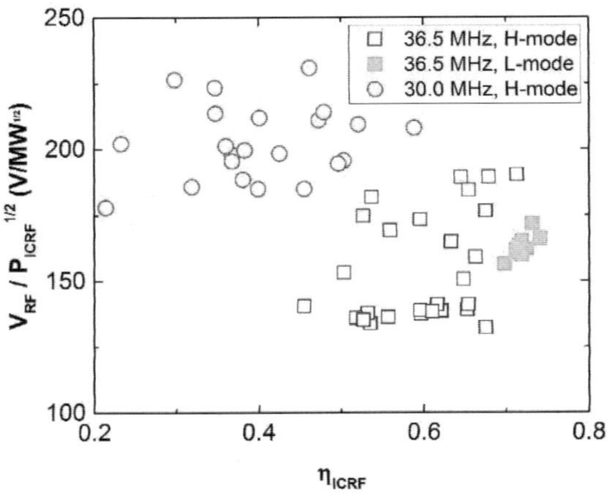


**Fig. 3.** The normalized fast wave amplitude $V_{RF}$ from the Sector 2 high field side probe, as a function of the estimated ICRF power absorption coefficient $\eta_{ICRF}$. $\eta_{ICRF}$ is estimated using the break-in-slope method [9]. Note the the break-in-slope method underestimates $\eta_{ICRF}$ for the case of H-minority heating [9].

system, it is now possible to measure ICE frequency spectra emanating from the magnetized tokamak plasma in the presence of energetic ions. An example of one ICE type is obtained from one of the shots that are used for the ICRF power absorption study (Fig. 4). The 10 Hz power "blips" of ICRF operated at 36.5 MHz are clearly visible, as well as the aliased second harmonic at 52.0 MHz. We,

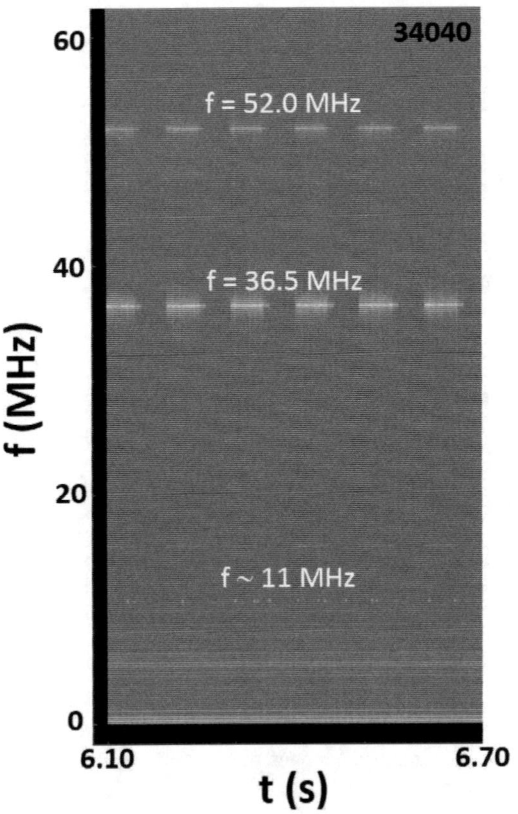


**Fig. 4.** An example of the ion cyclotron emission (ICE) spectrum on ASDEX Upgrade. The ICE signal at ~11 MHz is attributed to the energetic deuterium ions, produced by neutral beam injections at 60 kV grid bias.

additionally, observe short bursts at ~11 MHz and these are generally attributed to the neutral beam deuterium ions exciting magnetoacoustic cyclotron wave emissions from the low field side [10]. The on/off bursts of the emission are correlated with the ELMs. We also see uninterrupted "emission" lines below 10 MHz, which are the internal electronics noise of the diagnostic.

H- and L-mode discharges shown so far are not the only tokamak plasma modes that are studied for the purpose to provide favourable fusion environment. The so-called I-modes, which have beneficial transport properties [11], are also of interest on ASDEX Upgrade. One of the defining features of the I-mode is the appearance of a mode in the plasma pedestal region, with a characteristic frequency of ~100 kHz. Such a feature is, indeed, observed with a newly installed MEM diagnostic head, equipped with a pair of HFB probes (as well as a pair of Langmuir probes) (Fig. 5). The manipulator was plunged for the duration of 2.8-3.2 s, to coincide with the appearance of an I-mode (Fig. 5 (a)). The high frequency mode is visible as a modulation of the ICRF wave amplitude with a broad local maximum at ~100 kHz (Fig. 5 (b)). The feature is commonly called the weakly coherent mode, or WCM [11]. Note, that the I-mode was "sampled" by the manipulator probe in the inserted position at the major radius of the limiter, while the L- and

DOI: 10.1051/epjconf/201715703038

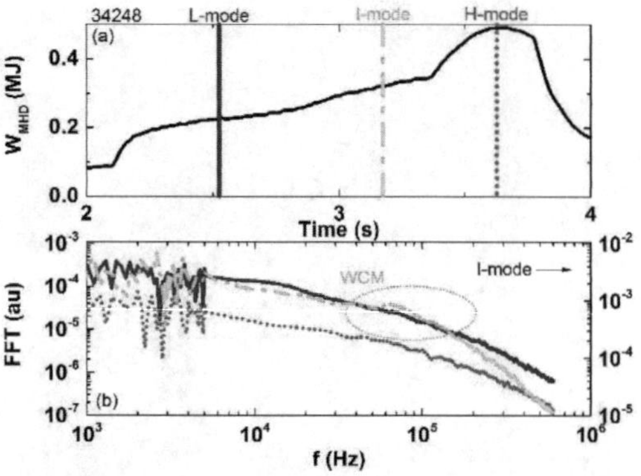


**Fig. 5.** (a) The stored plasma energy $W_{MHD}$ is used to identify three different plasma modes, the L-, the I-, and the H-modes. (b) The frequency spectra of the ICRF fast wave amplitude measured with the MEM diagnostic head in the three plasma modes. The WCM refers to the broad weakly coherent mode observed near 100 kHz in the I-mode part of the discharge.

H-mode data were obtained in the retracted position of the manipulator (10 cm behind the limiter).

## 4 Conclusion

The HFB diagnostic on ASDEX Upgrade has been greatly expanded during the 2016 opening. The expansions include additional probes that increase the toroidal and radial diagnostic coverage, as well as the additional ability to measure frequency spectra emitted by the plasma in the ICRF range. Possible studies with the new HFB diagnostic suite include the efficiency of core ICRF wave absorption in different heating schemes (Fig. 3) or identification of ion cyclotron emissions under various plasma conditions (Fig. 4). It is also capable of providing information on ICRF wave-plasma turbulence interactions in the SOL and pedestal plasma regions with the new midplane manipulator head (Fig. 5).

This work has been carried out within the framework of the EUROfusion Consortium and has received funding from the Euratom research and training programme 2014-2018 under grant agreement No 633053. The views and opinions expressed herein do not necessarily reflect those of the European Commission.

## References

1. V. Bobkov, F. Braun, R. Dux, et al., Nucl. Fusion Nucl. Fusion **56** (2016) 084001.
2. S. J. Wukitch, M. L. Garrett, R. Ochoukov, et al., Physics of Plasmas **20**, 056117 (2013).
3. B. Soutic, et al., Plasma Phys. Control. Fusion **36** (1994) B123-BI32.
4. F. Durodié, M. Nightingale, A. Argouarch, et al., Fusion Engineering and Design **84** (2009) 279–283.
5. D. Milanesio and R. Maggiora, Nucl. Fusion **50** (2010) 025007.
6. A. Bader, G. Bosia, A. Messiaen, et. al., Fusion Engineering and Design, **in press** (2017).
7. W. Zhang, V. Bobkov, T. Lunt, et. al., Nucl. Fusion **56** (2016) 036007.
8. R. Ochoukov, V. Bobkov, H. Faugel, H. Fünfgelder, et. al., Review of Scientific Instruments **86**, 115112 (2015).
9. E. A. Lerche, D. Van Eester and JET EFDA Contributors, Plasma Phys. Control. Fusion **50** (2008) 035003.
10. R. O. Dendy, K. G. McClements, C. N. Lashmore-Davies, et al., Phys. Plasmas **1** (10), 1994.
11. D. G. Whyte, A. E. Hubbard, J. W. Hughes, et al., Nucl. Fusion **50** (2010) 105005.

EPJ Web of Conferences **157**, 03039 (2017) DOI: 10.1051/epjconf/201715703039
*22 Topical Conference on Radio-Frequency Power in Plasmas*

# ICRF-Induced Changes in Floating Potential and Ion Saturation Current in the EAST Divertor

*Rory* Perkins[1,⋆], *Joel* Hosea[1], *Gary* Taylor[1], *Nicola* Bertelli[1], *Gerrit* Kramer[1], *Chengming* Qin[2], *Liang* Wang[2], *Jichan* Yang[2], and *Xinjun* Zhang[2]

[1] *Princeton Plasma Physics Laboratory, Princeton, New Jersey, U.S.A*
[2] *Institute for Plasma Physics, Chinese Academy of Science, Hefei, China*

**Abstract.** Injection of waves in the ion cyclotron range of frequencies (ICRF) into a tokamak can potentially raise the plasma potential via RF rectification. Probes are affected both by changes in plasma potential and also by RF-averaging of the probe characteristic, with the latter tending to drop the floating potential. We present the effect of ICRF heating on divertor Langmuir probes in the EAST experiment. Over a scan of the outer gap, probes connected to the antennas have increases in floating potential with ICRF, but probes in between the outer-vessel strike point and flux surface tangent to the antenna have decreased floating potential. This behaviour is investigated using field-line mapping. Preliminary results show that mdiplane gas puffing can suppress the strong influence of ICRF on the probes' floating potential.

## 1 Introduction

RF rectification is a sheath phenomenon that is important for impurity production via ion bombardment, hot-spot formation on antennas, modifications to the scrape-off layer (SOL), and dissipation of wave power (far-field sheaths). Many aspects of RF rectification focus on the possible increase in plasma potential resulting from relatively strong RF fields driving electrons into plasma-facing components. However, in the divertor of NSTX, the floating potential of certain swept probes decreases and the currents to certain tiles increases with the application of ICRF heating. This is as expected for relatively weaker RF fields for which the oscillating plasma potential drives a net (RF-averaged) electron current without depleting the linking flux tube of electrons or modifying the I-V characteristic observed in the absence of RF fields [1]. The averaged electron current induced by this form of RF rectification can heat the tiles locally and is the leading hypothesis for the heat flux observed in the spirals found in the divertors of NSTX during fast-wave heating [2]. This effect is observed for probes attached to field lines that pass through the SOL in front of the fast-wave antenna but are not linking the antenna. On the Alcator C-Mod device, when the field lines intercept the antenna, the RF rectification in the large electric field there does result in large positive increases in the plasma potential, presumably to reduce the electron current at the antenna [3].

The Experimental Advanced Superconducting Tokamak (EAST) provides an excellent opportunity to study RF rectification in the SOL in finer detail. EAST is designed for high-power long-pulse discharges, and a major component is an ICRF system consisting of two antennas and 12 MW of source power [4]. One antenna is located at I port and consists of four toroidally-spaced folded straps. The second antenna is at B port and consists for a two-by-two array. The nominal parallel wavenumber for the I port array is around 14 $m^{-1}$ for dipole phasing while for B-port $k_{\|} \approx 12$ $m^{-1}$. Current levels of coupled ICRF power are sufficient to modify both the SOL and the core plasma. EAST has extensive arrays of triple probes in both the carbon lower divertor [5] and the tungsten upper divertor [6]. The upper divertor has two arrays spaced apart by 112 degrees. The probes do not have RF compensation and are thus sensitive to RF rectification. While RF rectification is commonly studied in the vicinity of the antenna where the RF fields are expected to be strongest, far-field effects are imporant not only for NSTX but perhaps also for Alcator C-Mod, where far-field sheaths on limiters or divertor tiles may also be contributing to impurity production [7].

This paper explores several different cases of RF rectifcation in the divertor. Section 2 continues studies of an previously observed effect where probes on flux surfaces that intercept the antenna have a positive increase in floating potential whereas those more inboard swing negative. Magnetic mapping of the probe location to the midplane helps assess the location of these probes relative to the antennas, the density profile measured by a reflectometer, and between upper- and lower-divertor probes. Second, in Sec. 3 we present preliminary findings regarding the effects of gas puffing at one of the antennas on the response of probes. Both the floating potential ($V_{fl}$) and ion saturation current ($I_{sat}$) can be dramatically changed. Interestingly, in some instances, the large modulation of floating potential with RF is greatly reduced.

⋆e-mail: rperkins@pppl.gov

DOI: 10.1051/epjconf/201715703039

## 2 The dependence of floating potential polarity with ICRF on magnetic connection in the SOL

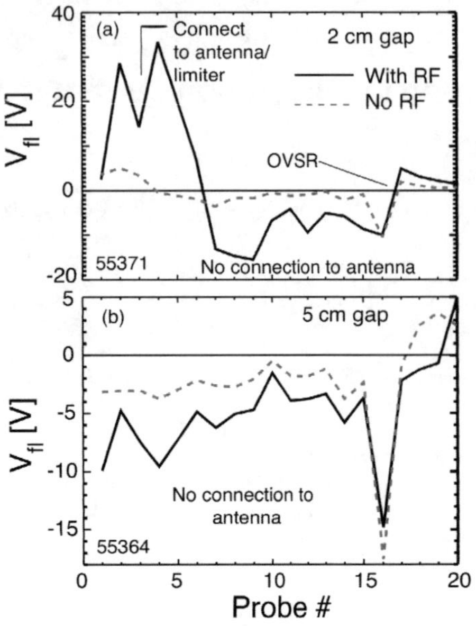


Figure 1: Profiles of floating potential for the lower-divertor array for two different outer-gap values. Black curves are profiles taken at $t$ = 3.525 s, during the RF pulse and averaged over 40 ms. Green curves are taken at $t$ = 3.190 s, before the RF pulse. OVSR indicates the outer-vessel strike radius.

Over a scan of the outer gap size, the lower-divertor probes tended to show a decrease in floating potential with ICRF unless the probe was further outboard than the flux surface tangent to the antenna, in which case the change in floating potential is positive [8] . These shots were lower-single null discharges. During shot 55364, 0.75 MW of ICRF power were coupled through both antennas; during shot 55371, the I-port antenna coupled 0.85 MW while the B-port antenna coupled 1.15 MW. In each case, the ICRF power was applied from 3.277 s to 4.248 s. With a nominal outer gap size of 2 cm (Fig. 1(a)), the probes outboard of the flux surface tangent to the antenna tended to rise in floating potential with ICRF whereas those inboard tended to fall in floating potental. The most inboard probes (probes numbered 17-20 in this plot) have a positive change in floating potential, but they also lie inboard of the outer-vessel strike radius (OVSR). At a 5 cm gap (Fig. 1(b)), the lower array shows a negative value for floating potential during ICRF, and for this outer-gap value none of the probes connect to the antenna. As an aside, the very negative floating potential for probe 16, present even without ICRF, may be due to the probe's location relative to the outer strike point [9].

These discharges are now analyzed using field-line mapping in Fig. 2, which maps both upper (D-port) and lower probes to the midplane. Also shown are the midplane major radii of the ICRF limiter faces for each antenna. The general trends found before are more or less reproduced and can be directly compared between upper and lower divertor probes. In the case of shot 55364 (Fig. 2a), all probes lie inboard of the antenna, and their floating potential are below zero or marginally above it. Strong positive floating potentials are only found for those probes with mapped midplane radii beyond the antenna limiters indicated in the figures (Fig. 2b).

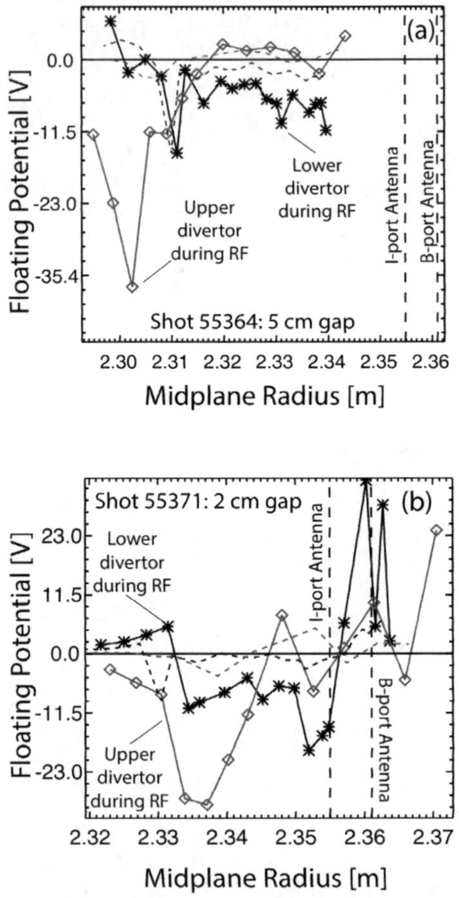


Figure 2: Floating potential for both the lower-divertor array (black) and the Bay-D upper-divertor array (red) as a function of major radius mapped to the midplane. Dashed lines show pre-ICRF profiles at $t$ = 3.19 s averaged over 16 ms. (a) Shot 55364, 5 cm outer gap, and (b) shot 55371, 2 cm outer gap.

Figure 3 shows the magnetic mapping for both sets of probes, plotting the probe major radius at the divertor against the major radius mapped to the midplane. This plotting scheme is chosen so that density profiles from the J-port edge reflectometer can be overlaid as a function of midplane radius. The midplane density is quite low compared to the righthand cutoff density for fast waves. For instance, taking an ICRF frequency of 27 MHz and launched

 DOI: 10.1051/epjconf/201715703039

$k_{\parallel}$ = 12.2 m, an on-axis toroidal field of 2 T (extrapolating to 1.6 T at the antenna location) gives a cutoff density of roughly $10^{19}$ $m^{-3}$, which greatly exceeds the measured density at the midplane anywhere in the SOL and places the cutoff layer well inside the separatrix according to Fig. 3. The relationship between this large evanescent layer for fast waves and the divertor probe signals will be the topic of future work.

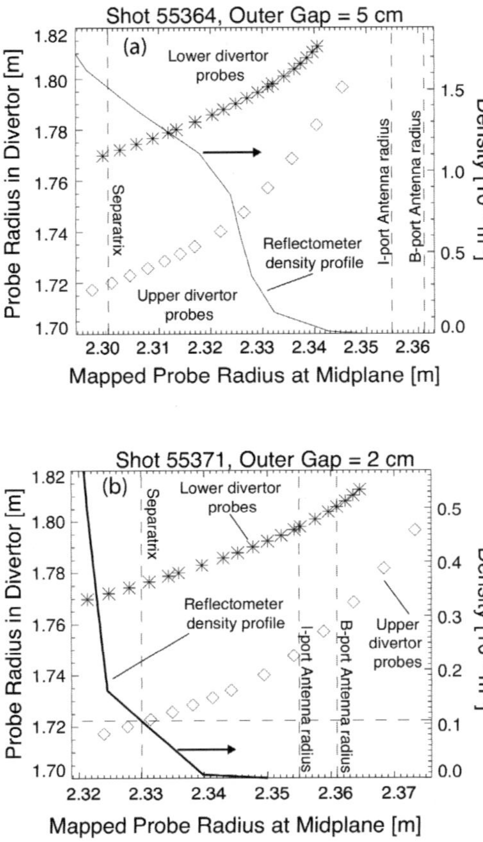


Figure 3: Magnetic mapping of the divertor probes to the midplane for comparison with the midplane density profile obtained by reflectometry. (a) Shot 55364 and (b) Shot 55371, both at $t$ = 3.525 s during the ICRF pulse.

## 3 The effect of gas puffing on divertor probes

This section shows preliminary results for a series of shots with increasing gas puffing in single upper null discharges driven by 1.4 MW of 4.6 GHz lower hybrid heating. The ICRF power is applied to each antenna separately; in all cases presented here, the I-port antenna is powered on first at $t = 3.2s$ until $t = 5.12s$, while B-port is powered from $t$ = 5.77 s to $t$ = 7.2 s, with each modulated between full power (1 MW) and half power. The gas injector, located near the B-port antenna, is active from $t$ = 2.5 s to 9 s, and increases from shot 69938 to 69939 and again to 69940. Further increase in gas puffing resulted in a prompt disruption. Shot 69942 is a fiducial shot without gas puffing. Note that the B-port antenna arced off around $t$ = 6.8 s for shot 69940. Also, changes in the outer gap, shown in Figure 4, must be taken into account. The outer gap is held rather constant during the modulation, varying 3 mm peak-to-peak. Shots 69938 and 69939 have a 0.5 cm increase in outer gap over the fiducial. The gap for shot 69940 increased steadily during the I-port pulse and dramatically during the B-port pulse.

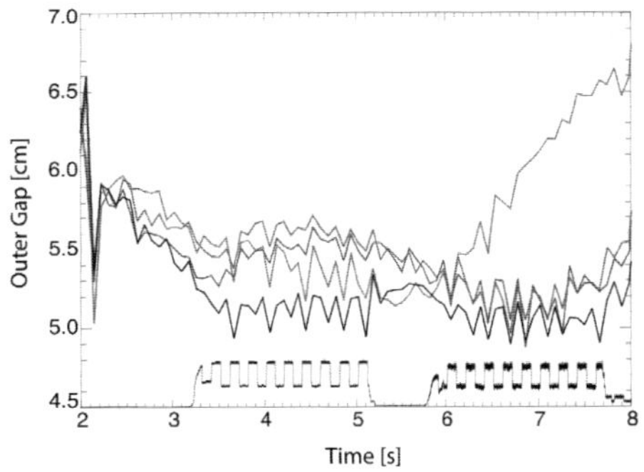


Figure 4: Outer gap values for no-gas-puff (black), smallest puff (blue), medium puff (red) and largest puff (green).

The most dramatic effect of the gas puffing on the divertor probes can be seen on Probe 6, Fig. 5a. Without gas puffing, this probe shows a strong negative response to both the I-port pulse and the second half of the B-port pulse. With gas puffing, this response is strongly suppressed at intermediate gas puffs. Similar behavior is seen on neighboring probes. For Probe 12 (Fig. 5b), though, the modulations in $V_{fl}$ become bigger with gas puffing.

The $I_{sat}$ signals response to ICRF is also modified by the gas puffing; Fig. 6 shows $I_{sat}$ for probe 4 and Fig. 7 shows $I_{sat}$ for Probe 5 in the lower divertor. Probe 4 is strongly positively correlated with the first half of the B-port pulse for 69938 and 69939, whereas Probe 5 is anti-correlated.

## 4 Discussion and Conclusions

Changes to divertor probe signals due to ICRF can be due to wave fields directly interacting with the sheaths, modifications to the SOL, and core heating. Part of this work is to determine the former in relationship to the other components for comparison with the NSTX results. Negative changes in floating potential are likely due to RF rectification. It is commonly reasoned that plasma potential rises with RF, so in principle the floating potential should also rise, being given by first-principle considerations as: $V_{fl} = V_{pl} - \alpha T_e$ for some $\alpha$ depending on geometry and gas species. However, in an RF field, the floating potential can drop when RF currents flow through the sheath and deliver a net electron current to the probe as per Ref. [1]. From

DOI: 10.1051/epjconf/201715703039

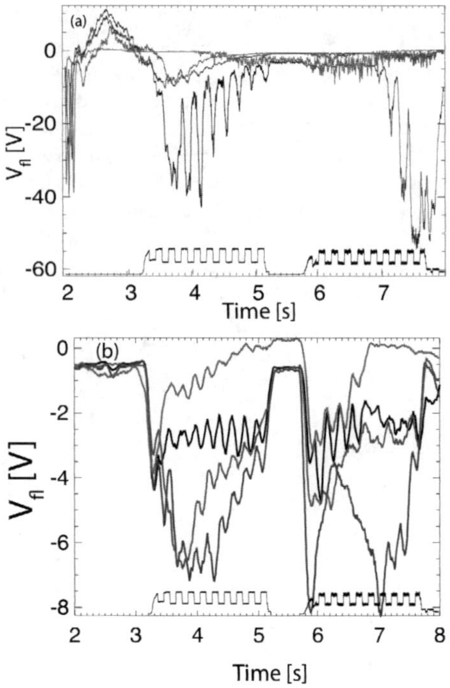


Figure 5: $V_{fl}$ for Probes 6 (a) and 12 (b) in the upper divertor. Black: shot 69942, blue: 69938, red: 69939, green: 69940.

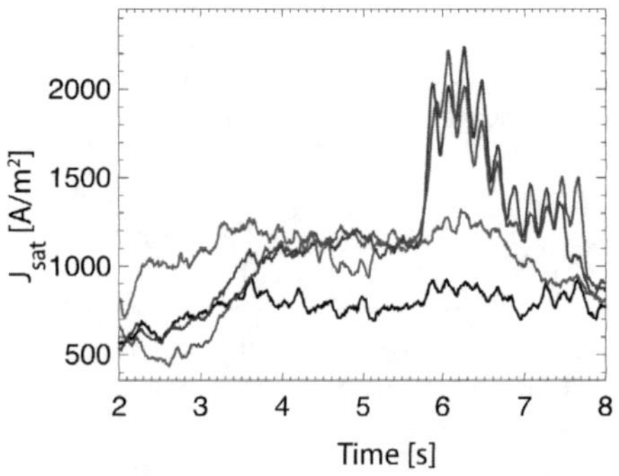


Figure 6: $I_{sat}$ for Probe 4 of the lower divertor during a gas-puff scan. Black: shot 69942, blue: 69938, red: 69939, green: 69940.

our observations, probes connected to the antenna see an increase in plasma potential in response to rectification at the antenna which then propagates along field lines. Inboard probes with prompt anti-correlated responses to RF, however, are likely subject to ICRF wave fields.

In conclusion, magnetic mapping of the EAST divertor probes back to the midplane allows comparison between upper and lower probe sets and also to diagnostics such as the edge reflectometer. A strong rise in floating potential is seen only for outboard probes in the case of a 2 cm outer gap when such probes lie outboard of the flux surface tangent to the antenna face. Gas puffing is seen to strongly modify the probe signals and in some instances substantially reduces their modulation with ICRF; this effect will be studied further in the future.

This research is partly supported by the National Key Research and Development Program No. 2016YFA0400601 and also by the China Magnetic Confinement Fusion Science Program No. 2015GB101001. This research is also supported by US DOE Contract No. DE-AC02-09CH11466.

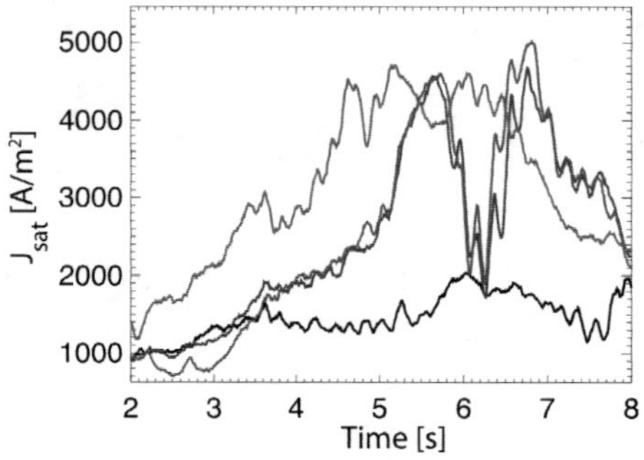


Figure 7: $I_{sat}$ for Probe 5 of the lower divertor during a gas-puff scan. Black: shot 69942, blue: 69938, red: 69939, green: 69940.

## References

[1] A. Boschi, F. Magistrelli, Il Nuovo Cimento **29**, 487 (1963)
[2] R. J. Perkins et al., Phys. Plasmas **22** (2015)
[3] R. Ochoukov et al., Plasma Physics and Controlled Fusion **56**, 015004 (2014)
[4] Y. Zhao et al., Fusion Engineering and Design **89**, 2642 (2014)
[5] T. Ming et al., Fusion Engineering and Design **84** (2009)
[6] J.C. Xu et al., Rev. Sci. Instrum. **87** (2016)
[7] S. Wukitch et al., Bull. APS **61** (2016)
[8] J. Hosea et al., Bull. APS **61** (2016)
[9] J.C. Hosea et al., AIP Conference Proceedings **1580**, 105 (2014)

# A Finite-Orbit-Width Fokker-Planck solver for modeling of energetic particle interactions with waves, with application to Helicons in ITER

*Yuri V.* Petrov[1,*] and *R.W.* Harvey[1]

[1]CompX, PO Box 2672, Del Mar, CA 92014, USA

**Abstract.** The bounce-average (BA) finite-difference Fokker-Planck (FP) code CQL3D [1,2] now includes the essential physics to describe the RF heating of Finite-Orbit-Width (FOW) ions in tokamaks. The FP equation is reformulated in terms of Constants-Of-Motion coordinates, which we select to be particle speed, pitch angle, and major radius on the equatorial plane thus obtaining the distribution function directly at this location. Full-orbit, low collisionality neoclassical radial transport emerges from averaging the local friction and diffusion coefficients along guiding center orbits. Similarly, the BA of local quasilinear RF diffusion terms gives rise to additional radial transport. The local RF electric field components needed for the BA operator are usually obtained by a ray-tracing code, such as GENRAY, or in conjunction with full-wave codes. As a new, practical application, the CQL3D-FOW version is used for simulation of alpha-particle heating by high-harmonic waves in ITER. Coupling of high harmonic or helicon fast waves power to electrons is a promising current drive (CD) scenario for high beta plasmas. However, the efficiency of current drive can be diminished by parasitic channeling of RF power into fast ions, such as alphas, through finite Larmor-radius effects. We investigate possibilities to reduce the fast ion heating in CD scenarios.

## 1 General description of FOW features

An accurate modeling of fast ion interaction with RF waves should include Finite-Orbit-Width (FOW) effects. The bounce-average (BA) finite-difference Fokker-Planck (FP) code CQL3D [1] has been recently upgraded to incorporate all necessary FOW physics [2]. The general formulation is based on writing the original 6D FP equation in canonical action-angle space, averaging over periodic angle variables, and making a transformation from the 3D canonical action space to a computationally convenient set of invariants $\mathbf{I} = (I_1, I_2, I_3)$. Our choice is dictated by the desire to keep computational grids as close as possible to the original Zero-Orbit-Width (ZOW) version of CQL3D, and to maintain direct visualization of the results in velocity and configuration space. Thus, we adopt $\mathbf{I} = (u_0, \theta_0, R_0)$, where $R_0$ is the major radius coordinate along the minimum-$B$ surface (the midplane of tokamak, in case of up-down symmetrical equilibrium). For each given $R_0$ point, the value of particle speed (relativistically, momentum-per-mass) $u_0$ and value of the pitch-angle $\theta_0$ at this point determine a unique orbit.

The transformation from canonical action space $\mathbf{J}$ to our $\mathbf{I} = (u_0, \theta_0, R_0)$ space results in the general form of the bounce-averaged FPE for the particle distribution function $f_0(\mathbf{I},t)$ [2]:

$$\frac{\partial}{\partial t} f_0(\mathbf{I},t) = \frac{1}{\Im}\frac{\partial}{\partial I^i}\left[\Im D^{ij}\frac{\partial}{\partial I^j} f_0 + \Im F^i f_0\right] + S_0(\mathbf{I},t), \quad (1)$$

where $\Im \equiv |\partial\mathbf{J}/\partial\mathbf{I}|$ is the Jacobian of the transformation, and the BA diffusion coefficients $D^{ij}$ and advection terms $F^i$ are expressed through the local diffusion $\mathbf{D}^{\mathbf{uu}}$ (from collisions and/or resonant RF heating) and through the local collisional friction $\mathbf{F}^{\mathbf{u}}$,

$$D^{ij} = \left\langle \frac{\partial I^i}{\partial \mathbf{u}} \cdot \mathbf{D}^{\mathbf{uu}} \cdot \frac{\partial I^j}{\partial \mathbf{u}} \right\rangle, \quad F^i = \left\langle \frac{\partial I^i}{\partial \mathbf{u}} \cdot \mathbf{F}^{\mathbf{u}} \right\rangle. \quad (2)$$

The bounce-averaging involves the transformation coefficients $\partial\mathbf{I}/\partial\mathbf{u}$, which effectively give rise to non-local neoclassical phenomena, such as the radial diffusion described by $D^{33} = \langle(\partial R_0/\partial\mathbf{u})\cdot\mathbf{D}^{\mathbf{uu}}\cdot(\partial R_0/\partial\mathbf{u})\rangle$. Also, Eq. (1) includes the BA particle source operator $S_0$, such as from NBI or fusion alphas. The initial benchmarking of the code is performed for calculation of the bootstrap current [2]. The results from other tests will be published elsewhere.

In this paper, we apply the FOW version of the code to simulation of alpha-particle heating by high-harmonic fast waves in ITER. Using high frequency fast waves for electron current drive (CD) is a promising scenario for high beta plasmas [3]. However, its efficiency can be diminished by parasitic channeling of RF power into fast ions, such as alphas, through wave-particle resonant Larmor-radius effects. Here, we consider two ITER scenarios [4, 5], one with a relatively low content of alphas, $P_{\alpha\text{-source}}$ = 58 MW, and another with a high content of alphas corresponding to $P_{\alpha\text{-source}}$ = 112 MW.

* Corresponding author: petrov@compxco.com

The parameters that we change in simulations are the wave frequency and the vertical position of antenna.

## 2 ITER scenario LH_60s_3

This scenario [4], which is based on the total auxiliary power of 73 MW (33 NB + 20 EC + 20 LH) has a high central temperature $T_{e0} \approx T_{i0} \approx 31$ keV and high density $n_{e0} \approx 8.5e19$ m$^{-3}$ ($n_D = n_T = 3.2e19$ m$^{-3}$), which results in a high power of fusion alphas $P_{\alpha\text{-source}} = 111.6$ MW. The other plasma parameters are $I_p = 12.5$ MA, $B_0 = 5.3$ T. The fast waves at 500 MHz (called "high harmonic" or "helicon" waves in this frequency range) can encounter a wide range of cyclotron harmonics for alphas, $\omega/\omega_{c\alpha}$ = 9-17. As the first step of the modeling we use the GENRAY ray-tracing code [6] to obtain linear absorption on ions and electrons, RF field components and other data along rays, which is further imported by the CQL3D code to form the bounce-average quasilinear RF diffusion operator for the heating of alpha-particles (the Fokker-Planck'd species). As the fusion-born alphas are slowing down by collisions on one hand, and absorbing RF power on the other, the remaining RF power (linear damping on electrons and ions, if any) is adjusted self-consistently. Results from a GENRAY run are shown in **Fig. 1**. The antenna is shifted down by 1 m, which allows a better access of the plasma center by rays. In this GENRAY run, almost all RF power is deposited to electrons. Other species in ray-tracing simulations are thermal $D^+$, $T^+$, thermalized He-4, and a small fraction of Be-9 and Ar-40 that make up $Z_{eff} = 1.7$. In all simulations discussed in this paper we use the total RF power of fast waves equal to 10 MW, and the initial parallel refractive index is set to $n_{\|} = 1$-5. For the rays shown in **Fig. 1** the profile of power absorption by electrons is quite broad from plasma center to the edge, with a peak of power being at $\rho \approx 0.63$ ($\rho$ is defined through the sqrt of toroidal magnetic flux) and another narrow peak present at $\rho \approx 0.05$.

The results from CQL3D runs are shown in **Fig. 2**. The profile of RF power absorbed by alphas in **Fig. 2(a)** is based on the solution of FP equation obtained by CQL3D-FOW, while the profile for deposition on electrons is based on self-consistent linear damping of the power flowing in rays, using the damping rate from GENRAY. It is seen that as the power deposition to alphas grows in time (thin lines converging to the bold line) the power absorbed by electrons is diminished – mainly at $\rho < 0.6$. The damping on alphas is mostly caused by resonance with their 16$^{th}$ IC harmonic. The parasitic absorption by alphas is very high, 36% of the total injected RF power. This is related to penetration of waves into the plasma core where the content of alphas is relatively large. In a modified launching condition, when RF antenna is not shifted from the midplane, the rays tend to miss the core. As a result, in the GENRAY run there is no deposition of power to electrons at $\rho < 0.2$. From the CQL3D calculations, the damping on alphas is reduced from 36% to 30%.

A better result in power partition is achieved with higher wave frequency $f_{RF}$ = 800 MHz. The profiles of power deposited to alphas and electrons are shown in **Fig. 3**. The antenna is positioned at the midplane in this run. The power absorbed by alphas is 14%. If the antenna is shifted by 1 m down off the midplane, it becomes somewhat larger, 19% of the total RF power. The improvement is due to two factors. First, at higher frequency, the helicon waves tend to propagate more poloidally than radially, which moves the region of power deposition away from the plasma core. Second, and more important factor is that the waves interact with higher harmonic numbers: $\omega/\omega_{c\alpha}$ = 21-28 along rays for the $f_{RF}$ = 800 MHz runs, which reduces the rate of alpha IC heating.

Another concern in parasitic power absorption is the damping on fast $D^+$ ions produced by NBI. In a separate run, using 33 MW NBI source, we obtained a 5% damping on $D^+$. It is located at $\rho \approx 0.65$-0.95, and is caused by resonances at 24$^{th}$-27$^{th}$ IC harmonics of $D^+$.

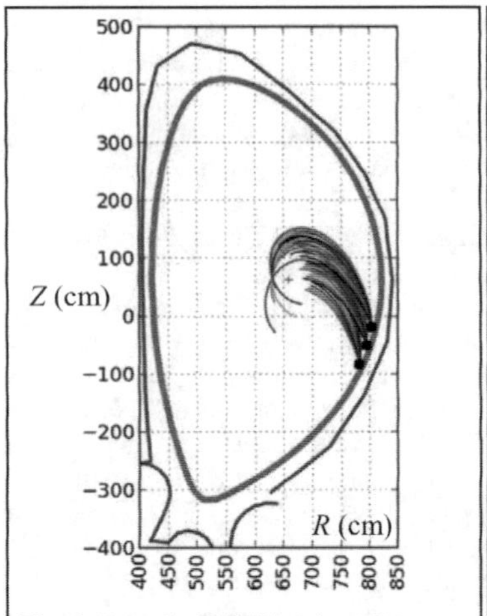


**Fig. 1.** Rays for HHFW at $f_{RF}$ = 500 MHz. Antenna is shifted by 1 m down from the midplane.

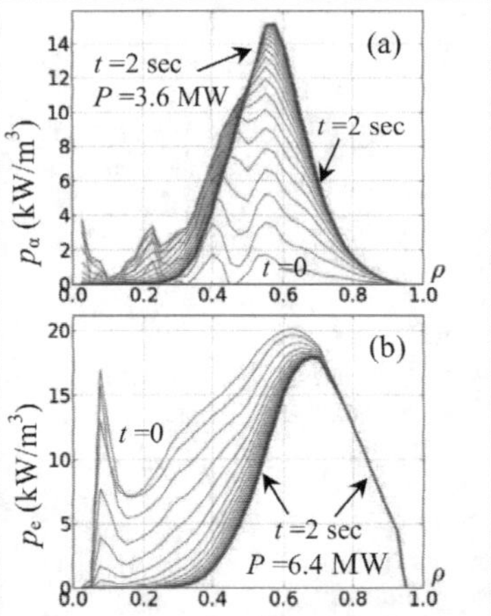


**Fig. 2.** Power deposition for 500 MHz waves: (a) to alphas, and (b) to electrons at different time steps.

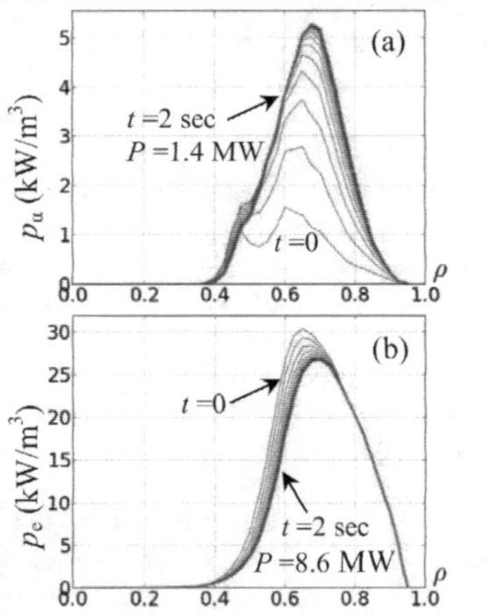


**Fig. 3.** Power deposition for 800 MHz waves: (a) to alphas, and (b) to electrons at different time steps.

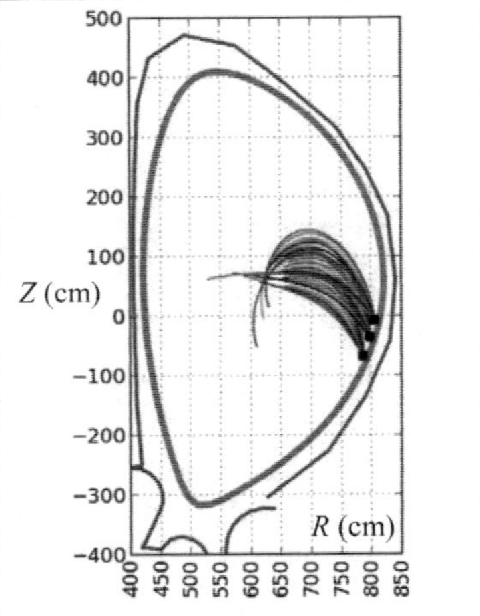


**Fig. 4.** Rays for HHFW at $f_{RF}$ = 500 MHz. Antenna is shifted by 1 m down from the midplane.

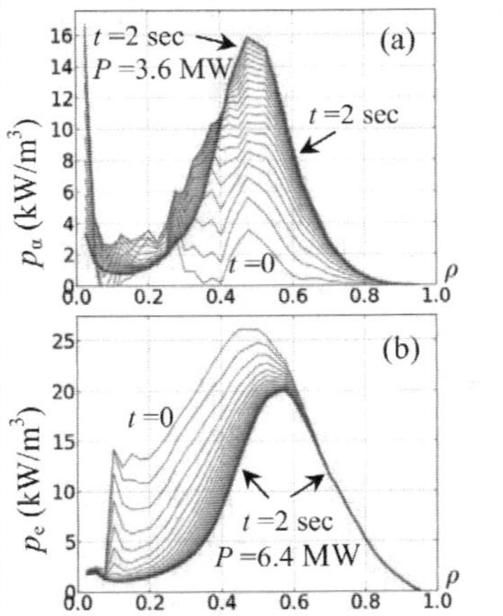


**Fig. 5.** Power deposition for 500 MHz waves: (a) to alphas, and (b) to electrons at different time steps.

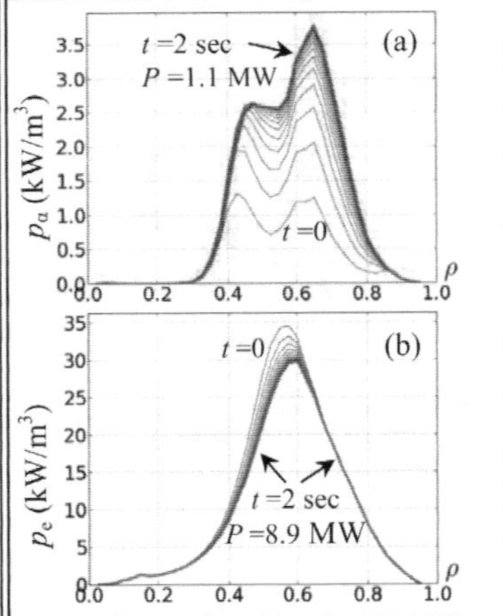


**Fig. 6.** Power deposition for 800 MHz waves: (a) to alphas, and (b) to electrons at different time steps.

## 3 ITER scenario EC_60s_1f

Different from the previous case, this scenario [5] has a lower auxiliary power of 56.5 MW (16.5 NB + 20 EC + 20 LH). Although the central magnetic field is unchanged, 5.3 T, the plasma current is lower, 9 MA. The central temperature is almost the same, $T_{e0} \approx T_{i0} \approx$ 36 keV, but the density is reduced to $n_{e0}$ = 6.0e19 m$^{-3}$, $n_D = n_T \approx$ 2.4e19 m$^{-3}$. Because of the lower density of $D^+$ and $T^+$, the power of fusion alphas is only $P_{\alpha\text{-source}}$ = 57.9 MW. One would expect a significantly smaller absorption of RF power on alphas, but it is not the case. The results from GENRAY run for a similar case of **Fig. 1** with $f_{RF}$ = 500 MHz and the vertically shifted antenna are shown in **Fig. 4**. The corresponding profiles of power deposition obtained with CQL3D-FOW are shown in **Fig. 5**. The absorption of RF power on alphas is the same as in the previous scenario, i.e., 36%. This can be explained by observation that because of the lower density, the rays' radial group velocity is higher, and the power is deposited into the interior region with the larger population of alphas. The change of the power deposition profile is noticeable from comparison of **Fig. 2** and **Fig. 5**. In the present case the peak of the $p_e(\rho, t{=}0)$ profile is at $\rho$ = 0.48 (**Fig. 5(b)**, the top curve), while in the previous scenario the peak is at $\rho$ = 0.63 (**Fig. 2(b)**). The peak position of the power deposition to alphas is also shifted from $\rho$ = 0.57 to $\rho$ = 0.48 in the present scenario, and there is an additional absorption at $\rho$ < 0.3, as seen in **Fig. 5(a)**. For the alpha power absorption, the most of damping occurs at $\omega/\omega_{c\alpha}$ = 16 resonance layer, which is shifted by 15 cm inward ($\Delta\rho \approx$ –0.1) for present equilibrium conditions, comparing to the previous scenario.

Switching to the higher frequency of $f_{RF}$ = 800 MHz yields a better power partition in favor of electrons. The results for the case of the midplane-centered antenna are shown in **Fig. 6**. The absorption on alphas is only 11%. This is the lowest fraction of RF power absorbed by alphas among the two ITER scenarios with different frequencies (500 or 800 MHz) and antenna positions. If the antenna is shifted by 1 m down off the midplane (in the present scenario with 800 MHz) the power deposited to alphas becomes 1.7 MW (17%). An additional CQL3D-FOW calculation for the power deposition to fast $D^+$ ions produced by 16.5 MW NBI source (half of that in the previous scenario), gives only a 1% of RF power damping on $D^+$ with two narrow peaks in the plasma edge region around $\rho \approx$ 0.8 and $\rho \approx$ 0.9. Thus, the result shown in **Fig. 6** appears to be optimal for electron CD by the helicon waves. The profile of current based on GENRAY calculations is shown in **Fig. 7**. The CD efficiency is 0.047 A/W (total RF current is 0.47 MA), based on asymptotic formula by Ehst-Karney [7]. The linear CD calculation is sufficient for present purposes, since the electron quasilinear diffusion is weak. This CD efficiency is comparable to that expected for LHCD in ITER (0.04-0.06 A/W).

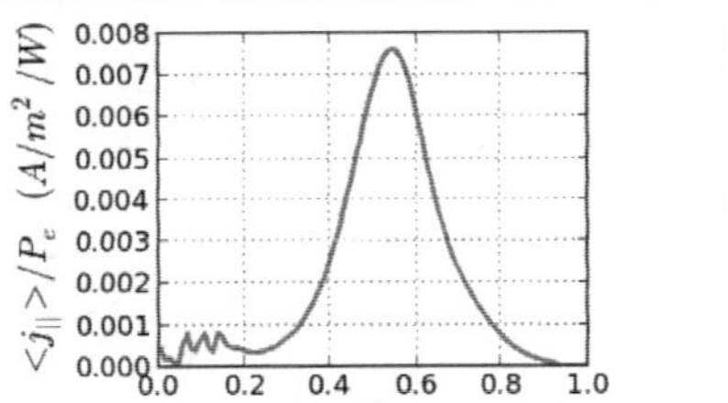


**Fig. 7.** Profile of electron current density generated by helicon waves at $f_{RF}$ = 800 MHz (normalized by the launched RF power 10 MW).

It should be mentioned though, that the distribution of alphas is generating a current as well, as they slow down and are transported in the radial direction. The profile of this current, derived from the solution of the FPE by CQL3D-FOW, is shown in **Fig. 8**. The value of this current reaches 0.37 MA at $t$ = 4 sec (0.24 MA if the electron screening current is taken into account). It has nothing to do with the RF power, but it is rather a result

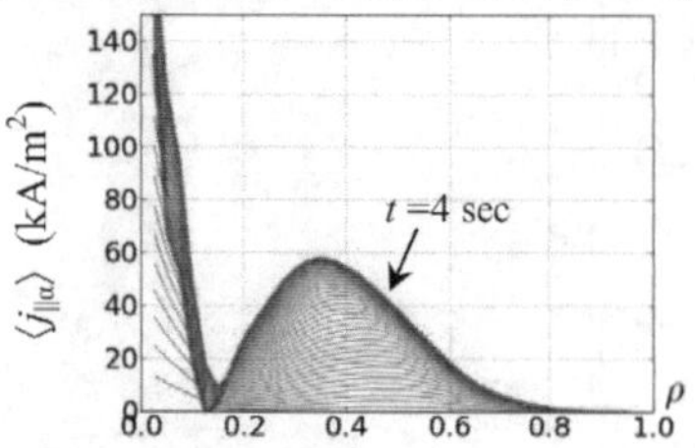


**Fig. 8.** Flux-surface-average parallel current density of alphas at different time steps, due to the neoclassical transport. Calculated by CQL3D-FOW.

of the neoclassical-type radial transport. It cannot be calculated by the model bootstrap formulas because they are based on a Maxwellian distribution. In the present FOW calculations, the most current is produced by alpha particles with energies 0.3-3.5 MeV, rather than by thermal part of the distribution. With the RF power turned on or off in calculations, this current remains almost unchanged, even in the case of $f_{RF}$ = 500 MHz when 3.6 MW is deposited to alphas.

Returning to the optimal CD scenario with $f_{RF}$ = 800 MHz, 1.1 MW deposited to alphas, it is instructive to look at the distribution function of alphas. For the peak position of power deposition at the midplane coordinate $R_0$ = 779 cm ($\rho$ = 0.65) the distribution function is shown in **Fig. 9**. It is the steady state solution of FPE at $t$ = 2 sec (except the thermalized population of alphas at $u/c$ < 5e-3 that keeps increasing). It is seen that the waves produce a high-energy tail beyond the birth energy of 3.52 MeV, especially at pitch angles close to $\pi/2$. At energies below 3.52 MeV, the distribution has a strong variation in pitch angle. This variation originates from the similar variation of the source operator: the co-passing and especially co-going barely trapped particles have excursions into the hot inner region of plasma which is well populated by alphas. In contrast, the counter-going particles come from a relatively cold edge region. Consequently, the source operator is over-populated at pitch angles less than $\pi/2$ (co-going side) and under-populated at pitch angles greater than $\pi/2$ [8]. This is another important feature (besides the neoclassical transport) that appears in FOW formulation.

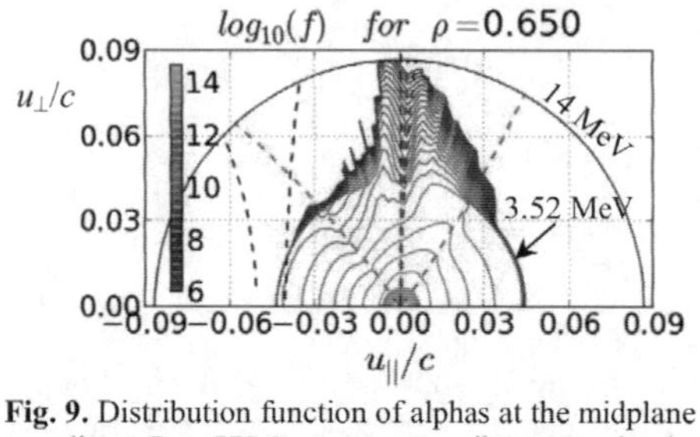


**Fig. 9.** Distribution function of alphas at the midplane coordinate $R_0$ = 778.8 cm corresponding to results shown in Fig. 6. It is the steady state solution of FPE at $t$ = 2 sec (except the thermalized population of alphas at $u/c$ < 5e-3 that keeps increasing). Dashed lines mark specific boundaries in velocity space [2].

## 4 Conclusion

Using high-harmonic or helicon fast waves for electron current drive in ITER is a promising CD method, but a big concern is the parasitic absorption by fusion-born alphas. The modeling with the CQL3D-FOW bounce-average Fokker-Planck code for the selected two ITER scenarios proves that the absorption of RF power on alphas can be as high as 36%. To reduce this absorption, the RF frequency has to be increased to at least 800 MHz. In the best case that we could achieve the absorption on alphas was 11% of the total RF power. The vertical displacement of RF antenna can serve as an additional tool for changing in some extend the power partition between electrons and alphas, at a cost of changing the localization of the absorption power profile.

Each CQL3D-FOW simulation described in the paper is done in a relatively short computational time: 1.5 hours using 192 cores. The steady-state solutions for a physical time $t$ = 2 sec are obtained in 20 time steps. Computer runs with the time step $dt$ = 0.1 sec (and larger) are stable because of the fully-implicit solver. The code can be applied for other ITER scenarios in future work. In general, it is demonstrated that the code is a valuable tool for modeling of fast ions' RF and transport physics in ITER and other tokamaks.

The authors acknowledge encouragement of the RF-SciDAC team, and support of the USDOE under grants ER54649, ER54744, and SC0006614. We would like to thank Sun-Hee Kim, Alexei Polevoi, and Mireille Schneider for providing data on ITER scenarios.

## References

1. R.W. Harvey and M.G. McCoy, "The CQL3D Fokker Planck Code", www.compxco.com/cql3d
2. Yu.V. Petrov and R.W. Harvey, Plasma Phys. Control. Fusion **58**, 115001 (2016)
3. V.L. Vdovin, Plasma Phys. Reports **39**, 95 (2013)
4. S.H. Kim, et al., Nucl. Fusion **56**, 126002 (2016)
5. S.H. Kim, et al., in Proc. 7th IAEA-TM on Steady-State Operation (Aix-en-Provence) O8 (2013)
6. A.P. Smirnov and R.W. Harvey, "The GENRAY Ray Tracing Code" (available at www.compxco.com/Genray_manual.pdf) (2000)
7. D.A. Ehst and C.F.F. Karney, Nucl. Fusion **31**, 1933 (1991)
8. Yu.V. Petrov and R.W. Harvey, 26$^{th}$ IAEA Fusion Energy Conference (Kyoto), TH/P4-24 (2016)

EPJ Web of Conferences **157**, 03041 (2017) DOI: 10.1051/epjconf/201715703041
*22 Topical Conference on Radio-Frequency Power in Plasmas*

# EC power management in ITER for NTM control: the path from the commissioning phase to demonstration discharges

*Francesca* M. Poli[1,*], *Eric* Fredrickson[1], *Mark* A. Henderson[2], *Nicola* Bertelli[1], *Daniela* Farina[3], *Lorenzo* Figini[3], *Emanuele* Poli[4]

[1]Princeton Plasma Physics Laboratory, Princeton, NJ 08543, USA
[2]ITER Organization, St Paul Les Durance, France
[3]Istituto di Fisica del Plasma, Milano, Italy
[4]Max-Planck-Institute fur Plasmaphysiks, 85748 Garching, Germany

**Abstract.** Time dependent simulations that evolve consistently the magnetic equilibrium and plasma pressure profiles and the width and frequency rotation of magnetic islands under the effect of the Electron Cyclotron feedback system are used to assess whether the control of NTMs on ITER is compatible with other simulataneous functionalities of the EC system, like core heating and current profile tailoring, or sawtooth control. Results indicate that the power needs for control can be reduced if the EC power is reserved and if pre-emptive control is used as opposed to an active search for an already developed island.

## 1 Introduction

Covering nearly the entire plasma radius, the Electron Cyclotron system planned on ITER has high flexibility, potentially allowing for combined central heating, current profile tailoring and MHD stability control of sawteeth and Neoclassical Tearing Modes [1-3]. Every application has to be balanced with the other heating and current drive sources for optimization of capabilities and resources. A critical application of the EC system is for NTM control and stabilization, for which the Upper Launcher (UL) has been specifically designed. The Upper Launcher can accommodate the entire 20MW available, with up to two thirds delivered to either the upper or the lower steering mirror. Depending on how much power is needed for NTM control and stabilization in each phase of the discharge, other applications might be restricted. This work aims at finding conditions for optimal usage of the EC in all phases of the discharge, going from operation at half-field to full field.

Similarly to previous work dedicated to the assessment of NTM stability and control on ITER, this work also relies on the use of a Modifed Rutherford Equation that includes the effects of the EC current and heating. Differently from previous work, the assessment is done by evolving consistently the plasma equilibrium and the pressure profiles and the magnetic island width and frequency under the action of the EC feedback control. The simulations confirm some of previous offline calculations, for example that an alignment within half the EC deposition width is needed for NTM stabilization [4-6]. In addition, they identify a need for optimization of the ramp-up and ramp-down phase, in particular of entry and exit from H-mode. They also point to limitations in the choice of the control schemes due to the fast growth rate of the (2,1)-NTM compared to the mechanical response of the steering and of the power switch between launchers.

## 2 Approach to control simulation

In order to assess the EC control system requirements, it is important to simulate the evolution of the NTM island in combination with the plasma magnetic equilibrium and the kinetic profiles, as they evolve in response to the external heating and current drive sources. In order to provide a simulated response of the plasma, a MRE has been interfaced with TRANSP [7] and coupled to a feedback control algorithm for the tracking of rational surfaces and for the management of the input power in response to the NTM stability. The MRE used here is based on the approach by Fredrickson [8], which was validated against the (3,2)-NTM on TFTR. The original MRE has been extended to include the effect of the EC heating and current drive [5,9] and the rotation of the island. The electron cyclotron calculations are performed with the beam tracing code TORBEAM [10]. The effect of the alignment (or misalignment) of the EC with the rational surfaces is consistently taken into account in the calculation of the ECCD and ECH terms and in the tearing stability term through the magnetic equilibrium and plasma current profile calculated in TRANSP. Only cases with continuous EC injection have been considered since power modulation is expected to be effective only when the deposition width is comparable to or wider than the magnetic island [4], which is not the case for the ITER simulations considered here.

*Corresponding author: fpoli@pppl.gov

DOI: 10.1051/epjconf/201715703041

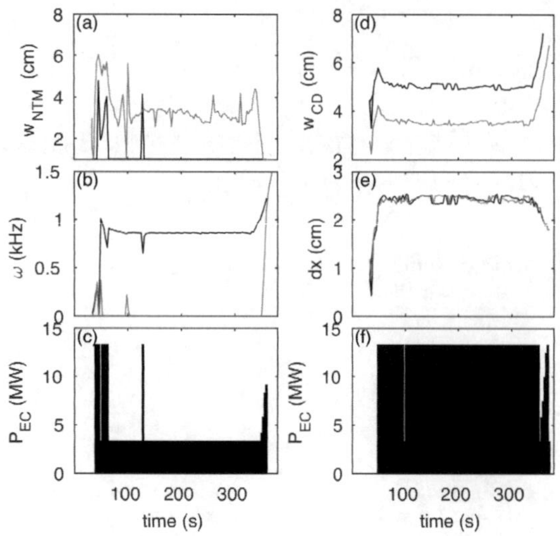


**Fig. 1.** EC feedback control calculations for a D-T discharge at 2.65T and 7.5MA. (a) width of the (2,1)-NTM (b) rotation frequency (d) EC deposition width (d) alignment between the EC and the q=2 surface (e) EC power for the case with broader $w_{EC}$ (f) EC power for the case with broader $w_{EC}$.

## 3 Assessment of control requirements

Approaches to NTM control can be divided into two categories: control of modes that have grown above the detection threshold size and prevention of the triggering of instabilities. Figure 1 shows two simulations of a plasma discharge with D-T gas mix at 2.65T and 7.5MA. The time step in the simulations is 3s and corresponds to an upper limit on the constraints imposed from the hardware, namely the power switch between waeguides and the poloidal steering between the q=1 and the q=2 surface.

The Upper Steering Mirror (USM) is used to track the q=2 rational surface while the Lower Steering Mirror (LSM) is reserved for control of the (3,2)-NTM. A tolerance of 2.5-3.0 cm is assumed on the precision of the EC alignment, based on the requirements from the Plasma Control System [11]. It is also assumed that the feedback control reacts only when the magnetic island has grown to at least 4cm, which is consistent with projections of the ECE diagnostic resolution designed for ITER [12-14]. It is noticed that increasing the threshold size to 6cm does not increase the power requirements, while increasing the tolerance on the alignment does. In particular, if the EC deposition is misaligned by more than half the EC deposition width, stabilization might not be possible even with the entire available power of 20MW [15].

The time-dependent simulations (not shown here) indicate that the (3,2) does not lock in these plasmas while the (2,1) can lock in less than 5 seconds both in plasmas at half-field and full field. Thus the analysis is here focussing on the stabilization of the (2,1)-NTM with the USM. The two cases shown in the figure refer to different hypotheses on the beam width. Broadening of the beam width can be obtained by selecting the launching waveguides to spread the superposition of the beamlets. Additional broadening will be inevitable and caused by scattering of the waves due to turbulence fluctuations, in particular from the region around the separatrix [16]. These simulations can be interpreted as representative of a general case, where the wider the beam the lower the peak current density, for a constant total integrated current. The temperature and plasma current profiles are calculated consistently with the EC heating and current profiles and are therefore consistent as opposed to rescaling width and current for fixed temperature and bootstrap current profiles.

As shown in the figure, a broadening of the profile up to about 6cm is needed to fully suppress the (2,1)-NTM. After an initial investment of 13.4MW, which is the maximum available on either of the two mirrors of the UL, maintaining a constant level of about 3.32MW on the q=2 rational surface would be sufficient to prevent the NTM from growing back. The two phases with the highest power requirement are typically the end of the ramp-up phase, when the plasma enters H-mode, and the start of the ramp-down phase, at the H-mode back transition. Tracking the NTMs in this phase is challenged by current and density decay, with current profile peaking and inward shift of rational surfaces, as well as a natural broadening of the EC deposition width because the deposition becomes more tangential to the rational surfaces. Optimization of this phase is needed to reduce the risk of NTM mode locking and subsequent disruptions. The simulation with narrower deposition width indicates that the NTM cannot be suppressed even with 13.4MW power on the rational surface. Although the other mirror could be added to have the full available power available for stabilization on the q=2, here we are taking a safety margin approach to account for the limitations of the model used and assume that this mirror is reserved for stabilization of the (3,2)-NTM, if needed.

Simulations of the baseline scenario come to similar conclusions on the maximum tolerance on the alignment and on the minimum threshold size for detection. One difference is that the nominal Gaussian beam width in the O1-mode polarization at full field is about 75% wider than the width in X2-mode polarization at half-field. The constraints on the minimum deposition width needed for stabilization are therefore less stringent. Similarly to operation at half-field, also at full field the most critical phases are the entry to H-mode and the first third of the current ramp-down phase. Depending on the conditions of access to H-mode the (2,1)-NTM is predicted to lock between one and 15 seconds after exceeding the threshold width. Despite these differences, these cases all require an initial investment of up to 13.4MW on the mirror for suppression of the first NTM, then stability can be sustained with constant power on the q=2 surface. Similarly to the cases analysed at half-field, with a deposition width of at least 5 cm and with EC alignment maintained within half of the EC

 DOI: 10.1051/epjconf/201715703041

deposition width, stability can be maintained with up to 6 MW continuously tracking the q=2 surface. Because in the flattop phase the plasma is stationary, the requested poloidal angle is constant and a continuous tracking does not challenge the fatigue limit of the system.

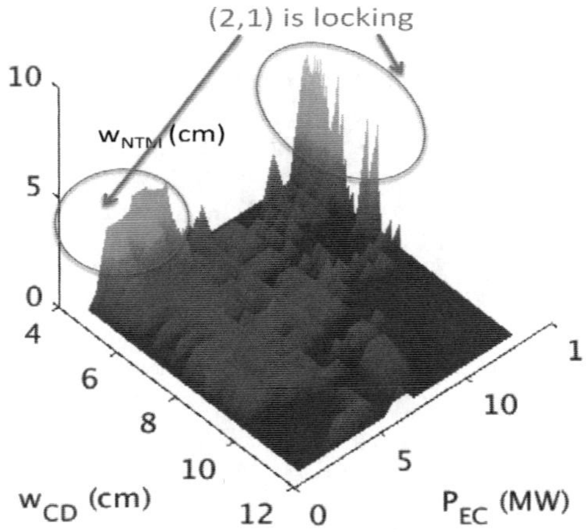


**Fig. 2.** Histogram of the (2,1)-NTM width as a function of the deposition width $w_{CD}$ and of the injected power $P_{EC}$.

Figure 2 summarizes a number of simulations run under dfferent assumptions: removing the power from the q=2 surface after mode suppression is achieved and maintaining a constant power of 2 to 8 gyrotrons. For all these cases the deposition width $w_{CD}$ has been varied between the nominal Gaussian width and 12cm. As shown in the figure, there is only one critical operational range at narrow deposition width. The two regions with the highest EC power correspond to simulations that drop the EC power to zero after suppression of the (2,1)-NTM. In those cases, the mode grows back soon after the power is removed from the rational surface. All cases with narrow $w_{CD}$ lock, even with maximum power injected. Tipically, the island needs to be maintained below 6 cm to avoid locking. For deposition width broader than 5cm an upper limit on the power needed for stabilization can be set to about 5-7MW depending on the deposition width.

## 4 How NTM control affects plasma performance in the baseline scenario

The target of the baseline scenario is to demonstrate operation at fusion gain Q=10, the challenges include doing so while ensuring MHD stability, ELM control, disruption avoidance and low power heat load to the divertor. Since Q is calculated as the tatio of the alpha power to the external power, $Q=5P_\alpha/P_{ext}$, there is concern that NTM stabilization would reduce Q by increasing the EC power. Control schemes should therefore minimize the reduction in Q by minimizing the power needs and at the same time maximize the plasma performance by avoiding the reduction in confinement that is subsequent to the presence of NTM. This requires careful discharge planning, design and a robust control system that manages complex applications. Using the belt model [17] as a simple approximation for the expected reduction in confinement time, $\Delta\tau_E/\tau_E\sim 4\rho_s 3w_{sat}/a$, and assuming that the (2,1)-NTM can be maintained at a width of about 5-6 cm to avoid locking, the reduction in confinement would be only 5%. If this target is achieved with minimum EC power investment then Q can still be maintained around 9-10.

Figure 3 summarizes the variations of Q when active control is compared with pre-emptive control. For each time step in the flattop phase we have grouped all data as a function of the EC power on the q=2 surface and calculated the average Q and its standard deviation for a given value of $P_{EC}$. The coloured area represent the variation of Q among all cases analysed. The cases dubbed as active control are simulations where the EC power is turned-on only when the NTM is above the detection threshold size, whivh has been varied between 4cm and 6cm. The cases dubbed as pre-emptive are maintaining a constant power on the q=2 surface. For these cases a scan on the misalignment has been performed. The simulations that achieve the largest Q are those with pre-emptive control and with alignment maintained within half the EC deposition width.

Figure 4 shows a simulation of the ITER baseline scenario with EC feedback control. The current ramp-up phase is 80s long, with the plasma being diverted at about 12s and the radio-frequency heating and current drive being turned-on shortly after.

The electron density is built-up rapidly within the first 20s to provide a background plasma for good absorption of both Electron and Ion Cyclotron waves. In the flattop phase the EC power is reserved for NTM control; after the stabilization of the first (2,1)-NTM the power is dropped to a constant 5 MW and turned-on only

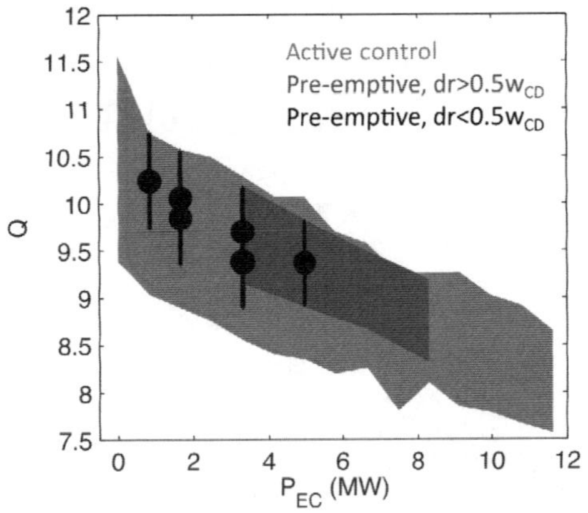


**Fig. 3.** Dependence of Q on the EC power on the q=2 surface for simulations with different control schemes and different assumptions on the EC alignment.

when needed. Since the power is reserved and does not need to be transferred from another application, the turn-on of the gyrotron is in practice instantaneous. The IC power is turned-off in the flattop under the assumption that it is reserved for other applications, for example sawtooth control. The apha power could be increased even further by turning-on the IC power, but this would also affect the sawtooth period and the fast ion stability. This simulation achieves Q=10 during the flattop phase. Since the (2,1)-NTM is maintained at a width below 5 cm, the reduction in the confinement is less than 5%. It is expected that turbulence effects would broaden the EC deposition profile. However, since it is assumed that the EC power is never completely removed from the q=2 surface, scattering effects are not expected to be an issue. In fact, while defocusing of the beam might be a problem when trying to stabilize an island that has already achieved a finite size that exceeds the EC deposition width, in the case of pre-emptive stabilization would instead play a favourable role by minimizing the negative effects of EC misalignment.

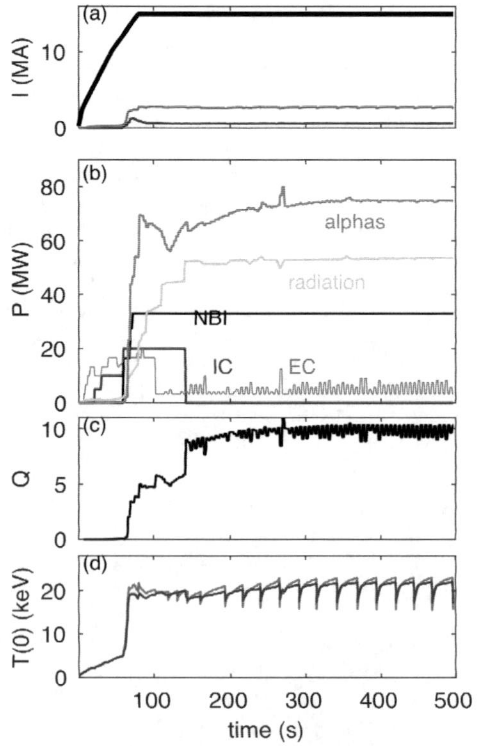


**Fig. 4.** TRANSP simulation of the ITER baseline scenario with EC feedback control for NTMs. (a) total current, bootstrap (red) and NBCD (blue) (b) external power, radiation and alpha power (c) fusion gain (d) central electron and ion temperature.

## 5. Conclusions

Controlling NTMs in ITER is critical for component protection, disruption avoidance and for optimization of the plasma performance and demonstration of fusion gain of Q. Time-dependent simulaitons that evolve consistently the plasma magnetic equilibrium and pressure profiles in response to EC feedback control of NTMs indicate that the EC power planned on ITER is sufficient to ensure control and that the main cosntraints come from the precision of the magnetic equilibrium reconstruction rather than on detection threshold of the magnetic island or time scales imposed by the hardware. While these results can inform on future steps to be taken on the design of a robust control on ITER, they depend on the assumptions in the simulations and on the limitations of the reduced models used for this assessment. In particular the NTM threshold and the effect of toroidal effects and mode coupling on the tearing stability term should be verified against nonlinear MHD codes. Validation of the MRE implemented in TRANSP is currently being undertaken against experiments on JET, ASDEX-U and DIII-D,both with and without EC feedback control.

Work supported by ITER under IO/RFQ/13/9550/JTR and by DOE under DEAC02-09CH11466. The views and opinions expressed herein do not necessarily reflect those of the ITER Organization.

## References

1. M.A. Henderson *et al*, Phys. Plasmas **22** 021808 (2015)
2. D. Farina *et al* Phys. Plasmas **21** 061504 (2014)
3. D. Farina *et al*, Nucl. Fusion **52** 033005 (2012)
4. E. Poli *et al* , Nucl. Fusion **55** 013023 (2015)
5. D. De Lazzari *et al*, Nucl. Fusion **50** 079201 (2010)
6. D.J. La Haye *et al*, Nucl. Fusion **48** 054004 (2008)
7. R. Hawryluk, Proceedings of the Varenna conference, **1** 1946 (1981) Elsevier Ltd.
8. E. Fredrickson *et al*, Phys, Plasmas **7** 4112 (2000), Phys. Plasmas **9** 548 (2002).
9. N. Bertelli *et al*, Nuclear Fusion **51** 103007 (2011)
10. E. Poli, Comput. Phys. Commun. **136** 90 (2001)
11. J.A. Snipes *et al*, Fus. Eng, Design **87** 1900 (2012)
12. V. Udintsev *et al*, Proc. 24$^{th}$ IAEA FEC 2012 ITR/P5-41
13. H. van den Brand *et al*, Nucl. Fusion **53** 013005 (2013)
14. H. van den Brand *et al*, Plasma Phys. Control. Fusion **54** 094003 (2012)
15. F.M. Poli *et al*, submitted to Nucl. Fusion
16. A. Snicker *et al*, submitted to Nucl. Fusion
17. Z. Chang and J. Callen Phys. Plasmas **5** 1076 (1998)

DOI: 10.1051/epjconf/201715703042

# Parametric Instabilities During High Power Helicon Wave Injection on DIII-D

*M.* Porkolab[1] and *R.I.* Pinsker[2]

[1]Massachusetts Institute of Technology, Plasma Science and Fusion Center, Cambridge, Massachusetts 02139, USA
[2]General Atomics, P.O. Box 85608, San Diego, California 92186-5608, USA

**Abstract.** High power helicon (whistler) waves at a frequency of 0.47 GHz are being considered for efficient off-axis current generation in high performance DIII-D plasmas and in K-Star [3]. The need for deploying helicon waves for current profile control has been noted in previous publications since penetration to the core of reactor grade plasmas is easier than with lower hybrid slow waves (LHCD) which suffer from accessibility limitations and strong electron Landau absorption in fusion grade high temperature plasmas. In this work we show that under typical experimental conditions in present day tokamaks with 1 MW of RF power coupled per antenna, the associated perpendicular electric fields of the order of 40 kV/m can drive strong parametric decay instabilities near the lower hybrid layer. The EXB and polarization drift velocities which are the dominant driver of the PDI can be comparable to the speed of sound in the outer plasma layers, a key measure of driving PDI instabilities. Here we calculate growth rates and convective thresholds for PDIs, and we find that decay waves into hot ion lower hybrid waves and ion cyclotron quasi modes dominate in the vicinity of the lower hybrid layer, possibly leading to pump depletion. Such instabilities in future reactor grade high temperature plasmas are less likely.

## 1 Introduction

We consider the possibility of parametric decay instabilities (PDI) in the proposed 1 MW helicon wave (whistler or fast wave at frequencies below the central lower hybrid frequency in a tokamak plasma) current drive experiments on DIII-D [1, 2]. The importance of introducing such a current drive method was already recognized in the ARIES reactor studies where it was necessary to introduce the fast lower hybrid wave (helicon wave in current use) in the 1 GHz range to provide some of the necessary driven current in the outer region of the plasma core, yet well inside the pedestal region [3]. Previous work postulated [4] that significant levels of PDI would be unlikely in such experiments due to the weaker RF electric fields of these predominantly electromagnetic waves, in contrast to those of the quasi-longitudinal (electrostatic) slow lower hybrid wave [5, 6, 7, 8]. In this work we show that this is not necessarily the case under present day experimental conditions in medium size tokamaks, e.g., DIII-D, where 1 MW of RF power at 476 MHz is proposed to be injected using a "comb-line" traveling wave antenna [2]. We note that similar experiments are also in the planning stage for the KSTAR tokamak [9]. Under such conditions perpendicular electric fields of 40 kV/m may be generated which then drive ExB velocities and also, comparable values of the polarization drift velocity. While these are weaker fields than those typical for the slow wave ($E_{//}$ = 300 kV/m and the perpendicular value of E is even larger) they are comparable to the local acoustic speed in the relatively low density and temperature scrape-off layer near the local lower hybrid frequency and may be sufficient to drive PDI [4, 5].

In what follows we will summarize the directly relevant theoretical work of the author from past years and proceed to calculate growth rates for typical DIII-D parameters. We will then estimate convective growth rates of the expected decay into ion cyclotron quasi-modes and the slow lower hybrid waves, in particular the hot ion plasma waves, under the greatly simplified mathematical formulation of a slab geometry [5,6]. We find that under the idealized conditions of our calculations, significant convective growth may occur near the lower hybrid resonance layer in the edge plasma. However, under reactor like conditions, where the electric fields may remain the same but the edge plasma parameters, in particular the temperatures will be much higher, such instabilities may not be important. However, this remains the subject of future studies.

## 2 Parametric Instabilities in the Lower Hybrid Range of Frequencies

We shall consider the theory of parametric decay instabilities (PDI) in the lower hybrid range of frequencies as in past works [5,6], i.e., assume that the

* Corresponding author: porkolab@psfc.mit.edu

injected high amplitude RF wave (pump wave) has a wavelength that is much longer than that of the decay waves, and hence we shall calculate growth rates in the so-called "dipole" approximation, namely $\mathbf{k_0} << \mathbf{k_j}$ , where j=1, 2 corresponds to the decay or daughter waves, and in fact we shall set $\mathbf{k_0} = \mathbf{0}$ (where j=1, 2 corresponds to the decay or daughter waves). The daughter waves in the present work will either be hot ion plasma waves at the lower sideband $\omega_2 = (\omega_0 - \omega)$, (an ensemble of ion Bernstein waves propagating through high cyclotron harmonic frequencies nearly perpendicular to the weakly inhomogeneous magnetic field), and ion cyclotron quasi-modes (heavily cyclotron damped electrostatic response of the plasma near the beat frequency, $(\omega_0 - \omega_2) = n\Omega_i$, where n is an integer and $\Omega_i$ is the angular ion cyclotron frequency ) [5, 10]. The selection rules in the dipole approximation reduce to $\mathbf{k_1} = - \mathbf{k_2}$ and this greatly simplifies the theory which then can be derived even for relatively large amplitude pump electric fields [ 5, 6, 11]. The growth rates for decay waves (of order 1 MHz ) are much larger than the bandwith of the pump wave (1 kHz range).

The relevant theoretical results for PDI will be summarized by the equations in the sections below.

## 2.1. Parametric Decay Instabilities in the Large Amplitude Limit

We start with the Vlasov equation in the presence of the large amplitude pump wave,

$$E_0(x,t) = (E_{0x}\hat{x} + E_{0z}\hat{z})\cos\omega_0 t, \qquad (1a)$$

$$\partial f/\partial t + \mathbf{v}\cdot\nabla f + (q/m)\left(\frac{\mathbf{v}\times\mathbf{B}}{c} + \mathbf{E} + \mathbf{E}_0\cos\omega_0 t\right)\cdot\nabla_v f = 0 \qquad (1b)$$

the fundamental parametric instability dispersion relationship in the dipole approximation is given by [5, 11]:

$$1+\frac{1}{\chi_i} = J_0^2(\mu)\frac{\chi_e}{1+\chi_e} + J_1^2(\mu)\left(\frac{\chi_e^+}{1+\chi_e^+} + \frac{\chi_e^-}{1+\chi_e^-}\right)$$
$$+\frac{J_0^2(\mu)J_1^2(\mu)\{[\chi_e/(1+\chi_e)]-[\chi_e^+/(1+\chi_e^+)]\}^2}{1+(1/\chi_i^+)-J_0^2(\mu)[\chi_e^+/(1+\chi_e^+)]-J_1^2(\mu)[\chi_e/(1+\chi_e} \qquad (2)$$
$$\frac{J_0^2(\mu)J_1^2(\mu)\{[\chi_e/(1+\chi_e)]-[\chi_e^-/(1+\chi_e^-)]\}^2}{1+(1/\chi_i^-)-J_0^2(\mu)[\chi_e^-/(1+\chi_e^-)]-J_1^2(\mu)[\chi_e/(1+\chi}$$

where

$$\mu = (e/m_e)\left[\left(\frac{E_{0\parallel}k_\parallel}{\omega_0^2} + \frac{\mathbf{E}_{0\perp}\cdot\mathbf{k}_\perp}{\omega_0^2 - \omega_{ce}^2}\right)^2 + \frac{|\mathbf{E}_{0\perp}\times\mathbf{k}_\perp|^2\omega_{ce}^2}{(\omega_0^2-\omega_{ce}^2)^2\omega_0^2}\right]^{1/} \qquad (3)$$

provides the coupling between the low frequency quasi-modes and the lower hybrid ion plasma waves at the sidebands. The parallel and perpendicular components of the pump wave electric field are indicated by the appropriate subscripts, and the parallel and perpendicular components of the wave numbers of the daughter waves are also indicate appropriately. We recognize the three terms in $\mu$, as the parallel oscillating drift (or quiver) velocity, the perpendicular polarization drift velocity and the EXB drift velocity, respectively. The appropriate electrostatic susceptibility of electrons and ions is designated by the Greek alphabet symbol $\chi_e$ or $\chi_i$, respectively [5] and the lower $(\omega_0 - \omega)$ or upper sideband $(\omega_0 + \omega)$ responses are designated by the super $(^-)$ or super$(^+)$ symbols, respectively, while the susceptibilities at the low frequency $\omega = n\Omega_i$ (ion cyclotron quasi-modes in the present case) response have no superscripts. This dispersion has been configured into a numerical computer code MPPDI in the 1970s, including all orders of the electron and ion Larmor radius through retention of the modified Bessel functions in the hot electrostatic linear dispersion relationships to at least 50 cyclotron harmonics, and electron and ion Landau and cyclotron damping are also accounted for [5]. This code has been tested numerous times in the last 4 decades and is still being use at MIT by various students and research scientists, including the author. In the second half of the paper we shall present numerical results of the growth rates and convective thresholds of the most unstable modes corresponding to the planned helicon experiments in DIII-D.

## 2.2 Parametric Decay into Two Weakly Damped Modes

For small values of $\mu << 1$, the ordinary Bessel function J can be expanded in a power series and various terms combined into a compact form, equivalent to the usual PDI dispersion relationships. Indeed, it was shown in our earlier papers [5, 6, 10] that the familiar PDI dispersion relationship for either resonant decay instability, the OTSI instability, and for decay into quasi modes (nonlinear Landau damping) can be readily deduced from Eq. (2) and are of the following form:

(a) Resonant decay into two weakly damped modes:

$$\epsilon(\omega) + \frac{\mu^2}{4}\chi_i(\omega)\,[\chi_e(\omega)]\left[\frac{1}{\epsilon(\omega-\omega_0)} + \frac{1}{\epsilon(\omega+\omega_0)}\right] = 0\,, \qquad (4)$$

with the threshold for temporal growth rates given by expansion about weakly damped normal modes,

$$(\gamma+\gamma_1)(\gamma+\gamma_2) = \frac{\mu^2}{4}\chi_i(\omega_1)\chi_e(\omega_1)\frac{\partial\epsilon_1}{\partial\omega}|_{\omega_1}\frac{\partial\epsilon_2}{\partial\omega}|_{\omega_2} \qquad (5)$$

(b) Decay into low frequency quasi-modes:

$$\gamma = -\gamma_2 + \mu^2\mathrm{Im}[\chi_e(\omega_1)]\,(|\,\chi_i(\omega_1)\,|^2 + \mathrm{Re}[\chi_i(\omega_1)])$$
$$+\mathrm{Im}[\chi_i(\omega_1)]\,(|\,\chi_e(\omega_1)\,|^2 + \mathrm{Re}[\chi_e(\omega_1)])\,/\,|\,\partial\epsilon/\partial\omega\,|_{\omega_2}|\,\epsilon(\omega_1)\,|^2\} \qquad (6)$$

where the quasi mode response is given by

$$(\omega_0 - \omega_2)/k_\parallel \simeq n\Omega/k_\parallel \simeq v_{ti} \qquad (7)$$

and the lower hybrid sideband wave in the straight hot ion orbit approximation (summation of high harmonic Bernstein waves) is given by :

$$\omega^2 = \omega_{LH}^2 \left[ 1 + \frac{k_\parallel^2}{k^2}\frac{m_i}{m_e} + \frac{3k_\perp^2 T_i}{\omega^2 m_i}\left(1 + \frac{1}{4}\frac{T_e}{T_i}\frac{\omega^4}{\omega_{ce}^2\omega_{ci}^2}\right)\right] \quad (8)$$

Once we have the growth rates from either Eqs. 2 or 6, we can obtain the convective threshold which has to be considered in all three dimensions, x, y or z [5,6,12]. For example, in the x direction (radial direction in our case) the threshold is given by

$$\frac{\gamma\Delta x}{|v_{2x}|} = \frac{\gamma L_x}{|v_{2x}(v_{2z}/v_{2x} - v_{0z}/v_{0x})|} = \pi \ , \quad (9)$$

or more generally we need to consider the diagram of convective losses in 3 dimension, as given in Ref. [12]. In fact, when we consider PDI near the lower hybrid frequency, we find that the minimum convective threshold is given when the perpendicular group velocity vanishes, and consequently convective losses in this direction vanish (ie, the second and third terms in Eq 8 balance), and therefore only convective losses in the z (toroidal) direction need to be considered. Additional stabilizing mechanisms in the radial direction include the density inhomogeneity, and in the poloidal direction the finite height of the antenna. Therefore in the relevant case the convective threshold is given by $\gamma L_z/V_{gz} = \pi$, where $\gamma$ is the growth rate obtained above from Eq 2 (or Eq 6) and $L_z$ is the effective antenna extent and $V_{gz}$ is the group velocity, both along the magnetic field, approximately in the z direction. Initial results for the helicon pump wave growth rates and convective losses are as follows. For 1.25 MW coupled power, GENRAY finds 45 kV/m or 450Volt/cm, see Ref.[1]. $E_{0r}$ is significantly smaller and $E_{0z}$ is negligible. Near the lower hybrid mode conversion layer near the plasma edge (where $n_e \approx 3\times10^{18}$ m$^{-3}$ , $T_e$ = 30 eV, $T_e = T_i$, and B= 1.3 Tesla) based on results from extensive growth rate calculations from Eq 2, we find maximum growth rates for $N_{2II}$ = 10.0, $k_x\lambda_D$ = 0.08, $\omega_2/\omega_0$ = 0.077, $\gamma/\omega_0$ = 0.0114. Here for the pump wave we assumed $N_{0II}$ = 3.0, and the appropriate selection rules for parallel wavenumbers were introduced into Eq.2 to properly calculate nonlinear Landau damping. For the presently considered DIII-D antenna length of $L_z$ = 160 cm, taking an "effective" length of 80 cm (since the pump wave amplitude decreases along the length of the antanna) we find that $\gamma L_z/V_{gz}$ = 24. Therefore the estimated power threshold is $\pi$/24 x 1.25 MW = 0.16 MW, (the growth rate for quasi mode decay is proportional to $E_0^2$) and at 1 MW input power level we expect complete pump depletion. It should be remarked, however, that the convective losses are based on the group velocity calculated from the approximate dispersion relationship of Eq. 8 and in addition, the resonance cone behaviour of helicon waves and impact of poloidal magnetic field has been neglected in this calculation. While thresholds are easily exceeded in the present experimental situation because of the relatively low electron temperatures near the lower hybrid layer, at higher edge electron temperatures in reactor relevant plasmas the effective growth rates would be reduced and it is not clear that PDI would be important. We shall carry out more extensive calculations in the near future for such candidates for burning plasmas, such as ITER or beyond.

## Acknowledgements

We wish to thank Dr. Ron Prater for providing the experimentally relevant pump wave electric fields in the early phases of this work, Dr. Charles Moeller for helpful discussions regarding the proposed antenna design, and Dr. Craig Petty for careful reading of this manuscript and helpful comments.

## References

[1] R. Prater, C. P. Moeller, R. I. Pinsker, *et al,* Nucl. Fusion **54**, 083024 (2014).
[2] R. I. Pinsker, C. P. Moeler, J. P. Anderson, et al, Paper presented at this conference.
[3] S. C. Jardin, *et al*, Fusin Eng. Design **38**, 27 (1997).
[4] V. L. Vdovin, Plasma Physics Reports **39** 95 (2013).
[5] M. Porkolab, Phys. Fluids 20, 2058 (1977).
[6]. M. Porkolab, Nucl . Fusion 18, 043009 (1978).
[7] S. G. Baek. R. R Parker, P. T. Bonoli. *et al,* Nucl. Fusion **55**, 043009 (2015).
[8] R. Cesario, A. Cardinali, C. Castaldo, et al, Nucl. Fusion **46,** 462 (2006).
[9] S. J. Wang, H. H. Wi, H. J. Kim, et al, Nucl. Fusion **57,** 046010 (2017)
[10] M. Porkolab and R. P. H. Chang, Rev. Modern Phys. **50,** 745 (1978).
[11] M. Porkolab, Phys Fluids **17**, 1432 (1974).
[12] Y. Takase and M. Porkolab, Phys. Fluids **26**, 2992 (1983)

Work supported by DOE contracts DE-FC02-04ER54698 at DIII-D, a DOE FES User Facility, and Grant DE-SC001-6154 at MIT.

# CNES – Chalmers – IAP – ONERA - XLIM activities in the domain of high RF power breakdown phenomena

*J.* Puech[a], *E.* Sorolla[a,b], *V.E.* Semenov[c], *E.I.* Rakova[c], *D.* Anderson[d], *M.* Belhaj[e], *J.* Hillairet[f]

[a] CNES (France), 18 Avenue Edouard Belin, 31401 Toulouse, Cédex 9, France
[b] XLIM (France), 123 Avenue Albert Thomas, 87060 Limoges Cedex, France
[c] IAP (Russia), 46 Ulyanov St., Nizhny Novgorod, 603950, Russia
[d] Chalmers (Sweden), Chalmers University of Technology, SE-412 96 Göteborg, Sweden
[e] ONERA (France), BP74025 – 2 Avenue Edouard Belin, 31055 toulouse, Cédex 4, France
[f] CEA, IRFM, F-13108 Saint Paul-lez-Durance, France

**Abstract.** Multipactor breakdown is an important potential failure mechanism in many different microwave devices working under close to vacuum conditions. Applications range from space borne RF equipment to high-power microwave generators. The basic physics involved in the multipactor phenomenon is well known for the case of two infinite pallel plates made of metal. However, most realistic RF device geometries involve inhomogeneous RF electric fields and curved field lines and sometimes also dielectric material. The purpose of this paper is to set up methodologies to determine the Multipactor threshold in such situations.

## 1 Introduction

The trend in the space telecommunication market is to increase the number of space segment users, these being spread over wide areas and asking for higher and higher data rate. At the payload level, more and more powerful microwave amplifiers are inserted in the chain. Components must handle very high RF electric fields in vacuum. Two kinds of discharges can occur depending on the pressure range:

- in a rarefied atmosphere, an avalanche-like increase of the electron density may occur due to secondary electron emission when electrons, accelerated by the RF electric field, hit the wall of the microwave device ( Multipactor effect)
- at finite pressures, a similar growth of the plasma density in the gas filling the device may occur due to impact ionization of neutral gas molecules by free electrons (Corona effect).

Risks of discharge such as Multipactor or Gaseous Discharge triggered by Multipactor can become critical and must be taken into account at payload level, as well as at equipment level. The consequences may be the link budget degradation or even damages to the equipment.

Different CNES projects involving the Institute of Applied Physics (Nizhny Novgorod – Russia), Chalmers University of Technology (Göteborg – Sweden), XLIM/University of Limoges, ONERA and Centre National d'Etudes Spatiales (Toulouse – France) were working together on these phenomena. The purpose of this article is to describe methodologies which were set-up to determine the Multipactor threshold in 2 cases: RF components with metal iris and RF components with dielectric material at the level of the critical gap.

## 2 RF components with iris

Iris geometry (fig. 1) is quite commonly used in microwave components.

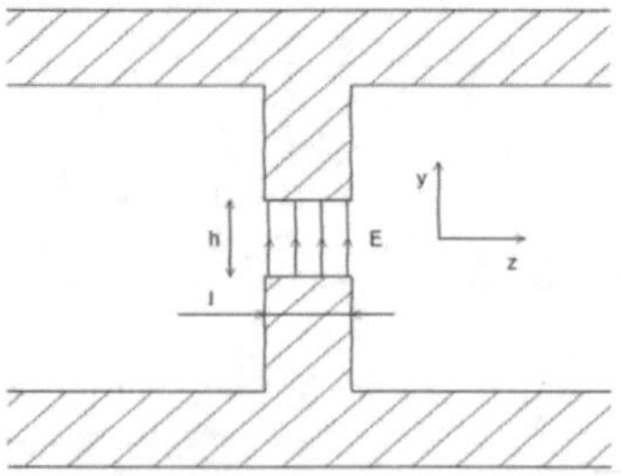


**Fig. 1.** Iris geometry.

In such situations, 3 physical phenomena have to be taken into account:

- random walk of the electrons along the propagation axis,
- Miller force (ponderomotive force) near the iris edge,
- RF electric field component along the propagation axis

In this case, the spread in emission angles leads to a random walk of the corresponding electron trajectories that ultimately carries them out of the finite iris region and thus constitutes a loss mechanism that tends to increase the multipactor breakdown threshold. The importance of this diffusion out of the iris is determined by the ratio of the width, a, and height, h, of the iris and it was shown in [1] (assuming that the electric field was homogeneous inside the iris). The model is based on the statistical approach to determine the probability of surviving a certain number of gap crossings and thus the number of electrons after k steps. It enables to draw the the Multipactor chart as in the case of the 2 parallel plate model but with the ratio height over width ratio. The Figure 2 below illustrates this methodology for the alodine material (both sides of the iris) :

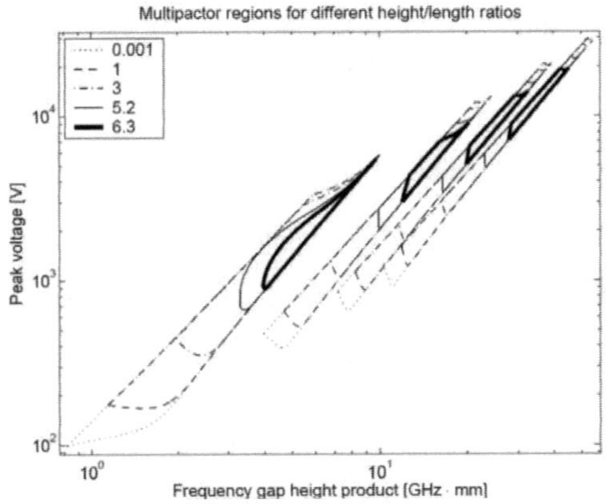


**Fig. 2.** Multipacor chart in the iris case.

A more rigorous and detailed analysis of the iris problem was made in [2] where the geometry of the iris was transformed into the classical 2 infinite parallel plate geometry using conformal mapping assuming that the electric field in the iris was approximately quasi-static. In parallel, a software called SEMA/MULSIM based on the Monte-Carlo approach was developed to determine the Multipactor threshold inside RF components with waveguide sections. Comparisons were drawn on the same RF component.

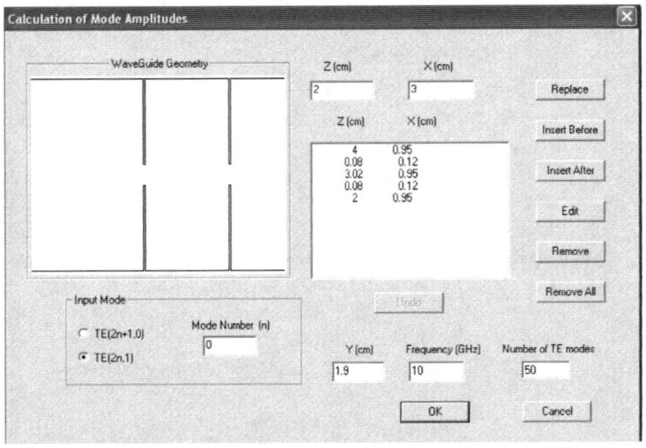


**Fig. 3.** RF component analysed with SEMA/MULSIM.

The figures 4 and figures 5 illustrates the number of electrons after T period versus the voltage U. For the figure 4, the calculations were realized with the SEMA/MULSIM software. For the figure 5, the calculations were realized with the conformal mapping approach.

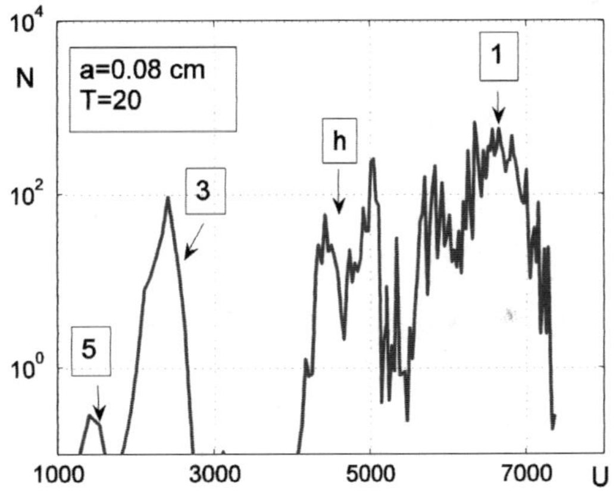


**Fig. 4.** Results obtained with the SEMA/MULSIM software.

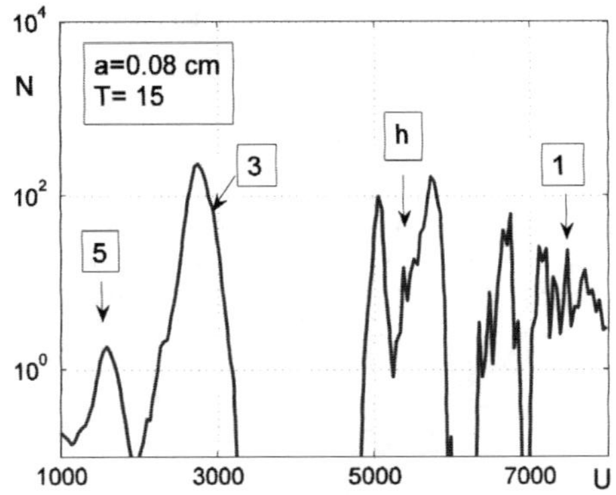


**Fig. 5.** Results obtained with the conformal approach.

The 2 graphs are similar which demonstrates a good agreement between the 2 approaches.

Cross-validations were also performed between the conformal mapping approach and Multipactor measurements. These comparisons were performed within the EVEREST project between the conformal mapping software using ONERA secondary emission yield measurements of "technical" silver and Multipactor measurements of a L band coaxial filter with fringing field geometries [3]. The figure 6 shows the good agreement between the threshold found by measurement and the one found by the conformal mapping approach.

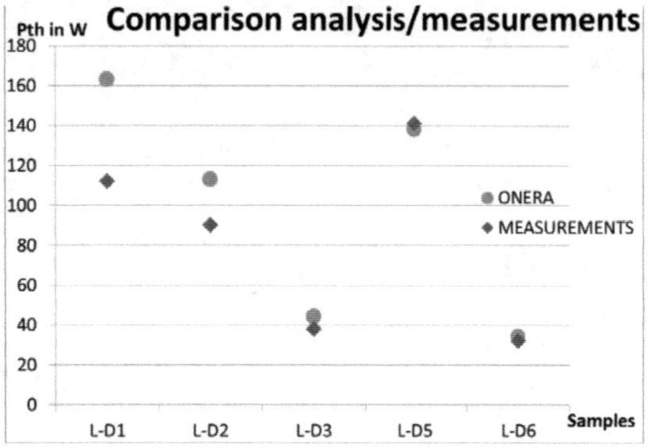


**Fig. 6.** Comparison between numerical simulations and measurements.

The agreement is excellent between the 2 approaches. It validates the conformal mapping methodology.

## 3 RF components with dielectrics

The trend in the space telecommunication payload is to reduce the footprint of the different high power RF components. One way to reach this goal is to use dielectric material inside these components, in order to lower the size while maintaining the overall RF performances. It is essential to determine the power capabilities of such components though Multipactor analysis. Thus it is necessary to study the electron dynamics in such situations.

Recent theories and secondary emission measurements at ONERA [4] have put in evidence the fact that the secondary emission properties of the dielectric varies with time leading to an equilibrium state. As a matter of fact, the secondary emission yield creates a positive or a negative charge which accelerates or decelerates the electrons before the collision. This charge reduces the SEY [5].

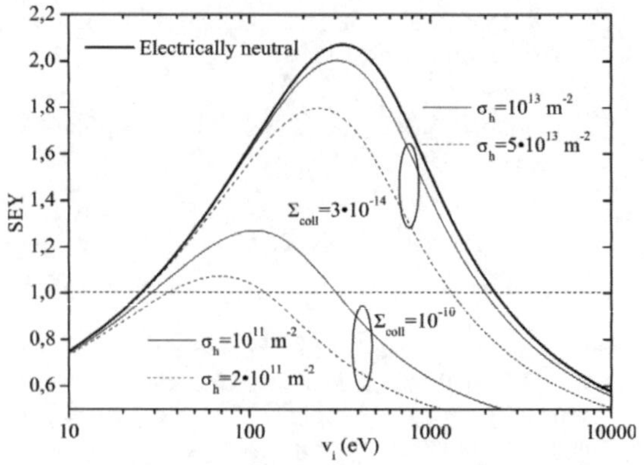


**Fig. 7. Variation of the SEY with the surface density of holes.**

Moreover, the charge created in the dielectric generates a DC field that modifies the force the electron cloud is subjected to. The electron dynamics within a 2 parallel-plate geometry partially filled with dielectrics has been analyzed using a mono-energetic secondary emission model. The approximation of the electron cloud by a thin sheet of charge was done, what is valid slightly above the Multipactor voltage threshold. Both the decrease of the SEY with the charge in the dielectric and the subsequent electrostatic field generated are taken into account to find the equation of motion of the electron sheet when the waveguide is excited with the fundamental TM-mode.

When the dielectric filling factor of the waveguide is close to one, both the position and velocity of the electron cloud can be found analytically. The dimensionless equation for the position is as follows:

$$\Lambda = \left[V_\omega - \left(\cos\alpha - \Omega\alpha\right)\right]\left(\varphi - \alpha\right) + \sin\varphi - \sin\alpha - \frac{\Omega}{2}\left(\varphi^2 - \alpha^2\right)$$

where the dimensionless position, $\Lambda \equiv m_e\omega^2\left(x' - x_0\right)/|e|E_0$, electric field, $\Omega \equiv E_d/E_0$, and velocity, $V_\omega \equiv m_e\omega v_0/|e|E_0$ and the parameter α (phase of the electric field when the electron begins the flight) have been defined. When the charge on the dielectric is close to zero, $\Omega = 0$. A mathematical expression to evaluate the Multipactor electric field breakdown as function of the dielectric charge has been obtained using the classical resonance theory.

$$E_0^{th} = \frac{2m_e\pi f|\gamma|}{|e|\cos\lambda}\sqrt{1 + \frac{\mu\sin\lambda}{\gamma} + \left(\frac{\mu}{2\gamma}\right)^2} \quad (1)$$

where $\gamma \equiv [N\pi(v_0 - N\pi|e|Ed/2me\omega) - \omega X]/S$, $\mu \equiv (vi - v_0 + N\pi|e|Ed/me\omega)$ and $S = ((N\pi)2 + 4)^{1/2}$, being $N$ the order of resonance. It must be reminded that this expression is valid when ($|Ed| \ll |E_0|$).

The comparison between this expression and the numerical simulations is illustrated in the figure 8. It shows the electric field threshold obtained with the simulations for different values of the initial density of holes together with the predictions of the classical resonance theory (eq. 1).

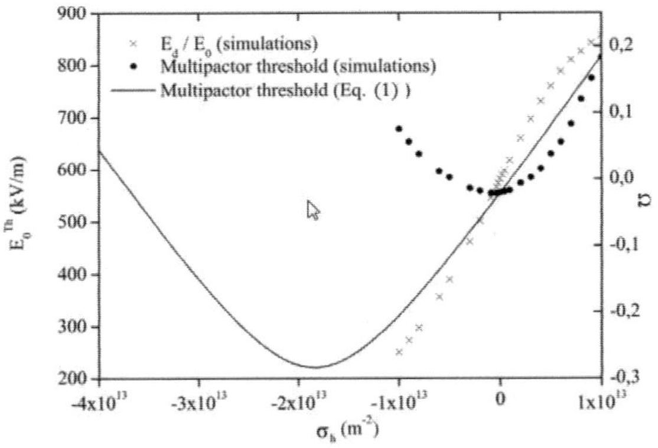


**Fig. 8. Comparison of the electric field threshold obtained with the classical resonance theory and the simulations**

When the dielectric charge is positive, the 2 methodologies do not differ too much. On the contrary when the dielectric charge is negative, the classicial theory underestimates the level of the Multipactor threshold. Some deeper studies need to be performed to understand such discrepancies. The mainpoint is that the simulations and the equation (1) predict a Multipactor power threshold that depends on the initial charge.

## Conclusion

The problem of the iris geometry was treated in this paper. A first approach based on the random walk was depicted and enabled to draw Multipactor chart similar to the charts drawn in the parallel plate case but with the ratio height over width as parameter. The main drawback of this method is that it is too conservative as the 2 other electron losses mechanism as the Miller force and the fringing field effect are not taken into account. An approach proposed by V.E. Semenov and E.I. Rakova was set up. It was based on the conformal mapping methodology, taking into account the 3 electron losses mechanism. Comparisons of the Multipactor threshold on the same cases were drawn between the conformal mapping approach with the use of SEY measured data and measurements.A very good agreement could be observed what validates the approach.

This paper also focuses on the Multipactor threshold estimation inside RF samples with dielectric. It was observed that the conductivity and the subsequent charging phenomena of dielectrics have a strong impact on the Multipactor threshold determination. Based on theoretical elements, the dependence of the breakdown threshold on the ratio $E_{DC}/E_{RF}$ was given. The results presented enable to give an idea of the conductivity domain in which the "metal like" analysis is still valid.

As perspectives, the next step is to study RF components such as isolators/circulators and dielectric windows. This activity is performed in common with CEA IRFM and ONERA through a PhD student who focus on the influence of a DC magnetic field on the Multipactor. In that frame, a first inovant study [6] was realized and was dealing with the design of a SEY test set-up to quantify the influence of a DC magnetic field on the SEY characteristics of a given material sample.

## Acknowledgments

The comparison between the conformal mapping simulations of iris and Multipactor measurements were realized in the frame of the TRP ESA funded activity EVEREST n°22452/09/NL/GLC

## References

1. R. Udiljak, D. Anderson, M. Lisak, J. Puech, and V. E. Semenov, "Multipactor in a waveguide iris", *IEEE Trans. Plasma Sci.* Vol. **35**, No. 2, pp. 388-395, 2007.
2. V. E. Semenov, E. Rakova, R. Udiljak, D. Anderson, M. Lisak, J. Puech, "Conformal mapping analysis of multipactor breakdown in waveguide irises", *Phys. Plasmas,* vol. 15, p. 033501, 2008.
3. J. Puech and EVEREST consortium, "Synthesis of the results of the EVEREST project", MULCOPIM 2014, Valencia, 17-19 Sept. 2014
4. M. Belhaj, D. Payan, J. Puech, "About the electron emission properties of dielectric materials", MULCOPIM 2017, Noordwijk, 5-7 April 2017.
5. E. Sorolla, M. Belhaj, J. Sombrin and J. Puech, *Phys. Plasmas* (unpublished).
6. N. Fil, M. Belhaj, J. Hillairet, J. Puech, R. Mathevet, "Electron Emission under Uniform Magnetic Field of Conductor Materials used for Spatial and Fusion Applications", MULCOPIM 2017, Noordwijk, 5-7 April 2017.

EPJ Web of Conferences **157**, 03044 (2017)
*22 Topical Conference on Radio-Frequency Power in Plasmas*
DOI: 10.1051/epjconf/201715703044

# Study of a distributed ICRF antenna system in DEMO

*Riccardo* Ragona[1,2,*] and *André* Messiaen[1]

[1]Laboratory for Plasma Physics, Royal Military Academy, (LPP-ERM/KMS), BE-1000, Brussels, Belgium
[2]Ghent University, Department of Applied Physics, 9000 Ghent, Belgium

**Abstract.** Different feeding options for distributed ICRF antenna systems in DEMO are analysed and compared, based on detailed circuit modelling. Voltages and currents (amplitude and phase) are computed for each element of the array. The power delivered by the feeding networks is evaluated characterizing each system. Although the results presented are for a small number of radiating elements, the procedure can be scaled to any arbitrary number of straps and feeding connections. Among the three options analysed (i.e. all-fed; distributed; resonant ring) the resonant ring feeding appears to be the most suitable configuration for a traveling wave array distributed antenna system in the DEMO reactor.

## 1 Introduction

An Ion Cyclotron Range of Frequency (ICRF) distributed antenna integrated in the blanket of the DEMO reactor is presently considered as a candidate system to heat the central plasma core without density cut-off. Lower power density is envisaged in order to not exceede the voltage standoff due to the large coupling evanescent length. A large number of radiating elements are used to couple the required power while achieving the low level of power density. The feeding of those elements plays an important role in the overall design of the ICRF system. While an independent feeding for each element allows the best control on the radiated spectrum, it poses some technical difficulties like the large number of intrusions in the reactor vessel and the intrinsic coupling between adjacent lines that complicates the external matching circuitry. A travelling wave array [1] is then a possible solution allowing a considerably smaller number of feeding lines and a simplified external circuitry. Integration in the DEMO design requires a detailed analysis of the possible system solutions for the launcher and the corresponding feeding mechanism. The aim of this contribution is to explore some of those possibilities.

Three different feeding cases are here analysed: (i) an array with all elements fed [2], (ii) an array with periodic conventional feeding [3] and (iii) an array with resonant ring recirculation feeding system [1]. We first analyse case (i), the case with the best possible performances [4] although with the drawback of requiring a large number of mutually coupled feeding lines. In the second case the dependency of the response of the structure with different relative phasing of the generators is analysed and the power capabilities are studied. In the last section an array fed with one resonant ring is compared to the traveling wave array (TWA) of [1] and a general case of two consecutive resonant rings is studied. Conclusions are then presented.

## 2 Analyses

### 2.1 Simulation parameters

All the three cases analysed are based on an array of 28 elements, i.e. $N_{str}$=28. In Fig. 1 the array is presented in one of the feeding configuration analysed (case (ii)). A predominantly diagonal matrix can describe this array with some elements at the upper right and lower left corners due to the symmetry of each radiating element w.r.t. the neighbouring ones. Here only the mutual coupling with the first two neighbours (on each side) are considered. Exploiting the symmetry of the system means that all values of the self-inductance $L_i$=$L$ and mutual couplings $M_{i+1}$=$M_1$, $M_{i+2}$=$M_2$, for $i$ = 1, …, $N_{str}$, are respectively the same and all other elements of the

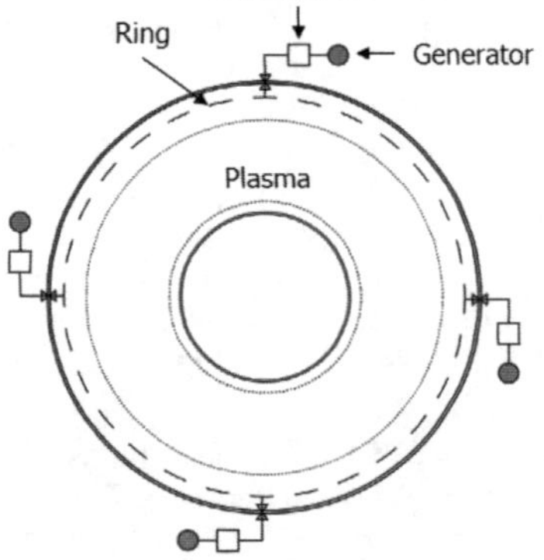


**Fig. 1.** Sketch of the array fed as described in section 2.3.

* Corresponding author: riccardo.ragona@rma.ac.be

strap impedance matrix $\mathbf{Z}_s$ are set equal to zero. At a frequency $f_0$=50MHz, $\omega_0 L$=56Ω, $\omega_0 M_1$=17.3Ω and $\omega_0 M_2$=7.25Ω, being $\omega_0 = 2\pi f_0$ the angular frequency. A resistance $R_i = R$=2.5Ω is added in series to each strap to account, in an appropriate way, for the coupling of power to the plasma. This arbitrary value is chosen as reference to compare the different cases and it is not consistently derived from a coupling code (e.g. ANTITER [5]). The array is configured as combline with a tuning reactance on each strap and can be described by an array impedance matrix $\mathbf{Z}_a$ such that $1/\mathbf{Z}_a = 1/\mathbf{Z}_s + 1/\mathbf{Z}_c$ where $1/\mathbf{Z}_c$ is the admittance matrix of the tuning capacitors. The details of the modelling procedure are explained in [4]. It is worth to remind that, in its pass band, the combline acts as transmission line with characteristic impedance $Z_0 = Z_{it} = (Z_{11}^2 - Z_{12}^2)^{0.5}$ [1] where: $Z_{it}$ is the iterative impedance of the structure [6] and $Z_{ii}$ ($i$=1,2) are the terms of the symmetric $\mathbf{Z}$ matrix that characterise the two-port network. In other words, $Z_{it}$ is the impedance seen at the input of a two-port network when the output is terminated on the same input impedance.

## 2.2 360° array – all straps fed

A continuous array distributed 360° all around the machine is fed at each element by an independent generator connected through its own matching-decoupling network [2]. The generators are feedback controlled to provide the same strap current amplitude $|I_s|$=178A with a phase difference $\Delta\phi$ between elements such that the peak of the spectrum is centred on $k_{z,M} = p\, k_{z,M0}$. The index $p$ represents the harmonic number of $k_{z,M0} = 2\pi/(N_{str} S_z)$ where the denominator is the perimeter of the array with $S_z$ the inter-strap distance, i.e. the distance between the centre of two consecutive elements. For $k_{z,M}$ = 3.74m$^{-1}$ ($S_z$=0.3m, $k_{z,M0}$ =0.748m$^{-1}$ harmonic $p$=5) each strap has the following values: maximum voltage $|V_{str}|$ = 11.09kV, input impedance and input current respectively $Z_{in}$ = (187+i503)Ω and $|I_{in}|$ = 20.67A seen and provided by the generators. The input current is given by the sum of the current on the strap and the one in the tuning capacitor: $I_{in} = I_s + I_c$.

In Fig. 2, the value of the input impedance of the array is

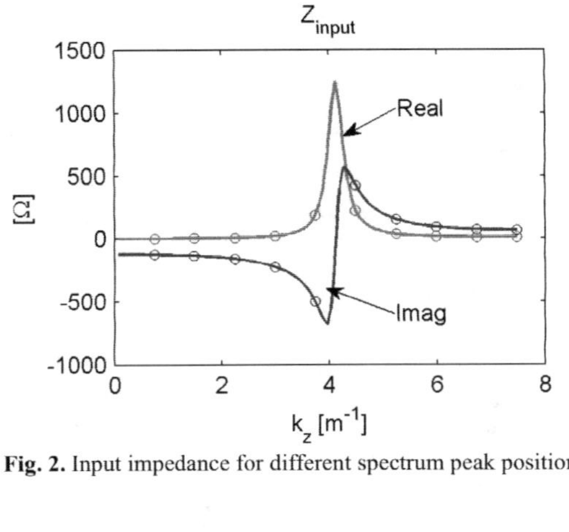


**Fig. 2.** Input impedance for different spectrum peak positions.

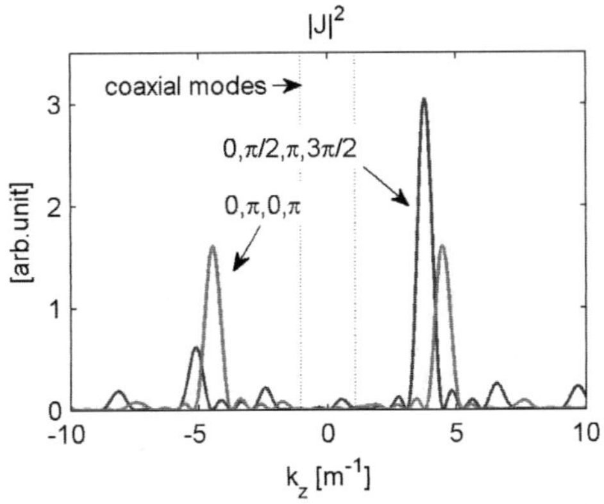


**Fig. 3.** Spectrum of the two cases for the array periodically fed in the conventional way with self-tuned elements.

computed for the discrete set of $k_{z,M}$, each of those is chosen by a different $\Delta\phi$ (represented by the dots). The computation of a continuous spectrum for $k_{z,M}$ (lines) reveals a resonance in the input impedance when $k_{z,M}$ is equal to the one obtained by the dispersion relation of the traveling wave structure at the tuned frequency $f_0$. For this value we have minimum amplitude of the input voltage required to obtain the requested strap current. This is valid only in the assumption that the loading resistance is the same for each element. A variation of the plasma loading on one (or more) radiating element(s), or a failure of the corresponding generator(s), will lead to a change in the input impedance seen by the neighbouring generators causing an important power imbalance. A complicated matching-decoupling network is needed to compensate the impedance mismatch. Then the system cannot be made easily load resilient.

## 2.3 360° array – periodic feeding

A possible way to reduce the number of feeding lines is to provide periodic feeding to the 360° array. Here we analyse a case where the array is fed with 4 independent generators evenly distributed (depicted in Fig. 1) and connected to the elements 1, 8, 15 and 22. Two different phasing cases for the generators are studied: (A) $\Delta\theta = \pi$

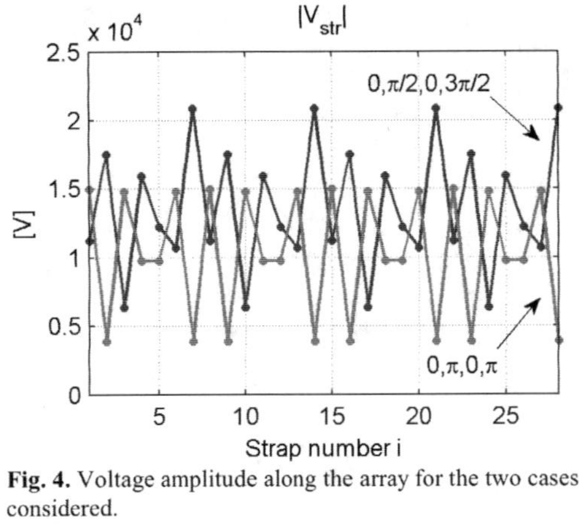


**Fig. 4.** Voltage amplitude along the array for the two cases considered.

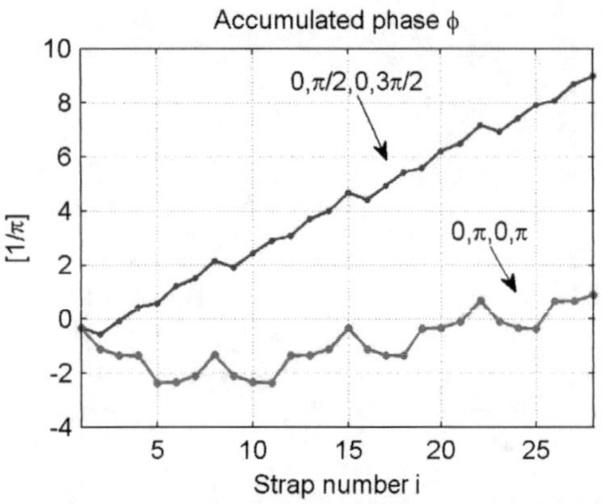


**Fig. 5.** Phase along the array for the two cases considered.

and (B) $\Delta\theta = \pi/2$. The resulting excitation spectra are shown in Fig. 3 where the reader will recognise the shape with two main peaks typical of case (A). From the point of view of traveling waves, (A) corresponds to the case in which two opposite directed traveling waves interfere leading to a standing wave pattern. To be noted the absence of coaxial modes ($k_z<k_0\approx1.05\text{m}^{-1}$) excitation. The voltage amplitude distribution on each strap is shown in Fig. 4, for an applied forward voltage of $V_+ =$ 10kV. It appears to vary considerably and every 7 straps a spatial symmetric pattern emerges from the graph in both cases.
Not surprisingly this is the same number of elements in between two adjacent generators.
The current amplitude distribution presents the same characteristics of the voltage one. The value of the phase at each strap is shown in Fig. 5.
The array impedance matrix $\mathbf{Z}_a$

$\mathbf{Z}_a$=

| | | | |
|---|---|---|---|
| 42.11 - i6.29 | -13.76 - i36.81 | 2.04 + i49.39 | -13.76 - i36.82 |
| -13.76 - i36.81 | 42.11 - i6.29 | -13.76 - i36.81 | 2.04 + i49.39 |
| 2.04 + i49.39 | -13.76 - i36.81 | 42.11 - i6.29 | -13.76- i36.81 |
| -13.76 - i36.81 | 2.04 + i49.39 | -13.76 - i36.81 | 42.11 - i6.29 |

reveals the strong coupling between the fed elements (i.e. #1, #8, #15 and #22) connected to the 4 generators. The characteristic impedance of the ring is $Z_{0,a} = 69\Omega$. The contribution of each generator to the active power provided by the 4 fed elements is given by the power matrix $\mathbf{P}_A$ and $\mathbf{P}_B$ for the two different cases analysed.
The real part of the two power matrices is:

Re($\mathbf{P}_A$)= (100kW)

| | | | |
|---|---|---|---|
| 2.521 | 0.824 | 0.122 | 0.824 |
| 0.824 | 2.521 | 0.824 | 0.122 |
| 0.122 | 0.824 | 2.521 | 0.824 |
| 0.824 | 0.122 | 0.824 | 2.521 |

and

Re($\mathbf{P}_B$)= (100kW)

| | | | |
|---|---|---|---|
| 5.616 | 4.911 | -0.271 | -4.911 |
| -4.911 | 5.616 | 4.911 | -0.271 |
| -0.271 | -4.911 | 5.616 | 4.911 |
| 4.911 | -0.271 | -4.911 | 5.616 |

where each generator is delivering respectively $P_{G,A}$ = 429kW or $P_{G,B}$ = 534kW giving rise to total power levels of $P_{tot,A}$ = 1.72MW for case (A) and $P_{tot,B}$ = 2.14MW for case (B).

## 2.4 360° array – consecutive resonant rings

The circular array of 28 elements used in the previous examples is now fed firstly by one resonant ring circuit and secondly by two consecutive one (Fig. 6). A TWA section fed by a resonant ring system was analysed in [1] and will be here compared with the first case where the resonant ring is connected to two arbitrary but consecutive elements, e.g. #28 and #1. The main difference with a normal TWA section is the mutual coupling between the inputs of the circuit due to the circular symmetry of the ring. This leads to a substantial differences in the array behaviour: (i) a different characteristic impedance of the array $Z_0 = Z_{it}$ and (ii) a much larger voltage and current amplitudes variation compared to the exponential decay of the single section.
An example of the amplitude of the voltage is shown in Fig. 7. The parameters are: $Z_{0,r} = 85\Omega$ for the ring, $Z_{0,s}$ = 148Ω for the section and $V_{G+}$ = 10kV. The forward power is then computed by $P_+ = 0.5|V_{G+}|^2/Z_0$, i.e. $P_{+r}$ = 588kW and $P_{+s}$ = 338kW. The linear phase variation

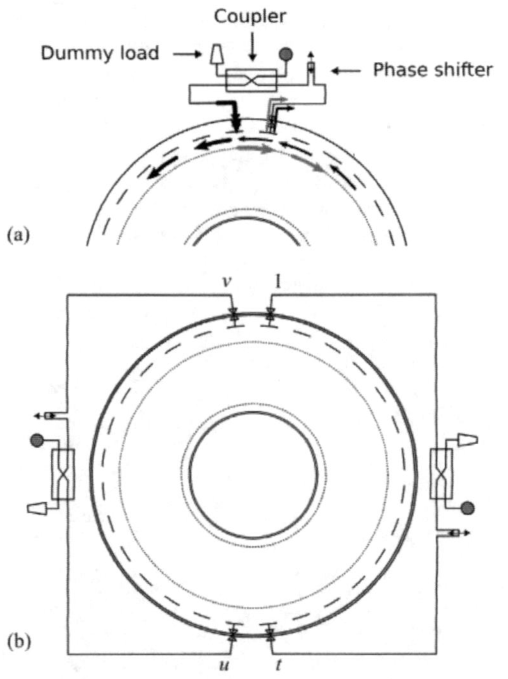


**Fig. 6.** Ring array fed by (a) one or (b) two consecutive resonant rings.

DOI: 10.1051/epjconf/201715703044

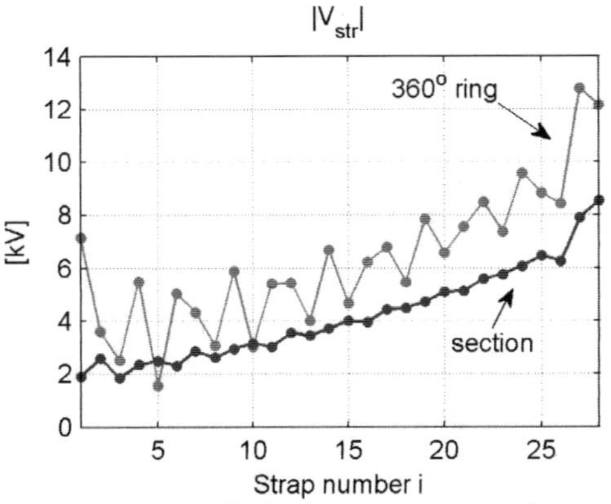


**Fig. 7.** Voltage amplitude along the array for the two cases considered and for a generator forward voltage $V_{G+}$=10kV.

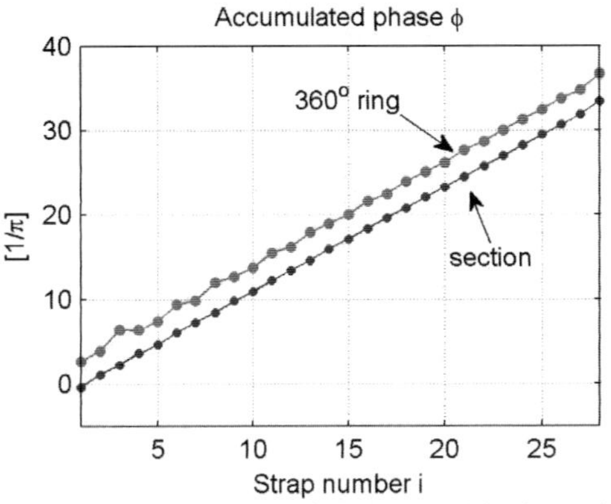


**Fig. 8.** Phase along the array for the ring and for the section.

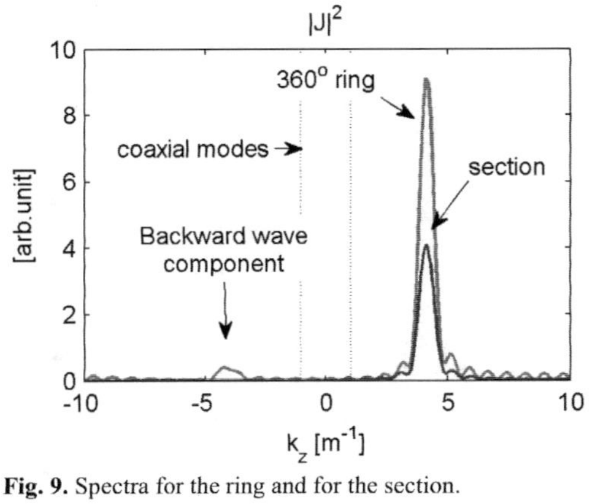


**Fig. 9.** Spectra for the ring and for the section.

along the array is however preserved (see Fig. 8). As a consequence, the spectra for the two cases assume the shape shown in Fig. 9 where in the case of the ring a

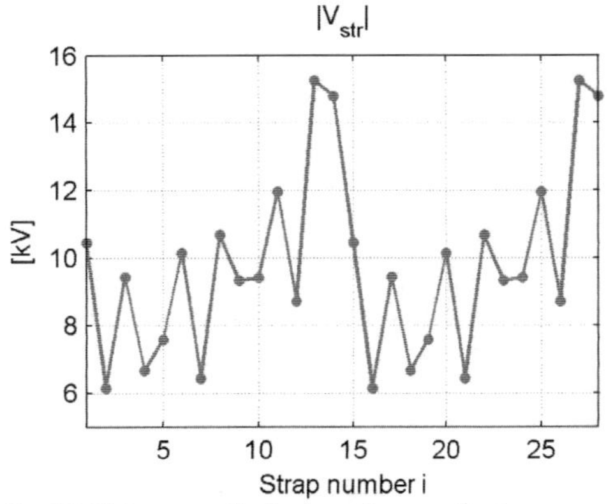


**Fig. 10.** Voltage amplitude along the array for the two cases considered and for a generator forward voltage $V_{G+}$=10kV.

backward component appears as a sign of a wave traveling in the opposite direction, injected in the line due to the mutual coupling between the inputs, as depicted in Fig. 6a by red arrows.

Both the section and the ring maintain the capability of controlling the position of the peak varying the frequency.

The second case consists of the same ring fed by two consecutive resonant rings (see Fig. 6b), connected symmetrically at the elements 1, $t$, $u$ and $v$. As seen above, there is mutual coupling between the output (e.g. 1, $u$) and the input (e.g. $t$, $v$) of the two resonant ring feeding systems due to the terms $\mathbf{Z}(1,v)$, $\mathbf{Z}(v,1)$, $\mathbf{Z}(t,u)$ and $\mathbf{Z}(u,t)$ of the array impedance matrix $\mathbf{Z}_a$. For our particular case: $v$ = 28, $u$ = 15 and $t$ = 14. Also in this case the voltage amplitude presents large fluctuations, shown in Fig. 10, due to the interference of opposite directed traveling waves that manifests itself again in the spectrum, like for the case in Fig. 9. But here, due to the strong modulation of the current on the straps, side peaks appear in the spectrum as shown in Fig. 11.

The real part of the power matrix for this case is:

$$\mathrm{Re}(\mathbf{P}_A)=\begin{vmatrix} 1.615 & -1.046 & -0.078 & -4.807 \\ 1.006 & 4.474 & 4.646 & -0.217 \\ -0.078 & -4.807 & 1.615 & -1.046 \\ 4.646 & -0.217 & 1.006 & 4.474 \end{vmatrix} \mathrm{100kW}$$

that gives $P_1 = P_{15}$ = -432kW, $P_{14} = P_{28}$ = 991kW. The negative values show that the power is extracted from the structure and recirculated in each resonant ring. The total power delivered by the array is $P_{tot}$ = 1.18MW.

## 3 Conclusions

 DOI: 10.1051/epjconf/201715703044

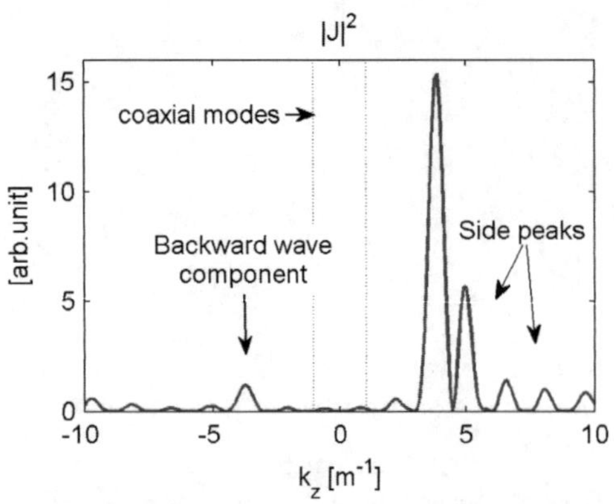


**Fig. 11.** Spectrum for the ring fed by two consecutive resonant ring systems.

The analysis based on the 28 elements array can be generalized to an arbitrary number of radiating elements and sections. The results presented, that are independent on that arbitrary number, shows that the preferred system for DEMO is the self-tuned strap array sections fed in resonant ring configuration (e.g. Fig. 6b), where the number of sections depends on the total number of elements. It has the advantage of: (i) best selective excitation spectrum, (ii) load resilience and (iii) simplicity of its adjustment. The load resilience is a characteristic of the resonant recirculation ring [1]. Moreover, a configuration with separate sections (e.g. Fig. 12), ideally one per each sector of the machine, provides better reliability than all other solution: if one element (or one generator) of one section fails, the other ones will not be significantly affected. In the example in Fig. 12 the sections are isolated by means of short blanket segments (blue rectangles).

An example of a DEMO system is given in [4].

The second layout that provides excellent RF performances is the 360° ring array with each strap fed independently. For this system, the problems of load resilience and mutual coupling between feeding networks are not yet solved and all solutions based on a complete 360° strap ring will suffer from mutual coupling between the feeding networks and from

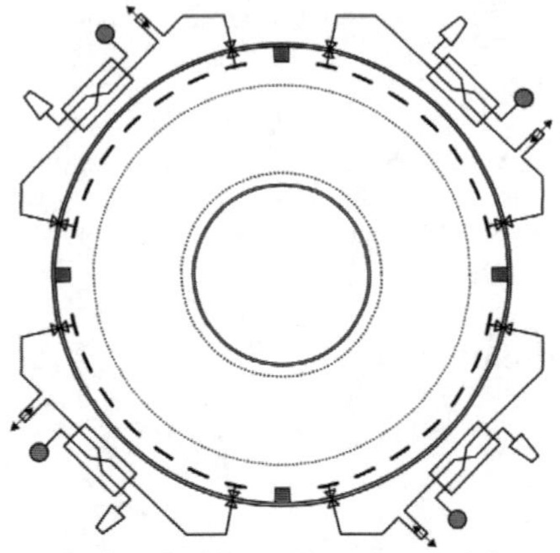

**Fig. 12.** Distributed TWA sections separated by septa or blanket modules. All sections are independent and isolated.

standing waves along the TWA structure that will lower the performance.

This work has been carried out within the framework of the EUROfusion Consortium and has received funding from the Euratom research and training programme 2014-2018 under grant agreement No 633053. The views and opinions expressed herein do not necessarily reflect those of the European Commission.

## References

1. R. Ragona and A. Messiaen, Nucl. Fusion **56** (2016) 076009.
2. G. Bosia and R. Ragona, this conference.
3. G. Bosia, Fus. Eng. and Design **92** (2015) 8-15.
4. A. Messiaen and R. Ragona, this conference.
5. A. Messiaen et al, Nucl. Fusion **50** (2010) 025026.
6. C. G. Montgomery, R. H. Dicke, and E. M. Purcell. *Principles of Microwave Circuits*, volume 8 of *MIT Radiation Laboratory Series*. McGraw-Hill, New York, 1948

EPJ Web of Conferences **157**, 03045 (2017) DOI: 10.1051/epjconf/201715703045
*22 Topical Conference on Radio-Frequency Power in Plasmas*

# Lower hybrid current drive in the presence of electric field

*Alexander* Saveliev[1,*], *and Vladimir* Zakharov[1]

[1]Ioffe Institute, Plasma Physics Laboratory, 194021 St. Petersburg, Russia

**Abstract.** A new one-dimensional approach to the lower hybrid current drive (LHCD) modelling in the presence of an inductive electric field is suggested in this paper. The approach is based on using time-dependent solutions of a well-known Fokker-Planck equation for the distribution function of fast electrons calculated concurrently with solving plasma transport equation in the Automated System for TRansport Analysis (ASTRA) [1]. A good agreement between experimental and modelling results is demonstrated for an FT-2 [2] plasma shot. Also new formulae for the steady-state solution of this kinetic equation are found.

## 1 Analytical analysis

The well-known steady-state solution for the one-dimensional distribution function of fast electrons in the absence of an electric field [3] is still used for the lower hybrid current drive modelling because of its simplicity combined with a rather good calculated driven current accuracy [4]. Actually, the electric field is only absent in a steady-state situation in which the plasma current is entirely driven by the waves. In most LHCD experiments the inductive electric field is present and can play a noticeable role in the LH current generation. In this paper we describe a steady-state solution of the 1D kinetic equation for the fast ($V_{||}/V_{T_e} \gg 1$) electrons and discuss application of the solution to the lower hybrid (LH) current drive simulations. This equation has the following form in standard notations [5].

$$\frac{\partial f}{\partial \tau} = E\frac{\partial f}{\partial v} + \frac{\partial}{\partial v}\left\{D(v)\frac{\partial f}{\partial v}\right\} + \beta(Z_{eff})\frac{\partial}{\partial v}\left\{\frac{1}{v^3}\frac{\partial f}{\partial v} + \frac{f}{v^2}\right\} \quad (1)$$

Here $D(v)$ is the quasilinear diffusion coefficient that represents the effect of the waves, $v = V_{||}/V_{T_e}$, $V_{\mathrm{Te}} = \sqrt{T_e/m_e}$, $E = E_{||}/E_c$, $E_c = 4\pi e^3 n_e L/T_e$ , $L$ is the Coulomb logarithm and time is normalized to the Spitzer collision time $\tau = t/\tau_0$ , $\tau_0 = m_e^2 V_{T_e}^3/4\pi e^3 n_e L$. To reconcile the 1D solution with the prediction of the 2D theory we introduce a parameter $\beta = (5 + Z_{eff})/5$, which actually means renormalization of the Spitzer collision frequency. The equation (1) includes the run-away problem leading to a divergence of the electron distribution function (EDF) wing $f_{acc}(v)$ where the electric field $E$ accelerates electrons. Nevertheless, for small normalized parallel electric field $|E| \ll 1$ the run-away problem can be solved as stationary [6] because a very large number of electrons in the Maxwellian body of the distribution function play the role of a large reservoir feeding a small stationary electron flux in the direction of higher energies. Following the methods given in [6] and [7], we found for the $f_{acc}(v)$ a steady-state solution $\left(\frac{\partial f}{\partial \tau} = 0,\ |E| \ll 1\right)$ of the equation (1) in the form of integrals of elementary functions.

$$f_{acc}(v) = \frac{1}{\sqrt{2\pi}\,C}\int_v^\infty \frac{x^3}{1+Dx^3} exp\left(\int_v^x \frac{t-|E|t^3}{1+Dt^3}dt\right)dx$$
$$C = \int_0^\infty \frac{x^3}{1+Dx^3} exp\left(\int_0^x \frac{t-|E|t^3}{1+Dt^3}dt\right)dx \quad (2)$$

For the wing of the distribution function $f_{dec}(v)$ where the electric field decelerates electrons we obtain the following solution of the equation (1) for $\frac{\partial f}{\partial \tau} = 0$ and an arbitrary absolute value of the electric field$|E|$:

$$f_{dec}(v) = \frac{1}{\sqrt{2\pi}} exp\left[-\int_0^v \left(\frac{t+|E|t^3}{1+Dt^3}\right)dt\right] \quad (3)$$

It should be noted that the solution (2) in contrast to the solution (3) has a nonzero flux of fast electrons propagating in the direction of higher energies. In the particular case $D(v) = 0$, integrals in (2) can be calculated analytically. This simplifies solution (2) to a combination of the probability integral and elementary functions.

$$f_{acc}(v) = \frac{e^{-\frac{1}{4|E|}}}{\sqrt{2\pi}}\left(\frac{e^{x^2}\Psi(x)+2\sqrt{\frac{|E|}{\pi}}}{2-\Psi\left(\frac{1}{2\sqrt{|E|}}\right)+2\sqrt{\frac{|E|}{\pi}}\,e^{-\frac{1}{4|E|}}}\right) \quad (4)$$

Here the following notations are used:

$$x = \frac{1}{2}\left(v^2\sqrt{|E|} - 1/\sqrt{|E|}\right)$$
$$\Psi(x) = 1 - \Phi(x); \quad \Phi(x) = \frac{2}{\sqrt{\pi}}\int_0^x e^{-t^2}dt$$

This formula (4) is a new one. Then, for the decelerating electrons and $D(v) = 0$ we easily obtain from (3) a simple formula (5).

$$f_{dec}(v) = \frac{1}{\sqrt{2\pi}} exp\left(-\frac{v^2}{2} - \frac{|E|v^4}{4}\right) \quad (5)$$

* Corresponding author: alexander.n.saveliev@mail.ioffe.ru

DOI: 10.1051/epjconf/201715703045

## 2 Numerical analysis

### 2.1. Zero electric field

We start with an example of the LH driven current profiles calculated for the case of E=0 by a stand-alone version of the Fast Ray-Tracing Code (FRTC) [8, 9] for the FT-2 plasma parameters of the shot #041514_35 at the time slice t=0.032 seconds and 32 kA of plasma current. The simulation result is given in Fig. 1.

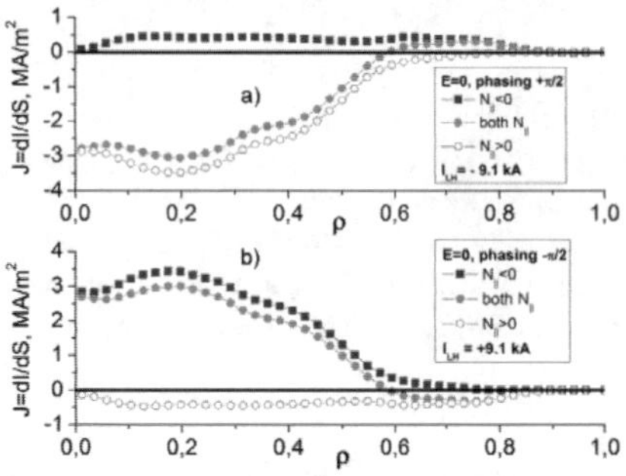


**Fig. 1.** Driven current profiles calculated by FRTC with E=0. a) antenna phasing is $+\pi/2$; b) antenna phasing $-\pi/2$.

Initial $N_{||}$ spectra for the FT-2 LH two-waveguide antenna noticeably differs from the standard LH spectra because its positive and negative lobes contain comparable amounts of the LH input power. Examples of the FT-2 spectra [10] calculated for opposite waveguide phasing is present in Fig. 2.

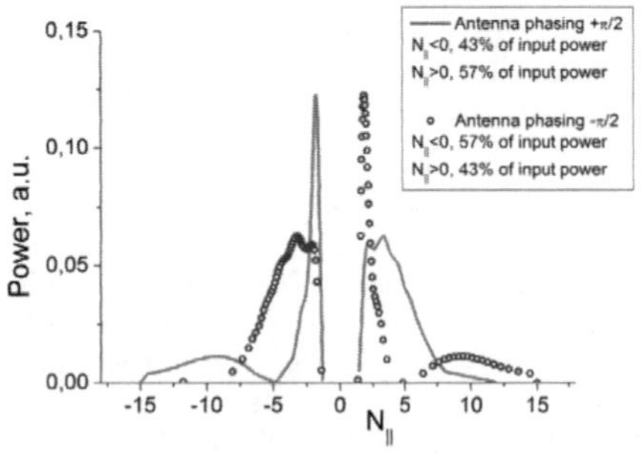


**Fig. 2.** Calculated initial FT-2 $N_{||}$ spectra for $+\pi/2$ and $-\pi/2$ phasing of the FT-2 antenna.

Predicted driven currents in Fig. 1 for the case of the inductive electric field absence turns out oppositely directed for the $+\pi/2$ and $-\pi/2$ antenna phasing's. It must considerably influence the loop voltage, which is never the case in the FT-2 LHCD experiments, where changing phasing from $+\pi/2$ to $-\pi/2$ never changed the sign of the loop voltage. This weak dependence of the loop voltage on the antenna phasing can be explained by the influence of the inductive electric field on the LH current generation. Namely, the electric field enhances $f_{acc}(v)$ and suppresses the $f_{dec}(v)$ wings of the distribution function, leading to an additional asymmetry of the EDF, which may strongly change driven current profile and value. The other part of the paper is devoted to the LHCD modelling with $|E| > 0$.

### 2.2. Nonzero electric field

Numerical solution of the Fokker-Planch equation (1) with initial Maxwellian distribution function and typical profile of the quasilinear diffusion coefficient $D(v)$ given in Fig. 3 showed that the steady-state distribution function of fast electrons (2) is reached at time $t_{asymp}$, which is approximately equal to ten collision times $t_{st(4V_{Te})}$ of fast electrons ($V_{||} = 4\,V_{Te}$, where the quasilinear plateau starts) with the thermal electrons.

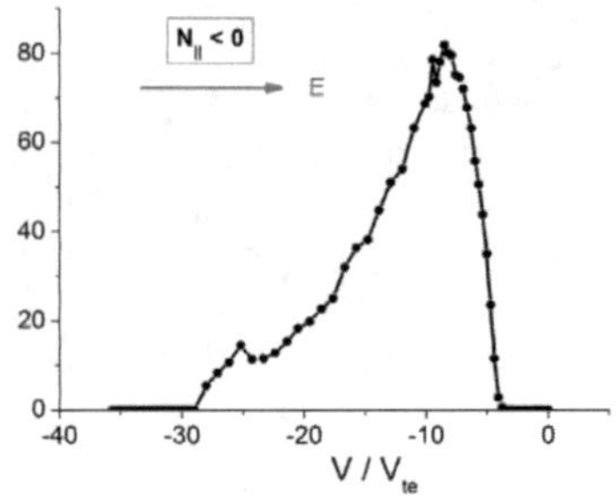


**Fig. 3.** Typical $D(v)$ profile obtained by the FRTC.

In Fig. 4 you can see evolution of the $f_{acc}(v)$ calculated for $|E| = 0.04$ and $D(v)$ given in Fig. 3.

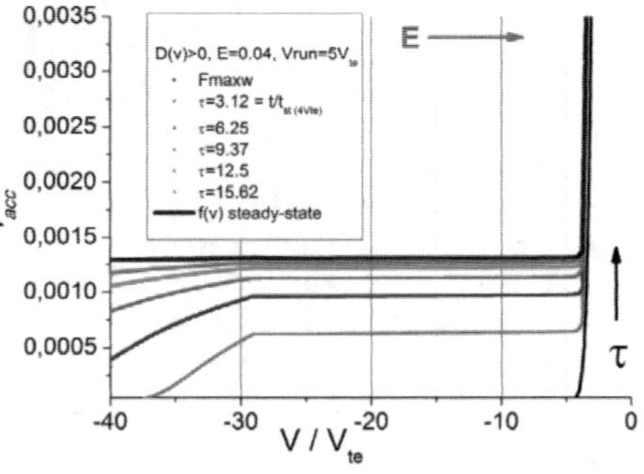


**Fig. 4.** $f_{acc}(v)$ evolution with time $\tau = t/t_{st(4V_{Te})}$

Note, that $t_{asymp}$ depends mainly on the quasilinear plateau's width and is almost independent of the electric field value. This statement is illustrated by the Fig. 5.

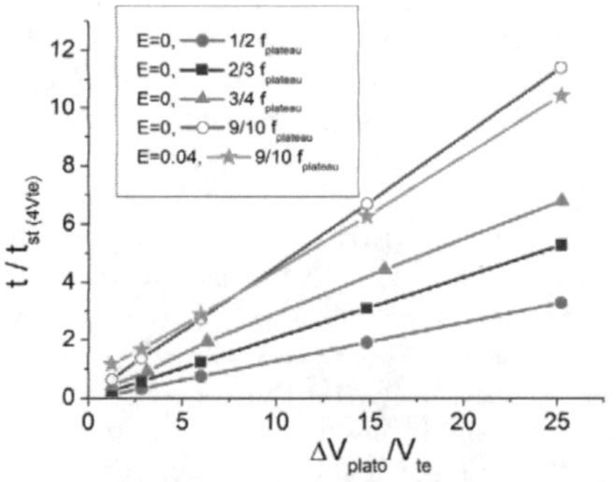


**Fig. 5.** Normalized times needed to reach different levels of the steady-state EDF $f_{acc}(v)$: (1/2 $f_{plateau}$, 2/3 $f_{plateau}$, etc.).

A few dependences of $t_{asymp}$ on plasma temperature and density for $\beta(Z_{eff}) = 1$ are given in Fig. 6. For $\beta \neq 1$ the $t_{asymp}$ should be divided by $\beta$.

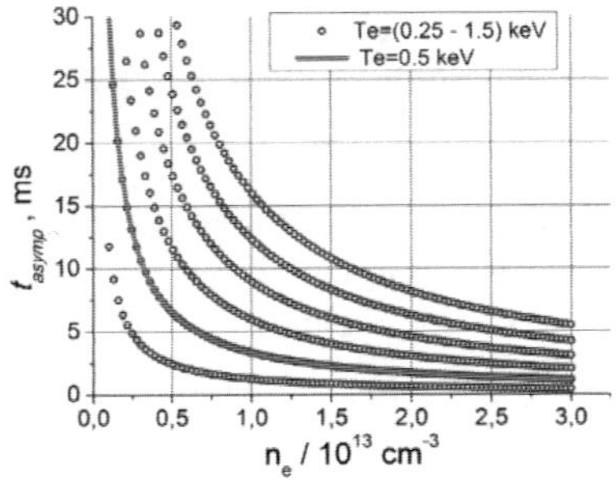


**Fig. 6.** $t_{asymp}$ as a function of plasma temperature and density. Each curve corresponds to a fixed temperature on the equidistant greed in the interval 0.25÷1.5 keV.

It is seen from the Fig. 6 that $t_{asymp}$ may vary across the minor tokamak radius from a few to a few dozen milliseconds, so the formulae (2-5) for the steady-state distribution function can be, in principle, used in the LHCD modelling of experiments with much longer LH impulse duration than the $t_{asymp}$. For small tokomaks Globus-M, Globus-M2 and FT-2 the LH impulse duration is comparable with the $t_{asymp}$ which makes the LHCD simulation task nonstationary for these tokamaks. Moreover, the application of the stationary solution (2-5) for the distribution function of fast electrons in modelling of the FT-2 LHCD experiments gives obviously wrong result in many cases, namely the overestimated driven current. An example of the steady-state formulae (2-5) application is given in Fig. 7, where you can see the LH driven current profiles calculated by a stand-alone FRTC version for the same plasma conditions as in Fig. 1 and the electric field radial profile taken from the ASTRA run.

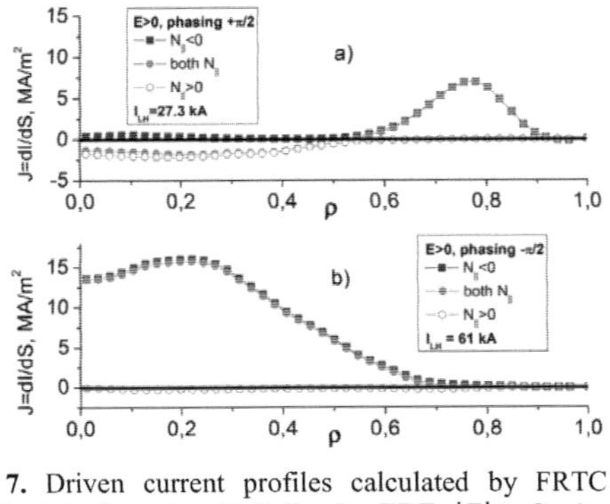


**Fig. 7.** Driven current profiles calculated by FRTC using steady-state formulae (2-5) for the EDF, $|E| > 0$. a) antenna phasing is $+\pi/2$, b) reverse antenna phasing $-\pi/2$.

It is seen from the Fig. 7 that calculated LH driven current for $+\pi/2$ antenna phasing is almost equal to the total plasma current, which should lead to almost zero loop voltage, but it is not the case in this experiment (see Fig. 8). Results of modelling for $-\pi/2$ antenna phasing are even more far from the all FT-2 experimental experience. So, we have to make a conclusion that formulae (2-5) lead to driven current overestimation for the FT-2 experimental conditions. To explain this disagreement between experiment and modelling we supposed that distribution function for the fast electrons does not have enough time to reach the steady-state level described by the formulae (2-5), because duration of the LH power impulse is only 6 milliseconds in the FT-2 experiments which is almost equal to the corresponding value of the $t_{asymp}$. Therefore the EDF time evolution must be taking into consideration for an adequate modelling of the FT-2 experiments.

To take into account the evolution of the EDF, plasma parameters and electric field radial profile we modified the FRTC code to include a relatively weak inductive electric field using a time-dependent solution of the Fokker-Planch equation (1) for the EDF. This version of the FRTC was coupled with the ASTRA. The whole scheme works as follows. At a given ASTRA time step the FRTC reads equilibrium, plasma parameters and electric field radial profile from ASTRA, calculates electron distribution function of fast electrons at this time step, finds the corresponding LH driven current and power absorption density profiles and returns these profiles into ASTRA as external current and heat source, which are then used by ASTRA for equilibrium and plasma parameters recalculation. So, the whole scheme provides self-consistent simulations of plasma shots, which can be compared with experimental results. The very similar scheme was used in the paper [11] where TRANSP and CQL3D were utilized instead of ASTRA and FRTC.

In Fig. 8 you can see the first modelling result for the FT-2 shot #041514_35 obtained by using the above scheme. It is seen from the Fig. 8 that the nonstationary approach gives a rather good agreement between experimental and modelling results. Calculated loop voltage generally follows the experimental time trace. Agreement between modelling and experiment can be improved, for instance, by introducing a small change to the power sharing in the initial $N_{||}$ spectrum lobes or a small variation to the spectrum form, because we actually know this spectrum only approximately.

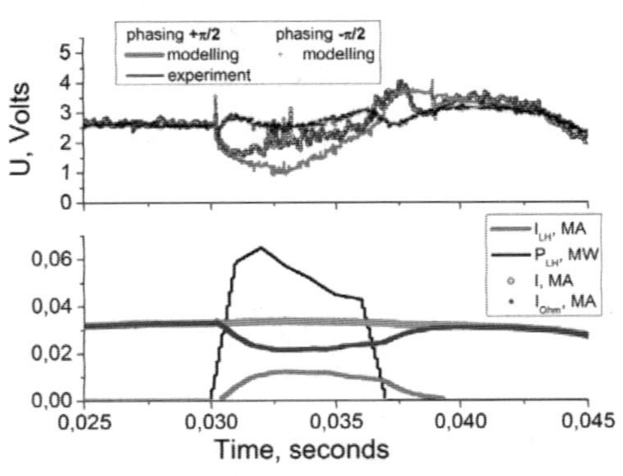


**Fig. 8.** Comparison of experimental and modelled loop voltage for FT-2 tokomak LHCD experiment, shot #041514_35.

DOI: 10.1051/epjconf/201715703045

The last figure Fig. 9 of this paper shows some radial profiles of the FT-2 shot#041514_35 at time slice t=0.032 seconds plus the LH driven current profile given in Fig. 7a. Comparing this radial profile (solid curve, $I_{LH}$=27.3 kA) with the corresponding self-consistently modelled LH current profile (circles, $I_{LH}$=11 kA), one can estimate the difference between the time-dependent approach and the steady-state EDF application for the LHCD simulations in the case when $t_{asymp}$ is comparable with the LH power impulse duration.

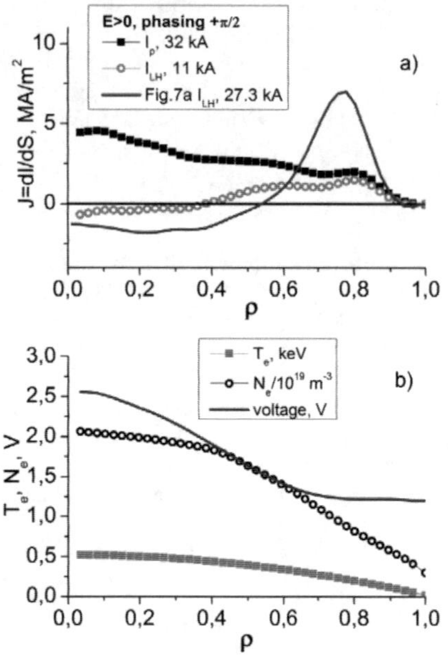


**Fig. 9.** Radial profiles for the shot #041514_35, LHCD experiment in the FT-2 tokomak.
a) Solid squares – plasma current, cycles – driven current, solid curve – driven current calculated by FRTC using the steady-state formulae (2-5) for the EDF.
b) Solid squares – plasma temperature, cycles – plasma density, solid curve – voltage radial profile.

## 3 Summary

Formulae (2-4) and the conclusion that nonstationary approach should be used in many cases in the LHCD modelling for adequate explanations of experimental results are the main results of this paper. Also a new scheme for the LHCD modelling is developed. A good applicability of the scheme is demonstrated using an example of an FT-2 shot.

The financial support of the RSF grant 16-12-10043 and the Ioffe Institute is acknowledged.

## References

1. G.V. Pereverzev and P.N. Yushmanov, ASTRA Automated System for TRansport Analysis IPP-Report IPP **5/98**, (2002)
2. S. I. Lashkul et al. Plasma Physics Reports, **36**(9), 751 (2010)
3. C.F.F. Karney and N.J. Fisch, Phys.Fluids, **22**(9), 1817 (1979)
4. N.J. Fisch, Rev. Mod. Phys. **59**(1), 175 (1987)
5. A.A. Vedenov, in *Reviews of Plasma Physics*, edited by M.A. Leontovich (Consultants Bureau, New York, 1967), **3**, 229
6. E.M. Lifshitz, L.P. Pitaevskii, *Physical Kinetics*, **10** (Pergamon Press., 1st ed., 1981)
7. A.V. Gurevich, Soviet Physics JETP, **12**(8), 904 (1960)
8. Esterkin A R and Piliya A D, Nucl. Fusion, **38**, 1501 (1996)
9. Piliya A D and Saveliev A N, Preprint JET-R, **01** (1998)
10. M. A. Irzak and O. N. Shcherbinin, Nucl. Fusion **35**, 1341–56 (1995)
11. F. Poli et al, Plasma Physics and Controlled Fusion **58**, 095001 (2016)

EPJ Web of Conferences **157**, 03046 (2017) DOI: 10.1051/epjconf/201715703046
*22 Topical Conference on Radio-Frequency Power in Plasmas*

# ICRF heating schemes for the ITER non-active phase

*Mireille* Schneider[1,*], *Jean-François* Artaud[2], *Paul* Bonoli[3], *Yevgen* Kazakov[4], *Philippe* Lamalle[1], *Ernesto* Lerche[4,5], *Dirk* Van Eester[4] and *John* Wright[3]

1 ITER Organization, Route de Vinon-sur-Verdon, CS 90 046, 13067 St. Paul-lez-Durance, France
2 CEA, IRFM, F-13108 Saint Paul-lez-Durance, France
3 Plasma Science and Fusion Center, MIT, Cambridge, MA 02139, USA
4 LPP-ERM/KMS, Association Eurofusion-Belgian State, TEC partner, Brussels, Belgium
5 Euratom/CCFE Fusion Association, Culham Science Centre, Abingdon, United Kingdom

**Abstract.** ITER plasma operation requires a non-active phase for tokamak initial commissioning, covering First Plasma and Pre-Fusion Power Operation phases, PFPO-1 and PFPO-2. Non-active operation consists of hydrogen and helium plasmas to minimize the neutron production rate. The present document describes some Ion Cyclotron Radio Frequency (ICRF) heating schemes in terms of their predicted performance for the main foreseen scenarios of the ITER non-active phase in hydrogen and helium. Emphasis is given on remaining issues and physics uncertainties to be addressed for successful ICRF heating in ITER.

## 1 Introduction

Auxiliary heating power is essential for future tokamaks to achieve and sustain fusion performance. In ITER, three auxiliary heating methods are foreseen: Electron Cyclotron Resonance Heating (ECRH), Ion Cyclotron Resonance Frequency heating (ICRF) and Neutral Beam Injection (NBI). The new ITER research plan is based on a staged approach including two Pre-Fusion Power Operation phases PFPO-1 and PFPO-2 consisting of hydrogen and helium plasmas. The ITER fusion operation phases, i.e. deuterium and deuterium-tritium plasmas, are not covered in the present analysis.

Efficient fusion performance relies on achieving an improved confinement regime, i.e. the so-called H-mode. The L-H power threshold is predicted by scaling laws derived from experimental database and mostly depends on the core density, the plasma surface area and the toroidal magnetic field [1].

Three main magnetic fields are now under consideration to improve H-mode access capabilities in the non-active phase: $B_0$ = 1.8T, 2.65T and 5.3T. The 5MA/1.8T operation is aimed at enabling H-mode access early in the plan, i.e. in the PFPO-1 phase where only ECRH and ICRF should be available (20 MW of ECRH and possibly 10 MW of ICRF if the PFPO-1 installation of one IC antenna is confirmed). The PFPO-2 phase will benefit from the full baseline heating capabilities, i.e. ECRH, ICRF and NBI heating providing a total auxiliary power of 73 MW. A number of ICRF heating schemes have been investigated for these magnetic fields, both in hydrogen and helium plasmas. The expected RF absorption efficiencies of each of these schemes, based on 1D RF wave modelling [2], is reported here.

* Corresponding author: mireille.schneider@iter.org

## 2 ICRF heating schemes

Transport simulations have been carried out using the METIS 0.5D transport code [3], which combines a 1D current diffusion solver, a 2D equilibrium solver and a hybrid 0D/1D calculation of heat transport dynamics, based on scaling laws. These simulations have provided a consistent initial set of temperatures and densities that are used for ICRF wave calculations [4], for which central core values are displayed in table 1 as a reference for scenarios covered in the present document.

**Table 1.** Central electron and ion temperatures and densities for all scenarios covered by this analysis.

| $B_0$ (T) | 5.3 | | 2.65 | | 1.8 | | 3 | 3.3 |
|---|---|---|---|---|---|---|---|---|
| $I_p$ (MA) | 15 | | 7.5 | | 5 | | 8.5 | 9.5 |
| $P_{AUX}$ (MW) | 30 | | | | | | 73 | |
| Main ion | H | He | H | He | H | He | H | H |
| $Te^0$ (keV) | 10 | 13 | 7.2 | 11 | 14 | 16 | 14 | 11 |
| $Ti^0$ (keV) | 12 | 16 | 6.0 | 12 | 7.0 | 8.0 | 10 | 9.3 |
| $ne^0$ ($10^{19}$ m$^{-3}$) | 5.6 | 5.5 | 2.8 | 3.0 | 1.9 | 1.9 | 3.2 | 3.4 |
| $ni^0$ ($10^{19}$ m$^{-3}$) | 5.4 | 2.9 | 2.7 | 1.7 | 1.8 | 1.0 | 2.2 | 2.3 |

ICRF wave calculations have been performed with the TOMCAT 1D and TORIC 2D ICRF codes [2, 5]. In this analysis, only the core Single Pass Absorption (SPA), summed over electrons and ion species, is included, since it has been demonstrated to reliably quantify the quality of a heating scheme when compared to experiment, see e.g. [6]. All concentrations are given with respect to the electron density.

The ability of the antenna system to couple ICRF power to these plasmas through the edge region is not discussed in the present paper.

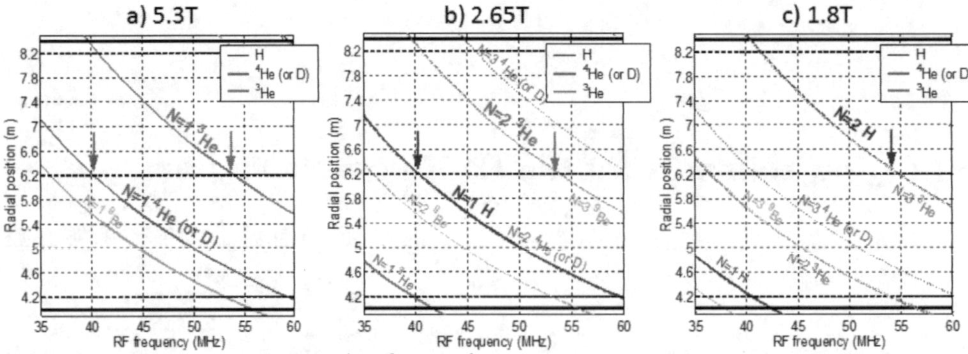


**Fig. 1.** Cyclotron resonance layers of H, $He^4$, $He^3$ and $Be^9$ ions as a function of the ICRF wave frequency at (a) 5.3T, (b) 2.65T and (c) 1.8T for left / middle / right figures respectively.

Various ion species have been taken into account to explore the best ICRF heating schemes in the ITER non-active phase: H, $He^4$, $He^3$ and $Be^9$. Majority ions can be heated at harmonics of the cyclotron resonance, while fundamental majority heating is inefficient due to the screening effect, i.e. when the wave electric field component rotating in the same sense as the ions vanishes at the resonance location of the majority ion [7]. Other ICRF heating schemes rely on the presence of a minority species with a charge-to-mass ratio different from the majority ion one. These are the two-ion minority heating schemes. Fundamental ion heating schemes typically work best at reduced concentration while harmonic heating requires preheating or a large ion population. H-$He^4$ mixtures have also been explored to investigate the three-ion minority heating schemes, based on the following fast wave property: if the L-cutoff layer of a two-ion plasma is located close to the fundamental cyclotron resonance of a third species, the latter, if present in small quantities, absorbs almost all the power [8]. Majority species, two-ion and three-ion heating schemes have been investigated at 5.3T, 2.65T and 1.8T, as described in the next paragraphs. Resonances of all considered ions are displayed in Figure 1 for the three values of the magnetic field as a function of the ICRF wave frequency, of which the accessible range in ITER is 40-55 MHz.

## 2.1 Heating schemes at a magnetic field of 5.3T

As illustrated in Figure 1.a, central heating can be achieved at 5.3T for fundamental resonance of $He^3$ at 53 MHz or fundamental resonance of $He^4$ at 40 MHz. Off-axis fundamental $Be^9$ heating is also possible at 40 MHz. Characteristics of associated heating schemes are summarized in Table 2 and subsequently discussed.

### *2.1.1 Hydrogen plasmas*

Fundamental minority $Be^9$ heating exhibits a low absorption (SPA < 0.4) with dominant electron heating for a $Be^9$ level of 0.5%, which degrades at higher $Be^9$ concentration, leading to poor ICRF heating at full field.

Inverted schemes such as $He^3$ and $He^4$ heating in hydrogen are more appropriate. They are sensitive to the level of background impurity species, as observed at JET with a C-wall [9,10] and Be-wall [11,12]. Hence, the presence of $Be^9$ impurities in hydrogen influences the $He^4$ and $He^3$ cyclotron absorption in a similar way as D or $He^4$ dilution helps the $He^3$ absorption in the D or $He^4$-($He^3$)-H three-ion scheme. An example is given in Figure 2, which shows that the higher the background $Be^9$ level, the lower the $He^4$ concentration can be, with equivalent absorption. There is no precise estimate of the $Be^9$ level in ITER, assumed to be ~2% in the core plasma transport modelling.

**Table 2.** Performance of various ICRF heating schemes at 15 MA / 5.3 T. Shaded cells highlight high SPA schemes.

| Main ion | Heating scheme | $f_{IC}$ (MHz) | SPA | Constraints and optimal conditions |
|---|---|---|---|---|
| H | n=1 min $He^3$ $He^4$-($He^3$)-H | 53 | ~ 0.95 | $X[He^3] < 1\%$ $X[He^4] \sim 10\text{-}15\%$ |
| | n=1 min $He^4$ ($He^4$)-H and $Be^9$-($He^4$)-H | 41 | 0.6-0.8 | $X[He^4] \sim 0.1\text{-}2\%$ Depends on $Be^9$ level (see Figure 2) |
| | n=1 min $He^3$ ($He^3$)-H | 53 | ~ 0.8 at $X[Be^9]$~1% | $X[He^3] \sim 2\text{-}4\%$ Depends on $Be^9$ level |
| | n=1 min $Be^9$ ($Be^9$)-H | 40 | < 0.4 at $X[Be^9]$~0.5% | Low damping rate |
| $He^4$ | n=1 min $He^3$ ($He^3$)-$He^4$ | 53 | ~ 0.8 | $X[He^3] \sim 4\text{-}5\%$ |
| | n=1 min $Be^9$ ($Be^9$)-$He^4$ | 40 | < 0.25 | Low damping rate |
| | n=1 maj $He^4$ | 40 | ~ 0.2 | Low damping rate ($E_+$ screening) |

The SPA for the ($^3$He)-H scheme is computed to maximize to ~ 0.8 for an $He^3$ concentration of ~ 2-4% assuming 1% of $Be^9$. Operating at lower $He^3$ concentrations would lead to a lower efficiency but may be desirable to limit the operation costs. As discussed above, the $He^3$ concentration can be reduced with equivalent efficiency if the $Be^9$ concentration is higher [10,12]. Similarly, heating $He^4$ as a minority in hydrogen is possible. While this regime was not accessible in JET-C wall operation [13] ($Z/A$ ratio for C is the same as $He^4$ and D), changing the main intrinsic low-Z impurity from C to $Be^9$ allows this scheme to work. Indeed, since $(Z/A)_{Be9} < (Z/A)_{He4} < (Z/A)_H$, $He^4$ ions can be effectively heated at very low concentrations (~0.1-0.2%) via the $Be^9$-($He^4$)-H scheme provided the core $Be^9$ concentration

* Corresponding author: mireille.schneider@iter.org

is ~2% (see Figure 2). If $Be^9$ level is low (<0.5%), then the absorption is maximized at higher $X[He^4]$ ~1%. Note that at $X[Be^9]$ > 3%, minority heating of $He^4$ ions becomes inefficient. Injecting $Ne^{22}$ isotope ($Z/A$ close to that for $Be^9$) could be an option to operate at very low $He^4$ levels, since controlling $^9Be$ is unlikely.

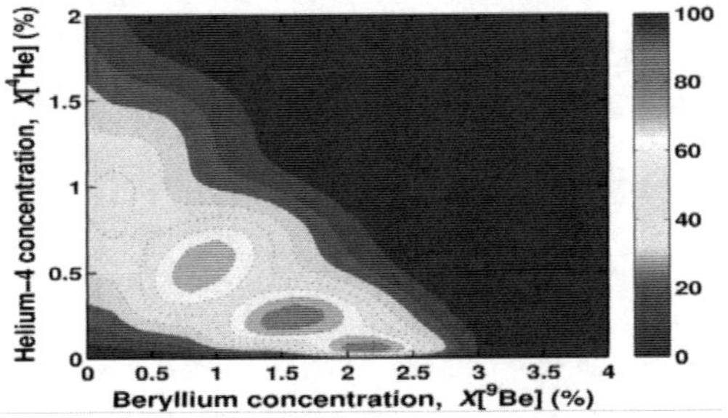


**Fig. 2.** Absorption by $He^4$ in H wrt $X[Be^9]$ and $X[He^4]$.

The three-ion heating scheme offers a more interesting option as its efficiency relies on very low minority concentration: heating $He^3$ ions with $X[He^3]$ < 1% in H-$He^4$ plasmas ($X[He^4]$ ~ 10-15%) provides a SPA of 0.95 making it the best absorption scheme, as experimentally demonstrated at JET and C-mod [14].

However, controlling the $He^3$ concentration in the presence of $He^4$ is challenging, since spectroscopy measurement can generally not distinguish them. They may however be discriminated by their different widths in wavelengths since $He^3$ ions are ICRF-accelerated while $He^4$ ions remain thermal [15]. An alternative is to inject a prepared $He^4$:$He^3$ mixture in the machine.

### 2.1.2 Helium plasmas

Fundamental majority $He^4$ heating exhibits a low absorption (SPA ~ 0.24) due to the $E_+$ screening effect, as discussed above. The situation may improve at higher temperatures due to enhanced Doppler-broadening, taking the vanishing $E_+$ point away from the resonance, as observed in JET-C in NBI preheated plasmas [16].

Fundamental minority $Be^9$ heating presents a SPA < 0.25 with dominant electron heating. Hence, the best heating scheme in helium at 5.3T is fundamental minority $He^3$ with a SPA of 0.8. It however requires $He^3$ concentration ~ 4-5 % which is scarce and costly.

## 2.2 Heating schemes at a magnetic field of 2.65T

Figure 1.b indicates that, at 2.65T, central deposition can be achieved with fundamental H or second harmonic $He^4$ heating at 40 MHz, and with second harmonic $He^3$ heating at 53 MHz. Characteristics and performance of investigated heating schemes are summarized in Table 3.

### 2.2.1 Hydrogen plasmas

Fundamental majority H heating exhibits a low absorption (SPA ~ 0.22) due to the screening effect. Again, the screening could be reduced by the Doppler-broadening at higher temperature but it has not been investigated yet. Second harmonic minority $He^3$ heating provides no significant improvement with a SPA < 0.25, even at high $He^3$ concentration. Both schemes have been tested in JET-C and the poor heating performance was confirmed [6]. Finally, second harmonic minority $He^4$ heating is a better heating scheme, providing a SPA still below 0.5 and with a dominant electron heating, leading to a very broad deposition. Thus, there is no efficient ICRF scheme at 2.65T in hydrogen. For this reason, optional schemes are envisaged at 3T / 3.3T to provide an operational space close to half field with an efficient ICRF heating scheme, as described in section 2.4.

**Table 3.** Performance of various ICRF heating schemes at 7.5 MA / 2.65T. Shaded cells highlight high SPA schemes.

| Main ion | Heating scheme | $f_{IC}$ (MHz) | SPA | Constraints and optimal conditions |
|---|---|---|---|---|
| H | n=2 min $He^4$ | 40 | < 0.5 | Even at high $X[He^4]$; dominant on electrons |
| | n=2 min $He^3$ | 53 | < 0.25 | Even at high $X[He^3]$ |
| | n=1 maj H | 40 | ~ 0.22 | Low damping rate ($E_+$ screening) |
| $He^4$ | n=1 min H | 40 | ~ 1 | X[H] ~ 2-8% |
| | n=2 maj $He^4$ | 40 | 0.78 | - |
| | n=2 min $He^3$ | 53 | > 0.6 | $X[He^3]$>5%, off-axis $He^4$ absorption |

### 2.2.2 Helium plasmas

Second harmonic minority $He^3$ heating displays a SPA of 0.6 with rather high concentration of $He^3$, here above 5%, hence this scheme should not be favoured. Second harmonic majority $He^4$ heating shows a SPA of 0.78, making it a suitable candidate. Finally, fundamental minority H heating at 40 MHz is the best candidate with a SPA close to 1 and $X[H]$~2-8%. The scheme could probably still be efficient at higher H concentration, which may not be avoidable due to the dilution effect when injecting hydrogen pellets into helium plasmas.

## 2.3 Heating schemes at a magnetic field of 1.8T

Operating at low magnetic field of 1.8T may enable H-mode operation during the PFPO-1 phase where auxiliary heating will be provided by 20 MW of ECRH and probably 10 MW of ICRF (one IC antenna may be installed for this phase but it is still under evaluation). Operating at low density is essential to minimize the L-H power threshold. 40% of the Greenwald density is considered as optimal as described in [17]. The threshold rollover at lower density can be due to the strong electron-ion decoupling, leading to $T_e \gg T_i$, where TEM and ETG modes may play a role and METIS scaling laws are beyond their validity domain. Hence, kinetic profiles used for this modelling need comparison with other transport codes. Operating at such low magnetic field induces an overcompensation of the magnetic field ripple by ferromagnetic inserts, leading to a ripple of the order of -1.3% which could enhance fast ion losses, still to be quantified. In addition, low density means low collisionality, further reducing the fast ion confinement, potentially leading to enhanced RF-induced plasma wall interaction [18].

As illustrated in Figure 1.c, central resonances are available at 1.8T for third harmonic $He^4$ at 40 MHz, and second harmonic H or third harmonic $He^3$ at 55 MHz. All these schemes are summarized in Table 4.

**Table 4.** Performance of various ICRF heating schemes at 5 MA / 1.8T. Shaded cells highlight high SPA schemes.

| Main ion | Heating scheme | $f_{IC}$ (MHz) | SPA | Constraints and optimal conditions |
|---|---|---|---|---|
| H | n=2 maj H | 53 | ~ 1 | - |
| | n=3 min $He^4$ | 40 | ~ 0.5 | Even at high X[$He^4$]; dominant on electrons |
| $He^4$ | n=2 min H | 53 | ~ 1 | $X$[H] = 5% |
| | n=3 maj $He^4$ | 40 | ~ 0.77 | Dominant el |
| | n=3 min $He^3$ | 53 | > 0.7 | Even at high X[$He^3$]; dominant on electrons |

### *2.3.1 Hydrogen plasmas*

Third harmonic minority $He^4$ heating exhibits a SPA of 0.5 with dominant electron heating, even at high $He^4$ concentration. The best scheme is second harmonic majority H heating with a very good SPA close to 1, making it an excellent scheme, provided the fast ions generated by harmonic heating are properly confined.

### *2.3.2 Helium plasmas*

Third harmonic minority $He^3$ heating shows a SPA ~ 0.7 with dominant electron heating (> 0.8), even at high $He^3$ concentration: in this case and more generally when there is no adequate ion resonance, one can rely on Fast Wave Electron Heating and avoid $He^3$ injection. This generates broader deposition profiles but with an overall good absorption. Third harmonic majority $He^4$ heating shows a SPA of 0.77, again with dominant electron heating. Finally, second harmonic minority H heating shows a SPA close to 1, making it the best candidate. Good minority heating at cyclotron harmonics is unusual but can be achieved thanks to the beneficial $1/B_0^2$ scaling of the Finite Larmor Radius absorption [19].

### 2.4 Optional schemes at 3T & 3.3T in hydrogen

To compensate the lack of efficient ICRF scheme in hydrogen at 2.65T, one option is to increase the magnetic field to 3T or 3.3T and heat a small amount of $He^3$ (~1%) at its fundamental resonance at 40 MHz. Minimizing the $He^3$ level to ~0.05% can be achieved in a H-$He^4$ mixture with ~10-15% of $He^4$ ions (three-ion scheme). Under these conditions, the SPA is ~0.8-0.9 shared between central electron heating and off-axis $He^3$ ion heating on the HFS as illustrated in Figure 3.

These optional heating schemes could offer the possibility to reach H-mode in hydrogen when operating close to half field, enabling long pulse operation and ELM control at $q_{95}$=3. Adding 10-15% of $He^4$ reduces the minimum allowable operational density with respect to NBI shinethrough, and could allow access to hydrogenic H-mode operation at lower input power, if observations from recent JET experiments are confirmed [20]. However, tungsten accumulation with reduced central heating needs to be considered and quantified. Furthermore, as discussed above, $He^3$ level control in the presence of $He^4$ may prove challenging. Finally, ion losses are still to be assessed. They can be significant due to off-axis fast trapped ion acceleration [21]. They can however always be reduced by increasing the $He^3$ concentration to reduce the fast ion tail.

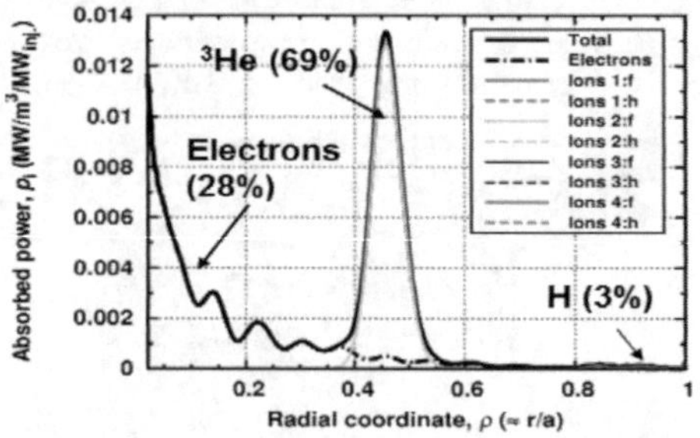


**Fig. 3.** Wave absorption for fundamental minority $He^3$ heating in an H:$He^4$ mixture at 3.3T (TORIC run).

## 3 Conclusion

Various ICRF heating schemes for the ITER non-active phase have been investigated. For the ITER ICRF frequency range 40-55 MHz, effective absorption schemes are available in hydrogen and helium at magnetic fields of 5.3T, 2.65T and 1.8T, except for hydrogen operation at 2.65T. This may be compensated by optional schemes at 3T and 3.3T with fundamental minority $He^3$ heating in a H-$He^4$ mixture (three-ion heating scheme). Results at 5MA/1.8T are preliminary as they imply a regime with strong electron / ion decoupling ($T_e >> T_i$) where TEM and ETG modes play a significant role, not covered by METIS scaling laws. Hence, kinetic profiles need to be crosschecked with other transport codes and made self-consistent with heating sources. Furthermore, most of the results presented in this paper have been produced by the TOMCAT 1D code and should be advanced with 2D full wave computations. Future antenna coupling studies will also contribute to complete the picture.


## Acknowledgments

ITER is a Nuclear Facility INB-174. The views and opinions expressed herein do not necessarily reflect those of the ITER Organization.


## References


1. Y.R. Martin et al, J. of Physics: Conf. Series **123** (2008) 012033
2. D. Van Eester and R. Koch, PPCF **40**, 1949–1975 (1998)
3. J.F. Artaud et al 2010 Nucl. Fusion **50** 043001
4. M. Schneider, 17th ITPA-IOS TG meeting – Naka, Japan (2016)
5. M. Brambilla, PPCF **41**, 1–34 (1999).
6. E. Lerche et al, PPCF 54 (2012) 069601 (6pp)
7. T. Stix, Waves in Plasmas, AIP Press (1992)
8. Ye.O. Kazakov et al, Nucl Fusion **55**, 032001 (2015)
9. M.-L. Mayoral et al, Nucl. Fusion 46 (2006) S550–S563
10. D. Van Eester et al, PPCF 51 (2009) 044007
11. D. Van Eester et al, EPS 2015 Lisbon (Portugal), P2.117
12. D. Van Eester et al, this conference
13. P. Lamalle et al. Nucl. Fusion 46, 391-400 (2006)
14. Ye.O. Kazakov et al., APS. 61, APS-DPP 2016, San Jose (USA)
15. M. von Hellerman et al., ITPA-EP, Seville (Spain), April 2017.
16. A. Krasilnikov et al, A V Krasilnikov et al 2009 PPCF 51 044005
17. F. Ryter et al, Nucl. Fusion **54** 083003 (2014)
18. E. Lerche et al, AIP Conference proceedings, 1689 (2015) 040003
19. M. Porkolab AIP Conference Proceedings 314, 99 (1994)
20. J. Hillesheim et al.,26th IAEA FEC, Kyoto, Japan, 2016., EX/5-2
21. J. Joly et al, this conference.

# Extension of electron cyclotron heating at ASDEX Upgrade with respect to high density operation

*Martin* Schubert[1,*], *Jörg* Stober[1], *Albrecht* Herrmann[1], *Walter* Kasparek[2], *Fritz* Leuterer[1], *Francesco* Monaco[1], *Bernhard* Petzold[1], *Burkhard* Plaum[2], *Stefan* Vorbrugg[1], *Dietmar* Wagner[1], *Hartmut* Zohm[1], and the ASDEX Upgrade Team[1]

[1] Max Planck Institute for Plasma Physics, Boltzmannstr. 2, 85748 Garching b. München, Germany
[2] Institut für Grenzflächenverfahrenstechnik und Plasmatechnologie, Pfaffenwaldring 31, D-70569 Stuttgart

**Abstract.** The ASDEX Upgrade electron cyclotron resonance heating operates at 105 GHz and 140 GHz with flexible launching geometry and polarization. In 2016 four Gyrotrons with 10 sec pulse length and output power close to 1 MW per unit were available. The system is presently being extended to eight similar units in total. High heating power and high plasma density operation will be a part of the future ASDEX Upgrade experiment program. For the electron cyclotron resonance heating, an O-2 mode scheme is proposed, which is compatible with the expected high plasma densities. It may, however, suffer from incomplete single-pass absorption. The situation can be improved significantly by installing holographic mirrors on the inner column, which allow for a second pass of the unabsorbed fraction of the millimetre wave beam. Since the beam path in the plasma is subject to refraction, the beam position on the holographic mirror has to be controlled. Thermocouples built into the mirror surface are used for this purpose. As a protective measure, the tiles of the heat shield on the inner column were modified in order to increase the shielding against unabsorbed millimetre wave power.

## 1 Introduction

Progress has been made at the ASDEX Upgrade tokamak in controlling high power ($P_{heat} \approx 20$ MW) discharges, where the power normalized to the major radius ($R_0$=1.65 m) is approaching ITER relevant values [1]. A solid tungsten divertor allows to gain experience in operation under ITER-like heat loads [2]. Accumulation of tungsten in the plasma center can be effectively avoided, if the central heating power density is high enough [3,4]. At ASDEX Upgrade (AUG) the electron cyclotron resonance heating (ECRH) can serve for this purpose. This system routinely operates at 105 and 140 GHz [5] and it is used for various wave heating schemes and current drive. Movable launcher mirrors and a broad-band polarization system allow for flexible launching geometries and heating schemes.

With respect to high density and high heating power operation, an O-2 mode scheme at 140 GHz has been developed [6]. This scheme extends the range of accessible plasma densities above the X-2 cut-off, however it may suffer from reduced single pass absorption of the order of 60-75% [7]. Installing reflectors into the heat shield on the inner column of AUG improves the situation significantly, since the unabsorbed fraction makes a second pass through the plasma. Several constraints on the reflector design had to be taken into account and this finally led to the holographic or phase grating mirror design [8].

This paper will give an overview of the current status of the AUG ECRH installations in the next section. The O-2 mode scheme is prepared by simulations, which are described in section 3. Some points of the holographic mirror design, manufacturing and instrumentation are considered in section 4. Protective measures for the AUG machine are explained in section 5 and with a few comments on ECRH operation with the holographic mirrors this paper shall close.

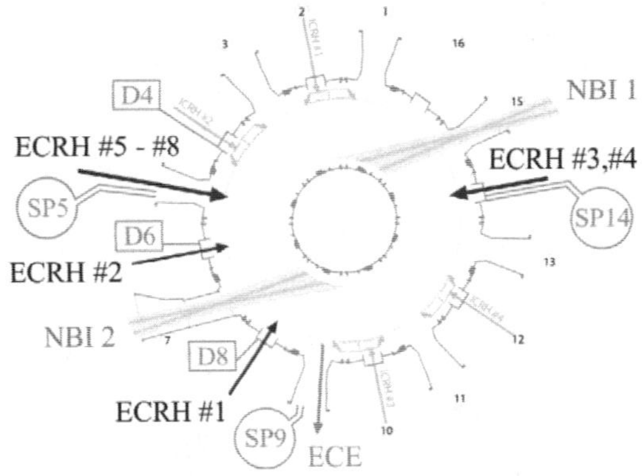


**Fig. 1.** Top view of equatorial cross section through the AUG vacuum vessel with launch positions of heating systems. ECRH relevant detectors for unabsorbed millimetre waves are also shown in light colour. 'SP' Sniffer sphere, 'D' D-band pick-up waveguide, mounted on quartz window [12,13].

* Corresponding author: Martin.Schubert@ipp.mpg.de

DOI: 10.1051/epjconf/201715703047

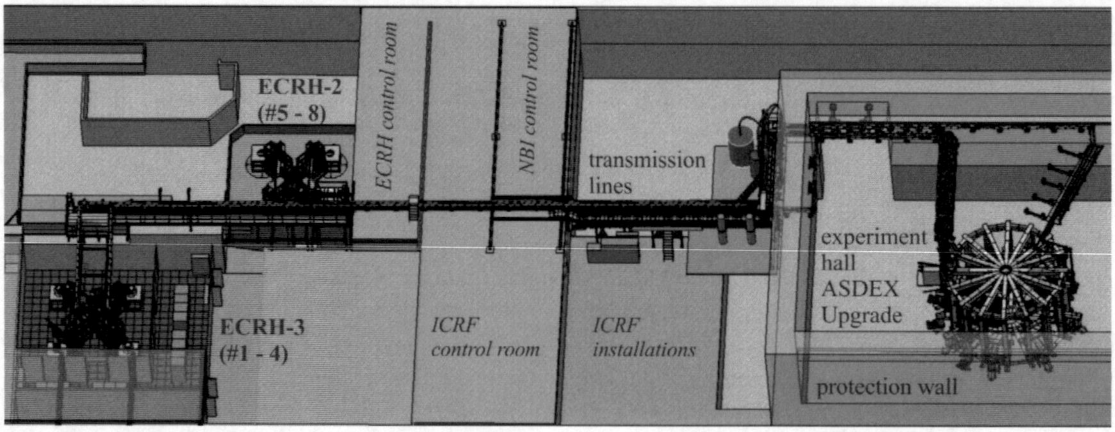


**Fig. 2.** Electron cyclotron resonance heating installations comprising three storeys in main ASDEX Upgrade building. Ground floor: blue, experiment hall. First floor: light brown, access to each group of four Gyrotrons ('ECRH-2','ECRH-3'). Second floor: grey, transmission lines.

## 2 Extension of ECRH installations

An overview of ECRH launchers at AUG is given in figure 1 and the installations are shown in figure 2. In 2016 four Gyrotrons with 10 s pulse length and output power close to 1 MW per unit were available. This group of four Gyrotrons belongs to an installation which was described in [9]. Following a decision which dates back to 2011, the system is being extended to eight similar units in total. The launching positions #1-4 (figure 1) stem from AUG's first ECRH system [10] and comments on operation experience are found in [11]. The periphery of this original ECRH system was dismantled and the launchers are now connected to the recent installations, cf. figure 2.

In order to detect unabsorbed millimeter waves, Sniffer spheres (figure 1) were originally installed only. This was with the intention to make the measurement insensitive to polarization and phase. From the measurements described in [12] and [13] it became clear that the dynamics of the main plasma is mainly responsible for mixing the phase and polarization of the stray radiation. At AUG a simple D-band pick-up wave guide, therefore, provides basically the same signal properties [13] as the measurement via Sniffer sphere.

## 3 Simulation of the O-2 mode scheme

An O-2 mode scheme exists for the launchers #5 and #6, as described in [6]. There are similar upgrades (figure 3) for #7 and #8. The reflector #8 has been manufactured, tested and is ready for operation, while #7 is intended to be prepared for the 2018 campaign. The criteria [6] for the selection of the beam geometry were basically:

(i) toroidal angle for maximum absorption [7],
(ii) poloidal/vertical angle to allow central heating,
(iii) beam should end on a robust surface after 2$^{nd}$ pass.

Both angles are adapted such that the reflector fits into the tile columns of the heat shield.

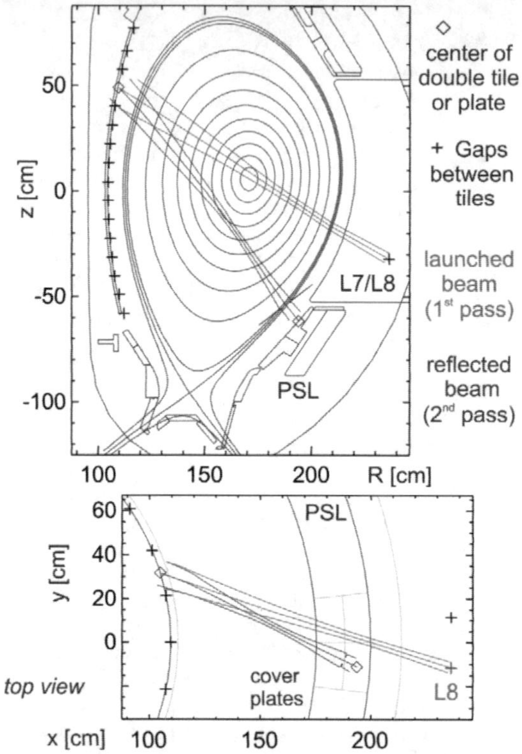


**Fig. 3.** Simulation of the O-2 mode beam propagation and absorption [14]. First pass: 75% absorption, second pass: 80% giving a total of 95% of the launched power. After the 2$^{nd}$ pass, spikes may occur in the simulated beam width. These are artefacts produced by insufficient smoothing in the experimental density profile, particularly in the pedestal region. Input data: AUG #30505, t=3.3 s, $I_p$= 1.2 MA, $B_t$= -2.48 T, central electron density $n_e$= 1.15·10$^{20}$ m$^{-3}$ and $T_e$= 3.6 keV.

The robust beam facing component after the 2$^{nd}$ pass is important, since the remaining millimeter wave power may contain up to 15% of the launched power. In this case the cover plates of the lower passive stabilization loop (PSL, figure 3) are considered large and robust enough. The remaining power is reflected on this plates and spreads out due to beam divergence.

The simulations shown in figure 3 are based on the Torbeam [14] beam tracing code, using data from AUG experiment #30505 at t= 3.3sec with a total heating power of approx. 19.5 MW. This is considered a model high power discharge with partial detachment.

All beamlines including #1-4 are to be equipped with appropriate reflectors on the central column by 2019. Individual constraints arise from the launching position, the position of the appropriate beam dump, and from material and instrumentation of the holographic mirrors.

## 4 Reflector design

The plasma facing surface of the central column inside AUG is composed of tiles (figure 4) and the surface contours of each tile are designed such, that leading edges are avoided. According to the simulation result, the reflector is large enough, if a double tile of approx. 210 x 180 mm is used.

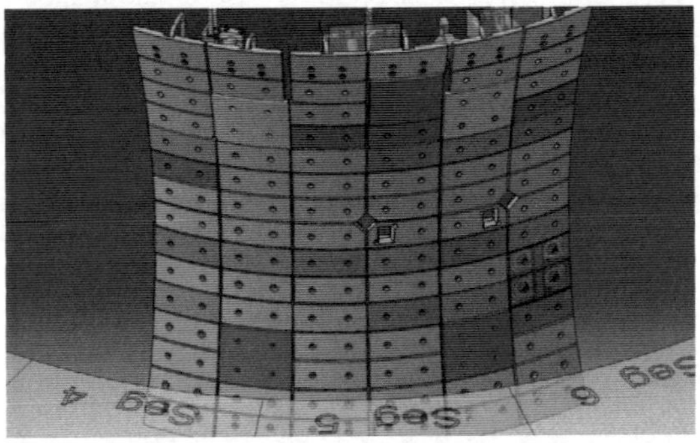

**Fig. 4.** Heat shield tiles in AUG vessel segments 4-6. The lower magenta double tiles are the reflectors #5 and #6 [6]. The upper green double tiles are the upgrades for #7 and #8 and correspond to figure 3. The upper magenta tile in the middle is a reflector that was mounted for measurement purposes [8].

### 4.1 Optimized gratings

The reflectors need optimized gratings in order to direct the beam in the direction of the second pass through the plasma. The requirements for the grating are

(i) the direction of the reflected beam,
(ii) possibly a refocusing characteristic,
(iii) polarization independence.

Furthermore, the tiles on the high field side have typically a saddle-like surface, into which the grooves are milled. Additional constraints are imposed by the available milling tools, which limit the difference of the minimum and maximum slope angles of a grating.

The optimization is done by calculating the local **k** vectors on the reflector and approximating the fields by plane waves, which results in a 2D field problem. The 2D groove profiles are then optimized in order to maximize the reflected intensities $I_n^{TM}$ and $I_n^{TE}$ for both polarizations in the desired diffraction order $n$ and minimize the phase difference $\phi^{TM-TE}$ between the TM and TE polarizations. The overall efficiency becomes:

$$\eta_\circ = \frac{1}{4} I_n^{TM} + \frac{1}{4} I_n^{TE} + \frac{1}{2}\sqrt{I_n^{TE} I_n^{TM}} \cos(\phi^{TM-TE}) \quad (1)$$

The remaining power will be radiated in other diffraction orders and contributes to the stray radiation level in the vessel. The 2D gratings are optimized separately, efficiencies are calculated using a BEM algorithm [8]. The complete 3D-grating is interpolated from the 2D profiles [6]. The separate optimization of the 2D profiles can be simply parallelized, because no data needs to be transferred between the optimization processes.

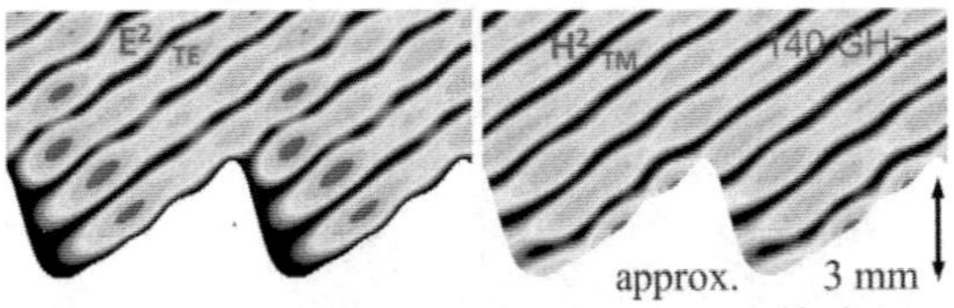


**Fig. 5.** Exemplary mirror grating and contours of the millimetre wave amplitude squared [8]. This 3rd order grating with $\theta_{in}$ = 60° and $\theta_{out}$ = 50° has an efficiency of 0.89 for TE and 0.97 for TM.

**Fig. 6.** Reflector #8, machined from graphite with optimized grating and coated with tungsten.

### 4.2 Thermocouple instrumentation

If the millimeter wave beam has an overlap with a gap between the tiles, part of it may penetrate into the gap and cause damage. The beam alignment on the reflector is, therefore, critical. Thermocouples built into the surface of the reflectors can be used to verify the beam position [6]. The fastest response times are achieved, if the thermocouples are flush mounted (figures 7 and 8).

During assembly (figure 9) special attention has been paid with respect to maintainability. The NiCr-Ni element is fixated in a single spot on the rear side and it is intended to be exchangeable.

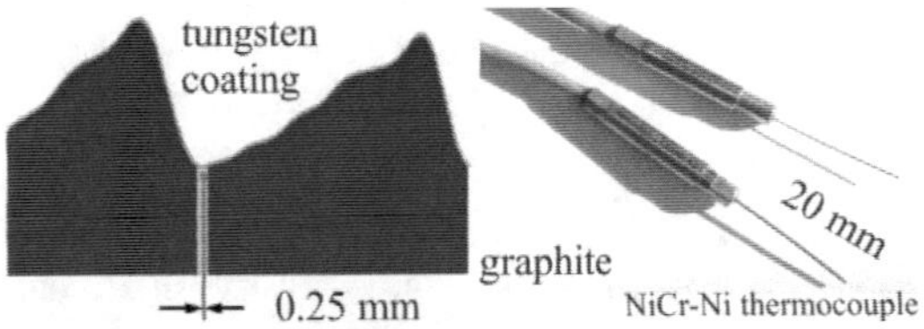


**Fig. 7.** Scheme of a thermocouple mount (left) in the solid body below the grating. The tip of the NiCr-Ni thermocouple (right) has to be adjusted flush with the plasma facing surface.

**Fig. 8.** Flush mounted NiCr-Ni thermocouple, its tip can be recognized as a small bright spot within the borehole. The borehole diameter of 0.5 mm is already in cutoff for the millimetre wave.

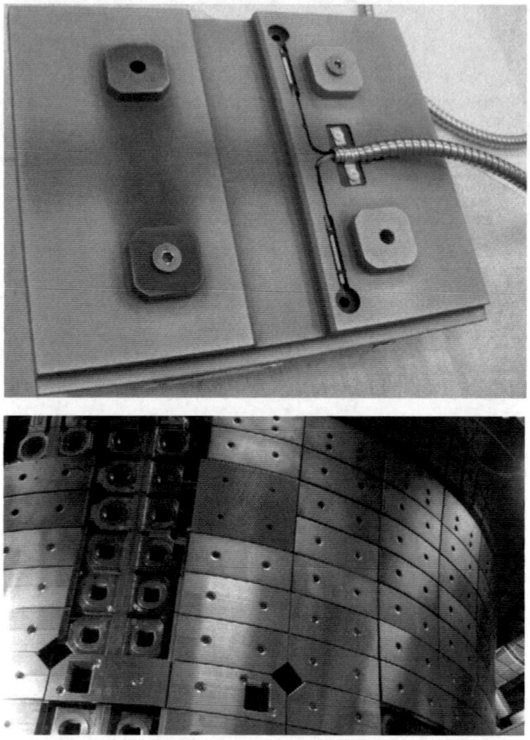

**Fig. 9.** Top: Grooves machined into the rear part increase the shielding of the thermocouples and their cables. In addition, cables are protected by flexible stainless steel tube. Bottom: Reflector #8 assembly completed inside AUG.

## 5 Machine protection and operation

Considerable effort has been spent on AUG in order to harden the components on the inner column against the millimetre waves from the ECRH. An overlapping design of the tile edges (figure 10) increases the shielding against mm-waves and reduces the risk of severe damage, if an error occurs. The shielding factor has been measured on a laboratory test mock-up and its worstcase value 0.036 or -14.4 dB (140 GHz) is considered sufficient for this purpose. Vital parts behind the tiles have received additional tight covers.

Launcher alignment can be accurately checked in vacuum, sweeping the beam over the thermocouple positions while the Gyrotron generates a pulse train with reduced microwave power [15].

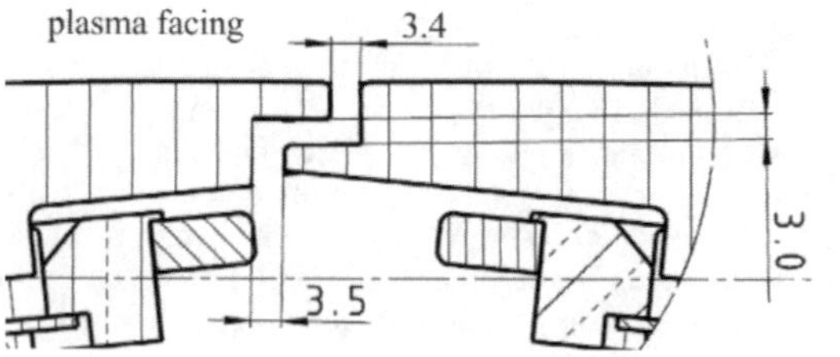


**Fig. 10.** The edges of the heat shield tiles on the inner column of AUG have been redesigned with a labyrinth-like dogleg structure in order to increase the shielding against millimetre waves.

During plasma operation it is found that the toroidal position of the beam on the reflector is very robust with respect to changes in the plasma configuration. The vertical position of the beam on the reflector may change in particular with plasma density [6] and an automated feedback of the launch angle, depending on the thermocouple signals was tested successfully. In the harsh environment of high power plasma operation and under certain conditions, the thermocouple signals are, however, subject to cross-talk from the AUG ICRF. It is to be tested, whether the improved shielding of the thermocouples and their cables (section 4.2) is sufficient.

Video observation is helpful in order to detect arcs, which are ignited if the beam crosses the tile edge. Automated real-time video analysis is being developed and will be discussed elsewhere.

The skillful and precise assembly by H. Schuetz is gratefully acknowledged.

This work has been carried out within the framework of the EUROfusion Consortium and has received funding from the Euratom research and training programme 2014-2018 under grant agreement No 633053. The views and opinions expressed herein do not necessarily reflect those of the European Commission.

## References

1. A. Kallenbach *et al*, Nucl. Fusion **55**, 053026 (2015)
2. A. Herrmann *et al*, Nucl. Fusion **55**, 063015 (2015)
3. A. Kallenbach *et al*, Nucl. Fusion **51**, 094012 (2011)
4. O. Gruber *et al*, Nucl. Fusion **49**, 115014 (2009)
5. D. Wagner *et al*, J Inf Milli Terahz **37**, 45-54 (2016)
6. H. Höhnle *et al*, Nucl. Fusion **51**, 083013 (2011)
7. V. Erckmann, U. Gasparino, Plasma Phys. Control. Fusion **36**, 1869-1962 (1994)
8. O. Mangold, PhD thesis (Stuttgart University, 2009)
9. F. Leuterer *et al*, Fus. Eng. Des. **66-68**, 537 (2003)
10. B. Streibl *et al*, Fusion Sci. Technol. **44**, 578 (2003)
11. F. Leuterer *et al*, Fusion Sci. Technol. **55**, 31 (2009)
12. M. Schubert *et al*, EPJ Web Conf **32**, 02013 (2012)
13. M. Schubert *et al*, EPJ Web Conf **87**, 02010 (2015)
14. E. Poli, A. Peeters, G. Pereverzev, Comput. Phys. Commun. **136**, 90 (2001)
15. M. Schubert *et al*, *43rd Conference on Plasma Physics*, **ECA 40A**, P1.026 (2016)

 DOI: 10.1051/epjconf/201715703048

# RF wave simulation for cold edge plasmas using the MFEM library

*S. Shiraiwa*[1,*], *J. C. Wright*[1], *P. T. Bonoli*[1], *T. Kolev*[2], *and M. Stowell*[2]

[1]MIT Plasma Science and Fusion Center, 190 Albany St Cambridge, 02139, USA
[2]Center for Applied Scientific Computing, Lawrence Livermore National Laboratory, Livermore, CA 94551, USA

**Abstract.** A newly developed generic electro-magnetic (EM) simulation tool for modeling RF wave propagation in SOL plasmas is presented. The primary motivation of this development is to extend the domain partitioning approach for incorporating arbitrarily shaped SOL plasmas and antenna to the TORIC core ICRF solver, which was previously demonstrated in the 2D geometry [S. Shiraiwa, et. al., "HIS-TORIC: extending core ICRF wave simulation to include realistic SOL plasmas", Nucl. Fusion in press], to larger and more complicated simulations by including a 3D realistic antenna and integrating RF rectified sheath potential model. Such an extension requires a scalable high fidelity 3D edge plasma wave simulation. We used the MFEM [http://mfem.org], open source scalable C++ finite element method library, and developed a Python wrapper for MFEM (PyMFEM), and then a radio frequency (RF) wave physics module in Python. This approach allows for building a physics layer rapidly, while separating the physics implementation being apart from the numerical FEM implementation. An interactive modeling interface was built on πScope [S Shiraiwa, et. al. Fusion Eng. Des. **112**, 835] to work with an RF simulation model in a complicated geometry.

## 1. Introduction

SOL plasmas affect both antenna coupling and core heating/current drive performance through various mechanisms, including RF rectified sheath potential for the ion cyclotron range of frequency (ICRF) problem and parasitic edge absorption for the lower hybrid (LH) waves. The need of incorporating global SOL plasmas to RF heating and current drive simulations has been widely recognized in order to better understand and assess roles of SOL plasmas. Such a simulation requires theoretical and computational advancement from the past RF simulations, where the majority of the works started from separating the plasma into core and SOL/antenna regions.

The separation has been made mainly motivated by the need of capturing rather different aspects of wave physics in the two regions. In the core, spatially dispersive hot plasma conductivity is important to properly model the wave propagation and absorption, whereas, in the SOL/antenna region, reproducing the detailed 3D geometry of antenna structure is important. The issue is that, after separating the core and SOL/antenna region, a rather simplified assumption on the interfacing layer was made to define a boundary condition. Several works [1, 2, 3, 4] in the literature have attempted to eliminate or at least soften the impact of the separation. These works oriented toward constructing a monolithic solver to handle the two regions together, and they were partially successful in expanding the existing RF simulation capability to include both core and edge. However, none of them represents a simulation approach universally applicable to all (from ICRF to EC) frequency ranges to include hot core plasma, surrounding cold SOL plasmas and a complicated 3D antenna structure all together.

Recently we tackled this issue with an alternative strategy, in which we solved core and edge regions separately, but stitch them afterwards. An initial work of this approach ("domain partitioning") was presented at the previous conference [5], and a successful simulation of H-minority ICRF heating on Alcator C-Mod was

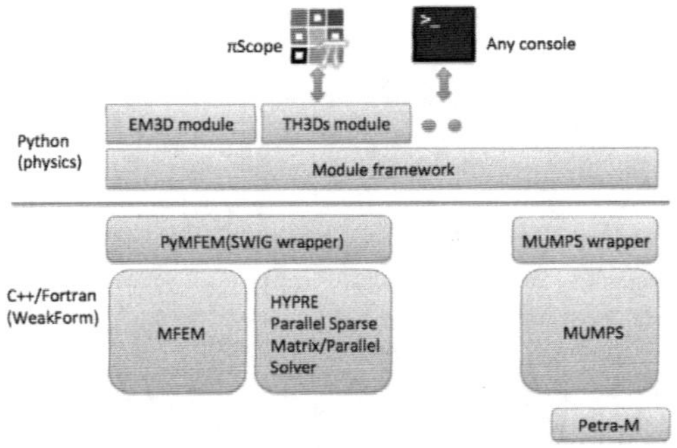


**Fig. 1.** Module structure developed for a new SOL RF wave simulation. The bottom layer (below the red line) is C++/Fortran libraries and their python wrappers. The modules shown in red and πScope are already available as open source software.

* Corresponding author: shiraiwa@psfc.mit.edu

recently published [6]. In ref 5 and 6, the wave in the hot core plasma was solved by the TORIC spectral solver [7], and the SOL and antenna regions were discretized by unstructured triangles (HIS-TORIC: Hybrid Integration of SOL to TORIC) and solved using a commercial FEM package, COMSOL. We showed in 2D axisymmetric simulation that the correct final solution is obtained from the superposition of "mode solutions".

While the formulation for connecting two regions derived in Ref 6 is not restricted to 2D geometry, extending the work to 3D [8] requires solving a much larger linear problem and better automation of the simulation workflow, both of which are difficult to realize when using commercial software. Furthermore, we are envisioning the need of integrating additional physics to improve the reality of simulation. Such physics includes the RF rectified sheath potential on the material surface [9] and the interaction of RF waves with edge turbulence. Introducing any hot conductivity effect in the SOL region would make a linear system matrix denser, which would potentially require massively parallel computation. Driven by this long-term research direction, we developed a new FEM framework and an edge RF wave simulation environment based on an open source scalable FEM library, MFEM [10].

In this paper, we discuss the software stack we developed in Sec 2. Initial result of applying it to a HIS-TORIC simulation is discussed in Sec 3, and Sec 4 summarizes the work.

## 2 Module structure

Figure 1 shows the module structure we developed to implement a new SOL wave simulation. We use the MFEM (modular finite element method) library for the finite element discretization. MFEM is a relatively compact C++ library developed in LLNL and is available as open source software. It uses the Hypre linear solver library [11] and allows for directly filling a distributed sparse matrix in a ParCSR (dual-CSR, compressed row storage) format. The library is written for a generic FEM analysis, not specifically for solving the Maxwell's equations. However, it is equipped with basis functions and linear/bilinear integrators required to formulate a linear system corresponding to a Maxwell problem.

## 2.1 PyMFEM wrapper

We developed a python wrapper layer (PyMFEM [12]) for MFEM and implemented physics modules discussed in the next section using Python. Such a hybrid approach is advantageous for rapid development and has been used in other open source FEM projects such as the FEniCS project [13].

The main role of the PyMFEM layer is to provide an access to the MFEM library objects from Python. The wrapper layer is generated mostly by SWIG (simple wrapper interface generator). A user can instantiate a C++ MFEM object and call its methods from Python. We designed the wrapper so that a user can perform most of the operations using Python's native objects such as List or Numpy array.

An important feature implemented in PyMFEM is a wrapper class of C++ function pointer used in the FunctionCofficient and other similar objects in MFEM. In MFEM, coefficients appearing in an inner product, such as $\frac{1}{\mu_0}$ in Eq. 1, is given as a C++ function pointer, where the function arguments are spatial coordinates. PyMFEM allows for using a regular Python class method instead of a C++ function. This way, we can define material properties such as a cold plasma tensor in Python.

The wrapper layer also provides an easy access to the ParCSR matrix. A user can fill the local portion of ParCSR matrix in each processor using a standard method available in the scipy.sparse module and combine them to make a final distributed ParCSR matrix. It is also possible to decompose a ParCSR matrix into scipy.sparse matrices distributed among processors.

## 2.2 Petra-M (Physics Equation Translator for MFEM)

In order to solve a physics problem using MFEM, a user needs to translate physics equations, often given in the form of partial differential equations (PDEs), into a weak form. It would be much more convenient if a user can use regular physics terminologies when configuring a simulation. For this purpose, we developed a physics equation translation layer (Petra-M), which corresponds to the blue part of Fig. 1. Petra-M consists of two layers. The lower layer is a generic infrastructure providing routines to define a multi-physics problem, where a solution is expressed using a combination of multiple finite element function spaces. The support for a complex valued problem is also realized in this layer. For a complex valued problem such as a frequency domain simulation, Petra-M fills the real part and imaginary part separately and generates a final linear system matrix either as a complex value matrix, or 2x2 real value block matrix. Differences between serial and parallel (MPI) environments are absorbed in this layer, too.

On the upper layer of Petra-M, we define the PDE system we want to solve. A module to solve a 3D frequency domain Maxwell problem is implemented here. This module fills a linear system for the following weak form of Maxwell's equation [14, 15]

$$\left(\nabla\times F, \frac{1}{\mu_0}\nabla\times E\right) - (F, (\omega^2\epsilon - j\sigma\omega)E) + \langle F, n\frac{1}{\mu_0}\nabla\times E\rangle = -j\omega(F, J_{\text{ext}}),$$

where $E$ and $F$ are test and trial functions in the H(curl) finite element space, and two inner products are defined as $(u,v) \stackrel{\text{def}}{=} \int uv\, d\Omega$ and $\langle u,v\rangle \stackrel{\text{def}}{=} \int uv\, d\partial\Omega$, corresponding to volumetric and surface contribution,

DOI: 10.1051/epjconf/201715703048

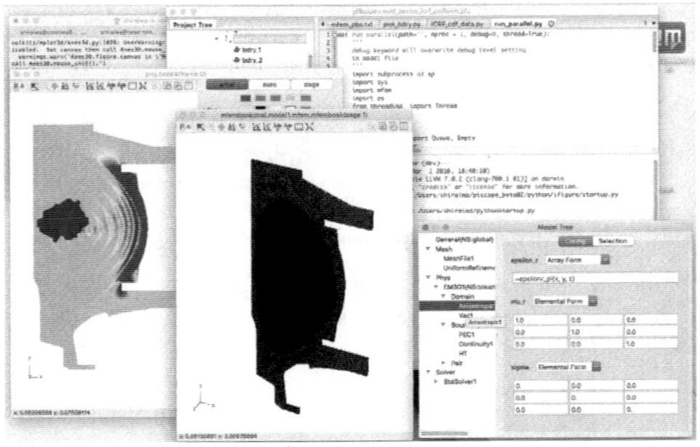

**Fig 2**. GUI developed in πScope. 3D mesh geometry, RF physics parameter setting, and a solution of RF field is shown. The right bottom panel shows how parameters for anisotropic media are specified, showing that the relative dielectric constant is given as "=epsilonr_pl(x,y,z)", where epsilonr_pl is a user defined python function. The dark region in the left most window indicates the typical size of domain solved by one CPU (total 16 CPU was used in this case).

respectively. The later term is yielded from a partial integration.

The first and second terms in the above form correspond to the curl-curl operator and the conductivity, while the last term is the contribution from volumetric external currents. Presently, the module supports boundary conditions (BCs) including, PEC (perfect electric conductor), PMC (perfect magnetic conductor), coax and waveguide ports BC, Floquet periodic BC, surface current BC, surface magnetic field BC, electric field BC, and impedance BC.

Optionally, a user can add a divergence-free constraint for the current. When using this constraint, the solution space is expanded to contain two finite element spaces, one for the electric field and the one for the time derivative of charge. Petra-M discretizes the second one using the $H^1$ element and formulates a linear system in the saddle point format. This constraint is used in the vacuum region in the simulations shown in the next section.

## 2.3 User interface layer

Setting up a 3D RF simulation in a realistic geometry is not a trivial task. A graphical user interface is prepared in πScope [16], an open-source python based data analysis environment, in which a user can set domain and boundary conditions as discussed above. A screen capture of the user interface is shown in Fig 2.

Additionally, Petra-M supports generating a python script from the model settings prepared by using the GUI. The generated script is easy to modify, allowing for a user to setup, for example, a parametric scan of simulation parameters quickly from an existing simulation.

# 3. Application to HIS-TORIC

The original HIS-TORIC simulation involves cumbersome steps including running multiple COMSOL simulations programmed by Matlab and executing separate TORIC simulations. We re-implemented the edge RF wave physics model in HIS-TORIC using Petra-M, which moved the most of simulation workflow under πScope control and allows for extending HIS-TORIC to the 3D geometry [8]. Here, as a verification case, we discuss the simulation of the H-minority heating scenario for an Alcator C-Mod discharge, the same simulation as shown in Ref 3, but ran with TORIC and Petra-M is shown in Fig. 3.

This model simulates an ICRF wave with the n=10 mode using the axis-symmetry. A single toroidal mode in 3D is imposed in a similar way as discussed in Ref. 16, namely by using extremely thin (0.1 deg) toroidal sector and imposing a Floquet boundary condition. On the left figure, the model geometry and boundary conditions are shown. The SOL region has 2.5 cm thickness surrounding the core TORIC region (not shown). Outside the SOL plasma is set to vacuum, and we imposed a divergence-free constraint in this vacuum region. The ICRF wave at 80MHz is excited using surface currents on the low field side curved surface. The left figure also shows the size of mesh used in the simulation. The FEM problem is assembled using Petra-M and solved using MUMPS direct sparse linear solver [18].

The right figure shows the η component ($\sim B\times\nabla\psi$) of the RF electric field after combining TORIC and Petra-M regions. It can be seen that the wave smoothly propagates through the interface between SOL and TORIC regions towards the core and is strongly absorbed in the core. Note that there is also observed a short wavelength ICW wave near $R = R_0$, where $R_0$ is the major radius, which wouldn't exist if a cold plasma approximation were used in the core.

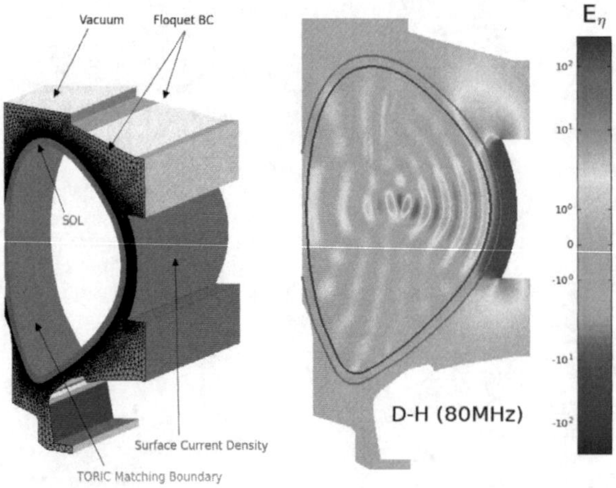


**Fig 3**: (Left) a simulation geometry to implement axis-symmetric simulation of ICRF wave propagation in Alcator C-Mod. (Right) a simulation result using the same input parameter presented in Ref. 3. Note that the actual geometry is extremely thin in the toroidal direction. The left figure is enlarged in the toroidal direction for better visualization.

## 4. Conclusions

In this paper, we discussed the implementation details of a new edge SOL plasma simulation environment. We built our simulation model on the top of the open source MFEM C++ FEM library. Petra-M is developed using the Python wrapper for MFEM (PyMFEM) to set up a MFEM simulation using physics (not weak from) terminologies, and a 3D frequency domain RF simulation model is then developed as a part of Petra-M. A graphical user interface is developed on πScope to configure an RF simulation with a complicated realistic geometry.

The cold plasma wave simulation model used in the previous HIS-TORIC simulation was re-implemented using Petra-M. The model was tested by reproducing the Alcator C-Mod H minority heating simulation presented in Ref. 3. With Petra-M, the model can be extended in 3D as presented at this conference [8], and the HIS-TORIC simulation workflow can be fully automated. This open-soruce software based platform will be used to integrate more advanced physics such as an RF rectified potential discussed in [9] in our future work.

Work supported by the U.S. Department of Energy, Office of Science, Office of Fusion Energy Sciences, using User Facility Alcator C-Mod, under Award Number DE-FC02-99ER54512 and by US DoE Contract DE-FC02-01ER54648 under a Scientific Discovery Through Advanced Computing Initiative. The work at Lawrence Livermore National Laboratory was performed under the auspices of DoE under Contract DE-AC52-07NA27344, LLNL-PROC-703397. The simulations presented in this paper were performed on the MIT-PSFC partition on the Engaging cluster at the MGHPCC (www.mghpcc.org).

## References

1. T. G. Jenkins, D.N. Smithe and R. I. Pinsker, 2015 AIP Conf. Proc. 030003
2. D. L. Green, L. Berry, G. Chen, P. Ryan, J. Canik and E. Jaeger, 2011 Phys. Rev. Lett. **107**, 145001
3. O. Meneghini, S. Shiraiwa and R. Parker, Phys. Plasmas **16**, 090701 (2009).
4. S. Shiraiwa, J. Ko, O. Meneghini, et. al, Phys. Plasmas **18**, 080705 (2011).
5. J. C. Wright and S. Shiraiwa, AIP Conf. Proc., **1689**, 050007 (2015).
6. S. Shiraiwa, J.C. Wright, J.P. Lee, and P.T. Bonoli, Nucl. Fusion **57** 086048 (2017).
7. M. Brambilla, 1999 Plasma Phys. Control. Fusion **41**, 1
8. J. C. Wright and S. Shiraiwa, "Antenna to Core: A New Approach to RF Modelling" in *22nd Topical Conference on Radio Frequency Power in Plasmas*, European Phys. J. Web of Conferences, **this issue**, Inv-13 (2017).
9. J. R. Myra and H. Kohno, "A post-processing method to simulate the generalized RF" in *22nd Topical Conference on Radio Frequency Power in Plasmas*, European Phys. J. Web of Conferences, **this issue**, Inv-13 (2017).
10. MFEM (http://mfem.org)
11. HYPRE : Scalable Linear Solvers and Multigrid (http://llnl.gov/casc/hypre)
12. PyMFEM : Python wrapper for MFEM (https://github.com/MFEM/PyMFEM)
13. FEniCS project (https://fenicsproject.org)
14. J. Jin, Finite Element Method in Electromagnetics (New York: Wiley) (1993)
15. A. Bondeson A., T. Rylander and P. Ingelström Computational Electromagnetics (Berlin: Springer) (2000)
16. S Shiraiwa, T Fredian, J Hillairet, J Stillerman, Fusion Engineering and Design **112**, 835-838
17. S Shiraiwa, O Meneghini, R Parker, G Wallace, J Wilson, 2009 AIP Conf. Proc. **1187**, 363
18. MUMPS : a parallel sparse direct solver (http://mumps.enseeiht.fr/)

 DOI: 10.1051/epjconf/201715703049

# Control of high power microwave radiation by electromagnetic band gap structures

*Leanid* Simonchik[1,*], *Vladislav* Babitski[1], *Thierry* Callegari[2], *Jerome* Sokoloff[2] and *Maxim* Usachonak[1]

[1] Institute of Physics of NAS of Belarus, 220072 Minsk, Belarus
[2] LAPLACE, UPS-CNRS, 31062 Toulouse, France

**Abstract.** A plasma control of powerful microwave propagation through the 1D and 2D EBG structures is investigated. Pulsed discharges in argon (or helium) at atmospheric pressure are applied as the plasma inhomogeneities. Temporal behavior of electron concentration in discharge is determined. The time-dependent transmission spectra of 1D EBG structure formed solely by plasma in the X-waveguide are measured. The amplitudes of short (~200 ns) powerful (50 kW) microwave pulses at frequency of 9.15 GHz are strongly suppressed (more than 40 dB) when they fall in the time interval of plasma structure existing. The propagation of these powerful microwave pulses through the triangle metallic 2D EBG structure with the plasma control elements is investigated too: when plasma acts as a compensator of defect in the front row of EBG structure the transmission in direction of 45° quickly ceases (during a few tenth of nanoseconds), or one quickly arises, when plasma is as an additional defect.

## 1 Introduction

During the last decade, there has been an increasing interest in the microwave (MW) reconfigurable devices the transmitted properties of whose are changed by variation of external electric or magnetic fields or other factors. To create electrically tunable devices based, for instance, on the use of electromagnetic band gap (EBG) structures, their some passive elements must be replaced by active ones. However some critical problems for high-speed, large scale or complex systems which use computers or automation servers are occured at that. It can be a significant time delay or an operation at low microwave power or an extra development cost. At the same time, the discharge plasma has a great potential for the application in microwave devices in the function of control elements due to its possibility for variations in size, geometry and density by changing the discharge current [1-3]. In this work, we demonstrate the ability to control the propagation of MW radiation at high (about 50 kW) power by one- and two-dimension (1D and 2D) EBG structures using the pulse discharges in argon or helium at atmospheric pressure.

## 2 Control by 1D EBG structure

### 2.1 Experimental arrangement

We developed an 1D EBG structure in a X-band waveguide (WR90) using three pulse discharges, which are arranged along waveguide with the period of 30 mm (Fig. 1). This period corresponds to the wavelength $\Lambda_0$ in

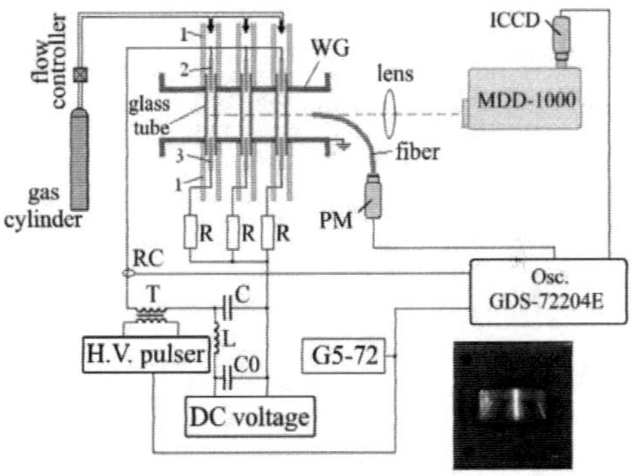


**Fig. 1.** Schematic of experimental setup. 1 – silicon tubes, 2 and 3 – copper electrodes, WG – waveguide, R – ballast resistors, T – transformer, L – throttle, C and C0 – capacitors, RC – Rogovskii coil. The image of discharge in waveguide is at the right inferior angle.

waveguide for frequency about 11.5 GHz determined by the equation:

$$\Lambda_0 = \lambda_0(1-(\lambda_0/2a)^2)^{-1/2}\ , \qquad (1)$$

where $a$ is the wide wall dimension of waveguide and $\lambda_0$ is the wavelength in air.

The discharges are ignited in quartz tubes with inner diameters of 1.6 mm, wherethrough argon or helium flows (~0.5 slm) at atmospheric pressure are provided. The electrodes (2, 3) with diameters about 1 mm are fixed inside tubes. In the experiments, the electrode gap is set at 11-12 mm.

* Corresponding author: l.simonchik@ifanbel.bas-net.by

The ignition of discharges is carried out by a short pulse power supply (H.V. pulser) through a pulse transformer T and capacitor C. Simultaneously, capacitor C0 (4 μF) charged from a DC voltage source (Fig. 1) begins to discharge through the choke L, the secondary winding of the pulse transformer T and the electrically broken down gaps in three tubes. The image of discharge in waveguide produced by Canon PowerShot A530 camera is shown in Fig. 1 (in the right inferior angle).

## 2.2 Results and discussion

Typical time evolution of single discharge current is shown in Fig. 2 (solid curve). It is shown that the discharge current rapidly increases and reaches a maximum value (about 10 A) at 20-25 μs. Then it slowly

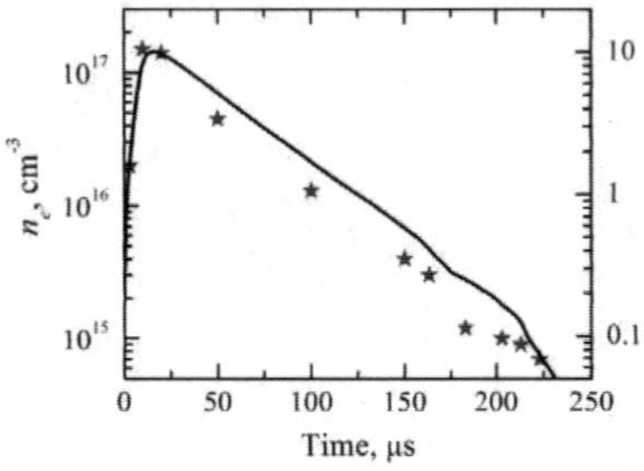


**Fig. 2.** Time evolutions of electron concentration (stars) and discharge current.

decreases and disappears after 250 μs.

Electron concentration in plasma columns is determined using the $H_\alpha$ line shapes registered at different discharge time. The hydrogen $H_\alpha$ line broadening is maximal (> 1 nm) around 20 μs at the discharge current maximum. Estimations of the Hα line half widths, which are caused by different broadening mechanisms (instrumental, Doppler and Van-der-Waals profiles) are shown to be less than experimental profile by one-two orders of magnitude. Therefore, the main broadening of the $H_\alpha$ line is due to the Stark broadening. According to [5], a half width Δλ of $H_\alpha$ line, broadened due to the electrical field of ions, depends on the plasma electron density $n_e$ (cm$^{-3}$) as follows:

$$\Delta\lambda = 2.5\cdot 10^{-9}\cdot \alpha_{H\alpha}\cdot n_e^{\ 2/3}, \qquad (2)$$

where $\alpha_{H\alpha}$ is the Stark broadening parameter. The result is shown in Fig. 2 (stars). It is seen that the electron concentration reaches its maximum (>$10^{17}$ cm$^{-3}$) at the 20$^{th}$ μs, corresponding to the discharge current maximum, and then slowly decreases to less than $10^{15}$ cm$^{-3}$ at 200 μs proportionaly to the current discharge

In order to obtain the transmission spectrum, we used the spectrum analyzer of pulse signals S4-80 at 3 μs gating. Measurements were taken step-by-step at different frequencies at the discharge current maximum (20 μs after ignition) and at the end of the discharge (200 μs), see Fig. 3. Transmission spectrum of 1D EBG structure formed by three inhomogeneities in X-waveguide is discussed in detail in [3]. It consists of the alternating pass- and stopbands. At the 20$^{th}$ μs the transmission spectrum is well contrasting. Thus in the bandpass (around 11.5 GHz) the attenuation of the MW signal is only about 4 dB, while it is reduced by more

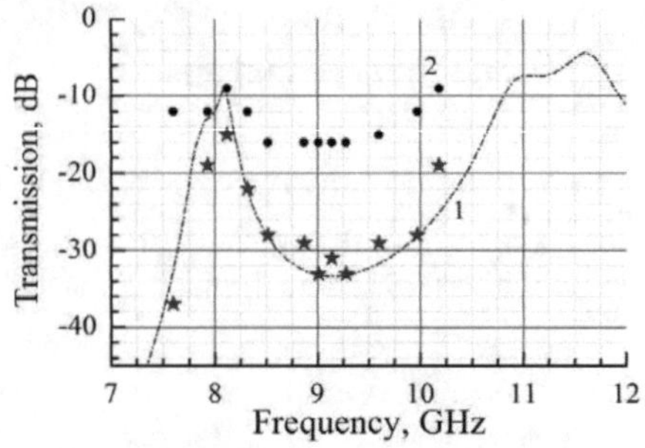


**Fig. 3.** Transmission spectra (experiment: stars – 20 μs, circles – 200 μs; simulation by HFSS [5]: dashed – electron concentration is $10^{16}$ см$^{-3}$, effective diameters – 1.4 mm).

than 30 dB in the stopband (around 9 GHz). There is a second bandpass at frequency of about 8 GHz, however, attenuation of the MW signal is significant (about -10 dB). At longer times, at 200 μs, the transmission signal attenuation is less expressed since the suppression of the MW signal in the stopband is only -20 dB. At the same time, MW signal in bandpass (around 11.5 GHz) decreases significantly less, only about 1 dB.

The 1D EBG structure including pulsed plasma discharges in tubes can be irradiated by two different MW signals at frequency 9.15 GHz: powerful pulses (about 50 kW, pulse duration of about 150 ns and repetition frequency of 2 kHz) and a continuous microwave (CW) signal at low power (5 mW). The transmited MW power normalized by the incident power is shown in Fig. 4 at low CW power (solid curve). It is seen that the transmitted signal is reduced by 3 orders of magnitude at the 25$^{th}$ μs. After the 150$^{th}$ μs the transmission begins to recover and fully recovered after 230$^{th}$ μs. Therefore, the duration of transmitted signal suppression correlates with time of plasma existing, when its concentration exceeds $10^{15}$ cm$^{-3}$ (Fig. 2).

After that, the transmittance behaviour at high MW power was investigated. The repetition frequency of high power pulses is constant and equals 2 kHz. At the same time, the discharge repetition frequency could be only a few Hz. We had not the opportunity to synchronize these two devices at this stage. Therefore, the following scheme for recording the transmitted MW signal was chosen. Pulse voltage source and the oscilloscope were

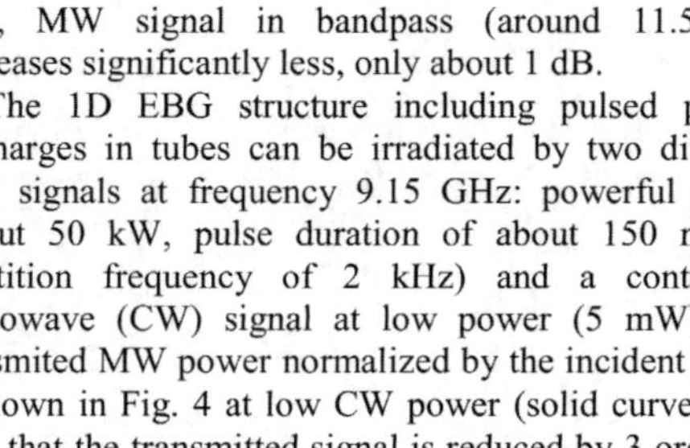


**Fig. 4.** Time dependence of 1D EBG plasma structure transmission at low (solid) and high (vertical dached lines) MW power.

started synchronously, and a oscillogram was recorded for given time interval. An envelope of short MW pulse is appeared randomly in all of these oscillograms. Having received a set of such oscillograms and combining them, we obtained a series of random pulses. Pulses amplitudes are corresponding to the transmittance in the moment of the pulse passing through a EBG plasma structure. The results of this procedure are shown in Fig. 4 (vertical strokes).

Therefore, the MW signal suppression at the level of -30 dB in forbidden band is demonstrated for low MW power of 5 mW and for high power of 50 kW. The transmission spectrum switching rate is evaluated as a function of the discharge parameters (type of gas, flow rate, applied pulse voltage). A switching time of 10-20 ns and a repetition frequency up to 20 kHz can be received.

## 3 Control by 2D EBG structure

### 3.1 Experimental arrangement

A triangle 2D EBG structure is formed by copper rods of 140 mm in length and 5 mm in diameter which are located in parallel to each other in square lattice points (Fig. 5). The lattice spacing is 22 mm. This arrangement allows a diagonal propagation of wave mode (around ±45°) for the frequency of 9.15 GHz when a defect is present in the front row, while it is forbidden for the principal propagating mode (0° direction) [2]. In order to excite the diagonal mode, it is necessary to put a defect in the front rod row of the EBG structure by removing one or two rods (Fig. 5, inset II) or by adding a rod (or plasma column) (Fig. 5, inset III).

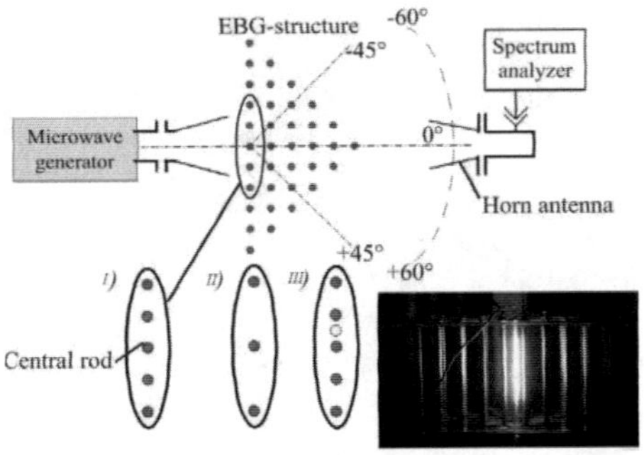


**Fig. 5.** Schematic of the experimental setup (I – perfect EBG structure, II – structure with tow vacations, III – structure with additional defect). The inset (bottom-right) – photo of extended discharge in 2D EBG metallic

The structure is irradiated by continuous MW radiation at frequencies in the range of 8.5-10.0 GHz and power of 5 mW using a horn antenna. MW radiation transmitted through the 2D EBG structure is received by another horn antenna. Receiving antenna with a waveguide-to-radio adapter is moved along a circular arc of 0.5 m radius, centered in the middle of the triangular periodic structure base. This movement is realized within an angle range of ± 60° relative to the axis of the radiating horn antenna (Fig. 5).

### 3.2 Results and discussion

In this study the directional pattern of this 2D EBG structure at frequency of 9.15 GHz using a low-power generator (5 mW) operating in continuous regime was investigated. When the structure has no defect (Fig. 6, solid), there is only weak MW radiation reaching the receiving antenna, where the signal level is less by 3-4 orders of magnitude than the MW radiation signal from

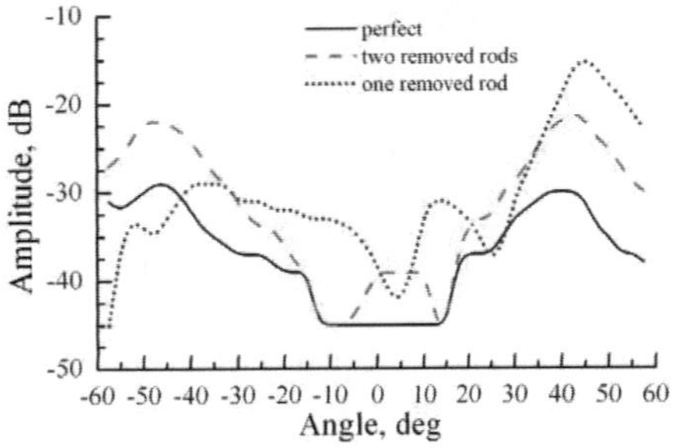


**Fig. 6**. Directional patterns of the 2D EBG structure.

the radiating horn antenna without 2D EBG structure. If the two nearest rods to the central rod in the first row are removed, MW radiation begins to propagate in directions ±45° and signal level grows up by 7 dB (dashed curve). If there is only one rod removed, a MW signal increases by 15 dB (dotted) at the removed rod side. The same behavior takes place in directional diagram in case of additional defect as well.

After that the plasma columns were placed as additional defects. A photo of a discharge placed in a two-dimensional electromagnetic crystal as an additional defect is shown in Fig. 5 (insert right below). Since the aperture of the irradiating antenna is 110×50 mm, the discharge gap is 50 mm. The inner diameter of the tube is 2.5 mm, the diameter of the electrodes is 2 mm. The discharge is powered by a switching power supply, which is shown schematically in Fig. 1 for discharges in a one-dimensional structure.

An estimation of the electron concentration in a pulsed discharge was also performed by the Stark broadening of the hydrogen line contour Hα, which was recorded experimentally with a time scan. The electron concentration was calculated according to formula (2) and its time dependence in a plasma of an extended pulsed discharge in argon at atmospheric pressure is demonstrated in Fig. 7. It is shown that the electron concentration maximum is at $20^{th}$ μs, and the maximum concentration is about $7\times10^{16}$ $cm^{-3}$. After that the concentration decreases exponentially to $6\times10^{15}$ $cm^{-3}$ at about $400^{th}$ μs.

When the crystal is irradiated with CW radiation of about 5 mW power, transmission of crystal in the direction of 45° changes during the first 300 μs when electron density in tubes is higher than $10^{16}$ $cm^{-3}$. This is

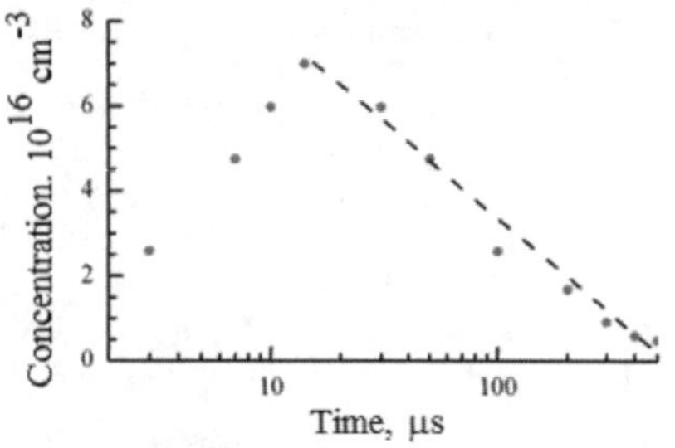


**Fig. 7.** Time dependence of the electron concentration in a plasma of an extended pulsed discharge (5 cm long and 2.5 mmdiameter) in argon at atmospheric pressure

the case in both configurations, when the plasma column is in place of the additional inhomogeneity (curve 2 in Fig. 8) or when the plasma is in vacancy place (curve 2 in Fig. 9). After 300 μs, the electron concentration in the discharge channel is less than $10^{16}$ $cm^{-3}$ (Fig. 6), indeed the plasma has no more effect on EBG structure.

Now, we investigate the case where the structure is irradiated by high power (about 50 kW) MW pulses at 9.15 GHz frequency and about 150 ns duration. As in case of 1D EBG structure, a waveform registration is synchronized with discharge ignition, and envelopes of MW pulses are randomly distributed during every registration time interval. Summarizing these waveforms we obtained a set of pulses distributed frequently in time scale (Fig. 8, vertical dashed lines). It is shown that the transmission change of an electromagnetic crystal for both types of defects is maximal in the discharge phase, when the electron concentration is higher than $10^{16}$ $cm^{-3}$, and is leveled to 400 μs, which is later than at low MW power. Probably, it is connected with the maintenance of

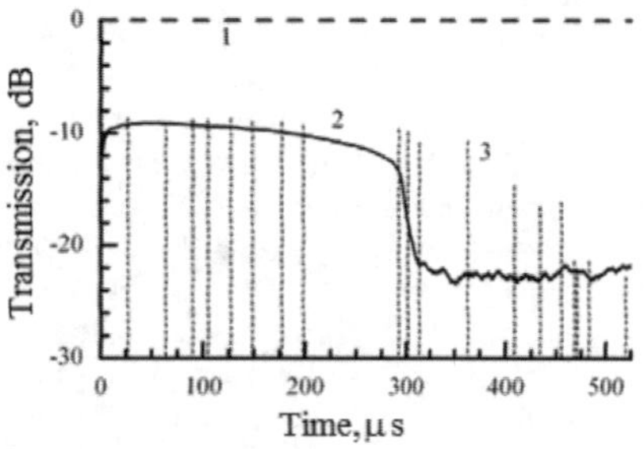


**Fig. 8.** The time dependence of the transmission at a 9.15 GHz frequency of a two-dimensional electromagnetic crystal in the direction of 45° with an extended pulse discharge as an additional defect. 1 – without EBG, 2 – 5 mW and 3 – 50 kW pulses.

the electron concentration in the plasma inhomogeneity due to the collision absorption in the plasma of the transmitted powerful MW radiation. It should be noted that the maximum change (14±2 dB) of the transmitted MW signal in the 45° direction from the defect side is observed approximately from 5 μs to 100 μs relative to the leading edge of the discharge current pulse, and it is the same as in the case of metallic inhomogeneities use.

## 4 Conclusions

The control possibilities of powerful MW by the 1D and 2D EBG plasma structures were demonstrated experimentally. 1D EBG plasma structure in the waveguide formed solely by pulse discharges at atmospheric pressure allows to obtain a MW signal suppression up to 40 dB in forbidden band and its decrease by about 4 dB in passband by variation of electron density in the range of $10^{14}$-$10^{16}$ $cm^{-3}$. 1D EBG plasma structure can be used as quickly switched (10-20 ns) attenuator, modulator, frequency switcher, filter and etc. for high power microwaves.

Also we demonstrated one plasma column as a key element, allowing to change the transmitted properties of regular metallic structure using 2D EBG structure.

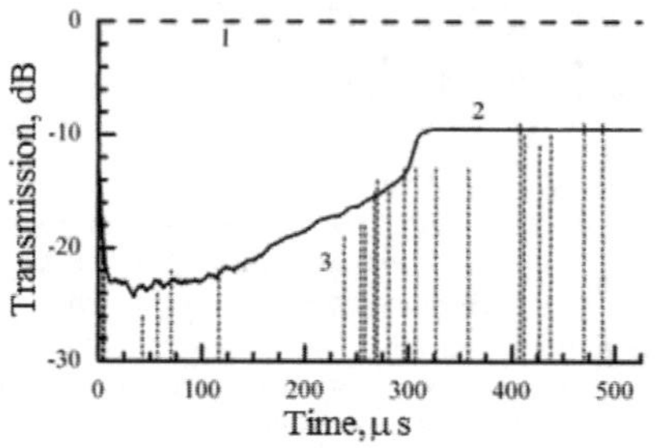


**Fig. 9.** The time dependence of the transmission at a 9.15 GHz frequency of a two-dimensional electromagnetic crystal in the direction of 45° with an extended pulse discharge as a defect compensator. 1 - without EBG, 2 – 5 mW and 3 - 50 kW pulses.

When the discharge is used as an additional defect, the radiation is transmitted through the structure in (45.0±2.5)° direction. On the contrary, transmission disappears in that direction when the discharge is used as a defect compensator. The transmission changes were about 15 dB in both cases.

The work was partially supported by the grants the number F15F-004 from BRFFR and PICS number 151539 from CNRS.

## References

1. O. Sakai, K. Tachibana, Plasma Sources Sci. Techn., **21(1)** (2012)
2. J. Lo, J.Sokoloff et al., Applied Physics Letters, **96** (2010)
3. V.I. Arkhipenko et al., Journal of applied physics, **116(12)** (2014)
4. H.R. Griem, "Spectral lines broadening" (Academic Press, New York and London, 1974).
5. Ansoft HFSS, http://www.ansys.com/Products/Simulation+Technology/Electromagnetics/High-Performance+Electronic+Design/ANSYS+HFSS

EPJ Web of Conferences **157**, 03050 (2017)
*22 Topical Conference on Radio-Frequency Power in Plasmas*

DOI: 10.1051/epjconf/201715703050

# Observation of the X-mode anomalous absorption in the plasma filament associated with the two upper-hybrid-plasmon decay

*Leanid* Simonchik[1,*], *Alexey* Altukhov[2], *Valery* Arkhipenko[1], *Alexey* Gurchenko[2], *Evgeniy* Gusakov[2], *Alexey* Popov[2] and *Maxim* Usachonak[1]

[1] Institute of Physics of NAS of Belarus, 220072 Minsk, Belarus
[2] Ioffe Physical-Technical Institute of RAS, 194021 St-Petersburg, Russia

**Abstract.** The strong anomalous absorption of the X-mode wave associated with the two upper-hybrid-plasmon decay in the plasma at density higher than the UH resonance value for the half frequency of the pump by means of optical and microwave diagnostics is observed. The threshold and growth rate of the anomalous phenomena are estimated and compared to the theory predictions. The low frequency waves excited in plasma are investigated using the enhanced scattering diagnostics.

## 1 Introduction

In the recent decade there has been accumulated a substantial number of observations in the ECR plasma heating experiments in tokamaks and stellarators that do not fit into a simple linear picture. The most distinct among them are the anomalous backscattering phenomena studied in detail in the second harmonic X-mode heating experiments at the Textor tokamak [1]. In these experiments it was shown that the radiation temperature of the observed frequency downshifted backscattering signal is thousand times larger than the electron temperature, and its amplitude is modulated at the magnetic island rotation frequency. The highest level of anomalous backscattering was achieved when the plasma density in the island coincides with the upper hybrid density for the frequency equal to the half value of the pump frequency. Based on the latter observation the theoretical model [2] was proposed recently allowing to explain the effect of anomalous backscattering as a result of the two upper-hybrid (UH) plasmon parametric decay instability (PDI) possessing very low threshold due to trapping of nonlinearly excited plasmons in the vicinity of the density maximum that accompanies the island. The theory [2] also predicts substantial (up to 25%) anomalous absorption due to this process.

In the present paper the low-threshold two-plasmon PDI leading to excitation of the trapped UH waves is modelled experimentally in the laboratory plasma.

## 2 Experimental setup

The decay occurs in a plasma filament extended in the direction of the magnetic field and produced by high-frequency discharge in long quartz tube with an inner diameter of 23.5 mm filled with argon at about 2 Pa pressure. Quartz tube passes through the waveguide (72 x 34 mm$^2$) in parallel to the wide walls (see Fig. 1). The RF power of about 100 W at frequency of 27 MHz is supplied to the ring electrodes placed outside of the tube and disposed on both sides of the waveguide at a distance of 30 cm.

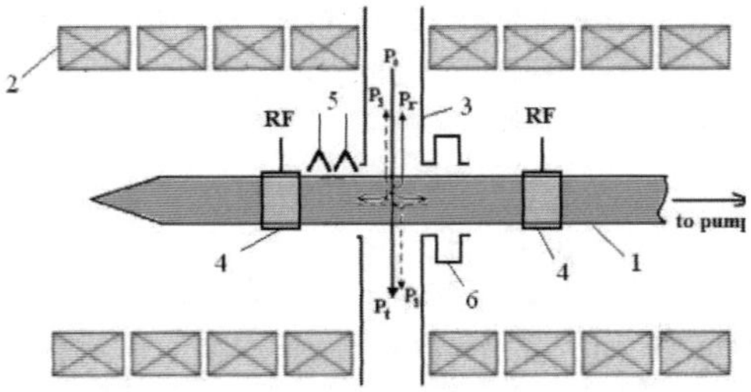


Fig. 1. Schematic of experimental setup. 1 – quartz tube; 2 – magnetic system; 3 – waveguide; 4 – RF antenna; 5 – microwave antenna; 6 – microwave cavity.

At the maximal RF power the volume averaged plasma density measured using the cavity diagnostics is about $10^{10}$ cm$^{-3}$ changing by 15-20% at the variation of the magnetic field from 0 to 450 Oe. The radial distribution of the plasma luminosity is approximated by expression $(1\text{-}r^2)^{1.6}$. Asuming it to be proportional to the plasma density we get the maximal density in the discharge about $2\times10^{10}$ cm$^{-3}$.

The X-mode microwave pulses (up to 160 W) at frequency $f_0$ = 2.35 GHz, substantially higher than the EC and UH values, were incident onto the plasma along the waveguide. The temporal behaviour of the transmitted and reflected microwave signal as well as the plasma luminosity were monitored in the experiment at the different pump power and magnetic field. The low-frequency plasma turbulence was investigated using the UHR enhanced microwave scattering diagnostics at frequency of about 1.2 GHz.

* Corresponding author: l.simonchik@ifanbel.bas-net.by

DOI: 10.1051/epjconf/201715703050

## 3 Theoretical predictions

The theoretical model of the low-threshold two-plasmon decay proposed in [2] consists of two parts shown schematically in Fig. 2. Firstly, it is a model of the PDI linear stage describing the absolute instability leading to extra-ordinary $t_X$-wave decay into two UH plasmons $l_{UH}(f_1)$ and $l_{UH}(f_2)$, which are exponentially growing from the thermal noise level $b_{therm}$. Secondly, it is a model of the absolute instability nonlinear saturation as a result of cascade of several low-threshold decays in which UH waves $l_{UH'}$ trapped in plasma and low frequency waves (ion sound $l_{IS}$ in the case of our experiment) are excited. The saturation level $b_m$ determines the efficiency of the anomalous absorption due to the PDI.

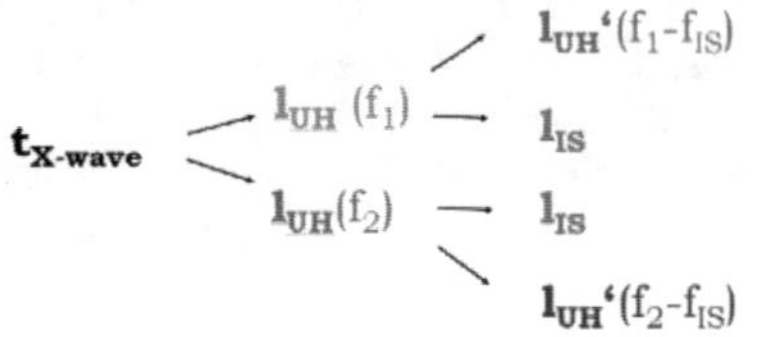


**Fig. 2.** The scheme of the two-plasmon decay and it's saturation.

In the theoretical analysis of the PDI linear stage in the specific "Granit" device conditions we assumed an extraordinary pump wave propagating across the magnetic field and possessing the $m$ = 1 mode azimuthal structure represented by expression

$$E_y(\mathbf{r})\mathbf{e}_y = \frac{E_0(z)}{2}(\mathbf{e}_\theta \cos\theta + \mathbf{e}_r \cos\theta)\exp(-i\omega_0 t) + c.c.,$$

$$E_0(z) = \cos\left(\pi \frac{z}{2w_z}\right)$$

where $\omega_0$ – is the pump frequency, $z$ – is the coordinate along the magnetic field and $\vartheta$ – is the azimuthal angle.

As a result of the PDI the UH waves are excited and described by their potentials

$$\begin{aligned}\phi_1(\mathbf{r},t) &= b_{s+1}(z,t)\varphi_{n_1,s+1}(\rho)\exp(i(s+1)\theta - i\omega_1 t)\\ \phi_2(\mathbf{r},t) &= b_s(z,t)\varphi_{n_2,s}(\rho)\exp(is\theta + i\omega_2 t)\end{aligned}$$ , where

$\varphi_{n_1,s+1}(\rho)\exp(i(s+1)\theta - i\omega_1 t)$ and $\varphi_{n_2,s}(\rho)\exp(is\theta + i\omega_2 t)$ stand for eigen functions describing the radially trapped

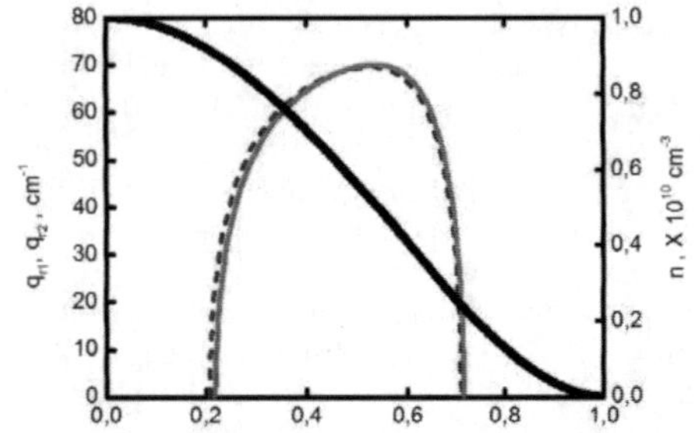


**Fig. 3** (left and bottom axes) Disppersion curves of the UH waves $q_{r2}$ ($f_2$ = 1.177 GHz, $s=15$, dashed line) and $q_{r1}$ ($f_1 = f_0 - f_2$, s+1, solid line); (right and bottom axes) – density profile (thick solid line); $H$ = 320 Oe, $T_e$ = 2 eV.

and azimuthally propagating UH eigen modes satisfying the decay condition $\omega_2 = \omega_0 - \omega_1$, possessing numbers $n$ and $s$. The dispersion curves (dependence of the WKB radial wavenumber of the UH wave on the coordinate) indicating the UH wave localization region are shown in Fig. 3 as well as the plasma density profile.

The temporal and spatial behavior of the plasmon amplitudes $b_{s+1}(z,t)$ and $b_s(z,t)$ satisfies the following equations taking into account plasmon nonlinear pumping $\Gamma_{s+1,s}$, $\Gamma_{s,s+1}(z)$ collisional damping $\nu_{d1,2}$ and their diffractive losses from the pump localization region:

$$\begin{aligned}\left(\frac{\partial}{\partial t} + iD_{z,s+1}\frac{\partial^2}{\partial z^2} + \nu_{d1}\right)b_{s+1} &= \Gamma_{s+1,s}(z)b_s,\\ \left(\frac{\partial}{\partial t} - iD_{z,s}\frac{\partial^2}{\partial z^2} + \nu_{d2}\right)b_s &= \Gamma_{s,s+1}(z)b_{s+1}.\end{aligned} \qquad (1)$$

where at s>>1

$$\left[\Gamma_{s+1,s}(z);\Gamma_{s,s+1}(z)\right] = \frac{1}{4\pi}\frac{\omega_0|\omega_{ce}|^2 c}{4\Delta_0}\left[\frac{1}{\omega_0} + \frac{s+1}{\omega_1} - \frac{s}{\omega_2}\right]\times$$

$$\int\limits_{\max(r_{1l},r_{2l})}^{\min(r_{1r},r_{2r})} \frac{d\rho}{\sqrt{|q_{r1}||q_{r2}|}}\frac{s^2}{\rho^3}\exp\left(i\int\limits_{\max(r_{1l},r_{2l})}^{\rho}(q_{r1}-q_{r2})d\rho'\right)\times$$

$$\left[\frac{E_0(z)}{H}\int\limits_{r_{1l}}^{r_{1r}}\frac{dr}{|q_{r1}|}\left(\frac{s}{\rho}\right)^2;\frac{E_0^*(z)}{H}\int\limits_{r_{2l}}^{r_{2r}}\frac{dr}{|q_{r2}|}\left(\frac{s}{\rho}\right)^2\right]; \text{ and } \Delta_0 = \omega_0^2 - \omega_{ce}^2$$

It should be noted that for the experimental conditions of the "Granit" device the inverse collisional losses time $\nu_{d1,2} = f(\nu_{ea}) \propto 3 \div 5\mu s^{-1}$ exceeds substantially the diffractive loss time $\tau_D^{-1} = D_{zs,s+1}/(\pi w_z^2) \propto 10^{-2}\mu s^{-1}$. Under these conditions we may neglect the diffractive losses in our analysis and obtain the following expression for the absolute PDI growth rate:

$$2\gamma_s = -(\nu_{d1}+\nu_{d2}) + \sqrt{(\nu_{d1}+\nu_{d2})^2 + 4\Gamma_{s,s+1}\Gamma_{s+1,s} - 4\nu_{d1}\nu_{d2}} \qquad (2)$$

This expression at $\gamma_s = 0$ is providing the PDI threshold, which is determined by the plasmon collisional damping according to the formula $\Gamma_{s,s+1}\Gamma_{s+1,s} = \nu_{d1}\nu_{d2}$ similar to the corresponding threshold in homogeneous plasma theory.

## 4 Experimental results and discussion

The first observations of the microwave anomalous absorption at the pump frequency higher than the second electron cyclotron harmonic were performed at the maximal available power of 160 W. The magnetic field was taken smaller than 42 mT, which corresponds to the cyclotron resonance for half the pump frequency of 1.175 GHz. At a small plasma density no distortions of the microwave pulse transmitted through the plasma or reflected from it were observed during the pulse duration, as it is seen in Fig. 4*a*. However at plasma density exceeding a critical value after a time delay dependent on the density value, a fast decrease of both transmitted and reflected power was observed thus indicating the turning on of the anomalous absorption (see Fig. 4*b*). This effect was accompanied by a sharp growth of the plasma luminocity also shown in this figure.

The time delay of the anomalous phenomena appearance was dependent on the pump power, as it is seen in Fig. 4*c*, however their main features look similar.

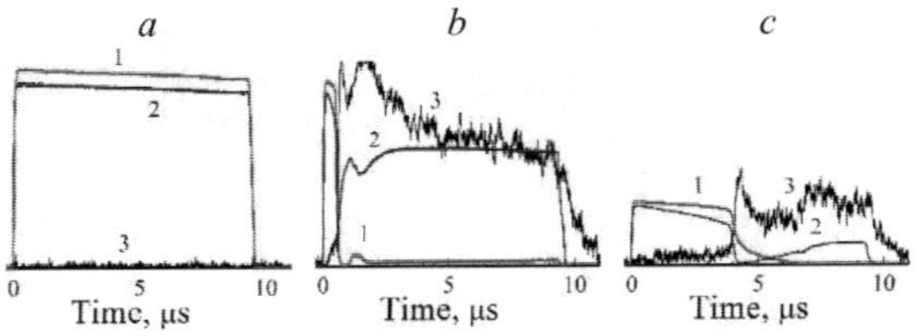


**Fig. 4.** Waveforms of transmitted (1), reflected (2) pulses and of the plasma luminosity (3). *a* –small plasma density, *b* – $P_0$ = 160 W, c – $P_0$ = 30 W

Starting from the microwave beginning a slow decrease of the reflected power is accompanied by a slow increase of the plasma luminocity, whereas the transmitted power is practically constant. Then an abrupt decrease of the transmitted power happens, which is followed by a fast fall of the reflected power and growth of the plasma luminosity. After that the reflected power increases, but doesn't reach the value it had before the onset of the anomalous absorption.

The critical density needed for switching on of the anomalous phenomena is growing with decreasing magnetic field, as it is shown by circles in Fig. 5. As it is

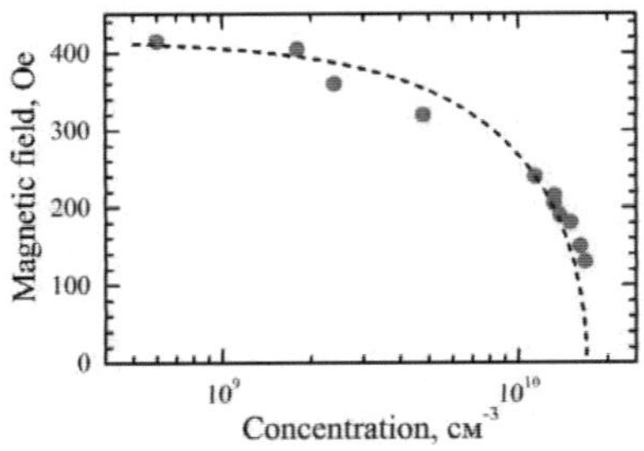


**Fig. 5.** Dependence of the density critical for onset of the anomalous absorption on the magnetic field. Broken line - UHR condition for 1.175 GHz; $P_0$ = 160 W.

seen there, this dependence is close to the theoretical UHR density dependence on the magnetic field plotted in the figure for the half pump frequency

$$n_e = \frac{m}{4\pi e^2}\left(\pi^2 f^2 - \left(\frac{eB}{m_e c}\right)^2\right) \qquad (3)$$

The temporal delay of the anomalous absorption onset is most likely associated with the time needed for growth of the UH wave noise from the thermal level to the saturation level. In order to check this hypothesis we plotted the dependences of inverse delay time against the magnetic field measured at the constant plasma density, but slightly different pump power and compare them to the expression (2) $\frac{1}{T} = 2\gamma_p / \ln\left(\frac{b_m^2}{b_{therm}^2}\right)$ provided by the theoretical analysis. As it is seen in Fig. 6 the experimental data fits the theoretical dependence of the PDI growth rate on the magnetic field at an assumption of of the UH noise amplification in saturation regime by 10 orders of magnitude.

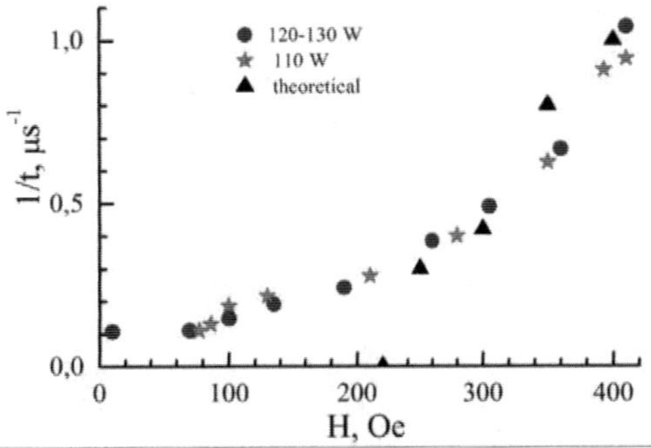


**Fig. 6.** Experimental (stars and circles) and theoretical (triangles) inverse time of the anomalous absorption onset versus magnetic field.

We have also studied the dependence of the two-plasmon instability growth rate on the pump power. The

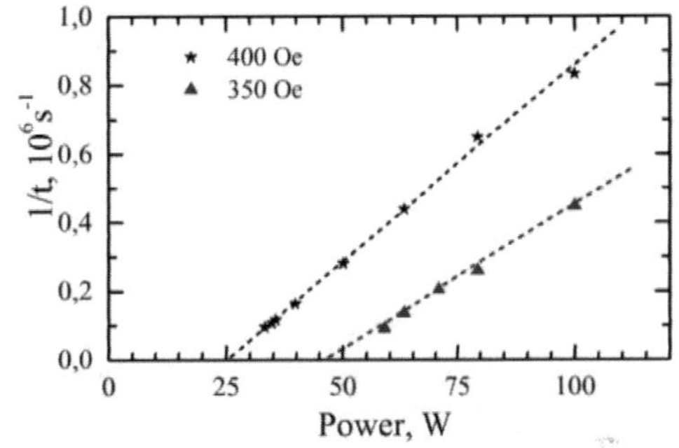


**Fig. 7.** Dependence of the inverse time of the anomalous absorption onset on the pump power for different magnetic fields.

measurements were performed for two magnetic fields 350 and 400 Oe, but at the constant averaged plasma density about $10^{10}$ см$^{-3}$ controlled by the plasma RF source power and Ar presure of about 1.5 Pa. The results for the inverse time of the anomalous absorption turning on shown in Fig. 7 appear to be in a qualitative agreement to the theory predictions for the growth rate

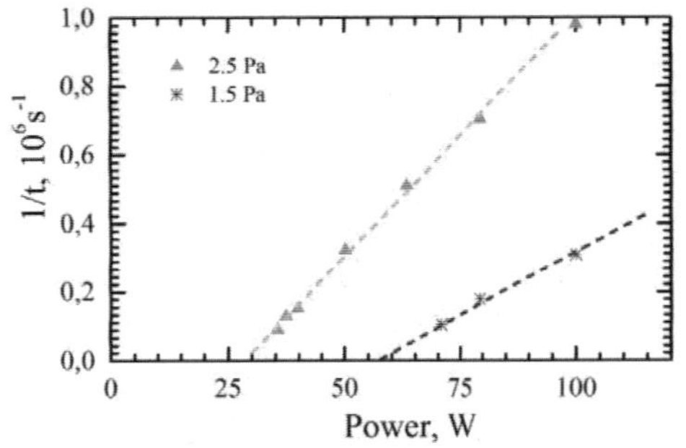


**Fig. 8.** Dependence of the inverse time of the anomalous absorption onset on the pump power for different values of Ar presure.

DOI: 10.1051/epjconf/201715703050

increase with increasing pump power and for the PDI power threshold decrease with growing magnetic field.

The dependence of the inverse time of the anomalous absorption onset on the pump power was also studied for two values of Ar pressure 1.5 Pa and 2.5 Pa. The obtained results shown in Fig. 8 qualitively confirm the theoretical predictions on the PDI threshold growth with the increasing electron-atom collision frequency and accordingly with the gas presure.

Using the spatially resolved plasma luminocity measurements we studied the localization of the anomalouse absorption, which appeared to be non-central in qualitative agreement with the theoretical expectations based on the UH eigen mode localization (see Fig. 3). Moreover the radial position of the plasma radiation drastic growth determined at the constant value of the averaged plasma volume density was shown to be dependent on the magnetic field. Figure 9 is demonstrating the dependence of the UHR plasma density value, determined using expression (3), values of

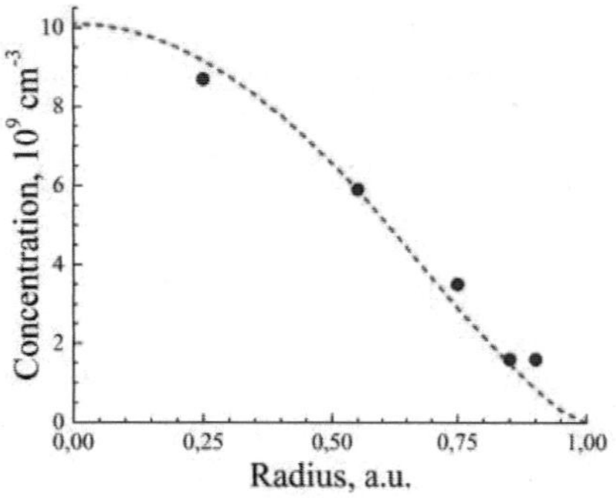


**Fig. 9.** Dependence of the UHR density value on the radius of the maximal anomalous absorption determined by variation of the magnetic field.

the magnetic field and the half pump frequency, versus the radius of the maximal plasma luminocity growth. As it is seen the obtained dependence fits closely the unperturbed plasma density profile $(1-(r/r_0)^2)^{1.6}$ reconstructed using the volume averaged density measurements and the Abelian transformed profile of plasma luminocity. This agreement, as well as those demonstrated above, confirms the validity of the two-plasmon decay mechanism for explanation of the strong anomalous absorption observed in our experiment.

In order to investigate the nonlinear saturation mechanism of the two-plasmon decay instability, the measurements of the low-frequency small-scale plasma waves were performed using the enhanced UHR backscattering (BS) diagnostics. For this purpose plasma was probed by microwaves in the frequency range 1.1-1.2 GHz, for which the UHR condition was fulfilled within the plasma volume. It is shown that an intensive BS signal is observed when the probing frequency is close to the half pump frequency.

The temporal variation of the BS spectrum is demonstrated in Fig. 10. The white line on the top of the figure shows the waveform of the transmitted signal. It should be stressed that the most intensive BS signal is observed only in the period of the strongest suppression of the transmitted signal, thus in the period of the strong anomalous absorption onset. During this period the BS

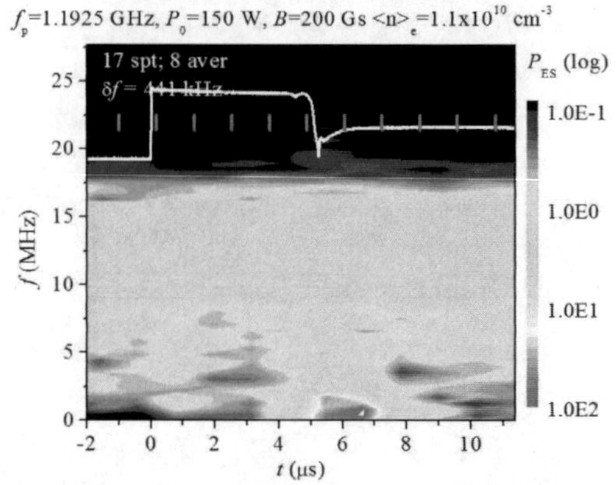


**Fig. 10.** The temporal evolution of the BS spectrum. The upper curve corresponds to the transmitted pulse.

signal is enhanced by two orders of magnitude. The frequency shift of the BS signal is close to 2-3 MHz, which corresponds to the ion acoustic frequency range along with the PDI saturation mechanism discussed in theory.

## 5 Conclusions

The strong anomalous absorption of the microwave power is observed in the plasma at density close to the UHR value for the half pump frequency by means of optical and microwave diagnostics. The effect localization and its dependence on magnetic field, plasma density, gas pressure and microwave power are shown to be close to the theoretical predictions for the two-plasmon decay. The threshold and growth rate of the anomalous phenomena are shown to agree with the theory results. The low frequency waves spectra are shown to be enhanced in the strong absorption period .

The work was performed at the support of grants RFFR Bel-a 16-52-00019 and BRFBR F16R-095.

## References

1. S.K. Nielsen, M. Salewski, et al., Plasma Phys. Control. Fusion, **55** (2013) 115003
2. E.Z. Gusakov and A.Yu. Popov, Physics of Plasmas, **23** (2016) 082503

DOI: 10.1051/epjconf/201715703051

# Investigation of the coupling properties of the ion cyclotron fast wave under applied magnetic perturbations and MHD phenomena in ASDEX Upgrade

*Guillermo* Suárez López[1,2,⋆], *Roman* Ochoukov[1], *Matthias* Willensdorfer[1], *Volodymyr* Bobkov[1], *Mike* Dunne[1], *Helmut* Faugel[1], *Helmut* Fünfgelder[1], *Jean-Marie* Noterdaeme[1,3], *Erika* Strumberger[1], *Wolfgang* Suttrop[1], *Hartmut* Zohm[1,2], the ASDEX Upgrade team, and the EUROfusion MST1 Team⋆⋆

[1]*Max Planck Institute for Plasmaphysics, Boltzmannstraße 2, 85748 Garching bei München, Germany*
[2]*Ludwig Maximillian University of Munich, Geschwister-Scholl-Platz 1, 80539 München, Germany*
[3]*Applied Physics Department, University of Ghent, Sint-Pietersnieuwstraat 41, 9000 Ghent, Belgium*

**Abstract.** The modulation of the ion cyclotron fast wave coupling to the plasma due to non-axisymmetric changes of the distance antenna-R-cutoff is studied. These changes can arise when magnetic perturbation (MP) fields are used, or when MHD activity is present. The application of MP fields can excite a low field side midplane plasma kink response that amplifies the vacuum perturbation field, leading to appreciable 3D plasma displacements. This effect is studied via NEMEC simulations. Rigid rotation of the MP field is found to produce a coherent antenna loading resistance modulation, suggesting an interplay between the non-axisymmetric magnetic field structure and the wave coupling properties. MHD modes are shown to introduce similar loading resistance oscillations, coherent with the mode rotation frequency. The case of a (2,1) mode is presented.

## 1 Introduction

The excitation of the fast magnetosonic wave in the ion cyclotron range of frequencies (few tens of MHz) is an effective technique to transport several megawatt of power up to the core plasma of a fusion device. Its excitation is produced by launchers, usually, antennas that are flushed with the first wall of the plasma device, such as a tokamak **[1]**, and the wave energy is deposited via damping mainly into the ion species.

The major drawback of the technique is the fact that the launched fast wave with $k_{\|} \neq 0$ is evanescent in low-density plasmas, such as the ones encountered in the so-called scrape-off layer or edge plasma region of a tokamak. The transported power into the non-evanescent, higher density, plasma can then be expressed as an exponential decay that depends on the optical thickness ($\eta$) of the evanescent region, i.e. $P_{trans} \approx P_0 e^{-2\eta}$ with $\eta = k_0 \int_l n(l) dl$. For a monotonically increasing perpendicular refractive index $n^2_{\perp,FW} \in \left[n^2_{0\perp,FW}, 0\right]$, the transported power can be approximated as $P_{trans} \approx P_0 e^{-1.1 k_{\|} d}$ **[2]** with d being the distance from the antenna up to the R-cutoff where the wave becomes propagative, i.e. $n^2_{\perp,FW} > 0$.

In this study, we address non-axisymmetric changes of the distance *«d»* arising from the application of magnetic perturbation (MP) fields from in-vessel saddle coils and naturally occurring MHD modes. Measurements of the antenna performance via loading resistance are used in order to assess their effect on the accessibility of the fast wave to the propagating region. Furthermore, simulations are performed with the ideal non-linear MHD code NEMEC (NESTOR code for the free boundary problem + VMEC code) **[3]** to predict the plasma kink response under different applied MP phasing.

The paper is organized as follows: In **section 2** the experimental setup is described, including a brief description of the ASDEX Upgrade saddle coils and the ion cyclotron system. In **section 3** the analyzed discharges are presented and NEMEC predictions are shown. In **section 4** the influence of MHD modes on the loading resistance of the antennas is described for the specific case of a (2,1) mode. Finally, in **section 5** conclusions are drawn and a future outlook is sketched.

## 2 Experimental setup

### 2.1 Magnetic perturbation saddle coils

The ASDEX Upgrade tokamak is complemented with two poloidally shifted toroidally-distributed arrays of eight in-vessel saddle coils. MP fields are applied from these coils for edge localized mode (ELM) mitigation and suppression studies, mode unlocking, and resistive wall mode control **[4]**. Individual power sources feeding each coil allow a wide range of configurations. The coil currents can be set to $0 < I_{coil} < 1.25$ kA/turn and the phasing between the upper toroidal row and lower row can be set to $-\pi < \Delta\varphi_{UL} < \pi$. The applied phase difference determines the vacuum poloidal spectrum of the MP field.

⋆e-mail: guillermo.suarez@ipp.mpg.de
⋆⋆See the author list of IAEA FEC 2016 OV/P-12 by H. Meyer et al. , to be published in Nuclear Fusion

DOI: 10.1051/epjconf/201715703051

When the MP field is applied, a plasma response is commonly observed, especially strong at low collisionality ($\nu^*$), that leads to pressure driven kink modes around the low field side (LFS) midplane and a peeling response around the X-Point and plasma top **[5, 6]**. Both responses are sensitive to the applied MP field poloidal and toroidal spectrum. Furthermore, the plasma response can amplify the vacuum MP field, leading to non-axisymmetric displacements that range from a few millimeters to the centimeter order **[7]**, as illustrated in **figure 1**.

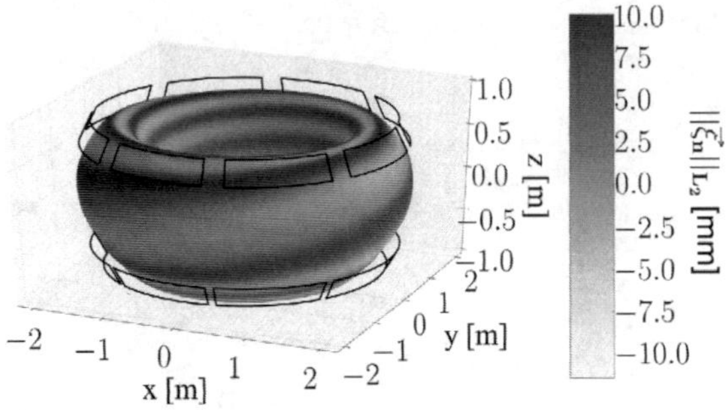


**Figure 1.** NEMEC 3D equilibrium of shot #32091 at t=2.9 s where the saddle coils have been artificially supplied with 5kAt and $\Delta\varphi_{UL} = -140°$ for this simulation. The normal displacement of the last closed flux surface $||\vec{\xi}_n||_{L_2}$ is color-coded.

### 2.2 Ion cyclotron resonance heating (ICRH) system and diagnostics

Four ICRH antennas are presently installed in ASDEX Upgrade: a pair of 2-strap antennas with boron coated limiters and a pair of 3-strap antennas with tungsten coated limiters **[8]**. The antennas are fed in pairs; the 2-strap antennas (a1 and a3) are fed by two RF generators connected together through two hybrid 3dB couplers. The transmission lines are then individually matched by a set of two stubs and a T-junction. The 3-strap antennas (a2 and a4) are also fed by two RF generators, one feeds the central strap of each antenna and the other feeds the side straps via suitable 3dB hybrid couplers. In this case, each individual transmission line is matched by two stubs without T-junction, see **figure 2**.

The measured loading resistance ($R_L$) is an experimentally accessible quantity that gives a good estimate of the coupled electromagnetic power to the plasma. It relates the power flowing through the coaxial line feeding a strap and the voltage maximum in that line, such that:

$$R_L = \frac{2P_{line}Z_0^2}{V_{line,max}^2} \tag{1}$$

Where $Z_0$ is the characteristic impedance of the line. Behind the matching system, a directional coupler measures the forward and reflected power, thus providing $P_{line}$. A voltage probe located at the voltage antinode of the resonant line provides $V_{line,max}$.

## 3 Experimental results. NEMEC simulations

### 3.1 Analyzed discharges

The three ASDEX Upgrade presented discharges, #32319, #32320 and #32321 were performed at a toroidal magnetic field of $B_t = -2$ T and a plasma current of $I_p = 0.8$ MA, resulting in an edge safety factor of $q_{95} \approx -4.3$. Rigid rotations of the MP fields with different phasing at $\Delta\varphi_{UL} = -30°, 40°, 150°$ respectively were performed with a maximum current of $I_{coil} = 6.25$ kAt at $\nu = 6$ Hz. The applied MP field is affected by the passive stabilization loop (PSL), a copper bulk holding the saddle coils that serves to increase the resistive time constant of the vertical instability. The PSL is induced with eddy currents whenever an AC current circulates the saddle coils, attenuating and lagging the MP field with respect to the applied coil current **[9]**. This results in an effective smaller MP field. At a frequency of $\nu = 6$ Hz, the attenuation can become as high as 50%.

The four ICRH antennas were switched on simultaneously with ranging powers, such that 0.6 MW< $P_{antenna}$ < 0.85 MW. The presented loading resistance time traces of **figure 3** are chosen so discharge profiles, such as the normalized plasma beta $\beta_N$, line-averaged density $\langle n_e \rangle$, ICRH power $P_{ICRH}$, $q_{95}$, etc, were kept as similar as possible during the rigid rotation. A clear modulation of the loading resistance, coherent with the rotating MP field can be observed. It is seen that for the case of $\Delta\varphi_{UL} = +150°$ the loading resistance changes as much as $\Delta R_L \sim 0.3\,\Omega$, value peak to peak. This loading variation accounts for ~ 20% of the base value, whereas for the case of $\Delta\varphi_{UL} = -30°$ the change in loading is moderated (~ 10%). The case of $\Delta\varphi_{UL} = 40°$ presents a loading resistance variation of about (~ 10% - 15%). This points to a dependence of the loading resistance variation with the applied MP phasing. The three discharges ended prematurely due to MHD activity deriving in a locked mode, hence difficulting a more profound analysis.

### 3.2 NEMEC simulations

A series of simulations were performed with the nonlinear spectral ideal MHD code NEMEC. NEMEC uses a variational principle to self-consistently minimize the plasma energy in the confined region, while the plasma boundary is deformed according to the external vacuum field (outside of the last closed flux surface) region. The saddle coils, as well as the other conductors (TF, PF and auxiliary coils), provide such vacuum field. The scrape-off layer is not included in the code due to its spectral description. Because of the lack of resistivity terms, the nested flux surfaces are preserved and the normal component of the magnetic field vector $\vec{B} \cdot \vec{u}_n$ vanishes. Hence, no magnetic islands can arise with the NEMEC code.

The experimental plasma pressure and toroidal current reconstructed from the CLISTE equilibrium code **[10]** serve as input to the NEMEC code. Plasma parameters from discharge #32091 at 2.9 s were provided

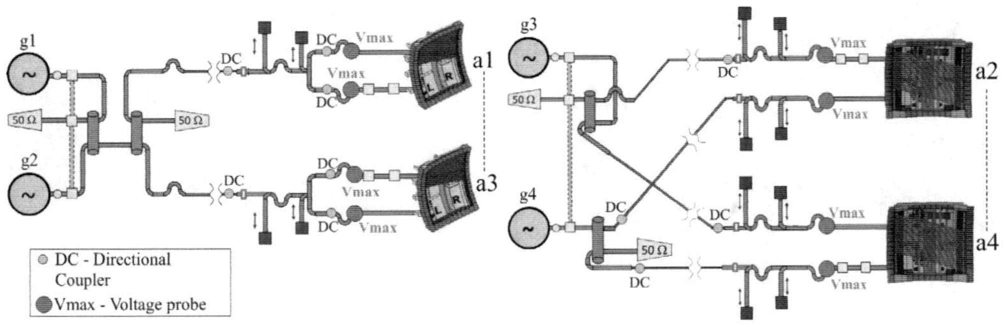


**Figure 2.** ASDEX Upgrade ICRH system. RF generators are marked from g1 to g4. Antenna 1 (a1) and antenna 3 (a3) are 2-strap antennas, antenna 2 (a2) and antenna 4 (a4) are 3-strap antennas.

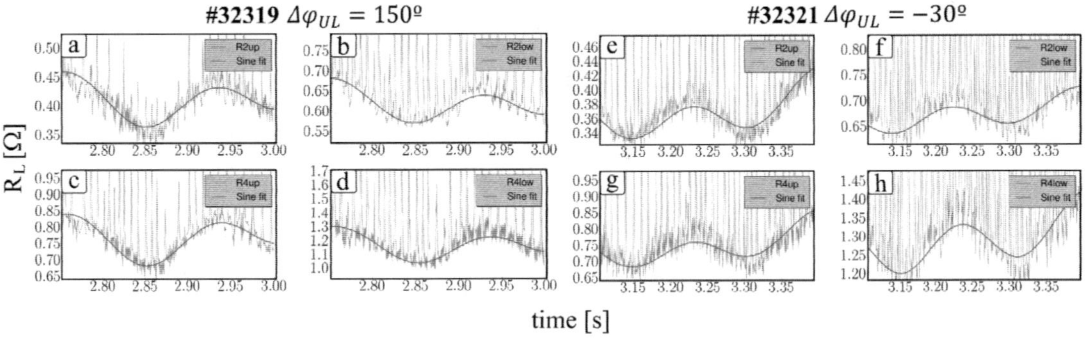


**Figure 3.** Loading resistances of the 3-strap antennas feeding lines in discharges #32319 (a-d) and #32321 (e-h). R2up designates a2, upper feeder and so on. A sine fit (green) is added to better visualize the base variation of the loading.

for the simulations. Discharge #32091 is very close in parameters to the presented discharges where $R_L$ measurements have been taken, $q_{95} = -4.4$, $\beta_N = 1.9$, $\langle n_e \rangle_{edge} = 1.95 * 10^{19}\,\mathrm{m}^{-3}$, $\langle n_e \rangle_{core} = 3.6 * 10^{19}\,\mathrm{m}^{-3}$. The 2D converged NEMEC equilibrium was compared with the CLISTE one and minimal adjustment to the external PF coil currents were performed to match the q-profile from both codes. Then, the 3D free boundary NEMEC code was run, with the previously converged 2D equilibrium parameters, adding this time the different $\Delta\varphi_{UL}$ saddle coil cases in order to obtain the final non-axisymmetric equilibria. The simulation parameters were chosen as 1000 flux surfaces, 26 poloidal mode numbers and 8 toroidal mode numbers for each period. For the saddle coils, n=2 perturbations with $I_{coil} = 5$ kAt were used. The LFS normal displacement at $\theta = 0$ (midplane) of the computed last closed flux surface ($||\vec{\xi}_n||_{L_2}$) is presented in **figure 4**. The normal displacement is defined as the $L^2$ norm of the normal displacement vector to the axisymmetric last closed flux surface.

It is predicted from the simulations that a LFS midplane normal displacement of about 0 mm $< ||\vec{\xi}_n||_{L_2} <$ 7 mm should occur for different applied $\Delta\varphi_{UL}$ and $I_{coil} = 5$ kAt. The ion cyclotron fast wave coupling to the plasma is very sensitive to the density profile in front of the launching antenna. A LFS midplane normal displacement of a few mm should influence this coupling. This phenomenon is indeed registered in the loading resistance measurements of **figure 3**, as a harmonic oscillation coherent with the rotating MP field. It is hence also expected from these simulations that the loading resistance variation should exhibit a $\Delta\varphi_{UL}$ dependence. A detailed comparison of loading resistance and density measurements with simulations is beyond the currently available data.

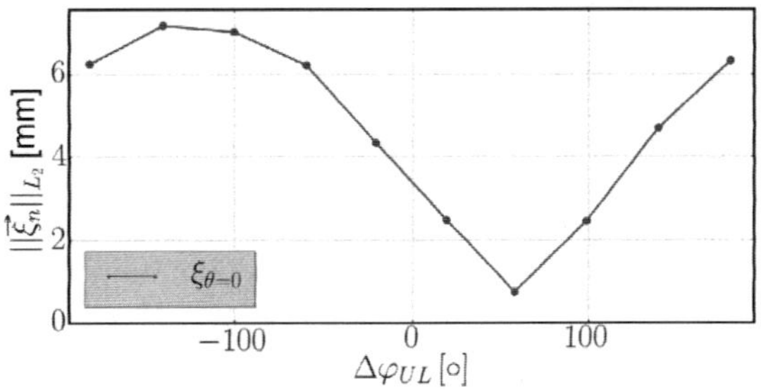


**Figure 4.** NEMEC LFS midplane normal displacement vector norm vs. different $\Delta\varphi_{UL}$.

 DOI: 10.1051/epjconf/201715703051

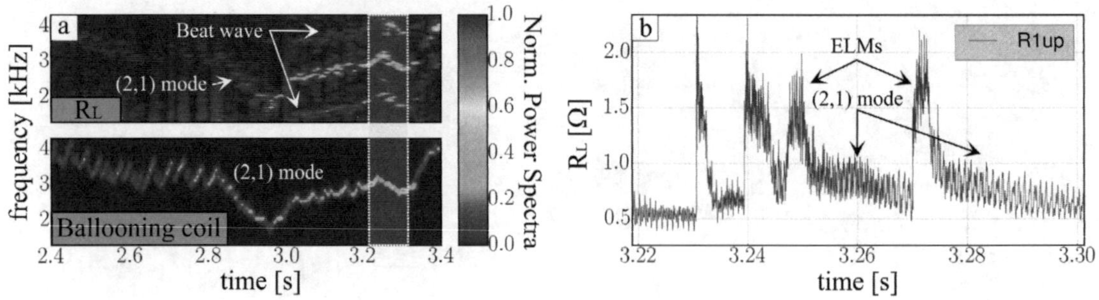


**Figure 5.** Discharge #32749. (a) Spectrograms of the loading resistance measured in the upper feeder of a1 and Ballooning coil for reference. The signature of an MHD mode is clearly visible on both signals. (b) Loading resistance during the highlighted time region.

## 4 Ion cyclotron fast wave coupling under MHD modes

MHD modes can also deform the plasma boundary in a non-axisymmetric fashion. The displacement they produce depends upon their radial profile ($\vec{\xi}(r)$) and growth rates ($\gamma$), supposing that the mode structure can be described as $\vec{\xi}(r,\theta,\phi,t) = \vec{\xi}(r)cos(m\theta - n\phi + wt)e^{\gamma t}$. We present here the case of a coherent modulation of the loading resistance in the ion cyclotron antennas with the rotation of a (2,1) mode. A spectrogram shows the same frequency components for the loading resistance measurements and a Ballooning coil, which registers the change of radial magnetic field due to the mode rotation (**figure 5**).

The change in loading resistance is hypothesized to be connected with a density modification in front of the ICRH antennas, due to the radial profile of the MHD mode. The loading resistance spectrogram in **figure 5** also displays a beat wave pattern from 3 s on. The reason is the activation of the 3-strap antennas, which are shifted by 1 kHz in their operational frequency with respect to the 2-strap antennas. The (2,1) mode numbers are identified via the poloidal and toroidal arrays of Mirnov and Ballooning coils present on the tokamak vessel.

A loading resistance oscillation an order of magnitude faster (few kHz) than the ELM cycle is observed during the appearance of this (2,1) mode, as seen in **figure 5**. For the ASDEX Upgrade case, the hybrid 3dB couplers close to the RF generators assure a steady load-resilient operation.

## 5 Conclusions and outlook

MP fields induce a LFS midplane non-axisymmetric plasma kink response that influences ion cyclotron antennas coupling properties. Rigid rotation of the MP field causes a harmonic oscillation of the antenna loading resistance. In this study, with relatively fast rotating MP fields ($\nu = 6$ Hz) and $I_{coil} = 6.25$ kAt, the loading resistance is found to change as much as $\Delta R_L \sim 0.3\,\Omega$. This loading variation accounts for ~ 20% of the base value. The 3dB hybrid system is able to assure normal operation of the ICRH system. Nevertheless, loading resistance oscillations such as the ones presented in this study need to be taken into account, especially, for future ion cyclotron systems. Higher loading oscillations could be encountered if the plasma displacement becomes larger (i.e. slow rotating MP field with higher coil currents). In future reactors such as ITER, this effect should be considered if MP fields are to be used for ELM control.

MHD modes are found to introduce a coherent loading resistance variation with mode rotation frequency, most likely due to their radial extent. This is illustrated by the specific case of a (2,1) mode. A deeper study is required in order to assess their effects on current and future ion cyclotron systems.

This work has been carried out within the framework of the EUROfusion Consortium and has received funding from the Euratom research and training programme 2014-2018 under grant agreement No 633053. The views and opinions expressed herein do not necessarily reflect those of the European Commission.

## References

[1] K. Steinmetz, Nuclear Fusion **29**, 277 (1989)
[2] R. Bilato, M. Brambilla, D.A. Hartmann, A. Parisot, Nuclear Fusion **45**, L5 (2005)
[3] S.P. Hirshmann et. al., Computer Physics Communications, **43**, 143 - 155, (1986)
[4] W. Suttrop et. al., Phys. Rev. Lett., **106**, 22 - 225004 (2011)
[5] R.A. Moyer et. al., Nuclear Fusion **52**, 123019 (2012)
[6] Y. Liu, A. Kirk, Y. Gribov et. al., Nuclear Fusion **51**, 8-083002 (2011)
[7] M. Willensdorfer et. al., Plasma Phys. Control Fusion **58**, 114004 (2016)
[8] V. Bobkov et al., Nuclear Fusion **53**, 093018 (2013)
[9] M. Rott, U. Seidel, B. Streibl et. al., Fusion Engineering and Design **84**, 1653–1657 (2009)
[10] P. J. Mc Carthy, Plasma Physics and Controlled Fusion, **54**, 015010, (2012)

EPJ Web of Conferences **157**, 03052 (2017) DOI: 10.1051/epjconf/201715703052
*22 Topical Conference on Radio-Frequency Power in Plasmas*

# Time-Dependent Simulations of Fast-Wave Heated High-Non-Inductive-Fraction H-Mode Plasmas in the National Spherical Torus Experiment Upgrade

*Gary* Taylor[1,*], *Nicola* Bertelli[1], *Stefan P.* Gerhardt[1], *Joel C.* Hosea[1], *Dennis* Mueller[1], *Rory J.* Perkins[1], *Francesca M.* Poli[1], *James R.* Wilson[1], and *Roger* Raman[2]

[1]Princeton Plasma Physics Laboratory, Princeton University, Princeton, New Jersey 08543, USA
[2]William E. Boeing Department of Aeronautics and Astronautics, University of Washington, Seattle, Washington 98195, USA

**Abstract.** 30 MHz fast-wave heating may be an effective tool for non-inductively ramping low-current plasmas to a level suitable for initiating up to 12 MW of neutral beam injection on the National Spherical Tokamak Experiment Upgrade (NSTX-U). Previously on NSTX 30 MHz fast wave heating was shown to efficiently and rapidly heat electrons; at the NSTX maximum axial toroidal magnetic field ($B_T(0)$) of 0.55 T, 1.4 MW of 30 MHz heating increased the central electron temperature from 0.2 to 2 keV in 30 ms and generated an H-mode plasma with a non-inductive fraction ($f_{NI}$) ~ 0.7 at a plasma current ($I_p$) of 300 kA. NSTX-U will operate at $B_T(0)$ up to 1 T, with up to 4 MW of 30 MHz power ($P_{rf}$). Predictive TRANSP free boundary transport simulations, using the TORIC full wave spectral code to calculate the fast-wave heating and current drive, have been run for NSTX-U $I_p$ = 300 kA H-mode plasmas. Favorable scaling of $f_{NI}$ with 30 MHz heating power is predicted, with $f_{NI} \geq 1$ for $P_{rf} \geq 2$ MW.

## 1 Introduction

A Fusion Nuclear Science Facility based on the spherical tokamak concept needs to operate with little or no inductive drive from a central solenoid [1]. A long-term research goal of the National Spherical Tokamak Experiment Upgrade (NSTX-U) [2] is to develop fully non-inductive plasmas, and the NSTX-U fast-wave heating system may be an effective tool for non-inductively ramping low-current plasmas to a level suitable for initiating up to 12 MW of neutral beam injection [3]. On NSTX, 30 MHz fast-wave heating efficiently and rapidly heated electrons; at the NSTX maximum axial toroidal field ($B_T(0)$) of 0.55 T, 1.4 MW of fast-wave heating increased the central electron temperature ($T_e(0)$) from 0.2 to 2 keV in 30 ms and generated an H-mode plasma (NSTX shot #138506) with a non-inductive fraction ($f_{NI}$) ~ 0.7 at a plasma current ($I_p$) = 300 kA [4]. The NSTX-U fast-wave heating system will couple up to 4 MW of RF power ($P_{rf}$) and will operate again at 30 MHz, which is the 4$^{th}$ harmonic of deuterium at the maximum NSTX-U $B_T(0)$ of 1 T. Predictive TRANSP free boundary transport simulations [5] have been run for NSTX-U $I_p$ = 300 kA H-mode plasmas in order to study the dependence of $f_{NI}$ on $B_T(0)$ and $P_{rf}$. The TRANSP simulations used the TORIC full wave spectral code [6] to calculate the fast-wave heating and current drive. The transport in the simulations was benchmarked to the experimental results obtained on NSTX. This paper presents the results of these simulations and the implications for using 30 MHz power on NSTX-U to ramp $I_p$ non-inductively.

## 2 Predictive modelling assumptions

The thermal transport model used in the predictive TRANSP simulations was the Multimode MMM7.1 model [7]. This model gave reasonably good agreement to plasma parameters (eg. $f_{NI}$, $T_e(0)$ and plasma pressure) obtained experimentally during NSTX shot #138506 for a NSTX-U simulation with similar $B_T(0)$ and $P_{rf}$. Effective charge, impurity, rotation and other profiles were taken largely from NSTX plasma shot #142305, a plasma that was part of an experimental campaign to support NSTX-U and next-step Spherical Tokamak devices [8].

All the simulations used a $k_{//}$ = 8 m$^{-1}$ antenna phasing, the current drive antenna phasing used for shot #138506. The RF pulse was ramped up to the flat top level in 30 ms, starting at 100 ms. This RF power ramp rate was typical for experiments on NSTX.

In the simulations the electron density profile and central density ($n_e(0)$) were initially chosen to be similar to shot #138506, which had $n_e(0)$ = 1.15x10$^{19}$ m$^{-3}$. The density in the simulation was ramped to this level in 100 ms, at that time a H-mode transition was imposed by flattening the density profile, as shown in figure 1. However when $B_T(0)$ > 0.75 T it was found that when using this relatively low density the plasma became unstable and the simulations terminated prematurely,

* Corresponding author: gtaylor@pppl.gov

typically only ~100 ms after the start of the RF pulse. Increasing $n_e(0)$ to $1.43\text{x}10^{19}$ $\text{m}^{-3}$ resulted in stable plasma simulations up to $P_{rf}$ = 4 MW and $B_T(0)$ = 1 T. Simulations were run for $P_{rf}$ = 1.4, 2, 3 and 4 MW, and for plasmas having $B_T(0)$ = 0.5, 0.65, 0.75, 0.89 and 1 T.

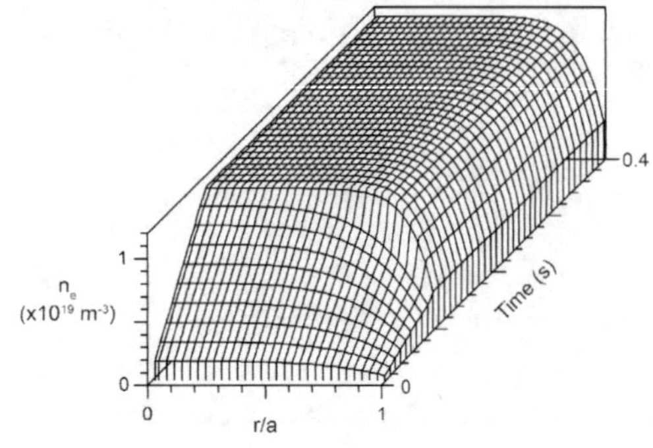


**Fig. 1.** Time evolution of the density profile used in the $n_e(0) = 1.15\text{x}10^{19}\text{m}^{-3}$ TRANSP simulation.

## 3 Results

Figure 2 summarizes the predictive TRANSP simulation results for a NSTX-U plasma with the same $P_{rf}$, $I_p$ and $n_e(0)$ as NSTX shot #138506, and $B_T(0)$ = 0.5 T, compared to $B_T(0)$ = 0.55 T for shot #138506. As shown in figure 2(a), $T_e(0)$ during the simulation was 2.2-2.4 keV, compared to 2.5-3 keV during shot# 138506. $f_{NI}$ during the simulated NSTX-U plasma reached 0.6, compared to a fluctuating $f_{NI}$ = 0.7±0.2 during shot #138506. One notable difference was that the bootstrap current was larger than the current driven directly by RF power in shot #138506. The fluctuation in $f_{NI}$ during shot #138506 is due entirely to large fluctuations in the plasma pressure gradient resulting in fairly rapid changes in bootstrap current. In contrast, in the NSTX-U plasma simulation shown in figure 2(b) the bootstrap current is comparable to the RF-driven current and neither non-inductive currents have significant fluctuations.

The best value of $f_{NI}$ reached during the lower density, $n_e(0) = 1.15\text{x}10^{19}$ $\text{m}^{-3}$ simulations was 1.7. This high value of $f_{NI}$ was achieved for a simulation with $B_T(0)$ = 0.65 T and $P_{rf}$ = 4 MW. The results from this simulation are summarized in figures 3 and 4. Figure 3(a) shows the time evolution of $T_e(0)$ reaches 5 keV. As shown in figure 3(b), the bootstrap current remains constant around 100 kA and the RF-driven current reaches 370 kA at 0.4 s. Figure 4(a) shows $T_e$ at the start of the RF pulse (dashed line) and at 0.4 s (solid line) for the case shown in figure 3. The $T_e$ profile becomes very peaked during the RF pulse (figure 4(a)). Figure 4(b)

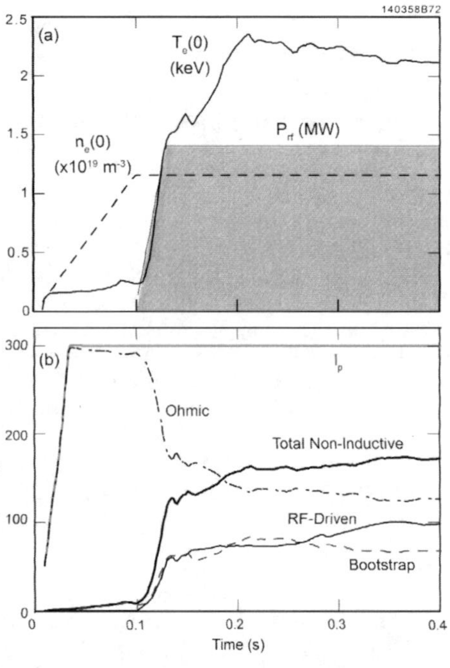


**Fig. 2.** Predictive TRANSP results for an $I_p$ = 300 kA, $B_T(0)$ = 0.5 T NSTX-U plasma with $P_{rf}$ = 1.4 MW. (a) Time evolution of $n_e(0)$, $T_e(0)$ and $P_{rf}$. (b) Time evolution of various currents predicted by TRANSP. The predicted $T_e(0)$ and $f_{NI}$ are similar to NSTX shot #138506.

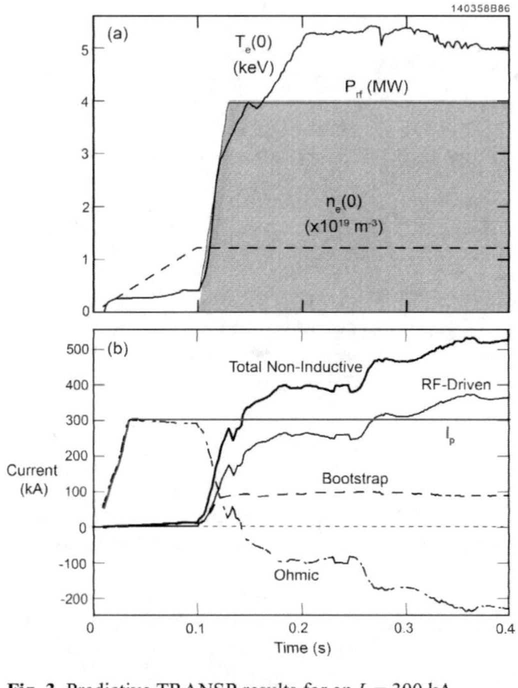


**Fig. 3.** Predictive TRANSP results for an $I_p$ = 300 kA, $B_T(0)$ = 0.65 T NSTX-U plasma with $P_{rf}$ = 4 MW. (a) Time evolution of $n_e(0)$, $T_e(0)$ and $P_{rf}$. (b) Time evolution of various currents predicted by TRANSP.

shows the area integrated current densities at 0.4 s. 90% of the RF-driven current is within $r/a$ = 0.2, while 90% of the bootstrap current is generated within $r/a$ = 0.5. Because the RF heating was strongly peaked near the axis this results in the development of a highly peaked $T_e(R)$ profile (figure 4(a)) which further increases the RF-driven current near the magnetic axis. This peaking

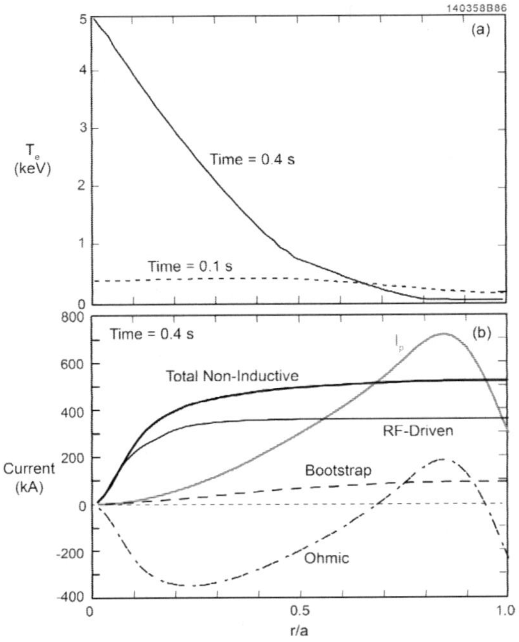


**Fig. 4.** (a) Electron temperature profile at 0.1 s, the start of the RF pulse (dashed line) and at 0.4 s (solid line). (b) Area integrated currents versus normalized minor radius for the case shown in figure 3 at 0.4 s. 90% of the RF-driven current is located inside $r/a$ = 0.3.

of $T_e(0)$ and the RF-driven current at higher $P_{rf}$, particularly at higher $B_T(0)$, caused the plasma simulation to become unstable in the first 100 ms of the RF pulse, prematurely terminating the simulation, as mentioned earlier. This behaviour was also seen in NSTX shot #138506. Although it didn't cause the plasma to disrupt it did result in a magneto-hydrodynamic reconnection in the plasma core that rapidly reduced the stored energy resulting in the plasma moving radially and briefly tripping the RF heating pulse.

All the lower density, $n_e(0)$ = 1.15x10$^{19}$ m$^{-3}$ simulations that didn't terminate by ~ 0.2 s reached a quasi-steady state condition by 0.4 s. Figure 5 shows the dependence of $f_{NI}$ at 0.4 s on $P_{rf}$ for the $n_e(0)$ = 1.15x10$^{19}$ m$^{-3}$ simulations. Also plotted in this figure is the $f_{NI}$ achieved during NSTX shot #138506. At $P_{rf}$ = 1.4 MW all the simulations have an $f_{NI}$ ~ 0.6, independent of $B_T(0)$, and in fairly good agreement with the $f_{NI}$ = 0.7±0.2 achieved experimentally during NSTX shot #138506. Increasing $P_{rf}$ from 1.4 to 4 MW for the $B_T(0)$ = 0.5 T simulation approximately doubled $f_{NI}$. For the $B_T(0)$ = 0.65 T and 0.75 T simulations $f_{NI}$ increased by 2.8 and 2.4, respectively. The $B_T(0)$ = 0.89 T simulation became unstable at ~ 0.2 s when $P_{rf}$ was

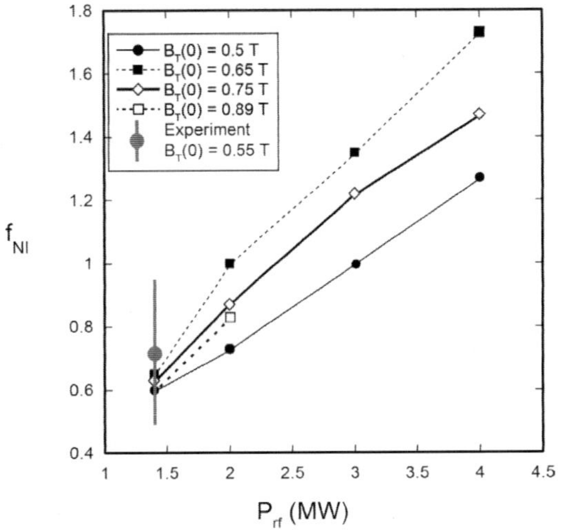


**Fig. 5.** $f_{NI}$ at 0.4 s versus $P_{rf}$ for plasma simulations with $n_e(0)$ = 1.15x10$^{19}$ m$^{-3}$. Also plotted in this figure is the $f_{NI}$ achieved during NSTX shot #138506 (grey filled circle with range during the shot indicated by the vertical bar).

increased above 2 MW and the $B_T(0)$ = 1 T simulations were all unstable early in the RF pulse.

Figure 6 shows the dependence of $f_{NI}$ at 0.4 s on $B_T(0)$ for the $n_e(0)$ = 1.15x10$^{19}$ m$^{-3}$ simulations. Increasing $B_T(0)$ from 0.5 to 0.65 T is predicted to significantly increase $f_{NI}$, particularly at the higher $P_{rf}$. However, $f_{NI}$

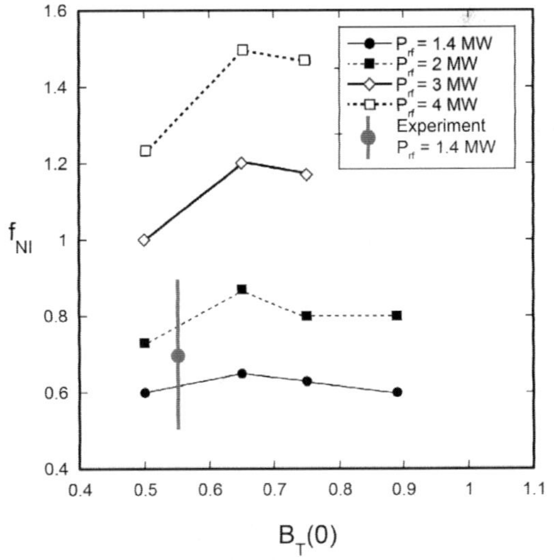


**Fig. 6.** $f_{NI}$ versus $B_T(0)$ for the $n_e(0)$ = 1.15x10$^{19}$ m$^{-3}$ plasma simulations. Also plotted in this figure is the $f_{NI}$ achieved during NSTX shot# 138506 (grey filled circle with range during the shot indicated by the vertical bar).

decreases above $B_T(0) = 0.65$ T at all values of $P_{rf}$.

For the $n_e(0) = 1.15\text{x}10^{19}$ m$^{-3}$ simulations fully non-inductive operation is predicted with as little as 2 MW of RF power for a $B_T(0) = 0.65$ T plasma,

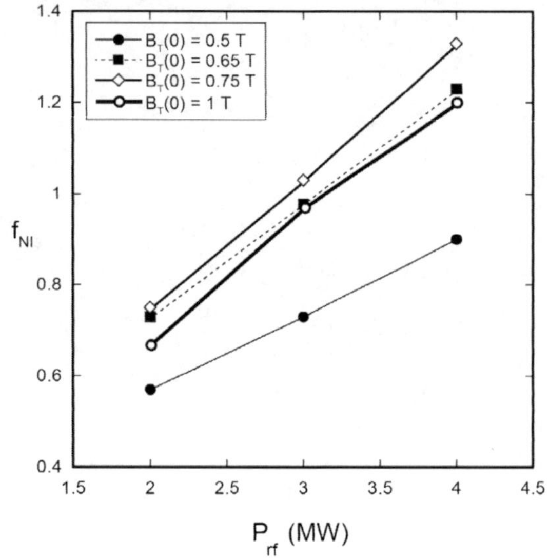


**Fig. 7.** $f_{NI}$ versus $P_{rf}$ for the plasma simulations with $n_e(0) = 1.43\text{x}10^{19}$ m$^{-3}$.

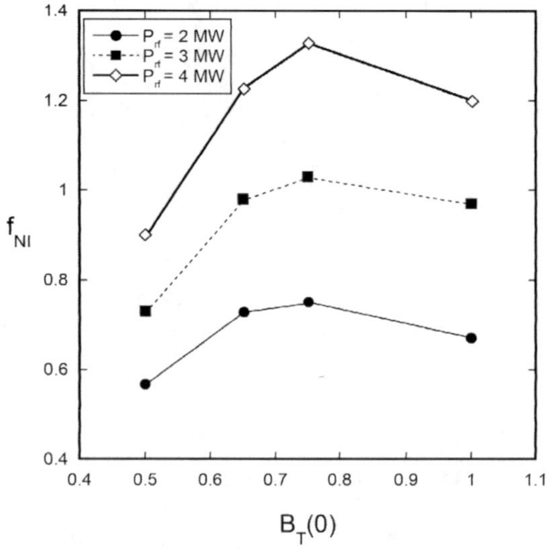


**Fig. 8.** $f_{NI}$ versus $B_T(0)$ for the plasma simulations with $n_e(0) = 1.43\text{x}10^{19}$ m$^{-3}$.

As mentioned earlier, increasing $n_e(0)$ to $1.43 \text{ x } 10^{19}$ m$^{-3}$ provided stable plasma simulations up to $P_{rf} = 4$ MW at $B_T(0) = 1$ T. Figure 7 plots $f_{NI}$ at 0.4 s versus $P_{rf}$ for this slightly higher density case. $f_{NI} \geq 1$ is predicted for $P_{rf} \geq 3$ MW for the plasma simulations with $B_T(0) = 0.75$ T. At $P_{rf} = 4$ MW all the plasma simulations with $B_T(0) \geq 0.65$ predict fully non-inductive operation. Increasing $P_{rf}$ from 2 to 4 MW increased $f_{NI}$ by 1.6, 1.7, 1.8 and 1.4 for the 0.5 T, 0.65 T, 0.75 T and 1 T simulations, respectively. As expected, the somewhat higher density reduced the $f_{NI}$ achieved at a given $P_{rf}$ and $B_T(0)$ compared to the lower density simulations.

Figure 8 plots $f_{NI}$ at 0.4 s versus $B_T(0)$ for the $n_e(0) = 1.43 \text{ x } 10^{19}$ m$^{-3}$ simulations. While $f_{NI}$ increases significantly when $B_T(0)$ is raised from 0.5 T to 0.65 T, $f_{NI}$ decreases when $B_T(0)$ increases from 0.75 to 1 T, as predicted for the $n_e(0) = 1.15 \text{ x } 10^{19}$ m$^{-3}$ simulations.

## 4 Conclusions

The TRANSP simulation results presented here strongly support the possibility of achieving stable $f_{NI} \geq 1$ plasmas with $P_{rf} = 2\text{-}3$ MW in an $I_p = 300$ kA NSTX-U plasma. However while increasing $P_{rf}$ to higher levels than were achieved on NSTX shot #138506 is predicted to increase $f_{NI}$ the simulations also predict that the plasma may become unstable as $P_{rf}$ is increased, particularly at the highest $B_T(0)$ values achievable on NSTX-U. The simulations predict that increasing $B_T(0)$ to 0.65 T from 0.55 T, the value used in NSTX experiment, may significantly increase the achievable $f_{NI}$. One caveat is that operationally it was difficult to control $I_p = 300$ kA NSTX plasmas when fast wave power was coupled into the $I_p$ flattop, even at $P_{rf} = 1.4$ MW. Coupling fast wave power at even higher levels will no doubt be challenging. But even more challenging will be coupling fast wave power immediately after start up in order to ramp $I_p$ up to 300-400 kA. These TRANSP simulation predictions must now await experimental validation in NSTX-U.

This work was supported by United States Department of Energy contract no. DE-AC02-09CH11466.

## References

1. Y.-K.M. Peng, et al., Fusion Sci. Technol. **56**, 957 (2009)
2. J. E. Menard, et al., Nucl. Fusion **52**, 083015 (2012)
3. F. M. Poli, et al., Nucl. Fusion **55**, 123011 (2015)
4. G. Taylor, et al., Phys. Plasmas **19**, 042501 (2012)
5. R. Andrei, Bull. Am. Phys. Soc. **57**, 12 (2012)
6. M. Brambilla, Plasma Phys. Control. Fusion **44**, 2423 (2002)
7. T. Rafiq, et al., Phys. Plasmas **20**, 032506 (2013)
8. S. P. Gerhardt, et al., Nucl. Fusion **51**, 073031 (2011)

DOI: 10.1051/epjconf/201715703053

# 3-Dimensional density profiles in edge plasma simulations for ICRF heating

*Wouter* Tierens[1,*], *Jonathan* Jacquot[1], *Wei* Zhang[1,2], *Jean-Marie* Noterdaeme[1,2], *Volodymyr* Bobkov[1], *Laurent* Colas[3], and The ASDEX Upgrade team[1]

[1] *Max-Planck-Institut für Plasmaphysik, Boltzmannstrasse 2, D-85748 Garching, Germany*
[2] *Ghent University, Sint-Pietersnieuwstraat 41, Gent, Belgium*
[3] *CEA, IRFM, F-13108 Saint Paul-lez-Durance, France*

**Abstract.** In this paper we discuss improvements made to two codes for the simulation of ICRF waves in edge plasmas: SSWICH-SW, which self-consistently models the interplay between sheath physics and radiofrequency waves (the slow wave), and RAPLICASOL, a Finite Element solver for Maxwell's equations in the cold plasma approximation. We have extended both to be able to handle 3D plasma density profiles. A comparison between a 1D and a 3D simulation reveals that the density profile dimensionality has a relatively small effect on $E_{\parallel}$ at the aperture, but a large effect on the sheath potential at the antenna limiters

## 1 Introduction

It is important to understand the behaviour of Ion Cyclotron Range of Frequencies (ICRF) waves in edge plasmas and near the antenna, in order to understand the physical mechanisms underlying experimental observations such as ion sputtering [1, 2], RF induced convection [3], and variation in power coupling to the plasma.

Several codes are commonly used to numerically simulate these electromagnetic waves [4, 5]. The vast majority, or perhaps all, of these codes until now use a 1D density profile, which depends only on the radial coordinate.

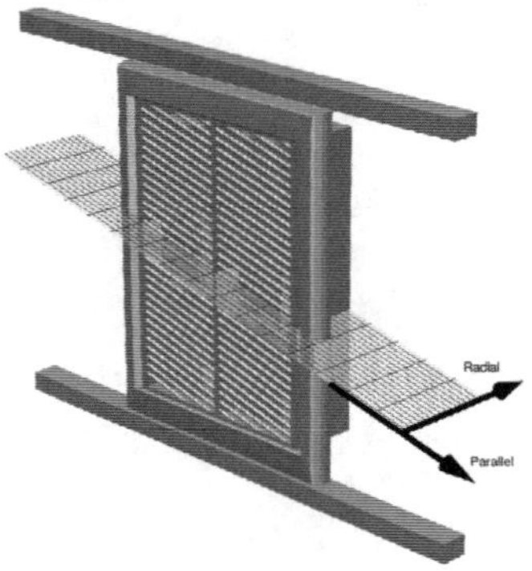


**Figure 1.** RAPLICASOL antenna geometry for the ASDEX Upgrade 2-strap antenna. A 2D radial-parallel slice used by the SSWICH-SW code is also shown. The tilt of this slice corresponds to the tilt of the background magnetic field: the SSWICH-SW slice follows the field lines.

In this paper we discuss two codes in particular:

- RAPLICASOL (Radiofrequency wAve couPLing for Ion Cyclotron. Antenna in Scrape-Off-Layer [6]) to solve Maxwell's equations in cold plasma.
- SSWICH-SW (Self-consistent Sheaths and Waves for Ion Cyclotron Heating (Slow Wave) [7–9]) to self-consistently determine the nonlinear plasma sheath potential and related quantities.

Both codes model physics in Cartesian coordinates, which neglect toroidal and poloidal curvature, but are nonetheless spatial coordinates, as opposed to codes such as TOPICA which operate in Fourier $\vec{k}$ space. Because the cold plasma dielectric tensor depends on quantities that depend on the spatial coordinates, it is relatively straightforward to extend them to handle 3D density profiles (or in fact 3D variation of other physical quantities such as the background magnetic field or the electron and ion temperature).

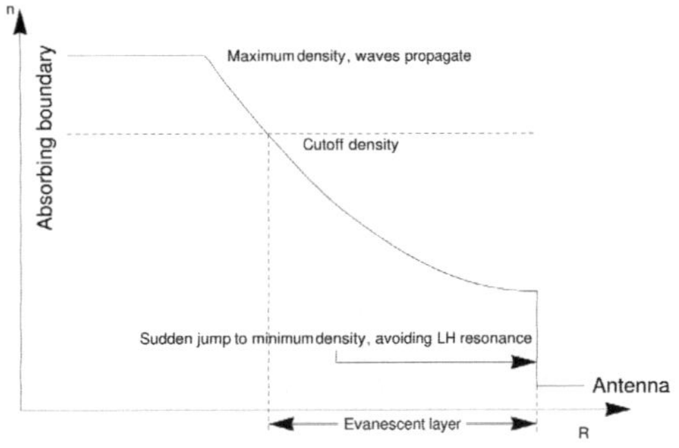


**Figure 2.** Summary of the density profile in RAPLICASOL.

*e-mail: wtt@ipp.mpg.de

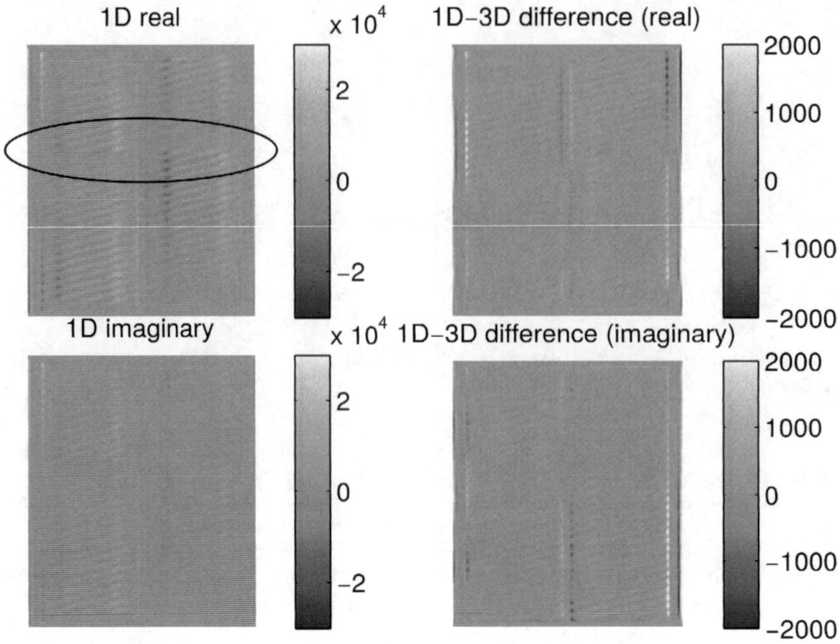


**Figure 3.** Comparison between $E_{\parallel}$ field maps $(V/m)$ at the aperture of ASDEX Upgrade antenna 1 (a 2-strap antenna), calculated by RAPLICASOL with 1D resp. 3D density profiles. The left column contains $\Re(E_{\parallel,1D})$ (top) and $\Im(E_{\parallel,1D})$ (bottom). The right column contains the absolute difference between the field maps for the 1D and 3D density profiles: $\Re(E_{\parallel,1D} - E_{\parallel,3D})$ and $\Im(E_{\parallel,1D} - E_{\parallel,3D})$. The apparent horizontal discontinuity in $E_{\parallel,1D}$, indicated with an ellipse, is due to the way the straps are folded.

## 2 RAPLICASOL

RAPLICASOL is a 3D finite-element solver based on COMSOL, which solves Maxwell's equations in the cold plasma approximation in the neighbourhood of a realistic 3D antenna geometry as shown in figure 1. RAPLICASOL does not model the core plasma: the simulation region is terminated by a layer of absorbing material (a so-called Perfectly Matched Layer or PML [11], adapted for cold magnetized plasma). By limitting the size of the simulation region, the PMLs also reduce the computational cost (RAM and CPU time) of the simulations.

There are two issues with allowing a completely arbitrary 3D density profile in RAPLICASOL. First, we should avoid the lower-hybrid resonance, which we do by imposing a discontinuous jump from vacuum to a certain minimum density (in SSWICH, we do not have to do this). Second, the density should be constant along the PML interface, which we ensure by imposing a maximum density. Although coupling and reflection depend on the radial gradient of the density, this maximum density does not have a large effect on the reflected waves, if it is chosen large enough for the waves to be propagating at that density (i.e. it should be larger than the cutoff density for all toroidal $k$ modes of interest). All this is summarized in figure 2.

## 3 SSWICH-SW

The SSWICH-SW code [7–9] self-consistently determines the nonlinear plasma sheath potential. Unlike RAPLICASOL, it is essentially 2D: it operates on consecutive 2D radial-parallel slices, as shown in figure 1, and neglects derivatives in the vertical direction. SSWICH-SW imports an $E_{\parallel}$ field map at the antenna aperture from RAPLICASOL, which excites the slow wave and the sheaths. SSWICH-SW solves three coupled fluid equations:

- The equation for the parallel electric field of the slow wave (SSWICH-SW models only the slow wave)

$$\left(\epsilon_{\parallel}\Delta_{\parallel} + \epsilon_{\perp}\Delta_{\perp} + \epsilon_{\parallel}\epsilon_{\perp}\frac{\omega_0}{c^2}\right)E_{\parallel} = 0 \qquad (1)$$

  where $\epsilon_{\parallel}, \epsilon_{\perp}$ are elements of the Stix tensor, $\Delta_{\parallel}, \Delta_{\perp}$ are the parallel resp. perpendicular terms in the laplacian, and $\omega_0$ is the antenna frequency. In SSWICH-SW's asymptotic mode [8], which we use in this paper, the boundary condition at the limiter surfaces is $E_{\parallel} = 0$.

- The radiofrequency potential $V_{RF}$ at the limiters

$$\epsilon_{\perp}\Delta_{\perp}V_{RF} = \pm\epsilon_{\parallel}\partial_{\parallel}E_{\parallel} \qquad (2)$$

- The DC current, given by a parallel and a perpencidular conductivity, is source-free in the bulk:

$$\nabla \cdot J_{DC} = (\sigma_{\parallel}\Delta_{\parallel} + \sigma_{\perp}\Delta_{\perp})\,V_{DC} = 0 \qquad (3)$$

The sheath boundary condition couples (2) and (3)

$$\begin{aligned} J_{DC\parallel} &= -\sigma_{\parallel}\nabla_{\parallel}V_{DC} \\ &= i^{+}\left(1 - \exp\left(\frac{V_f + V_b - V_{DC}}{T_e}\right)\right) \end{aligned} \qquad (4)$$

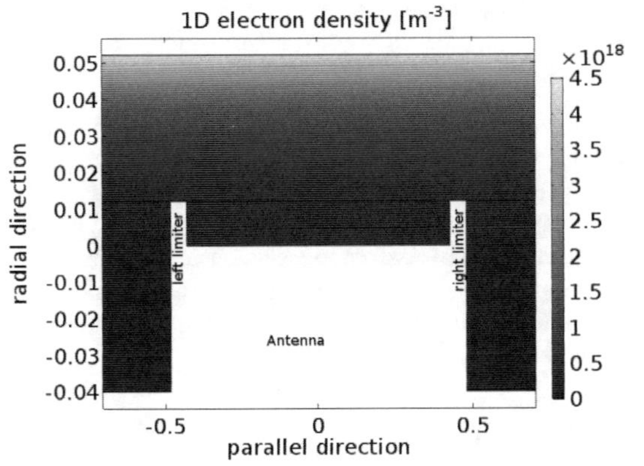


**Figure 4.** 1D and 3D electron density plotted along a radial-parallel SSWICH-SW slice at vertical position 0m. The density variation in the parallel direction is mostly due to poloidal, rather than toroidal, variation of the actual density profile, because the parallel direction is slanted.

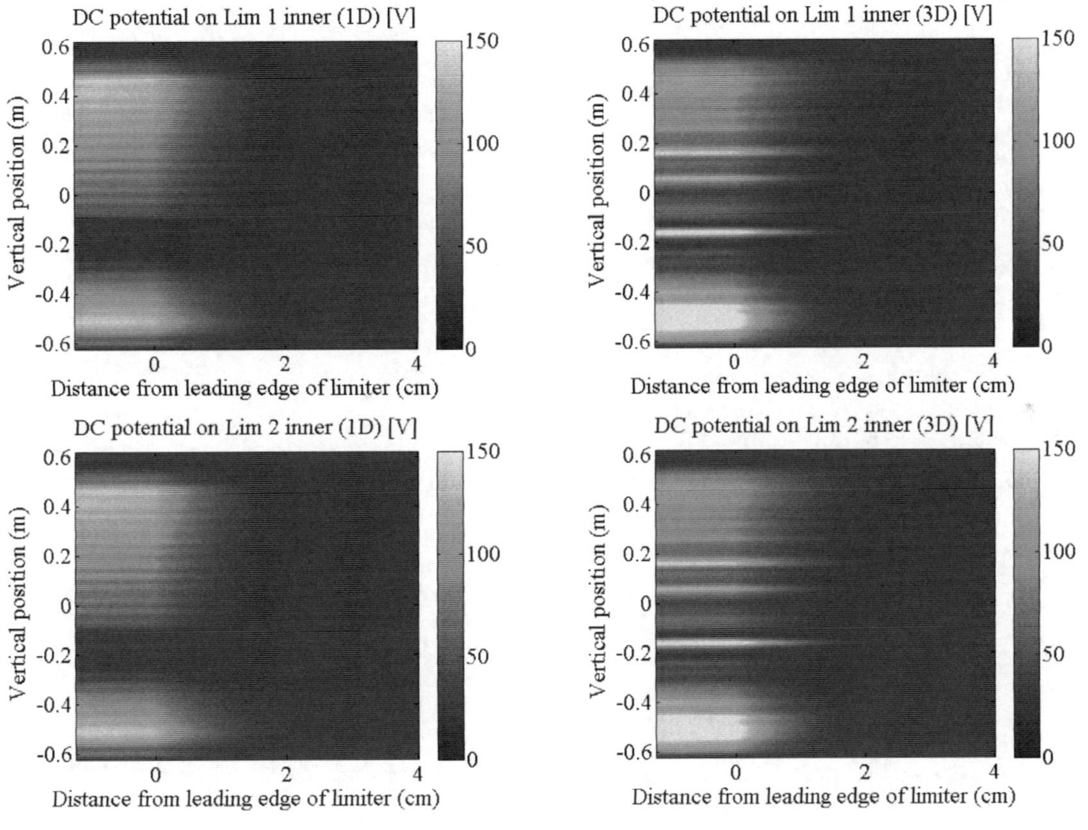


**Figure 5.** DC potential on the inside of limiters 1 (left limiter) and 2 (right limiter) vs. radial and vertical coordinate, for 1D (left) and 3D (right) density profiles.

where $i^+$ is the local ion saturation current, $V_f$ is the floating potential and $V_b$ is the biasing voltage which is a function of $V_{RF}$. In practice, (4) is limited by a saturation condition when the electron saturation current is reached. For further details see [7, 12].

These equations are density-dependent through the Stix tensor elements $\epsilon_{\parallel}$ and $\epsilon_{\perp}$, but not through the DC conductivity $\sigma_{\parallel}$ and $\sigma_{\perp}$, which are assumed to be proportional to the Spitzer conductivity (in reality, perpendicular DC currents are likely due to turbulence. The description in terms of a conductivity is purely phenomenological).

## 4 Comparison between simulation results with 1D and 3D density

In what follows we will use 3D density data determined using EMC3-Eirene for the "Mid-3" gas-puffing case in ASDEX Upgrade [13]. Let us call this $n_{\text{Mid3}}(R, \theta, z)$. We will compare it with results obtained with the 1D density profile obtained by holding $\theta$ and $z$ constant: $n_{\text{Mid3}}(R, \theta_0, z_0)$, where $\theta_0$ and $z_0$ are the coordinates of the center of the antenna.

In figure 3, we compare RAPLICASOL results with the 1D and 3D density profiles. The difference in $E_{\parallel}$ field strength is the largest close to the right limiter, with a maximum difference in field strength of about $6kV/m$. This difference in field strength is comparable to the typical field strength in either the 1D or the 3D case, although it is about an order of magnitude smaller than the maximum field strength. From the point of view of SSWICH-SW, which uses these field maps as input, the most important differences are those at or near the limiters [9]. The per-port coupling resistance is calculated from the standing wave ratio at the ports with dipole phasing and 1MW of coupled power. In the 3D case it is $[2.08\Omega, 2.20\Omega]$ and in the 1D case $[2.14\Omega, 2.18\Omega]$, a difference of a few %.

The SSWICH-SW code, as discussed before, operates on radial-parallel 2D slices (with the "parallel" direction being parallel to the background magnetic field). We have plotted the 1D and 3D density profile on one such slice in figure 4. Note that there is a poloidal variation as well: in addition to depending on the parallel coordinate as seen in figure 4, the 3D density also changes from slice to slice.

SSWICH-SW calculates the nonlinear plasma DC potential along the limiters, and outputs 2D radial-vertical $V_{DC}$ maps on the inner and outer edge of each limiter (in SSWICH-SW, the limiter geometry is somewhat simplified: walls are either parallel or perpendicular to the magnetic field. This is necesary to satisfy the assumption that there is only slow wave propagation). It is these $V_{DC}$ maps which are shown figure 5. The short-wavelength vertical oscillations which appear even in the 1D case are due to the Faraday screen (the oscillations change when we change the number of Faraday screen bars, and disappear alltogether if there is no Faraday screen). The 3D density profile has a much larger effect on the DC potentials than on the $E_{\parallel}$ field maps. The increase in $V_{DC}$ at the lower half of the antenna in the 3D case is likely due to the density which is lower there in the 3D density profile than in the 1D density profile.

## 5 Conclusion

We have extended the SSWICH-SW and RAPLICASOL codes to be able to handle 3D plasma density profiles. A comparison between results with 1D and 3D density profiles, where the 3D density profile was obtained using EMC3-Eirene and data from an ASDEX Upgrade gas puffing experiment, reveals that the change in density profile (at least for the specific parameters considered here) has a relatively small effect on the $E_{\parallel}$ field map at the aperture and on the coupling resistance, but a large effect on the sheath potential at the antenna limiters.

## References

[1] V. Bobkov, D. Aguiam, R. Bilato, S. Brezinsek, L. Colas, H. Faugel, H. Fünfgelder, A. Herrmann, J. Jacquot, A. Kallenbach et al., Plasma Physics and Controlled Fusion **59**, 014022 (2016)

[2] V. Bobkov, M. Balden, R. Bilato, F. Braun, R. Dux, A. Herrmann, H. Faugel, H. Fünfgelder, L. Giannone, A. Kallenbach et al., Nuclear Fusion **53**, 093018 (2013)

[3] W. Zhang, W. Tierens, J. Noterdaeme, V. Bobkov, D. Aguiam, D. Coster, H. Fünfgelder, J. Jacquot, R. Ochoukov, A. Silva et al., Submitted to Nuclear Fusion (2017)

[4] V. Lancellotti, D. Milanesio, R. Maggiora, G. Vecchi, V. Kyrytsya, Nuclear Fusion **46**, S476 (2006)

[5] M. Brambilla, Plasma Physics and Controlled Fusion **35**, A141 (1993)

[6] J. Jacquot, V. Bobkov, L. Colas, S. Heuraux, A. Křivská, L. Lu, J.M. Noterdaeme, R.I. Pinsker, *Full wave propagation modelling in view to integrated ICRH wave coupling/RF sheaths modelling*, in *AIP Conference Proceedings* (AIP Publishing, 2015), Vol. 1689, p. 050008

[7] J. Jacquot, Ph.D. thesis, Université de Lorraine (2013)

[8] L. Colas, J. Jacquot, S. Heuraux, E. Faudot, K. Crombé, V. Kyrytsya, J. Hillairet, M. Goniche, Physics of Plasmas **19**, 092505 (2012)

[9] L. Colas, L. Lu, A. Křivská, J. Jacquot, J. Hillairet, W. Helou, M. Goniche, S. Heuraux, E. Faudot, Plasma Physics and Controlled Fusion **59**, 025014 (2017)

[10] W. Zhang, V. Bobkov, J. Noterdaeme (2017), this conference

[11] J.P. Berenger, Journal of computational physics **114**, 185 (1994)

[12] J. Jacquot, D. Milanesio, L. Colas, Y. Corre, M. Goniche, J. Gunn, S. Heuraux, M. Kubič, Physics of Plasmas **21**, 061509 (2014)

[13] W. Zhang, V. Bobkov, J.M. Noterdaeme, W. Tierens, R. Bilato, D. Carralero, D. Coster, J. Jacquot, P. Jacquet, T. Lunt et al., Plasma Physics and Controlled Fusion **59**, 075004 (2017)

DOI: 10.1051/epjconf/201715703054

# Modelling RF-plasma interaction in ECR ion sources

David *Mascali*[1], Giuseppe *Torrisi*[1,*], Alessio *Galatà*[2], Gino *Sorbello*[1,3], Giuseppe *Castro*[1], Luigi *Celona*[1], Lorenzo *Lega*[1], Ornella *Leonardi*[1], Maria *Mazzaglia*[1], Eugenia *Naselli*[1,4], Lorenzo *Neri*[1], and Santo *Gammino*[1]

[1]Laboratori Nazionali Del Sud, INFN, via S. Sofia 62, Catania, 95123, Italy
[2]Laboratori Nazionali di Legnaro, INFN, Viale dell'Università 2, 35020 Legnaro (Padova), Italy
[3]Università degli studi di Catania, Viale Andrea Doria 6, Catania, 95125, Italy
[4]Dipartimento di Fisica e Astronomia, Università of Catania,Via S. Sofia, 64 I-95123 Catania

**Abstract.** This paper describes three-dimensional self-consistent numerical simulations of wave propagation in magnetoplasmas of Electron cyclotron resonance ion sources (ECRIS). Numerical results can give useful information on the distribution of the absorbed RF power and/or efficiency of RF heating, especially in the case of alternative schemes such as mode-conversion based heating scenarios. Ray-tracing approximation is allowed only for small wavelength compared to the system scale lengths: as a consequence, full-wave solutions of Maxwell-Vlasov equation must be taken into account in compact and strongly inhomogeneous ECRIS plasmas. This contribution presents a multi-scale temporal domains approach for simultaneously including RF dynamics and plasma kinetics in a "cold-plasma", and some perspectives for "hot-plasma" implementation. The presented results rely with the attempt to establish a modal-conversion scenario of OXB-type in double frequency heating inside an ECRIS testbench.

## 1 Introduction

Three-dimensional self-consistent numerical simulation of full-wave RF fields in inhomogeneous non-uniformly magnetized ECRIS plasma is a big challenge. Simulations are necessary to improve the comprehension of microwave-to-plasma coupling. Numerical results about electron heating, ionization, ion confinement and ion beam formation in an ECRIS can be very useful in order to increase the performances in terms of output beams currents, charge states, beam quality (emittance minimization, beam halos suppression, etc.) .

Several full-wave numerical simulation codes have been developed to describe waves in fusion toroidal devices [1-3], exploiting symmetry or ray-tracing technique; at the same time, only few methods have been developed to obtain selfconsistent solutions for non-axisymmetric ECRIS plasma [4-6].

In ECRIS devices (having magnetic field of the order of few Tesla, electron density in the order of $10^{18}$ $m^{-3}$ and frequencies in the range 3-40 GHz) the Wentzel, Kramers and Brillouin (WKB) method is not applicable because the wavelength is comparable with the plasma gradient and dielectric tensor scale lengths. Only full-wave methods are suitable to capture waves' propagation features, strongly affected by multiple reflections at the cavity-plasma chamber.

During the last years, several numerical simulations of the microwave heating process in ECR plasmas has been carried out at INFN-LNS and LNL [7-9]. They allowed to give a reasonable explanation of the "Frequency Tuning Effect" affecting the ECRIS plasmas as depending on the excited modes into the plasma chamber.

However, a self-consistent calculation of wave fields and plasma distribution requires a closed loop joining three physics models: Wave-solver, Kinetic solver and collisions. Previous approaches to simulate RF fields in ECRIS using the conductivity kernel of 1-D mirror magnetic field are described in [6].

In this paper we present numerical results of a set of 3D simulations based on the quasi-self consistent approach described in [10, 11], including RF wave propagation in a "cold-plasma" solved by a finite element (FEM) method, and a single species (electrons) kinetic code. Our code-suite conjugates FEM solutions of Maxwell equations including the resonator (metallic plasma chamber) effects and the "cold" dielectric tensor of the plasma, with a "particle mover", i.e. a kinetic code developed in MATLAB for solving the single particle relativistic equation of motion of thousands of electrons according to the Boris method [12].

A criticality in microwave-heated magnetoplasmas occurs near the resonance regions, where the plasma becomes a spatial dispersive medium, and the hot magnetized plasmas response has to be considered. In this case, the dielectric tensor depends on the wave vector and the wave solution needs to be calculated by means of iterations. In our FEM code it is proposed a method to take into account the hot plasma response in this paper.

* Corresponding author: giuseppe.torrisi@lns.infn.it

 DOI: 10.1051/epjconf/201715703054

## 2 Coupled Electromagnetic and Particle motion model

Assuming a quasi-stationary equilibrium, particles can be followed until they "die" (i.e., they impinge on the chamber walls), while their trajectories are stored in a 3D mesh which finally produces a charge density distribution: this distribution is then included in the following calculation step to compute the electromagnetic field distribution.
In a magnetized plasma modelled as a cold magneto-fluid with collisions, the field-plasma interaction is described by the tensorial relation **ε·E**. Hereby we will apply the "cold plasma" approximation, (i.e. $V_{\square} >> V_{th}$, being $V_{\square}$ the wave's phase speed and $V_{th}$ the electron thermal speed). In the case of an ECRIS, due to the shape of its magneto-static structure lacking of uniformity and axial symmetry, **ε** depends in a complex way from the magneto-static field and the local electron density $n_e(x,y,z)$.
Full-wave simulations were carried out through the COMSOL Multiphysics FEM solver, considering the geometry shown in Fig. 1 with an initial electron density $n_e(x,y,z)$ and the above described "cold plasma" model. The simulations reproduce the main features of the Flexible Plasma Trap that recently came into operations at INFN-LNS [13]. The Flexible Plasma Trap (FPT allows axial and radial launching of the waves and it is based on a simple-mirror configuration. The magnetic field is however flexible, allowing "flat" and "beach" magnetic configurations.
Since in proximity of the resonance surface (individuated from the iso-surface $\mathbf{B_0}=B_{ECR}=m_e/e\omega$) the permittivity varies strongly, the discretization of such a narrow region needed a very fine mesh: to achieve this, we adopted an adaptive mesh procedure (that was allowed by a specific feature of the solver). The obtained solution for the electromagnetic fields {E(r), H(r)} was then coupled to the 3D kinetic Matlab® code, initially developed for particles-in-cell calculations of ion dynamics in a plasma. Such a code describes the motion of N macro-particles (being N much smaller than real plasma particles), calculating 3D spatial coordinates and velocities described by the functions $r_i(t)$ and $v_i(t)$ for $i^{th}$ particle. The calculations take into account only electrons motion, including the so called electro-static (Spitzer) 90° collisions and the Lorentz force:

$$d\boldsymbol{v}_i(t)/dt=q/m[\boldsymbol{E}(\boldsymbol{r}_i(t),t)+\boldsymbol{v}_i(t)\times\boldsymbol{B_0}(\boldsymbol{r}_i(t),t)] \quad (1)$$

The kinetic code solves equation (2) for each simulated particle applying the relativistic Boris scheme [10, 14]: the calculation stops as soon as all the particles hit the chamber walls, while at each time step particles positions are stored in a matrix forming a 3D density map. This map is the output of the kinetic code, and, once rescaled to real values, is used as local electron density $n_e(x,y,z)$ to compute again the electromagnetic field {E(r), H(r)}. This "self-consistent loop" runs iteratively, until a convergence is reached.
Hereby we show results up to step 1, i.e. after the evaluation of the magnetized plasma action on RF and viceversa.

### 2.1 Full-Wave simulation strategy

The following calculation uses FPT as simulation domain in order to explore the possibility to obtain overdense plasmas, generating electron Bernstein waves (EBW) by mode-conversion, from the incident O-mode (via OXB conversion) or directly through an X-mode.
The presence of the resonant cavity causes a polarization scrambling, thus strongly complicating the OXB modeling: this makes unpractical some straightforward solutions, already adopted for fusion plasma applications [15]. We hereinafter describe our approach:

1) *Preparatory virtual experiment*: we assume $n_e$ and B profiles, then verifying the consistency of the assumed scenarios with modal conversion constraints;
2) *Numerical simulations*: they allows to check step-1, allowing to find cut-offs and resonances displacements into the plasma; the full-wave FEM code has been applied to study mode conversion mechanism features of O-X-conversion in FPT;
3) Starting from point 2) *a microwave launcher* has been designed, suitable for modal-conversion in FPT.

FPT has a cylindrical plasma chamber 260 mm long and 41 mm in radius, fed by microwaves in the range 4-7 GHz generated by a TWT, while the magnetic field is obtained by means of three coils.

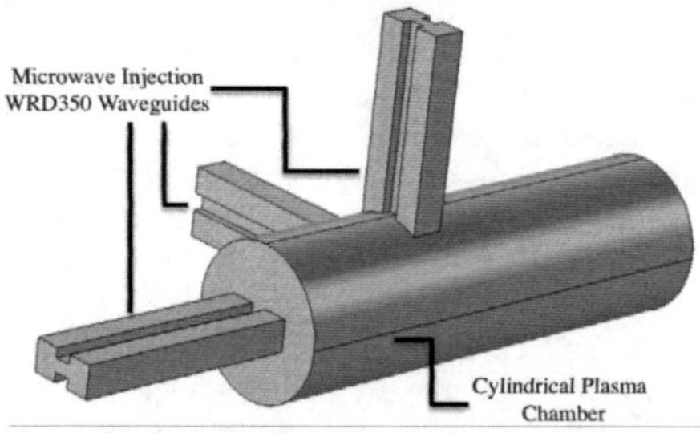


**Fig. 1** Simulated Geometry: Cylindrical plasma chamber cavity and microwave waveguide injections (axial and radial ports).

Fig. 2 and 3 display the electron density $n_e$ and magnetostatic field $B_0$ , used as input parameters for the simulations. In particular, we considered two magnetic field profiles (Fig. 2) as input parameters to compute the "cold" permittivity tensor for 3D full-wave simulations:

1. *Simple mirror configuration with double ECR: Single-ECR configuration.*
2. *"Simple Mirror- single ECR" configuration.*

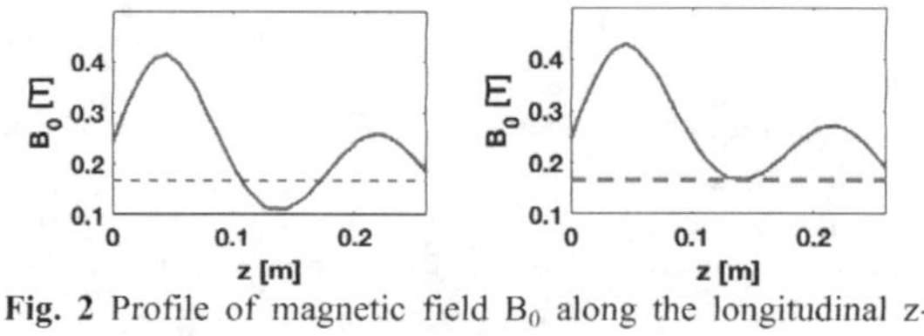


**Fig. 2** Profile of magnetic field $B_0$ along the longitudinal z-

DOI: 10.1051/epjconf/201715703054

axis. (x = 0, y = 0, z): on the left the *"Simple mirror – double ECR"* configuration; on the right the *"Simple Mirror- single ECR"* configuration. The horizontal red lines show the ECR magnetic field value.

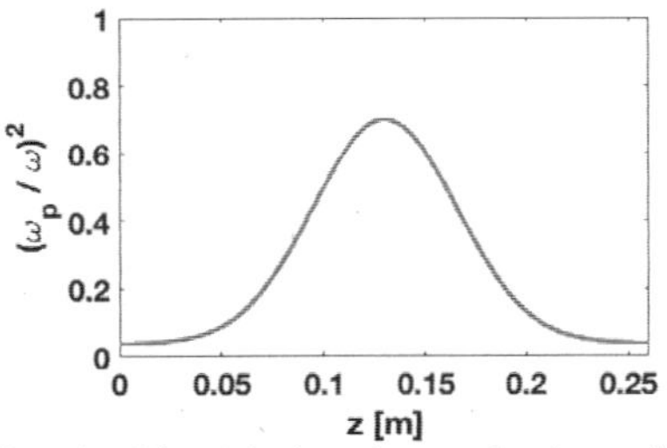


**Fig. 3** Gaussian Dimensionless electron density profile along longitudinal z-axis used as input parameters to compute the cut-off and resonance zones.

The virtual experiment consists in assuming the desired density profile, then selfconsistently checking that the resonance/cut-off layers displacements are compatible with the assumed profile. If the desired profile is a single peaked gaussian with maximum in the central part of the plasma chamber (because of energy transport issues, center-peaked profile is greatly preferable in order to have a hot plasma core), the compatibility check consists in verifying that the energy deposition, occurring at the ECR harmonics, is placed in proximity of the assumed density peak. The cut-off layers where the conversion may happen, as well as the resonance layers for absorption of the electromagnetic waves, can be predicted thanks to our full-wave code as it can be seen in Fig. 4 for the *"Simple Mirror- single ECR" configuration:* in this figure the chacteristic cutoffs and resonances frequencies $\omega_{Ch}$ are plotted along the longitudinal z-axis considering the electron cyclotron frequency $\omega_c$, the plasma frequency $\omega_p$, the Upper Hybrid Resonance $\omega_{UHR}$ and the L and R-cutoffs $\omega_L$ and $\omega_R$. It could be controlled by proper adjustment of the magnetic induction inside the plasma during the experimental phase. The self-consistency check has been performed by means of these type of plots (like Fig. 4). It seems the reciprocal displacement of cutoffs and resonances is suitable for RF power conversion through OXB mechanism.

## 3 Numerical Results and Discussion

The following step consisted in the full-wave simulation of the field structure into the plasma chamber. Then, the full system of Maxwell's equations, including the cold magnetized plasma already described in section 2, has been solved for the geometry of the FPT plasma chamber.

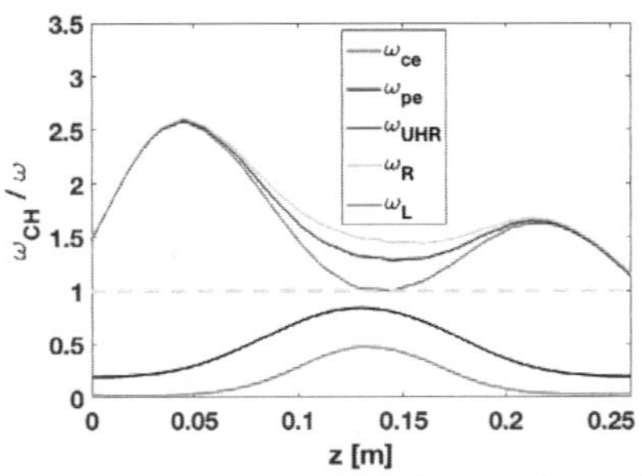


**Fig. 4** Cutoffs and resonances in FPT linear plasma along 1D direction, assuming the density of Fig. 3 and the case 2 magnetic profile for frequency normalized to heating wave angular frequency 4.65 GHz.

We performed a mesh refinement study, exploiting the adaptive mesh tool based on an error estimate during the solution stage. An extremely fine mesh on the ECR and UHR surface has been obtained, thanks to "functional evaluation" based on the electric field gradient.

Figures 5 and 6 illustrate the COMSOL outputs for electric field distribution in the two magnetic fields cases: for the *Simple mirror configuration* the electric field has a strong gradient near the ECR layer, that corresponds to resonance-induced wavelength shortening. The electric field strength inside the cavity appears enhanced by the presence of the plasma and, as expected, the largest fraction of input power is absorbed at ECR.

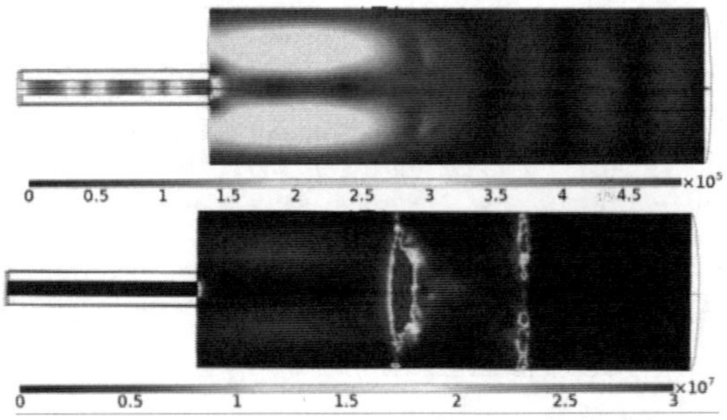


**Fig. 5** *Simple mirror magnetic configuration 1 of* Simulated electric field |E| [V/m] (on the Top) distribution and dissipated Power $P_d$ [W/m^-3] (on the bottom) in proper colours scales range that emphasize the strong gradient near the resonance layers.

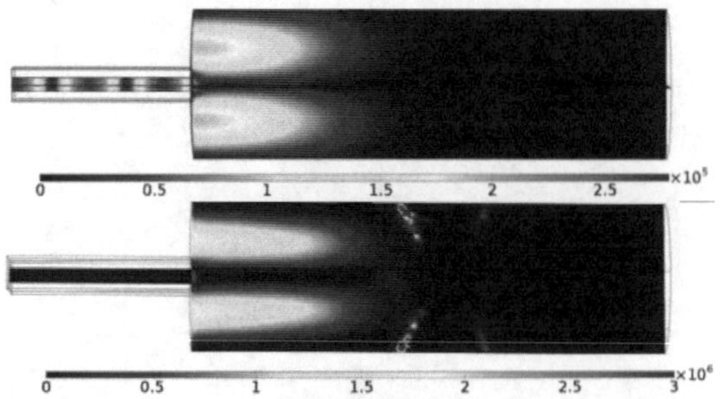

**Fig. 6** Simulated electric field amplitude |E| [V/m] distribution and dissipated Power $P_d$ [W/m^-3] for the *Simple Mirror-single ECR configuration* in proper scale to emphasize the strong gradient near the resonance layer.

The situation in the case 2 (single-ECR point) is somehow different since most of the power seems to be absorbed in the peripheral region of the plasma. The two different displacements of power deposition regions are consistent with visual inspection of the plasma, as it appears from images of figure 7. Experimentally measured X-ray emission is instead a factor 10 larger in case 2, probably due to the smaller volume of the hot plasma, which causes some generation of suprathermal electrons.

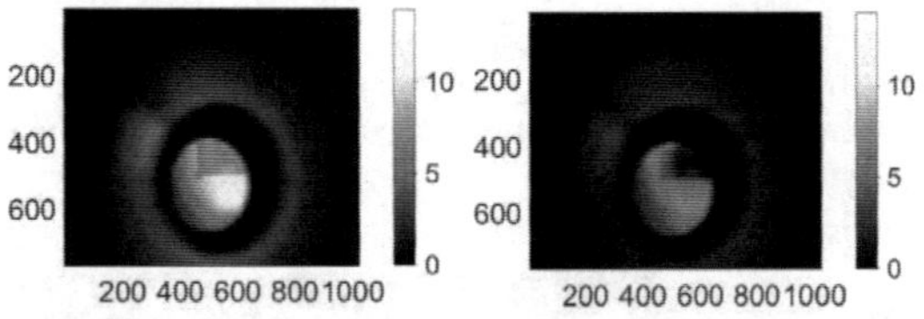

**Fig. 7**. Imaging of the plasma of FPT in the optical range. Case 1 is depicted on the left: most of the plasma is visible in the central part of the trap. Case 2 is shown on the right: the emission is much weaker and the plasma assumes an annular shape. The rectangular shadow is the axial waveguide.

The fields obtained for both magnetic configurations were used to calculate the electrons motion as described in the following section.

### 3.1 Electromagnetic field applied to the particles

The Particle Mover uses the calculated full-wave RF field in order to compute the phase-space stationary electron distribution by integrating equation of motion according to the Boris method. The simulation time was set to 40 µs, while the time step of each iteration was fixed to $10^{-12}$ s. By dividing the simulation domain in small cells of 1 $mm^3$, the code provides two important outputs: the occupation number of each cell (resembling the spatial plasma electron density $n_e$) and the average energy deposited in each cell, from which a local plasma temperature can be derived. Moreover, the code is able to store positions and velocities of lost electrons for post-processing. Figure 8 shows the energy distribution of both lost (a) and survived (b) electrons for the case 1. Lost electrons reach energies in the order of tens of keV. Experimentally observed strong X-ray emission (SDD and HpGe X-ray detectors collected photons up to several tens of keVs) is evidently due to such lost energetic particles. Figure 9 shows a comparison between the two magnetic configurations. In particular, figures 9-a and 9-b show the transversal electron density distribution: case 1 is characterized by higher values if compared to case 2, with a more localized density distribution in the center of the trap, in good agreement with the higher light emission rate observed experimentally as shown in figure 7. Figure 9-c and 9-d show the longitudinal (xz plane) energy dnsity distribution: the different displacement of ECR layers (i.e., of the regions of power deposition) is clearly visible; the effect of the magnetic field on the volumetric density distribution is finally shown in figures 9-e and 9-f.

### 3.2 "Hot-plasma" model

The next step would be the coupling of a second frequency in presence of a self-consistently established plasma. The idea is to sustain the peaked-profile plasma thanks to the axial microwave injection at frequency $f_1$; then using the radial microwave port in order to inject an O-mode wave at the second harmonic frequency $f_2=2f_1$. A dedicate launcher – with orientable emission lobe [16, 17] – has been developed on purpose in order to establish proper OXB scenarios. However, in terms of modelling, we need to go beyond cold-plasma approximation. A method to solve the wave-plasma interaction in the "hot plasma" scenario consists in post-processing to the full-wave fields by means of a Continuous-Wavelet-Transform (CWT). We opted for a complex-Morlet CWT, decomposing the spectral local components along a (magnetostatic) field line.
We have started the modeling of well-known systems, possibly giving the opportunity of analytical solutions. We formerly performed the 1D-Wavelet Spectrogram of the $E_z(z)$ component on the electric field **E**, that is shown in figure 10. It is evident – despite the fact the method must be still improved – the analysis was able to "capture" the displacement of the two resonances along the chamber axis. Due to the afore-mentioned lack of symmetries in ECRIS machines, 1D-CWT must be extended to a fully- 3D system. This upgrade requires many computational efforts which are still ongoing.

## 4 Conclusions

Power deposition and plasma density profiles in the FPT scenario have been investigated in order to establish a proper OXB mode conversion scenario. The simulations have been performed with a numerical suite by coupling the full-wave solutions of the field together with particle kinetics. The simulation will support the double-frequency launching inside the FPT based on a phase-variable waveguides array.

 DOI: 10.1051/epjconf/201715703054

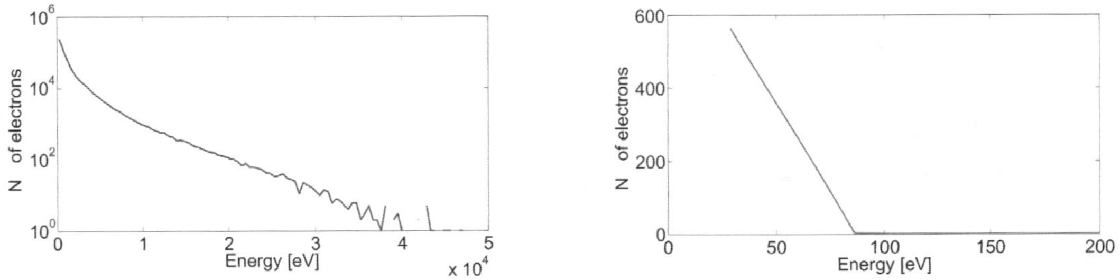


a) b)

**Fig. 8** a) Energy distribution of lost electrons and b) of survived electrons (log scale) in case 1

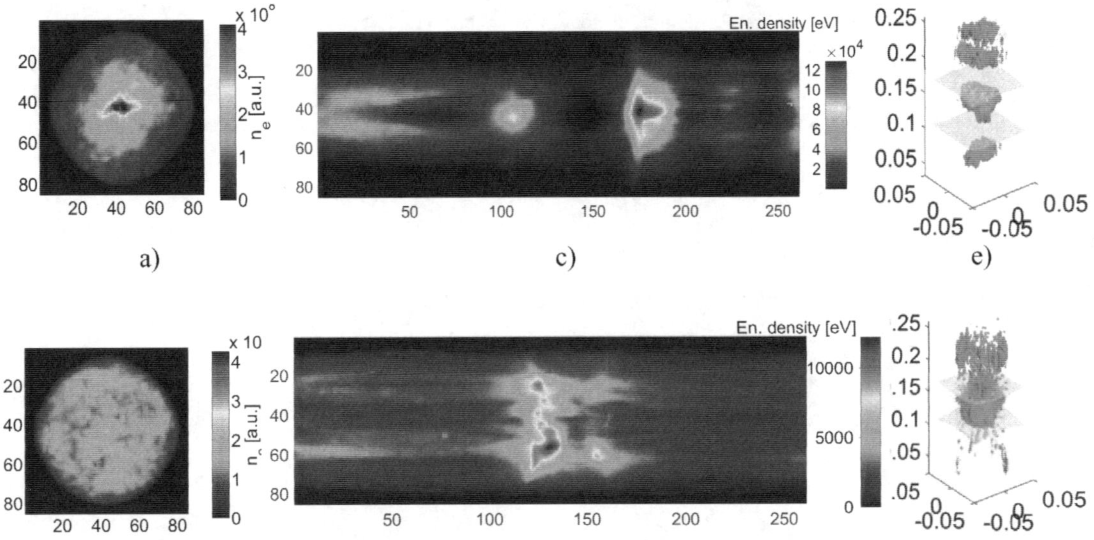


a) c) e)

b) d) f)

**Fig. 9** a) and b): Simulated transverse (xy plane) electron density distribution; c) and d) longitudinal (xz plane) energy density distribution; e) and f) volumetric electron density distribution (in a.u.) for case 1 (upper row) and in case 2 (lower row)

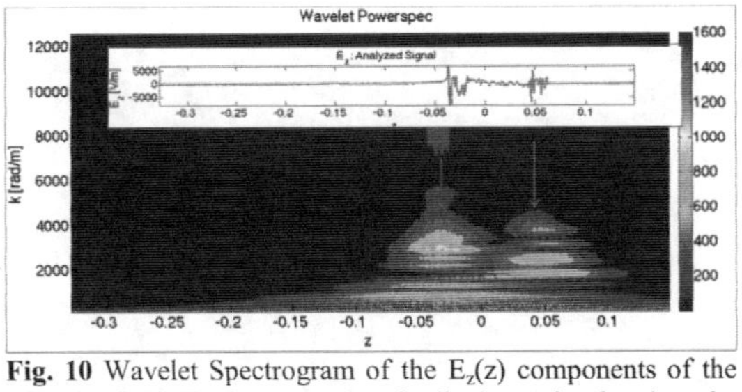


**Fig. 10** Wavelet Spectrogram of the $E_z(z)$ components of the Electric field **E** along the longitudinal z-axis showing the increase of the wavenumber close to the ECR regions.

## Acknowledgement

The authors wish to thank the 5th Nat. Comm. of INFN, under the Grant VESPRI, UTOPIA, RDH, VESPRI, PANDORA for the financial support.

## References

1. M. Brambilla and T. Kreucken, Nucl. Fusion **28**, 1813 (1988).
2. Brambilla, M., Plasma Phys. Controlled Fusion 41 (1999)
3. E. F. Jaeger, L. A. Berry, E. D'Azevedo, D. B. Batchelor, and M. D. Carter, Phys. Plasmas **8**, 1573 (2001).
4. E. G. Evstatiev, V. A. Svidzinski, J. A. Spencer, and J. S. Kim, Rev. Sci. Instrum. **85**, 02A503 (2014).
5. B. P. Clugghish and J. S. Kim, Nucl. Instrum. Methods Phys. Res., Sect. A **664**, 84–97 (2012).
6. V. A. Svidzinski, J. S. Kim, J. A. Spencer, L. Zhao, S. A. Galkin, and E. G. Evstatiev Physics of Plasmas **23** , 112101 (2016); doi: 10.1063/ 1.4966638
7. S. Gammino, G. Ciavola, L. G. Celona, D. Mascali and F. Maimone, IEEE Transactions on Plasma Science, vol. **36** , 4 1552-1568, Aug. (2008).

DOI: 10.1051/epjconf/201715703054

8. S. Gammino, G. Ciavola, L. G. Celona, D. Mascali and F. Maimone, **36**,. 4, pp. 1552-1568, Aug. 2008.
9. A. Galatà et al., RSI DOI 10.1063/1.4934209
10. G. Torrisi, D. Mascali, G. Sorbello, L. Neri, L. Celona, G. Castro, T. Isernia and S. Gammino, , J. Electromagnet, Wave. , **28** (9), (2014) pp. 1085-1099.
11. D. Mascali, G. Torrisi, L. Neri, G. Sorbello, G. Castro, L. Celona, and S. Gammino, Eur. Phys. J. D **69** (2015): 27 DOI: 10.1140/epjd/e2014-50168-5
12. C. K. Birdsall, A. B. Langdon, "Plasma physics via computer simulation" (Taylor and Francis, New York 2005) p. 356-35.
13. S. Gammino et al; "The flexible plasma trap for the production of overdense plasmas", submitted for publication on JINST (to be published).
14. A. Galatà et al., PSST DOI 10.1088/0963-0252/25/4/045007
15. A Kohn, G Birkenmeier, E Holzhauer, M Ramisch and U Stroth ,Plasma Phys. Control. Fusion **52** (2010) 035003 (13pp) doi:10.1088/0741-3335/52/3/035003
16. G. Torrisi, G. Sorbello, O. Leonardi, D. Mascali, L. Celona, and S. Gammino, Microw. Opt. Technol. Lett., **58**: 2629–2634. doi:10.1002/mop.30117
17. G. Castro et al., "Overdense plasma generation in a compact ion source", Plasma Sources Sci. Technol. **26**, 055019, 2017.

EPJ Web of Conferences 157, 03055 (2017)
22 Topical Conference on Radio-Frequency Power in Plasmas
DOI: 10.1051/epjconf/201715703055

# Innovative launching schemes in ECR ion sources

Giuseppe *Torrisi*[1,*], David *Mascali*[1], Gino *Sorbello*[1,2], Giuseppe *Castro*[1], Eugenia *Naselli*[1,3], Ornella *Leonardi*[1], Luigi *Celona*[1] and Santo *Gammino*[1]

[1]INFN-Laboratori Nazionali del Sud, via S. Sofia 62, Catania, 95123, Italy
[2]Università degli studi di Catania, Viale Andrea Doria 6, Catania, 95125, Italy
[3]Dipartimento di Fisica e Astronomia, Università of Catania,Via S. Sofia, 64 I-95123 Catania

**Abstract.** An optimised RF power delivery in magnetoplasmas of ECR ion sources is crucial in order to provide a cost effective upgrade (plasma density and temperature increase and multiply charged ions production) of these machines without recurring to higher and higher magnetic fields, power level and pumping wave frequency. This can be done by following two strategies: a) in a pure ECR-heating scenario, by multiplexing different frequencies; b) in a modal-conversion scenario, by multiple-launching at different frequencies, controllable angles and polarization. The paper will show two typical cases in both the aforementioned scenarios, as developed at INFN-LNS. Test-benches have been developed on purpose, such as the "Plasma Reactor" and "Flexible Plasma Trap", and solutions have been proposed also for ion beams current boosting in the injectors of the Superconducting Cyclotron.

## 1 Introduction

There is great interest in the ECRIS community in increasing the efficiency of ECR heating as a way to maximize the current of extracted ions. Experiments have identified a number of useful techniques for maximizing the extracted currents, such as increasing the frequency scaling and exploiting the multiple frequency heating with the "frequency tuning" [1-4].
Proper coupling of the RF power to the plasma in an electron-cyclotron resonance (ECR) ion source is probably the most important factor in achieving satisfactory performances. Poor coupler design in fact can lead to a variety of undesirable effects such as high reflected power, low plasma density, unstable operation, and, eventually, poor performance.
Several efforts have been paid to RF coupling to ECRIS plasmas at INFN-LNS in the last decade. Most of them were devoted to Frequency Tuning Effect [5, 6] , which is now routinely adopted for beam optimisation (both in terms of currents and phase-space parameters) in many ECRIS worldwide. Before the use of frequency tuning technique, the two/multiple frequency method proved to be a reliable way to increase the produced ion beam currents in ECRIS. The positive effect of this technique was observed for the first time already in '90s [3]. The impact was observed to be higher on the ions at higher charge states, and this was a proof that the effect produced a more efficient heating of the ionizing electrons.
Output currents and charge states are in fact related to plasma density and lifetime by the well known relations: $\langle q\rangle \propto n_e\tau_i$, provided that $T_e$ is high enough to allow deep ionisation of atoms, and $I\propto n_e/\tau_i$, where I is the output current and $\langle q\rangle$ the average charge state. More recently, some authors [7] have shown the use of two frequencies instead than one, even when they carry the same (but shared) amount of RF power, contributes to damping of instabilities.
The above mentioned methods can be further optimized, and some recent results relying to a very compact setup for summing up Frequency Tuning Effect (FTE) to Two Frequency Heating (TFH) in small-size ECRIS will be briefly described in the following.
Another challenging research in the ECRIS field concerns the intrinsic currents limitations affecting these machines. These rely with the well-known "omega-squared" law (due to the electromagnetic cutoff and RF scattering into the [8] cone). The only way for any performance boost, in the recent past, has been the increase of the pumping wave frequency, which in turn implies a scaling of the confining magnetic field as well. One possibility to overcome this problem is to implement in ECRIS also some modal-conversion scheme, such as XB or OXB conversion. Some attempts have been already done [9].
The optimization of the OXB conversion helps to maximize the amount of the absorbed RF power.
It is well known in the field of fusion reactors that modal-conversion require high control of RF launching into the plasma [10]: not only the power, but launching angle and wave polarisation are critical too.
The latter point is deemed as crucial for any future developments of ECRIS. Up to now, in fact, RF launching in these machines has been based on the very simple scheme of rectangular (or, more rarely, circular) waveguides mechanically matched to cylindrical plasma

* Corresponding author: giuseppe.torrisi@lns.infn.it

DOI: 10.1051/epjconf/201715703055

chambers. This scheme, despite its simplicity, cannot allow any mastering of the RF coupling with the plasma. R&D on RF launchers for modal conversions is therefore deemed to become more and more important for any further boost of their performances.

## 2 Multiplexing Two Different Frequencies in an ECR-H Scenario

The “Two Frequency Heating" (TFH) has been implemented many times in ECRIS equipped by a double microwave-port at the injection endplate. We tried to implement the same method in a compact ECRIS named CAESAR, and operating since 2002 at INFN-LNS as injector of the Superconducting Cyclotron. Due to its compactness, the plasma chamber has only one microwave port allowing microwave launching. Therefore TFH requires a proper setup to be correctly implemented.

The TFH experiment involved the use of a "diplexer Hybrid coupler" Waveguide Combiner in the frequency band [13.5-15.6] GHz, up to 1.5 kW, (see Fig. 1).

This 4-ports passive device implements a "multiplexing" in the frequency domain, i.e. it allows to combine the power coming out from the two 500 W TWTs amplifiers at two different frequencies, combined on an output port which goes towards the plasma chamber.

The use of two TWT allows the contemporaneous tuning of both the frequencies, in separate phases. Once optimized both, in terms of relative maximization of the source’s performances, they can be summed up by the diplexer, and launched at the same time into the plasma chamber. The remaining fourth port of the device, called “insulated port”, is terminated on an adapted high-power waveguide load (up to 1 kW of Power Handling) in order to minimize and absorb unwanted reflections.

The first TWT amplifier works at 13.75 - 14.5 GHz at maximum 700 W.

The second TWT amplifier allows to produce 500-Watts maximum power, in the 8.0 - 18.0 GHz frequency range.

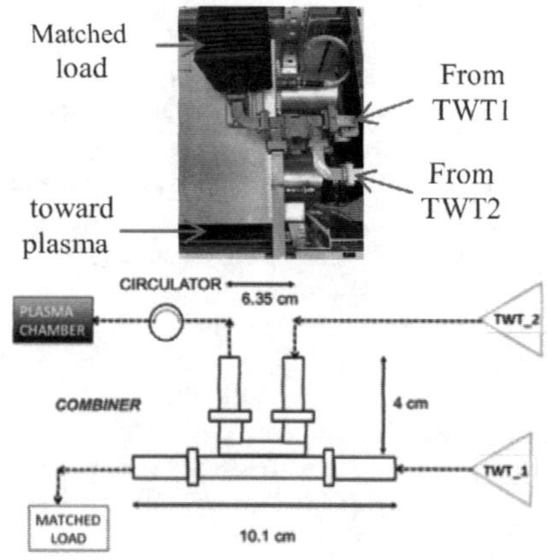


**Fig. 1** Microwave scheme for implementing TFH on the CAESAR source at LNS. The same scheme can be used for any ECRIS equipped by only one microwave port for RF matching with the plasma chamber.

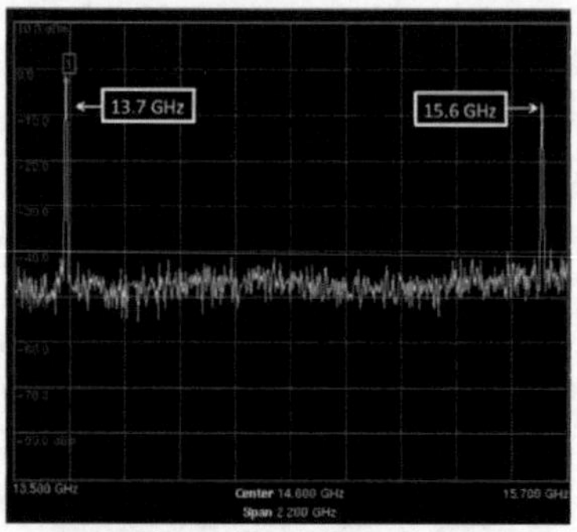


**Fig. 2** Spectrum Analyzer Acquisition: $f_1$ =13.5 GHz $f_{stop}$=15.7 GHz $f_{center}$=14.6 GHz Span=2.2 GHz.

Before to proceed with the plasma heating, we tested the Experimental setup by using a directional coupler terminated on a waveguide matched load. One port of the directional coupler has been connected to a Spectrum Analyzer Anritsu MS2720T-0743 9 kHz-43 GHz working at Resolution Bandwidth RBW= 3 MHz and Sweep Time=200 ms in Continuous sweep mode.

As it is possible to see in Fig. 2, thanks to the diplexer we are able to simultaneously transmit two frequencies (13.7 GHz and 15.6 GHz) without interference and power loss.

### 2.1 Comparison of the injector of the Superconducting Cyclotron in double-frequency heating mode with respect to the single frequency mode operation

The CAESAR source is equipped by two room-temperature solenoids providing axial confinement, and an hexapole for the radial one. The coils are able to produce a maximum field of 1.2 T when both are energized by 1100 A of currents. In order to allow a proper beam extraction, the second solenoid is usually operated at currents not larger than 800 A, so that the mirror configuration is asymmetric. In the specific case of TFH operations, we used Coil 1: 1021 A, Coil 2: 745 A. The extraction potential was fixed at $V_{extr}$= 23726 kV; while $^{32}S$ charge states were monitored through a magnetic mass spectrometer. The plasma was formed by $SF_6$ gas injected into the plasma chamber at injection pressure of 1.73E-6 mbar. The obtained results are shown in table 1.

**Table 1.** $^{32}S$ charge states: double-frequency heating mode vs single frequency mode operation.

| Charge state | RF power [W] | Frequency [GHz] | Current [μA] |
|---|---|---|---|

 DOI: 10.1051/epjconf/201715703055

| | | | |
|---|---|---|---|
| $^{32}S^{11+}$ | 305 | 14.360 | 1.0 |
| $^{32}S^{11+}$ | 380+170 =550 | 14.450+ 15.434 | 2.8 |

Each beam current was optimized at the given frequency (or double frequency) operating mode in order to minimize the reflected power and the beam ripple. Frequency tuning method was adopted for the single frequencies. Results show that when operating in "double frequency" mode, we were able to obtain I=2.8 μA of $^{32}S^{11+}$ , i.e. a beam current that was almost tripled even though the power was less than doubled. This result is even more relevant if one considers that in ECRIS the beam currents are less than linear dependent on the RF power in single frequency operations.

## 3. Innovative launching scheme

A novel microwave launcher has been electromagnetically characterized to search for the possibly of exciting Electrostatic Bernstein Waves (EBW) inside the prototype installed at INFN-LNS called FPT – Flexible Plasma Trap [11]. In FPT, the plasma can be generated by using two or three different microwave systems, for parallel and perpendicular microwave injections, as figure 3 shows. In particular, we are now working with two TWTs: one operates at a central frequency of 5 GHz, in the range 3-7 GHz, and the other around 14 GHz (range 13.75 – 14.5 GHz).

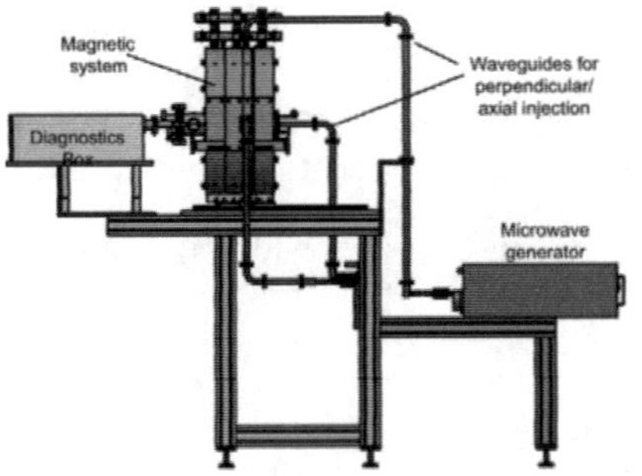

**Fig. 3** General layout of the FPT setup.

FPT is equipped by three solenoids providing three different possible profiles of the magnetic field: simple mirror; quasi-flat; and magnetic beach. No beam extraction systems have been coupled to the machine. One of the chamber endplates is dedicated to microwave launching and gas-injection, the other one to plasma diagnostics.

### 3.1 Modal-conversion scenario

The EB waves are electrostatic modes and need to be excited via a inner plasma mode-conversion process, starting from electromagnetic waves (the X mode) at the upper hybrid resonance (UHR) layer, where their generation may involve one (XB) or two mode conversion scenarios (OXB) that are efficient only in a narrow window of the following parameters: frequency, polarization and angle of incidence.
Numerous experiments were performed in fusion plasma toroidal devices where ray tracing approximation allows to compute the electromagnetic waves traveling inside the plasma [12, 13]. In these "big-size" scenarios, well-established optical techniques (e. g. grooved mirrors, lenses and grating polycromator), were adopted to optimize the O-X-B conversion, in terms of wave polarization and incidence angle.
On the contrary, in ECRIS, the pumping wavelength is comparable with the plasma radius [14] and density scale-length, thus prohibiting any ray tracing approach. Full-wave calculations must be applied to simulate the wave behaviour in the plasma [15, 16]. The presence of resonant cavity additionally causes the polarization scrambling, and it further complicates the OXB modeling. The mechanical constraints also make very difficult the positioning of complex launching systems.
The solution hereby proposed consists of a phased waveguide "array" of two elements: using a relative phase shift between them, a steering of the radiated beam can be produced, in order to obtain an optimum incidence angle. The system is described in details in [17].

### 3.2 The new microwave launcher based on a two-waveguides-array

This section is devoted to the microwave transmission line of the heating antenna. Simulations showed O-X-B-like scenarios are possible by properly launching the pumping O-wave from the radial port. As mentioned in the previous section, the antenna should be optimized in order to emit the heating power with an optimal **k**(r) angle with respect to the external magnetic field $B_0$. The absorption of the waves in the vicinity of the resonance layer should be almost complete in a single "pass", if the proper mode of irradiation is chosen.

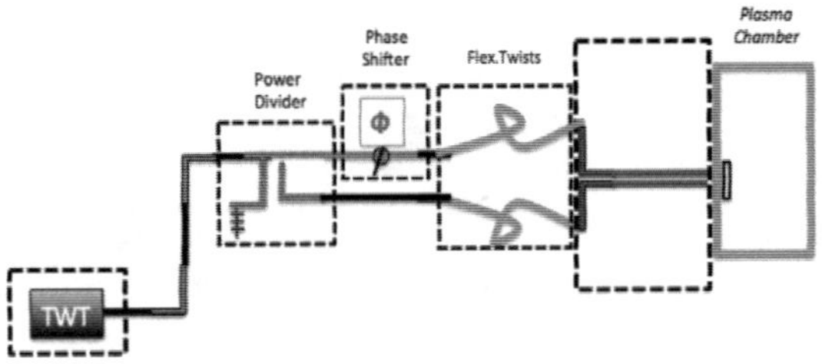

**Fig. 4** Microwave transmission line

Taking inspiration from phased multi-waveguide structures (nicknamed waveguide "grill") [17, 18] designed to launch microwaves at the lower hybrid resonance to heat large toroidal plasmas, a double-waveguide antenna system has been proposed for the

FPT. The microwave transmission line is schematically shown in fig. 4.

The launcher consists of two waveguide driven in $TE_{10}$ mode with relative phase controlled by a calibrate phase shifter, a loaded four-port power divider and two fexible and twistable WR62 waveguides.
The **k** direction for different phase shift can be simulated, as shown in figure 5. For a $\Delta\phi=0°$ (Fig. 5a), the lobe points in forward direction, meaning that the O-wave propagates as purely perpendicular to the magnetic field, while for delta $\Delta\phi=180°$ ((Fig. 5b) a two lobes structure is generated, with an angle of 60° with respect to the magnetostatic field direction.

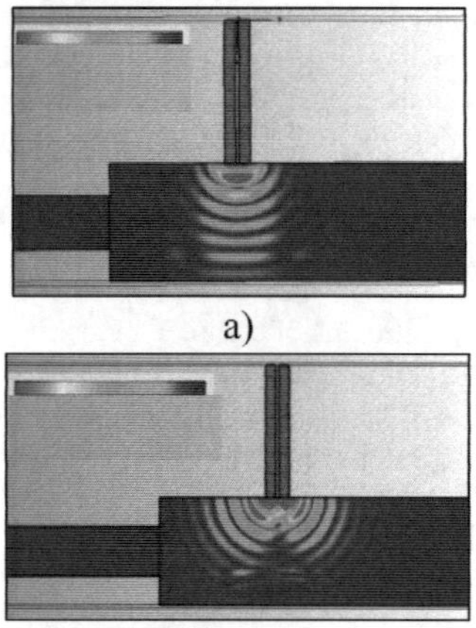

a)

b)

**Fig. 5** CST microwave studio "Free-space" vs "In-cavity" simulation of microwave radiated Electric field |E| from the double waveguide array launcher (plots are given with pseudocolor representation of the electric field intensity) a) when the phase shift=0°; b) when the phase shift=180°

## Conclusions

The paper has commented the schemes of microwave-to-plasma coupling in ECRIS-type machines. Other than in the framework of ECR-heating of plasmas generated in such machines, efforts are now devoted to new launching schemes, possibly suitable for modal-conversion. In the first case, frequency tuning and multiple-frequency heating techniques can be used also in combination, and even in case of very compact and simple machines. For the modal-conversion scenario, difficulties arise due to the critical scale-lengths involved in such compact machines. Nevertheless a phased-waveguide-array has been designed and tested on a testbench, and it is now going to be implemented for OXB modal-conversion experiments at INFN-LNS.

## Acknowledgement

The authors wish to thank the 5th Nat. Comm. of INFN, under the Grant VESPRI, UTOPIA, RDH, VESPRI, PANDORA for the financial support. M. Castro, S. Marletta, S. Passarello, G. Pastore, S. Vinciguerra, E. Zappalà, G. Messina. and the INFN-LNS mechanical workshop people are warmly acknowledged for the support given.

## References

1. Geller R., et al. Proc. 8th Int. Workshop ECR Ion Sources, Vol. 1 (1987), p. 22; East Lansing, MI-USA NSCL report MSUCP-47.
2. S. Gammino and G. Ciavola , Plasma Source Sci. Technol., **5** 19 (1996).
3. Z. Q. Xie, Rev. Sci. Instrum. **69**, 625 (1998);
4. D. Mascali et al., Rev. Sci. Instrum. **83**(2):02A336. (2012)
5. L. Celona, G. Ciavola, F. Consoli, S. Gammino, F. Maimone, D. Mascali, P. Spädtke, K. Tinschert, R. Lang, J. Mäder, J. Roßbach, S. Barbarino, R. S. Catalano, Rev. Sci. Instrum. **79**, 2,023305, 2008, 10.1063/1.2841694
6. Z. Q. Xie and C. M. Lyneis, in Proceedings of the 12th International Workshop on ECRIS, Wako, INS-J-182 (Institute for Nuclear Study, University of Tokyo, Japan, 1995), p. 24.
7. V. Skalyga, I. Izotov, T. Kalvas, H. Koivisto, J. Komppula, R. Kronholm, J. Laulainen, D. Mansfeld, and O. Tarvainen, Physics of Plasmas **22** , 083509 (2015); doi: 10.1063/ 1.4928428
8. A. Girard et al., Phys. Rev. E, **62**(1), 1182, (2000).
9. G. Castro et al., "Overdense plasma generation in a compact ion source", Plasma Sources Sci. Technol. **26**, 055019, (2017).
10. A Kohn, G Birkenmeier, E Holzhauer, M Ramisch and U Stroth, Plasma Phys. Control. Fusion **52** (2010) 035003 (13pp).
11. S. Gammino et al Journal of Instrumentation, **12**, 07, P07027, July 2017.
12. Petrov, Yu. V. and Becoulet and A. Monakhov, , American Physical Society, 41st Annual Meeting of the Division of Plasma Physics, (1999).
13. Van Eester, Plasma Phys. Controlled Fusion. **34**:1539-1562, (1992).
14. R Rácz et al, Plasma Sources Sci. Technol. **26** (2017) 075011 (14pp).
15. G. Torrisi, D. Mascali, G. Sorbello, L. Neri, L. Celona, G. Castro, T. Isernia and S. Gammino, J. Electromagnet, Wave. , **28** (9), (2014) pp. 1085-1099.
16. D. Mascali, G. Torrisi, L. Neri, G. Sorbello, G. Castro, L. Celona, and S. Gammino, "Eur. Phys. J. D **69** (2015): 27 DOI: 10.1140/epjd/e2014-50168-5
17. G. Torrisi, G. Sorbello, O. Leonardi, D. Mascali, L. Celona, and S. Gammino, Microw. Opt. Technol. Lett., **58**: 2629–2634. doi:10.1002/mop.30117
18. M. Brambilla, NUCLEAR FUSION 16, (1976)
19. S. Bernabei, M A Heald, M Hooke, R W Motley and F J Paoloni and M. Brambilla and W D Getty, NUCLEAR FUSION **17** 5 (1977),

DOI: 10.1051/epjconf/201715703056

# Discharge initiation by ICRF antenna in IShTAR

*Matěj* Tripský[1,2,⋆], *Tom* Wauters[1], *Rodolphe* D'Inca[3], *Kristel* Crombé[1,2], *Jonathan* Jacquot[3], *Roman* Ochoukov[3], *Fabrice* Louche[1], *Mariia* Usoltceva[2,3,4], *Ana* Kostic[2,3], *Anatoli* Lyssoivan[1], *Jean-Marie* Noterdaeme[2,3], and *Michael* Van Schoor[1]

[1] *Laboratory for Plasma Physics, ERM/KMS, 1000 Brussels, Belgium*
[2] *Ghent University, Department of Applied Physics, 9000 Ghent, Belgium*
[3] *Max-Planck-Institut für Plasmaphysik, Boltzmannstr. 2, 85748 Garching, Germany*
[4] *Université de Lorraine, 54506 Vandœuvre-lés-Nancy, France*

**Abstract.** IShTAR is a linear magnetized plasma test facility dedicated to the investigation of RF wave/plasma interaction. The IShTAR ICRF system consists of a single strap RF antenna. When using the antenna for plasma production without an external plasma source, it is shown that the plasma is either produced in front of the antenna strap or inside the antenna box depending on the antenna parameters. Here, we present experimental and numerical investigation of the plasma initiation parametric dependencies. Detailed pressure and RF power scans were performed in helium at $f = 5.22$ MHz and $f = 42.06$ MHz. The experiment shows the parameter ranges for which the plasma is produced in front of the strap, or inside the antenna box. These ranges are validated by simulations with the RFdinity model, and by theoretical predictions.

## 1 Introduction

Using the ICRF antennas for discharge initiation brings the problem that the antenna load before plasma breakdown is weak. Therefore, until the plasma is created in the vessel, a high voltage is located on the antenna straps. When the plasma is formed, this voltage decreases together with an increase in the coupled power. This moment indicates the plasma breakdown. For antenna safety reasons, it is important to set the antenna parameters (power, frequency, and phasing) and the neutral pressure to minimize the breakdown time. Such optimization reduces the likelihood for the sparking inside of the antenna box.

The dependency of the breakdown time (or ionization rate) on the antenna parameters and the neutral gas pressure on Tokamaks is subject of ongoing research [1]. Simulation results have shown qualitative agreement with the experimental observations for the neutral pressure and antenna power dependency. They indicate that above a certain value of the antenna power, the breakdown times does not change with increasing power. Therefore, a first optimization consist of lowering the antenna power to avoid very high voltages. On the other hand, the neutral pressure scans demonstrate a decreasing dependency of the breakdown time with increasing pressure. For that reason, a maximum allowed pressure should be set for ICRF plasma initiation. This maximum allowed pressure is typically around $5 \cdot 10^{-2}$ Pa. The simulations in [1] show that the ionization rate decreases rapidly with antenna frequency and circumference length. These dependencies may be of concern for plasma production in ITER. Its ICRF system operates at high frequencies ($f = 40-55$MHz), and the circumference of the ITER torus is much longer than any present-day fusion machine. Nevertheless, simulations with ITER ICRF vacuum electric fields demonstrated that plasma initiation is possible even at these high frequencies and large circumference [2]. The simulations also showed that the plasma first appears close to the antenna straps. Therefore, it is crucial to understand better our set of safety conditions for the pressure, frequency and antenna power in order to avoid any possibility of plasma formation inside the ITER ICRF antenna.

The IShTAR device provides an excellent opportunity to focus on specific aspects of ICRF plasma production. It is hereto important to highlight that the IShTAR is a linear device. It can therefor be used to investigate the plasma formation in ponderomotive wells in front of the antenna box, or inside of the antenna. In toroidal devices, electrons leaving the antenna region can ionize the neutral gas along the magnetic field line and afterward they can again reenter the antenna region.

## 2 Experiment on IShTAR

The IShTAR (Ion cyclotron Sheath Test ARrangement) project focuses on studying antenna near-fields and RF sheath effects in the presence of a plasma and magnetic field [3]. It is a linear device with a cylindrical main vessel of 1.1 m length and 1 m diameter. The magnetic field is created by two coils around the main vessel. The experiment was operated with the current in the coils of

⋆e-mail: matej.tripsky@rma.ac.be

DOI: 10.1051/epjconf/201715703056

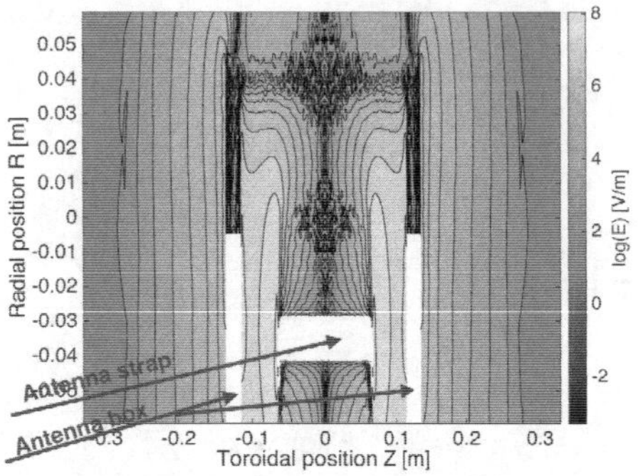


Figure 1: Contour plot of the absolute value of the parallel $E_z$-field in (R-Z)-coordinates for one-strap IShTAR antenna obtained in the horizontal plane using the commercial electromagnetic software CST Microwave Studio®(MWS) for $f$ = 5 MHz, $P$ = 1 W.

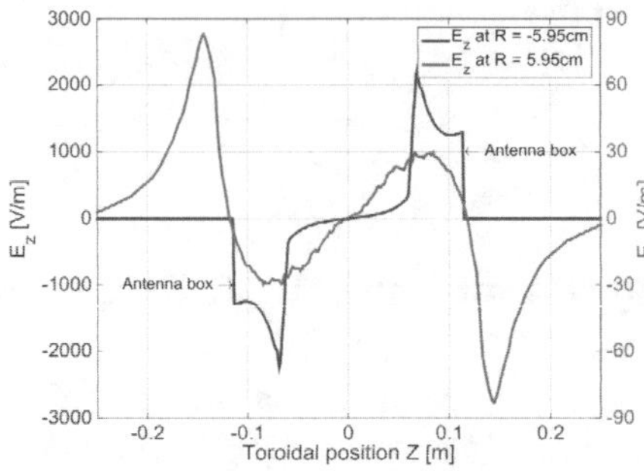


Figure 2: Vacuum $E_z$-field at $R = -5.95$ cm (inside the antenna box) and $R = 5.95$ cm (in front of the antenna strap) used by the model ($f$ = 5 MHz, $P$ = 1 W).

2 kA, corresponding to the magnetic field of 0.06T. For the presented experiment plasma was initiated solely by the IShTAR ICRF antenna system. The antenna consists of a single strap fed by transmission lines connected to a generator with a power up to 1 kW and with a frequency range [0.1 – 100 MHz]. We operated the IShTAR antenna at two frequencies, $f$ = 5.22 MHz and $f$ = 42.06 MHz.

## 3 Simulations

The Monte Carlo collision model `RFdinity1d3v` [1] simulates the motion of electrons along a parallel magnetic field line. The model includes inelastic collisions (*excitation* and *ionization*) and elastic scattering between electrons and helium atoms. Electrons interact with the vacuum parallel electric field $E_z$. Depending on the electric field and frequency, electrons can be trapped in the potential well in front of the antenna strap [4]. If electrons remain trapped for a sufficient time compared to the ionisation characteristic time, the electron density may increase. The trapped electrons must be accelerated by the electric field (the generator input power) above the ionization potential ($\Delta E^{\text{ion}} \approx 24.6$ eV for helium). Furthermore, for trapping the electrons, the electric field must have the shape (depending on the radial position) to minimize the electron losses. The performed simulations use the parallel electric field profiles as obtained by the commercial electromagnetic software CST Microwave Studio® [5]. Figure 1 shows the absolute values of the $E_z$-field in the horizontal plane 0.5 cm below the antenna top for $f$ = 5 MHz, $P$ = 1 W. The figure demonstrates that not only the amplitude of the electric field changes radially but also the shape. We investigate plasma initiation at two radial locations: $R = -5.95$ cm (inside the antenna box) and $R = 5.95$ cm (in front of the antenna strap). The vacuum parallel electric field for these two radial positions are visualized in figure 2 for a coupled power of $P$ = 1 W and frequency $f$ = 5 MHz. Two arrows indicate locations of the antenna box walls marked by the abrupt change of the $E_z$-field profile for the electric field inside the antenna box ($R = -5.95$ cm).

### 3.1 Plasma initiation at $f$ = 5 MHz

Simulations were launched at $f$ = 5 MHz with varying neutral gas pressure and coupled power. The simulations show that the ionization rate has a linear dependency on the neutral gas pressure as predicted by Schüller [6], $\nu_{\text{ion}}^{\text{RF}} \sim n_{He}$. By fitting this linear dependence, we identify the minimum pressure above which the ionization rate is higher than zero. In the experiment, the discharges at $f$ = 5.22 MHz were initiated in front of the strap and never in the antenna box. The experiment demonstrated that with the increasing generator power the minimum neutral gas pressure for plasma initiation decreases. Figure 3 indicates that the minimum pressure decreases steeply at lower generator power (100 – 150 W) whereafter the minimum pressure decreases more slowly up to the maximum operated power on IShTAR ($P_G$ = 400 W).

In the figure, the coupled power range used in the simulations is very different from the generator power range. The electric fields obtained from the MWS do not take into account eventual power losses in the matching system or transmission lines. Therefore, it is hard to link the generator power in the experiment and the coupled power from the MWS. No measurements of the voltages on the antenna strap were available. The coupled power $P$ = 100 kW corresponds to the electric field amplitude $E_0 \approx 24$ kV/m at $R$ = 5.95 cm for $f$ = 5MHz, which from experience appears to be a realistic field strength in IShTAR device. The minimum pressures in our simulations are slightly higher than the experimental ones. This shift towards the higher minimum pressures in the simulations may also be caused by the fact that our model contains only one neutral gas type, i.e. helium atom. However, the base pressure in IShTAR is $10^{-4}$ Pa. It suggests that a high level of impurities can be present in the vessel ($H_2O$) with significantly lower ionization potentials than for the helium atom.

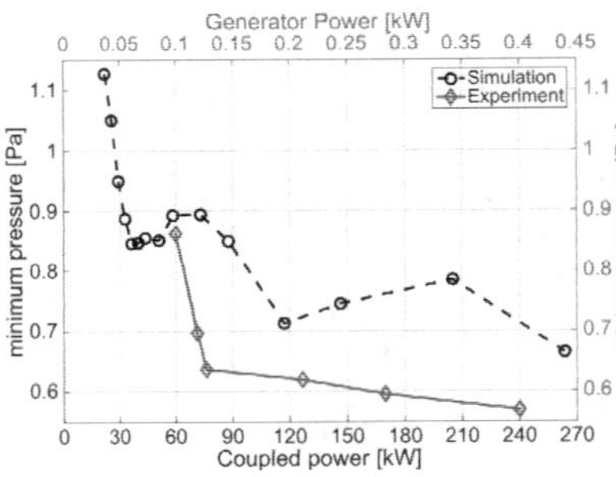


Figure 3: Dependence of the minimum pressure on the power in the experiment at $f$ = 5.22 MHz (red line) with the generator power and the simulations at $f$ = 5 MHz (black line) with the coupled power.

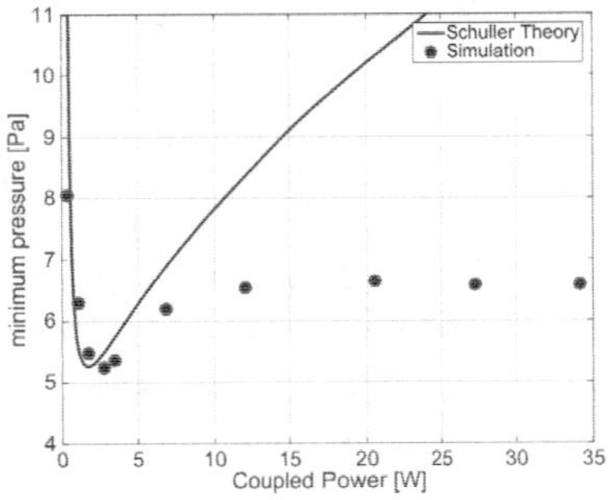


Figure 4: Comparison of the minimum pressure dependency on the power in the simulations and theoretical predictions (*Schüller* theory) for $f$ = 42 MHz, and ($W_{ANT}$ + $L_z$) = 0.48 m.

### 3.2 Plasma initiation at $f$ = 42 MHz

Unlike at $f$ = 5.22 MHz, at $f$ = 42.06 MHz the plasma was initiated inside the antenna box. This difference between the plasma column and its location is shown in figure 6. The figure illustrates two images from a camera on IShTAR facing the ICRF antenna for identical generator power ($P_G$ = 290W). At $f$ = 5.22 MHz the plasma column is homogeneous and located in front of the antenna, whereas at $f$ = 42.06 MHz the plasma is also created inside the box near the feeding point (the brightest light on the top). To understand this occurrence we reinvestigate the frequency and pressure dependency for trapping electrons in ponderomotive well to initiate the electron density build-up. The first approximation of the plasma initiation by ICRF antenna was introduced by *Lyssoivan* and *Carter* [7]. They derived conditions for the electric field and frequency to accelerate electrons above the ionization potential in the RF field. The model was later on improved by *Schüller* to include also the neutral pressure dependency [6]. The comparison of the theoretical predictions and simulations results for the minimum pressure dependency on the coupled power for the plasma at $f$ = 42 MHz are plotted in figure 4. According to the theoretical predictions and simulations results the plasma should not exist in the antenna box below 5 Pa at $f$ = 42.06 MHz. Nevertheless, we observed in the experiment the plasma formation in the antenna box for the neutral gas pressures $p_{He} > 0.2$ Pa for the medium generator power $P$ = 290 W.

This discrepancy between the experiment and theoretical or simulations predictions shows that we neglected in the model and theory an effect contributing in the electron multiplication in the antenna box. One of the candidates was the secondary electron emission due to electrons hitting the antenna box walls. We have implemented the secondary electron emission into our model using the yield dependency on the incoming energy from [8]. The secondary electron yield (SEY) depends strongly on the surface properties (roughness, cleanness, etc.) rather than on the material type. Therefore, we studied in more details two cases: (i) low SEY with the maximum yield of 0.65, and (ii) high SEY with the maximum yield of 2.16. The SEY energy dependencies for these two scenarios are plotted in Figure 5(C). Figures 5(A-B) show the electron density evolution for simulations inside the antenna box for these two scenarios (High and Low SEY) with the same electric field strength for the experimental frequencies $f$ = 5MHz, and $f$ = 42MHz at different neutral pressure levels. Figure 5(A) for the case with the high SEY confirms the experimental observations that the plasma is created in the antenna box only for $f$ = 42MHz even at $p_{He}$ = 0.2Pa. Whereas for $f$ = 5MHz, the electron density is not increasing even at the extremely high-pressure $p$ = 8Pa. For the low SEY scenario, we didn't observe the electron density increase for both frequencies. The results

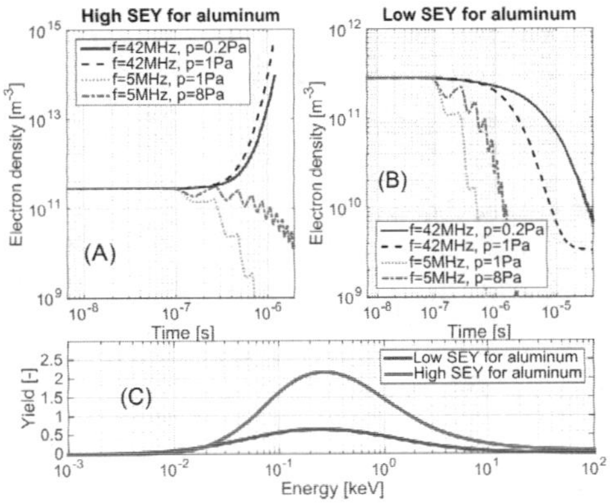


Figure 5: Electron density evolution in time for simulations with the same electric field strength at $f$ = 5MHz, $f$ = 42MHz, and various pressure levels for two scenarios with different SEY profile: (A)-High SEY, (B)-Low SEY for aluminum. (C) The SEY energy dependency for aluminum.

DOI: 10.1051/epjconf/201715703056

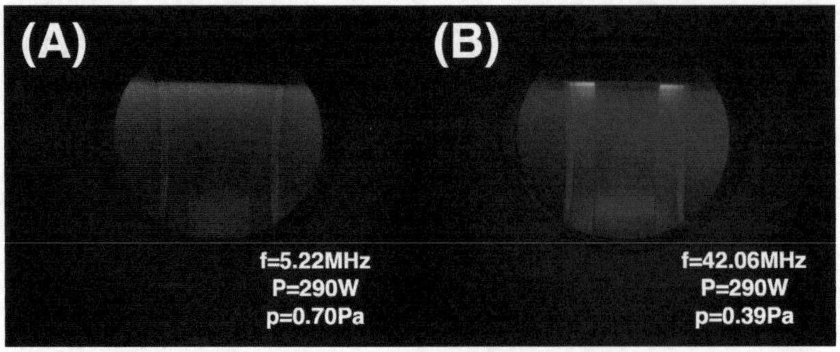


Figure 6: Camera views in IShTAR on the front side of the ICRF antenna for two frequencies at the generator power $P_G$ = 290W: (A) plasma is located in front of the strap for $f$ = 5.22MHz, $p_{He}$ = 0.7Pa, (B) plasma is created in antenna box for $f$ = 42.06MHz, $p_{He}$ = 0.39Pa.

suggest a strong dependency not only on the frequency but the SEY energy profile as well. By implementing high but, realistic energy dependent yields from literature one easily can evoke enough multiplications to overcome particle losses at the antenna box wall.

## 4 Conclusion

A first experimental study of ICRF plasma initiation in Helium in IShTAR for two frequencies is presented. The experimental observations were compared with modeling and theoretical predictions. In the experiment, the homogeneous plasma column in front of the antenna was formed only at $f$ = 5.22 MHz. The minimum pressures above which the plasma was created depend on the generator power and are in qualitative agreement with the modeling predictions. However, the experimental values of the minimum pressures are slightly shifted towards lower pressures. This shift may be caused by insufficient vacuum conditions or significant power losses in transmission lines and a matching unit. Impurities presented in the vessel with lower ionization potential than for the helium could increase the ionization rate. Our model (`RFdinity1d3v`) in the current version is not able to simulate a gas mixture.

For experiments at $f$ = 42.06 MHz, the plasma is always created in the antenna box instead of in front of the antenna strap. The plasma was observed in the antenna box starting from pressures above 0.2 Pa for the generator power $P_G$ = 290 W. This observation meet the modeling predictions only after including the secondary electron emission into the model. The simulations show that the density increase in the antenna box for the given input power is possible only for $f$ = 42MHz and not for $f$ = 5MHz with the high SEY. It demonstrates that the secondary emission plays a crucial role in the discharge initiation inside the antenna box. The plasma inside the box was observed in the ITER ICRF frequency range. Thus more accurate model and further investigations are needed to carefully set the operation pressure limit in order to avoid plasma initiation in the antenna box in ITER.

## Acknowledgments

We acknowledge the CINECA award under the ISCRA initiative, for the availability of high performance computing resources and support. This work has been carried out within the framework of the EUROfusion Consortium and has received funding from the Euratom research and training program 2014-2018 under grant agreement No 633053. The views and opinions expressed herein do not necessarily reflect those of the European Commission.

## References

[1] M. Tripsky et al., European Conference Abstracts (ECA), **38F**, P1.133, Berlin, Germany, (2014)

[2] M. Tripsky et al., AIP Conference Proceedings, **1689**, 060009, California, USA (2015)

[3] K. Crombé et al., AIP Conference Proceedings, **1689**, 030006, California, USA (2015)

[4] T. Wauters, M. Tripsky, Proc. 22$^{nd}$RF topical conference, this conference (2017)

[5] CST AG, *http://www.cst.com*, CST STUDIO SUITE (R) (2015)

[6] F. Schüller, *Report on Applications of ICWC on ITER*, Technical Report IO/2009/ADM-014 report version 3, p. 16-23, ITER - IO, 13115, St. Paul-lez-Durance, France, (2009)

[7] M.D. Carter, A.I. Lyssoivan and et al. *Nuclear Fusion*, **30**, 723, (1990)

[8] D.C. Joy *A database of electron-solid interactions*, EM Facility, University of Tennessee, and Oak Ridge National Laboratory, (2008)

 DOI: 10.1051/epjconf/201715703057

# Characterization of the mutual influence of Ion Cyclotron and Lower Hybrid Range of frequencies systems on EAST

*Guillaume* Urbanczyk[1, 2, 3, 4,*], *Xinjun* Zhang[1], *Chengming* Qin[1], *Yanping* Zhao[1], *Tao* Zhang[1], *Ling* Zhang[1], *Jiangang* Li[1], *Shuai* Yuan[1], *Liang* Chen[1], *Heng* Zhang[1], *Jiahui* Zhang[1], *Jianhua* Wang[1], *Xiuda* Yang[1], and *Jinping* Qian[1]

[1]Institute of Plasma Physics, Chinese Academy of Sciences, Hefei 230031, People'sRepublic of China
[2]Université de Lorraine-CNRS UMR 7198 BP 70239 FST_IJL F-54506 Vandoeuvre
[3]University of Science and Technology of China, Hefei 230026, People's Republic of China
[4]CEA, IRFM, F-13108 Saint Paul-lez-Durance, France

**Abstract.** Waves in the Ion Cyclotron (ICRF) and Lower Hybrid (LH) Range of Frequencies are efficient techniques respectively to heat the plasma and drive current. Main difficulties come from a trade-off between good RF coupling and acceptable level of impurities release. The mutual influence of both systems makes such equilibrium often hard to reach [1]. In order to investigate those interactions based on Scrape-Off Layer (SOL) plasma parameters, a new reciprocating probe was designed allying a three tips Langmuir probe with an emissive wire. The emissive filament provides a precise measure of plasma potential [2], which can be used to calibrate Langmuir probe's results. This paper reports on experimental results obtained on EAST, where there are two ICRF antennas and two LH launchers. Among others diagnostics, the new reciprocating probe enabled to evidence the deleterious influence of ICRF power on LHWs coupling in L-mode plasmas. In areas connected with an active ICRF antenna, SOL potentials increase while densities tend to decrease, respectively enhancing impurities release and deteriorating LHWs coupling. This phenomenon has mostly been attributed to RF sheath; the one that forms on top of Plasma Facing Components (PFCs) and causes ExB density convections [3]. From those experiments it seems ICRF has a strong influence on magnetically connected areas, both in the near field – influencing ICRF waves coupling – and in farther locations such as in front of LH grills. Moreover, influence of ICRF on LH system was observed both in L and H modes. Those results are consistent with RF sheath rectification process. Concerning the influence of LHWs on ICRF coupling, nothing was observed in L-mode. Besides during H-mode experiments, LHWs have been identified as having a mitigating effect on ELMs [4], which on average lowers the pedestal, increasing edge densities to the profit of ICRF waves coupling.

## 1 Introduction

Lower Hybrid Current Drive (LHCD) consists in launching waves at a frequency lying in the GHz range of frequencies. Excited by klystrons, waves are transported into waveguides and launched in the plasma from a grill. Since LH Waves (LHWs) are evanescent at too low densities (bellow $7.10^{17}m^{-3}$ at 4.6GHz), gas puff is often used to increase electron densities locally in front of the grill to improve coupling efficiency. Thanks to their large parallel electric field, LHWs have the interesting property of damping very efficiently at high parallel phase velocities to the electron thermal speed. They are consequently able to drive off-axis current, allowing good control of plasma current profile, enabling to build internal transport barriers and access attractive steady-state operations with high bootstrap current. On ITER, LHCD system should enable to reach more than 70% of bootstrap current, with high $\beta_N$ (~3) and good confinement ($H_{ITER-89}$~2.5) [5].
Ion Cyclotron Range of Frequencies (ICRF) goes from 10 to 100MHz. ICRF waves are mostly being used for heating plasmas on tokamaks. Unlike LHWs, ICRF waves are excited from an antenna composed of current straps as close as possible to the plasma. ICRF waves are also evanescent at low densities and must tunnel through a thin evanescent layer before reaching higher density plasma where they can propagate normally until reaching a resonance layer and essentially exchange their energy with ions before being spread to the whole plasma by collisions. Unfortunately, single-pass absorption and coupling efficiency are never perfect, so that waves undergo several reflections in the Scrape-Off Layer (SOL), enhancing dramatically interactions with Plasma Facing Components (PFCs). Those interactions are generally the result of RF sheaths which characterization from a global point of view can be done through an increase of impurities production and consequently an increase in radiations. From a local point of view, ExB convections induce strong modifications on density profiles at the edge, often degrading both ICRF and LH waves coupling efficiencies.
By presenting experiments recently conducted on Experimental Advanced Superconducting Tokamak, we aimed at characterizing interaction processes during ICRH and LH operations. A brief overview of EAST and the position of objects and diagnostics of interests for this study are given in section 2. Further details on the design of a newly designed reciprocating probe which enabled to collect key information

* Guillaume Urbanczyk: Guillaume.urbanczyk@cea.fr

for this study are provided in section 3. Finally experimental results are addressed in section 4.

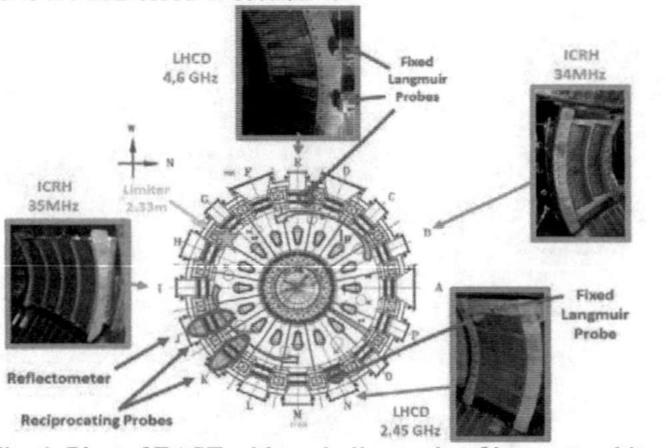


**Fig. 1.** Plan of EAST with each diagnostic of interest and its toroidal position

## 2 EAST presentation

Two ICRF antennas and two LH grills are available on EAST. In order to minimize interactions between each launcher, their positions were changed after the 2014 campaign to maximize distances between them (**Fig. 1.**). B and I ports ICRF antennas such as E and N ports LH grills are almost diametrically opposite in the torus to minimize mutual influences. While 6MW power is available on each ICRF antenna roughly working at same frequencies (34 and 35 MHz), each LH grill has different power, frequency, and wave number range; respectively 4MW for N-port at 2.45GHz with $2< n_{\|} <3$, and 6MW for E-port at 4.6GHz with $1.8< n_{\|} <2.5$. Langmuir probes were also fixed on each LH grill, respectively on top of the N-port and on the right side of the E-port. Reciprocating probes can be inserted from J and K ports, with an X-mode reflectometer in J-port [7] providing density profiles potentially useful to benchmark both probes results. However, the reflectometer's frequency range (from 4 to 6 GHz) is not appropriate to measurements in the SOL; it can only provide good results at densities above $10^{18}m^{-3}$. On the C-port, an Extreme UV spectrometer enables measurement of high-Z impurities in the plasma core [7].

## 3 Reciprocating Emissive and Langmuir probe

This probe (**Fig. 2.**) was specifically designed for the study of ICRF waves influence on the SOL. It consists in a combination of a 2cm tungsten wire of 150μm diameter, doped with thorium

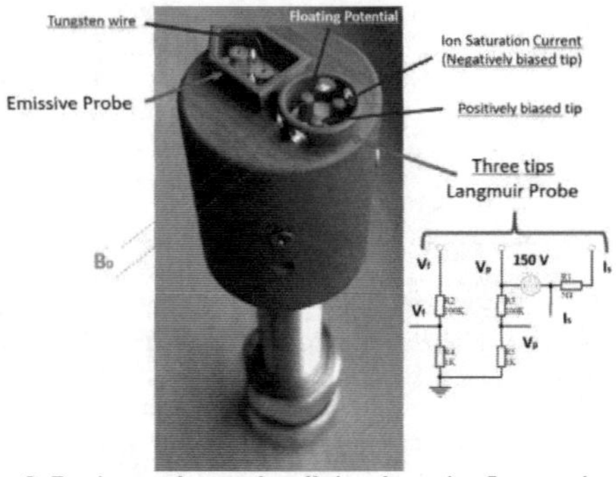


**Fig. 2.** Reciprocating probe allying three tips Langmuir and emissive techniques

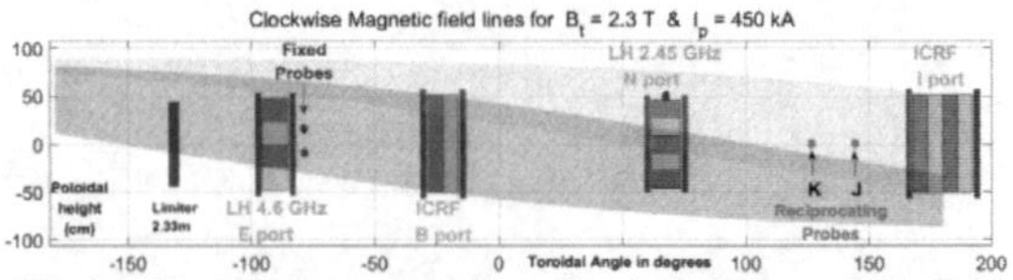


**Fig. 3.** Magnetic connection lines for clockwise magnetic field

to boost its emissive properties and a three tips Langmuir probes. Because emissive probe is hot, it enables to measure the potential of the plasma with much better precision than cold Langmuir probes. The idea is then to use emissive wire floating potential instead of Langmuir tips one, to calculate more precise values of plasma potential, electrons temperature and density. Reciprocations are fully controlled shot by shot from the main control room. Plunges timing, deepness, stationary laps inside the plasma are all set before each shot. Installation of the probe can conveniently be done in a small chamber which can be isolated from tokamak vacuum. Once the probe is ready, vacuum is restored after about two days pumping, and the probe can enter in the tokamak. This was particularly useful for experiments with the emissive probe, because unlike in machines with short discharges, EAST provides several seconds plasmas so that tungsten filament rapidly breaks after few plunges. Probes can be inserted into the middle plane of J and K ports, both well connected magnetically to I-port and not connected to B-port ICRF antennas (**Fig. 3.**).

## 4 Experimental results

Series of experiments on EAST aimed at characterizing the influence of ICRH on the SOL, by comparing the values of parameters measured by the reciprocating probes with and without ICRH, on L-mode plasma with currents of 450kA (**Fig. 4**). The probe was inserted two times per shot materialized by red and blue areas on **Fig. 5**, and remained deep in the SOL, behing the ICRF antenna limiter not to break the filament by plunging too deep. This explains why parameters measured are so small. Both ICRF antennas were also alternatively activated (**Fig. 5**), so that their respective influence can be compared. Figure 3 clearly shows that the K-port reciprocating probe was close and connected to the I-port antenna (red curves on **Fig. 4**), but far and not connected to the B-port antenna (blue curves on **Fig. 4**). Profiles in **Fig. 4** show that ICRH significantly enhances potentials on connected regions where RF fields are more intense (red curve gets above the blue one as the probe approaches the plasma), presumably leading to ExB density convections consistent with pump-out effects observed (red curves). SOL rectifications in regions far and not connected to the active antenna are much smaller than in connected cases, yet those rectifications are significant as can be seen comparing blue (non-connected ICRF antenna) with dashed black curves (no ICRH).

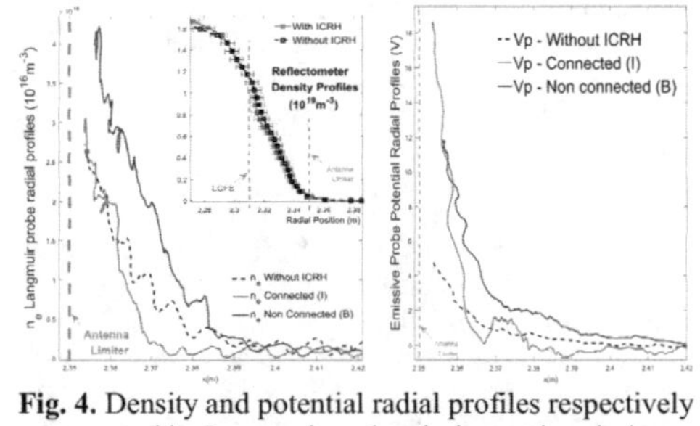


**Fig. 4.** Density and potential radial profiles respectively measured by Langmuir and emissive probes during reciprocations in the SOL in shots 64619 and 64623. Black

dashed curves represent cases without ICRH while red curves correspond to connected I-port antenna active and blue curves to non-connected B-port antenna active.

A density profile from the reflectometer was also added

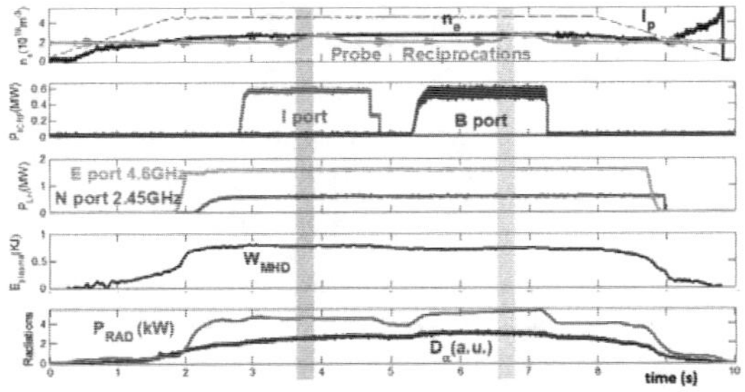


**Fig. 5.** Overview signals of the shot N°64619 similar to 64623. To the top are represented the line integrated density in blue, plasma current in dashed black and probe reciprocations in orange. Underneath we find each ICRF antenna power, then each LH grill power. Plasma energy is represented in in brown. Purple curve corresponds to plasma oscillations $D_\alpha$and radiations are in magenta.

During another series of experiments (**Fig. 6**), we tried to characterize the sheath formation at proximity and faraway of ICRF antennas for different power levels using a Langmuir reciprocating probe on the J-port and plunging until the LCFS. **Fig. 7** shows floating potentials radial profiles measured for each reciprocation. Blue curves correspond to B-port launcher, which was launching relatively low power, was not magnetically connected and too far to significantly influence SOL parameters of the J-port. Red curves corresponding to I-port antenna launching much higher power and being very close and connected show much different trends. In the far SOL, oscillations between positive and negative values over one to two centimeters are characteristics of strong electric fields and convections. A typical point of the graph is the hump around the antenna limiter. This zone is subject to significantly higher potential gradients than others as it has been observed on most devices equipped with ICRH. Amplitude of oscillations also increases with ICRF power, indicating good correlation with RF sheath-induced mechanism.

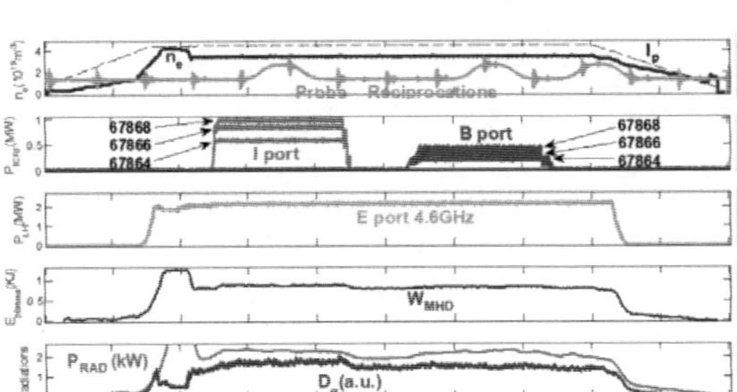


**Fig. 6.** Overview of the shots N°67864 / 6 / 8 with respectively three level of ICRF power. Represented signals are the same as in figure 5.

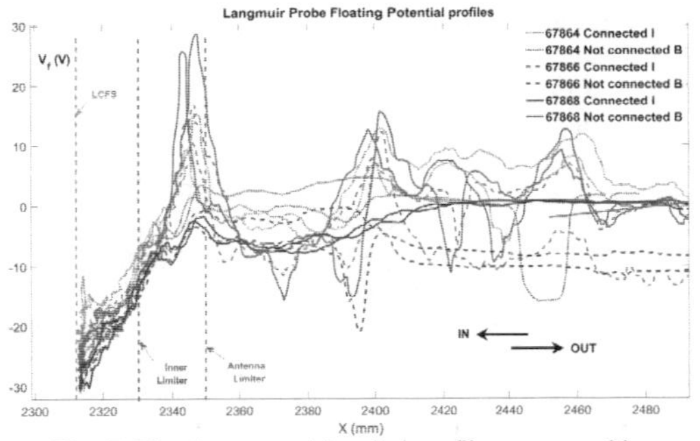


**Fig. 7.** Floating potential radial profiles measured by reciprocating Langmuir probe during shots N°67864 (dotted), 67866 (dashed) and 67868 (full). Red curves (respectively blue) correspond to I-port ICRF antenna active (B-port).

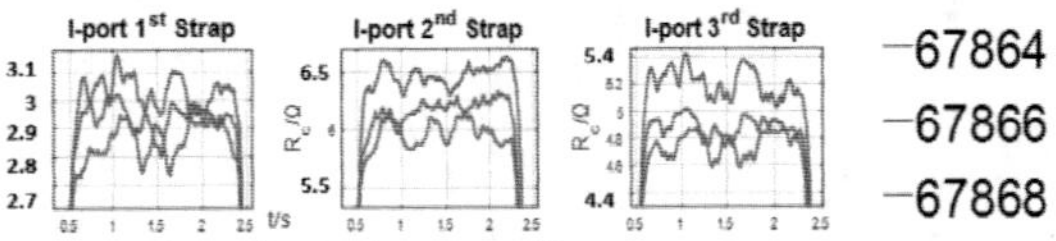


**Fig. 8.** Coupling resistances of I-port antenna current straps during shots N°67864 67866 and 67868, showing that coupling efficiency decreases with power.

As a consequence, potential differences result in electric fields and ExB density convections. Also known as convective cells, those may locally increase evanescent layers width and have deleterious effect on the coupling efficiency of surrounding current straps as can be seen in **Fig. 8.**

In order to investigate even more the role of ICRF power, series of experiments with power modulations (**Fig. 9**) not only enabled to evidence the role of the power on near field sheath looking at ICRF coupling resistances (bottom right graph), but also on farther regions looking at LH reflection coefficients [7] (bottom left graph). ICRF-induced sheath basically enhances density convections in the SOL making waves coupling more difficult. Those observations remain true both in L and H modes.

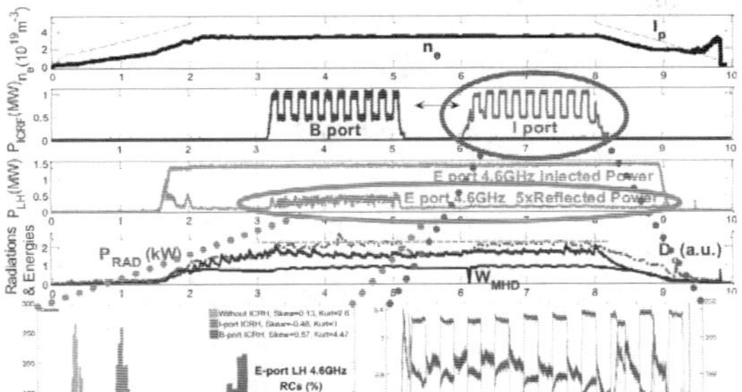


**Fig. 9.** Overview of the shot N°69949 on top (same signals as in figure 5).

Bottom left bar chart shows reflection coefficients of the E-port 4.6GHz LH launcher at three different times, respectively without ICRH and both in non-connected and connected cases. Bottom right graph finally shows the coupling resistance of an I-port ICRF antenna current strap. All show deleterious impact of ICRH power.

While no influence of LHWs on ICRF was noticed in L-mode, on the contrary in H-mode, LHWs were found to have some positive effects [8, 4]. Density gradient being stiff, in H-mode, pedestal tends to become sharper, density increases in the center more than at the edge and ICRF waves coupling becomes more complex. Fortunately, LHWs have been found to have a mitigating effect on ELMs [4] - changing type I into type III - diminishing their intensity and increasing their frequency. As can be seen in **Fig. 10**, this changes pedestal shape, it becomes softer, regulating edge densities, shortening ICRH waves evanescent length and improving the coupling. Such process could be very useful on ITER where it is planned to achieve high power H-mode plasmas.

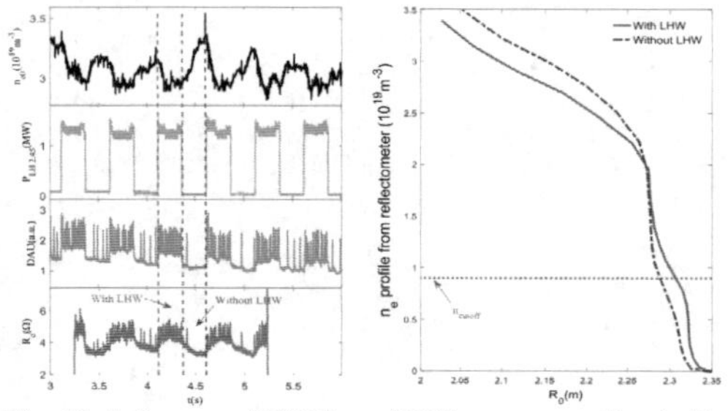


**Fig. 10.** Influence of LHWs on ICRF waves coupling in H-mode plasma. LH power (in green) mitigates ELMs (in orange), softening pedestal (right graph), decreasing core density at the benefit of the edge (black curve), improving ICRF waves coupling efficiency (coupling resistance in red)

## 5 Conclusions and Prospects

Over the EAST 2016 campaign, series of experiments in L-mode were carried out with the aim to characterize the mutual influence of ICRF and LH systems. In areas both magnetically connected and disconnected to active ICRF antennas, RF sheath was evidenced using a reciprocating probe allying both three tips Langmuir and emissive wire methods. No obvious influence was observed in non-connected regions, whereas in connected areas, SOL potential and temperature (cf formulas in Appendix) tend to increase with RF power. Electric fields appear at the edge and seem to create density convections, consistent with both global density decrease and ICRF deleterious influence on LHWs coupling. Reciprocally, whereas no influence of LHWs on SOL parameters was observed in L-mode, they can mitigate ELMs in H-mode, which influences the plasma edge by softening the pedestal and improving ICRF coupling.

## Appendix

We mentioned in the third part of this chapter that main advantages of the emissive probe are to provide a much more precise measurement of plasma potential, which can be written:

$$\Phi = V_{EP} + \alpha_{hot}.T_e / e \approx V_{EP} + T_e / e \qquad (1)$$

with the hot probe correction coefficient:

$$\alpha_{hot} = \ln\left(\frac{I_{es}}{I_s + I_{em}}\right) \approx 1 \qquad (2)$$

With $I_{em}$ the emitted current, $I_{es}$ and $I_s$ the electron and ion saturation currents respectively. Langmuir tips are in graphite; two of them are biased at ±150V and we assume it is sufficient to reach regimes of saturation in electrons ($V_p$) and ions($I_s$). The third tip is measuring the floating potential $V_f$. Using the following formulas we calculate the electron temperature $T_e$ (equ 3), the electron density $n_e$ (equ 4) and the plasma potential Φ (equ 5); with e the electron charge, $m_i$ the ion mass and $A_{pr}$ the tip effective area:

$$- T_e = (V_p - V_f) / \ln 2 \qquad (3)$$

$$- n_e = 2I_s / (eA_{tip}\sqrt{2eT_e / m_i}) \qquad (4)$$

$$- \Phi = V_f + \alpha_{cold}.T_e / e \approx V_f + 2.8T_e / e \qquad (5)$$

With the cold probe rectification coefficient

$$\alpha_{cold} = \ln(I_{es} / I_s) \approx \ln(\sqrt{\pi m_i / m_e}) = 2.8 \qquad (6)$$

Advantages of such triple probe are its cheap cost (simple electronic circuit), and very high time resolution, convenient for observing fast variation of plasma parameters due to turbulences. Nevertheless its sensitivity to RF sheath is its main drawback. The RF→DC rectification induced by ICRF waves shifts the I-V characteristic so that it is hard to determine if the fluctuations observed really reflect plasma dynamics or are just a mirage from DC rectification. For this reason, it is interesting to combine it with the emissive technique to benchmark the results with a better measure of the plasma potential (equ 5). For instance, combining relations 1 and 5 lead to another expression of the temperature, no longer depending on $V_p$– often leading to aberrant values due to electron drifts towards the edges in presence of RF E field – but only as a function of the floating potentials $V_f$measured by the cold Langmuir probe and $V_{EP}$ measured by the hot emissive wire:

$$T_e = (V_{EP} - V_f) / (\alpha_{cold} - \alpha_{hot}) \approx (V_{EP} - V_f) / 1.8 \qquad (7)$$

## Acknowledgements

The whole EAST ICRH team, colleagues in charge of reciprocating probes and French CEA colleagues; *Laurent Colas* and *Annika Ekedahl* are warmly acknowledged.
This work has been carried out within the framework of the EUROfusion Consortium and has received funding from the Euratom research and training program 2014-2018 under grant agreement No 633053. Views and opinions expressed herein do not necessarily reflect those of the European Commission.
This work was supported partly by National key research and development program (grant nos 2016YFA0400600 and 2016YFA0400601). This work was supported partly by National Magnetic confinement Fusion Science Programme (grant nos 2015GB101001). This work was also supported partly by the National Natural Science Foundation of China under grant nos 11675213, 11375235 and 11375236.

## References

1. L. Colas et al., PPCF **49** B35–B45 (2007)
2. R. Schrittwieser et al PPCF **44** 567-578 (2002)
3. K.K Kirov et al PPCF **51** 044003 (2009)
4. A. Ekedahl et al, 26th IAEA FEC 2016 (2016) EX/P7-34
5. Y. Peysson 2009 *5th Korean Fusion Plasma Summer School*
6. E.H.Kong & al. *Plasma Physics and Controlled Fusion* **54** (2012) 105003
7. Zhang Shoubiao *et al* 2014 *Plasma Sci. and Technol.* 16 311
8. J. H. Zhang et al, 2017 *Nucl. Fusion* **57** (2017) 066030

EPJ Web of Conferences **157**, 03058 (2017) DOI: 10.1051/epjconf/201715703058
*22 Topical Conference on Radio-Frequency Power in Plasmas*

# IShTAR ICRF antenna field characterization in vacuum and plasma by using probe diagnostic

*Mariia* Usoltceva[1,2,3,*], *Roman* Ochoukov[2], *Rodolphe* D'Inca[2], *Jonathan* Jacquot[2], *Kristel* Crombé[1,4], *Ana* Kostic[1,2], *Stéphane* Heuraux[3], *Eric* Faudot[3], *Jean-Marie* Noterdaeme[1,2], and the IShTAR team[2]

[1]Department of Applied Physics, Gent University, 9000 Gent, Belgium
[2]Max-Planck-Institut für Plasmaphysik, Boltzmannstr. 2, 85748 Garching, Germany
[3]Université de Lorraine, 54506 Vandœuvre-lès-Nancy, France
[4]LPP-ERM-KMS, TEC partner, 1000 Brussels, Belgium

**Abstract.** RF sheath physics is one of the key topics relevant for improvements of ICRF heating systems, which are present on nearly all modern magnetic fusion machines. This paper introduces developement and validation of a new approach to understanding general RF sheath physics. The presumed reason of enhanced plasma-antenna interactions, parallel electric field, is not measured directly, but proposed to be obtained from simulations in COMSOL Multiphysics® Modeling Software. Measurements of RF magnetic field components with B-dot probes are done on a linear device IShTAR (Ion cyclotron Sheath Test ARrangement) and then compared to simulations. Good resulting accordance is suggested to be the criterion for trustworthiness of parallel electric field estimation as a component of electromagnetic field in modeling. A comparison between simulation and experiment for one magnetic field component in vacuum has demonstrated a close match. An additional complication to this ICRF antenna field characterization study is imposed by the helicon antenna which is used as a plasma ignition tool in the test arrangement. The plasma case, in contrast to the vacuum case, must be approached carefully, since the overlapping of ICRF antenna and helicon antenna fields occurs. Distinguishing of the two fields is done by an analysis of correlation between measurements with both antennas together and with each one separately.

## 1 Introduction

Interaction between ICRF antenna fields and plasma of a tokamak edge has been studied for decades. RF sheath-accelerated ions have been shown to be the main reason of antenna limiters sputtering [1]. Attempts done on different experimental machines to reduce negative effects accompanying plasma-wall interactions are usually heuristic, with conclusions drawn basing on consequential parameters like impurities concentration, RF currents or temperature of limiting structures [2,3], not on the electrical field. Improvements are typically achieved for specific machine parameters, magnetic field angle, antenna geometry, toroidal phasing, etc. Beside that, modelling tools have been developed for SOL ICRF physics simulations. They are reviewed in details in [4].

In this paper, an approach to understand general RF sheath physics is described and first results are presented. Parallel electric field, supposedly responsible for the enhancement of interactions between antenna and plasma, is inherently connected to the magnetic field of an RF wave. Experimental results of magnetic field components measurements in the linear device IShTAR are used as a reference for simulations in COMSOL Multiphysics® Modeling Software in order to obtain target parallel electric field values.

## 2 Field characterization with probe diagnostic

The object of interest, a thin layer in the vicinity of an antenna, is hardly approachable by diagnostic tools, especially since the most common (and so far the only one applicable in SOL plasma) diagnostic for potential measurements, electrical probes, is known to introduce significant changes by its presence. That is why indirect measurements are preferred, though their results should be necessarily linked to the E-field values.

The electromagnetic field of an ICRF antenna can be simulated and values for each component of the field can be calculated separately. A relatively simple B-dot probe diagnostic is used for magnetic field measurements in our experimental device. A B-dot probe is able to provide local measurement of one component of the magnetic field. Assuming that agreement of experimental and numerical results for magnetic field components would mean the equality of electrical field values, we can expect to deduce the sought parallel electric field from simulation, as soon as we confirm the match of experiment and simulation results for the magnetic field.

* Corresponding author: mariia.usoltceva@ipp.mpg.de

## 3 Experimental setup

The objectives of IShTAR (Ion cyclotron Sheath Test ARrangement) are described in [5,6]. Fig. 1 illustrates all important features of the device: main vacuum chamber of 0.5 meter radius with an ICRF antenna (5.22 MHz, 1 kW of power) and a manipulator in front of it; small vacuum chamber attached on a side, with helicon antenna for plasma ignition (11.76 MHz, up to 3kW); two sets of magnetic coils of up to 0.24 T and 0.03 T, surrounding correspondingly big and small vacuum chambers. This setup allows around 10 second of plasma confinement, with a high density (~$10^{17}$ $m^{-3}$) plasma column of 15-20 cm in radius being in close proximity to the specially designed one-strap ICRF antenna. The working gas is He (Ar can be also used).

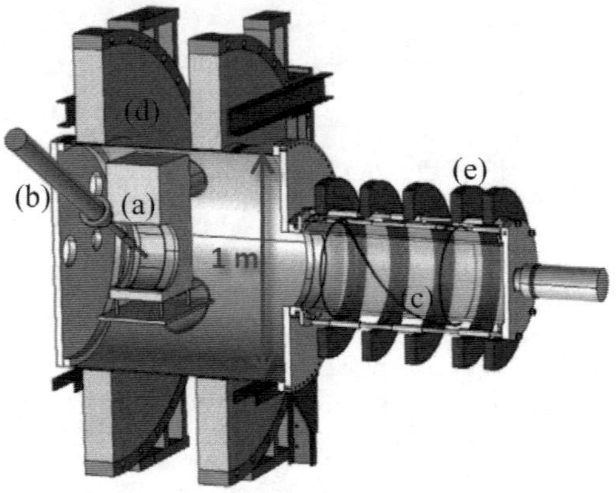


**Fig. 1.** IShTAR scheme: (a) ICRF antenna, (b) manipulator, (c) helicon antenna, (d) big magnetic coils, (e) small magnetic coils.

ICRF antenna magnetic field components are measured by B-dot probes. An array of 4 probes (Fig. 2) is installed on a radially movable manipulator in parallel to the main magnetic field (z-direction) with coordinates of -15, -5, 5 and 15 cm counting from the ICRF antenna center.

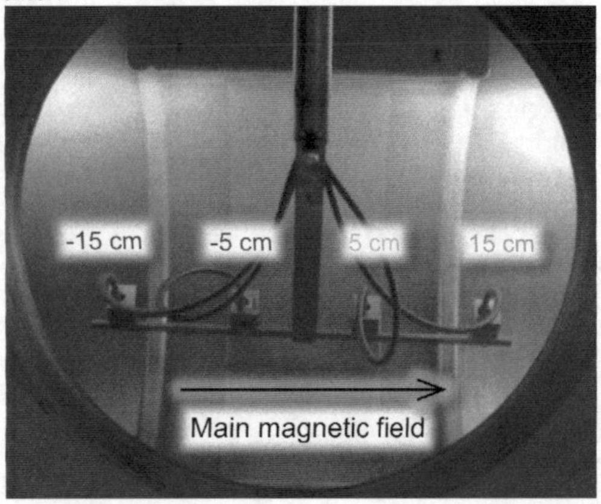


**Fig. 2.** Probe array on a manipulator in front of an ICRF antenna inside IShTAR.

## 4 Experimental results

The first thing that should be noted before proceeding to the experimental results, is that there is more than one radio-frequency antenna in the experimental setup. The helicon antenna situated in the small chamber and dedicated to start a discharge also emits RF waves. Its presence complicates the task of ICRF antenna magnetic field measurements and its contribution must always be taken into account.

In order to distinguish the magnetic fields of the two antennas, following scenarios have been studied:
(a) Plasma with only helicon antenna.
(b) Plasma with only ICRF antenna (since this antenna is not designed to ignite plasma, only weak plasma is produced; no wave propagation is registered, evanescent wave signal decays as in vacuum).
(c) Plasma with both antennas simultaneously.

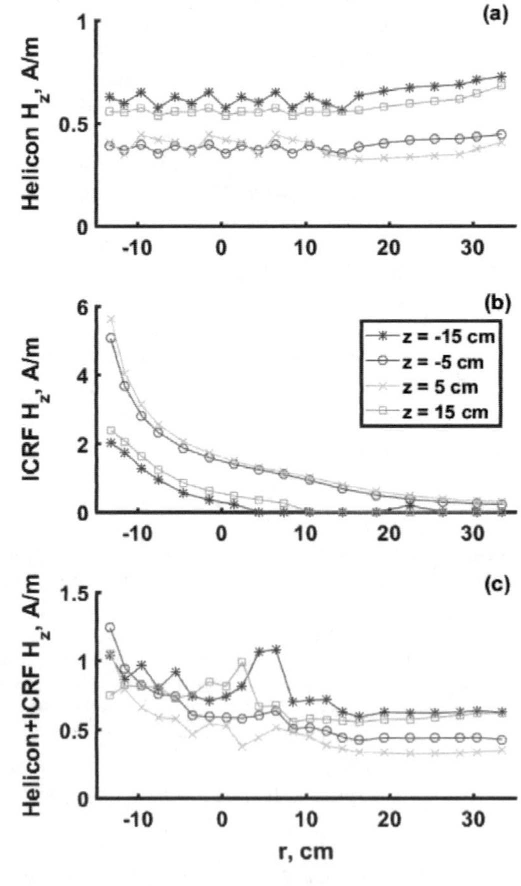


**Fig. 3.** Experimental $H_z(r)$ of: (a) helicon antenna, (b) ICRF antenna, (c) helicon and ICRF antennas together.

Experimental results (Fig. 3) reveal evident superposition of the two antenna fields. In the range of positive values of radius (farther away from the ICRF antenna) profiles of $H_z$ for all 4 probes have similar shape on two plots: Fig. 3a which shows the case with helicon antenna only, and Fig. 3c which is for the case with two antennas. In the vicinity of the ICRF antenna

DOI: 10.1051/epjconf/201715703058

(negative r) measurements demonstrate the impact of the ICRF antenna field on the resulting signal. A significant rise of signal towards the ICRF antenna, similar in shape to that observed on Fig. 3b, is especially well seen for the two central probes.

It can be noticed, however, that absolute values of ICRF antenna magnetic field on Fig 3b are significantly higher than those measured in plasma with both antennas. The reasons is the B-dot probes detectors, which have a constant attenuation factor for frequencies greater than 10 MHz, thus aiming to cut low-frequency noise. It is still possible to measure the 5.22 MHz signal of ICRF antenna with those detectors, using a different calibration (as it was done for results on Fig. 3b). But in combination with the 11,76 MHz frequency of the helicon antenna the signal of ICRF antenna becomes weak, being attenuated much stronger than the helicon antenna signal. For a complete analysis of the superimposed magnetic fields of the two antennas new detectors need to be made, with equal attenuation factor for both frequencies. Nevertheless, it can be already concluded that it is feasible to distinguish the ICRF antenna magnetic field from the sum of the two fields.

## 5 Simulation

Realistic geometry of all in-vessel parts (including ICRF antenna, manipulator and probes) has been built in COMSOL Multiphysics®. Fig. 4 is a view on the ICRF antenna without a side wall: the antenna strap, the limiting (supporting) structure and the feeding coaxial transmission line can be seen. On the background the small chamber is visible.

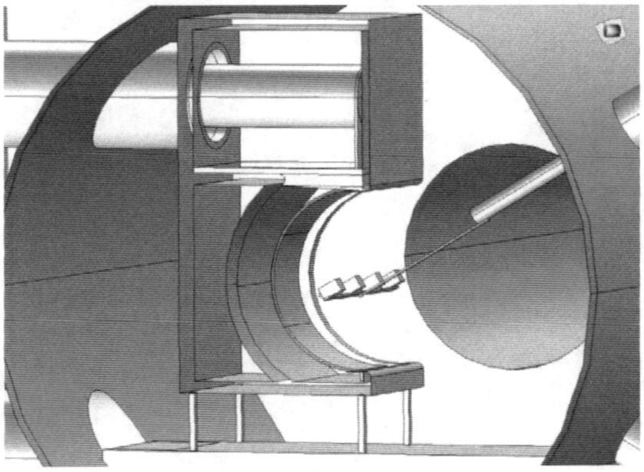

**Fig. 4.** IShTAR model with all main components: ICRF antenna (side wall removed), manipulator and probes.

### 5.1 Vacuum case

The standard Comsol Electromagnetics Module makes it possible to model the vacuum electromagnetic field of the IShTAR ICRF antenna. In vacuum, an RF wave of 5.22 MHz propagates in the coaxial transmission line and then becomes evanescent, causing very small fields inside the IShTAR chamber. Magnetic field distribution in logarithmic scale (log of A/m) is plotted on Fig. 5.

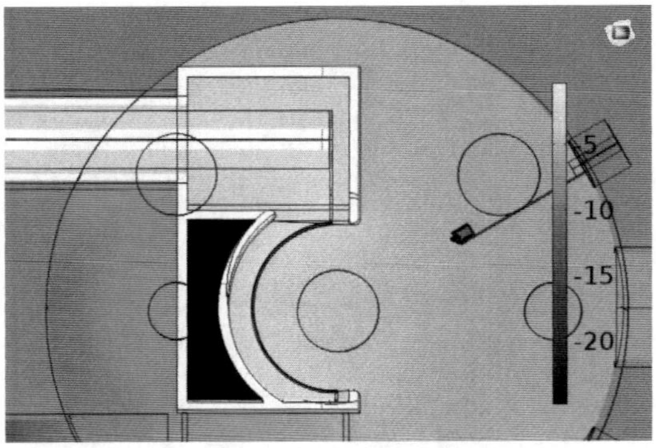


**Fig. 5.** Magnetic field of IShTAR ICRF antenna in vacuum in logarithmic scale.

Magnetic field measured by a B-dot probe is represented in simulation as an average field inside a cylindrical volume at the position and of the size of a real inductor used in a probe. Radial profiles of $H_z$ at 5.22 MHz for the 4 probes (Fig. 6) have the same shape in the experiment and in the modeling. Due to a perfect symmetry of the ICRF antenna and probes in the model results for two outer and two central probes are identical, which is well reproduced in the experiment. Absolute values do not play any role here, since they are fully dependent on the amount of injected power and the matching system parameters. This can be always varied in the simulation, if needed.

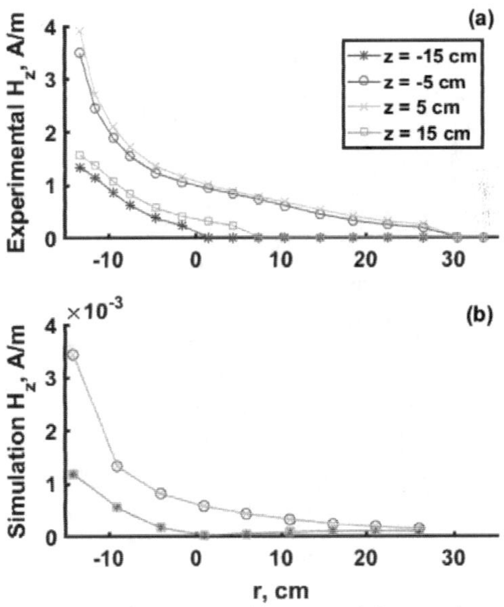


**Fig. 6.** $H_z(r)$ in vacuum: (a) measured in experiment, (b) calculated in simulation.

Four radial profiles obviously do not provide a full 2D plot of the magnetic field parameters. Here we are making an assumption that if a comparison of 4 radial experimental and simulation profiles is successful, an electromagnetic field distribution from the simulation is considered to closely depict the real field in our device.

DOI: 10.1051/epjconf/201715703058

For more certainty not only one component of magnetic field can be compared, but at least 2.

Since a simulation of electromagnetic field in COMSOL gives all components of the field at any geometrical point, parallel electric field is thus known at any required position. Calculated $H_z$ and $E_z$ in front of the antenna are provided as examples in Fig. 7 and 8.

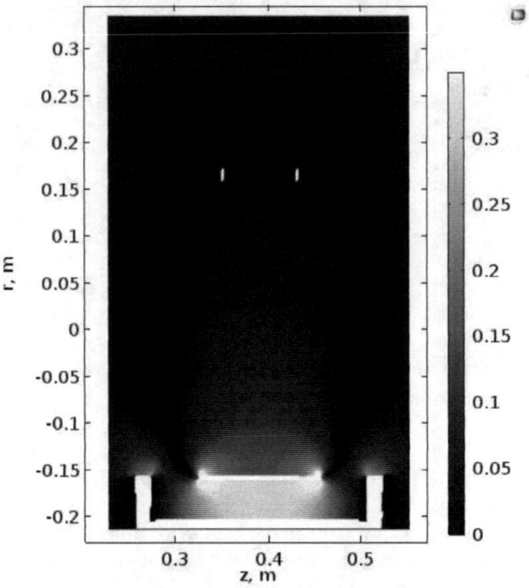


**Fig. 7.** Calculated 2D distribution of $H_z$ (in A/m) in front of the ICRF antenna in vacuum.

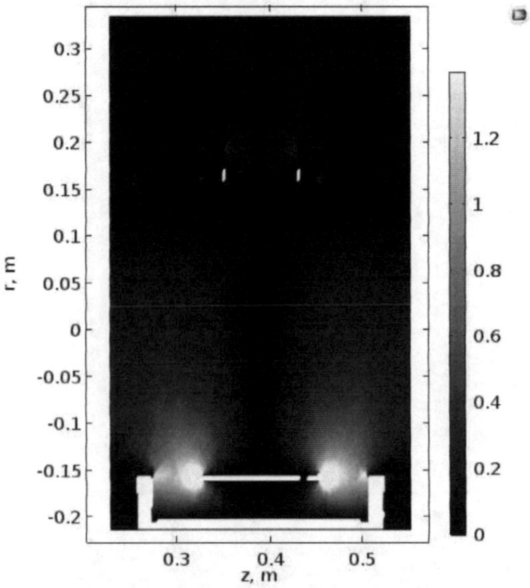


**Fig. 8.** Calculated 2D distribution of $E_z$ (in V/m) in front of the ICRF antenna in vacuum.

## 5.2 Plasma case

Plasma can be programmed in COMSOL's Electromagnetics Module as a material with manually assigned physical properties. The next step of the validation of the approach presented in this paper is planned to be a plasma simulation in IShTAR geometry. Ideally, each case (ICRF antenna, helicon antenna, both antennas together) needs to be addressed separately and comparisons with experiment have to be made accordingly.

# 6 Conclusion

An approach for obtaining parallel electric field of an ICRF antenna using a combination of experiment and simulation is presented. The obtained agreement of experimental and modelling results for one of the magnetic field components in vacuum case supports the proposed assumption of feasibility of the explained approach. Further comparisons, for other magnetic field components in vacuum and for the plasma case will provide more confident and detailed summary of the applicability of the discussed approach.

*This work has been carried out within the framework of the EUROfusion Consortium and has received funding from the Euratom research and training programme 2014-2018 under grant agreement No 633053. The views and opinions expressed herein do not necessarily reflect those of the European Commission.*

### References

1. D.A. D'Ippolito and J.R. Myra, Plasma Phys. Control. Fusion **33**, 607 (1991)
2. V.Bobkov et al., Nucl. Mater. Energy (2016)
3. W. Zhang et al., *22nd RFPPC* (2017)
4. L. Colas et al., *22nd RFPPC* (2017)
5. K. Crombé et al., *26th IAEA Fusion Energy Conference* (2016)
6. R. D'Inca et al., *AIP Conference* (2015)

# Modeling RF waves in spatially dispersive inhomogeneus plasma using an iterative wavelet spectral method

*Pablo* Vallejos[1,*], *Thomas* Johnson[1], and *Torbjörn* Hellsten[1]

[1]KTH Royal Institute of Technology, Teknikringen 29, 100 44 Stockholm, Sweden

**Abstract.** The wave equation for a spatially dispersive inhomogeneous magnetized plasma is given by an integro-differential equation. The effects caused by spatial dispersion in the directions perpendicular and parallel to the magnetic field are quite different. In this study, we show how to solve the wave equation using a newly developed iterative wavelet spectral method for two cases. In the first case, the method is applied to a propagating kinetic Alfvén wave in the perpendicular direction and solved to all orders in FLR. To conserve the kinetic energy flux, first order corrections in equilibrium gradients are used in the dielectric response tensor. In the second case, we verify the method for a fast wave minority heating scenario and study the up- and downshift in the parallel wave number.

## 1 Introduction

Magnetized plasma exhibit spatial dispersive effects both in the perpendicular and parallel directions. Perpendicular spatial dispersion arises due to finite Larmor radius (FLR) effects and becomes important when the perpendicular wavelength is comparable to the Larmor radius of a plasma species. Parallel spatial dispersion is related to the Doppler shifts caused by the parallel thermal velocity and is important near cyclotron resonances. The response of a spatially dispersive medium is of non-local character. If the medium is inhomogeneous the wave equation becomes an integro-differential equation [1-6].

Common numerical methods to solve the wave equation include the finite element method (FEM), spectral methods, or combinations of the two. By expansion of the FLR effects with respect to the inhomogeneity the integro-differential equation can be approximated with a higher order differential equation that can be solved with FEM. The up- and downshift in the parallel wave vector is more difficult to account for since the resonant interactions take place over several wavelengths. Methods based on FEM typically neglect the up- and downshift in the parallel wave vector [7-8]. Fourier spectral methods can be used for solving problems with spatial dispersion. However, these methods tend to produce dense and large matrices which are time consuming to solve.

Green et al proposed a method where the dielectric response is splitted into a dispersive and non-dispersive part [9]. By using an iterative scheme, the induced current is added by means of iteration.

More recently, a new method has been proposed based on operator splitting, an iterative scheme and wavelets [3,4]. Operator splitting is applied to separate dispersive and non-dispersive terms in the wave equation. The dispersive terms are added by means of iteration using Anderson Acceleration scheme [10] and the spatially dispersive response is evaluated using a Morlet wavelet representation of the electric field and dielectric response [11].

The purpose of this study is to verify the method by applying it to two cases. In the first case, the method is applied to a propagating kinetic Alfvén wave in the perpendicular direction and solved to all orders in FLR. To conserve the kinetic energy flux, first order corrections in equilibrium gradients are included [5,6]. In the second case, we apply the method to a fast wave minority heating scenario and study the up- and downshift in the parallel wave number.

## 2 Spatially dispersive wave equation for Maxwellian plasma

The spatially dispersive wave equation for time harmonic waves in inhomogeneous media is given by

$$\nabla \times \nabla \times \mathbf{E} - \frac{\omega^2}{c^2} K[\mathbf{E}](\mathbf{r}) = i\omega\mu_0 \mathbf{J}_{ant}(\mathbf{r}), \qquad (1)$$

where $K$ is an operator defined as [5,6]

$$K[\mathbf{E}](\mathbf{r}) = \int \frac{d\mathbf{k}}{(2\pi)^3} \int d\mathbf{r}' \mathbf{K}(\mathbf{r},\mathbf{k}) \mathbf{E}(\mathbf{r}') e^{i\mathbf{k}\cdot(\mathbf{r}-\mathbf{r}')} \qquad (2)$$

$$= L[\mathbf{K}(\mathbf{r},\mathbf{k}), \mathbf{E}(\mathbf{r})]$$

where $\mathbf{K}$ is the dielectric kernel. The $L$ operator has been defined for simplicity in writing. Note that if the response is not spatially dispersive, then

$$L[\mathbf{K}(\mathbf{r}), \mathbf{E}(\mathbf{r})] = \mathbf{K}(\mathbf{r})\mathbf{E}(\mathbf{r}). \qquad (3)$$

* Corresponding author: pablova@kth.se

To simplify our analysis, we have taken the limit for the parallel dielectric response, such that $E_{||} \to 0$.

### 2.1 Kinetic Alfvén wave

For a plasma consisting of electrons and deuterium ions, the equation for a propagating kinetic Alfvén is given by

$$n_{||}^2 E_x(x) - L[K_{xx}(x,k_x), E_x(x)] = 0, \quad (4)$$

$$K_{xx}(x,k_x) = 1 + \chi_{xx}(x,k_x),$$

where $n_{||}$ is the parallel refractive index. To include first order corrections in equilibrium gradients, we use the susceptibility given by [3]

$$\chi_{xx}(x,k_x) = \left(1 - \frac{i}{2}\frac{\partial}{\partial k_x}\frac{\partial}{\partial x}\right)\chi_{xx}^{HOMO}(x,k_x) \quad (5)$$

where the second term in the bracket is the first order correction term and

$$\chi_{xx}^{HOMO}(x,k_x) = \sum_{j=e,D} \frac{\omega_{p,j}^2}{\omega}\left(\frac{1}{\omega+\Omega_j} + \frac{1}{\omega-\Omega_j}\right)e^{-\lambda_j}\frac{I_1(\lambda_j)}{\lambda_j},$$

is the hot susceptibility tensor for homogeneous media given by Stix [1,2], $\omega_{p,j}$ is the plasma frequency, $\Omega_j$ is cyclotron frequency, $I_1$ is the modified Bessel function of order 1 and $\lambda_j = 0.5k_x^2\rho_{L,j}^2$ where $\rho_{L,j}$ is the Larmor radius. We have assumed that $\omega \ll \Omega_D, \Omega_e$ so that $K_{xy} \approx 0$. The tensor is Hermitian (no absorption) and the wave does not propagate in the $y$ direction. The kinetic energy flux is given by [1,2]

$$T_x(x) = -\frac{\omega\epsilon_0}{4}E_x^*\left(\frac{d}{dk_x}K_{xx}\right)E_x. \quad (6)$$

### 2.2 Fast wave minority heating

The up- and downshift in the parallel wave number is here studied for a fast wave minority heating scenario in a plasma of electrons, deuterium and hydrogen minority ions. The geometry can be described using a Cartesian coordinate system, $(x,y,z)$, where the plasma is homogeneous in the $y$ and $z$ directions. The magnetic field is described by

$$\mathbf{B}(x) = B(x)(\cos\alpha\, \boldsymbol{e}_z + \sin\alpha\, \boldsymbol{e}_x),$$

where $\alpha$ is the angle between the $z$-axis and parallel direction. Neglecting the parallel electric field, the wave equation for the fast wave in a Maxwellian plasma is given by [1,2]

$$\mathbf{ME}_\perp - \frac{\omega^2}{c^2}L[\mathbf{K}_0(x,k_x), \mathbf{E}_\perp(x)] = \mathbf{0}, \quad (7)$$

where

$$\mathbf{M} = \begin{bmatrix} k_z^2\cos^2\alpha & 0 \\ 0 & k_z^2 - \frac{\partial^2}{\partial x^2} \end{bmatrix},$$

$$\mathbf{K_0} = \mathbf{K}_{cold} + \chi_H,$$

$$\mathbf{K}_{cold} = \mathbf{I} - \sum_{j=e,D}\frac{\omega_{p,j}}{\omega(\omega^2-\Omega_j^2)}\begin{bmatrix}\omega & i\Omega_j \\ -i\Omega_j & \omega\end{bmatrix},$$

$$\chi_H = \frac{1}{2}\frac{\omega_{p,H}^2}{\omega}\left(\frac{Z(\zeta)}{k_{||}v_{th}}\begin{bmatrix}1 & i \\ -i & 1\end{bmatrix} - \frac{1}{\omega+\Omega_H}\begin{bmatrix}1 & -i \\ i & 1\end{bmatrix}\right),$$

$$\zeta = \frac{\omega - \Omega_H + k_{||}v_p}{k_{||}v_{th}},$$

$$k_{||} = k_z\cos\alpha + k_x\sin\alpha, \quad (8)$$

where $Z(\zeta)$ is the plasma dispersion function [1,2], $v_{th}$is the thermal velocity and $v_p$ is the mean plasma fluid velocity. The local absorption is given by [1,2]

$$P(x) = \omega\epsilon_0\mathbf{E}_\perp{}^*\mathbf{K}_0^A\mathbf{E}_\perp \quad (9)$$

where $\mathbf{K}_0^A$ is the anti-Hermitian part of the dielectric tensor.

# 3 Wavelet representation of the wave equation

The dielectric response tensors in Eqs. (2), (4) and (7) are Fourier transformed in space. To represent the dielectric response using a complex Morlet wavelet representation, we use the following equation [3,4]

$$L[\mathbf{K}(x,k), \mathbf{E}(x)] = \mathbf{WT}^{-1}[\mathbf{K}_{a,b}(x)\mathbf{E}_{a,b}](x) \quad (10)$$

$$\mathbf{K}_{a,b}(x) = \mathbf{K}\left(x, \frac{\sigma}{a}\right) + iX\left.\frac{\partial\mathbf{K}}{\partial k}\right|_{\frac{\sigma}{a}} + \frac{1}{2a^2}(1-X^2)\left.\frac{\partial^2\mathbf{K}}{\partial k^2}\right|_{\frac{\sigma}{a}}$$

where $\mathbf{WT}^{-1}$ is the inverse wavelet transform, $a$ and $b$ are the wavelet scale and dilatation parameters, $\sigma = 6$ is a constant, $X = (x-b)/a$ and

$$\mathbf{E}_{a,b} = \mathbf{WT}[\mathbf{E}(x)](a,b), \quad (11)$$

is the wavelet transform of the electric field [11].

### 3.1 Wavelet representation of the kinetic Alfvén wave equation

To construct an iterative solution for Eq. (4) we move the dispersive terms to the right hand side and use Eq. (10) to obtain

$$(n_{||}^2 - 1 - \chi_e)E_x = \mathbf{WT}^{-1}[\chi_{a,b}^i(x)E_{a,b}](x).$$

Since the left hand side is algebraic, a fix-point formulation can be constructed

$$E_x^{(n+1)} = \frac{1}{n_{||}^2 - 1 - \chi_e}\mathbf{WT}^{-1}[\chi_{a,b}^i(x)E_{a,b}^{(n)}](x). \quad (12)$$

This equation tend to be unstable. However, stable solutions can be found using the Anderson Acceleration scheme [10].

### 3.2 Wavelet representation of the fast wave minority heating equation

In this study, the curl-curl operator in Eq. (7) is solved using the iterative wavelet spectral method. By Fourier transforming the curl-curl operator in the $x$ direction, we can rewrite Eq. (7) as

$$L\left[\frac{c^2}{\omega^2}\mathbf{M}_k(k_x) + \mathbf{K}_0(x, k_x), \mathbf{E}_\perp(x)\right] = \mathbf{0}. \quad (13)$$

By using Eq. (10) and Anderson Acceleration scheme, the solution is given by

$$\mathbf{E}_\perp^{(n+1)} = -\mathbf{K}_{cold}^{-1} \cdot \mathbf{WT}^{-1}\left[\left(\frac{c^2}{\omega^2}\mathbf{M}_{a,b}^k + \boldsymbol{\chi}_{a,b}^H\right)\mathbf{E}_{\perp,a,b}^{(n)}\right](x). \quad (14)$$

## 4 Results

### 4.1. Kinetic Alfvén wave

The solution to the kinetic Alfvén wave was obtained on an interval $x = [2,8]$ using the following parameters; $B = 1.32 + 0.1cos\,(-\pi(x-2)/6)$ T, $T_e = 1$ keV, $n_e = 5e19$ m$^{-3}$, $k_z = 1$ m$^{-1}$ and a wave frequency of 500 kHz. The wave is excited on the left boundary and propagates to the right. The wave is absorped at the right boundary (no reflection) and reflections due to inhomogeneities are neglected.

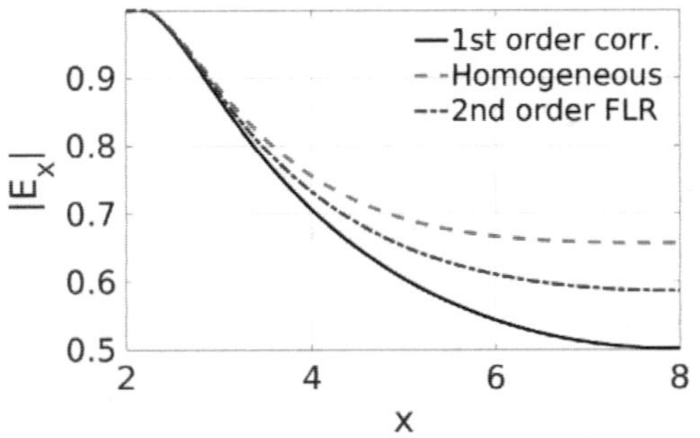


**Fig. 1.** Amplitude of the electric field $E_x$. The solid black curve is the solution to Eq. (4). The red dashed and blue dashed-dotted solutions are obtained using $K_{xx} = 1 + \chi_{xx}^{HOMO}$. FLR effects have been expanded to order 2 for the blue dashed-dotted solution.

The amplitude of the solution to Eq. (4) is shown in Figure 1 (black solid line). The solution is compared to the red dashed and blue dashed-dotted solutions, which are obtained assuming a homogeneous dielectric response tensor, i.e $K_{xx} = 1 + \chi_{xx}^{HOMO}$. The FLR effects are expanded to order 2 in the blue dashed-dotted solution.

As the wave propagates, the rate of amplitude decay depends on the dielectric tensor model used in the wave equation. The solution to Eq. (4) has the fastest decay in amplitude, while the solutions using a homogeneous response tensor have a weaker amplitude decay.

Figure 2 shows the kinetic energy flux of the three wave solutions. The solution to Eq. (4) show a constant energy flux, i.e. conserves the energy (energy flux difference between left and right boundary is ~1%). The other two solutions show a strongly increasing trend, i.e. energy is not conserved.

Figure 3 shows the wavelet spectrum of the wave solution. The wavelet scale parameter and wave number are related through $a = \sigma/k_x$. The figure shows how the spectrum is localized near values that correspond to the dispersion relation of the kinetic Alfvén wave.

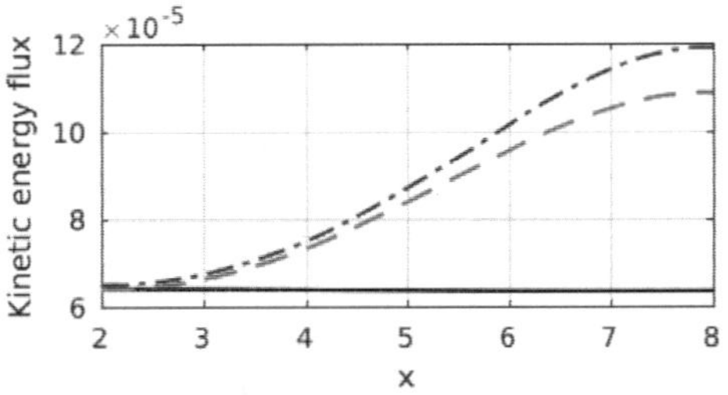


**Fig. 2.** Kinetic energy flux as a function of $x$. Same color coding as Figure 1. Conservation of the energy is only acquired using first order corrections in the dielectric response tensor.

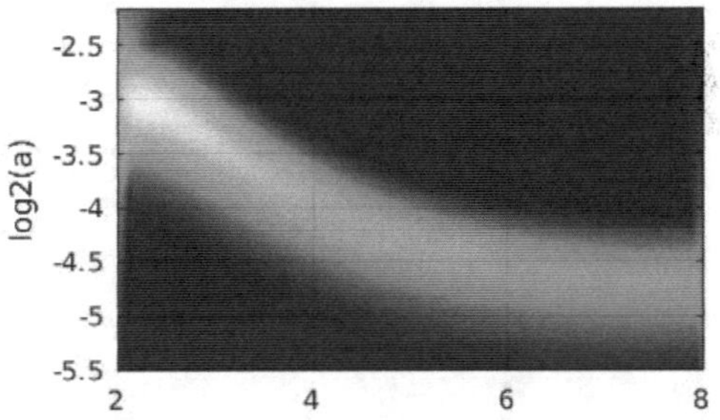


**Fig 3.** Wavelet spectra of $E_x$ component of the kinetic Alfven wave. The scale vector is related to the wave number through $a = \sigma/k_x$.

### 4.2 Fast wave minority heating

The solution to the fast wave was obtained on an interval $x = [2,4]$ using the following parameters; magnetic field $B = B_0R_0/x$, $B_0 = 3.8$ T, $R_0 = 3$ m, $T_e = 8\,keV$, $n_e = 4e19$ m$^{-3}$, $n_\phi = 25$, $\alpha = 0.051$ rad with a wave frequency of 51.1 MHz and hydrogen minority concentration of 1.5%. To separate the location of the absorption of the incident and reflected waves, a finite equilibrium velocity is included, $v_p = 8e6$ m/s. The cold hydrogen cyclotron resonance (i.e. $\omega = \Omega_H$) is located at $r = 3.43$ m, while the Doppler shifted cyclotron resonance (i.e. $\omega = \Omega_H + k_\| v_p$) is located near $r = 2.82$ m when excluding the up- and downshift in the parallel wave number.

The solution to Eq. (7) is shown in Figure 4. The fast wave is excited using a boundary condition on the right side and propagates to the left (see blue curve). The wave is reflected at the left boundary and propagates back to the right boundary (see red curve), were it is absorbed. The fast wave will therefore pass the ion cyclotron resonance twice, but with different $k_\|$.

The deuterium resonance is located near $r$ = 1.7 m, i.e. outside the simulated interval. As the wave approaches the deuterium resonance, the solution in Figure 4 verifies that the amplitude for $E_+$ decays due to screening by the deutrons.

Similarly, the hydrogen ions tend to screen $E_+$ near the Doppler shifted resonance, causing a local minimum. Figure 4 shows that the minima for the incident and reflected waves are located at 3.0 and 2.8 m respectively. The difference in location is due to the up- and downshift in $k_{||}$ of the two waves.

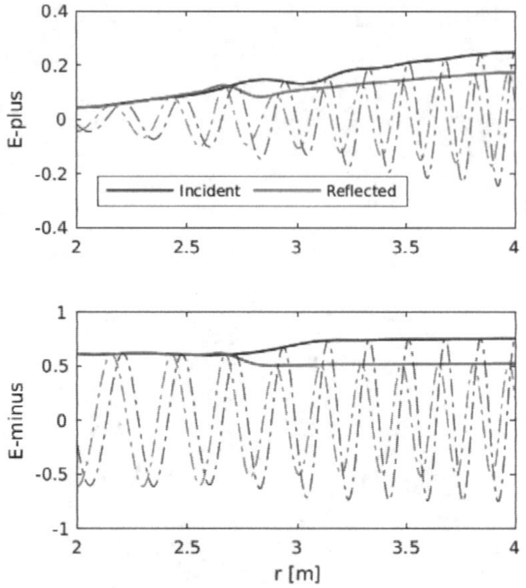


**Fig 4.** Left and right hand circular wave components of the fast wave. Solid and dashed-dotted curves show the amplitude and real part of the wave solutions. The wave is excited at the right boundary (blue) and propagates to the left boundary where it is reflected (red). The reflected wave propagates back to the right boundary, were it is absorbed.

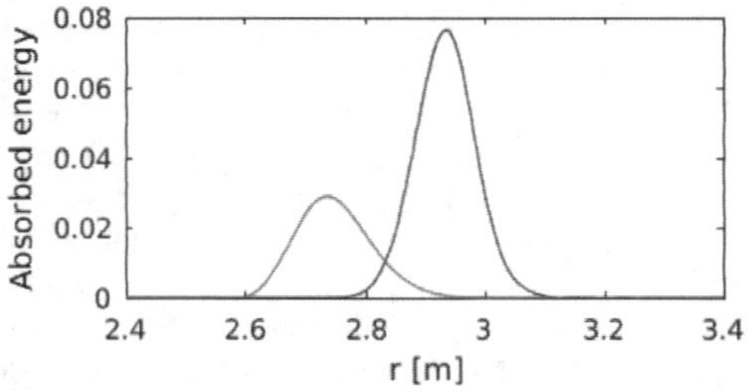


**Fig 5.** Absorbed energy of the incident (blue) and reflected (red) wave.

The absorption of energy is shown in Figure 5. The absorption peak of the incident wave is located at 2.93 m, while the reflected wave have the peak at 2.74 m. This shift in the absorption is a direct consequence of the up- and downshift of $k_{||}$. Another effect of is the different full width at half maximum (FWHM) values. The FWHM values are 0.12 m and 0.14 m for the incident and reflected waves respectively.

## 5 Discussion and conclusions

In this study we have shown that this new iterative scheme is capable of solving spatially dispersive wave equation for inhomogeneous media.

The method has been used to study the energy flux of a propagating kinetic Alfvén wave to all orders in FLR. The results show that in order to conserve the kinetic energy flux, first order corrections in equilibrium gradients must be included. The solution using the homogeneous dielectric tensor (no first order corrections) did not conserve the energy. Note that the solution to the wave equation using a second order expansion in FLR effects gives a better amplitude solution compared to the all-order solution without first order corrections. To conclude, first order corrections in the equilibrium gradients are essential for accurate solutions and energy conservation in inhomogeneous plasma.

In the second case, the method has been successfully applied to a fast wave minority scenario to study the up- and downshift of the parallel wave number. The method is capable of separating the responses for the incident and reflected waves and account for the up- and downshift of $k_{||}$, which is difficult to do with FEM.

The calculation of the fast wave heating model took about ~200 iterations. The reason is that the curl-curl operator is included iteratively. The performance is expected to improve by inverting the curl-curl operator with a finite element scheme and use the iterative wavelet spectral scheme to include the spatial dispersion.

## References

1. T.H. Stix, *Waves in plasmas* (AIP, New York, 1992)
2. D.G. Swanson, *Plasma Waves* $2^{nd}$ ed., (IOP Publishing, 2009)
3. T. Hellsten, T. Johnson, P. Vallejos, Journal of Physics: Conference Series **561** 012010 (2014)
4. P. Vallejos, T. Johnson, T. Hellsten, Journal of Physics: Conference Series **775** 012016 (2016)
5. R.J. Dumont, C.K. Phillips, D.N. Smithe, Phys. of Plasmas **12** 042508 (2005)
6. D.N. Smithe, Plasma Phys. Control. Fusion **31** 1105 (1989)
7. J.Hedin et al, Nuclear Fusion **42** 527-540 (2002)
8. L. Villard et al, Comp. Phys. Reports **4** 95 (1986)
9. D.L. Green, L.A. Berry, Comput. Phys. Commun. **185** 736 (2014)
10. H.F. Walker, P. Ni, SIAM J Numer. Anal. **49** 1715 (2011)
11. I. Daubechies Ten lectures on wavelets (Philadelphia, PA, USA: Society for Industrial and Applied Mathematics) (1992)

EPJ Web of Conferences **157**, 03060 (2017) DOI: 10.1051/epjconf/201715703060
*22 Topical Conference on Radio-Frequency Power in Plasmas*

# On all-FLR order numerical modelling of RF wave propagation and damping

*Dirk* Van Eester[1,*], *Ernesto* Lerche[1,2]

[1]LPP-ERM/KMS, EUROfusion Consortium Member - Trilateral Euregio Cluster, Brussels, Belgium
[2]CCFE, Culham Science Centre, Abingdon, OX14 3DB, UK

**Abstract.** Determining the fate of waves in hot magnetized plasmas in magnetic confinement fusion machines close to arbitrary cyclotron harmonics retaining finite Larmor radius effects requires solving the relevant integro-differential wave equation for waves of arbitrary wavelength. Anticipating exploitation in as realistic geometry as possible this requires massive computer resources. Work is ongoing to attempt reducing the required CPU memory and time. As Morlet wavelets are localised in x-space as well as k-space, they potentially offer a means to solve the relevant wave equation at reduced CPU requirement cost. Encouraging first results are obtained.

## 1 Introduction

Ion cyclotron resonance heating (ICRH) is a routinely used method to bring plasmas to fusion relevant temperatures in magnetic confinement fusion machines [1]. Properly modelling the dielectric response is a challenge both from the physics and the computational point of view. More often than not simplifications are made, first of all by truncating the dielectric tensor at the leading order terms in finite Larmor radius (FLR) corrections and secondly by simplifying the geometry. A new solver is under development to help understand the ICRH wave propagation and damping in multiple dimensions. It is inspired on the basic philosophy pioneered by Jaeger [2] for the AORSA code, and intends to provide a rigorous computation of the wave polarisation at arbitrary cyclotron harmonics. It will be useful both to study hot plasma wave propagation and damping effects in fusion relevant plasmas, and to add kinetic effects to the description contained in codes describing the electric field close to the antenna. The key idea at this initial exploratory stage is to assess the importance of higher order finite Larmor radius corrections while keeping the zero order distribution functions simple (Maxwellian) and the geometry prescribed.

## 2 Finite Larmor radius effects

A hot plasma admits an infinity of wave modes. Most of the key physics is captured by the "FLR2" models (retaining up to second order FLR kinetic terms i.e. corrections in $k_p\rho$ where $k_p$ is the perpendicular wavelength and $\rho$ the Larmor radius) routinely used to describe the wave propagation and damping. Some physics aspects are outside this approach's scope, however. Examples are heating at higher cyclotron harmonics (N>2 where N is the cyclotron harmonic), wave absorption by very energetic populations and the fate of power carried by short wavelength waves (in particular the waves excited by the fast wave at the ion-ion hybrid layer).

The FLR2 model resting on the smallness of the $k_p\rho$, upgrading the model to avoid errors arising from violations of this assumption necessitates solving the wave equation in its integro-differential form. Solving such an equation requires dedicated tools.

Jaeger solves the relevant equation directly in **k**-space, relying on the collocation method [2]. In his approach, the continuous integral in **k**-space allowing to account for the dielectric response of the hot plasma to the wave is replaced by a discrete set of Fourier integrals. In order to be rigorous, the considered **k**-spectrum should cover all physically important modes. Although proven to be extremely powerful, Fourier analysis has some drawbacks: since all Fourier modes are coupled, the system matrix of the corresponding linear problem is full, requiring vast computer resources. Another drawback is that Fourier analysis is ill-suited to describe evanescent modes since its building blocks (although forming a complete base and thus mathematically perfectly suited) represent propagative i.e. oscillating waves. It takes a lot of oscillatory base functions to approximate exponential growth or decay sufficiently accurately.

The ultimate goal of this work is to solve the hot plasma wave equation in 2 or 3 dimensions accounting for the realistic geometries of tokamaks and stellarators. Two roads are being pursued in parallel: (i) a "brute force" Fourier technique that puts the computational effort on supercomputers as it requires inversion of massive, full matrices, and (ii) a technique that explores the possibility of reducing the computational effort by semi-analytically solving some of the required integrals by hand and exploiting the fact that the net effect of very fast oscillations potentially offers a justification for reducing the full matrix to a sparse(r) one. Whereas the first is directed at exploring available numerical techniques to solve large linear systems swiftly and accurately, the present paper focusses on the reformulation of the equation in a suitable way attempting to avoid the need to invert a large, full matrix by transforming it in a sparse(r) one. Prior to exploring multi-dimensional application, tackling the integrodifferential equation in 1D is thought to be a first necessary step.

Recently, the "influence matrix" technique was explored [3, 4]. This technique is similar to the finite element technique in that it locally approximates the solution by a

* *Corresponding author: d.van.eester@fz-juelich.de*

DOI: 10.1051/epjconf/201715703060

low order polynomial. The equation at hand is solved in each subdomain for a complete set of "base solutions" consistent with the equation. Hence it represents all possible linearly independent solutions for the degree of accuracy imposed by the chosen base polynomials. The remaining degrees of freedom are exploited to impose the connection of the physical solution across interfaces. Although this technique was demonstrated to be very versatile and fairly easy to implement, it proved unable to provide converged solutions for the wave equation at hand.

A wavelet approach has now been attempted. Wavelets have the advantage that they are local, a characteristic they share with finite element base functions. In an elegant way, this potentially yields sufficiently sparse rather than full matrices for the non-local wave equation at hand. Moreover, certain classes of wavelets resemble wave packets and are therefore particularly suited to describe waves. In the magnetic confinement fusion domain, this technique was successfully pioneered by Vallejos et al. [5].

## 3 Wavelet approach

Morlet wavelets are a much adopted kind of wavelets. They are a family of functions

$$\varphi_{a,b} = \frac{1}{a^{1/2}}\varphi((x-b)/a)$$

where $\varphi_{1,0=}\varphi$ is the mother wavelet

$$\varphi\,(x) = \frac{1}{\pi^{1/4}}\exp\left[-\frac{x^2}{2}\right][exp[i\sigma x] - \kappa]$$

with $\kappa = \exp\,[-\sigma^2/2]$ and $\sigma$ a suitable constant. For wave applications, it is more logical to use $K = 1/a$ rather than $a$ as a variable. This has the supplementary advantage that it removes a singular factor $1/a^2$ from the integrand of the inverse wavelet transform, which is defined as

$$f(x) = \int_{-\infty}^{+\infty} dK \int_{-\infty}^{+\infty} db\, \tilde{f}(K,b)\;\varphi_{K,b}(x)$$

for a function $f$. The Morlet wavelet is localised in space around $x = b$ and in k-space around $\sigma/a$. An example is shown in Fig. 1. The top graph depicts the wavelet and the bottom graph its Fourier transform.

The method can be illustrated via a simple example: the solution of the wave equation

$$\frac{d^2\mathrm{E}(x)}{dx^2} + Q(x)\mathrm{E}(x) = 0$$

with 2 boundary conditions of the type

$$\alpha_0\mathrm{E} + \alpha_1\frac{d\mathrm{E}}{dx} = \beta$$

at the extremities of the interval of interest can be found by introducing the definition of the wave transform in the equation and the boundary conditions. The coefficients $\tilde{\mathrm{E}}(K,b)$ are unknown while the base functions and their derivatives are. Similar to the technique adopted by Jaeger, the continuous Fourier integrals are approximated by discrete Fourier sums on a grid $(K_i, b_j)$ and the equation is imposed at as many grid points as there are coefficients $\tilde{\mathrm{E}}(K_i, b_j)$. At the extremities of the interval of interest, the boundary conditions are either imposed via 2 Lagrange multipliers or – simpler – substituted for the differential equation. The resulting linear system can be resolved by standard matrix inversion techniques. Since the wavelet base is localised, the system matrix is sparse: Introducing a suitable clipping criterium, contributions from $b$'s too far from the reference position $x$ can safely be neglected. As an illustration, the solution of the Airy equation $\mathrm{E}'' + x\mathrm{E} = 0$ for wave incidence from the right is depicted ($' = d/dx$). Both regions of propagation and evanescence are treated to satisfaction.

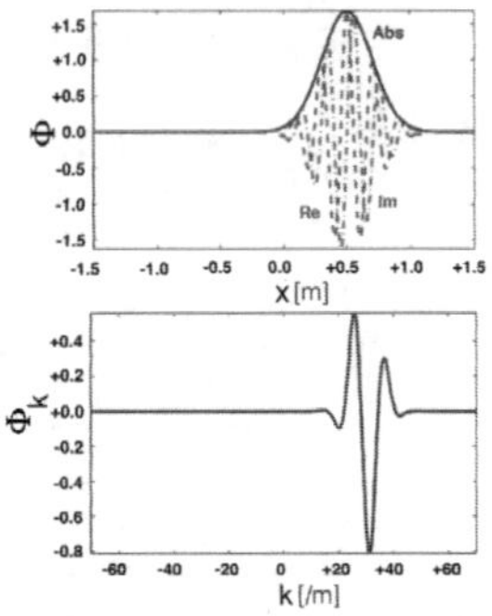


**Fig. 1**: Example of Morlet wavelet with $\sigma = 6$, $a = 1/K = 0.2$ and $b = 0.5$ yielding a maximum in physical space around $x \approx 0.5m$ and in k-space around $k = 30/m$.

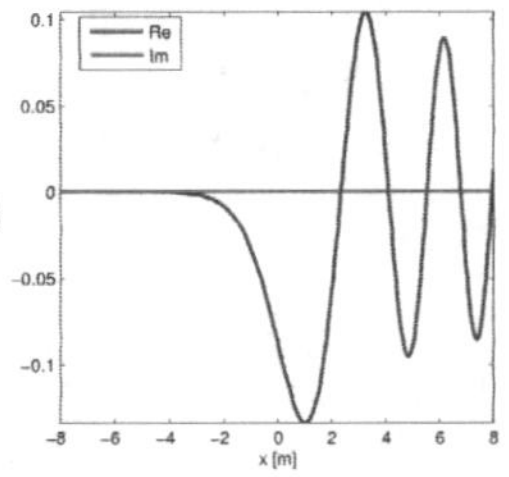


**Fig. 2**: Solution of the Airy equation imposing $E = 10^{-4}$ at the left and $E = 10^{-2}$ at the right edge.

## 4 Hot plasma wave equation

The *homogenous* plasma dielectric response of a hot plasma to an electromagnetic wave is well known [6,7]. In *inhomogenous* plasma the wave spectrum contains many coupled modes, each of which have a different response. It is customary to account for the inhomogeneity in an approximate way locally using a "quasi-homogeneous" assumption and writing the total RF current density as an integral composed out of contributions belonging to the various modes of the spectrum but assuming the dielectric response for a given

DOI: 10.1051/epjconf/201715703060

**k** can simply be written as the response in a *uniform* plasma. The relevant equations are then given in the above given references. The hidden subtlety is the fact that the equation of motion is solved locally pretending the confining magnetic field is constant. Inside the plasma, the wave equation can then be written as

$$\nabla\times\nabla\times\bar{E}(\bar{x}) = k_0^2 \int d\bar{k}\, \bar{\bar{K}}(\bar{x},\bar{k}).\, \bar{E}_{\bar{k}} \exp\,[i\bar{k}.\bar{x}]$$

in which $\bar{E}(\bar{x})$ is the electric field vector, $\bar{E}_{\bar{k}}$ its Fourier transform and $\bar{\bar{K}}(\bar{x},\bar{k})$ the homogeneous plasma dielectric tensor; $k_o=\omega/c$, in which $\omega$ is the angular frequency of the driver and c the speed of light. Thanks to the available expressions for this tensor, the formulation allows to study wave propagation and damping at arbitrary harmonics and for any wave type. The expressions for the dielectric tensor are known explicitly for bi-Maxwellian distributions with a parallel drift. Relying on numerically evaluated derivatives of the derivatives the corresponding expressions for the dielectric response are equally available for exploitation for non-Maxwellian distributions, as was e.g. demonstrated by recent work involving AORSA [8].

Substituting the Fourier transform of the electric field by its definition one obtains the non-local wave equation

$$\nabla\times\nabla\times\bar{E}(\bar{x}) = \frac{k_0^2}{(2\pi)^3} \int d\bar{k}\, \bar{\bar{K}}(\bar{x},\bar{k}). \int d\bar{x}'\, \bar{E}(\bar{x}') \exp\,[i\bar{k}.(\bar{x}-\bar{x}')]$$

which can - be it at the price of more bookkeeping – be solved in the same way as described in the simple example. For 1D application in the $x$-direction, the equation becomes

$$\int dK \int db[\bar{\bar{C}}_0 \varphi_{K,b}(x) + \bar{\bar{C}}_1 \varphi'_{K,b}(x) + \bar{\bar{C}}_2 \varphi''_{K,b}(x)] = \frac{k_0^2}{2\pi} \int dK \int db\, [\, I_0(K,\xi) \bar{\bar{K}}(\bar{x},\bar{k}) + I_1(K,\xi) \bar{\bar{K'}}(\bar{x},\bar{k}) + I_2(K,\xi) \bar{\bar{K}}''(\bar{x},\bar{k})].\, \bar{E}(K,b)$$

in which $k = k_x$. A truncated Taylor series expansion of the dielectric tensor has been made in $k=k-K$ and

$$I_m(K,\xi) = \int \mathrm{d}k\, k^m \int \mathrm{dx} \exp[\mathrm{i}(\mathrm{k+K})(\xi\text{-x})] \varphi_{K,0}(\mathrm{x})$$

while $\mathrm{x} = x' - b$ and $\xi = x - b$. Although the k and x integrals run from $-\infty$ to $+\infty$, the integration interval can be truncated in practice since the wavelets are localised. This is necessary to ensure the adopted Taylor series expansion is justified. The $I_m$ integrals can be integrated analytically. An example is shown in Fig. 3. The $\bar{\bar{C}}_{0-2}$ coefficients are the coefficients obtained when rewriting $\nabla\times\nabla\times\bar{E}(\bar{x})$ in terms of the derivatives of the field components.

To ensure the system matrix becomes sparse, negligible contributions to it should be omitted. It can easily be seen that if $|K|$ is small the contribution should be retained. But when it is large, looking at the shape of the base function and only retaining contributions larger than $\varepsilon$ the inequality $(-2ln\varepsilon)^{1/2} < |K\xi|$ can be adopted as clipping requirement.

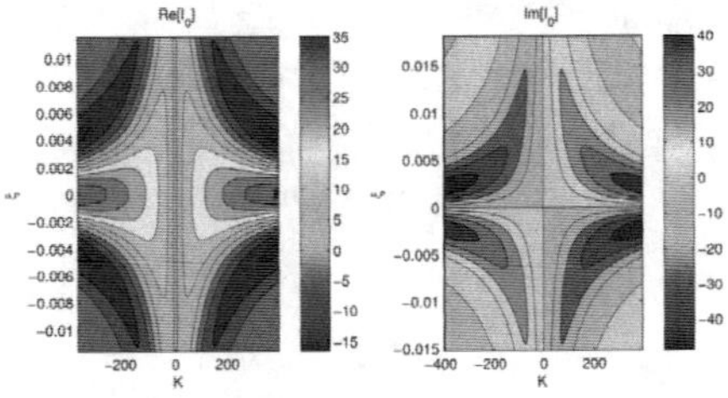


**Fig. 3**: Real and imaginary part of the $I_0$ integral.

## 5 Numerical example

Although the testing of the method is still fully ongoing, a preliminary example is provided. The case examined is one of the key RF heating schemes in JET: fundamental cyclotron hydrogen minority heating in a deuterium plasma. The following parameters were taken: The magnetic field strength is $B_o$=3.45T, which allows for core cyclotron heating when adopting f=51MHz as the driver frequency. Dipole phasing was considered but only the dominant component in the antenna spectrum ($n_{tor}$=27) was retained. The central density is $3\times10^{19}/m^3$ and the core temperature 3keV. The minority concentration is assumed to be 5%. The JET major radius is $R_o$=2.97m and a minor radius of $a_p$=0.9m was considered. The equation is solved in slab geometry but retaining the toroidal curvature; poloidal field effects are ignored. The poloidal (y) and toroidal (z) directions are assumed to be ignorable; $k_y$=0 was chosen. The relevant fast wave dispersion roots is depicted in Fig. 4; x=R-$R_o$.

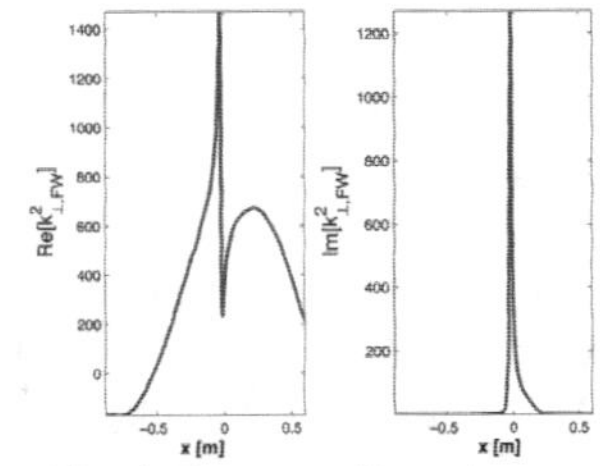


**Fig. 4**: Fast wave dispersion equation root.

The fast wave becomes evanescent at the far high field side (left in the picture) and has a confluence with the Bernstein wave in the plasma centre. The dominant damping is expected close to the minority cyclotron layer ($\omega=\Omega_H$ at x=0.14m). The electric field components are depicted in Fig. 5. A fast wave is incident from the low field side. Since the Bernstein wave (excited near the ion-ion hybrid layer) is electrostatic, no trace of short wavelength behaviour is seen on the $E_y$ field component but it is visible in the $E_x$ component. The component of the electric field parallel to the static magnetic field is small in the whole interval. At the far high field side the wave is reflected from the cutoff across which its amplitude decreases exponentially. The presence of the Berstein wave can also be seen in the Fourier spectrum of the field; see Fig. 6. The biggest peak lies at $k_x$=-25/m and corresponds to the fast wave incoming from the low field (propagating towards negative x); the slightly smaller peak with the opposite sign is the reflected fast wave. After being born at the mode conversion layer the Bernstein wave propagates leftward. This wave being a backward wave with a wavelength shorther than that of the fast wave, its signature is seen to the right of the fast wave root at $k_x$>25/m.

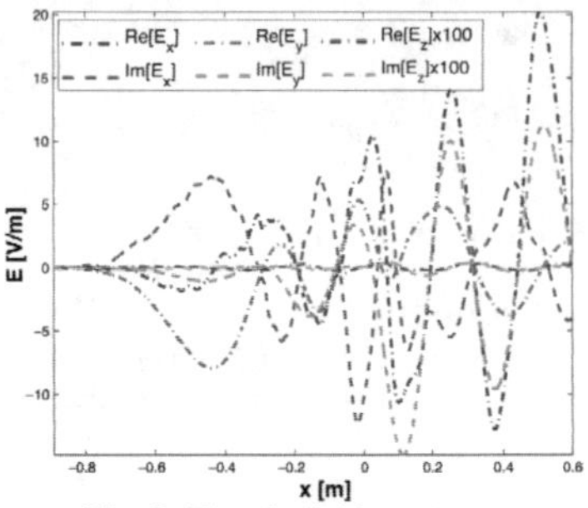

**Fig. 5**: Electric field components.

Fig. 7 depicts the corresponding power deposition. The hydrogen minority absorbs most of the power while the deuterons absorb some of the power at their second harmonic layer, coinciding with the fundamental cyclotron layer of the minority. To the scale of the plot, the electrons do not absorb any wave power.

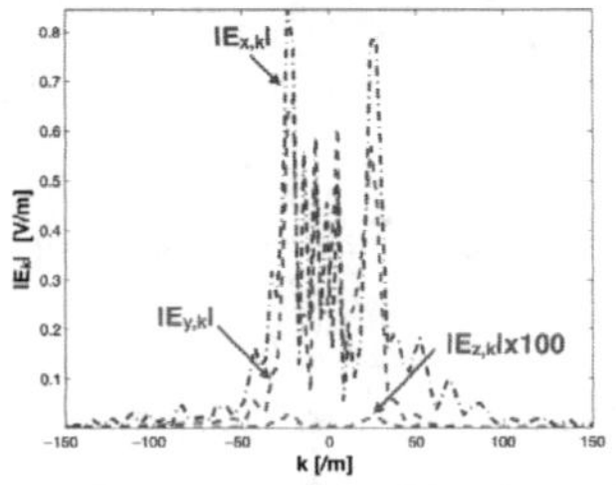

**Fig. 6**: Fourier spectrum of the electric field.

Although the obtained solution contains many features that are in line with expectations, there are various clear hints that the solution is *not* yet converged. A closer look at the power balance reveals that the majority absorption dips slightly below zero, which is unphysical for a plasma in which all distributions are Maxwellian. Also the fact that the electrons do not take part at all in the absorption is highly unusual. Electron absorption scaling with $k_\perp^2$ when adding the electron Landau and transit time magnetic pumping contributions [9, 10], the absence of electron absorption suggests that the considered spectrum not containing high enough $k_x$-modes. More in depth analysis confirms this is indeed the case: see the inset for the corresponding evaluation using the TOMCAT differential equation solver [11] which shows a small but finite electron absorption exhibiting both long (fast wave) and short (Bernstein wave) wavelength structure.

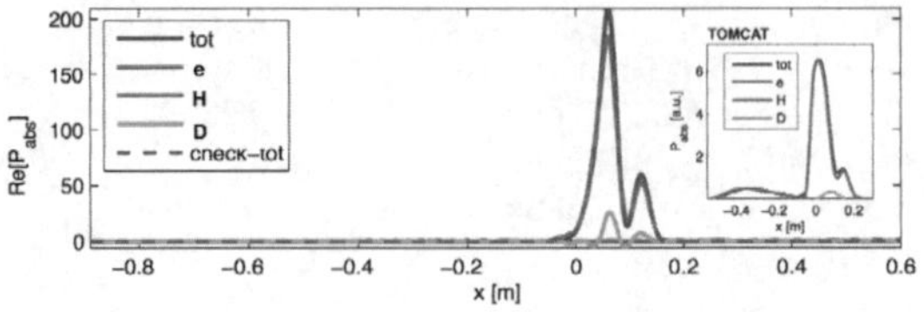


**Fig. 7**: Power deposition profile; the inset is the corresponding TOMCAT profile (different normalisation).

## 6 Final remarks

In this paper the wavelet technique was adopted to solve the hot magnetized plasma integro-differential wave equation. As is done by Jaeger for AORSA, it maximally exploits the dielectric tensor expressions available in the literature. The approach proves promising but though more localised and thus yielding sparse matrices, the double integral on $K$ and $b$ is somewhat more CPU demanding than was anticipated and necessitates further optimization. Among the techniques under examination are the exploitation of pushing the analytical evaluation of the coefficients needed in the wave equation further and attempts to push down the CPU requirements by avoiding to invert the system matrix adopting e.g. iterative solvers for the linear system.

## Acknowledgements

This work has been carried out within the framework of the EUROfusion Consortium and has received funding from the Euratom research and training programme 2014-2018 under grant agreement No 633053. The views and opinions expressed herein do not necessarily reflect those of the European Commission.

## References

1. E.A. Lerche et al., Nuclear Fusion **54**, 073006 (2014)
2. E.F. Jaeger et al., Physics of Plasmas **8** (2001) 1573
3. D. Van Eester et al., *Proc. 43rd EPS Conference on Plasma Physics*, Leuven (2016), P2.0151
4. E.M. Rønquistand and A.T. Patera, Journal of Scientific Computing **2**, 389 (1987)
5. P. Vallejos et al., *Proc. Theory of Fusion Plasmas* (Varenna, 2016), P-18
6. T.H. Stix, Waves in Plasmas, AIP, New York (1992)
7. D.G. Swanson, Plasma Waves, Academic Press, Boston (1989)
8. J. Lee et al., Physics of Plasmas **24**, 052502 (2017)
9. T.H. Stix, Nuclear Fusion 15, 737 (1975)
10. F.W. Perkins, Nuclear Fusion **17**, 1197 (1977)
11. D. Van Eester and R. Koch, Plasma Phys. Contr. Fusion **40**, 1949 (1998)

# Recent H majority inverted radio frequency heating scheme experiments in JET-ILW

*Dirk* Van Eester[1,*], *Ernesto* Lerche[1,2], *Yevgen* Kazakov[1], *Philippe* Jacquet[2], *Yuri* Baranov[2], *Volodymyr* Bobkov[3], *Kristel* Crombé[1,4], *Agata* Czarnecka[5], *Remi* Dumont[6], *Pierre* Dumortier[1,2], *Jacob* Eriksson[7], *Luca* Giacomelli[8], *Carine* Giroud[2], *Marc* Goniche[6], *Carl* Hellesen[7], *Vasily* Kiptily[2], *Natalia* Krawczyk[5], *Tuomas* Koskela[9], *Filomena* Nave[10,2], *Massimo* Nocente[11], *Jef* Ongena[1], *Marko* Santala[2,12], *Mirko* Salewski[13], *Mireille* Schneider[14], *Henri* Weisen[15] and JET contributors[x]

[1]LPP-ERM/KMS, EUROfusion Consortium Member - Trilateral Euregio Cluster, Brussels, Belgium
[2]CCFE, Culham Science Centre, Abingdon, OX14 3DB, UK
[3]Max-Planck-Institut für Plasmaphysik, Garching, Germany
[4]Department of Applied Physics, Ghent University, Ghent, Belgium
[5]Institute of Plasma Physics and Laser Microfusion, Hery 23, 01-497 Warsaw, Poland
[6]CEA, IRFM, F-13108 Saint-Paul-lez-Durance, France
[7]Department of Physics and Astronomy, Uppsala University, SE-75120 Uppsala, Sweden
[8]Istituto di Fisica del Plasma "Piero Caldirola", IFP-CNR, via R. Cozzi 53, 20125 Milano, Italy
[9]NERSC, MS 59R4010A Berkeley, CA 94720, USA
[10]Associação EURATOM/IST, Instituto de Plasmas e Fusão Nuclear, Instituto Superior
[11]Dipartimento di Fisica "G. Occhialini", Università di Milano-Bicocca, 20126 Milano, Italy
[12]VTT Technical Research Centre, FI-02044 Espoo, Finland
[13]Technical University of Denmark, Department of Physics, Kgs. Lyngby, Denmark
[14]ITER Organization, Route de Vinon-sur-Verdon, CS 90 046, 13067 St. Paul-lez-Durance Cedex, France
[15]Ecole Polytechnique Fédérale de Lausanne (EPFL), SPC, CH-1015 Lausanne, Switzerland

[x]See the author list of "Overview of the JET results in support to ITER" by X. Litaudon et al. to be published in Nuclear Fusion special issue: "Overview and summary reports from the 26[th] Fusion Energy Conference (Kyoto, Japan, 17-22 October 2016)"

**Abstract.** Inverted $^3$He and D ion cyclotron minority heating scenarios were recently tested in JET-ILW. They confirm the good heating efficiency at low concentrations of ~3%. The $^3$He *minority* heating scheme is only modestly affected by the change from a carbon (JET-C) to a Beryllium (JET-ILW) wall but unlike what was the case in JET-C, the intrinsic Be ions D-like particles in terms of charge-over-mass ratio do not prevent the D (or $^4$He) minority regime from being exploited. Direct and indirect evidence of the existence of fast particle subpopulations was found in both cases.

## 1 Introduction

Lacking other ICRF (Ion Cyclotron Resonance Frequency) heating options at full field, $^3$He and $^4$He minority fundamental cyclotron radio frequency heating schemes will be crucial in ITER's non-activated H majority phase.

Both these so-called "inverted" schemes can be accessed in JET. They were recently tested in presence of the ITER-like wall (ILW). For "inverted" schemes the charge-over-mass ratio of the majority is larger than that of the minority so that the confluence layer is in between the antenna and the minority ion cyclotron layer where the dominant heating occurs. In these schemes, the mode conversion regime is reached at much lower minority concentration (2-3%) than is the case of standard schemes, for which $(Z/A)_{min}>(Z/A)_{maj}$; Z is the charge and A the atomic number. Moreover, increasing the minority concentration shifts the mode conversion layer where the Bernstein or ion cyclotron wave are born towards the low rather than the high magnetic field side with respect to the ion-cyclotron absorption layer.

Both in the past with the C wall and more recently with the ITER-like wall, fundamental $^3$He heating has been successfully exploited in JET [1-3]. It was used in the 2016 experimental campaign for L-H threshold experiments [4] and transport studies [5]. For the former, the available power was marginally sufficient to reach H mode at high field. For the latter, exploiting the minority heating regime at modest concentration and the mode conversion regime at high concentration allowed providing localized core ion and electron heat sources, respectively.

Equipped with a C wall in the past, JET's plasmas typically contained ~2% of C, the presence of which excluded fundamental minority $^4$He or D heating in H plasmas, the mode conversion regime being reached immediately as $(Z/A)_D=(Z/A)_{^4He}=(Z/A)_C$ [1]. Because JET's walls are now made of Be rather than C, these inverted minority heating regimes can now be accessed and heating schemes re-assessed. The fundamental D minority ICRH case was briefly revisited in the last JET-ILW hydrogen campaign; some representative findings will be presented.

## 2 Experimental results

The $^3$He minority experiments were conducted at $B_o$=3.2T to absorb waves at the fundamental cyclotron resonance layer of $^3$He. This layer resides in the hot plasma core when using f=32.5MHz, a frequency in reach both for the A2 [6] and the ILA antenna [7,8]. Up to 7.6MW was launched into the plasma. Dipole phasing

* *Corresponding author: d.van.eester@fz-juelich.de*

DOI: 10.1051/epjconf/201715703061

was imposed. The plasma was in L-mode and the plasma current was $I_p$=1.8MA; the strike point was on divertor tile 5.

Figure 1 shows the temporal evolution of some key parameters while Fig. 2 gives an example of the density and temperature profiles. Hydrogen was the majority ion species and the central electron density was $N_{eo}=4x10^{19}m^{-3}$. The (hydrogen) neutral beam power was modest, $P_{NBI}$~3MW. About 30% of the total power is lost through radiation. Temperatures of up to 4keV were reached, the electron temperature typically exceeding that of the ions in spite of the presence of sawteeth; opposite to past experiments performed at higher current ($I_p$=2.45MA), the recovery from sawtooth crashes was prompt and significant [9]. Relying on break-in-slope analysis of the plasma energy, optimal overall heating was observed at $^3$He minority concentrations of 2-4%; see Fig. 3. Heating efficiencies of $P_{abs}/P_{in}$~70% were reached. The large spread in the efficiency at a given $^3$He concentration is due to the variation of the parameters (sawtooth behavior, density profile, current, ...). The plasma energy crudely increases by ~0.1MJ/MW, which is a factor of ~2 lower than the performance for the frequently used H minority in a D majority JET ICRH scheme [10]; this lower efficiency is largely brought about by the different confinement of D plasmas but is also to a lesser extent due to the actual heating performance. At too small minority concentration the heating degrades while at larger concentrations the mode conversion regime is entered and direct electron heating comes more significantly into play. The actual mode conversion regime – where direct electron damping clearly dominates ion damping - was not reached. At the small minority concentrations typical for guaranteeing minority heating in inverted scenarios, the fraction of power indirectly flowing to the electrons via Coulomb collisional relaxation of the minority is significant. For the parameters relevant for the experiment, the tipping point where bulk ions collisionally receive as much energy as the electrons is at about $X[^3He]=N_{^3He}/N_e=1.5\%$ when $P_{RF}=0.5MW/m^3$. This explains why the experimentally observed core electron heating efficiency electron response is significant except when going to too high concentrations at which point the mode conversion efficiency response for significant direct core electron heating is lost and waves are sloshing over the plasma multiple times before being fully absorbed [3].

At the low concentrations optimal for heating ions when relying on an inverted scheme, energetic tails are formed. Consistent with the presence of the formation of fast particle ion populations, sawtooth stabilization is observed. A clear response of the sawtooth frequency to a step in ICRH power is observed. At $X[^3He]$~2% the sawteeth are fully stabilised.

Aside from assessing the overall heating performance, JET is equipped with diagnostics allowing to diagnose in detail the response of both electrons and bulk ions. Figure 4 shows an example of the experimental power deposition profiles obtained by modulating the ICRH power and analysing it performing a Fourier analysis of the response of the electron temperature to the changes of the power. When focusing on electron dynamics a modulation frequency $f_{mod}$ is preferred ($f_{mod}$=25Hz for the depicted data); focussing on the slower ion dynamics (not done here) necessitates accounting for the nonlinear saturation of the ion temperature response and requires operating at smaller modulation frequencies (see e.g. [11] with $f_{mod}$~4Hz). At low $^3$He concentrations, the deposition is close to the centre and highly peaked. When the concentration increases, the peak of the power deposition moves outward, reflecting the moving of the ion-ion hybrid layer whose position critically depends on the minority concentration and at which efficient localised absorption of the Bernstein wave by electrons takes place [12]. When $X[^3He]$~10%, the maximum of the power deposition has dropped by a factor 6 and its location has shifted to half-radius.

Adding a small minority of $^4$He to a H plasma allows to study a second inverted heating scheme relevant for the initial phase of ITER. $^4$He and D sharing the same mass-to-charge ratio, ICRH waves respond similarly to them. When JET was equipped with a C wall, D minority heating experiments were unsuccesful since C – equally sharing the deuteron's Z/A – eroded from the wall contaminated the main plasma and increased the D-like ion population concentration well beyond the limit tolerable for promising minority heating [1]. The C being replaced by Be in the ITER-like wall [13], this problem no longer poses and hence the scheme was re-addressed when the D concentration was very low (X[D]~0.7%); X[Be] was estimated to be of order 0.5%. Reaching the D minority heating regime at $B_o$=3.33T requires lowering the driver frequency to 25.3MHz, beyond the reach of the ILA. The A2 antenna's – on the other hand – can be operated at that frequency, be it that the system is not designed for it so that only limited power ($P_{RF}$~3MW) can be launched. D minority heating was briefly tested in the recent campaign; $^4$He minority heating experiments were postponed to later.

The obtained heating efficiencies were somewhat lower than those obtained for $^3$He ($P_{abs}/P_{in}$~50%); the D levels remained below the level X[D]~2% required for optimal heating. Energy and temperature increases of 0.08MJ/MW and 0.3keV/MW, respectively, were seen. Evidence of fast D was clearly observed by various diagnostics while such evidence was not found in presence of the C wall. Tail temperatures of 200keV were confirmed by the neutral particle analyser as well the time-of-flight diagnostics relying on neutrons born from the D-D reaction. The neutron rate clearly – but nonlinearly – correlates with the ICRH power. And gamma ray detectors [14] testify for nuclear reactions necessitating a fast D population. More details can be found in [15].

Figure 5 depicts the counts per energy bin of the neutrons detected by the time-of-flight neutron detector ; velocity being inversely proportional to time, evidence for fast particles is on the left in the plot. The black dots are the data while the lines are the corresponding error bars. The blue line is a synthetic diagnostic fit based on a Maxwellian deuterium distribution, which has an anisotropic pitch angle distribution centered closely around 90°, as expected for the RF accelerated ions. The dotted vertical line delimits the position beyond which JET's beams do not have any counts, illustrating that the RF wave's capacity to produce high energy deuterons; neutral beam injection is absent in the depicted shot.

Dedicated experiments were also performed to study D minority NBI heating [16]. Aside from operating at somewhat higher concentrations, reducing the plasma's collisionality and widening the cyclotron damping region is known to be beneficial (see e.g. [17] on the synergistic effects of combining ICRH and NBI). On top of that « classical » effect most often exploited for boosting harmonic heating schemes [18] and referred to as « NBI preheating », a supplementary heating scenario has been proposed more recently and was experimentally tested on JET and Alcator C-Mod [19-21]. It profits from the presence of 2 ion species (H and D in this case) to align the cyclotron layer of a third ion species ($^3$He) with a region where the polarisation is locally modified to make it well suited for ion heating. Combining assets from various effects makes RF heating of a minority beam – playing the role of the third type of ions species as its dominant cyclotron damping region is Doppler shifted away from the cold plasma cyclotron layer - an efficient heating scheme.

## 3 Summary and discussion

Inverted $^3$He and D ICRF heating scenarios in hydrogen majority plasmas were recently tested on JET-ILW at $B_o$~3.3T. Both are viable options for the non-active full-field operation of ITER ($B_o$=5.3T) at respectively 53 and 40MHz [22].

Heating efficiencies of ~70% have been achieved with $^3$He minority heating and and the experimental findings are in line with theoretical expectations. The change from JET-C to JET-ILW has limited impact on this minority heating scenario. Due to the optimal concentration being small (~3%), fast ion tails are created and Coulomb collisional electron heating is dominant in these conditions. This was confirmed by fast particle measurements, evidence of sawtooth stabilization and by the study of the electron response to ICRF power modulation. Real time control of X[$^3$He] is important in these conditions to keep the RF wave absorption in the optimal regime.

Whereas in JET-C carbon eroded from the JET walls prevented exploiting the D or $^4$He minority heating schemes (C being present at ~2% levels and $(Z/A)_C$ being identical to that of D and $^4$He so that the effective D-like concentration was too high for minority heating), proof-of-principle experiments in JET-ILW demonstrated successful D ion heating although optimal D minority concentrations were not reached. Evidence of the formation of fast particle subpopulations is available for this scenario as well and suggests that D ion tails with energies ~200keV have been created.

N=1 $^3$He minority heating in H proves to be a promising candidate for ITER's full field non-activated phase. Dedicated experiments are required to document the potential of the N=1 $^4$He minority heating scheme but the first evidence using a D minority ($Z/A_D=Z/A_{^4He}$) is encouraging.

## Acknowledgements

This work has been carried out within the framework of the EUROfusion Consortium and has received funding from the Euratom research and training programme 2014-2018 under grant agreement No 633053. The views and opinions expressed herein do not necessarily reflect those of the European Commission. The views and opinions expressed herein do not necessarily reflect those of the ITER Organization.

## References

1. M.-L. Mayoral et al., Nucl. Fusion **46**, S550–S563 (2006)
2. P.U. Lamalle et al, Nucl. Fusion **46**, 391 (2006)
3. D. Van Eester et al, Plasma Phys. Contr. Fusion **54**, 074009 (2012)
4. J. Hillesheim et al., *Proc. 26th IAEA-FEC Conference*, Kyoto (2016), EX/5-2
5. P. Mantica et al., *Proc. 26th IAEA-FEC Conference*, Kyoto (2016), EX/P6-14
6. A. Kaye et al., Fusion Engineering and Design, **24**, 1-21 (1994)
7. F. Durodié et al. , Fusion Engineering and Design **84**, Issues 2-6, 279-283 (2009)
8. F. Durodié et al., *Proc. 29th SOFT conference*, Prague, P4.037 (2016)
9. D. Van Eester et al., Proc. 42nd EPS Conf. on Plasma Physics (Lisbon, 2015), P2.117
10. V. P. Bhatnagar et al., Plasma Phys. Contr. Fusion **22**, 99 (1991)
11. F. Ryter et al., Plasma Phys. Contr. Fusion **51**, 113016 (2011)
12. R. Perkins, Nuclear Fusion **17**, 1197 (1977)
13. R. Neu et al., Plasma Phys. Control. Fusion **44**, 811 (2002)
14. Kiptily V G et al 2002 Nucl. Fusion 42 999-1007
15. E.A. Lerche et al., in preparation for Plasma Phys. Contr. Fusion

DOI: 10.1051/epjconf/201715703061

16. J. Ongena et al., "*Observations of Synergetic Acceleration of D-NBI Ions in the Vicinity of the Mode Conversion Layer in H-D Plasmas*", invited talk at this conference.
17. R. Koch et al., Fus. Science Technology **47**, 97 (2005)
18. M. Schneider et al., Nucl. Fusion **56**, 112022 (2016)
19. Ye. O. Kazakov et al., Nucl. Fusion **55**, 032001 (2015)
20. Ye. O. Kazakov et al., *Efficient generation of energetic ions in multi-ion plasmas by radio-frequency heating*, Nature Physics (2017); http://dx.doi.org/10.1038/nphys4167
21. J. Wright et al., *Proc. 26th IAEA Fusion Energy Conference*, Kyoto, EX/P3-5 (2016)
22. M. Schneider et al., "*ICRH scenarios for the ITER non-active phase*", paper C-19 at this conference.
23. M. G. Johnson et al., Nucl. Instrum. and Methods in Physics Research A, **591**, 417-430 (2008)

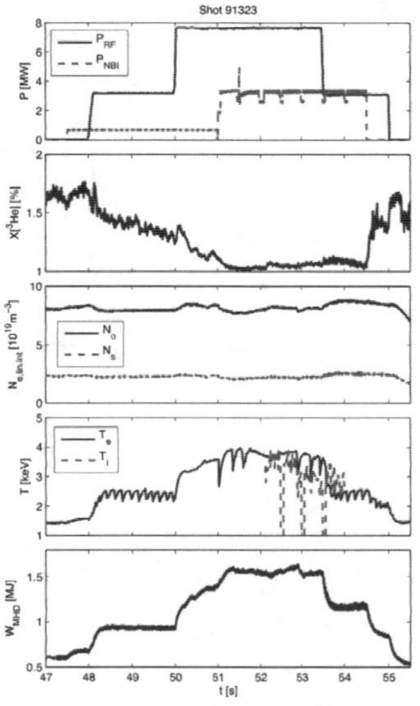


**Fig. 1** : RF and NBI power, minority concentration, core and edge line integrated densities, electron and ion temperature and plasma energy for the $^3$He minority heating study.

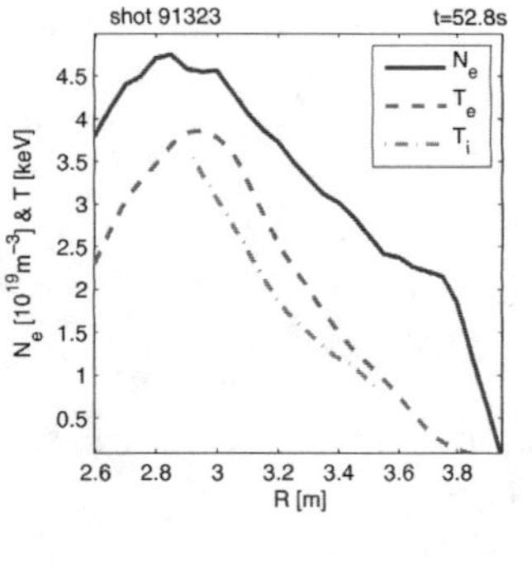


**Fig. 2** : Density and temperature profiles of the 91323 $^3$He minority heating shot.

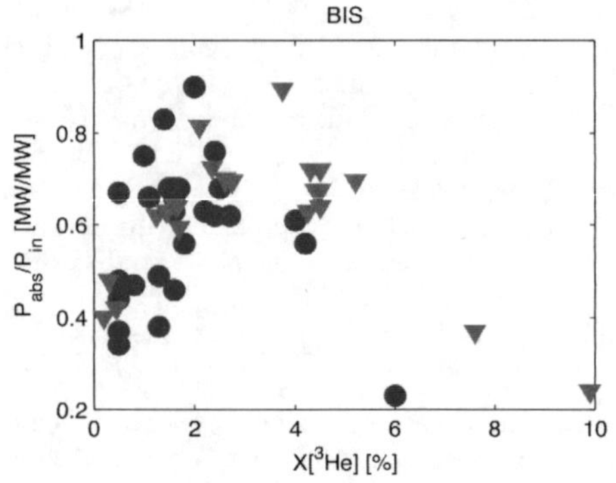


**Fig. 3** : Experimental heating efficiency as function of the $^3$He concentration as estimated during the experiment. The blue circles represent the recent results at $I_p$=1.8MA ; the green triangles correspond to earlier ($^3$He)-H experiments performed at $I_p$=2.45MA [9].

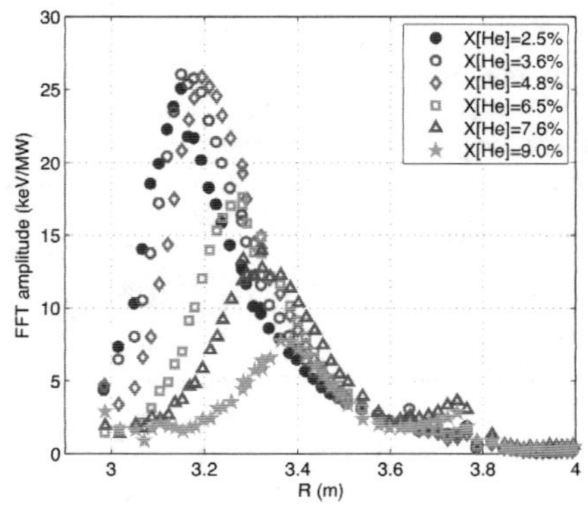


**Fig. 4** : Experimental electron power deposition profiles as function of the experimentally estimated $^3$He concentration.

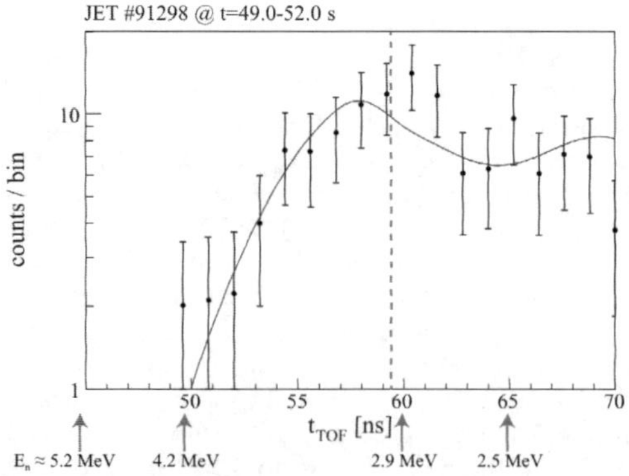


**Fig. 5** : TOFOR time-of-flight neutron detector's [23] signature of RF heated deutrium minority ions.

EPJ Web of Conferences **157**, 03062 (2017) DOI: 10.1051/epjconf/201715703062
*22 Topical Conference on Radio-Frequency Power in Plasmas*

# RF plasma generation in the chamber with the conducting walls

*Andey* Gavrikov[1, 2], *Sergey* Kuzmichev[1, 2], *Gennadiy* Lizyakin[2], *Valentin* Smirnov[2], *Rinat* Timirkhanov[1, 2], *Ravil* Usmanov[1,2] and *Nazar* Vorona[1,2,*]

[1]Joint Institute for High Temperatures of the Russian Academy of Sciences (JIHT RAS), 125412, Moscow, Russia
[2]Moscow Institute of Physics and Technology (MIPT), 141701, Dolgoprudny, Moscow Region, Russia

**Abstract.** This work is devoted to study of inductively coupled RF discharge within a metal chamber with a diameter of about 1 m and a length of about 2 m in argon gas (RF frequency is 5.28 MHz). The spatial distributions of electron temperature and density depending on the magnetic induction (from 0 to 0.2 T), plasma-forming gas pressure ($0.1 - 10$ mTorr) and RF power absorbed in the plasma are presented. For fixed gas pressure (6 mTorr) electron temperature decrease with increasing of magnetic field was found.

## 1 Introduction

Plasma separation method, proposed in [1] for spent nuclear fuel and radioactive wastes reprocessing, can to work in practice at about 1 m typical size of plasma volume, $0.1 - 0.2$ T magnetic induction and about 1 $kV$ electrical potential. The presence of buffer plasma with magnetized electrons allows creation conditions for compensation of space charge [2] and reproduction of necessary electric field profile providing efficient spatial separation of ion flows. This circumstance removes a limitation on ion current value and allows obtaining the reprocessing productivity attractive from industrial point of view.

The paper is devoted to experiments with plasma generation in the presence of magnetic field in the volume limited by conductive walls. Conditions are observed where plasma separation method of elements can be realized (plasma density is about $10^{12} - 10^{13}$ cm$^{-3}$, electron temperature is about units of eV) [3]. The properties of RF plasma source observed. Also the results of the measurements of argon plasma parameters (electron temperature and density) generated by the RF source are presented.

## 2 Plasma source: features of generation conditions and construction

The buffer plasma source for the SNF plasma separation method should satisfy a number of conditions:

1. The density of a plasma forming gas should not exceed $10^{13}$ cm$^{-3}$, otherwise collisions with separated ions become significant and their flows disintegrate and mix. On the other hand plasma density should be as high as possible, that increase productivity of the method.

2. Electron energy $T_e$ of buffer plasma should not exceed the second ionization potential of separated ions ($\leq 10$ eV).

3. The buffer plasma should provide a reproduction of required profile of electrical potential for efficient separation process. The physical mechanism of creation of electrical potential in plasma with magnetized electrons can be described as follows: the electron mobility along magnetic field ($B_0 \approx 0.1$ T) is several orders of magnitude higher than across it, therefore if a magnetic line of force lean by its ends on two electrodes immersed in plasma and biased at the same electrical potentials then this potential will be reproduced in a plasma volume. Thus a buffer plasma source should take into account a necessity of additional electrodes placement, which provide a potential of electric field.

4. As one of the important purposes of this work is a creation of engineering and physical fundamentals for the technology of SNF plasma separation, a significant part of the chamber must be metallic. This requirement is dedicated by conditions of radiation loads, manufacturability and large dimensions of the operating volume (typical size is more than 1 m).

According to the authors opinion, the most perspective is a helicon RF discharge [4], because it allows entering the energy into a plasma volume across a magnetic field, and at power up to 10 kV it is possible to generate plasma with density of about $10^{13}$ cm$^{-3}$ in a large volume of 1 m$^3$. The possibility of practical use of the helicon discharge for plasma generation in close experimental conditions was demonstrated in papers [5-7].

Given the above configuration of plasma source for SNF plasma separation method was proposed and it is presented on the Fig. 1. The vacuum chamber 1 is a metallic cylinder with internal diameter of 86 cm and length of 220 cm. The chamber 2 is hermetically separated from vacuum chamber 1 by dielectric wall passing RF radiation and by water-cooled Faraday shield. In the chamber 2 the antenna for RF plasma pumping is located. On the end faces of the chamber 1

* Corresponding author: raraavis@ihed.ras.ru

the electrodes able to provide electrical potentials into plasma are placed. The Helmholtz coils generate a magnetic field up to 0.2 T in the chamber 1. Operating frequency of RF generator is 5.28 MHz. A Rogowski coil is used for antenna current diagnostics. The power absorbed by a load is measured by directional couplers (sensors of incident and reflected waves).

Plasma density and electron temperature were measured by RF compensated double probe with cylindrical shape of the tips, each 0,35 mm in diameter and 8 mm long. Linear motion feedthrough is used to move the probe from the top of the plasma to the axis. The probe was located in the area between magnetic coils (Fig. 1).

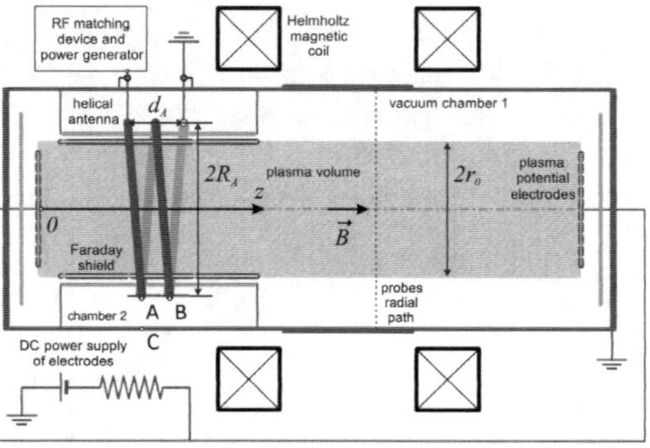


**Fig. 1.** Experimental setup layout and geometric parameters for the estimation of the plasma density.

For the estimation of antenna characteristics able to induce and maintain the helicon discharge with parameters required for plasma separation method (plasma density is up to $10^{13}$ cm$^{-3}$, plasma volume is ~1 m$^3$, electron temperature is about units of eV) an approach [8] was used. Estimations were performed in approximation of cold collision-free plasma with direct antenna placement within the single volume of gas-discharge chamber (GDC) (without chamber 2 and Faraday shield). All necessary parameters are shown in the Fig. 1.

In order the antenna radiation penetrates the argon plasma in homogenous magnetic field it is necessary for frequency $f$ to be in range of:

$$f_{c,i} = \frac{eB_0}{2\pi M_{Ar}} \ll f \ll f_{c,e} = \frac{eB_0}{2\pi m_e} \ll f_{p,e} = \frac{1}{2\pi}\sqrt{\frac{n_e e^2}{m_e \varepsilon_0}},$$

where $e$ is an elementary charge, $B_0$ is a magnetic field induction; $m_e$ and $M_{Ar}$ are masses of electron and argon respectively; $n_e$ is an electron density ($\varepsilon_0$ — vacuum permittivity). For $B_0 \approx 0.1$ T and $n_e = 10^{13}$ cm$^{-3}$ values of edge frequencies are $f_{c,i} = 0.038$ MHz, $f_{c,e} =$ 2830 MHz, $f_{p,e} = 28.5$ GHz. Chosen by the technical considerations operating frequency $f = 5.28$ MHz belongs to an obtained range.

For helicon waves with the frequency $\omega$ and wave vector $\vec{\boldsymbol{k}}(k_r, k_z)$ spreading in unbound plasma with electron density $n_e$ at an angle of $\theta$ with respect to a magnetic field direction the dispersion relation is

$$\frac{k^2}{\omega^2/c^2} = \frac{\omega_{pe}^2}{\omega(\omega_{ce}\cos\theta - \omega)},$$

$$k^2 = k_z^2 + k_r^2, \qquad k_z = k \cdot \cos\theta,$$

and waves are described by the expression like

$$E, B \sim \exp j(\omega t - k_z z - m\varphi),$$

where $\varphi$ is an azimuthal angle, $m$ is an azimuthal number ($m = 0, \pm1, \pm2, ...$) and $c$ is a speed of light in vacuum.

For the helicon wave with $m = 0$ spreading in plasma cylinder with radius $r_0$ minimal value of radial component $k_r$ of vector $\vec{\boldsymbol{k}}$ is $k_{r,\min} = 3{,}83/r_0$ [8].

The projection of vector $k_z$ on the direction of vector $\vec{\boldsymbol{B}}_0$ of magnetic induction is connected with the length of antenna $d_A$ (Fig. 1) as follows:

$$k_z = \pi(2l + 1)/d_A, l = 0,1,2,3, \ldots .$$

For electron density $n_e$ in a cylindrical area with radius $r_0 = 25$ cm corresponding to the helicon wave with frequency of $f = 5.28$ MHz and to the lowest value of wave vector induced by the antenna with length of $d_A = 0.2$ m we have ($\mu_0$ — vacuum permeability):

$$n_e = \frac{k_{\min}\ k_{z,\min}\ B_0}{e\mu_0 \cdot 2\pi f} \approx 5 \cdot 10^{12}\ \text{cm}^{-3}.$$

Thus such antenna can be used for a plasma source for realization of SNF plasma separation method.

Required power for RF pumping can be estimated from the relation [4]:

$$P \approx 3500 \cdot \left(\frac{r_0^2}{d_A}\right) \cdot \frac{n_e[\text{cm}^{-3}]}{10^{13}[\text{cm}^{-3}]}\,[\text{W}].$$

That gives $P \approx 55$ kW for plasma density $n_e = 10^{12}$ cm$^{-3}$.

Let us analyze electric field distribution occurring in the GDC before the breakdown and plasma formation (Fig. 1). In this case there are three components of electric field can be specified:

1. Electric field occurring due to the potential difference between antenna and GDC wall (points $A$ and $C$ Fig. 2). The magnitude of field $E_{1,\max}$ can be estimated as a ratio of antenna voltage $U_{ant}$ to a distance $h$ between antenna and GDC wall. For $h \approx 0.16$ m we have $E_{ant-wall} \approx U_{ant}/h \approx 7U_{ant}$.

2. Electric field occurring due to the potential difference between nearby antenna turns (points $A$ and $B$ Fig. 2). In this case the magnitude of field $E_{2,\max}$ can be estimated as a ratio of potential differences between the closest points on different antenna turns (in our case this value is about $U_{ant}/2$) to a distance between turns. For $d = 0.1$ m we have $E_{turn} \approx U_{ant}/2d \approx 5U_{ant}$.

3. Vortex electric field magnitude near the antenna is estimated as $E_{vort} \approx U_{ant}/2\pi R_A$. . For $R_A = 0.3$ m we have $E_{vort} \approx 0.5U_{ant}$.

Given analysis shows that in the case of direct antenna placement within the single volume of GDC filed intensities in the central part of GDC will be more than an order of magnitude lower than field intensities near walls. Therefore the initial breakdown and emerging discharge plasma as well will be concentrated near the walls and the main part of the operating volume, where plasma presence is necessary, will be free of it. This conclusion is also proved by results of preliminary experiments (Fig. 2).

DOI: 10.1051/epjconf/201715703062

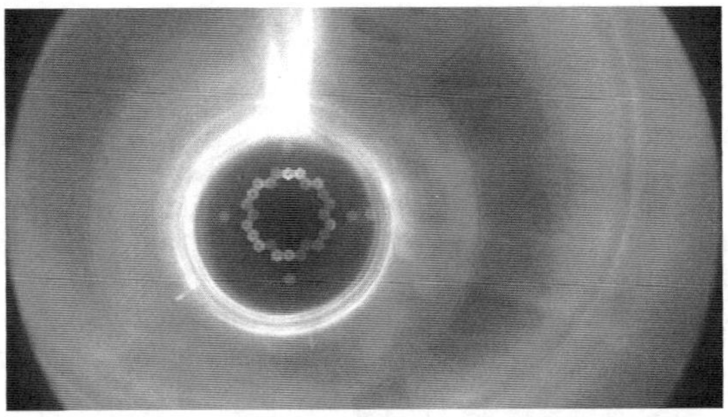

**Fig. 2.** Typicall discharge with direct antenna placement within the single volume (without chamber 2 and Faraday shield) ($P \approx 1.5$ kW, $p = 9$ mTorr (Ar), $B_0 = 0.14$ T).

To solve the problem the following actions were taken. Firstly, the antenna was placed in a separate volume hermetically isolated by dielectric divider (chamber 2). This volume is filled by air at atmospheric pressure that provides better electrical resistivity than in the main part of the chamber [9] and prevents a breakdown development between the antenna and the chamber. Secondly the Faraday shield made of conductive slits parallel to the magnetic lines was used [10-12]. The presence of the shield minimizes a capacitive coupling between the antenna and the plasma and increase the proportion of power transferred to the plasma through the inductive channel [13]. For inductive plasma a higher and more homogeneous electron density at relatively low electron temperature is typical [14] in contrast with capacitive plasma. Besides that inductive decoupling of plasma volume and antennas tract leads to a stabilization of the discharge and impedance of antennas tract as well [15] that allows efficiently calculate and perform adjusting independently of discharge parameters. It should be noted that Faraday shield also partially protects the dielectric divider surface, facing the plasma, from ion bombardment.

## 3 The experiment

With the help of RF plasma sources presented in previous section, plasma generation in different conditions were carried out.

Argon was used as a plasma forming gas and its pressure was in range of $2 - 6$ mTorr. Stable operation of RF discharge was obtained at magnetic induction of $0 - 0.12$ T. The photo of typical obtained discharge is shown on the Fig. 3.

On the Fig. 4(a) the obtained distributions of electron temperature $T_e$ and density $n_e$ at magnetic induction $B_0 = 0.026$ T and the buffer gas pressure $p = 5.9$ mTorr are shown. Electron temperature in this regime changes non monotonically and has a local maximum at $r = 5$ cm of from the discharge axis where its value reaches ~9 eV. Closer to the chamber periphery $T_e$ decreases to the value of 4 eV. Electron density $n_e$ is also distributed non monotonically on radius and has a fall in the region of $T_e$ peak, where it is equal to $1.7 \cdot 10^{11}$ cm$^{-3}$. On the discharge axis a density oscillations were observed that probably connected with instability of RF discharge at specified experiment parameters.

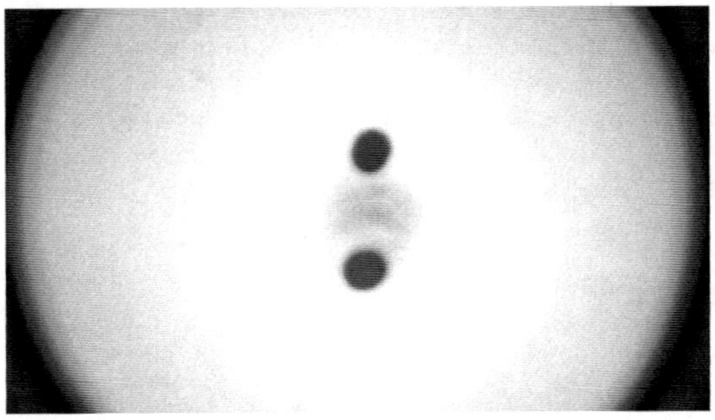

**Fig. 3.** Typicall discharge with Faraday shield ($P \approx 1$ kW, $p = 5.9$ mTorr (Ar), $B_0 = 0.026$ T. Spots in the center are the diagnostic elements in front of the glass).

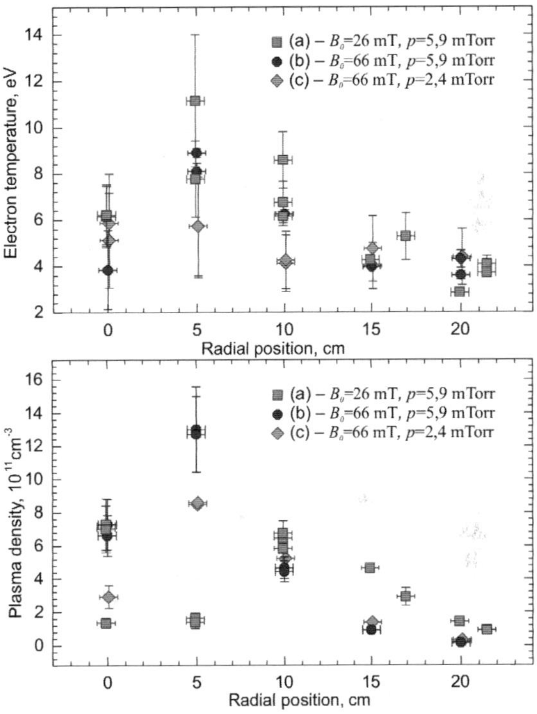


**Fig. 4.** Plasma density and electron temperature profiles of RF discharge at mentioned parameters (a-c) and $P \approx 1$ kW.

Electron temperature and density profiles were also measured at higher magnetic induction $B_0 = 0.066$ T and the same argon pressure $p = 5.9$ mTorr (Fig. 4(b)). The graphs show that increase of magnetic field causes a smoothing of electron temperature profile. The average value of $T_e$ in this regime decreases from the center to the periphery from 6 to 4 eV. Radial profile of electron density has a local maximum of $1.2 \cdot 10^{12}$ cm$^{-3}$ at $r = 5$ cm from the discharge axis and density decreases farther on radius. At $r = 20$ cm from the axis density falls to a value of about $1.2 \cdot 10^{10}$ cm$^{-3}$.

Electron temperature and density profiles at the same magnetic field $B_0 = 0.066$ T but at a lower argon pressure $p = 2.4$ mTorr are presented on the Fig. 4(c).

The decrease of the buffer gas pressure in contrast with previous regime leads to a non monotonically distribution of electron temperature along the radius of the chamber. The temperature has a local maximum of ~9 eV at $r = 5$ cm from the discharge axis. At the remaining points it stays in range of $4 - 6$ eV. The shape of plasma density distribution is not changed but the ratio of maximum density at $r = 5$ cm to a density on the axis is altered. If at the argon pressure of 5.9 mTorr the ratio is 2 then at the pressure of 2.4 mTorr it already becomes 3.

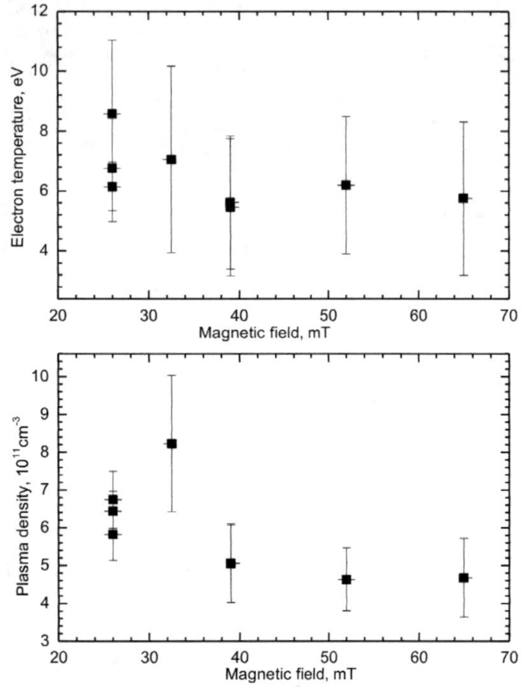


**Fig. 5.** The dependencies of electron temperature and density on magnetic field in the center of the chamber at the buffer gas pressure $p = 5.9$ mTorr.

In presence of RF plasma electrical potential on the chamber axis depending on magnetic field was measured by the method of floating probe [16]. Argon pressure was 6 mTorr. The measurement results are shown on the Fig. 6.

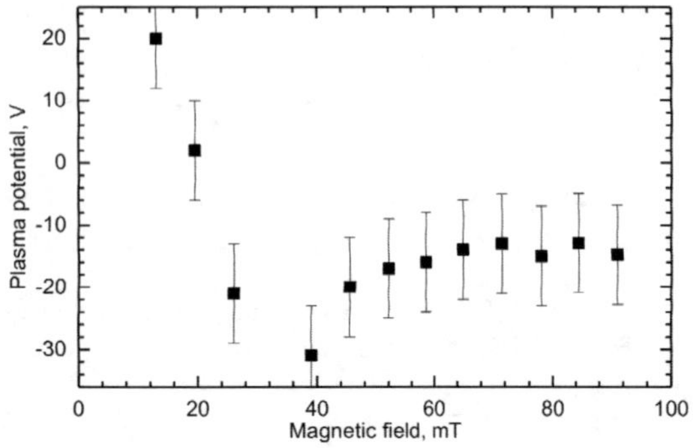


**Fig. 6.** Plasma potential versus magnetic field on the axis of the chamber.

Obtained relation demonstrating the transition of electrical potential through zero point can be explained in the following way. With the increase of magnetic field from 0 to 0.02 T electrons become more magnetized and accumulate on the axis, and ions recombine mainly on the wall, due to this processes the potential on the axis take negative values. Further increase of magnetic induction leads to partial ion magnetization and electrical potential becomes more compensated.

The study was supported by a grant of the Russian Science Foundation (project No. 14-29-00231).

## References

1. V. P. Smirnov, A. A. Samokhin, N. A. Vorona et al. Plasma Phys. Rep., **39**, 456 (2013)
2. A. I. Morozov, *Introduction to Plasma Dynamics* (CRC Press, 2012)
3. N. A. Vorona, A. V. Gavrikov, A. A. Samokhin et al., Phys. Atom. Nuclei, **78**, 1624 (2015)
4. F. F. Chen, in High Density Plasma Sources, edited by O. A. Popov (Noyes, New York, 1995)
5. R. Freeman, S. Agnew, F. Anderegg et al., AIP Conf. Proc, **694**, 403 (2003)
6. C. E. Ahfeld, J. D. Wagoner, D. L. Sevier et al., Fusion Engineering 2005, Twenty-First IEEE/NPS Symposium, 1 (2005)
7. A. Litvak, S. Agnew, F. Anderegg, B. Cluggish et al., 30th EPS Conf. Contr. Fusion and Plasma Phys,. **27A**, O-1.6A (2003)
8. P. Chabert, N. Braithwaite, *Physics of Radio-Frequency Plasmas* (Cambridge University Press, 2011)
9. V. A. Lisovskiy, V. D. Yegorenkov, J. Phys. D: Appl. Phys., 31, 3349 (1998)
10. B. P. Cluggish, F. A. Anderegg, R. L. Freeman et. al., Phys. Plasmas, 12, 057101 (2005)
11. P. Y. Nabhiraj, Ranjini Menon, G. Mohan Rao et. al, Nucl. Instrum. Methods Phys. Res. A, **621**, 57 (2010)
12. B. W. Longmier, A. D. Gallimore, N. Hershkowitz, Plasma Sources Sci. Technol., **21**, 015007 (2012)
13. E. A. Kralkina, P. A. Nekliudova, V. B. Pavlov et al., Plasma Sources Sci. Technol., **26**, 055006 (2017)
14. E. A. Kral'kina, Phys. Usp., **51**, 493 (2008)
15. M. H. Khater, L. J. Overzet, J. Vac. Sci. Technol. A, **19**, 785 (2001)
16. F. F. Chen, Lecture Notes on Principles of Plasma Processing (Plenum/Kluwer Publishers, 2002)

EPJ Web of Conferences **157**, 03063 (2017) DOI: 10.1051/epjconf/201715703063
*22 Topical Conference on Radio-Frequency Power in Plasmas*

# LHCD during current ramp experiments on Alcator C-Mod

*G.M.* Wallace[1], *F.* Poli[2], *M.A.* Chilenski[1], *J.W.* Hughes[1], *R.T.* Mumgaard[1], *S.D.* Scott[2], *S.* Shiraiwa[1], and *S.J.* Wukitch[1]

[1]MIT Plasma Science and Fusion Center, Cambridge, MA 02139 USA
[2]Princeton Plasma Physics Laboratory, Princeton, NJ 08543 USA

**Abstract.** The lower hybrid current drive (LHCD) system on Alcator C-Mod is capable of sustaining fully non-inductive discharges for multiple current relaxation times ($\tau_{cr} \sim 200$ ms) at line averaged densities in the range of $5x10^{19}$ m$^{-3}$. Some of these non-inductive discharges develop unstable MHD modes that can greatly reduce current drive performance, particularly in discharges with plasma current of 0.5 MA or less [1,2]. Avoiding these unstable MHD modes motivated an experiment to test if the stable current profile shape of a higher current non-inductive discharge could be achieved in a lower current discharge. Starting from a discharge at 0.8 MA, the plasma current was ramped down to 0.5 MA over 200 ms. The surface voltage of the plasma swings negative during the ramp, with the loop voltage reversal impacting the edge fast electron measurements immediately. Little change can be seen during the $I_p$ ramp in the core fast electron measurements, indicating that the loop voltage reversal does not penetrate fully to the magnetic axis on the timescale of the current ramp. The resulting discharge did not exhibit deleterious MHD instabilities, however the existence of this one discharge does not necessarily represent a robust solution to the problem.

## 1 Introduction

Lower hybrid current drive (LHCD) is capable of sustaining fully non-inductive discharges for multiple current relaxation times ($\tau_{cr} \sim 200$ ms) at line averaged densities in the range of $5x10^{19}$ m$^{-3}$ on Alcator C-Mod. C-Mod discharges with strong off-axis current drive ($\frac{\Delta V}{V} > 0.5$) sometimes develop MHD instabilities as the current profile evolves. These instabilities can be benign, but often result in loss of current drive efficiency and degraded transport [1]. Figures 1 and 2 show an example of a discharge with MHD instability developing around 1.15 s.

A database of existing discharges with lower hybrid current drive (LHCD) has been created. The database includes information on net LH power, launched $n_{\parallel}$, plasma current, loop voltage, electron density and temperature, $Z_{eff}$, driven current, 0-D LHCD efficiency, and the presence of deleterious MHD instabilities. The range of parameters for which MHD is problematic is shown in Figure 3. MHD instabilities also appear to be responsible for termination of $T_e$ ITBs in reverse shear discharges [2]. Although the MHD activity typically causes a decrease in current drive efficiency, some discharges with MHD continue to show good current drive efficiency as shown in Figure 3.

The goal of this experiment was to better understand how to avoid these instabilities in discharges with significant off-axis current drive. To that end, upgrades to the motional stark effect (MSE) diagnostic improve accuracy and time resolution of current profile reconstructions for 2015 [3] for input to MHD stability calculations.

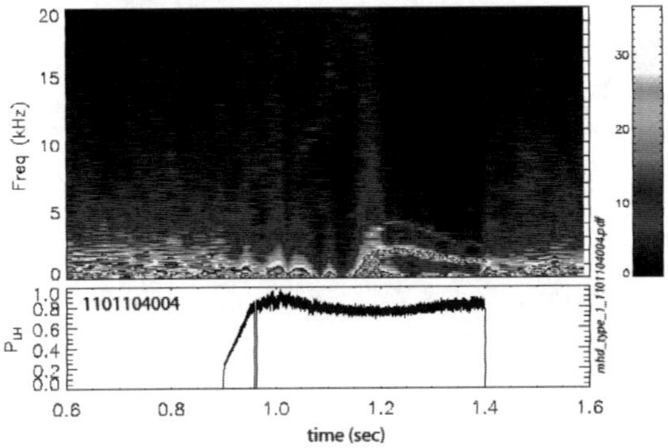


**Fig. 1.** Low-frequency (<~5 kHz) MHD is sometimes observed on magnetic pickup loops (shown above) and soft x-ray arrays during high-power LHRF in Alcator C-Mod. The onset occurs just before 1.2 s in the discharge shown.

## 2 Experimental approach

Experiments on Tore Supra [4] and ASDEX [5] have identified certain "pre-forming" techniques to eliminate the occurrence of MHD instabilities during LHCD. Some of the techniques identified are: step changes in $P_{LH}$ during the discharge; ramping down $I_p$ after turning on the LHRF; and intentional distortion of the $n_{\parallel}$ spectrum by

* Corresponding author: wallaceg@mit.edu

 DOI: 10.1051/epjconf/201715703063

reducing the number of waveguide columns energized. The experiments on C-Mod focused on ramping $I_p$ during the LHRF pulse.

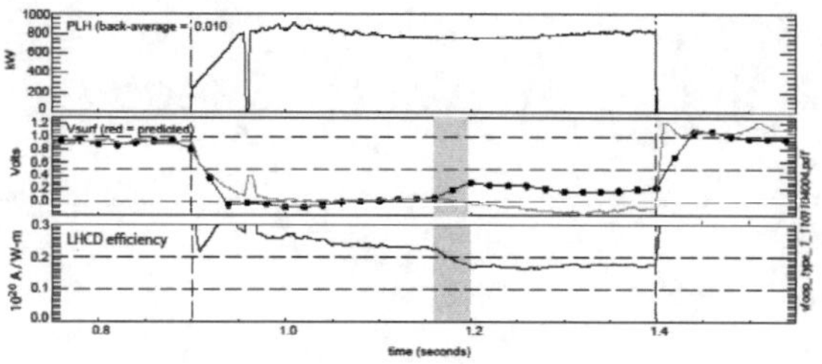


**Fig. 2.** Low-frequency MHD instabilities sometimes, but not always, results in a reduction in LH current-drive efficiency. The shaded region indicates the onset of low frequency MHD instability in this discharge.

Discharges at higher $I_p$ exhibit higher current drive efficiency and more off-axis current drive on C-Mod, while also less commonly suffering from low frequency MHD instabilties. Starting at higher $I_p$, then ramping the current down while LH is energized may help to keep a more stable, off-axis current drive profile while also increasing the fraction of power that is driven non-inductively.

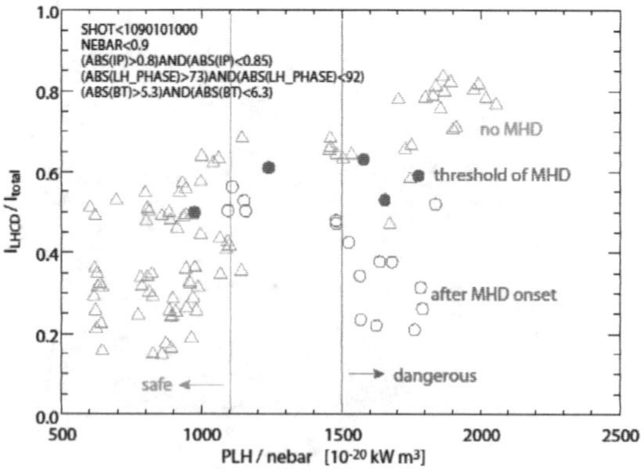


**Fig. 3.** The probability of deleterious MHD generally increases as the LHRF power normalized to line average density increases. Green triangles represent stable discharges. Solid red circles represent unstable discharges prior to onset of instability, and open red circles represent unstable discharges following onset of instability.

## 3 Experimental results

Experiments conducted in the final days of C-Mod operation explored the use of current ramps during the LHRF pulse to improve current drive performance and MHD stability. This paper will focus on shot 1160929029, an upper single null discharge with a current ramp from 0.8 MA to 0.5 MA during the LHRF pulse. Figure 4 shows some key plasma parameters from this discharge. The LHRF pulse began at 0.7 s, with the $I_p$ ramping down between 0.85 and 1.05 s. The LHRF pulse ended at 1.4 s. LHRF power was 650 kW at a launched peak $n_{\parallel}$ of 1.92. Plasma kinetic profiles measured before and after the current rampdown (Figure 5) show that the density and temperature both drop slightly across most of the plasma profile during the rampdown. The timing of the LHRF pulse and $I_p$ ramp allow for a current relaxation timescale ($\tau_{cr}$~0.2 s) to elapse following the end of the $I_p$ ramp.

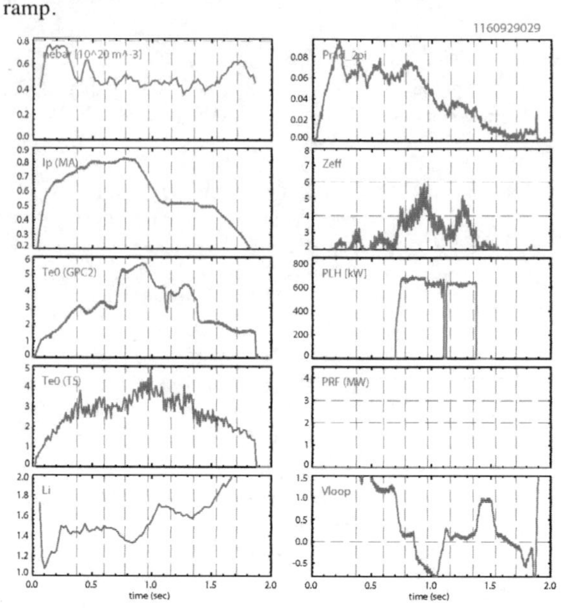


**Fig. 4.** Plasma parameters for discharge 1160929029. The plasma current ramps down from 0.8 to 0.5 MA during the LHRF pulse. Vertical dashed lines indicate MSE current profile measurement timepoints.

Prior to the current ramp, the loop voltage dropped from ~1.2 V to ~0.2 V during LHRF. As the current began to drop, the loop voltage swung to ~-0.7 V at ~1.0 s, then recovered back towards ~0.1 V at the end of the $I_p$ ramp. Although the loop voltage was negative at the boundary of the plasma during this interval, the diffusion of the negative DC electric field into the plasma, and its effect on LH wave damping and current drive, must be considered through time-dependent simulations as the current diffusion time of ~0.2 s is similar to the timescale of the loop voltage changes.

Figure 6 shows the MSE constrained EFIT current profile reconstructions from time slices during the Ohmic period, LHRF period before the $I_p$ rampdown, during the $I_p$ rampdown, and following the $I_p$ rampdown. The pre-LHRF current profile ($t$ = 0.6 s) was peaked with a central current density of 14 kA/cm$^2$. As LHRF was applied to the 800 kA segment of the discharge ($t$ = 0.78 s), the central current density dropped to ~12.5 kA/cm$^2$, with a corresponding increase in current density outside the mid-radius. As the current ramped down during the LHRF pulse ($t$ = 0.96 s), the central current density dropped sharply to slightly under 10 kA/cm$^2$, with the current profile peaked just off-axis. The final MSE measurement occurred after the current ramp finished ($t$ = 1.34 s) and the discharge reached a new equilibrium. The final state of the current profile did not exhibit the off-axis peak seen during the current ramp, but does show some flattening of the current profile near the magnetic axis.

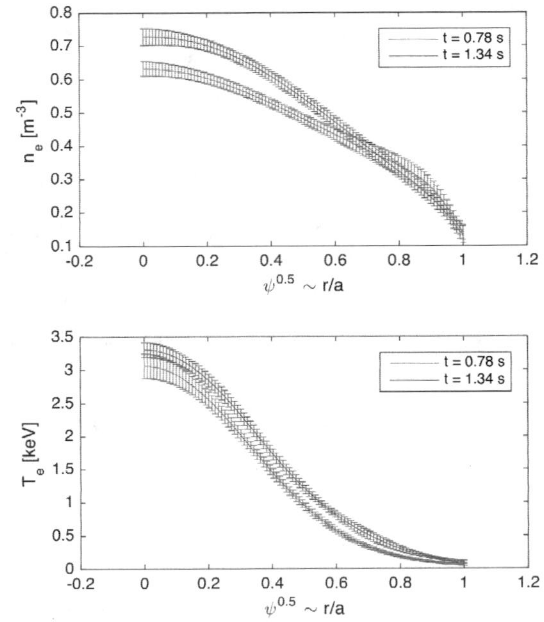


**Fig. 5.** Kinetic profiles measured by Thomson scattering for discharge 1160929029 before and after the current rampdown.

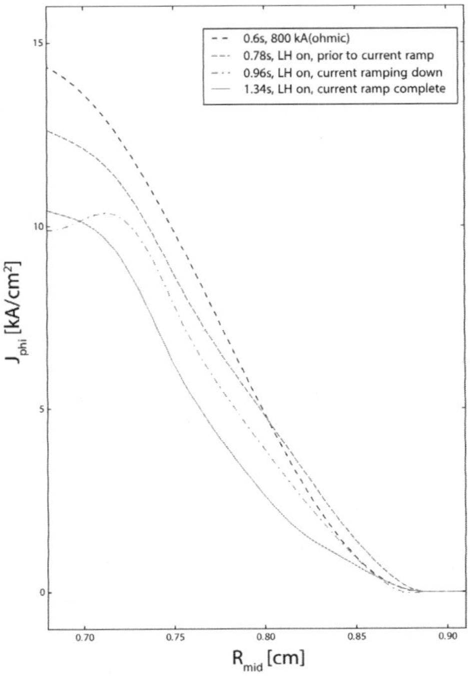


**Fig. 6.** MSE constrained EFIT current profile reconstructions of shot 1160929029.

Sawteeth in this discharge appear during the Ohmic period starting around 0.27 s, and are suppressed shortly after the beginning of the LH pulse (before the start of the $I_p$ ramp). The safety factor profiles in Figure 7 shows weak shear reversal during the current rampdown, with a flat safety factor profile in the core during the low $I_p$ portion of the discharge.

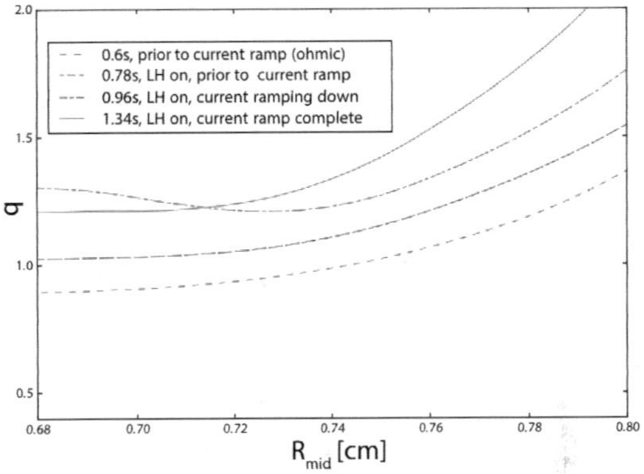


**Fig. 7.** MSE constrained EFIT safety factor profile reconstructions of shot 1160929029.

A short burst (0.99 to 1.02 s) of MHD activity was observed during the current rampdown, but otherwise the discharge was free of the deleterious low frequency MHD instabilities shown in Figure 1.

A poloidally viewing hard x-ray (HXR) camera is used to diagnose the non-thermal fast electron population for this discharge. Figure 8 shows the line integrated HXR emission profiles at the MSE measurement time points during the LHRF pulse. The HXR profile is considerably broader before the current rampdown as compared to after. The peak count rate in the plasma core does not change during the rampdown relative to the pre-rampdown value, indicating that the effect of the large negative loop voltage imposed at the edge of the plasma does not reach the core on the timescale of the ramp. Furthermore, the break in slop of the HXR emission plotted in Figure 9 occurs approximately 150 ms after the loop voltage at the plasma boundary changes sign, followed by a relatively modest reduction in HXR emission. This observation suggests that the externally applied toroidal electric field diffuses only to the outer regions of the plasma (HXR chords <10 and >22) on the timescale of the current rampdown.

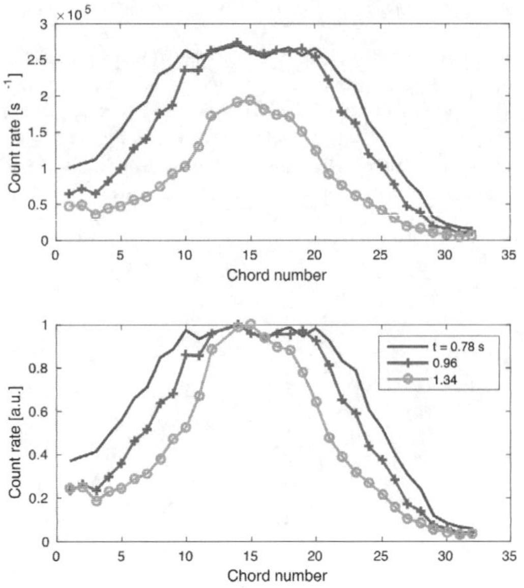


**Fig. 8.** Absolute (top) and normalized (bottom) hard x-ray emission profiles before, during, and after the current ramp.

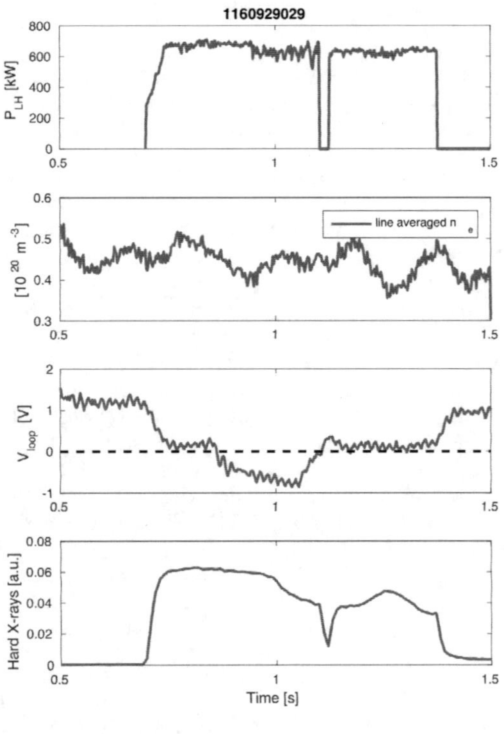


**Fig. 9.** Time history of net LHRF power, line averaged density, loop voltage, and hard x-ray count rate (chords 9-24).

## 4 Discussion and conclusions

The absence of MHD activity in this one discharge does not indicate that a solution to the problem of MHD instabilities reducing current drive efficiency has been identified, and more discharges are needed to increase statistical confidence. As Figure 3 shows, not every shot near full non-inductive current drive developed these instabilities, and additional discharges would be needed to determine if the technique is robust relative to 500 kA discharges without the $I_p$ ramp. Further stability analysis on this discharge following the method presented in [6] will yield additional insight into whether the current ramp technique results in a more stable current profile.

Time-dependent analysis of this discharge using TRANSP [7] in conjunction with the GENRAY/CQL3D ray-tracing/Fokker-Planck simulation suite [8,9] will be needed to determine the effect of the spatially and temporally varying toroidal electric field on wave damping and current drive. The use of GENRAY/CQL3D within TRANSP is described in [10], and a similar procedure will be followed in modelling this discharge.

This work was conducted on the Alcator C-Mod tokamak, an Office of Science user facility supported by US Department of Energy cooperative agreements DE-FC02-99ER54512 and DE-AC02-76CH03073.

## References

1. Scott, S., P. Bonoli, R. Mumgaard, S. Shiraiwa, G. Wallace, and D. Whyte. "Scaling of Global LHCD Efficiency in Alcator C-Mod." In APS Meeting Abstracts, vol. 1, p. 8021P. (2014).
2. S. Shiraiwa, et al. Nuc. Fus., Vol. 53, No. 11 (2013)
3. Mumgaard, R. T., S. D. Scott, and M. Khoury. "The multi-spectral line-polarization MSE system on Alcator C-Mod." Review of Scientific Instruments 87, no. 11 (2016): 11E527.
4. Zabiégo et al, Plasma Phys. Control. Fusion, 43 (2001) 1625–1639; Maget et al, Nucl. Fusion 47 (2007) 233–243
5. Zohm et al, PPCF, Vol. 33, No. 12, 1423-1441, 1991; Söldner et al, Nuc. Fus., Vol. 34, No. 7, 985-1016, (1994).
6. Ebrahimi, Fatima, A. Bhattacharjee, L. Delgado, S. Scott, J. R. Wilson, G. M. Wallace, S. Shiraiwa, and R. T. Mumgaard. "C-Mod MHD stability analysis with LHCD." In APS Meeting Abstracts. 2016.
7. Hawryluk, R. J. "An empirical approach to tokamak transport." Physics of plasmas close to thermonuclear conditions 1 (1980): 19-46.
8. A. P. Smirnov, R. Harvey, Bull. Am. Phys. Soc. 40 (1995) 1837.
9. R. W. Harvey, M. McCoy, The CQL3D Fokker-Planck Code, in: Pro- ceedings of the IAEA Technical Committee Meeting on Simulation and Modeling of Thermonuclear Plasmas, 1992, pp. 489–526.
10. F.M. Poli, et al. Plasma Physics and Controlled Fusion, Volume 58, Number 9 (2016).

DOI: 10.1051/epjconf/201715703064

# Advanced ponderomotive description of electron acceleration in ICRF discharge initiation

*Tom* Wauters[1,⋆], *Matej* Tripský[1,2], *Anatoli* Lyssoivan[1], *Fabrice* Louche[1], *Sören* Möller[3], *Riccardo* Ragona[1], and *Dirk* Van Eester[1]

[1]*Laboratory for Plasma Physics, ERM/KMS, 1000 Brussels, Belgium, TEC partner*
[2]*Ghent University, Department of Applied Physics, 9000 Ghent, Belgium*
[3]*Institute of Energy and Climate Research-Plasma Physics, Forschungszentrum Juelich, Germany - TEC Partner*

**Abstract.** This contribution proposes a new approach for the ponderomotive description of electron acceleration in ICRF discharge initiation. The motion of electrons in the parallel electric field $E_z$ is separated into a fast oscillation and a slower drift around the oscillation centre. Three terms are maintained in the Taylor expansion of the electric field (0$^{th}$, 1$^{st}$ and 2$^{nd}$ order). The efficiency for electron acceleration by $E_z(z,t)$ is then assessed by comparing the values of these terms at the slow varying coordinate $z_0$. When (i) the 0$^{th}$ order term is not significantly larger than 1$^{st}$ order term at the reflection point, or when (ii) the 2$^{nd}$ order term is negative and not sufficiently small compared to the 1$^{st}$ order term at the reflection point, then the electron will gain energy in the reflection.
An example for plasma production by the TOMAS ICRF system is given. Following the described conditions it can be derived that plasma production is (i) most efficient close to the antenna straps (few cm's) where the field gradient and amplitude are large, and (ii) that the lower frequency field accelerates electrons more easily for a given antenna voltage.

## 1 Introduction

According to Tripský et at. [1], the density evolution in the initial phase of an ICRF discharge can be subdivided into 4 phases characterised by their dominant physics mechanisms. While in later stage (phase III and IV) collective plasma effects are playing an important role, in phases I and II the ionisation rate can be assessed from single electron motions in the antenna vacuum electric field component parallel to the toroidal magnetic field, $E_z(z,t)$. The electron acceleration by $E_z(z,t)$ is described by *Lyssoivan* [2]. The theory separates the motion of charged particles into a fast oscillation and a slower drift around the oscillation centre. Rewriting the velocity and position as $v_z = v_0(t) + v_1(t)$ and $z = z_0(t) + z_1(t)$, the antenna electric field acting on charged particles around the slowly varying coordinate $z_0$ can be written in a Taylor series expansion:

$$E_z(z = z_0(t) + z_1(t)) = \sum_{n=0}^{\infty} \frac{z_1^n}{n!}\frac{\mathrm{d}^n E_z}{\mathrm{d}z^n}(z_0). \tag{1}$$

In the work of Lyssoivan [2], the Taylor series expansion retains the zeroth- and first-order term. Adding to this theory, for the purpose of this contribution, we introduce as well the quadratic term. Under the condition that $z_0$ varies slowly, we write two equations of motion representing the fast (eq. 2) and the slow (eq. 3) dynamics

$$\frac{\mathrm{d}v_1}{\mathrm{d}t} = -\frac{q_e}{m_e}\left(E_z(z_0) + \frac{z_1^2}{2}\frac{\mathrm{d}^2E_z}{\mathrm{d}z^2}(z_0)\right)\cos\omega t \tag{2}$$

$$\frac{\mathrm{d}v_0}{\mathrm{d}t} = -\left\langle \frac{q_e}{m_e} z_1 \frac{\mathrm{d}E_z}{\mathrm{d}z}(z_0)\cos\omega t \right\rangle, \tag{3}$$

where $\langle\cdots\rangle$ stands for the averaging over the fast oscillation period. The second order term is introduced in the equation of motion of the fast oscillation (eq. 2) as its sign will determine whether the oscillation is focussing around $z_0$ or defocussing (unstable). From eq. (2) the spatial oscillation amplitude can be derived $\hat{z}_1 = E_z(z_0)q_e/m_e\omega^2$ as well as the minimum electric field value for which the kinetic energy of the oscillating electron overcomes the ionization potential $\epsilon^{ion}$ [2]:

$$E_{z,\min} = \frac{\omega}{q_e}\sqrt{2m_e\epsilon^{ion}}. \tag{4}$$

From eq. (3) follows the ponderomotive force responsible for the slow drift of the oscillation centre $z_0$ with potential [2]:

$$\Phi_p = \left[\frac{q_e E_z(z_0)}{2\sqrt{m_e}\omega}\right]^2. \tag{5}$$

In order to represent with accuracy the electron motion using eq. (2) and (3), the first and second order term need to be smaller than the zeroth and first order term by factor $\xi_{v,1}$ (eq. 6) and $\xi_{v,2}$ (eq. 7) respectively,

⋆e-mail: t.wauters@fz-juelich.de

$$z_1 \frac{\mathrm{d}E_z}{\mathrm{d}z}(z_0) < \xi_{v,1} E_z(z_0) \tag{6}$$

$$\frac{z_1}{2}\frac{\mathrm{d}^2E_z}{\mathrm{d}z^2}(z_0) < \xi_{v,2}\frac{\mathrm{d}E_z}{\mathrm{d}z}(z_0)\,, \tag{7}$$

where excursion $z_1$ can be taken as the amplitude $\hat{z}_1$ of the zeroth order oscillation. The latter validity conditions are then summarised as

$$\xi_{v,1} > \xi_1 = \frac{q_e}{m_e\omega^2}\frac{\mathrm{d}E_z}{\mathrm{d}z} \tag{8}$$

$$\xi_{v,2} > \xi_2 = \left(\frac{\mathrm{d}E_z}{\mathrm{d}z}\right)^{-1}\frac{q_eE_z}{2m_e\omega^2}\frac{\mathrm{d}^2E_z}{\mathrm{d}z^2} \tag{9}$$

In the following, single electron simulations will be interpreted in light of the above validity limits.

## 2 Single electron simulations and interpretations

It is observed in PIC-MCC simulations with code RFdinity1D [1] that upon powering the ICRF antenna, initial electrons present in the torus are quickly expelled from the antenna area. Plasma production by ICRF antennas at typical ICWC pressure levels below $5 \cdot 10^{-4}$ mbar [3] relies therefore on the ability of the $E_z$ profile to change the energy (initial vs. final) of the electrons encountering the antenna area (passing electrons or reflected electrons).

Using a single electron model we investigate the threshold initial energy level $\epsilon_{tr}$ above which an electron is able to gain energy from the $E_z$ field. Parametric dependencies on field amplitude $E_0$, frequency $f$ and shape $\sigma$ are investigated. The electric field profile is parametrised as

$$E(z,t) = E_0\left(\mathrm{e}^{\left(-(z+7.5\,\mathrm{cm})^2/\sigma^2\right)} - \mathrm{e}^{\left(-(z-7.5\,\mathrm{cm})^2/\sigma^2\right)}\right)\cos(2\pi f t)\,. \tag{10}$$

The single electrons are launched at position $z_0$ nearby the antenna where $E_z^{RF}(z_0) \approx 0$ with initial energy sampled from a uniform distribution in energy spectrum $\epsilon_0 \in \left[10^{-1}, 10^2\right]$ eV and phase $\phi_0 \in [0, 2\pi[$. Each electron is tracked until it has encountered the antenna area 3 times, which corresponds for a 0.5 eV electron in an ASDEX Upgrade size of torus to a duration of about 0.2 ms. The model neglects all collisions. The threshold initial energy $\epsilon_{tr}$ is easily obtained by noting the substantial increase in the energy difference between the initial and the final electron energy in the simulation from certain initial energy level on.

Parametric scans for the electric field amplitude ($E_0$), frequency ($f$) and shape ($\sigma$) are plotted in Figure 1, resp. subfigures a), b) and c). The observed tendencies in these numerical experiments can be interpreted in light of the electric field threshold Eq. (4) and validity limits Eq. (8) and (9).

1. At high electric field amplitude ($E_0$), low frequency ($f$) and steep profiles ($\sigma$) it is observed that the threshold energy is closely related to the breaking of validity condition eq. (8). The oscillation centre

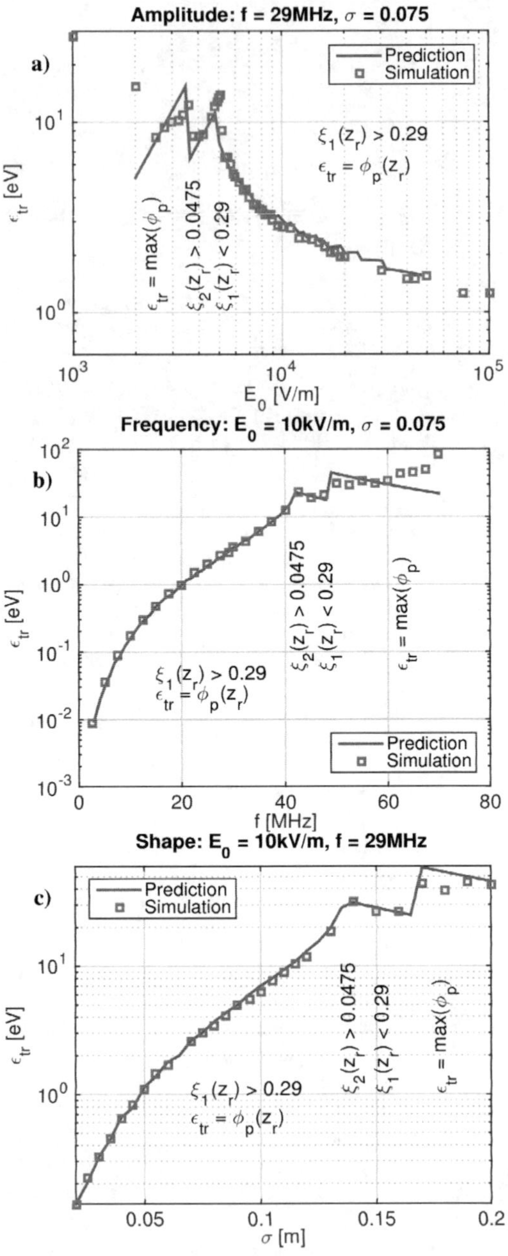


**Figure 1.** Threshold initial electron energy as function of a) the electric field amplitude ($E_0$), b) frequency ($f$) and c) shape ($\sigma$) of the parallel electric field (eq. 10). The observed tendencies in the numerical experiments (red rectangles) are result from the electric field threshold Eq. (4) and validity limits Eq. (8) and (9) (blue line)

of an electron that approaches the antenna climbs up the ponderomotive potential (eq. 5). In case the initial electron energy is smaller than the maximum ponderomotive potential $\epsilon_0 < \max(\Phi_p)$, it will be reflected. The initial energy equals the final energy ($\epsilon_0 \approx \epsilon_f$) as long as $\xi_1 < \xi_{v,1} \approx 0.29$ at the reflection point $z_r$. The electron was found to acquire energy from the vacuum electric field in case $\xi_1 > \xi_{v,1}$ at the reflection point. The reflection point corresponds to the location where the initial energy equals the ponderomotive potential ($\Phi_p(z_r) = \epsilon_0$). The threshold energy $\epsilon_{tr}$ can therefore be taken as the ponderomotive potential at the location where $\xi_1 = \xi_{v,1}$.

2. When validity limit eq. (8) is fulfilled for the entire electric field shape at the antenna edge ($\xi_1 < \xi_{v,1}$), i.e. at intermediate $E_0$, $f$ and $\sigma$, then the threshold energy $\epsilon_{tr}$ is determined by validity limit eq. (9). The single electron simulations showed that the final energy after reflection may differ from the initial energy as soon as the oscillation centre enters the area with negative second order term. This is understood from eq. (3); a positive second order term focusses the oscillating electron around the slowly varying coordinate, while a negative second order term leads to an unstable oscillation (defocussing). The initial energy equals the final energy ($\epsilon_0 \approx \epsilon_f$) as long as $\xi_2 < \xi_{v,2} \approx 0.0475$ at the reflection point $z_r$. The threshold energy $\epsilon_{tr}$ approximates the ponderomotive potential at the location where the second derivative of the electric field profile turns negative.

3. When both conditions (eq. 8 and 9) are fulfilled, then $\epsilon_{tr}$ is determined by the maximum ponderomotive potential. Only electrons with initial energy larger than the ponderomotive potential $\epsilon_0 > \max(\Phi_p)$ are able to enter into the antenna area and eventually to gain energy.

4. At low $E_0$, high $f$ or high $\sigma$, the electrons are unlikely to gain sufficient energy for ionisation in interaction with the vacuum electric field regardless of their initial energy (eq. 4).

## 3 Assessment of TOMAS ICRF vacuum electric field for plasma production

The above derived and studied set of conditions for electron acceleration by the parallel vacuum electric field in toroidal devices allows to pre-assess the ability for any ICRF antenna geometry, phasing, input power and frequency to initiate plasma as function of, e.g., the radial coordinate $R$. An example for the TOMAS ICRF system [4] is shown in figure 2 for an input voltage of 1.5 kV, corresponding to ~ 3 kW of forward RF power. Sub-plot 2(a) shows the maximum amplitude of the $E_z$ field for RF frequencies in range $f$ = 15 MHz to 45 MHz, together with the minimum amplitude required for accelerating electrons above the ionisation potential (horizontal

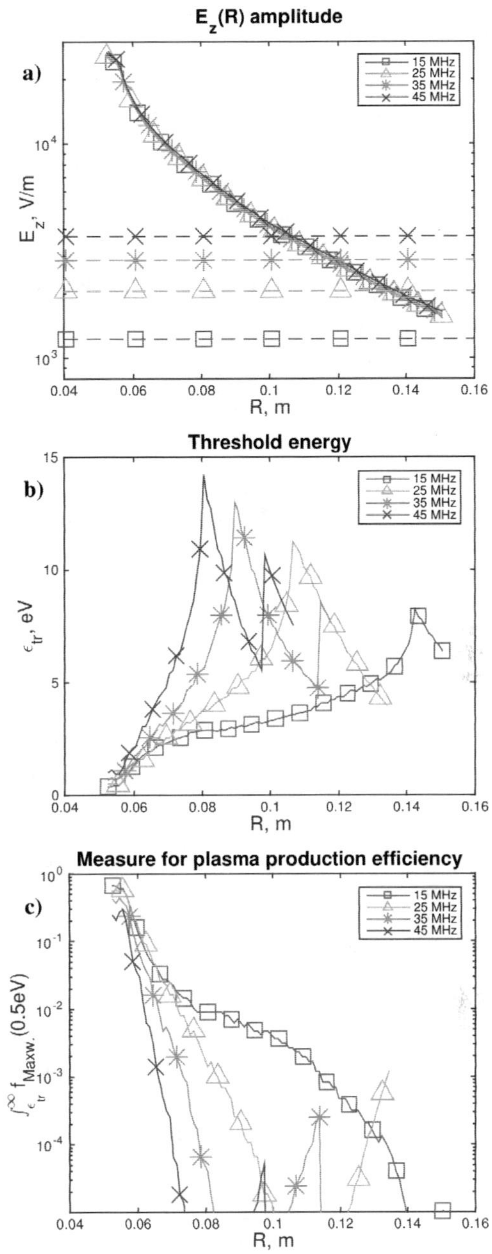


**Figure 2.** Illustration of dependencies as function of radius $R$ for TOMAS ICRF antenna. $E_z$ fields are obtained in the horizontal midplane using the commercial electromagnetic software CST Microwave Studio® .

lines as given by eq. (4)). The curves in sub-plot 2(b) and 2(c) are discontinued at radial positions where the $E_z$ amplitude becomes lower than this minimum. The antenna box face is located at 5.5 cm.

Sub-plot (b) shows the threshold energy as derived from the validity conditions. The threshold energy is minimum close to the antenna straps where the field gradient is steepest and largest in amplitude. The threshold energy, at first determined by validity condition eq. (8) ($\xi_1 < \xi_{v,1}$), increases to reach a maximum further in the torus (e.g. at 8 cm for 45 MHz). Beyond this point the threshold energy is determined by condition eq. (9) ($\xi_2 < \xi_{v,2}$) and decreases until a second discontinuity occurs (e.g. at 9.9 cm for 45 MHz). Beyond this point both validity limits are fulfilled and the threshold energy is determined by the maximum ponderomotive potential.

Sub-plot (c) shows the fraction of electrons in a Maxwellian energy distribution of 0.5 eV that has energy above the threshold energy $\epsilon_{tr}$. It provides a measure for the efficiency for accelerating a population of low temperature electrons that are initially present in the vessel. From this picture it is clear that the efficiency drops dramatically by 4 orders of magnitude over a distance where the threshold energy increases by only one order of magnitude. From this analysis it can therefore be concluded that plasma production in the vacuum $E_z$-field of poloidal ICRF antenna straps inside an antenna box occurs in the vicinity (few cm's) of the antenna box. Moreover it can be concluded that plasma production is more efficient at low frequencies than at high frequency for a given strap voltage (1.5 kV).

## 4 Conclusion

The majority of the electrons in the initial phase of an ICRF discharge reside outside of the antenna area and, in order to overcome the ionisation potential, need to gain energy via interaction with the vacuum parallel electric field that locates in the vicinity of the antenna. It is found that electrons with initial energy smaller than the maximum ponderomotive potential may gain energy in reflections at the sides of the antenna. This contribution proposes hereto an advanced ponderomotive description of electron acceleration. The motion of the electrons in the parallel electric field $E_z$ is separated into a fast oscillation and a slower drift around the oscillation centre. Three terms are maintained in the Taylor expansion of the electric field, namely the $0^{th}$, $1^{st}$ and $2^{nd}$ order term. The efficiency for electron acceleration by $E_z(z, t)$ is then assessed by comparing the values of these terms at the slow varying coordinate $z_0$ at the antenna outer edges. When (i) the $0^{th}$ order term is not significantly larger than $1^{st}$ order term at the reflection point, or when (ii) the $2^{nd}$ order term is negative and not sufficiently small compared to the $1^{st}$ order term at the reflection point, then the electron will gain energy in the reflection. (iii) Electrons with initial energy larger than the maximum ponderomotive potential will enter into the antenna area. Finally (iv) the $0^{th}$ order term gives the minimum electric field amplitude that accelerates an electron above the ionisation potential within a fast oscillation period.

An example for plasma production by the TOMAS ICRF system [4] is given. Following the described conditions it can be derived that plasma production is (i) most efficient close to the antenna straps (few cm's) where the field gradient and amplitude are large, and (ii) that the lower frequency field accelerates electrons more easily for a given antenna voltage.

## Acknowledgements

This work has been carried out within the framework of the EUROfusion Consortium and has received funding from the European research and training programme under grant agreement N 633053. The views and opinions expressed herein do not necessarily reflect those of the European Commission.

## References

[1] Tripský M et al, 2017, *Nucl. Fusion* at press: https://doi.org/10.1088/1741-4326/aa8446

[2] Lyssoivan A et al, 2012 *Plasma Phys. Control. Fusion* **54**, stacks.iop.org/PPCF/54/074014

[3] Wauters T et al, 2015, *J. Nucl. Mater.* **463**, http://dx.doi.org/10.1016/j.jnucmat.2014.12.097

[4] Louche F et al, 2017, *Fusion Eng. Des.*, in press: https://doi.org/10.1016/j.fusengdes.2017.04.123

DOI: 10.1051/epjconf/201715703065

# Simulations of High Harmonic Fast Wave Heating on the C-2U Advanced Beam-Driven Field-Reversed Configuration Device

*Xiaokang* Yang[1,*], *Yuri* Petrov[2], *Francesco* Ceccherini[1], *Alf* Koehn[3], *Laura* Galeotti[1], *Sean* Dettrick[1], *Michl* Binderbauer[1], and the TAE Team

[1]Tri Alpha Energy, Inc., P.O. Box 7010, Rancho Santa Margarita, CA 92688, USA
[2]CompX, P.O. Box 2672, Del Mar, CA 92014, USA
[3]IGVP, University of Stuttgart, Pfaffenwaldring 31, 70569 Stuttgart, Germany

**Abstract.** Numerous efforts have been made at Tri-Alpha Energy (TAE) to theoretically explore the physics of microwave electron heating in field-reversed configuration (FRC) plasmas. For the fixed 2D profiles of plasma density and temperature for both electrons and thermal ions and equilibrium field of the C-2U machine, simulations with GENRAY-C ray-tracing code have been conducted for the ratios of $\omega/\omega_{ci[D]}$ in the range of 6 - 20. Launch angles and antenna radial and axial positions have been optimized in order to simultaneously achieve good wave penetration into the core of FRC plasmas and efficient power damping on electrons. It is found that in an optimal regime, single pass absorption efficiency is 100% and most of the power is deposited inside the separatrix of FRC plasmas, with power damping efficiency of about 72% on electrons and less than 19% on ions. Calculations have clearly demonstrated that substantial power absorption on electrons is mainly attributed to high beta enhancement of magnetic pumping; complete power damping occurs before Landau damping has a significant effect on power absorption.

## 1 Introduction

The C-2U advanced beam-driven FRC device [1] is a simple compact toroid magnetic confinement system, that is, one without toroidal coils linking the plasma, and thus with predominantly poloidal fields. The attractions of such a configuration for a potential fusion reactor are its very high $<\beta_e>$ (near unity) thus allowing for efficient use of magnetic hardware, simple and linear geometry for ease of construction and maintenance, as well as a natural, unrestricted divertor configuration for facilitating energy extraction and fusion ash removal [2].

However, the unique characteristics of C-2U FRC plasmas, for example, the plasma being unusually over-dense ($\omega_{pe} > 30\omega_{ce}$ inside the separatrix) and the magnetic field dropping quickly to zero in the plasma core, make it extremely challenging to heat electrons in the core of FRC plasmas. Conventional electron heating scenarios such as electron cyclotron resonant frequency (or its second or third harmonics) heating which is widely utilized in tokamaks, stellarators, and mirror machines, cannot be adapted to FRC plasmas due to the issue of poor wave accessibility into the plasma core. Other electron heating scenarios, such as electron Bernstein waves, upper-hybrid resonant waves, and whistler waves, encounter similar problem or have low heating efficiency when they are applied to FRC plasmas [3, 4]. Fortunately, our recent survey indicated that the conflict between good wave accessibility and efficient power damping on electrons may be solved by using high harmonic fast wave (HHFW) heating, which has been successfully adapted to high beta, overdense spherical tokamak (ST) plasmas such as NSTX for the experiments of core electron heating and off-axis current drive [5 - 7].

It is well known that in the regime of fast wave direct electron heating, there are two collisionless damping mechanisms: Landau damping (LD), where the force acting on electrons is $\boldsymbol{F}_{\rm LD} = e\boldsymbol{E}_{//}$; and transit-time magnetic pumping (TTMP or MP), in which the force is $\boldsymbol{F}_{\rm MP} = -\nabla_{//}(\mu B_{//})$. Here $e$ and $\mu$ are electron's charge and magnetic moment, and $\boldsymbol{E}_{//}$ and $B_{//}$ are the parallel components of the fast wave electric and magnetic field, respectively. Conventional fast wave electron heating in tokamak plasmas requires wave parallel phase velocity $V_{\rm ph//} \equiv \omega/k_{//} \approx V_{\rm Te}$ (electron thermal velocity) for any significant absorption via a dominated LD; MP makes no significant contribution to electron damping and often it can be neglected. Moreover, the absorption of fast wave in tokamak plasmas is weak and therefore it is usually required to have a strong electron preheating by microwaves at electron cyclotron resonant frequency in order to enhance multiple-pass power absorption. However, in high beta, ST plasmas like NSTX, it was found that the MP significantly increases power absorption on electrons over the electron LD alone, and it becomes substantially large at a higher range of phase velocity, $\omega/k_{//} \leq 2.5V_{\rm Te}$ [5]. The combination of MP and LD can lead to 100% single pass absorption.

* Corresponding author: xyang@trialphaenergy.com

DOI: 10.1051/epjconf/201715703065

## 2 Parameter settings

Ray-tracing code Genray-C is a special version of Genray [8] adapted for plasmas like FRC or mirror machine where toroidal magnetic field can be absent. The C-2U machine configuration can be found in [1]; pure deuterium gas is routinely used in experiments. In our simulations, 5 rays are launched from one point but with different angles, each ray carrying 200 kW power. In this report, the launcher's z-axial position is fixed at the midplane (z = 0), which is a preferable position for HHFW antenna-plasma coupling and wave penetration; moreover, launching HHFW at the midplane can make antenna physics and engineering design easier. The 2D equilibrium field profile reconstructed from experimental measurements is utilized; Fig. 1 shows its radial profile at the midplane. For feasibility study, the 2D model density profiles are used, matching those from C-2U experimental measurements. Both electrons and ions (D) have the same peak density of $2.4\times10^{19}$m$^{-3}$ at the field null point and the same density of $1.5\times10^{19}$m$^{-3}$ at the separatrix. In this case, the plasma has an unusually large (comparing to tokamaks) dielectric constant $\varepsilon$ ($\equiv\omega_{pe}^2/\omega_{ce}^2$ > 4000 inside the separatrix layer). The $T_e$ and $T_i$ (thermal D ions) in the plasma core are 150 eV and 800 eV, respectively. For simplification the data (density and temperature) on fast ions generated by neutral-beam injection is not included in present calculations. The simulations with fast ions will be performed in future work.

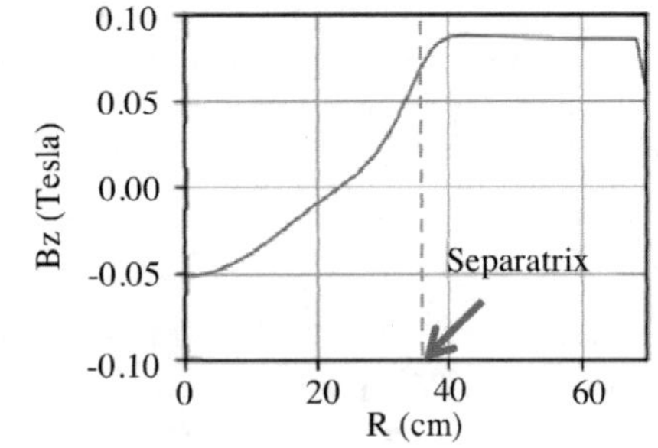


**Fig. 1.** Radial profile of C-2U equilibrium field $B_z$ at midplane.

## 3 Simulation results

Extensive simulations have been performed in order to find optimal regimes, for which the HHFWs have both good penetration and favourable power damping partition between electrons and ions. The parameters that have been scanned in the process of optimization include frequencies (thus the ratio of $\omega/\omega_{ci[D]}$ at rays' starting point since equilibrium field profile is fixed), antenna radial positions (the distance between antenna surface and plasma edge), and spectrum of parallel refractive index $n_{//}$ at the antenna position (launch angles). Those parameters can have significant effects on wave coupling, propagation, and penetration, as well as power damping partition between electrons and ions.

As an example of optimal regimes, Fig. 2 shows HHFW propagation trajectory in C-2U FRC plasmas. The HHFW frequency is at 7 MHz, and the initial $n_{//}$ is in a narrow range between 5 and 7 for the five rays. The background contours show the 2D profile of electron density, which is peaked at magnetic field null or local minimum of $|B|$. Rays launched at the possible antenna position near the wall (z = 0, r = 55 cm) can penetrate through the separatrix layer and propagate into the FRC plasma core. Figure 3 shows the change of power in each ray channel with the distance of wave propagation. The total distance of the rays' propagation is about 200 cm, but most of their power is deposited by non-collisional damping in the area between 40 cm to 100 cm (the region marked with shadow); even though rays can continue propagating another 100 cm into the plasma core, the remaining power is very small.

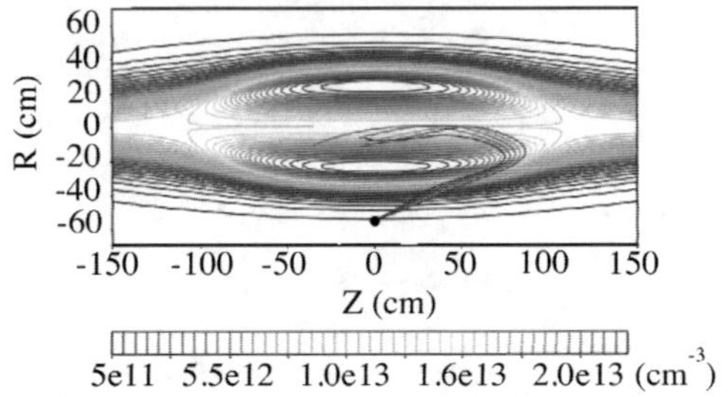


**Fig. 2.** The propogation trajectory of HHFW (7 MHz) in C-2U plasmas. Background contours show electron density profile.

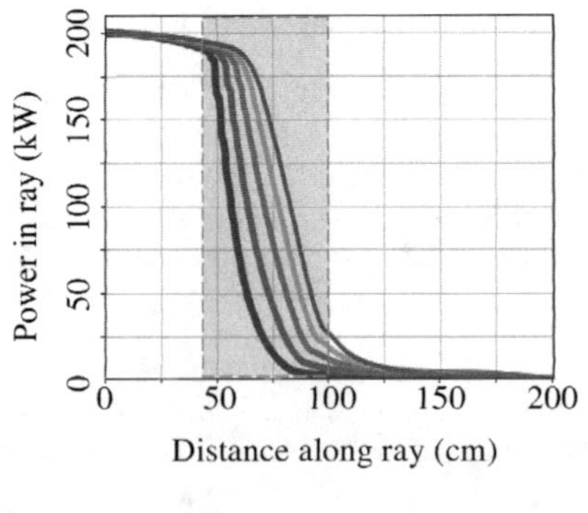


**Fig. 3.** The HHFW (7 MHz) power in ray channel along the distance of wave propagation. The initial $n_{//}$ = 5 for blue ray and $n_{//}$ = 7 for purple ray.

The radial profiles of power density for total power absorption, power damping on electrons and ions, and collisional power damping, are shown in Figs. 4(a)-4(d), respectively. The position of rho = 1 corresponds to the location of separatrix layer of FRC plasma, where rho is defined through the square root of poloidal flux. In this report, the area of rho <1 (with closed field lines) defines "plasma core" while the area of rho >1 outlines the region of scrape-off layer (SOL). As shown clearly in Fig. 4(a), single pass absorption efficiency of HHFW power is 100% (total launched power is 1000 kW), and most power (around 60%) is damped inside the separatrix. Also, Figs. 4(b) and 4(c) indicate that the power partition is 72% on electrons and about 19% on ions. The remaining power (about 9%) is damped through collisions (Fig. 4(d)) due to a relatively low electron temperature at plasma edge. The "sawteeth"-like features seen in profiles in Figs. 4(a) and 4(c) represent ion cyclotron harmonic resonance damping.

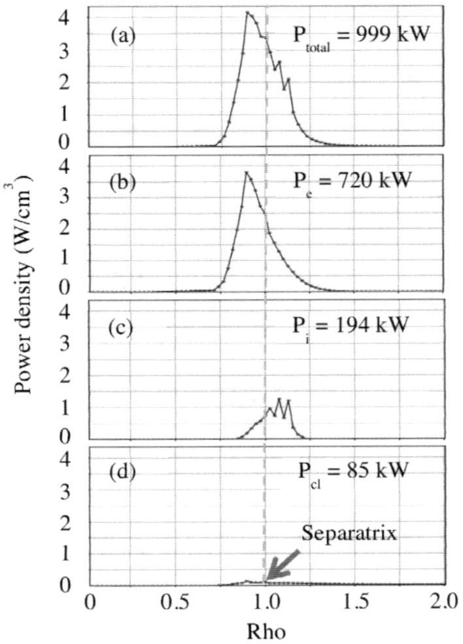


**Fig. 4.** Radial profiles of power density for (a) total absorption, (b) damping on electrons, (c) damping on ions, and (d) collisional damping.

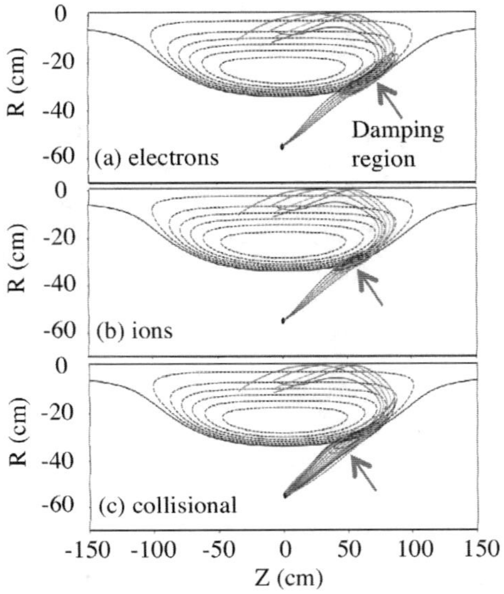


**Fig. 5.** The 2D profiles of power density for (a) damping on electrons, (b) damping on ions, and (c) collisional damping. Background contours show magnetic flux 2D profile.

The 2D power density profiles of damping on electrons, ions, and collisional damping are plotted in Fig. 5 with the marked area showing power damping region. As can be seen clearly in Fig. 5, the power damping regions for electrons and ions are overlapped, thus there is a competition of power partition between electrons and ions when HHFW propagates into this overlapped area.

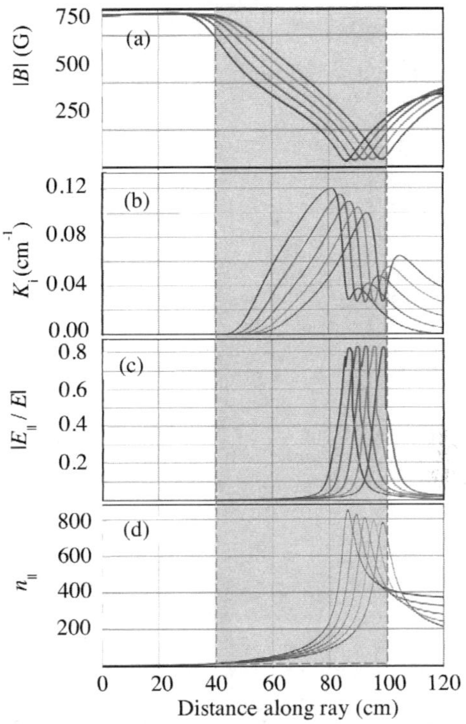


**Fig. 6.** Changes of (a) local $|B(r, z)|$, (b) imaginary part of perpendicular wave number $k_i$, (c) ratio of $|E_{//}/E|$, and (d) parallel refractive index $n_{//}$, with the distance along wave propagation.

Figure 6 can help to better understand the damping mechanisms of HHFW in FRC plasmas, which shows the changes of local magnetic field ($|B(r, z)| = (B_r^2 + B_z^2)^{1/2}$), imaginary part of perpendicular wave number $k_i$, the normalized parallel component of wave electric field $|E_{//}/E|$, and parallel refractive index $n_{//}$, with the distance along wave propagation. The value of $k_i$ in Fig. 6(b) describes power damping on electrons by both MP and LD; although we are presently unable to separate the contribution of MP and LD, we can use Fig. 6(c) to conclude where LD can have an effect of power damping on electrons because $\boldsymbol{F}_{LD} = e\boldsymbol{E}_{//}$. Clearly, if $|E_{//}/E|$ is small, there can be no pronounced LD. Meanwhile, according to Ref. [5], MP is strongly related to the local magnetic field, and it is found to scale as $\propto B^{-3}$.

As can be seen in Figs. 6(b) and 6(c), there is no collisionless damping on electrons at a distance of wave propagation from 0 to 40 cm because the equilibrium magnetic field remains large and at the same time the value of $|E_{//}/E|$ is nearly zero. A significant power

DOI: 10.1051/epjconf/201715703065

damping occurs in the area between 40 cm and 100 cm (also see the marked area in Fig. 3) due to a rapid drop of magnetic field (Fig. 6(a)), which leads to fast increase of $k_i$ (Fig. 6(b)). For the ray with blue color ($n_{//}$ = 5) in Fig. 6(c), the value of $|E_{//}/E|$ is almost zero from 40 cm to 70 cm; this value is still less than 0.2 even when $k_i$ reaches a maximum value at the distance of 82 cm, where $\omega/k_{//} \leq 0.5V_{Te}$ (this condition is different from either fast wave in tokamak plasmas or HHFW in ST plasmas). The value of $k_i$ drops to less than 1/3 of its maximum value when $|E_{//}/E|$ reaches a maximum at 87 cm, where $|B(r, z)|$ has a minimum and $n_{//}$ has a maximum. Therefore, conclusion can be made that the quick increase of $k_i$ is mainly attributed to high beta enhancement of MP, and power damping on electrons is almost completed before LD has a significant effect on power absorption.

## 4 Discussions

Through extensive calculations, it is found that efficient HHFW electron heating in C-2U plasmas is more difficult to realize than the case of HHFW heating in ST plasmas. The unique magnetic configuration of FRC plasmas can cause the conflict between good wave penetration and favorable power partition between electrons and ions. In C-2U FRC plasmas, the parameters such as the ratio of $\omega/\omega_{ci[D]}$, the initial value of $n_{//}$, the distance between antenna and plasma edge, and the launcher position along z axis, have significant impact not only on wave coupling and penetration, but also on power partition between electrons and ions. While in spherical tokamaks like NSTX, there is no issue of wave penetration into plasma core, thus one can only focus on how to avoid a parasitic ion heating or mitigate hot ion damping.

In C-2U plasmas, on the one hand, in order to mitigate the competition from ion heating, one can maximize power damping on electrons by increasing (1) the ratio of $\omega/\omega_{ci[D]}$, or (2) the initial value of $n_{//}$, or (3) the distance between antenna and plasma edge. However, the changes of those parameters may shift the area of power damping from inside the separatrix layer into the SOL. As a result, electrons at open field lines will be heated and therefore fast wave power will eventually be lost (at end divertors). Those results are somehow similar to the regime of Whistler wave heating.

On the other hand, in order to enhance HHFW penetration into the core of FRC plasmas, one can reduce (1) the ratio of $\omega/\omega_{ci[D]}$, or (2) the initial value of $n_{//}$, or (3) the distance between antenna and plasma edge. However, although single pass absorption efficiency is still 100% in those cases, power absorption by electrons decreases substantially and most of the HHFW power is deposited to the ions through ion (D) cyclotron harmonic resonance absorption. By comparing with the case shown in Fig. 2 (f = 7 MHz and $n_{//}$ = 5 - 7 ), one can find that the case shown in Fig. 7 (f = 5 MHz and $n_{//}$ = 3 - 5 ) has a better wave penetration (getting more closely to the region of magnetic field null) and stronger power absorption (shorter distance along wave propagation). Calculations indicate that for the case plotted in Fig. 7, power absorption partition is around 15% on electrons and 75% on ions, and the remainder of 10% power is deposited through collisional damping.

In summary, our simulation results have demonstrated for the first time that the HHFW heating scenario can offer a promising path to efficient heating of electrons in the core of FRC plasmas. In a wide range of plasma parameters, single pass power absorption efficiency is 100%; power damping on electrons increases quickly with the fast drop of magnetic field strength in the core of FRC plasma, and it reaches a maximum value before LD has any significant effect on power absorption. It is obvious that MP is the dominated power absorption mechanism for HHFW electron heating in FRC plasmas.

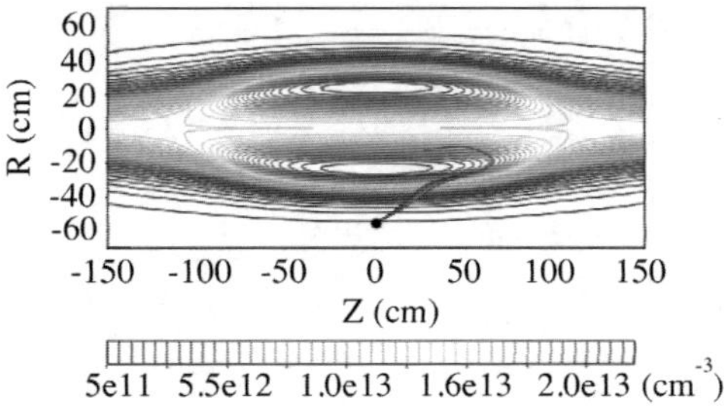


**Fig. 7.** The propagation trajectory of HHFW (5 MHz) in C-2U plasmas. Background contours show electron density profile.

We would like to acknowledge the support and contributions from the rest of the TAE team and collaborators. A debt of gratitude is also due to our shareholders for their continuous support.

## References

1. M. Binderbauer, et al., Phys. Plasmas, **22**, 056110 (2015).
2. M. Tuszewski, Nucl. Fusion, **28**, 2033 (1988).
3. X. Yang, et al., *21st Topical Conf. On RF Power in Plasmas*, AIP Conference Proceedings **1689**, 09008-1 (2015).
4. X. Yang, et al., *Bull. Am. Phys. Soc.* **58**, PP10.00013 (2016).
5. M. Ono, Phys. Plasmas, **11**, 4075(1995).
6. J. Hosea, et al., Phys. Plasmas, **15**, 056104 (2008)
7. G. Taylor, et al., Phys. Plasmas, **17**, 056114 (2010)
8. A. Smirnov and R. Harvey, *CompX report*, CompX-2000-01 (2001).

# Plasma edge modelling with ICRF coupling

*Wei Zhang*[1,2,3*], *David Coster*[1], *Yuhe Feng*[4], *Tilmann Lunt*[1], *Diogo Aguiam*[5], *Roberto Bilato*[1], *Volodymyr Bobkov*[1], *Jonathan Jacquot*[1], *Philippe Jacquet*[6], *Ernesto Lerche*[6], *Jean-Marie Noterdaeme*[1,2], *Wouter Tierens*[1], *the ASDEX Upgrade Team*[1], *the EUROfusion MST1 Team*[7]

[1]Max-Planck-Institut für Plasmaphysik, D-85748 Garching, Germany
[2]Applied Physics Department, Ghent University, B-9000 Ghent, Belgium
[3]Institute of Plasma Physics, Chinese Academy of Sciences, 230031 Hefei, P. R. China
[4]Max-Planck-Institut für Plasmaphysik, D-17491 Greifswald, Germany
[5]Instituto de Plasmas e Fusão Nuclear, Universidade de Lisboa, 1049-001 Lisboa, Portugal
[6]CCFE, Culham Science Centre, Abingdon, OX14 3DB, UK

**Abstract.** The physics of Radio-Frequency (RF) wave heating in the Ion Cyclotron Range of Frequencies (ICRF) in the core plasmas of fusion devices are relatively well understood while those in the Scrape-Off Layer (SOL) remain still unresolved. This paper is dedicated to study the ICRF interactions with the plasma edge, mainly from the theoretical and numerical point of view, in particular with the 3D edge plasma fluid and neutral transport code EMC3-EIRENE and various wave codes. Here emphasis is given to the improvement of ICRF coupling with local gas puffing and to the ICRF induced density convection in the SOL.

## 1 Introduction

The coupling of Radio-Frequency power in the Ion Cyclotron Range of Frequencies (ICRF) to the plasma depends sensitively on the plasma density in front of the antennas because the launched fast wave is evanescent below the cut-off density, usually of the order of $10^{18}$ m$^{-3}$ for typical frequencies and antenna spectrum in nowadays tokamak. Recent experiments on AUG [1] and JET [2, 3] indicate that by injecting the fueling gas from the main chamber instead of the divertor, the local density in front of the antennas and thus the ICRF power coupling can be greatly increased. The corresponding EMC3-EIRENE simulations [4-6] successfully reproduce these experiments and deepen our understanding on the associated physics. This paper will review these results, summarize the mechanisms and make extrapolations to ITER and DEMO.

While generating the fast wave ($E_\perp \gg E_\parallel$, parallel direction along $\vec{B}$) to heat the core plasmas, the slow wave is also excited parasitically. The parallel electric field of the slow wave, especially close to the limiters of the antennas, is expected to be the source of the nonlinearly enhanced sheath potential in front of the antennas by accelerating the more mobile electrons out of the plasma. The inhomogenous sheath potential drives E×B drifts in the Scrape-Off Layer (SOL) and modifies the plasma density locally. To understand this ICRF induced SOL density convection, two numerical methods have been developed: (a) simulations based on experimental data; (b) self-consistent simulations with several codes including EMC3-EIRENE [7] (extended to include prescribed drifts [8] and plasma convection), RAPLICASOL [9] (ICRF fields) and SSWICH [10] (rectified sheath potentials). This paper will review these two numerical methods and the corresponding results.

## 2 ICRF coupling

The local density in front of the ICRF antennas can be tailored by changing the deuterium (D2) gas puff locations. While changing the gas puff locations, no clear change of the energy confinement time and no plasma degration are seen [3]. To understand the effects of localized gas sources on the SOL density, comprehensive 3D simulations with the EMC3-EIRENE code on AUG [4], JET [5] and ITER-relevant [6] scenarios have been carried out.

In our simulations, toroidal 360° grids are used and almost all relevant the plasma facing components, including those toroidal nonsymmetrical (especially the limiters), are taken into account. An example of the computational grids (poloidal cross-section) and the gas puffing positions for AUG and JET is shown in Fig. 1. Note that both in AUG and JET, the mid-plane gas souce is essentially a poloidal and toroidal extended source filling the whole A-port. The divertor gas puffing is used as a reference case. Firstly the plasma of this reference case is validated with various experimental data by choosing the appropriate particle and energy transport coefficients. The separatrix density and total input power are set as the same as those in the experiments. Then the

* Corresponding author: wei.zhang@ipp.mpg.de
[7]See the author list of IAEA FEC 2016 OV/P-12 by H. Meyer et al., to be published in Nucl. Fusion

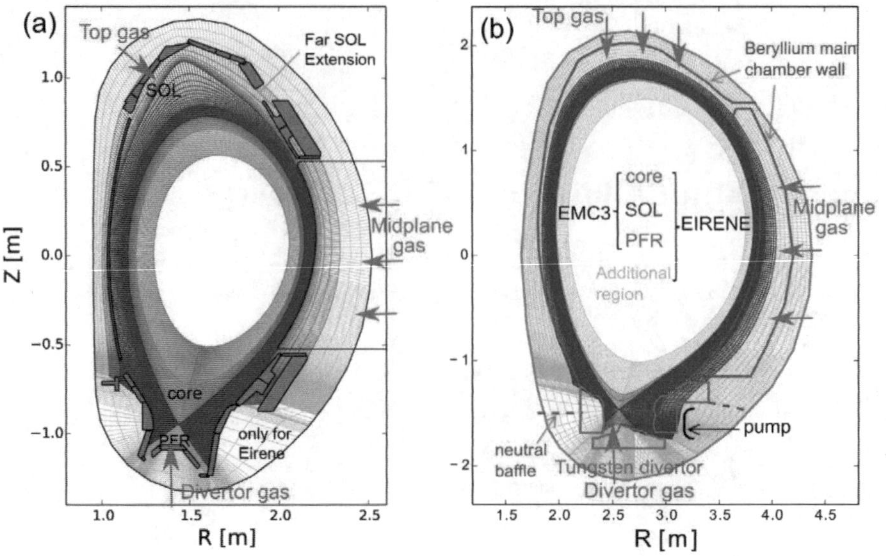


**Fig. 1.** Gas puffing positions and poloidal cross-section of the EMC3-EIRENE computational grid for (a) AUG; (b) JET. Figure reproduced with permission from [4, 5].

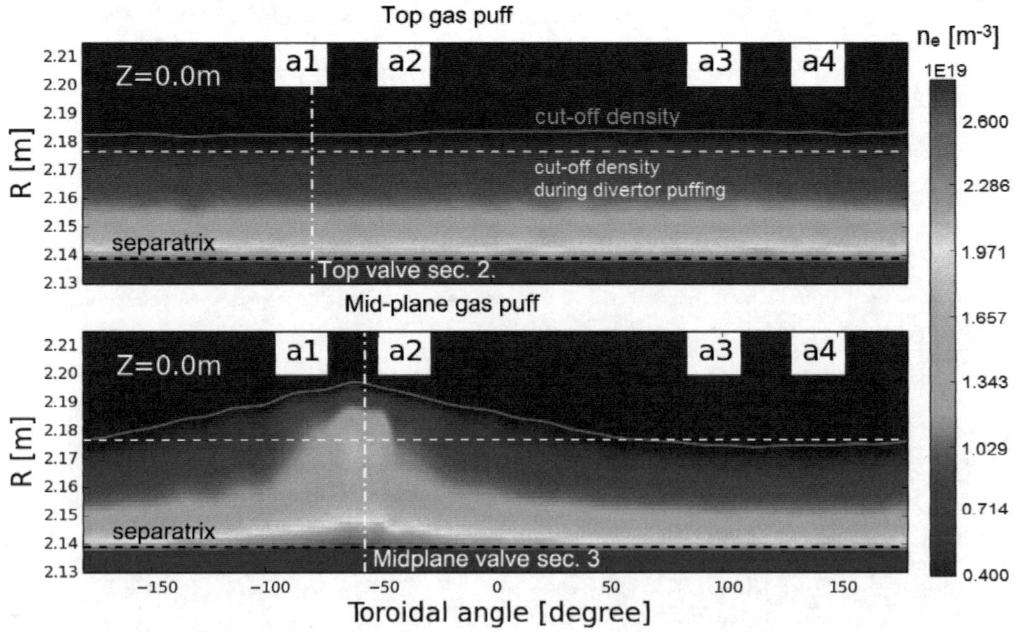


**Fig. 2.** Toroidal cross-sections (in the outer mid-plane) of the edge electron densities simulated by EMC3-EIRENE for top and mid-plane gas puffing in AUG H-mode plasmas. The white dashed line is the cut-off density during divertor gas puffing and is used as reference; the green line is the cut-off density during top or mid-plane gas puffing; the vertical dash-dotted lines represent the toroidal positions of the gas valves. The four ICRF antenna positions are represented by a1, a2, a3 and a4. Figure reproduced with permission from [4].

gas source is switched from the divertor to other locations of the machine to investigate other gas puffing cases.

Gas puffing from different locations of the machine ultimately lead to different SOL densities in front of the antenna. Such an example in AUG H-mode plasmas is shown in Fig. 2. Compared to divertor gas puffing, top gas puffing increases the SOL density near the outer mid-plane, where the antenna are located, almost toroidal uniformly but to a small extent, and this density increase is independent of the toroidal position of the top gas valve. Mid-plane gas puffing increases the SOL density very significantly but locally, and this density increase depends sensitively on the toroidal position of the mid-plane gas valve. The largest density increase is found in locations close to the gas valve and this increase decays in the toroidal direction. Same phenomena are observed also in JET.

DOI: 10.1051/epjconf/201715703066

The mechanisms of the local density increase with local gas puffing are summarized as follows. Gas puffing provides a local particle source. In a steady state, the ionized particles (ions+electrons) leave this ionization zone mainly via convection along the field lines. The associated convective energy flux and the power losses into the ionization processes provide an energy sink to the background plasma and thereby lower the plasma temperatures in the ionization region due to the finite classical heat conductivity, resulting in a local density rise as a consequence of the parallel pressure conservation.

The 3D plasma density obtained from the EMC3-EIRENE simulations are then used in the antenna codes such as FELICE [11] and RAPLICASOL [9] to calculate the antenna resistance $R_c$. Good agreements are found between the calculated and measured values. For top gas puffing, the increase of $R_c$, ~25% in AUG and ~40% in JET H-mode plasmas, is almost the same for all the antennas. For mid-plane gas puffing, the largest increase of $R_c$, ~120% both in AUG [1] and JET [3] H-mode plasmas, is found for the antenna closest to gas valve. This value decays with the distance between the antenna and valve becomes larger: the decay length is ~3.7m in AUG [1, 4] and ~3.0m in JET [3]. The difference of the decay lengths in various machines can be attributed to the peculiarity of the wall structures (for instance JET has wide poloidal limiters), different magnetic field geometries and plasma transport properties. Nevertheless, the results in AUG and JET strongly suggest that mid-plane gas valves close to the antenna is the best option to maximize the ICRF power coupling.

## 3 ICRF induced plasma convection

The interaction between the ICRF waves and SOL plasma is a complex nonlinear process: (1) the plasma density influences the $E_\parallel$ fields; (2) the $E_\parallel$ fields lead to an inhomogenous Direct Current (DC) sheath potential, $V_{DC}$; (3) the inhomogenous $V_{DC}$ drives E×B drifts in the SOL and modify the plasma density. The new plasma density will change the $E_\parallel$ fields, i.e. the loop comes to step (1) again.

A first simulation strategy is to use the measurements of the sheath potential of the reciprocating Retarding Field Analyzer (RFA) to estimate the E×B drifts, recently implemented in EMC3-EIRENE [8]. This makes possible to compare the impact of the E×B drifts on the density, predicted by EMC3-EIRENE, with the measurements of RFA or reflectometries. In our calculations, the averaged parallel energy $< W_\parallel(t) >$ measured by the RFA is equal to the Bohm energy plus the time-averaged (DC) sheath potential when assuming mono-energetic ions at sheath entrance [12], i.e.

$$< W_\parallel(t) >=< M_i u_{Bohm}^2/2 > +< eV_{sh}(t) >$$

in which $u_{Bohm}$ is the ion sound speed, $< M_i u_{Bohm}^2/2 >$ is the energy gained in the pre-sheath and $< eV_{sh}(t) >$ is the energy gained in the sheath. Thus, the whole plasma potential ($\Phi_{pl}$) which equals to the potential drop in the pre-sheath and sheath ($\Phi_{pre}$ and $\Phi_{sh}$) can be calculated with $\Phi_{pl} = \Phi_{pre} + \Phi_{sh} =< W_\parallel(t) >/q$. In addition, the sheath potential is assumed to be nearly constant along the magnetic field lines, and its value in other toroidal positions is extrapolated by tracing magnetic field lines. Our simulation results are in good agreement with the experimental ones, Fig. 3. The results indicate that convective cells are developed where high potential blobs exist. Density depletions are found in the center of the convective cells while density accumulations are found where the plasma interacts with the wall. The plasma flows driven to the wall by the convective drifts enhance the interaction between the plasma and wall.

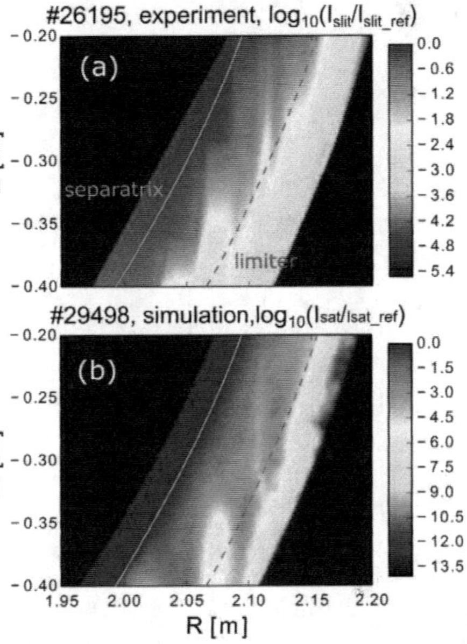


**Fig. 3.** Comparisons between experimental and simulated ion saturation current (in logarithmic scale). The ion saturation current can be expressed as $I_{sat}=qn_eC_s$, in which q is the ion charge number, $n_e$ is the electron density and $C_s$ is the plasma sound speed. Figure reproduced with permission from [8].

The second simulation strategy is of the ab-initio type, and it requires to interface several codes in a consistency loop [13]. Before starting the iterative loop, EMC3-EIRENE simulations are done for a phase of the plasma discharge without RF, and thus without E×B drifts induced by ICRF. This is necessary to set the particle and transport coefficients of EMC3-EIRENE consistently with the experimental plasma profiles. These coefficients are assumed unaffected by ICRF heating and thus kept constant in the loop. When this setting is completed, the consistency loop for the ICRF heating phase starts by calling RAPLICASOL code to evaluate the RF $E_\parallel$ fields. These fields are, then, used in SSWICH code to calculate $V_{DC}$. In the SSWICH calculations, the perpendicular conductivity in the

 DOI: 10.1051/epjconf/201715703066

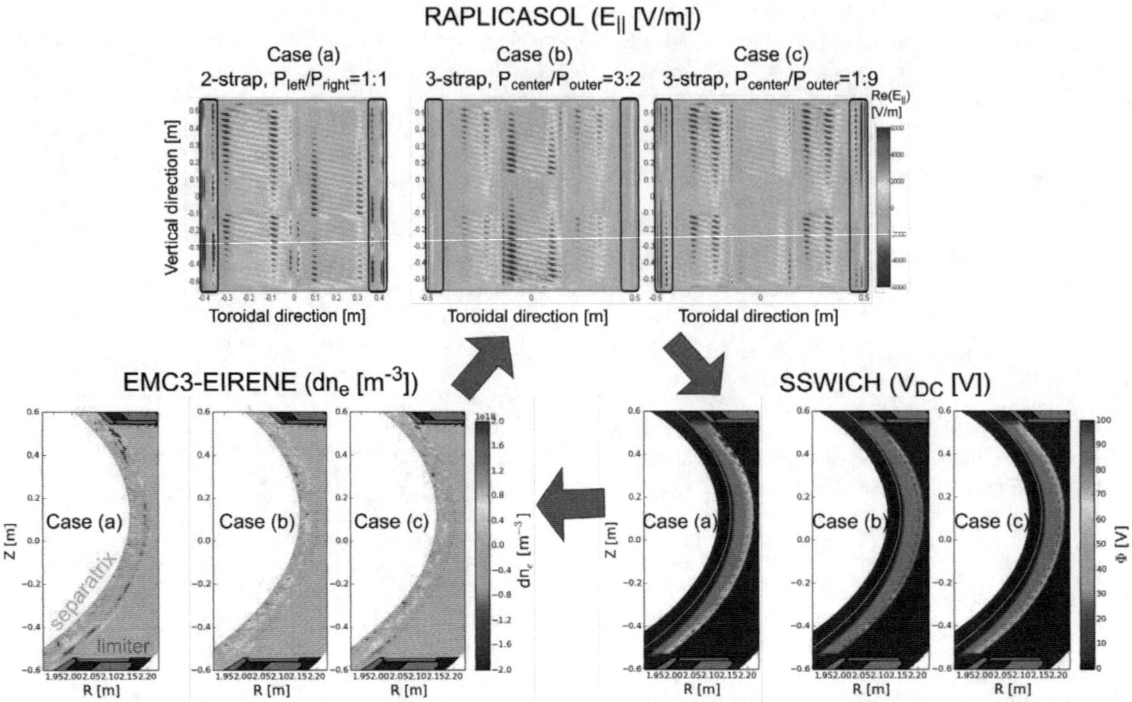


**Fig. 4.** Results of the self-consistent loop, including $E_{\parallel}$ calculated with the RAPLICASOLcode, $V_{DC}$ calculated with the SSWICH code and $n_e$ calculated with the EMC3-EIRENE code (here $dn_e = n_e - n_{e\_ref}$, in which $n_{e\_ref}$ is the unperturbed density). Figure reproduced with permission from [12].

private scrape-off layer ($\sigma_{\perp,p}$) and in the free scrape-off layer ($\sigma_{\perp,f}$) are assumed to be proportional to the parallel (Spitzer) conductivity: $\sigma_{\perp,p} = R_{s,pp} * \sigma_{\parallel}$, $\sigma_{\perp,f} = R_{s,fp} * \sigma_{\parallel}$. Here $R_{s,pp} = 4 \times 10^{-6}$ and $R_{s,fp} = 8 \times 10^{-8}$, values at which the sheath is near saturation and that seem to give reasonable behavior of $V_{DC}$. The calculated $V_{DC}$ covers the antenna and these values in other toroidal positions are extrapolated starting from the two sides of the antenna based on the assumption that $V_{DC}$ is almost constant along the magnetic field lines. The final $V_{DC}$, in turn, defines the E×B drifts for the call of EMC3-EIRENE. The iteration loop stops when the difference of a certain parameter (such as density) in one loop with that in the previous loop is lower than 5%. An example of this self-consistent simulation loop is shown in Fig.4. Three cases have been investigated: (a) 2-strap antenna with power balance between the left and right straps $P_{left}/P_{right}$=1:1 (the most typical settings for 2-strap antenna); (b) 3-strap antenna with power balance between the central and outer straps $P_{center}/P_{outer}$=3:2 (optimized power balance); (c) 3-strap antenna with $P_{center}/P_{outer}$=1:9 (non-optimized power balance). All the studied cases are with dipole phasing. The results (Fig. 4) indicate that the largest density convection happens in the top and bottom of the antennas. The largest sheath potential and density convection are induced by the 2-strap antenna (Case (a)) while the lowest values are induced by the 3-strap antenna with optimized power balance (Case (b)).

## 4 Conclusions

Through edge plasma modelling with ICRF coupling, significant progresses have been made on understanding (1) the effects of local gas puffing on edge plasma density and ICRF coupling; (2) the ICRF induced density convection.

The 3D SOL plasma density during different gas puffing are simulated with the EMC3-EIRENE code, which are then used in the antenna codes to calculate the coupling resistances. Different gas puffing cases in AUG and JET have been extensively studied and extensive comparisons between the simulations and experimental results have been made. It is shown that the gas vavles close to the antennas increase the local density and thus the ICRF power coupling most significantly. This conclusion is expected to be also true for other and future machines. Similar numerical studies for ITER and DEMO are being carried out to figure out optimized gas valve postions for the best ICRF power coupling.

The ICRF induced density convection has been studied either by EMC3-EIRENE simulations based on experimental data or by self-consistent simulations with several codes (EMC3-EIRENE + RAPLICASOL + SSWICH). Our simulation results are in qualitative agreement with the measurements. It is indicated that convective cells are often developed in the top and bottom of the antennas. The convective drifts change the density in front of the antenna locally, influence the

ICRF power coupling and enhance the plasma-wall interactions. Among the studied cases, the 3-strap antenna with optimized feeding configurations leads to the lowest rectified sheath potential and density convection. The developed numerical methods can be further used to predict and optimize the experiments.

*This work has been carried out within the framework of the EUROfusion Consortium and has received funding from the Euratom research and training programme 2014-2018 under grant agreement No 633053. The views and opinions expressed herein do not necessarily reflect those of the European Commission.*

## References

[1] Bobkov V. et al 2015 AIP Conference Proceedings **1689** 030004
[2] Lerche E. et al 2015 Journal of Nuclear Materials **463** 634-9
[3] Jacquet P. et al 2016 Nuclear Fusion **56** 046001
[4] Zhang W. et al 2016 Nuclear Fusion **56** 036007
[5] Zhang W. et al 2017 Nuclear Fusion **57** 056042
[6] Zhang W. et al 2017 Plasma Phys. Control. Fusion **59** 075004
[7] Feng Y. et al 2004 Contributions to Plasma Physics **44** 57-69
[8] Zhang W. et al 2016 Plasma Phys. Control. Fusion **58** 095005
[9] Jacquot J. et al 2015 AIP Conference Proceedings **1689** 050008
[10] Jacquot J. et al 2014 Physics of Plasmas **21** 061509
[11] Brambilla M. 1989 Plasma Physics and Controlled Fusion **31** 723-57
[12] Soucek J. et al 2005 Journal of Geophysical Research-Space Physics **110** A08102
[13] Zhang W. et al 2017 Nuclear Fusion **57** 116048

# AUTHOR INDEX

# AUTHOR INDEX

# AUTHOR INDEX

# AUTHOR INDEX

# AUTHOR INDEX